Bistatic Radar

Bistatic Radar: Principles and Practice

Edited by

Mikhail CHERNIAKOV

The University of Birmingham, UK

Authors:

David V. NEZLIN

Vladimir I. KOSTYLEV

Alexander B. BLYAKHMAN

Alexander G. RYNDYK

Alexander V. MYAKINKOV

John Wiley & Sons, Ltd

Telephone (+44) 1243 779777

Email (for orders and customer service enquiries): cs-books@wiley.co.uk
Visit our Home Page on www.wiley.com

Other Wiley Editorial Offices

John Wiley & Sons Inc., 111 River Street, Hoboken, NJ 07030, USA

Jossey-Bass, 989 Market Street, San Francisco, CA 94103-1741, USA

Wiley-VCH Verlag GmbH, Boschstr. 12, D-69469 Weinheim, Germany

John Wiley & Sons Australia Ltd, 42 McDougall Street, Milton, Queensland 4064, Australia

John Wiley & Sons (Asia) Pte Ltd, 2 Clementi Loop #02-01, Jin Xing Distripark, Singapore 129809

John Wiley & Sons Canada Ltd, 6045 Freemont Blvd, Mississauga, ONT, L5R 4J3

Wiley also publishes its books in a variety of electronic formats. Some content that appears in print may not be available is electronic books.

Anniversary Logo Design: Richard J. Pacifico

British Library Cataloguing in Publication Data

A catalogue record for this book is available from the British Library

ISBN 978-0-470-02630-4 (HB)

Typeset in 9/11pt Times by TechBooks International, New Delhi, India
This book is printed on acid-free paper responsibly manufactured from sustainable forestry in which at least two trees are planted for each one used for paper production.

Contents

PART III: Forward-scattering Radars

A.B. Blyakhman, A.G. Ryndyk, A.V. Myakinkov

List of Contributors

Mikhail CHERNIAKOV
The University of Birmingham,
United Kingdom
School of Engineering, EECE, University of Birmingham, Edgbaston,
Birmingham B15 2TT, UK
Email: m.cherniakov@bham.ac.uk
Phone: +44(0)121 4144286
Fax: +44(0)121 4144291

Vladimir I. KOSTYLEV
Voronezh State University,
Russia
Physics Faculty,
Voronezh State University,
Universitetskaya pl., 1,
Voronezh, Russia 394006
Email: kostylev@phys.vsu.ru
Phone: +74732208284
Fax: +74732208755

Alexander G. RYNDYK
Nizhny Novgorod State Technical University, Russia
Nizhny Novgorod State Technical University, 24, Minin Street,
Nizhny Novgorod, Russia 603600
E-mail: ryndyk@nntu.nnov.ru
Phone: (8312) 36-78-80
Fax: (8312) 36-23-11

David V. NEZLIN
Moscow Institute of Electronic Engineering, Russia
Moscow Institute of Electronic Engineering, Zelenograd, Moscow,
Russia 103498
Email: nezlin@rinet.ru
Phone: +74956875639
Fax: +74956874518

Alexander B. BLYAKHMAN
Nizhny Novgorod Research Radiotechnical Institute, Russia
Nizhny Novgorod Research Radiotechnical Institute, 5, Shaposhnikov Street, Nizhny,
Novgorod, Russia 603950
Email: blkhmn@pop.sci-nnov.ru
Phone: (8312)64-00-93
Fax: (8312)64-02-83

Alexander V. MYAKINKOV
Nizhny Novgorod State Technical University, Russia
Nizhny Novgorod State Technical University, 24, Minin Street,
Nizhny Novgorod, Russia 603600
E-mail: rs@nntu.nnov.ru
Phone: (8312)36-78-80
Fax: (8312)36-23-11

Preface

BISTATIC RADAR: PRINCIPLES AND PRACTICE

The second half of the last century is often referred to as the 'Electronic Age'. Audio and TV broadcasting, computers and communication networks, mobile phones and satellite navigation are amongst some of the most obvious electronic applications which have become part of our everyday lives.

Nevertheless, for many engineers and scientists the doyen of electronics is radar technology with its rich past and promising future. Radar provides the navigation for vessels, aircrafts and vehicles and is used in a wide spectrum of applications; from its use for the exploration of planetary structures to its role in medicine to diagnose illnesses. Furthermore, radar is the key component of numerous defence, military, security and traffic control systems. This list of radar applications is by no means exhaustive and radar's potential as the solution to a wide range of problems generates a suitably high level of interest. It is now difficult to count the number of companies, institutions, universities and research teams involved in radar related activity.

Bistatic radars, which are a subclass of radar systems and the main subject of this book, have a more complicated history. Although known of since the very early days of radar development, bistatic radar have some unique peculiarities that have great potential, but have found little practical application. The reasons for this are secondary to the central question of how to transfer bistatic radar's potential into a practical domain. A vital means in meeting this objective is to "promote" this promising technology by systematizing the knowledge recently acquired about it; the results of which are the writing of this book dedicated to bistatic radar.

The starting point of this book is the classic manuscript *Bistatic Radar* written by N.J. Willis, published more than 15 years ago (with 1995 Technology Service Corporation, second edition). This is the only source of systematic knowledge to have been collated in this field and a number of essential studies have been undertaken since this significant publication. Bearing in mind the constant growing interest of the research community in bistatic radar technology, the time has now come for a new updated publication.

An initial survey of the subject made it apparent that to cover all the complex issues of bistatic radar in one volume would prove an impossible task: to cover the key aspects alone would require a publication of more than one thousand pages. My experience of producing a publication of this kind on my own led me to approach the project as a collective monograph. I approached a number of scientists working in the bistatic radar and related fields and received a welcome reaction to the proposal of taking part in this project.

The book begins with a comprehensive, tutorial style part dedicated to radar Principles. This part is not intended to substitute the numerous and Excellent books entirely dedicated to radar systems. Nevertheless, it covers the main aspects of monostatic radar and specifically those to be discussed for the case of bistatic radar: radar systems base; signal, interference and signal processing; power budget analysis; target tracking; antennas and Aperture synthesis. This part may be omitted if readers are confident about Their knowledge in conventional radar. On the other hand, this part could be recommended to postgraduate students as its contents and complexity correspond to MEng or MSc level courses at traditional European universities. It is Also important to remember that, effectively, both monostatic and bistatic radar seamlessly represent the same technology.

The second part is dedicated to bistatic radar analysis and is the core of the book. This part contains a relatively concise introduction to the traditional bistatic radar problems: bistatic radar geometry, range, Doppler and angle resolutions, as well as the detection range. Material corresponding to this section may also be found in more detail in the N.J. Willis book[1]. The part mainly refers to the problem of electromagnetic waves analysis in bistatic systems, which is the fundamentally distinctive feature of bistatic radar; and yet has not been presented in a systematic manner in the literature. Here readers will find the fundamentals of electromagnetic waves bistatic scattering and, based on it, radar cross section analysis and signal models. The biggest section in the chapter presents a state of the art study in bistatic scattering based on more than 100 scientific publications in this area of analysis.

The final, third part of the book is entirely dedicated to forward scattering radar (FSR) for air target detection and tracking. The material presented in this part is unique and will be of specific interest to engineers working in this area. Major problems discussed here at a theoretical level are confirmed by real life FSR testing. The major topics covered here are: the basic principles of FSR forward scatter radar cross section; coverage area of FSR; target coordinate measurement; FSR with an array antenna and FSR design and experimental investigation.

Once again I want to express my gratitude to the authors who have made a tremendous effort in contributing to this book and who have effectively trained me in the art of bistatic radar technology. It is also a good place to thank all the reviewers for their wise advice and input on how to improve the books.

I hope that readers will find this book timely and useful. Of course, a number of problems presented in the books are far from their final solutions, but *omne ignotum pro magnifico est*[2] – and so, welcome to the bistatic radar wonderland.

Last but not least, behind any book there is a hidden job in manuscript preparation. I want to thank Mrs. Emma Makarova who has helped me with this.

Mikhail Cherniakov 2006

[1] N.J. willis (1995) *Bistatic Radar*, Technology Service Corporation.
[2] All unknown look great (Latin)

Part I

Radar Principles

D.V. Nezlin

1

Radar Systems

1.1 GENERAL PROPERTIES OF RADAR SYSTEMS

Radar systems and radar stations are intended for detecting various objects in space and establishing their current position, as well as determining velocities and trajectories for moving objects. To perform these tasks, the radar transmits a signal (also known as a 'bang') into the surrounding space and receives signals reflected by objects (Figure 1.1).

Strictly speaking, the output of a radar transmitter is not a signal since it does not carry any information. It is the returned signal that contains information about the target. However, in accordance with the established terminology, the electromagnetic waves generated by a radar transmitter will be reffered to as the transmitted 'signal'.

Radar detection and position measurement techniques are based on three physical phenomena, namely:

- the virtually rectilinear propagation of radio waves of the band(s) used a range of millimetres to metres at a practically constant speed $c \approx 3 \times 10^8$ m/s[1];
- the property inherent in objects of various natures to scatter electromagnetic waves that fall on them;
- the Doppler effect.

Because of the stated properties of radio waves, the echo signal contains certain information about the target. The fact of signal detection is indicative of the presence of an object in the volume of space covered by the radar. The time delay between the transmission of the radar signal and the reception of the echo signal, expressed as

$$t_s = \frac{2R}{c} \tag{1.1}$$

allows the object's range R to be measured. When the target is moving relative to the radar at a radial velocity $V_R = V \cos \alpha$ (Figure 1.2), the carrier frequency of the echo signal differs from the frequency,

[1] Over-the-horizon radars operating at the short-wave band (wavelengths of 10 m to 100 m) are not considered here.

Bistatic Radar: Principles and Practice Edited by M. Cherniakov

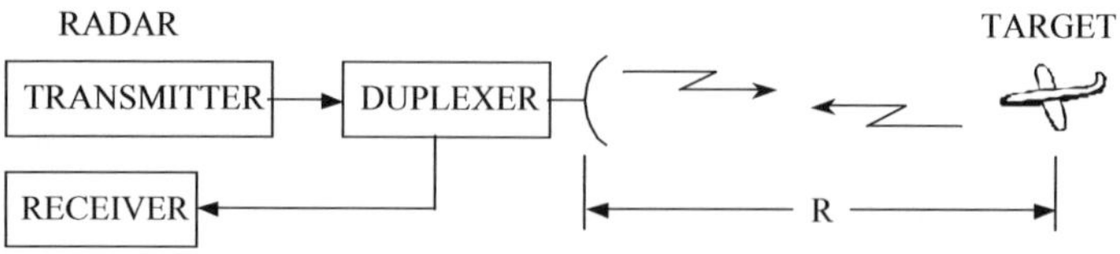

Figure 1.1 Radar principle of operation

f_0, of the transmitted signal on account of the Doppler effect. This difference is described by the equation

$$f_d = \frac{2V_R}{\lambda}, \tag{1.2}$$

where λ is the radar's wavelength and f_d is the Doppler shift in frequency. The target velocity component towards the radar, V_R, can be determined through the measurement of f_d.

The following example will make formula (1.2) clear. Take a radar with a harmonic transmitter output $U \cos 2\pi f_0 t$. The returned signal is given by

$$u_r = U_r \cos 2\pi f_0 \left(t - \frac{2R}{c} \right).$$

The frequency of this signal is

$$f_r = \frac{1}{2\pi} \frac{d}{dt} \left(2\pi f_0 t - 2\pi f_0 \frac{2R}{c} \right) = f_0 - \frac{2 f_0}{c} \frac{dR}{dt} = f_0 + \frac{2V_R}{\lambda} \tag{1.3}$$

or

$$f_r = f_0 + f_d.$$

In formula (1.3), the value of V_R is assumed to be positive if the target is approaching the radar and negative if it is moving away from the radar. This also applies to the sign of f_d.

The use of sharp-beam antennas for the transmission and reception of signals makes it possible to establish the angular position of the target by the position of the antenna system corresponding to the maximum peak of the received signal. Angular position measurement methods applied in practice are dealt with in Chapter 4.

The beamwidth of an antenna system with linear dimension L_A is expressed by the approximation

$$\theta \approx \frac{\lambda}{L_A} \quad \text{(radian)}. \tag{1.4}$$

Therefore, the practicability of narrow-beam designs with an acceptable antenna size (L_A measuring from fractions of a metre to several metres) is limited to S-band and X-band. Less common solutions are based on millimetre band radars (due to the atmospheric absorption of radio waves) and VHF (very high frequency) radars.

The angular position of a target is defined in terms of two angles in two mutually perpendicular planes: for example, the horizontal plane (azimuth β) and the vertical plane (elevation ε). As seen from the illustration in Figure 1.3, this corresponds to a system of spherical coordinates R, β, ε,

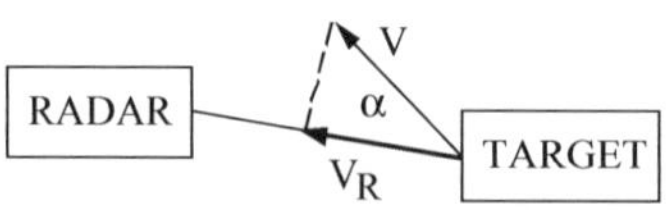

Figure 1.2 Target velocity

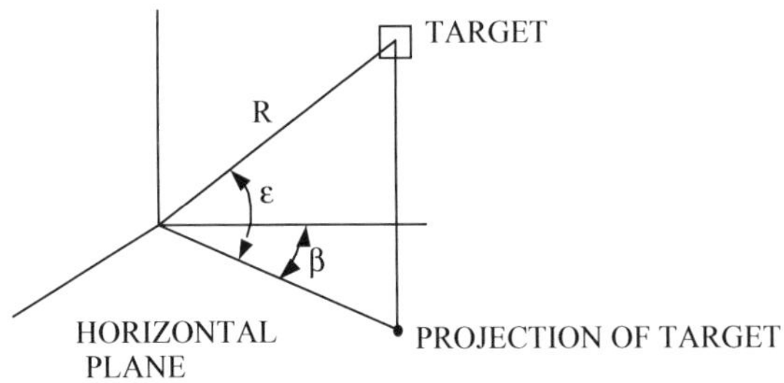

Figure 1.3 Coordinates of target

with the parameter of radial velocity (commonly referred to as the 'range rate') added in a number of cases.

The basic performance characteristics of a radar comprise range, position measurement accuracy and resolution, which are specified for operation both in the clear and in the preassigned interference conditions.

1.2 BLOCK DIAGRAM OF A RADAR

The block diagram of a pulse radar is given in Figure 1.4. The carrier generator produces a sinusoidal waveform with a stable frequency f_0 (Figure 1.5(a)). The synchronizer generates a train of rectangular pulses of duration τ_p and period T_p (Figure 1.5(b)). In this case, the modulator is an electronic switch triggered by the pulses of the synchronizer. At the output of the modulator there is a train of radio frequency (RF) pulses (Figure 1.5(c)). When there is a pulse at the output of the synchronizer, the duplexer connects the antenna to the power amplifier. As a result, the antenna emits into space an RF pulse of the bang. In the interval between the pulses of the synchronizer, the antenna is connected to the input of the receiver. The receiver picks up the pulses of echo signals. Amplified by the low-noise amplifier (LNA), the echo signal enters the frequency converter (the mixer in the diagram). The latter

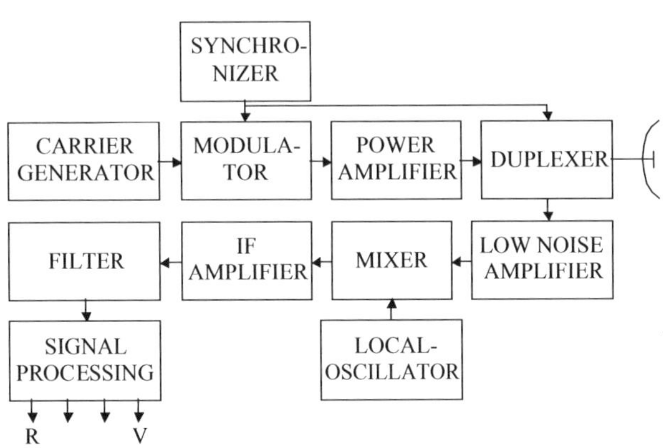

Figure 1.4 Block diagram of a pulsed radar

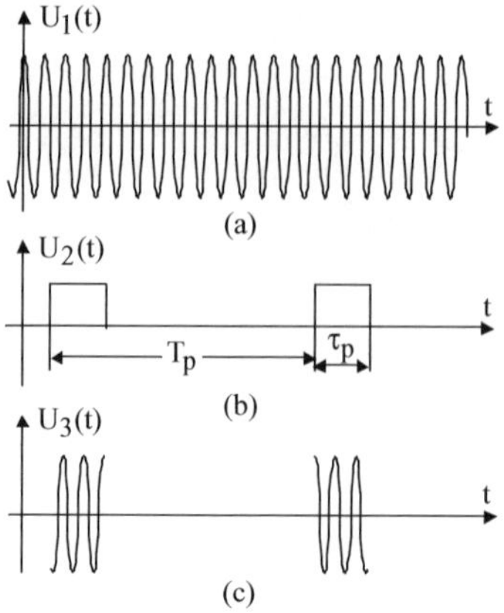

Figure 1.5 Flowcharts of transmitter voltages: (a) carrier generator; (b) sinchronizer; (c) modulator

puts out a rectangular pulse of intermediate frequency

$$f_{\text{int}} = f_0 - f_{\text{LO}}.$$

This voltage is further amplified and filtered. It is then digitized and fed to the digital signal processing unit. The processing unit, in turn, yields a target detection as well as the target's current coordinates R, β, ε and velocity component V_R.

The pulse repetition period is then chosen. To ensure an unambiguous and simple measurement of the target range, the transmitter's pulse repetition period should be chosen so as to satisfy the relation

$$T_p > \frac{2R_{max}}{c}. \tag{1.5}$$

Here, R_{max} is the range coverage of the radar. Violation of condition (1.5) brings about ambiguity in the target range measurement, which is illustrated by Figure 1.6. The figure shows the relative position of the

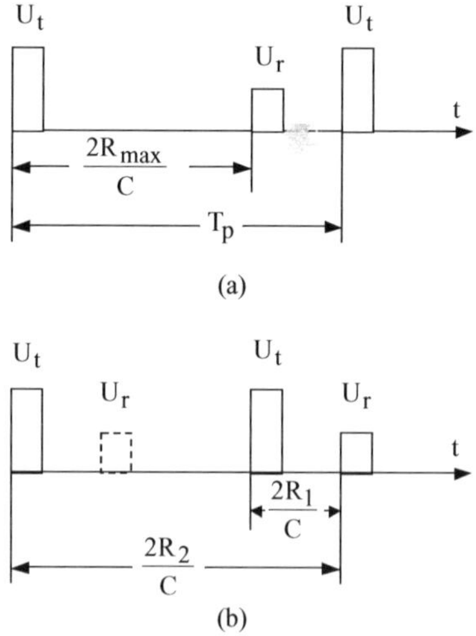

Figure 1.6 Reflected pulse disposition in relation to those transmitted: (a) $T_p > 2R_{max}/c$; (b) $T_p < 2R_{max}/c$

echo pulses and the transmitted pulses. Figure 1.6(a) refers to the case when $T_p > 2R_{max}/c$ and Figure 1.6(b) to the case when $T_p < 2R_{max}/c$. The dashed bar in Figure 1.6(b) represents the echo response to the pulse of the transmitter preceding the first transmitted pulse shown. It follows from Figure 1.6(b) that the measured target range is

$$R_1 = \frac{ct_1}{2}.$$

The true target range is $R_2 = ct_2/2$.

1.3 SIGNAL DETECTION

The detection of signals in radars is based on the thresholding method. A mixture of the received signal and interference (the receiver's own noise plus external noise) is compared to a certain threshold. When the magnitude of the above mentioned mixture crosses the threshold, a decision is made about the presence of a target signal. Otherwise, it is presumed that there is no target signal. However, this decision may be erroneous because of the random nature of interference and fluctuations of signal amplitudes (see Section 3.3).

There are two kinds of detection errors: miss detection and false alarm. The rate of occurrence of the former and the latter is evaluated in terms of the probability of missed detections P_{md} and the probability of false alarms P_{fa} respectively. P_{md} is regarded as a measure of the likelihood that the level of the signal–interference mixture will not reach the threshold level. P_{fa} is a measure of the likelihood that a surge of interference will cross the threshold in the absence of a target signal. Normally, the detection probability parameter P_d is used instead of P_{md}, the two being related thus:

$$P_d = 1 - P_{md}.$$

The required value of P_d and the tolerable value of P_{fa} are indicated in the requirements specification for any radar design. The threshold value is set on the basis of a known value P_{fa}. The desired P_d can be achieved if the signal-to-interference power ratio at the input of the receiver is not less than the acceptable minimum of ρ_{min}. An advisable method for ρ_{min} calculation is given in Sections 3.1 and 3.2.

It is shown in Section 1.6 that the lower the desired signal-to-interference ratio ρ_{min} is at the input of the receiver, the lower the transmitter power requirement, with other conditions being equal. To reduce the ρ_{min} value, it is practical to integrate the signal during the time allotted for detection. This procedure is implemented in most radars (see Sections 2.1 to 2.3). Such a detection device can be represented as a cascade connection of two elements (Figure 1.7).

The first element performs integration of the signal received. The second is a threshold detector. The best results are achieved using the so-called optimum filter in the first cascade, such as a matched filter, which is a specific case of the former. These matters are considered in detail in Section 2.3.

Targets are detected in a volume of space limited by a given solid angle Ψ and a range span from R_{min} to R_{max}. As shown in Section 3.3, it is practical to scan all the range cells simultaneously, making use of a multichannel device. As to the angular scan, it is advantageous to perform it sequentially, cell by cell. This is accomplished by sequentially moving the antenna beam within the limits of the angle Ψ.

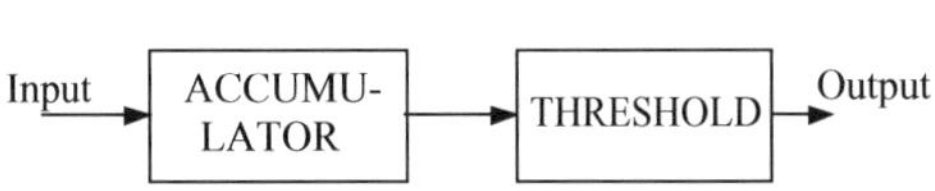

Figure 1.7 Block diagram of the detection unit

The number of angular cells is

$$N_a = \frac{\Psi}{\theta_{t.w.}}.$$

Here, $\theta_{t.w.}$ is the solid beam angle of the two-way directional pattern of a transmit–receive antenna. A two-way directional pattern is the product of the patterns of the transmitting and the receiving antennas.

Let us adopt the symbol T_s to designate the time allotted for one scan. Then, the target remains covered by the antenna beam for as long as

$$T_1 = \frac{T_s}{N_a} = \frac{T_s\theta_{t.w.}}{\Psi}. \tag{1.6}$$

The set of pulses received during the time T_1 is the referred to as a pulse burst. Therefore, when the radar operates in the angular scan mode, the echo signal received has the form of a burst, and T_1 is the burst duration. During each scan, one burst of the echo signal comes from each target.

The number of pulses in a burst is

$$N = T_1 f_p = \frac{T_s\theta_{t.w.} f_p}{\Psi}. \tag{1.7}$$

In this equation, $f_p = 1/T_p$ is the pulse repetition frequency expressed in hertz.

1.4 RADAR RESOLUTION

The resolution is the ability of a radar to detect and distinguish closely situated targets and to measure their coordinates and velocities separately. The term 'resolution' also refers to the quantitative characteristic of a radar's resolving power in a specific coordinate of the target. In the latter case, the resolution in a coordinate (range, azimuth, elevation) or velocity is the minimal difference in the values of that coordinate (velocity) for two targets that allows these two targets to be detected separately. It is of course assumed here that the other coordinates (velocity) of the targets are equal or, if anything, the targets cannot be resolved in them.

In the subsections to follow, a brief overview is given of the factors affecting the radar resolution of targets in range, angular position and velocity. Furthermore, the subject of signal resolution in range and velocity will be addressed in Section 2.9. Angular resolution is covered in more detail in Chapter 6. In the current section, definitions will be introduced for the volume resolution cell and the surface resolution cell of a radar.

1.4.1 Range Resolution

Let us estimate the range resolution of a pulsed radar as an example. Suppose there are two targets at ranges R_1 and R_2 (Figure 1.8(a)). The time gap between the pulses of the returned signals at the output of the processing circuitry (Figure 1.8(b)) is

$$t_c = \frac{2R_2}{c} - \frac{2R_1}{c} - \tau_{out} = \frac{2}{c}(R_2 - R_1) - \tau_{out}. \tag{1.8}$$

Here, τ_{out} is the duration of the echo pulse at the output of the processing circuitry. It follows from Equation (1.8) that the closer the targets, the shorter the t_c interval.

Let us assume the range resolution ΔR to be the difference in target ranges $R_2 - R_1$ when t_c turns to zero and the echo signals start to merge. Then, from Equation (1.8),

$$\frac{2\Delta R}{c} = \tau_{out}$$

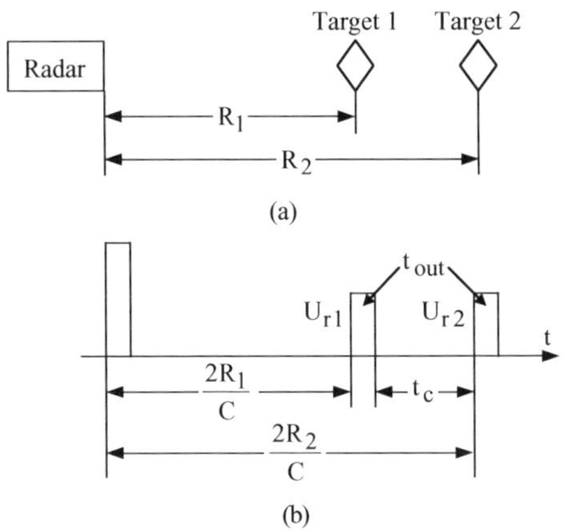

Figure 1.8 Radar range resolution: (a) target range to radar; (b) time intervals

and

$$\Delta R = \frac{c\tau_{\text{out}}}{2}. \tag{1.9}$$

1.4.2 Angular Resolution

Let us demonstrate that the angular resolution of a radar is driven by the antenna system beamwidth θ in the relevant plane. To this end, take two targets located at points A and B where the directions to these points from the radar form an angle β (Figure 1.9). Now imagine that the antenna beam turns through an angle α. Should the echo signals from the targets arrive at the input of the receiver in turn, the received signal powers $P_A(\alpha)$ and $P_B(\alpha)$ in the function of the angle α would be described by the antenna pattern (Figure 1.10).

Under real conditions, the total signal power at the input of the receiver is

$$P(\alpha) = P_A(\alpha) + P_B(\alpha).$$

In this case, it is the power levels that are summed and not the voltages since the signals have independent initial phases. Figure 1.11 shows the $P(\alpha)$ function as a set of curves plotted for different values of the parameter β/θ. Figure 1.11 permits the following conclusions to be made:

- When $\beta = \theta/2$, the curve $P(\alpha)$ has a single peak whose abscissa corresponds to the direction to the midpoint of segment AB. Hence, the two targets are perceived as one, i.e. there is no angular resolution.

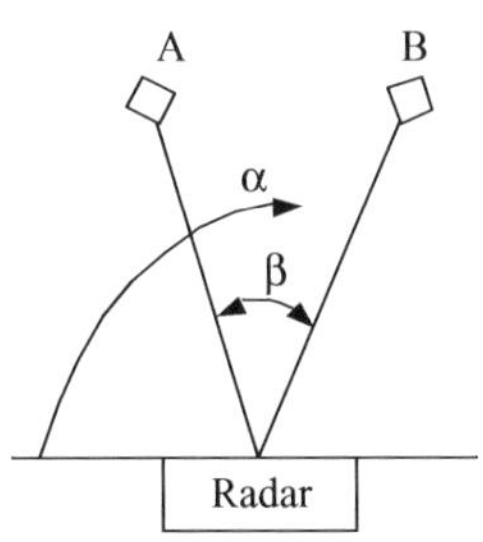

Figure 1.9 Disposition of objects A and B

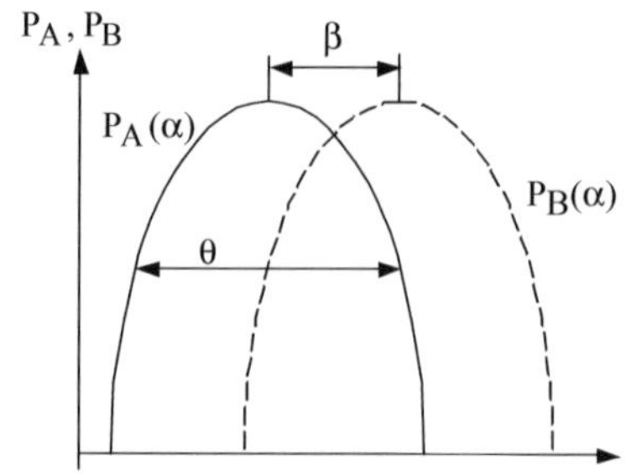

Figure 1.10 The values of signals reflected from objects A and B

- When $\beta = 1.5\theta$, the curve $P(\alpha)$ has two clearly cut peaks whose abscissas roughly correspond to the directions to targets A and B. Hence, the radar definitely perceives two targets whose angular positions can be estimated.
- When $\beta = \theta$, the curve $P(\alpha)$ has an intermediate shape. In addition to the peak in the direction of the AB midpoint, there are two side peaks indicative of the presence of two targets. Therefore θ is conventionally accepted as a measure of angular resolution.

1.4.3 Range Rate Resolution

According to Equation (1.2), the range rate (velocity component towards the radar) of a target is expressed in terms of a Doppler shift in frequency:

$$V_R = \frac{\lambda}{2} f_d.$$

Hence, the radar resolution in velocity is

$$\Delta V_R = \frac{\lambda}{2} \Delta f_d. \tag{1.10}$$

Here, Δf_d is the resolution in the Doppler-shifted frequency. The calculation procedure for Δf_d is elaborated in Section 2.8. Now, let us estimate this value for a single RF pulse of duration τ with a rectangular envelope and a constant value of carrier frequency.

We shall write the transmitted signal on the time interval from $-\tau/2$ to $\tau/2$ as

$$u(t) = U_0 \cos 2\pi f_0 t.$$

Suppose two targets are moving towards the radar. The first target has a velocity V_1 and the second target V_2. The echo signals have the same amplitude U. Then, the expressions for the two signals take the

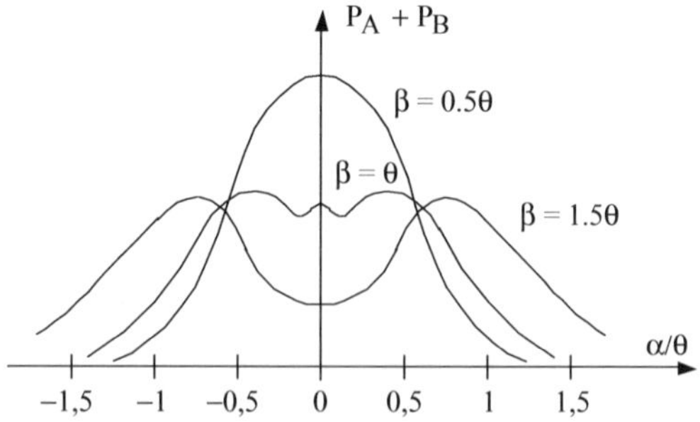

Figure 1.11 Summed power of signals reflected from objects A and B

following form:

$$\text{First target:} \quad u_1(t) = U \cos 2\pi(f_0 + f_{d1})t; \qquad f_{d1} = \frac{2V_1}{\lambda};$$

$$\text{Second target:} \quad u_2(t) = U \cos 2\pi(f_0 + f_{d2})t; \qquad f_{d2} = \frac{2V_2}{\lambda}.$$

Taking the Fourier transform of $u_1(t)$ and $u_2(t)$ gives the spectra of the echo signals:

$$S_1(f) = \frac{U\tau}{2}\frac{\sin[2\pi(f - f_0 - f_{d1})(\tau/2)]}{2\pi(f - f_0 - f_{d1})(\tau/2)} + \frac{U\tau}{2}\frac{\sin[2\pi(f + f_0 + f_{d1})(\tau/2)]}{2\pi(f + f_0 + f_{d1})(\tau/2)}; \tag{1.11}$$

$$S_2(f) = \frac{U\tau}{2}\frac{\sin[2\pi(f - f_0 - f_{d2})(\tau/2)]}{2\pi(f - f_0 - f_{d2})(\tau/2)} + \frac{U\tau}{2}\frac{\sin[2\pi(f + f_0 + f_{d2})(\tau/2)]}{2\pi(f + f_0 + f_{d2})(\tau/2)}. \tag{1.12}$$

The first terms in Equations (1.11) and (1.12) represent the spectra on the positive semi-axis of frequency. The second terms in these expressions give the signal spectra on the negative semi-axis of frequency.

Confining ourselves to the main lobes of the spectra and $f > 0$, let us plot the sum-of-squares curves for spectra $S_1(f)$ and $S_2(f)$, for the cases when $f_{d1} - f_{d2} = 1/\tau$ and $f_{d1} - f_{d2} = 0.7/\tau$ (Figure 1.12). It is not the spectra that are summed but the squares, since the phases of signals returned by different targets are independent.

Figure 1.12 enables the following conclusion to be drawn. If $f_{d1} - f_{d2} = 1/\tau$, the sum-of-squares curve has two peaks in the points corresponding to the peaks of spectra $S_1(f)$ and $S_2(f)$. Consequently, with the given difference in Doppler shifts f_{d1} and f_{d2}, the signals are well resolved and their frequencies can be measured independently. Therefore, the radar resolution in frequency is expressed in terms of

$$\Delta f_d = \frac{1}{\tau}. \tag{1.13}$$

Taking Δf_d from Equation (1.13) and substituting it into formula (1.10), the following expression for the radar resolution in target velocity is obtained:

$$\Delta V_R = \frac{\lambda}{2\tau}. \tag{1.14}$$

1.4.4 Volume Resolution Cell and Surface Resolution Cell

The terms 'volume resolution cell' and 'surface resolution cell' are used to give a comprehensive description to the spatial resolution of a radar. The volume resolution cell is a volume of space within which no points can be resolved. In other words, the signals returned by scatterers situated inside this

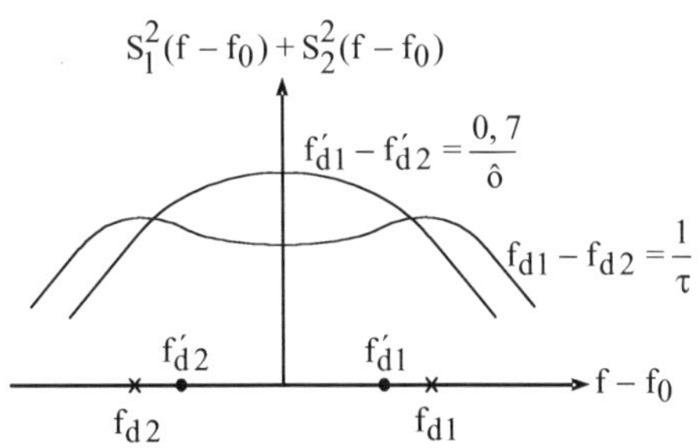

Figure 1.12 Sum of spectra of two signals with different Doppler shifts

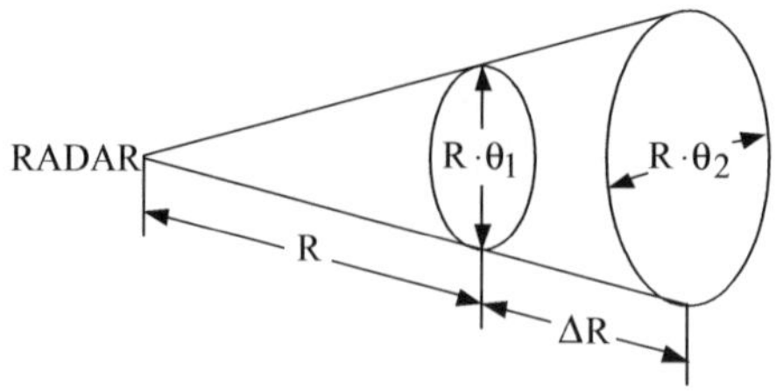

Figure 1.13 Volume resolution cell

volume reach the input of the radar receiver simultaneously. The resultant signal is the sum of their powers.

The volume resolution cell has the shape of a blunted cone with a generatrix length equal to ΔR and a solid angle at the apex equal to a half-power two-way beamwidth (Figure 1.13). The size of the volume resolution cell may be approximately estimated as

$$\Delta W = \frac{\pi}{4} \Delta R\, R^2\, \theta_1\, \theta_2. \tag{1.15}$$

In formula (1.15), R is the distance from the radar to the volume resolution cell and θ_1 and θ_2 are the beamwidths in two perpendicular planes (expressed in radians).

If scatterers on a surface are considered, the notion of the surface resolution cell comes into play as a measure of the spatial resolution of the radar. The surface resolution cell is represented by the trapezium ABCD in Figure 1.14, where no points within can be resolved by the radar.

The echoes from all the scatterers within the surface resolution cell make up a resultant signal at the input of the receiver. The size of the surface resolution cell is approximately estimated as

$$\Delta S = \frac{\Delta R}{\cos \varepsilon} R\theta_1, \tag{1.16}$$

where θ_1 is the beamwidth (in radians) and ε is the elevation angle of the beam.

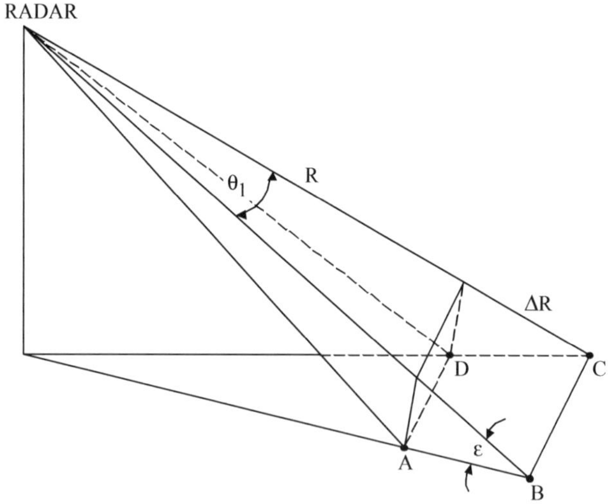

Figure 1.14 Surface resolution cell

1.5 RADAR MEASUREMENTS

Measurements produced by radars comprise target range, angular position and radial velocity (range rate). The target range is established from the time delay of the echo signal peak relative to the transmitted signal, t_R. Each of the angular coordinates is established from the position of antenna beam β_1 at the moment when the echo signal is at its peak. The radial velocity of the target, V_R, is established from the Doppler frequency shift of the echo signal f_d by the maximum of its spectrum.

Since the measurement methods employed in radar are much the same, let us consider the measurement of a generic coordinate α_0, which will be written for the true value of the coordinate. Then, let α denote the estimated value of the coordinate. The term 'generic coordinate' may be used to refer to any coordinate or velocity of the target.

The signal (spectrum) peak position is determined by the differential method. This method is based on the comparison of two signal (spectrum) amplitude samples separated by a fixed interval expressed in units of the measured coordinate. The value corresponding to the midpoint of this interval is assumed to be the coordinate (velocity) of the target. If the samples are equal in amplitude (Figure 1.15(a)), the midpoint of the interval coincides with the position of peak signal (spectrum).

Otherwise (Figure 1.15(b)), the midpoint of the interval is displaced relative to the peak signal by the magnitude of measurement error. This error is related to the difference in sampled amplitudes. If the samples are not equal in amplitude, their positions on the axis of the measured coordinate will be adjusted until their amplitudes are level.

When handling a moving target, the positions of samples are adjusted continuously to keep track of the varying coordinates (velocity) of the target. Hence, the instruments in question are termed 'trackers'. The two operations that would be normally performed in this kind of measuring device are the comparison of two samples of the echo signal in amplitude (spectrum) and the adjustment of the position of the samples according to the criterion of their equality. The subunit responsible for the first operation is called the discriminator. The second operation is carried out by what is called the 'controller', or the smoothing circuit.

Instruments for the measurement of target position and velocity are covered in sufficient detail in Chapter 4. This section only briefly describes the method of and the device for measuring the generic coordinate α.

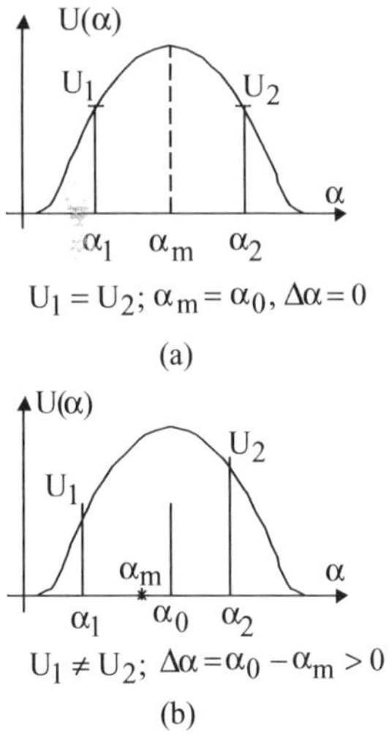

Figure 1.15 Differential method of signal maximum determination

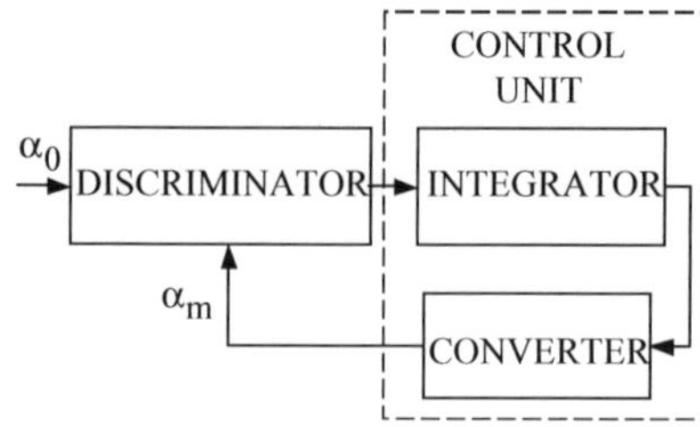

Figure 1.16 Block diagram of a tracker

The block diagram of a typical tracker is given in Figure 1.16. The discriminator output voltage U_{d}, which is proportional to the difference in the two samples of the signal, ultimately is a function of the current measurement error $\Delta\alpha$. Figure 1.17 shows the plot of $U_{\mathrm{d}} = f(\Delta\alpha)$, known as discrimination characteristic. For the sake of definiteness, assume that

$$\Delta\alpha = \alpha_0 - \alpha.$$

Suppose, at a certain instant of time, that the current error $\Delta\alpha_1 > 0$ (Figure 1.17). Then, $U_{\mathrm{d}} > 0$, and the integrator output voltage U_{i} is on the rise. This voltage governs the position of signal samples, which are formed in the converter. Consequently, the estimated coordinate, α, increases while $\Delta\alpha$ decreases, i.e. the estimated value, α, approaches the true value, α_0.

A distinction is drawn between discriminators (and trackers) employing a simultaneous and sequential comparison of signal samples. In the former case, both samples are taken simultaneously and compared in each signal period. The results of this comparison are accumulated during the time $2KT_{\mathrm{p}}$. In the latter case, K periods of signal are assigned for accumulating the values of the first sample and the next K periods are assigned for accumulating the values of the second sample. Only after that are the accumulated values of the first sample and the second sample compared. A device of the first type considered ensures higher measurement accuracy but requires a significant hardware input when used for angular position measurement. A device of the second type would be more compact in the same role, but it can yield a specific error due to signal level variations over the time KT_{p}.

Range and velocity measurements are typically performed with simultaneous comparison of signals. Both this and the other method are employed for angular position measurement.

By way of example, let us consider the typical structure of devices for measuring a single angular coordinate. First, take a tracker with simultaneous comparison of signals. The structure of such a device is shown in Figure 1.18. The discriminator is made up of a double-beam parabolic antenna with two feed elements, receivers 1 and 2, amplitude detectors and a voltage comparator. Since the feed elements are out of focus of the antenna dish, the antenna beams are offset relative to each other (see Figure 1.19). The controller comprises a direct current amplifier (DCA) and an antenna drive functioning as an integrator.

The straight line passing through the intersection point of antenna patterns 1 and 2 in Figure 1.19 is termed the equisignal line (ESL), or radar boresight. If the target signal arrives along the ESL, the receiver's output voltages U_1 and U_2 are equal. Thus, there is zero voltage at the output of the discriminator and the antenna drive is at a standstill. The signal's direction of arrival coincides with the direction of the maximum of the sum antenna pattern (designated by the dashed line in Figure 1.19).

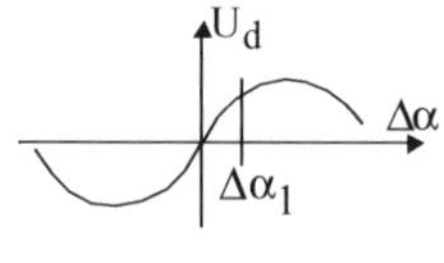

Figure 1.17 Discriminator characteristic

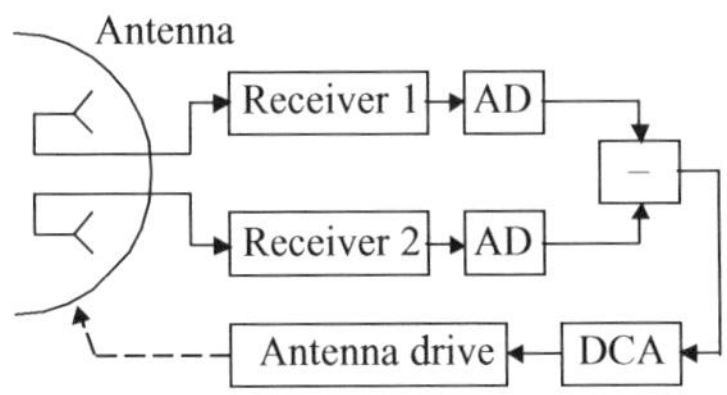

Abbreviation: AD – amplitude detector,

DCA – direct current amplifier.

Figure 1.18 Block diagram of a monopulse tracker

If the signal arrives from the direction α_1, then $U_1 - U_2 > 0$, and the drive turns the antenna to the left until the equisignal line matches the direction α_1. As soon as $U_1 - U_2 = 0$ the drive comes to a stop. The maximum of the sum antenna pattern coincides with the signal's direction of arrival.

The tracker functions in the same manner if the signal arrives from the direction α_2. In this case, as follows from Figure 1.19,

$$U_1 - U_2 < 0,$$

and the drive turns the antenna to the right until the equisignal line matches the direction α_2. Note that in both these cases, the discriminator output voltage furnishes the value and sign of the angular error almost instantly (even from a single pulse). Therefore, trackers with a simultaneous comparison of signals are called 'monopulse trackers'.

The structure of a tracker with a sequential comparison of signals is shown in Figure 1.20. The discriminator is made up of an antenna with a periodically varied position of the feed element (one of the positions is designated by a dashed line), a single-channel receiver, an amplitude detector (AD) and a phase detector (PhD). The voltage controlling the position of the feed element is used as the reference voltage for the phase detector. The controller of the tracker is composed of a direct current amplifier and an antenna drive.

To see into the operation of this device, we can refer again to Figure 1.19. Only imagine that the patterns $F_1(\alpha)$ and $F_2(\alpha)$ do not exist simultaneously, but appear in turn. Then, the above description of the operation of a monopulse tracker will be just as applicable.

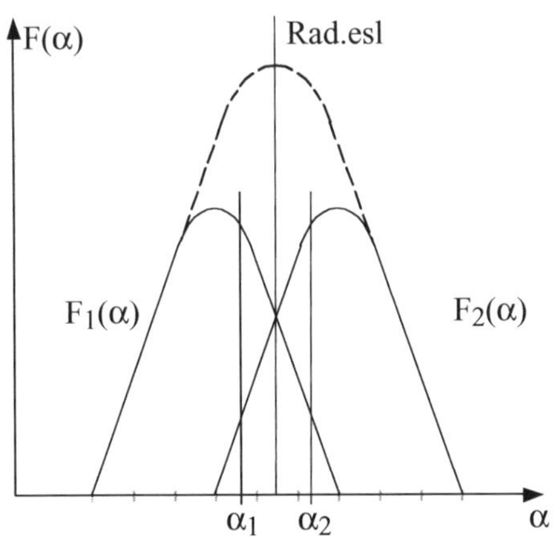

Figure 1.19 Three antennas diagrams in monopulse tracker. Rad. esl – radar equal signal line

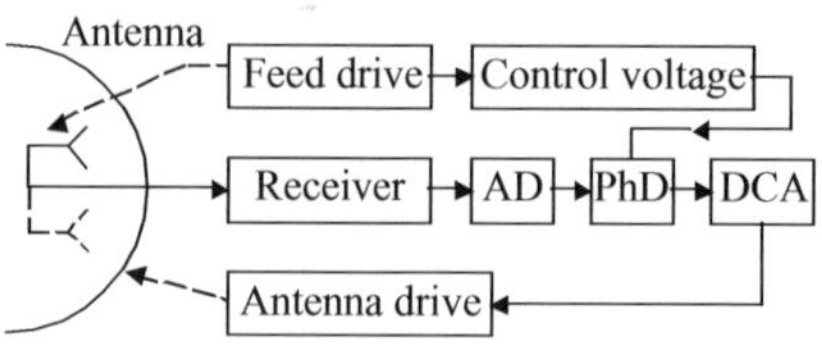

Figure 1.20 Block diagram of a linear scan tracker

1.6 RADAR EQUATION AND RANGE COVERAGE; TARGET RCS

1.6.1 Classification of Radars by the Relative Position of the Receiver and the Transmitter

Figure 1.4 shows the block diagram of an active radar. It is characteristic of this class of radars to have the transmitter collocated with the receiver. In a semi-active (bistatic) radar system, the transmitter and the receiver are spaced a great distance apart. One example of such a system is the semi-active homer in missiles.

For homing on to a target, the missile is fitted with a receiver, an antenna and a steering command generator mounted in its head. A powerful transmitter illuminating the target (target illumination transmitter) is collocated with the launcher (on the ground or on board a ship or an aircraft). The signal returned by the target is detected by the receiver built in the missile. The missile uses this signal to home on to the target.

The above example can also be used to illustrate the operation of a passive radar. If the target carries active radio equipment, the missile fitted with a homing device can home on to the target by the RF signal emitted by it. This eliminates the need for a target illumination transmitter.

The final class of radar systems is called the active radar with an active response (Figure 1.21). One typical application of these kinds of radar is the air traffic control system deployed at every airport. The repeater installed on board the aircraft responds to the transmitted signal, thus not only extending the range of the ground-based radar but also allowing the aircraft to be identified by the coded response.

Further, methods will be considered of range calculations for the four classes of radar system outlined above.

1.6.2 Range Coverage of Active and Semi-active Radar Systems in Free Space

To begin, let us consider the semi-active radar system of Figure 1.22. The power flux density of the transmitted signal at the target location is expressed by the following equation:

$$\mathrm{PD}_1 = \frac{P_t G_t}{4\pi R_1^2}.$$

Here, P_t and G_t are the transmitter power and the transmitting antenna gain. The power of the echo signal in the direction towards the receiver is proportional to the value of PD_1. The proportionality factor, σ,

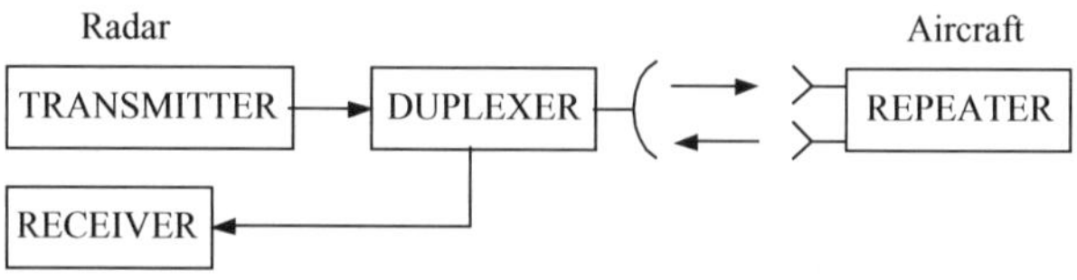

Figure 1.21 Radar system with an active reply

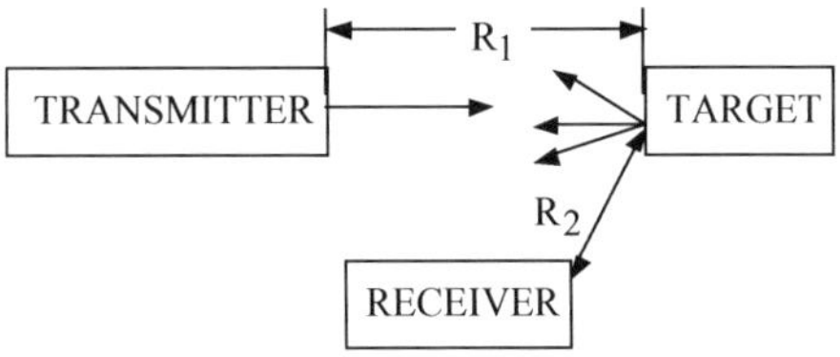

Figure 1.22 Semi-active radar system

which has the dimensionality of a flat area, is referred to as the target's radar cross-section. Consequently, the power flux density of the echo signal at the receiver location is

$$\mathrm{PD}_2 = \frac{\sigma \mathrm{PD}_1}{4\pi R_2^2} = \frac{P_\mathrm{t} G_\mathrm{t} \sigma}{16\pi^2 R_1^2 R_2^2}. \tag{1.17}$$

The echo signal power at the output of the receiving antenna is

$$P_\mathrm{r} = P_2 A = \frac{P_\mathrm{t} G_\mathrm{t} A \sigma}{16\pi^2 R_1^2 R_2^2} \tag{1.18}$$

where A is the effective aperture of the receiving antenna. Now, a dimensionless hardware loss factor $L_\mathrm{a} < 1$ will be introduced into Equation (1.18). Equation (1.18) then takes the form

$$P_\mathrm{r} = \frac{P_\mathrm{t} G_\mathrm{t} A \sigma L_\mathrm{a}}{16\pi^2 R_1^2 R_2^2}. \tag{1.19}$$

Therefore, the relationship between the antenna gain and the antenna effective aperture is

$$G = \frac{4\pi A}{\lambda^2}.$$

The next step is to formulate the signal-to-noise ratio referred to the receiver input. This should be done in view of two sources of noise, namely:

- environmental noise received by the antenna and
- the receiver's inherent noise.

Both of the noise components will be characterized by the noise temperature. Thus, in accordance with the results obtained in Reference [I.1], the total noise power is

$$N = kT_\mathrm{s} B_\mathrm{n}, \tag{1.20}$$

where $k = 1.38 \times 10^{-23}$ J/K is the Boltzmann constant, T_s is the system noise temperature expressed in Kelvin and B_n is the receiver's noise band, which approximately equals the normal –3 dB band. In its turn,

$$T_\mathrm{s} = (F_\mathrm{n} - 1)T_0 + T_\mathrm{a}.$$

Where F_n is the receiver noise factor, T_0 is the environmental temperature and $(F_\mathrm{n} - 1)T_0 = T_\mathrm{e}$ is the receiver effective noise temperature. If no special cooling is provided for the first stages of the receiver, then it is assumed that $T_0 = 290$ K. T_a is the antenna noise temperature, which is mainly due to emissions from the Earth's surface and solar radiation.

In the specific case when $T_\mathrm{a} = T_0$,

$$T_\mathrm{s} = F_\mathrm{n} T_0. \tag{1.21}$$

By combining Equations (1.18) and (1.20), the expression for the signal-to-noise ratio is obtained:

$$\rho = \frac{P_r}{N} = \frac{P_t G_t A \sigma L_a}{16\pi^2 R_1^2 R_2^2 k T_s B_n}. \tag{1.22}$$

For the desired target detection performance (probability of detection $\geq D_r$, probability of false alarm $\leq F_r$), ρ should be not less than a certain value of ρ_r. By substituting this value for ρ in Equation (1.22), the maximum product $R_1 R_2$ is obtained:

$$(R_1 R_2)_{max} = \sqrt{\frac{P_t G_t A \sigma L_a}{16\pi^2 \rho_r k T_s B_n}} \tag{1.23}$$

For an active radar, $R_2 = R_1 = R$. Thus, Equation (1.23) grades into a range equation for this class of radar systems:

$$R_{max} = \sqrt[4]{\frac{P_t G_t A \sigma L_a}{16\pi^2 \rho_r k T_s B_n}}. \tag{1.24}$$

Note that Equations (1.23) and (1.24) were derived with the assumption that the target is in the direction of maximum radiation of the transmitting antenna and the direction of maximum reception of the receiving antenna.

In Equations (1.23) and (1.24), the transmitter power is expressed in watts, the antenna effective aperture and target radar cross-section (RCS) in square metres, G_t, L_a, ρ_r in per unit, k in J/deg, T_s in Kelvin and B_n in hertz. R_1, R_2 and R will then be expressed in metres.

The sun does not always penetrate into the main lobe of the receiving antenna. If, in this case, the sidelobes of the receiving antenna are small enough (thermal radiation from the Earth's surface at the antenna output is small), antenna noise temperature, T_a, may turn out to be significantly less than T_e. Then,

$$T_s = T_e = (F_n - 1)T_0.$$

In this case, Equation (1.24) transforms into

$$R_{max} = \sqrt[4]{\frac{P_t G_t A \sigma L_a}{16\pi^2 \rho_r k T_0 (F_n - 1) B_n}}. \tag{1.25}$$

A single manipulation of Equation (1.25) can render it applicable to signals of a different structure. The signal-to-noise ratio at the output of the optimum processing circuitry increases by a factor of $\tau_p B_s$. Here, τ_p is the pulse duration of the transmitted signal and B_s is its spectral bandwidth (see Section 2.2). Incidentally, in the Russian technical literature, this bandwidth–duration product is usually referred to as the 'signal base' and is designated $\tau_p B_s = B$.

Granting the above, the numerator of the radicand in Equation (1.25) is multiplied by B. Since $B_n \approx B_s$, these quantities may be cancelled out in Equation (1.25), thus obtaining the final range equation for an active radar:

$$R_{max} = \sqrt[4]{\frac{P_t \tau_p G_t A \sigma L_a}{16\pi^2 \rho_r k T_0 (F_n - 1)}}. \tag{1.26}$$

The quantity E_R is given by

$$E_R = \frac{P_t \tau_p G_t A L_a}{\rho_r (F_n - 1)}, \tag{1.27}$$

which has a considerable influence on the range coverage of the radar, and is sometimes referred to as the 'radar power budget'.

1.6.3 Range Coverage of an Active Radar System with an Active Response

There are two equations for a given radar range R_{max}:

$$R_{max} = R_1; \quad R_{max} = R_2. \tag{1.28}$$

Where R_1 is the maximum range at which the repeater detects the transmitted signal and R_2 is the maximum range at which the radar detects the response signal from the repeater. Ranges R_1 and R_2 are calculated by a method similar to that used in the derivation of Equations (1.23) and (1.24). Therefore, the final expressions for R_{max} will be presented without proof:

$$R_{max} = \sqrt{\frac{P_1 G_1 A_2 L_{1.2}}{4\pi \rho_2 k T_{s2} B_{n2}}}; \tag{1.29}$$

$$R_{max} = \sqrt{\frac{P_2 G_2 A_1 L_{2.1}}{4\pi \rho_1 k T_{s1} B_{n1}}}. \tag{1.30}$$

Radar parameters on the right side of Equations (1.29) and (1.30) are labelled by the index '1', while repeater parameters are labelled by the index '2'. The coefficient $L_{1.2}$ represents the product of loss factors of the radar's transmitter and the repeater's receiver. The coefficient $L_{2.1}$ is the product of loss factors of the repeater's transmitter and the radar's receiver. Power budget parameters of the radar and the repeater are calculated by formulae (1.29) and (1.30).

1.6.4 Range Coverage of a Passive Radar

The range covered by a passive radar can be calculated from formula (1.30), provided that the receiver bandwidth is not less than the spectral bandwidth of the signal generated by the equipment borne by the target. Otherwise, the numerator of the radicand in Equation (1.30) is subject to multiplication by the ratio of the radar receiver bandwidth to the target signal bandwidth. Now, in Equation (1.30), R_{max} is the radar range coverage, '1' is the index designating radar parameters and '2' is the index designating the parameters of the target-borne equipment. $L_{2.1}$ is the loss factor in the receiving part of the radar.

1.6.5 Target Radar Cross-section; RCS of Selected Objects

Partial reflection of electromagnetic energy of the transmitted signal by an object is a corollary to the scattering of electromagnetic waves at the interface of two media with different electrophysical properties (conductivity, dielectric constant, magnetic permeability). According to Equation (1.17), the target RCS is described by the following equation:

$$\sigma = 4\pi R^2 \frac{\mathrm{PD}_2}{\mathrm{PD}_1}. \tag{1.31}$$

Where R is the range of the target relative to the radar receiver. This formula allows the RCS of objects with a simple geometry to be calculated. For most radar targets, this method of RCS evaluation is ineffective. Therefore, in this section, in addition to Equation (1.31), RCS characteristics will be given for various objects based on their physical notions.

First, all objects will be classified into two categories according to the ratio of their maximum linear dimension, L , to radar wavelength:

- small-sized objects, for which $L/\lambda \leq 1$;
- large-sized objects, for which $L/\lambda \gg 1$.

Hydrometeors (drops of water in the air) and half-wave chaff dipoles used to create radar clutter are typical representatives of the first class of object. This class also includes small elements of the Earth's surface. It is characteristic of these kinds of object to form accumulations occupying a certain volume of space or a certain area of surface. Hence, such objects are known as volume distributed or surface distributed.

The RCS of a volume-distributed object is estimated using the notion of the volume resolution cell ΔW by the formula

$$\sigma_{\mathrm{v}} = \sigma_{0\mathrm{v}}\, \Delta W. \tag{1.32}$$

Where, $\sigma_{0\mathrm{v}}$ is the specific RCS of a volume-distributed object of the kind at issue and is equal to the total RCS of the scatterers in unit volume (usually 1 m^3). This quantity has the dimension of m^2/m^3.

Similarly, the RCS of a surface-distributed object is calculated using the notion of the surface resolution cell ΔS by the formula

$$\sigma_{\mathrm{s}} = \sigma_{0\mathrm{s}} \Delta S. \tag{1.33}$$

Where, $\sigma_{0\mathrm{s}}$ is the specific RCS of a surface-distributed object and is equal to the total RCS of an object of this kind within the bounds of an area of 1 m^2. It has the unit of m^2/m^2, i.e. $\sigma_{0\mathrm{s}}$ is a dimensionless quantity. It is sometimes referred to as the reflectivity of a given type of surface.

Surface-distributed objects represent sources of clutter background, against which the target has to be detected. Therefore, when calculating radar cross-sections, σ_{v} and σ_{s}, the quantities ΔW and ΔS are determined by formulae (1.15) and (1.16).

Analysis of Equations (1.15) and (1.16) permits the following conclusion to be drawn: if a distributed object presents a source of clutter for the radar, it is desirable to improve radar resolution in range and angular position in order to have the object's RCS reduced (which, in turn, leads to a reduction in clutter power). When moving targets are to be detected against a background of distributed objects, improvement of the radar resolution in velocity is also to be sought.

More detailed information on the specific radar cross-sections, $\sigma_{0\mathrm{s}}$ and $\sigma_{0\mathrm{v}}$, can be found in the papers by Nathanson [I.2] and Skolnik [I.3].

The majority of radar targets belong to the second class of objects ($L/\lambda \gg 1$). Let us limit ourselves to the consideration of objects with a smooth, highly conductive surface (airplanes, helicopters, road vehicles, projectiles). For such objects, there is the following connection between radiated field intensity (E_1) and secondary field intensity (E_2) [I.4–I.6]:

$$\frac{E_2}{E_1} = \frac{1}{\lambda R_0} \left| \int_s \exp\left(-\mathrm{j}\frac{4\pi R}{\lambda}\right) \cos\gamma \,\mathrm{d}s \right|. \tag{1.34}$$

This formula reflects the fact that the echo signal field intensity is an integral sum of fields induced by the currents in ds elements of the object's surface taking into account their phase shifts, $\varphi = 4\pi R/\lambda$. The designations in formula (1.34) are explained by Figure 1.23. Here and further on in this section, the case in point is back-reflection towards the radar. This case is valid for a monostatic radar system with a collocated transmitter and receiver.

Let us substitute E_2/E_1 from formula (1.34) into Equation (1.31). It is assumed that $R = R_0 + r$, where R_0 is the distance from the radar to the nearest point of the target's surface. It is also taken into account that

$$\left| \exp\left(-\mathrm{j}\frac{4\pi R_0}{\lambda}\right) \right| = 1.$$

Thus,

$$\sigma = \frac{4\pi}{\lambda^2} \left| \int_s \exp\left(-\mathrm{j}\frac{4\pi r}{\lambda}\right) \cos\gamma \,\mathrm{d}s \right|^2. \tag{1.35}$$

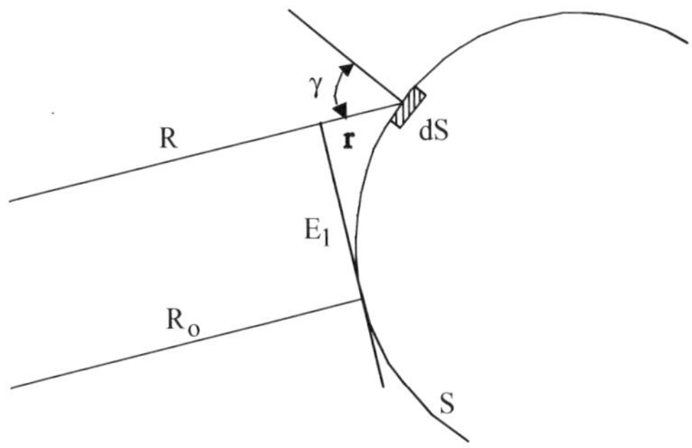

Figure 1.23 Illustration of the RCS of an arbitrary target calculation. $R = R_0 + r$

Using formula (1.35) it is not difficult to calculate the RCS for certain simple objects.

The RCS of a sphere with radius a, for $a/\lambda \gg 1$, is approximately equal to the area of its physical section:

$$\sigma \approx \pi a^2. \tag{1.36}$$

Most of the signal illuminating a smooth surface is scattered in accordance with the specular reflection law. Therefore, the main contribution to the RCS of a sphere comes from the neighbourhood of the so-called bright point (or a dominant scattering point). This is a point where the illumination direction is normal to the surface.

The RCS of a rectangular plate with dimensions a and b, when illumination falls at an angle γ (Figure 1.24), is determined by the formula given below:

$$\sigma = \frac{4\pi S^2}{\lambda^2} \cos^2\gamma \left(\frac{\sin 2\pi a \sin\gamma/\lambda}{2\pi a \sin\gamma/\lambda} \right). \tag{1.37}$$

Where $S = ab$ is the area of the plate. Specifically, when $\gamma = 0$ (the case of normal illumination),

$$\sigma = \frac{4\pi S^2}{\lambda^2} = S\frac{4\pi S}{\lambda^2}. \tag{1.38}$$

The plate behaves like a planar antenna that derives power $\mathrm{PD}_1 S$ from the primary field and re-emits it in the direction of the radar with a gain $G = 4\pi S/\lambda^2$, S being the area of the antenna aperture. Ordinarily, $4\pi S/\lambda^2 \gg 1$. Therefore, under normal illumination conditions, the RCS of a plate is significantly greater than its area. However, when the grazing angle is different from a right angle ($|\gamma| > 0$), there is a sharp decrease in plate RCS. *Corner reflectors* are made of three (or two) mutually perpendicular plates of square or triangular shape. Since the signal bounces against the plates of the reflector two or three times,

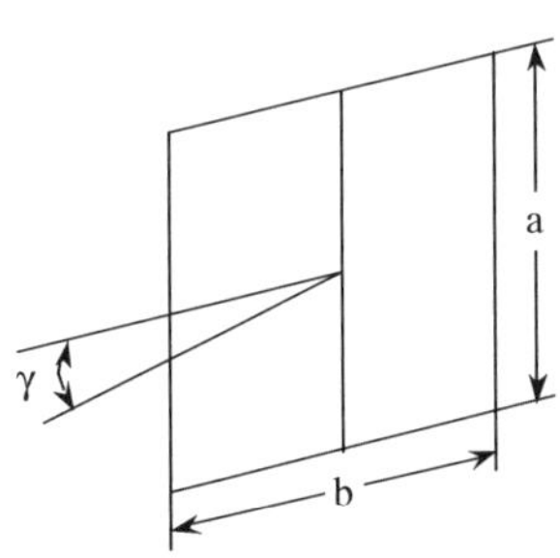

Figure 1.24 Illustration of the RCS of a plate calculation

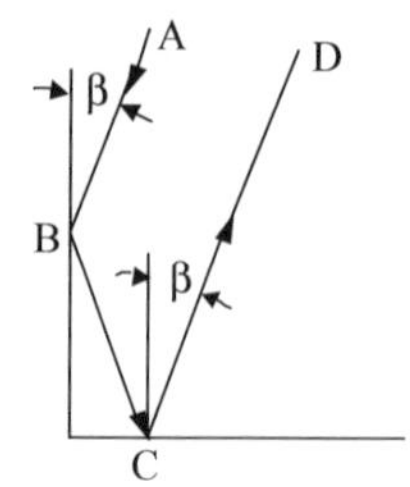

Figure 1.25 Corner reflector

most of its energy returns in the direction of the radar. The process of signal reflection is illustrated by Figure 1.25 for the specific case of twofold reflection.

Examination of this figure allows the following conclusions to be drawn

- Most of the energy is reflected in the same direction as that of the illumination.
- The length of the reflected signal path (ABCD) does not depend on the point of beam incidence on the reflector (point B).

Consequently, the echo signal is a co-phased beam similar to the return from a plate under normal illumination. Therefore, the RCS of an corner reflector remains practically unchanged over a wide angular domain. The RCS of an corner reflector approximates the RCS of one of its planes under normal illumination.

For a model of a complex target in the form of a cluster of bright points formula (1.36) can be generalized to the case of an arbitrary bicurvature surface. If such a surface has one bright point, its RCS can be expressed by the approximation

$$\sigma \approx \pi a_1 a_2.$$

Where a_1 and a_2 are the main (minimum and maximum) radii of the surface curvature at the bright point. The concept of a bright point gives an insight into the scattering of radio waves by objects with complex geometry. An object having two or more bright points on its surface is referred to as a 'complex target'.

The main contribution to the RCS of a complex target is constituted by echoes from bright points, which can be numerous. For instance, when an aircraft is illuminated, the main contribution to its RCS is given by the bright points on the nose or on the pods. Sometimes, there are bright points on the edges of the wing and empennage group. The mathematical expectation of the resultant RCS can be approximately evaluated as the sum of the RCSs of individual bright points:

$$\overline{\sigma} = \sum_{i=1}^{n} \overline{\sigma_i}.$$

Generally, a real object can be represented by a cluster of bright points. Such a representation of the complex target will be used in Sections 2.2 and 2.3 to obtain temporal and spectral characteristics of signals returned by complex targets.

Table 1.1, borrowed from Reference [I.3], gives mean RCS values for selected objects.

1.7 ATMOSPHERIC ATTENUATION OF RF SIGNALS

When a radio signal propagates in the atmosphere, its power flux density decreases due to oxygen and water vapour absorption and scattering by hydrometeors (particles of water). Signal attenuation is

Table 1.1 Mean RCS values for selected objects. (After Skolnik [I.3]. Reproduced by permission)

Object	RCS (m^2)
Cruise missile	0.1
Small-sized single-engine airplane	1
Large-sized fighter aircraft	6
Medium-sized jet airliner	20
Large-sized jet airliner	40
Helicopter	3
Motor vehicle	100
Human	1
Large bird	10^{-2}
Fly	10^{-5}

evaluated in terms of the attenuation coefficient α (expressed in dB/km). Numerically, this factor equals the reduction of signal power (in decibels) on a path of 1 km length.

If the target is at a range of R kilometres from an active radar, then the power of the echo signal is reduced by $\alpha\, 2R$ dB. Attenuation loss per unit is

$$L_{\mathrm{at}} = 10^{-0.2\alpha R}. \tag{1.39}$$

Normally, the radar range, $R_{\max}$, is a preset value. Therefore, for calculating a power-budget parameter of the radar, hardware loss, L_{a}, in Equation (1.26) should be replaced by the product $L_{\mathrm{a}}L_{\mathrm{at}}$. The quantity L_{at} is given by formula (1.39), where $R = R_{\max}$.

In the same fashion, for a semi-active radar system, L_{a} in formula (1.23) should be substituted by the following product:

$$L_{\mathrm{a}}L_{\mathrm{at}} = L_a \times 10^{-0.1\alpha(R_{1\max}+R_{2\max})}.$$

Here, $R_{1\max}$ and $R_{2\max}$ are the preset maximum target ranges for the system's transmitter and receiver.

For a passive radar, the losses are determined by the formula

$$L_{\mathrm{a}}L_{\mathrm{at}} = L_{\mathrm{a}} \times 10^{-0.1\alpha R_{2\max}}.$$

For an active radar with an active response, it is necessary to introduce the following multiplication factor in the numerators of formulae (1.29) and (1.30):

$$L_{\mathrm{at}} = 10^{-0.1\alpha R_{\max}}.$$

The attenuation coefficient, α, depends on the condition of the atmosphere and signal carrier frequency, f_0. Figure 1.26 [I.3] shows the relation $\alpha = \alpha(f_0)$ for a dry atmosphere (free of hydrometeors). Figure 1.27 presents the attenuation coefficient versus signal carrier frequency and rain intensity, which is measured as the amount of precipitation in mm/h. The plots have been borrowed from Reference [I.6].

Atmospheric loss calculations should be performed in view of the following considerations [I.2]. The curves for $\alpha(f_0)$ in a dry atmosphere have been obtained under an atmospheric pressure of 760 mm of mercury. In addition, if the target is at a considerable height, the radio beam path passes in part through the rarefied layers of the atmosphere. This results in lesser atmospheric losses. By way of example, Figure 1.28 [I.1] shows two-way attenuation curves for different values of the target elevation angle ε. From a comparison of Figures 1.26 and 1.28, it is obvious that atmospheric losses are notably less in the latter case.

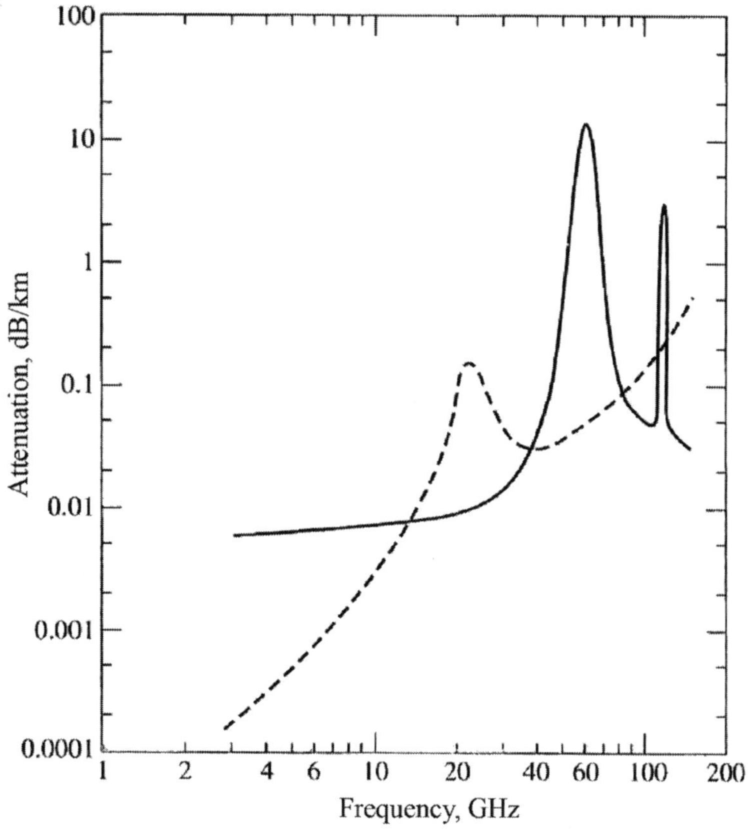

Figure 1.26 Signal attenuation in atmospheric gases: ——oxygen, - - - - -vapour. (After Skolnik [I.3], with permission)

Factors to be taken into account when calculating the attenuation loss in rain are given in Reference [I.2]. Statistical data point towards a low probability of heavy rain (with an intensity exceeding 4–5 mm/h), a small coverage area of heavy rain (the rain zone diameter is < 50 km for rain falling at 5 mm/h) and low rain intensity rates (≤ 4 mm/h) at heights over 4 km.

With all other conditions being equal, atmospheric losses increase as the wavelength of the transmitted signal decreases. Therefore, long-range radars do not typically use centimetric and millimetric wavebands, with the exception of high-altitude radars and space-based radars.

There is yet another observation to be made within this section. When it is necessary to estimate radar range taking account of atmospheric loss, with all the power-budget parameters of the active radar and the target RCS known, this range (designated by $R'_{\max}$) can be calculated using the formula

$$R'_{\max} = R_{\max} \times 10^{-0.05\alpha R'_{\max}}. \tag{1.40}$$

Formula (1.40) is derived from Equation (1.26) by adding the atmospheric loss coefficient $10^{-0.2\alpha R'_{\max}}$ to the numerator of the radicand with subsequent extraction of the fourth root. Formula (1.40) is a transcendental equation for $R'_{\max}$. An abacus is given in Reference [I.7] to simplify the estimation of $R'_{\max}$ when the values of $R_{\max}$ and α are known.

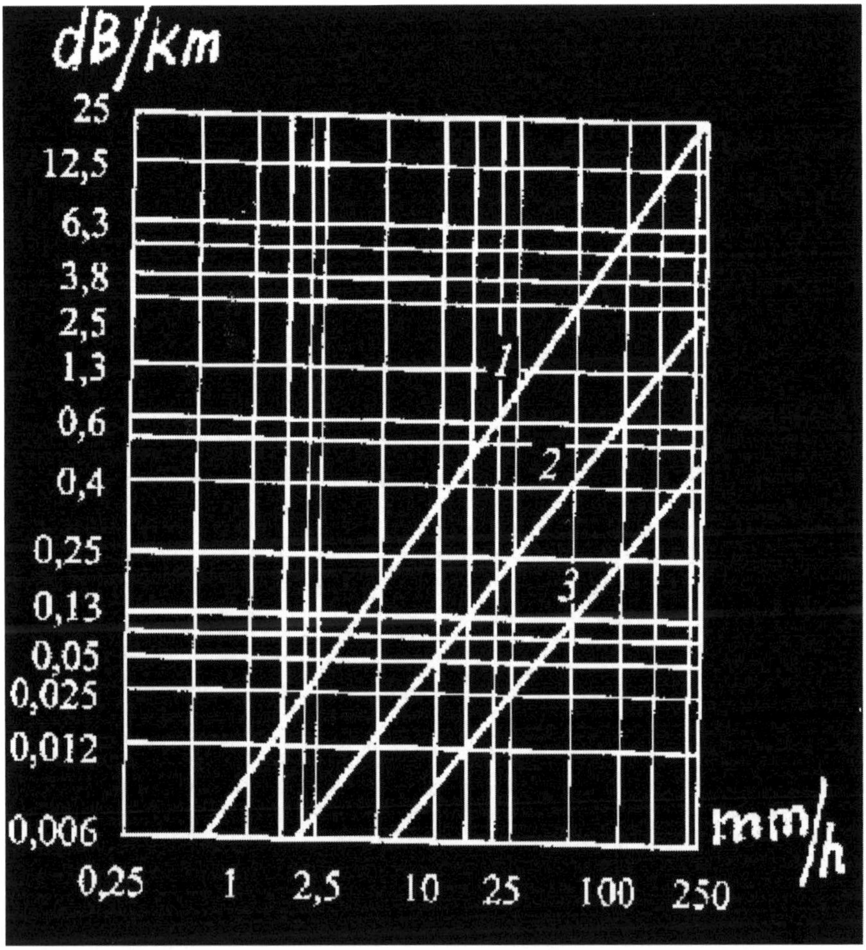

Figure 1.27 Signal attenuation in rain. (After Nathanson [I.2], with permission)

1.8 MAXIMUM RADAR RANGE; LINE-OF-SIGHT LIMITATION OF THE RADAR RANGE; TARGET ELEVATION MEASUREMENT

In Section 1.6, a method was discussed for estimating the radar range with no respect to the Earth's surface. In the current section and the sections to follow, we will also consider the effects of the Earth's surface on the radar range and target elevation measurement accuracy.

The Earth's surface has a dual effect on radar range coverage. With a transmitted signal frequency $f >$ (15–30) MHz ($\lambda <$ 10–20 m), the Earth's surface curvature confines radar range coverage to the line-of-sight range (R_d). The second effect of the Earth's surface relates to the fact that the transmitted signal can travel to the target along two or more paths: it can either follow the direct path 'transmitter to target' or be reflected from the Earth's surface. The power of the resultant signal does not only depend on the quantities that appear in Equations (1.23) to (1.30) but also on the interrelation of amplitudes and phases of the signals in question. The same applies to the echo signal travelling from the target to the receiver.

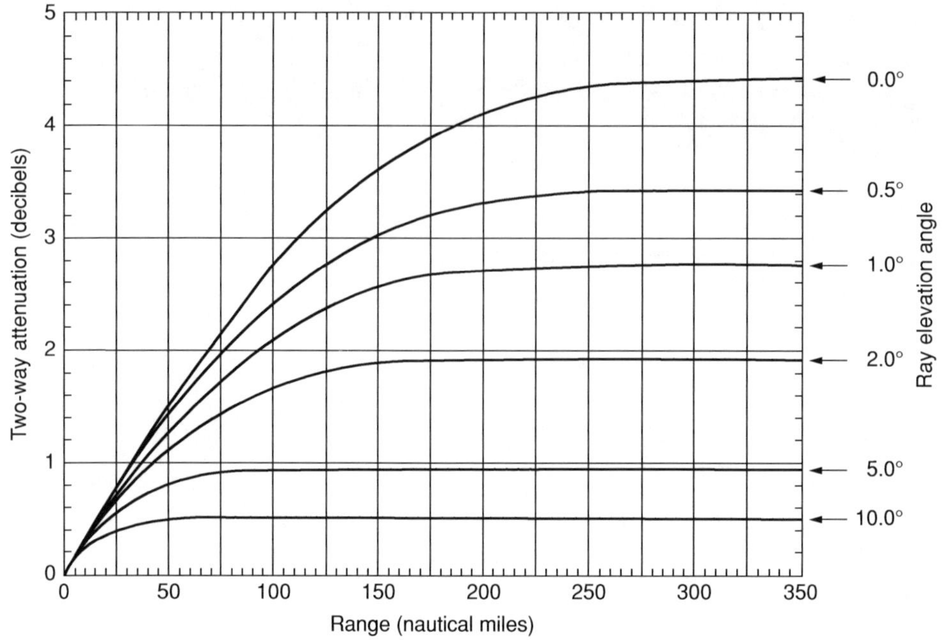

Figure 1.28 Two-way attenuation curves for different values of target elevation angles. (After Blake [I.1], with permission)

The quantity R_d is determined using the radar antenna height (h) and target height (H) above the Earth's surface, as well as the equivalent Earth radius (R_e). It follows from Figure 1.29 that

$$R_d = \sqrt{(R_e + h)^2 - R_e^2} + \sqrt{(R_e + H)^2 - R_e^2} \approx \sqrt{2R_e}\left(\sqrt{h} + \sqrt{H}\right). \tag{1.41}$$

Equation (1.41) was deduced on the assumption that $h \ll R_e$ and $H \ll R_e$. Referring to Reference [I.8] and taking into account the refraction of radio waves gives

$$R_e = \frac{1}{1/R_1 + dn/dH}. \tag{1.42}$$

Where R_1 is the physical radius of the Earth and n is the atmospheric refraction coefficient for the height H. The derivative dn/dH depends on the condition of the atmosphere. For the so-called *standard*

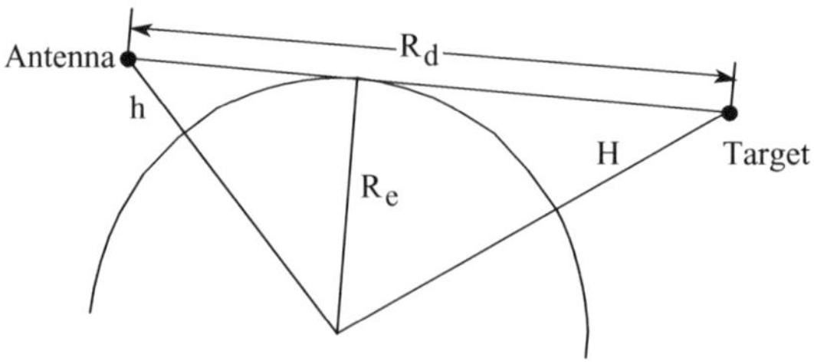

Figure 1.29 Direct visibility calculation

atmosphere,

$$\frac{dn}{dH} \approx -3.8 \times 10^{-5}\ \text{km}^{-1}.$$

Substituting this value of dn/dH into formula (1.42) gives the value of R_e for standard atmosphere: $R_e = 8500$ km.

Hence, the line-of-sight range under standard atmosphere conditions is

$$R_d(\text{km}) = 4.1\left(\sqrt{h(\text{m})} + \sqrt{H(\text{m})}\right). \tag{1.43}$$

The final expression for the radar range is

$$R_{max} = \min\left(R'_{max}, R_d\right).$$

Let us determine the relationship of the target elevation, ε, versus its height, H, and range, R. By definition, this is the angle between the target sight line and the local horizon. Analysis of Figure 1.30 gives

$$\varepsilon \approx \arcsin\frac{H - x}{R} \approx \frac{H - x}{R}.$$

Length x can be found using this approximation:

$$(R_e + x)^2 \approx R_e^{\,2} + R^2.$$

Hence,

$$x \approx \frac{R^2}{2R_e}.$$

Consequently,

$$\varepsilon \approx \frac{H}{R} - \frac{1}{2}\frac{R}{R_e}. \tag{1.44}$$

If the formula (1.44) is adopted to express the ε versus H and R relationship, as a sort of approximation the flat Earth model can be used.

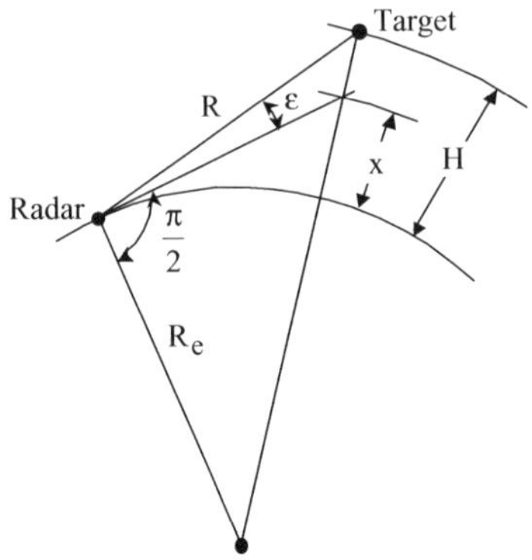

Figure 1.30 Value of the elevation angle calculation

1.9 THE IMPACT OF EARTH SURFACE REFLECTIONS ON THE RADAR RANGE AND ELEVATION MEASUREMENT ACCURACY

To solve the problem formulated in the heading, we will refer to a simplified Earth surface model. Assume that the Earth is flat and smooth. This first assumption is well justified because reflection points (for instance, point S in Figure 1.31) are not placed far from the antenna of the radar. With these assumptions, all surface reflections can be treated as specular.

The transmit–receive antenna is located at point N at a height h, while the target is located at point M at a height H (Figure 1.31). Field intensity of the transmitted signal at point M is equal to the sum of intensities of the direct wave E_1 and the reflected wave E_2. Suppose the target is at the maximum of the antenna beam. Then, in view of the designations in Figure 1.31, the amplitudes of those intensities can be expressed thus:

$$E_1 = E_{\mathrm{m}} F(0) = E_{\mathrm{m}};$$
$$E_2 = p E_{\mathrm{m}} F(\varepsilon + \alpha).$$

Where p is the modulus of the wave reflection coefficient and $F(\cdot)$ is the normalized beamwidth. The phase shift, Ψ, between the component fields E_1 and E_2 can be expressed as the sum:

$$\Psi = \Psi_1 + \Psi_2.$$

The term Ψ_1 is the phase shift due to reflection by the Earth's surface while Ψ_2 denotes the phase shift due to the difference of path lengths of the direct wave and the reflected wave. As seen from Figure 1.31,

$$\Psi_2 = \frac{2\pi}{\lambda} 2h \sin\varepsilon = \frac{4\pi h \sin\varepsilon}{\lambda}. \tag{1.45}$$

Hence, the resultant intensity E is

$$\begin{aligned} E &= \sqrt{E_1^2 + E_2^2 + 2E_1E_2\cos\Psi} \\ &= E_1\sqrt{1 + p^2F^2(\varepsilon+\alpha) + 2pF(\varepsilon+\alpha)\cos\Psi} \\ &= E_1\,\mathrm{PPF}(\varepsilon+\alpha). \end{aligned}$$

Where, PPF$(\varepsilon + \alpha)$ is the pattern propagation factor [I.1] or interference factor.

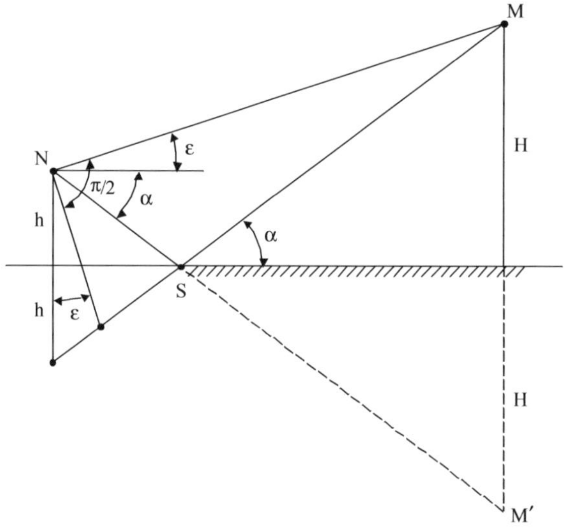

Figure 1.31 Phase difference of E_1 and E_2 estimations

The power flux density is proportional to the square of field intensity. Therefore, the ratio of transmitted signal power flux densities at point M with and without regard to reflections from the Earth's surface (PD'_1 and PD_1 respectively) takes the form

$$PD'_1 = PD_1\, PPF^2(\varepsilon + \alpha).$$

Similar phenomena occur when the target echo travels towards the receiving antenna. Therefore, the ratio of echo signal power flux densities at the radar location with and without regard to the influence of the Earth's surface (PD'_2 and PD_2) takes the form

$$PD'_2 = PD_2\, PPF^4(\varepsilon + \alpha).$$

Consequently, the connection between the radar range in view of reflections from the Earth's surface, R''_{max}, and radar range in free space, R_{max}, is given by the expression

$$\begin{aligned} R''_{max} &= R_{max} PPF(\varepsilon + \alpha) \\ &= R_{max}\sqrt{1 + p^2F^2(\varepsilon + \alpha) + 2pF(\varepsilon + \alpha)\cos(\Psi_1 + \Psi_2)}. \end{aligned} \tag{1.46}$$

The reflection coefficient modulus p and phase shift Ψ_1 depend on the properties of the Earth's surface at the reflection point, the grazing angle α, the signal frequency and the incident wave polarization.

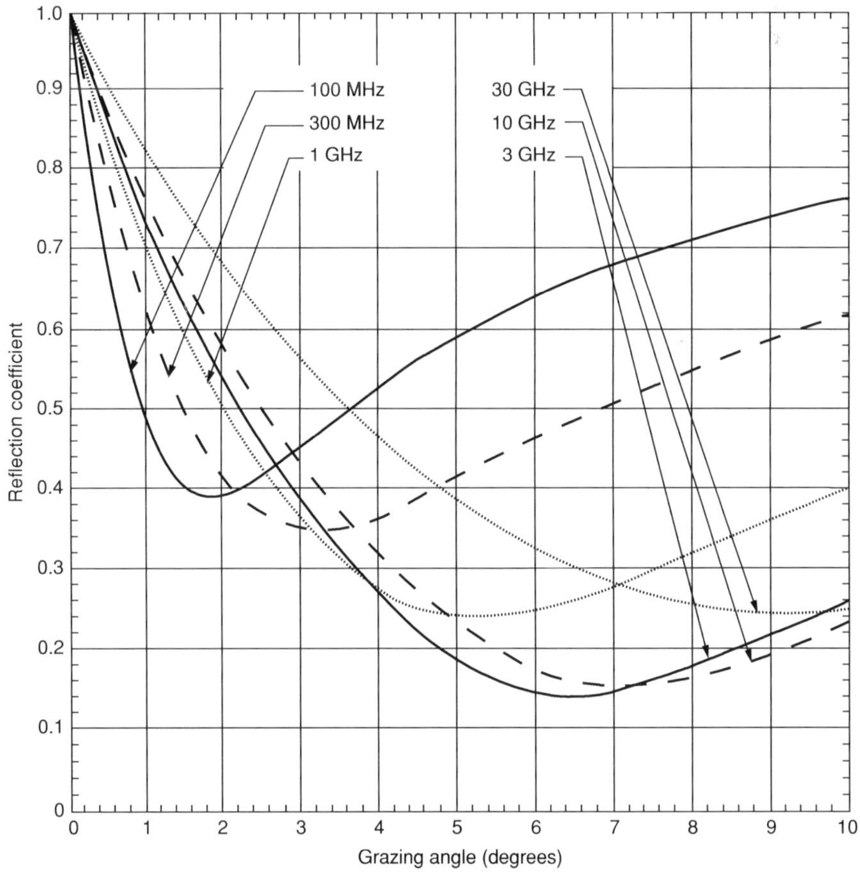

Figure 1.32 The behaviour of the reflection coefficient in the function of grazing angle and frequency. (After Blake [I.1], with permission)

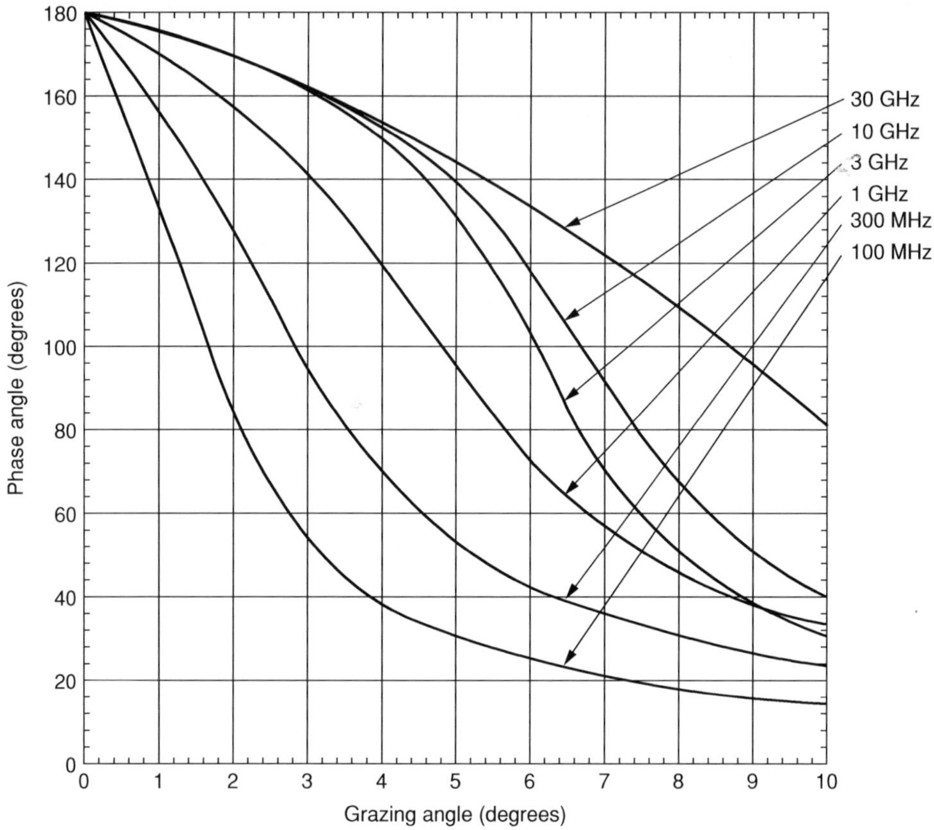

Figure 1.33 The behaviour of phase angle Ψ_1 in the function of grazing angle and frequency. (After Blake [I.1], with permission)

For a horizontal polarization, $p \approx 1$ and $\Psi_1 \approx \pi$ over a wide range of frequencies and grazing angles.

The behaviour of p and Ψ_1 in the function of the grazing angle, α, and frequency for vertical polarization is illustrated in Figures 1.32 and 1.33 borrowed from Reference [I.1]. What is implied here is RF signal propagation over a smooth water surface.

Analysis of formula (1.46) allows us to conclude that R''_{max} oscillates with low values of the angle ε. To make sure of this, it is assumed that

$$\alpha \approx \varepsilon.$$

When the value of the angle α is small, $\Psi_1 \approx \pi$ for both kinds of signal polarization. Furthermore, when $2\varepsilon < \theta/2$, $F(2\varepsilon) \approx 1$, where θ is the antenna beamwidth. In this case,

$$R''_{max} \approx R_{max}\sqrt{1 + p^2 - 2p\cos\Psi_2}. \tag{1.47}$$

It follows from formula (1.47) that, when Ψ_2 varies, the variation of R''_{max} lies within the limits $R_{max}(1 - p)$ and $R_{max}(1 + p)$. Specifically, when $\varepsilon = 0$, then $\Psi_2 = 0$, $p = 1$ and $R''_{max} = 0$. When $\varepsilon = \lambda/(4h)$, then $\Psi_2 = \pi$ and $R''_{max} \approx R_{max}$. When $\varepsilon \geq \theta/2$, then $F(\varepsilon + \alpha) \ll 1$ and $R''_{max} \approx R_{max}$.

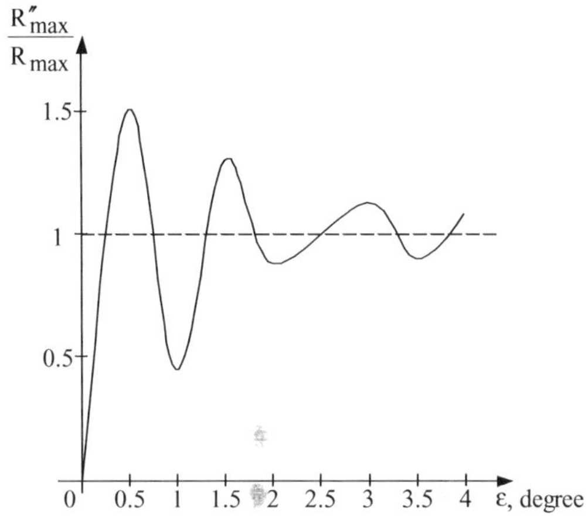

Figure 1.34 Function $R''_{\max}/R_{\max} = f(\varepsilon)$

Figure 1.34 illustrates the plot of $R''_{\max}(\varepsilon)$ for $\theta = 5^\circ$, a frequency of 3 GHz, and vertical polarization of the transmitted signal.

1.9.1 Target Elevation Measurement Errors

Returning to Figure 1.31, let us examine the dashed line connecting the target antipode (point M′) with the reflection point S. This line is an extension of the segment NS. Consequently, the signal reflected from the Earth's surface can be replaced by a new signal coming from the antipode. Thus, the presence of reflections from the Earth's surface brings about a second object (in addition to the target) – the antipode. If the antipode is within the bounds of the antenna beam in the vertical plane, it is not the target elevation that the angle tracker will measure but the elevation of a certain radar centre of a pair of objects, namely the target and the antipode. This results in a target elevation measurement error. The subject will be discussed in more detail in Chapter 4.

2

Radar Signals and Signal Processing

Radar signals are classified by several criteria. We distinguish between pulsed and continuous signals, coherent and noncoherent signals, signals with carrier frequency and baseband signals, and simple and complicated signals (including so-called pulse compressed signals).

Signal processing (filtering) devices are also classified by a number of criteria. These include: coherent (pre-detection) and noncoherent (post-detection) processors, optimum and matched filters and correlators, time-domain and spectral-domain filters, devices for processing signals with carrier frequency and baseband signal processors.

These matters, as well as those related to them, will be discussed in this chapter.

2.1 COHERENT AND NONCOHERENT SIGNAL SEQUENCES

Radar transmission often takes the form of a periodic sequence of RF pulses shaped as segments of a sinusoid. When the radar operates in the search mode, the signal returned by the target has the structure of a pulse burst.

A signal is called a coherent waveform (or a train of radio pulses or a burst of pulses) if its phase is a determinate function of time. By way of example, let us take a train of radio pulses made up of 'slices' of a continuous sinusoid (Figure 1.5). Given the initial phase, φ_0, in the first pulse, the initial phase in the ith pulse can be expressed as a determinate function of discrete time $(i-1)T_p$:

$$\varphi_i = \varphi_0 + (i-1)T_p\omega_0,$$

where ω_0 is the carrier in circular frequency units.

This kind of radar output is implemented in the radar represented by the block diagram in Figure 1.4. The 'slicing' of the sinusoidal voltage of the driving generator is carried out in the modulator. The waveform obtained in this fashion is referred to as a coherent signal of the first type of coherence. The second type of coherence implies that the train of pulses has the same initial phase in each pulse. A train, or burst, of RF pulses is called a noncoherent waveform if the initial phase of the carrier in each pulse is a random value. Separately, each pulse is coherent.

Bistatic Radar: Principles and Practice Edited by M. Cherniakov

There are cases of partial coherence when, in a burst of N pulses, the neighbouring m pulses are rather strongly correlated. Such pulse trains can be regarded as consisting of N/m noncoherent parts. Within them, each part composed of m pulses is coherent.

To compare the characteristics of coherent and noncoherent pulsed signals, let us analyse their spectra. To begin with, take a coherent burst containing N pulses. Let $\overline{S_1(\omega)}$ denote the complex spectrum of the first pulse. Here and hereafter, the bar symbol is used to denote complex quantities and complex functions. The time intervals between the first and the ith pulses of the burst are multiples of the pulse repetition period T_p. The complex spectrum of the ith pulse delayed from the first pulse by $(i-1)T_p$ is given by [I.9]

$$\overline{S_i(\omega)} = \overline{S_1(\omega)}\,\exp\left[-j\omega(i-1)T_p\right]. \tag{2.1}$$

The coherent burst spectrum $\overline{S(\omega)}$ is equal to the spectra sum of all the N pulses.

Hence,

$$\overline{S(\omega)} = \sum_{i=1}^{N} \overline{S_1(\omega)}\,\exp\left[-j\omega(i-1)T_p\right] = \overline{S_1(\omega)} \sum_{i=1}^{N} \exp\left[-j\omega(i-1)T_p\right]. \tag{2.2}$$

The second term on the right side of Equation (2.2) is the sum of a geometric progression with a common ratio $\exp(-j\omega T_p)$. Therefore,

$$\sum_{i=1}^{N} \exp\left[-j\omega(i-1)T_p\right] = \frac{1-\exp(-j\omega N T_p)}{1-\exp(-j\omega T_p)}. \tag{2.3}$$

Thus, the complex spectrum of the pulse burst is

$$\overline{S(\omega)} = \overline{S_1(\omega)}\frac{1-\exp(-j\omega N T_p)}{1-\exp(-j\omega T_p)}.$$

The modulus of this expression, i.e. the amplitude spectrum of the burst, is equal to the product of the terms modulo:

$$S(\omega) = \left|\overline{S(\omega)}\right| = S_1(\omega)\frac{\left|\sin(\omega N T_p/2)\right|}{\left|\sin(\omega T_p/2)\right|}. \tag{2.4}$$

Translating the circular frequency in Equation (2.4) to the frequency in hertz gives

$$S(2\pi f) = S_1(2\pi f)\left|\frac{\sin(\pi f N T_p)}{\sin(\pi f T_p)}\right|. \tag{2.5}$$

It follows from Equation (1.11) that the normalized spectrum of a single pulse on the positive semi-axis of frequency is

$$S_1(2\pi f) = \frac{\sin\left[\pi(f-f_c)\,\tau_p\right]}{\pi(f-f_c)\,\tau_p}, \tag{2.6}$$

where f_c is the carrier frequency taking account of the Doppler shift, i.e.

$$f_c = f_0 + f_d.$$

Substituting Equation (2.6) into Equation (2.5) and noting that the frequency f possesses the same value in both terms of Equation (2.5), the following expression is obtained for the amplitude spectrum of an echo signal burst:

$$S(f) = \frac{\sin\left[\pi(f-f_c)\,\tau_p\right]}{\pi(f-f_c)\,\tau_p}\left|\frac{\sin\pi(f-f_c)N T_p}{\sin\pi(f-f_c)T_p}\right|. \tag{2.7}$$

The first term in Equation (2.7) is a ratio of the form $\sin x/x$. The second term is a ratio of the form $\sin Nx/\sin x$. Both these ratios are maximal when $x \to 0$. Furthermore, $\lim \sin x/x|_{x\to 0} = 1$ and $\lim \sin Nx/\sin x|_{x\to 0} = N$.

Consequently, the maximum of the spectrum for a coherent pulse burst equals N and takes place at $f = f_c$. As the absolute value of the difference $f - f_c$ increases, both terms in Equation (2.7) decrease, with the second term decreasing much faster than the first term. The first term reaches zero for the first time when

$$|f - f_c| = \frac{1}{\tau_p}. \tag{2.8}$$

The second term reaches zero for the first time when

$$|f - f_c| = \frac{1}{NT_p}, \tag{2.9}$$

i.e. much earlier than the first term. Therefore, the spectrum bandwidth of a coherent burst is determined by the second term in Equation (2.7). With that, the null-to-null bandwidth is

$$\Delta f_0 \approx \frac{2}{NT_p}, \tag{2.10}$$

and the spectrum bandwidth referred to the level of $0.5N$ is

$$\Delta f_{0.5} \approx \frac{1.2}{NT_p}. \tag{2.11}$$

In formula (2.7), the $\sin x/x$ (first) term describes the spectrum of a radio pulse with duration τ_p. The $\sin Nx/\sin x$ (second) term can be called the multiplier of the pulse burst. The spectrum of a pulse burst on the positive semi-axis of frequency is illustrated in Figure 2.1. As seen from the graph, the nulls of the burst spectrum correspond to frequencies given by $f = f_c + k_1/(NT_p)$, where $k_1 = 1, 2, 3, \ldots$. The peaks of the spectrum are at frequencies given by

$$f = f_c + \frac{k_2}{T_p},$$

where $k_2 = 0, 1, 2, 3, \ldots$. The peak value decreases in accordance with the values of the first term in Equation (2.7).

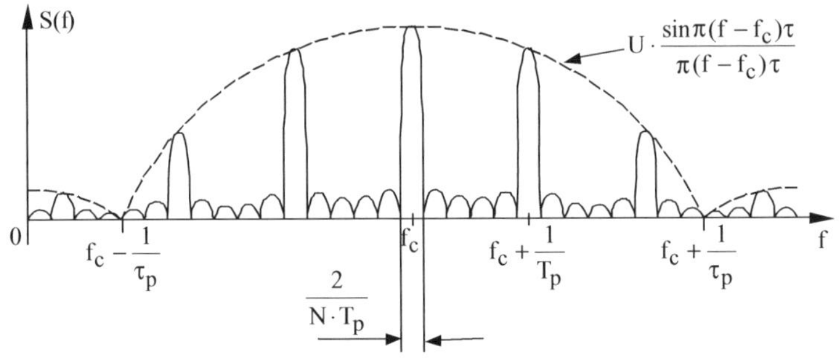

Figure 2.1 Amplitude spectrum of a coherent burst

If the frequency difference between two signals

$$\delta f < \frac{1}{T_p},$$

then the radar resolution in frequency is equal to the half-width of a coherent burst spectrum:

$$\Delta f = \frac{1}{NT_p}, \tag{2.12}$$

or the reciprocal of burst duration. Hence, the resolution in the target radial velocity is

$$\Delta V = \frac{\lambda}{2}\frac{1}{NT_p}.$$

We will now go on to determine the spectrum of a noncoherent burst of N pulses. Let us write the expression for the ith pulse in the form

$$a_i U(t)\cos(\omega_0 t - \varphi_i),$$

where a_i is the random (in the general case) amplitude, $U(t)$ is the rectangular envelope of a pulse with a duration τ_p, ω_0 is the carrier frequency and φ_i is the random initial phase. The sum of N pulses of the burst is

$$\begin{aligned} U_\Sigma &= \sum_{i=1}^{N} a_i U(t)\cos(\omega_0 t - \varphi_i) = U(t)\sum_{i=1}^{N} a_i\left(\cos\varphi_i \cos\omega_0 t + \sin\varphi_i \sin\omega_0 t\right) \\ &= U(t)\cos\omega_0 t \sum_{i=1}^{N} a_i \cos\varphi_i + U(t)\sin\omega_0 t \sum_{i=1}^{N} a_i \sin\varphi_i. \end{aligned} \tag{2.13}$$

Adopting the symbols

$$A = \sum_{i=1}^{N} a_i \cos\varphi_i \quad \text{and} \quad B = \sum_{i=1}^{N} a_i \sin\varphi_i,$$

expression (2.13) can be rewritten as

$$U_\Sigma = AU(t)\cos\omega_0 t + BU(t)\sin\omega_0 t.$$

Using the well-known trigonometric transformation, this can be changed further to

$$U_\Sigma = \sqrt{A^2 + B^2}\, U(t)\cos(\omega_0 t + \Psi), \tag{2.14}$$

where $\Psi = \mathrm{arctg}\, B/A$.

It follows from Equation (2.14) that the sum of pulses in a noncoherent burst can be likened to a single RF pulse of the same duration, τ_p, and the same carrier frequency, ω_0, as the pulses of the burst. Consequently, the normalized spectrum of the burst is given by formula (2.6). According to Equation (2.8), the half-width of the null-to-null spectrum is equal to $1/\tau_p$. Therefore, the same quantity can be used to estimate the frequency resolution for a noncoherent signal.

Comparison of the frequency resolutions of a coherent and a noncoherent pulse burst leads to the following conclusions:

- In the first case, the resolution is NT_p/τ_p times higher, where N stands for the number of pulses in the burst.
- The frequency resolution equals the reciprocal of the coherent interval, t_c, of the signal used. Indeed, $t_c = NT_p$ and $\Delta f \approx 1/(NT_p)$ for a coherent burst. For a noncoherent burst, $t_c = \tau_p$ and $\Delta f \approx 1/\tau_p$.
- For a partially coherent burst, $t_c = mT_p$, and it can be taken for granted that the frequency resolution $\Delta f = 1/\left(mT_p\right)$.

Figure 2.2 Block diagram of a noncoherent transmitter

As will be shown in Section 4.4, the error in the frequency estimate as measured against a noise background is inversely proportional to the length of the interval, t_c. In this regard also, it is preferable to use coherent signals.

Lastly, it is demonstrated in Section 3.3 that a coherent burst of pulses is more detectable. It was noted in Section 1.3 that the detection circuitry can be represented as a cascade connection of a signal integrator and a threshold detector. It will be shown later that it is more practical to accumulate a coherent signal than a noncoherent signal.

Under certain circumstances, however, the benefits of coherent waveforms are impossible or problematical to realize. One circumstance bringing about a noncoherent burst of echo signal in response to a coherent transmission is the scattering of radio waves by moving targets (see Section 3.4). Another such circumstance manifests itself when there are stringent requirements for weight, size, power consumption and cost of hardware, which compel the designer to employ relatively simple noncoherent waveform shapers and processors. Shown in Figure 2.2 is the block diagram of a transmitter with a noncoherent output. Modulator pulses constitute the driving voltage for a powerful carrier generator.

When a pulse is supplied to the generator (e.g. in the form of a magnetron), it drives the latter to produce a powerful RF pulse. Since the generator's self-excitation processes are independent, the initial phase of oscillations varies randomly from pulse to pulse, resulting in a noncoherent output.

A coherent transmitter output is represented in the block diagram of Figure 1.4. Here, the modulator functions as a switch, the output pulses taking the form of segments of a sinusoid. Such a waveform is naturally a coherent one. This coherent pulse-burst generator is structurally more complicated than the one represented in Figure 2.2. The same is true of coherent processing devices as compared to noncoherent waveform processors.

A comparison of coherent and noncoherent echo signal processors is given in Section 3.2.

2.2 OPTIMUM AND MATCHED FILTERS

The detection of a target echo and the evaluation of its parameters (target coordinates) are carried out against a background of noise. The detection probability characteristics, D and F, and target position accuracy improve as the signal-to-noise ratio increases at the output of the processing circuitry. Therefore, it is desirable to employ processing of such a structure that will allow the signal-to-noise ratio to be maximized. Such a processing device will be referred to as the optimum filter.

In most cases, interference, including the receiver's thermal noise, may be regarded as a random Gaussian process with a certain spectral power density $S_n(\omega)$. Granting this, the optimum filter belongs to the class of linear filters [I.10].

Let us find the complex frequency response of the optimum filter, $\overline{K(\omega)}$, by the criterion of the maximum signal-to-noise ratio at the output. The signal spectrum at the output of the filter is

$$\overline{S_{\text{out}}(\omega)} = \overline{S(\omega)}\,\overline{K(\omega)}.$$

The current value of the signal's output voltage is defined as a Fourier transform of its spectrum:

$$U_{\text{out}}(t) = \frac{1}{2\pi}\int_{-\infty}^{+\infty} \overline{S(\omega)}\,\overline{K(\omega)}\,\exp(\mathrm{j}\omega t)\,\mathrm{d}\omega.$$

At a certain instant of time, t_0, the output signal reaches its maximum, which is given by the equation

$$U_{\max} = U_{\text{out}}(t_0) = \frac{1}{2\pi} \int_{-\infty}^{+\infty} \overline{S(\omega)}\, \overline{K(\omega)} \exp(\mathrm{j}\omega t_0)\, \mathrm{d}\omega.$$

The noise spectral density at the output is

$$S_{\text{n.out}}(\omega) = S_{\text{n}}(\omega)\, K^2(\omega),$$

where $K(\omega) = \left|\overline{K(\omega)}\right|$ is the amplitude–frequency response of the filter.

The noise power at the output is

$$P_{\text{n.out}} = \frac{1}{2\pi} \int_{-\infty}^{+\infty} S_{\text{i}}(\omega)\, K^2(\omega)\, \mathrm{d}\omega.$$

At the instant t_0, the signal-to-noise power ratio at the output is

$$\frac{P_{\text{s.out}}}{P_{\text{n.out}}} = \frac{1}{2\pi} \frac{\left| \int_{-\infty}^{+\infty} \left(\overline{S(\omega)}/\sqrt{S_{\text{n}}(\omega)}\right) \sqrt{S_{\text{n}}(\omega)}\, \overline{K(\omega)} \exp(\mathrm{j}\omega t_0)\, \mathrm{d}\omega \right|^2}{\int_{-\infty}^{+\infty} S_{\text{n}}(\omega)\, K^2(\omega)\, \mathrm{d}\omega}. \tag{2.15}$$

Making use of the Cauchy–Schwarz inequality [I.11], gives

$$\left| \int_a^b \overline{Y(x)}\, \overline{Z(x)}\, \mathrm{d}x \right|^2 \leq \int_a^b \left|\overline{Y(x)}\right|^2 \mathrm{d}x \int_a^b \left|\overline{Z(x)}\right|^2 \mathrm{d}x \tag{2.16}$$

where $\overline{Y(x)}$ and $\overline{Z(x)}$ are arbitrary complex functions. It is not difficult to prove that the inequality (2.16) turns into an equality only subject to

$$\overline{Z(x)} = A\overline{Y(x)}^*, \tag{2.17}$$

where $A =$ constant.

To apply the inequality (2.16) to the numerator of expression (2.15), we will adopt the following designations:

$$\frac{\overline{S(\omega)}}{\sqrt{S_{\text{n}}(\omega)}} = \overline{Y(\omega)} \quad \text{and} \quad \sqrt{S_{\text{n}}(\omega)}\, \overline{K(\omega)} \exp(\mathrm{j}\omega t_0) = \overline{Z(\omega)} \tag{2.18}$$

Taking into account that $|\exp(\mathrm{j}\omega t_0)| = 1$, the expression (2.15) can be written in the form

$$\frac{P_{\text{s.out}}}{P_{\text{n.out}}} \leq \frac{1}{2\pi} \frac{\int_{-\infty}^{+\infty} \left|\overline{S(\omega)}\right|^2 / S_{\text{n}}(\omega) \mathrm{d}\omega \int_{-\infty}^{+\infty} S_{\text{n}}(\omega)\, K^2(\omega)\, \mathrm{d}\omega}{\int_{-\infty}^{+\infty} S_{\text{n}}(\omega)\, K^2(\omega)\, \mathrm{d}\omega}. \tag{2.19}$$

Cancellation of identical integrals on the right side of expression (2.19) simplifies its form to

$$\frac{P_{\text{s.out}}}{P_{\text{n.out}}} \leq \frac{1}{2\pi} \int_{-\infty}^{+\infty} \frac{\left|\overline{S(\omega)}\right|^2}{S_{\text{n}}(\omega)} \mathrm{d}\omega. \tag{2.20}$$

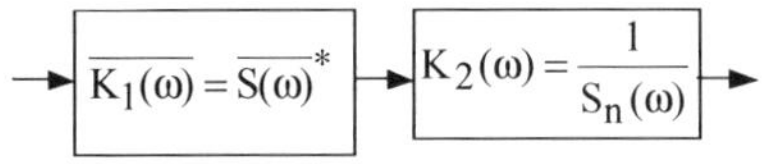

Figure 2.3 Block diagram of an optimum filter

Consequently, the ratio $P_{s.out}/P_{n.out}$ is at a maximum when the inequality (2.20) becomes equality. According to Equations (2.17) and (2.18), this occurs when

$$\sqrt{S_n(\omega)}\,\overline{K(\omega)}\,\exp(j\omega t_0) = \frac{A\,\overline{S(\omega)}^*}{\sqrt{S_n(\omega)}}.$$

Hence, the sought-for response of the optimum filter is

$$\overline{K(\omega)} = \frac{A\,\overline{S(\omega)}^*}{S_n(\omega)}\exp(-j\omega t_0). \tag{2.21}$$

Since we are interested to know the shape of the frequency response, the multiplier A is omitted. The term $\exp(-j\omega t_0)$ denoting the allowable time delay in the filter does not define the structure of the filter. Essentially, Equation (2.21) indicates that an optimum filter can be represented by a cascade connection of two elements with frequency responses $\overline{S(\omega)}^*$ and $1/S_n(\omega)$ (Figure 2.3).

When an optimum filter is used, the lenient inequality (2.17) turns into an equality, the latter defining the maximum signal-to-noise ratio at the output of the filter:

$$\frac{P_{s.out}}{P_{n.out}} = \frac{1}{2\pi}\int_{-\infty}^{+\infty}\frac{\left|\overline{S(\omega)}\right|^2}{S_n(\omega)}d\omega. \tag{2.22}$$

Let us examine the frequent case when noise has a constant spectrum density (e.g. white noise):

$$S_n(\omega) = \frac{N_0}{2}.$$

Substituting this expression into formula (2.21) gives the frequency response of the filter:

$$\overline{K(\omega)} = \frac{A\,\overline{S(\omega)}^* \times 2}{N_0}\exp(-j\omega t_0).$$

Cancelling out the constant factors, $2A$ and N_0, gives

$$\overline{K(\omega)} = \overline{S(\omega)}^*\exp(-j\omega t_0). \tag{2.23}$$

A filter with a frequency response given by Equation (2.23) is referred to as matched with a signal that has a spectrum $\overline{S(\omega)}$. Substituting

$$S_n(\omega) = \frac{N_0}{2}$$

into Equation (2.22), the expression is obtained for the signal-to-noise ratio at the output of the matched filter:

$$\frac{P_{s.out.m}}{P_{n.out}} = \frac{1}{2\pi}\int_{-\infty}^{+\infty}\frac{\left|\overline{S(\omega)}\right|^2}{N_0/2}d\omega = \frac{2E}{N_0}. \tag{2.24}$$

In Equation (2.24), the signal energy is

$$E = \frac{1}{2\pi}\int_{-\infty}^{\infty} \left|\overline{S(\omega)}\right|^2 \mathrm{d}\omega.$$

Therefore, when it is white noise that interferes with the signal, the signal-to-noise ratio at the output of the matched filter does not depend on the signal waveform. According to Equation (2.24), this ratio is only governed by the signal energy and noise spectral density.

Incidentally, for a matched filter to be an optimum filter, the noise spectral density is not required to be constant on the entire frequency axis. The only condition to be satisfied is that $S_n = N_0/2$ over a range that somewhat exceeds the spectral bandwidth of the signal. Inconstancy of the noise spectral components outside the signal spectrum will not have any significant bearing on the ratio, $P_{s.out.}/P_{n.out}$, owing to a decrease in the numerator in the integrand of Equation (2.22).

Let us define the pulse response for a filter matched with a signal, $u(t)$, that can generally be a complex signal. For any linear filter, the pulse response, $h(t)$, is equal to the Fourier transform of its frequency response:

$$h(t) = \frac{1}{2\pi}\int_{-\infty}^{\infty} \overline{K(\omega)}\ \exp(\mathrm{j}\omega t)\,\mathrm{d}\omega.$$

Substituting expression (2.23) into this relation, gives

$$\begin{aligned} h(t) &= \frac{1}{2\pi}\int_{-\infty}^{\infty} \overline{S(\omega)}^{*} \exp[-\mathrm{j}\omega(t_0 - t)]\,\mathrm{d}\omega \\ &= \frac{1}{2\pi}\left\{\int_{-\infty}^{\infty} \overline{S(\omega)} \exp[\mathrm{j}\omega(t_0 - t)]\,\mathrm{d}\omega\right\}^{*} = \overline{u(t_0 - t)}^{*}. \end{aligned} \tag{2.25}$$

Thus, in general, the pulse response of a matched filter is a complex conjugate of the signal's mirror image on the axis of time.

If the signal, $u(t)$, is a real-valued function, then

$$h(t) = u(t_0 - t) \qquad \text{at} \qquad t_0 - t \geq 0. \tag{2.26}$$

In other words, $h(t)$ is equal to the mirror image of the signal. Designating the signal duration by t_p gives

$$t_0 - t_p \geq 0. \tag{2.27}$$

Relation (2.26) is illustrated by Figure 2.4. It follows from Figure 2.4, that the physical meaning of (2.27) is that $h(t) > 0$ only when the values of the argument are not negative. This implies that in calculations any $t_0 \geq t_p$ can be chosen.

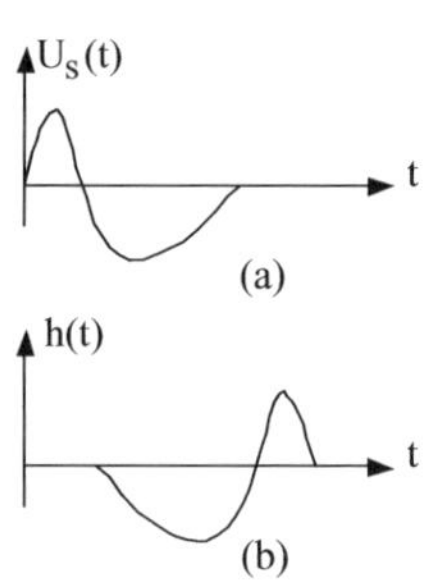

Figure 2.4 (a) Signal and (b) pulse response characteristic of the matched filter

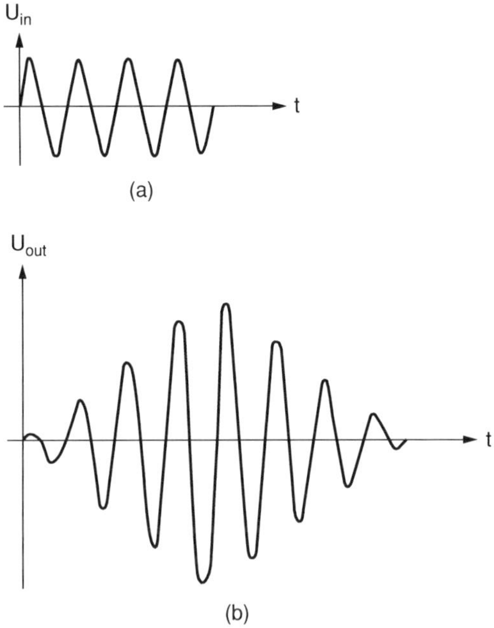

Figure 2.5 (a) Input and (b) output waveforms of a matched filter

Now let us derive the expression for the output voltage of a filter matched with a signal, $u_s(t)$, with an arbitrary voltage, $u(t)$, fed to its input. The output voltage of any linear circuit is equal to the convolution product of the input voltage and the circuit's pulse response:

$$u_{out}(t) = \int_0^t u(x)h(t-x)\,dx. \tag{2.28}$$

In this case, assuming that $u_s(t)$ is a real-valued function and referring to Equation (2.26) and (2.28), we obtain

$$u_{out}(t) = \int_0^t u(x)u_s(t_0 - t + x)\,dx. \tag{2.29}$$

As an example, consider a filter matched with an RF pulse, $u_s(t)$, having the form of a sinusoid segment of a duration, τ_p (see Figure 2.5 (a)). The output voltage, $u_{out}(t)$, calculated by formula (2.29) is represented in Figure 2.5 (b). It is assumed here that $t_0 = \tau_p$.

Judging by the form of the RF pulse spectrum (1.11), it is rather difficult to implement a filter accurately matched with this signal. Therefore, a quasi-optimal filter is frequently used in practice, whose frequency response roughly follows the shape of the main lobe of the signal spectrum. One case of such a filter is the bandpass filter (for instance, a resonant circuit) with a pass bandwidth

$$\Delta f \approx \frac{1}{\tau_p}.$$

The amplitude–frequency response of such a filter is shown by a dashed line in Figure 2.6. The signal-to-noise ratio at the output of the bandpass filter is approximately 1 dB less than that for the matched filter.

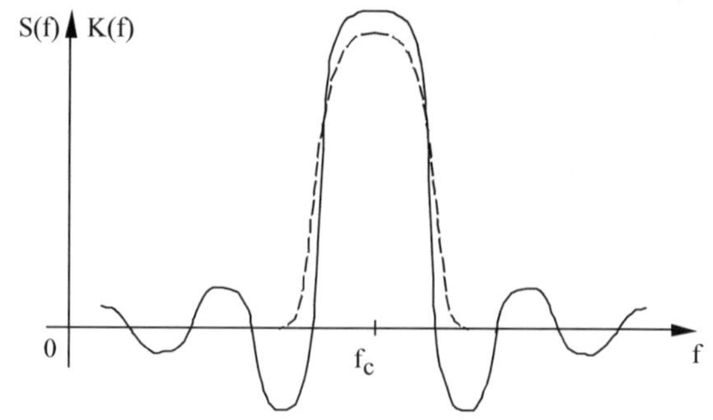

Figure 2.6 Signal spectrum (solid curve) and frequency response of a quasi-optimal filter (dashed curve)

Let us highlight three features here. First, the matched filter integrates the input signal. With the input signal amplitude being constant, the output voltage amplitude rises linearly as a function of time. Second, the peak of the output voltage falls on the final point of the input pulse. The physical meaning is obvious: when $u_s(t)$ ends, signal accumulation stops. The fall of $u_{out}(t)$ is not instantaneous because of a response lag in the filter. Third, as follows from formula (2.24), the peak output voltage is proportional to the signal energy.

Note two other characteristic properties of the optimum filter and the matched filter:

- Both these filters are invariant with respect to the signal time of arrival.
- The filters in question with such characteristics as obtained above are realizable for coherent signals only. Noncoherent signals are not described by determinate functions of time, which makes it impossible to evaluate (and to implement) the required pulse response. The structure of an optimum processor for a noncoherent pulse burst is discussed in Chapter 3.

2.3 TRANSVERSAL MATCHED FILTER

Figure 2.7 shows the block diagram of a transversal filter. This device pertains to the class of discrete nonrecursive filters, i.e. feedback-free filters. The filter is based on a tapped delay line with multiplication circuits (multipliers) and an adding circuit (adder). The multipliers are designated by triangles in Figure 2.7. Factors $b_0, b_1, b_2, \ldots, b_N$, which are known as filter coefficients, as well as the delay, T, in each element govern the filter's pulse and frequency response characteristics. The number of delay elements, N, is termed the order of the filter.

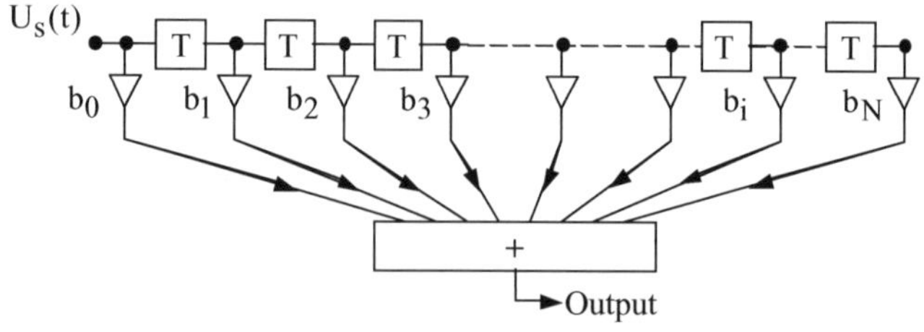

Figure 2.7 Block diagram of transversal filter. T is the T delay element

Transversal filters can be composed of either analogue or digital elements. In the first case, the tapped delay line with multipliers and adder usually takes the form of a surface acoustic wave (SAW) device. This class of devices is discussed in Section 2.8. In the second case, a shift register with digital multipliers and adder is used as a delay line. Delay elements are, as a rule, constructed in random-access main storage cells.

In this section the digital transversal filter will be considered. To evaluate the impulse response of the filter, it is necessary to determine its response to a single discrete pulse. Let us assume that, before the pulse comes to the input of the filter, there are zero voltages (numbers) at all its taps, b_i, and consequently at the output of the adder.

The arrival of a single pulse causes the number b_0 to appear at the output of the filter. After the time interval, T, the single pulse appears at the output of the first delay element and is multiplied by the factor b_1. Hence, there is a number equal to b_1 at the output. At the instant, iT, the single pulse appears at the output of the ith delay element and is multiplied by the factor b_i. As a result, there is a number equal to b_i at the output of the filter. Finally, at the instant NT, there is a number equal to b_N at the output of the filter, after which the output value turns to zero.

Thus, the pulse response of the transversal filter is constituted by a sequence of numbers b_0, b_1, $b_2, \ldots, b_i, \ldots b_N$, i.e. it coincides with the set of filter coefficients

$$h(iT) = b_i. \tag{2.30}$$

Since T is a known constant, formula (2.30) can be simplified to

$$h(i) = b_i. \tag{2.31}$$

Transversal filters are well suited for use as filters matched with discrete signals. Suppose, at the instant iT a discrete signal assumes values $U(iT)$, where $i = 0 - N$. Then, the pulse response of the matched filter can generally be expressed as

$$h(i) = U(N - i)^*.$$

This relationship is deduced from formula (2.25). If $U(iT)$ is a real-valued function, then

$$h(i) = U(N - i). \tag{2.32}$$

Filter coefficients are selected from the following relationship:

$$b_i = U(N - i). \tag{2.33}$$

It follows from formula (2.33) that, the number of delay elements, as well as the number of multipliers, is equal to the number of discrete samples of the input signal.

The frequency response of a discrete linear system is equal to the Fourier transform of its pulse response:

$$\overline{K(\omega)} = \sum_{n=0}^{N-1} h(n) \exp(-j\omega n). \tag{2.34}$$

A major advantage of nonrecursive filters is that their linear phase-response characteristic, $\phi(\omega)$, can be implemented easily and accurately. The delay of this or that spectral component of signal $S(\omega)$ in the filter equals

$$t_d(\omega) = -\frac{d\varphi(\omega)}{d\omega}.$$

When $\phi(\omega) = k\omega$ (the coefficient k does not depend on frequency), $t_d(\omega) = k$. Hence, all the spectral components undergo the same delay passing through the filter. This condition is essential for precluding distortion of the signal waveform.

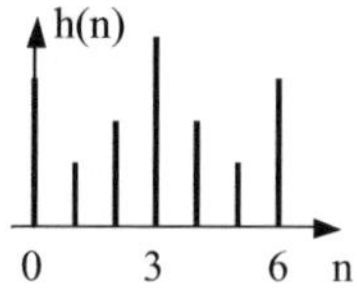

Figure 2.8 Symmetric pulse response

Let us show that meeting the condition of symmetry or antisymmetry of the filter's pulse response relative to its middle is sufficient for linearity of the phase response of the filter [I.12]. To avoid ambiguity, it will be assumed that there are an odd number of samples of the pulse response, which is equal to N. In this case, the numbers of the samples lie within the limits of 0–(N – 1), and the median sample is numbered $(N-1)/2$.

Figures 2.8 and 2.9 give examples of a symmetric and an antisymmetric pulse responses, where $N = 7$, sample numbers are 0–6 and the median sample number is three. It is obvious that the samples of the symmetric pulse response (except for the median) are connected in pairs by the following relationship:

$$h\left(\frac{N-1}{2}-n\right)=h\left(\frac{N-1}{2}+n\right), \tag{2.35}$$

where

$$n = 1 \text{ to } \frac{N-1}{2}.$$

For an antisymmetric pulse response of the same length, the relationship takes the form

$$h\left(\frac{N-1}{2}-n\right)=-h\left(\frac{N-1}{2}+n\right). \tag{2.36}$$

One of the conditions for an antisymmetric $h(n)$ is

$$h\left(\frac{N-1}{2}\right)=0. \tag{2.37}$$

Let us determine the frequency response for a filter with a symmetric $h(n)$. Using formulae (2.34) and (2.35) gives

$$\begin{aligned}\overline{K(\omega)} &= \sum_{n=1}^{(N-1)/2} h\left(\frac{N-1}{2}-n\right)\mathrm{e}^{-\mathrm{j}\omega[(N-1)/2-n]} \\ &\quad + \sum_{(n=1}^{(N-1)/2} h\left(\frac{N-1}{2}+n\right)\mathrm{e}^{-\mathrm{j}\omega[(N-1)/2+n]} \\ &= \mathrm{e}^{-\mathrm{j}\omega[(N-1)/2]}K(\omega).\end{aligned} \tag{2.38}$$

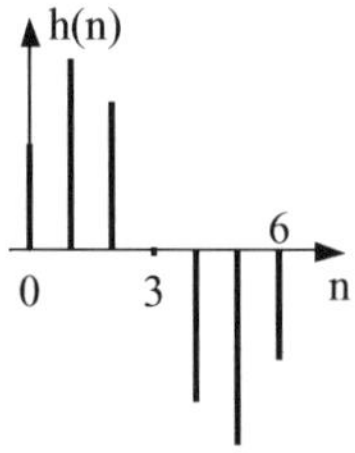

Figure 2.9 Antisymmetric pulse response

In Equation (2.38), $K(\omega)$ is the amplitude–frequency response of the filter:

$$K(\omega) = 2\sum_{n=1}^{(N-1)/2} h\left(\frac{N-1}{2} - n\right)\cos\omega n + h\left(\frac{N-1}{2}\right).$$

It follows from Equation (2.38) that the phase response of the filter is

$$\psi(\omega) = -\omega\left(\frac{N-1}{2}\right).$$

Hence, the filter's phase response is a linear function of frequency.

Similarly, in view of Equations (2.34), (2.36) and (2.37), the following expression is obtained for an antisymmetric pulse response:

$$\begin{aligned}\overline{K(\omega)} &= \sum_{n=1}^{(N-1)/2} h\left(\frac{N-1}{2} - n\right) e^{-j\omega[(N-1)/2-n]} \\ &\quad - \sum_{n=1}^{(N-1)/2} h\left(\frac{N-1}{2} - n\right) e^{-j\omega[(N-1)/2+n]} \\ &= e^{-j[\omega(N-1)/2-\pi/2]} 2\sum_{n=1}^{(N-1)/2} h\left(\frac{N-1}{2} - n\right)\sin\omega n.\end{aligned}$$

Now, it is obvious that in this case the phase response is also a linear function of frequency:

$$\psi(\omega) = -\omega\frac{N-1}{2} + \frac{\pi}{2}.$$

2.4 CORRELATION PROCESSING OF SIGNALS

Suppose an arbitrary voltage, $u(t)$, is applied to the input of a filter matched with the signal, $u_s(t)$. Then, according to formula (2.29), the output voltage is

$$u(t)_{\text{out}} = \int_0^t u(x)\, u_s(t_0 - t + x)\, dx.$$

At the instant $t = t_0$,

$$u(t)_{\text{out}} = \int_0^t u(x)\, u_s(x)\, dx. \tag{2.39}$$

Expression (2.39) is known as a 'correlation integral'. The voltage described by formula (2.39) is obtainable not only by means of a filter with the pulse response

$$h(t) = u_s(t_0 - t)$$

but also with the so-called correlation device (correlator). In this device (see Figure 2.10), the input voltage is multiplied by the reference voltage, $u_s(x)$, with a subsequent integration of the product, which is a direct implementation of expression (2.39). This gives us grounds to assert that the correlator, like the matched filter, is an optimum device for signal processing against a background of white Gaussian noise.

As an example, consider an RF pulse of a duration, τ_p, passing through a correlator:

$$u_s(t) = U\cos\omega_0 t.$$

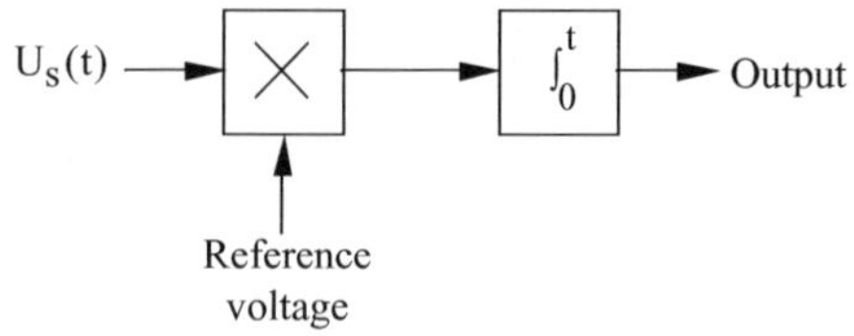

Figure 2.10 Block diagram of correlator

The same voltage is used as a reference. Hence,

$$u_{\mathrm{s.out}}(t) = \int_0^t u_s^2(x)\,\mathrm{d}x = \int_0^t U^2\cos^2(\omega_0 x)\,\mathrm{d}x$$
$$= \frac{U^2}{2}t + \frac{U^2}{4\omega_0}\sin 2\omega_0 t = \frac{U^2 t}{2} + \frac{U^2 T_0}{8\pi}\sin 2\omega_0 t. \tag{2.40}$$

Here, $T_0 = 2\pi/\omega_0$ is the period of the carrier wave. The second term of the sum in Equation (2.40) is a fast-oscillating quantity. Its maximum value does not exceed $U^2 T_0(8\pi)$, while the mean value over a period is equal to zero. Therefore, when $t > T_0$, this term can be neglected and expression (2.40) can be written as

$$u_{\mathrm{s.out}}(t) = \frac{U^2 t}{2}. \tag{2.41}$$

It follows from Equation (2.41) that in the example concerned, the output voltage of the correlator rises linearly throughout the duration of the input signal and thereupon remains equal to

$$u_{\mathrm{s.\,max}} = \frac{U^2}{2}\tau_{\mathrm{p}}. \tag{2.42}$$

The input and output waveforms of the correlator are shown in Figure 2.11. Comparing Figure 2.11 with Figure 2.5, it can be seen that the output voltage waveforms of the correlator and the matched filter differ, but the peak voltage values are the same. This validates the conclusion that the devices under consideration are equally effective.

Both the correlator and the matched filter have their merits and demerits. The key advantage of a matched filter is its invariance to the signal time of arrival. The drawback of the filter is that hardware

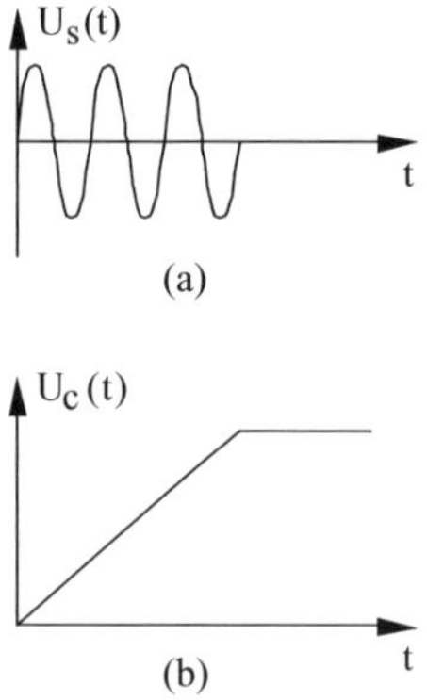

Figure 2.11 (a) Input and (b) output voltages of the correlator

input grows with greater signal durations (see Section 2.3). Conversely, input signal duration has almost no influence on hardware input in a correlator. It follows from Figure 2.10 that a correlator is made up of three subunits: a reference voltage source, a multiplier and an integrator.

The disadvantage of the correlator is its noninvariance to the signal time of arrival. It is related to the necessity of forming a reference voltage coinciding in time with the signal subject to processing. Such concurrence can be achieved in two ways: either using a multichannel correlator that implements a time shift in the reference voltage in each channel or searching for an appropriate time position of the reference voltage until it coincides with the input signal.

The above comparison brings us to the following conclusions. The matched filter is preferable when the signals to be processed are not very long. If the duration of a signal is considerable, then it is more expedient to use the correlator. One of the two processing devices can be chosen if the permissible hardware input for filter implementation and the permissible search time in the correlator are known.

2.5 COMPLEX ENVELOPE PROCESSING

Generally, radar signal can be expressed by the formula

$$u(t) = U(t)\ \cos[\omega_0 t + \varphi(t)]. \tag{2.43}$$

Where $U(t)$ is the amplitude modulation function and $\varphi\,(t)$ is the phase modulation function. Since

$$\cos x = \mathrm{Re}\,(\exp\ \mathrm{j}x)$$

and $U(t)$ is a real-valued function, Equation (2.43) can be rewritten as

$$u(t) = \mathrm{Re}\left[U(t)\ \exp\ \mathrm{j}\varphi(t)\ \exp \mathrm{j}\omega_0 t\right].$$

The product

$$\overline{U(t)} = U(t)\ \exp \mathrm{j}\varphi(t) \tag{2.44}$$

is referred to as the complex envelope of the waveform. Using this definition, the RF signal expression (2.43) can be written in the form

$$u(t) = \mathrm{Re}\left[\overline{U(t)}\ \exp \mathrm{j}\omega_0 t\right]. \tag{2.45}$$

The complex envelope, as a function of the signal amplitude and phase modulation, contains all the information the signal carries. Therefore, $\overline{U(t)}$ can be used in the receiving part of the radar instead of a signal with a carrier frequency. This substitution is practical in many cases because the complex envelope does not contain a carrier frequency and its processing does not require the use of high-speed devices. Moreover, the complex envelope can be readily employed to derive amplitude and phase modulation functions of the signal:

$$U(t) = \left|\overline{U(t)}\right|, \qquad \varphi(t) = \arg \overline{U(t)}. \tag{2.46}$$

In practice, the complex envelope is used as a combination of its real and imaginary parts. It follows from Equation (2.44) that

$$\overline{U(t)} = U(t)\ \cos\varphi(t) + \mathrm{j}U(t)\ \sin\varphi(t). \tag{2.47}$$

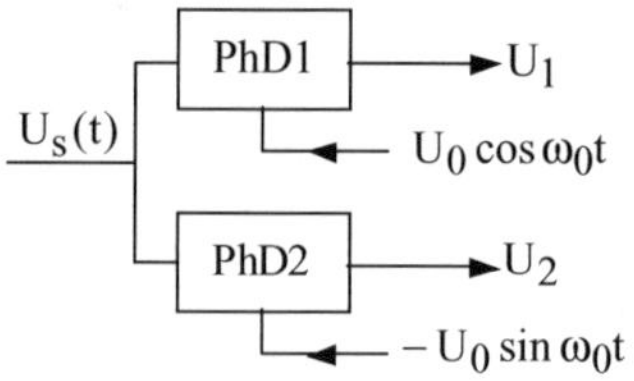

Figure 2.12 Block diagram of the device of a complex amplitude extraction

The block diagram of a device for complex envelope extraction from a signal with carrier frequency is given in Figure 2.12. This device includes phase detectors PhD1 and PhD2, a reference frequency generator and a phase shifter by $\pi/2$. The phase detector is known to consist of a multiplier and a lowpass filter (LPF).

At the output of the multiplier in PhD1,

$$\begin{aligned} u(t)_1 &= U(t)\cos[\omega_0 t + \varphi(t)]\cos\omega_0(t) \\ &= \tfrac{1}{2}U(t)\cos[2\omega_0 t + \varphi(t)] + \tfrac{1}{2}U(t)\cos\varphi(t). \end{aligned} \tag{2.48}$$

In the LPF, which is the next stage, the first term of the sum of Equation (2.48) is suppressed and the PhD1 output voltage is

$$u_{\text{PhD.1}} = \tfrac{1}{2}U(t)\ \cos\varphi(t). \tag{2.49}$$

Similarly, the PhD2 output voltage is

$$u_{\text{PhD.2}} = \tfrac{1}{2}U(t)\sin\varphi(t). \tag{2.50}$$

Comparing formulae (2.49) and (2.50) with formula (2.47), it can be seen that the output voltages of the extractor shown in Figure 2.12 are proportional to the real and imaginary parts of the complex envelope of the input.

Let us now determine the relationship between the spectra of a signal and of the signal complex envelope $\overline{S_c(\omega)}$. The original signal can be written as

$$u(t) = \tfrac{1}{2}\overline{U(t)}\exp \mathrm{j}\omega_0 t + \tfrac{1}{2}\overline{U(t)}^*\exp(-\mathrm{j}\omega_0 t). \tag{2.51}$$

The spectra of the first and the second terms of the sum in the right part of Equation (2.51) can be designated as $\overline{S_1(\omega)}$ and $\overline{S_2(\omega)}$ respectively. Then, the signal spectrum is given by the sum

$$\overline{S(\omega)} = \overline{S_1(\omega)} + \overline{S_2(\omega)}.$$

It is known [I.9] that the spectrum of the product is proportional to the convolution of the spectra of the multipliers. The spectrum of the complex exponent $\exp \mathrm{j}\omega_0 t$ is equal to $S_e = 2\pi\,\delta(\omega - \omega_0)$. Therefore,

$$\begin{aligned} \overline{S_1(\omega)} &= \frac{1}{4\pi}\int_{-\infty}^{\infty} \overline{S_c(x)}\ \overline{S_e(\omega - x)}\,\mathrm{d}x \\ &= \frac{1}{2}\int_{-\infty}^{\infty} \overline{S_c(x)}\ \delta(\omega - \omega_0 - x)\ \mathrm{d}x\ = \tfrac{1}{2}\overline{S_c(\omega - \omega_0)}. \end{aligned}$$

Similarly,

$$\overline{S_2(\omega)} = \tfrac{1}{2}\overline{S_c(-\ \omega - \omega_0)}^*.$$

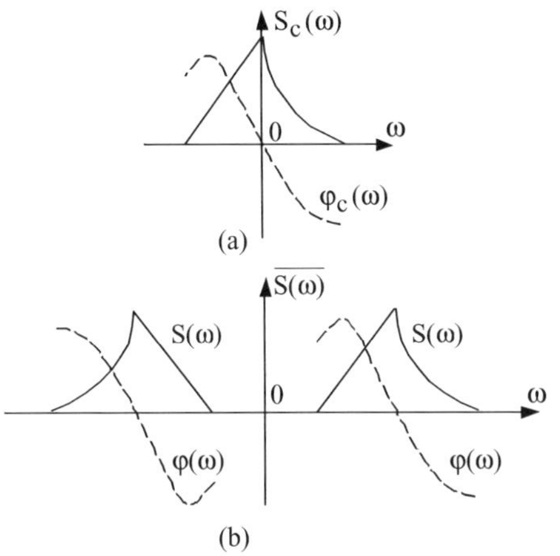

Figure 2.13 (a) Complex envelope spectrum and (b) signal spectrum

Hence,

$$\overline{S(\omega)} = \tfrac{1}{2}\left[\overline{S_c(\omega - \omega_0)} + \overline{S_c(-\,\omega - \omega_0)}^*\right]. \qquad (2.52)$$

The relationship given by Equation (2.52) is illustrated graphically in Figure 2.13.

If the signal contains a Doppler frequency shift f_d,

$$u(t) = U(t)\cos[2\pi(f_0 + f_d)t + \varphi(t)];$$

then the following voltages exist at the outputs of the two-channel device shown in Figure 2.12:

$$u_1 = U(t)\cos[2\pi f_d t + \varphi(t)],$$
$$u_2 = U(t)\sin[2\pi f_d t + \varphi(t)].$$

The frequency, f_d, is not a carrier; as f_d is proportional to the rate of closure of the target and the radar, it pertains to the signal's information parameters.

Let us determine the structure of a filter intended for processing the complex envelope of an arbitrary waveform signal

$$u(t) = \mathrm{Re}\left[\overline{U(t)}\ \exp(\mathrm{j}\omega_0 t)\right].$$

Suppose this signal is to be processed in a filter with a given pulse response

$$h(t) = a(t)\ \cos[\omega_0 t + \psi(t)] = \mathrm{Re}\left[\overline{H(t)}\exp(\mathrm{j}\omega_0 t)\right].$$

Here, $\overline{H(t)} = a(t)\ \exp \mathrm{j}\psi(t) = H_1(t) + H_2(t)$ is the complex envelope of the pulse response;

$$H_1(t) = a(t)\cos\psi, \qquad H_2(t) = a(t)\sin\psi(t).$$

The output voltage of the filter is

$$u_{\mathrm{out}}(t) = \int_0^t u(x)\,h(t-x)\,\mathrm{d}x = \int_0^t \mathrm{Re}\left[U(x)\ \exp(\mathrm{j}\omega_0 x)\right]\ \mathrm{Re}\left[H(t-x)\ \exp(\mathrm{j}\omega_0(t-x))\right]\mathrm{d}x.$$

The following relationship is valid for any complex quantities $\bar{A}$ and $\bar{B}$:

$$\mathrm{Re}\bar{A}\ \mathrm{Re}\bar{B} = \frac{\mathrm{Re}(\bar{A}\ \bar{B}) + \mathrm{Re}(\bar{A}\ \bar{B})^*}{2}. \tag{2.53}$$

Performing computations on the right side of Equation (2.53) and neglecting the integral of the fast-oscillating function that includes the term $\exp(\mathrm{j}2\omega_0 x)$, gives

$$u_{\mathrm{out}}(t) = \mathrm{Re}\left[\tfrac{1}{2}\int_0^t \overline{U(x)}\ \overline{H(t-x)}\ \mathrm{d}x\ \exp(\mathrm{j}\omega_0 t)\right].$$

By comparing this expression with formula (2.45), an expression can be deduced for the complex envelope of the output voltage:

$$\overline{U_{\mathrm{out}}(t)} = \tfrac{1}{2}\int_0^t \overline{U(x)}\ \overline{H(t-x)}\ \mathrm{d}x \tag{2.54}$$

Thus, $\overline{U_{\mathrm{out}}(t)}$ is connected with the complex envelopes of the input voltage and pulse response by the same relationship as the instantaneous values of the voltages and pulse response.

Let us transform expression (2.54):

$$\begin{aligned}
\overline{U_{\mathrm{out}}(t)} &= \tfrac{1}{2}\int_0^t U(x)[\cos\varphi(x) + \mathrm{j}\sin\varphi(x)]a(t-x)[\cos\psi(t-x) + \mathrm{j}\sin\psi(t-x)]\,\mathrm{d}x \\
&= \tfrac{1}{2}\left[\int_0^t U(x)\cos\varphi(x)a(t-x)\cos\psi(t-x)\,\mathrm{d}x - \int_0^t U(x)\sin\varphi(x)a(t-x)\sin\psi(t-x)\,\mathrm{d}x\right] \\
&\quad + \frac{\mathrm{j}}{2}\left[\int_0^t U(x)\cos\varphi(x)a(t-x)\sin\psi(t-x)\,\mathrm{d}x + \int_0^t U(x)\sin\varphi(x)a(t-x)\cos\psi(t-x)\,\mathrm{d}x\right] \\
&= U_1(t) + \mathrm{j}U_2(t).
\end{aligned} \tag{2.55}$$

Formula (2.55) describes the desired structure of the complex envelope filtering device. The functional diagram of the filter is given in Figure 2.14 (to the left of the dashed line).

PhD1 and PhD2 output voltages represent components of the complex envelope of the input signal:

$$U(t)\ \cos\varphi(t) \qquad \text{and} \qquad U(t)\ \sin\varphi(t).$$

These voltages are processed in filters, H_1 and H_2, having pulse responses $H_1(t)$ and $H_2(t)$. At the outputs of the subtraction and adding circuits, components of the output voltage complex envelope

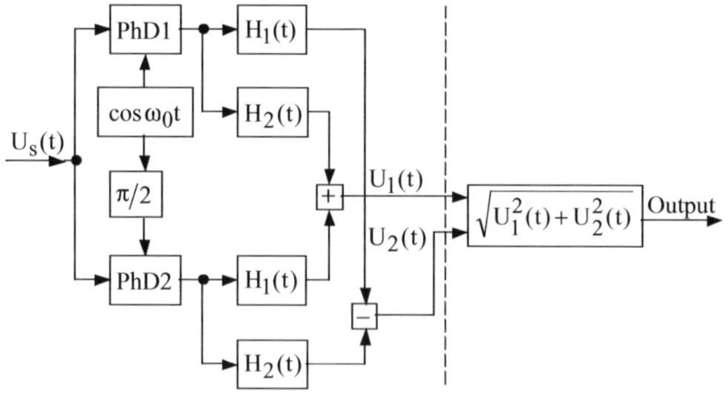

Figure 2.14 Block diagram of a filter of complex amplitude

are

$$U_1(t) = \tfrac{1}{2}\int_0^t U(x)\cos\varphi(x)a(t-x)\cos\psi(t-x)\,\mathrm{d}x$$
$$-\tfrac{1}{2}\int_0^t U(x)\sin\varphi(x)a(t-x)\sin\psi(t-x)\,\mathrm{d}x,$$
$$U_2(t) = \tfrac{1}{2}\int_0^t U(x)\cos\varphi(x)a(t-x)\sin\psi(t-x)\,\mathrm{d}x$$
$$+\tfrac{1}{2}\int_0^t U(x)\sin\varphi(x)a(t-x)\cos\psi(t-x)\,\mathrm{d}x.$$

In the case of matched filtering of the signal $\mathrm{Re}[\overline{U(t)}\ \exp(\mathrm{j}\omega_0 t)]$, the pulse response of the carrier frequency filter is

$$h(t) = U(t_0 - t) = \mathrm{Re}[\overline{U(t_0 - t)}^*\ \exp \mathrm{j}\omega_0 t\ \exp(-\mathrm{j}\omega_0 t_0)].$$

After rejecting the negligible multiplier $\exp(-\mathrm{j}\omega_0 t_0)$, we obtain

$$H(t) = \overline{U(t_0 - t)}^*. \tag{2.56}$$

Consequently, in accordance with Equation (2.56),

$$H_1(t) = \mathrm{Re}\left[\overline{U(t_0 - t)}\right] = U(t_0 - t)\ \cos\varphi(t_0 - t),$$
$$H_2(t) = -\ \mathrm{Im}\left[\overline{U(t_0 - t)}\right] = -\ U(t_0 - t)\sin\varphi(t_0 - t).$$

Let us compare the frequency response of the carrier-frequency filter, $\overline{K(\omega)}$, with that of the device for complex envelope processing, $\overline{K_{\mathrm{ce}}(\omega)}$. Since $h(t)$ can be represented as

$$h(t) = \tfrac{1}{2}\left[\overline{H(t)}\ \exp(\mathrm{j}\omega_0 t) + \overline{H(t)}^*\ \exp(-\mathrm{j}\omega_0 t)\right],$$

the frequency response

$$\overline{K(\omega)} = \int_0^\infty h(t)\ \exp(-\mathrm{j}\omega_0 t)\,\mathrm{d}t = \tfrac{1}{2}\int_0^\infty \overline{H(t)}\ \exp\left[-\mathrm{j}(\omega - \omega_0)t\right]\mathrm{d}t$$
$$+\tfrac{1}{2}\int_0^\infty \overline{H(t)}^*\ \exp[\mathrm{j}(-\omega - \omega_0)t]\,\mathrm{d}t = \tfrac{1}{2}\left[\overline{K_{\mathrm{ce}}(\omega - \omega_0)} + \overline{K_{\mathrm{ce}}(-\omega - \omega_0)}^*\right]. \tag{2.57}$$

In Equation (2.57), $\overline{K_{\mathrm{ce}}(\omega)} = \int_0^\infty \overline{H(t)}\ \exp(-\mathrm{j}\omega t)\,\mathrm{d}t$.

Analysis of formula (2.57) brings us to the conclusion that the shape of the frequency response is determined by the structure of the complex envelope of the pulse response. Furthermore, the frequency response of the complex envelope filter resides in the neighbourhood of zero, whereas the frequency response of the carrier-frequency filter – in the neighbourhood of frequencies ω_0 and $-\omega_0$. Therefore, when constructed in elements with a limited speed, the complex envelope filter is preferable to the carrier-frequency filter.

Expressions for the frequency responses of filters, H_1 and H_2, are defined by Fourier transforms:

$$\overline{K_1(\omega)} = \tfrac{1}{2}\int_0^\infty a(t)\ \cos\psi(t)\ \exp(-\mathrm{j}\omega t)\,\mathrm{d}t,$$
$$\overline{K_2(\omega)} = \tfrac{1}{2}\int_0^\infty a(t)\ \sin\psi(t)\ \exp(-\mathrm{j}\omega t)\,\mathrm{d}t.$$

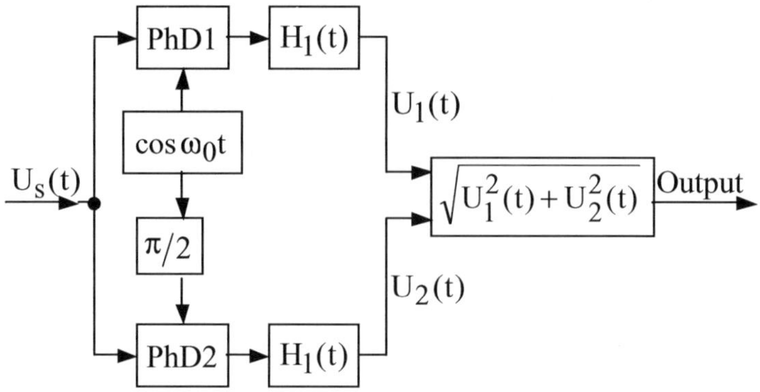

Figure 2.15 Block diagram of a filter: $\psi(t) = 0, \pi$.

Let us consider some important special cases:

1. Information about the target is embedded solely in the amplitude modulation of the signal. In this case, an additional circuit is introduced after the filter to generate a voltage proportional to the real amplitude of the signal.

$$U_{\text{out}}(t) = |U_{\text{out}}(t)| = \sqrt{\left[\text{Re}\overline{U_{\text{out}}(t)}\right]^2 + \left[\text{Im}\overline{U_{\text{out}}(t)}\right]^2}.$$

This estimator modulo circuit is shown in Figure 2.14 to the right of the dashed line.

2. Function $\psi(t)$ in the pulse response assumes values either 0 or π. In this case, $H_2(t) = 0$ and $K_2(\omega) = 0$, and there are two quadrature channels preserved in the processing device (Figure 2.15), in which identical filters are used, both having a pulse response $H_1(t)$.

2.6 FFT-BASED DIGITAL SIGNAL PROCESSING

2.6.1 Types of FFT Processor-Based Filters

The fast Fourier transform (FFT) is an economical algorithm for computing the discrete Fourier transform (DFT). The latter is the discrete spectrum, $\overline{S(i\Omega)}$, of a signal, u_n, specified by N time samples. Here, n and i are integers ranging from 0 to $N-1$. The DFT is calculated by the formula

$$\overline{S(i\Omega)} = \sum_{n=0}^{N-1} u_n \exp\left(-\text{j}\frac{2\pi}{N}\ in\right), \tag{2.58}$$

where $\Omega = 2\pi/(NT)$ is a frequency sample and T is the sampling interval.

If u_n is a periodic signal with a period, NT (in other words, one period contains N samples), the discrete spectrum of the signal is expressed by formula (2.58) accurately. However, for a nonperiodic input signal, the spectrum is a function of the continuous frequency, $\overline{S(\omega)}$, and $\overline{S(i\Omega)}$ is the result of $\overline{S(\omega)}$ sampled over the frequency axis. It is well known that, $\overline{S(i\Omega)}$ can be used to determine the original spectrum, $\overline{S(\omega)}$, which is discussed in Reference [1.9].

All of the filtering devices that can be constructed on the basis of FFT processor will be devided into two groups:

- comb filters (filter banks) with a frequency response dependent on FFT parameters;
- filters with an arbitrary frequency response, including matched filters.

2.6.2 FFT Processor as a Filter Bank

Let us determine the frequency response of an FFT processor. To this end, a discrete complex exponent is applied at its input. The FFT processor has N outputs; therefore the output effect will depend both on the frequency of the input exponent and on the channel number, i. For the ith channel, the frequency response is given by

$$\overline{K(i,\omega)} = \sum_{n=0}^{N-1} \exp \mathrm{j}\omega nT \exp\left(-\mathrm{j}\frac{2\pi}{N}ni\right). \tag{2.59}$$

Since $\Omega = 2\pi/(NT)$, the formula (2.59) can be rewritten as

$$\overline{K(i,\omega)} = \sum_{n=0}^{N-1} \exp(\mathrm{j}\omega nT)\exp(-\mathrm{j}i\Omega nT) = \sum_{n=0}^{N-1} \exp\left[\mathrm{j}(\omega - i\Omega)nT\right].$$

This expression is the sum of a geometrical progression with a ratio of $\exp\left[\mathrm{j}(\omega - i\Omega)T\right]$. Hence,

$$\overline{K(i,\omega)} = \frac{1-\exp\left[\mathrm{j}(\omega - i\Omega)NT\right]}{1-\exp\left[\mathrm{j}(\omega - i\Omega)T\right]}.$$

Going over to the modulus of the frequency response, we obtain

$$K(i,\omega) = \left|\frac{\sin(\omega - i\Omega)NT/2}{\sin(\omega - i\Omega)T/2}\right|.$$

The amplitude–frequency response as a function of frequency in hertz is expressed by the equation

$$K(i,f) = \left|\frac{\sin(f - iF)\pi NT}{\sin(f - iF)\pi T}\right|, \tag{2.60}$$

where $F = \Omega/(2\pi) = 1/(NT)$. Fixing the quantity i in formula (2.60) at $i = i_1$, the shape of the amplitude–frequency response for this particular channel can be determined as

$$K(i_1,f) = \left|\frac{\sin(f - i_1F)\pi NT}{\sin(f - i_1F)\pi T}\right|. \tag{2.61}$$

It follows from expression (2.61) that the maximum of the amplitude–frequency response occurs at frequency $f = i_1F$ and equals N. When $f = (i_1 \pm 0.5)F$,

$$K((i_1 \pm 0.5)F, i_1) = \left|\frac{\sin(\pm 0.5\pi)}{\sin(\pm 0.5\pi/N)}\right| \approx \frac{2}{\pi}N \approx 0.64N.$$

Consequently, $f = (i_1 \pm 0.5)F$ describes the limits of the filter passband referred to the level of –4 dB. When $f = iF$, given that $i \neq i_1$,

$$K(i_1, if) = \left|\frac{\sin(i - i_1)\pi}{\sin(i - i_1)\pi/N}\right| = 0.$$

Lastly, at the points $f = (i \pm 0.5)F$, given that $i \neq i_1$, we have

$$K(i_1,(i \pm 0.5)F) = \left|\frac{\sin(i - i_1 \pm 0.5)\pi}{\sin(i - i_1 \pm 0.5)\pi/N}\right| = \frac{1}{|\sin(i - i_1 \pm 0.5)\pi/N|}.$$

The amplitude–frequency response graph for a single channel of an FFT processor is presented in Figure 2.16, while Figure 2.17 shows the family of frequency responses for all the channels of an eight-point FFT processor. According to the figures, the FFT processor has a comb-shaped amplitude–frequency response with resonant frequencies of individual filters given by $f_i = iF$. Furthermore, the

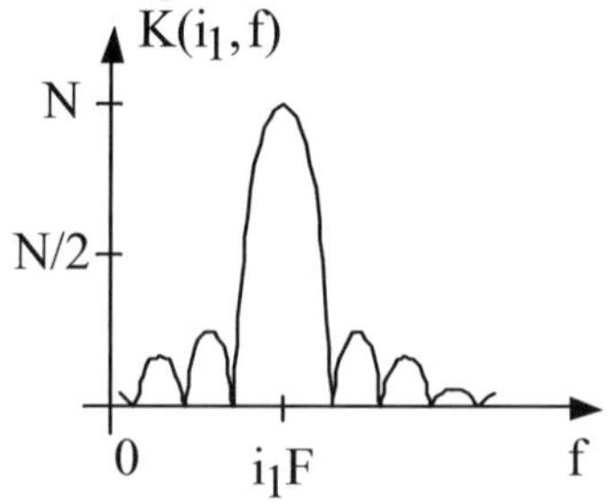

Figure 2.16 Amplitude–frequency characteristic of the ith channel

–4 dB passbandwidth of each filter is $\Delta F = F$. At points offset from the central frequency by a value other than iF, a rather high level of amplitude–frequency response exists. This shortcoming can be significantly reduced by weighting the input signal in the time domain or weighting the output signal in the frequency domain. In the first case, the input signal is multiplied by a weighting function, e.g. the Hamming window (cosine on a pedestal with certain coefficients). In the second case, the output of the channel considered is summed with the half-sum of the outputs of the adjacent channels [I.13].

2.6.3 FFT Processor-Based Filter with an Arbitrary Frequency Response

Figure 2.18 shows the block diagram for a filter with an arbitrary frequency response, based on an FFT processor (fast convolution device). At the output of the FFT processor, a discrete spectrum of the input signal, $\overline{S(i\Omega)}$, is formed. Then, all of the N components of the spectrum are multiplied by the corresponding samples of a desired frequency response, $\overline{K(i\Omega)}$. As a result, the output signal spectrum for a filter with a desired frequency response is obtained:

$$\overline{S_{\text{out}}(i\Omega)} = \overline{S(i\Omega)}\,\overline{K(i\Omega)}.$$

The next operation is the Fourier inversion (IFFT). The resultant signal is similar to the output signal of the filter with a desired frequency response, should the given signal be applied at its input. The filtering method in question is termed fast convolution as it employs two fast algorithms, namely the FFT and the IFFT.

Let us show that the choice of reference point for an array of N samples is of no importance for filtering a periodic signal in the spectral domain. To this end, an array of input voltage samples will be

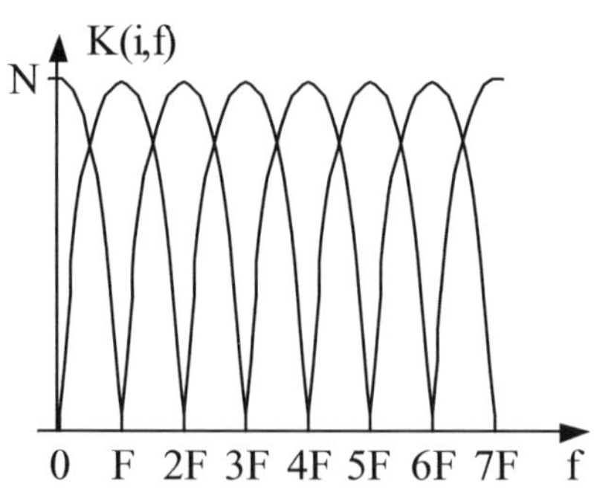

Figure 2.17 Amplitude–frequency characteristics of all channels (main lobes)

Figure 2.18 Block diagram of the fast convolution device

formed, beginning with the sample u_1 (Figure 2.19):

$$u_c(t) = u_1, u_2, u_3, \ldots, u_7.$$

The designation $\overline{S(i\Omega)}$ is adopted for the discreted spectrum of this signal and the designation $\overline{K(i\Omega)}$ for the frequency response of a matched filter. After performing all the fast convolution operations, the following output signal is obtained:

$$u_{\text{out}}(nT) = \sum_{i=0}^{N-1} \overline{S(i\Omega)}\ \overline{K(i\Omega)} \exp(\text{j}i\Omega nT). \tag{2.62}$$

Let us choose another reference point for the array of samples subject to processing, for instance the sample u_4.The new signal is given by

$$u_c'(t) = u_4,\ u_5,\ u_6,\ u_7,\ u_1,\ u_2,\ u_3 = u_c(t - mT).$$

In this example, $m = 3$. The spectrum of this signal is

$$\overline{S'(i\Omega)} = \overline{S(i\Omega)} \exp(-\text{j}i\Omega mT).$$

Hence, the voltage at the output of the processing device is described by

$$u_{\text{out}}'(nT) = \sum_{i=0}^{N-1} \overline{S(i\Omega)}\ \overline{K(i\Omega)} \exp\left[\text{j}i\Omega(n - m)T\right]. \tag{2.63}$$

Comparison of this expression with expression (2.62) leads to an obvious conclusion:

$$u_{\text{out}}'(nT) = u_{\text{out}}\left((n - m)T\right).$$

Thus, changing the sampling reference point for an array of samples brings about the same time shift in the output voltage, with its shape remaining unchanged.

In the special case when the filter is to be matched with the input signal, $\overline{K(i\Omega)} = \overline{S(i\Omega)}^*$. Consequently, $\overline{S(i\Omega)}\ \overline{K(i\Omega)} = \left|\overline{S(i\Omega)}\right|^2$ and expression (2.63) takes the form

$$u_{\text{out}}'(nT) = \sum_{i=0}^{N-1} \left|\overline{S(i\Omega)}\right|^2 \exp\left[\text{j}i\Omega(n - m)T\right]. \tag{2.64}$$

It follows from formula (2.64) that the output signal is at its maximum when $nT = mT$. This maximum is equal to the sum of squares of its spectral components or (in accordance with the Perceval theorem) the sum of squares of its time samples.

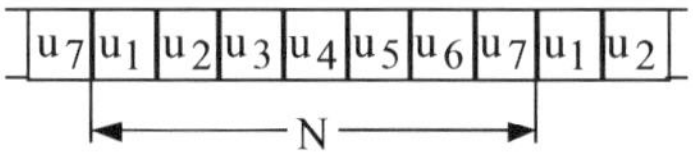

Figure 2.19 Fragment of a periodic seven-element signal

2.7 SIMPLE AND COMPLICATED WAVEFORMS; SIGNAL BASE

The product of the signal's coherent part duration and its spectrum bandwidth is termed the 'signal base' (B). Simple waveforms are those with $B \approx 1$. Signals with $B \gg 1$ (pulse compression waveforms) will be referred to as complicated waveforms. Simple waveforms are single pulses and pulse trains (bursts) with a nonmodulated carrier frequency.

A single RF pulse with an arbitrary duration, τ_p, has a spectrum bandwidth of

$$\Delta f_s \approx \frac{1}{\tau_p}.$$

Hence, the base of this pulse is

$$B = \tau_p \, \Delta_{fs} \approx 1.$$

The spectrum bandwidth of a noncoherent pulse burst is also given by $\Delta f_s = 1/\tau_p$ (see Section 2.1). The duration of the coherent interval is equal to τ_p. Hence, base $B = 1$.

The spectrum bandwidth of a coherent pulse burst is $\Delta f_s \approx 1.2/(NT_p)$ according to Equation (2.11). The duration of the coherent part is NT_p. Hence, the base of a coherent burst is

$$B = NT_p \frac{1.2}{NT_p} = 1.2 \approx 1.$$

Therefore, a coherent pulse burst is also a simple waveform.

Before we proceed to the characteristics of complicated waveforms, let us determine the connection between the range resolution and signal spectrum bandwidth. It follows from formula (1.9) that $\Delta R = c\tau_{out}/2$, where τ_{out} is the pulse duration at the output of the matched filter. Consequently, calculation of the range resolution requires knowledge of the signal duration at the output of the matched filter. For the sake of commonality of the duration estimate, τ_{out}, suppose that the signal amplitude spectrum, $S(\omega)$, has a rectangular envelope of width $\Delta\omega$ (Figure 2.20). This approximation is quite valid for the linear FM (chirp) waveform, which is widely used in radar. Although concerning signals of a different structure, the rectangular spectrum approximation is an assumption. The phase spectrum of the signal is assumed to be arbitrary.

The complex spectrum of the signal and the complex frequency response of the matched filter will be written as

$$\overline{S(\omega)} = S(\omega)\,\exp\left[j\varphi(\omega)\right]$$

and

$$\overline{K(\omega)} = S(\omega)\exp\left\{-j\left[\varphi(\omega) + \omega t_0\right]\right\}.$$

The voltage spectrum at the output of the matched filter is

$$\overline{S_{out}(\omega)} = \overline{S(\omega)}\;\overline{K(\omega)} = S^2(\omega)\,\exp(-j\omega t_0).$$

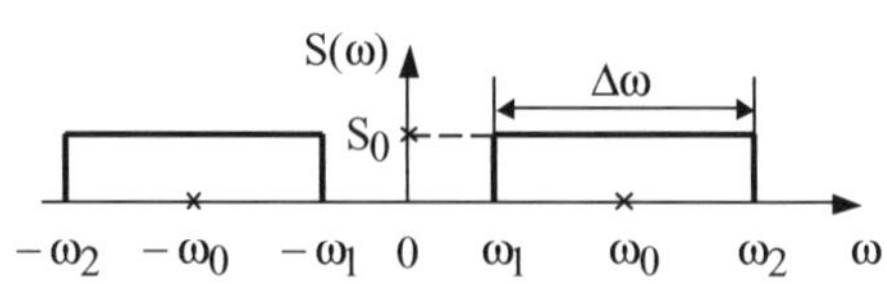

Figure 2.20 Rectangular amplitude spectrum approximation

The output voltage shape is obtained by means of the inverse Fourier transform:

$$u_{\text{out}}(t) = \frac{1}{2\pi}\int_{-\infty}^{\infty} S^2(\omega)\ \cos\omega(t-t_0)\ \mathrm{d}\omega + \frac{\mathrm{j}}{2\pi}\int_{-\infty}^{\infty} S^2(\omega)\sin\omega(t-t_0)\ \mathrm{d}\omega.$$

The second integral is equal to zero as the integrand is odd. Since the integrand in the first integral is even,

$$\begin{aligned} u_{\text{out}}(t) &= \frac{1}{\pi}\int_{\omega_1}^{\omega_2} S^2(\omega)\cos\omega(t-t_0)\ \mathrm{d}\omega = \frac{S_0^2}{\pi(t-t_0)}\left[\sin\omega_2(t-t_0) - \sin\omega_1(t-t_0)\right] \\ &= \frac{2S_0^2}{\pi(t-t_0)}\sin\frac{\omega_2-\omega_1}{2}(t-t_0)\cos\frac{\omega_2+\omega_1}{2}(t-t_0) \\ &= \frac{2S_0^2\Delta f_{\text{s}}}{\pi\Delta f_{\text{s}}(t-t_0)}\sin\pi\Delta f_{\text{s}}(t-t_0)\cos\omega_0(t-t_0). \end{aligned}$$

Here, $\Delta f_{\text{s}} = \frac{\Delta\omega}{(2\pi)}$ is the spectrum bandwidth of the input signal in hertz.

Obviously, the amplitude of this voltage is

$$u_{\text{out}}(t) = 2S_0^2\Delta f_{\text{c}}\frac{\sin\pi\Delta f_{\text{s}}(t-t_0)}{\pi\Delta f_{\text{s}}(t-t_0)}. \tag{2.65}$$

The voltage envelope as expressed by Equation (2.65) is depicted in Figure 2.21. The figure shows that the zero-level width of the signal's main lobe is $2/\Delta f_{\text{s}}$, while its –3 dB width is

$$\tau_{\text{out}} \approx \frac{1}{\Delta f_{\text{s}}}. \tag{2.66}$$

Hence

$$\Delta R = \frac{c}{2}\Delta t = \frac{c}{2\Delta f_{\text{s}}}.$$

For example, for $\Delta f_{\text{s}} = 1$ MHz,

$$\Delta R = \frac{3\times 10^8}{2\times 10^6} = 150\ \text{m}.$$

The use of complicated waveforms is a viable alternative when the radar is expected to provide a considerable range coverage, R_{max}, and a good range resolution, ΔR, at the same time. These quantities are closely related to the parameters of the transmitted signal. It follows from Equation (1.26) that the range coverage varies in proportion to $\sqrt[4]{P_{\text{t}}\tau_{\text{p}}}$. This means that to attain a specified range coverage with a limited transmitter power, P_{t}, it is practical to increase τ_{p}.

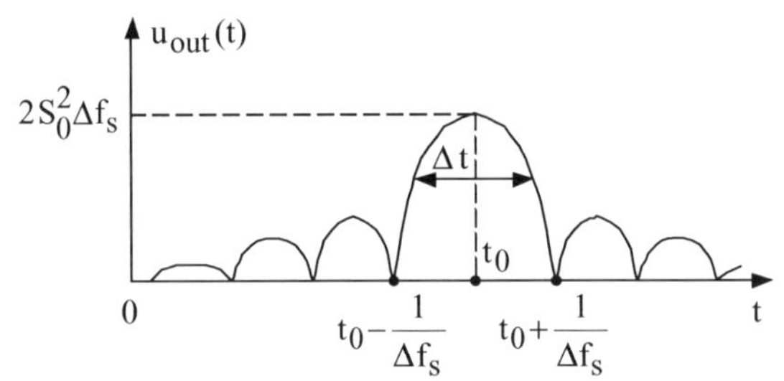

Figure 2.21 MF output voltage envelope

It can also be seen from Equation (2.66) that

$$\Delta R = \frac{c}{2\Delta f_s}.$$

Since

$$\Delta f_s = \frac{B}{\tau_p},$$

the resolution in range is

$$\Delta R = \frac{c}{2}\frac{\tau_p}{B}. \tag{2.67}$$

When simple waveforms are used, $B = 1$ and

$$\Delta R = \frac{c}{2}\tau_p.$$

It follows from this formula that if a good range resolution is to be achieved the value of τ_p should be small. Therefore, the use of simple waveforms can lead to conflicting requirements as to the duration of the transmitter's pulse. This discrepancy is eliminated by the employment of complicated waveforms. In this case, the duration of the transmitter pulse is chosen in order to attain the specified range coverage. To achieve the desired range resolution, it is adequate to use a signal with a base determined by Equation (2.67).

2.8 LINEAR FM AND PHASE-CODED WAVEFORMS

2.8.1 Linear Frequency Modulation Pulses

The expression for an RF pulse with a linear frequency modulation (also referred to as a chirp pulse) of duration τ_p (Figure 2.22) has the form

$$u_s(t) = U(t)\cos(\omega_0 t + bt^2), \qquad U(t) = \begin{cases} U_0, & \text{if } 0 \le t \le \tau_p \\ 0, & \text{if } t < 0, t > \tau_p \end{cases}. \tag{2.68}$$

The instantaneous frequency of this signal given by

$$f = \frac{1}{2\pi}\frac{\mathrm{d}}{\mathrm{d}t}\left(\omega_0 t + bt^2\right) = f_0 + \frac{b}{\pi}t$$

is a linear function of time. The variation of frequency within the pulse duration (see Figure 2.23) is

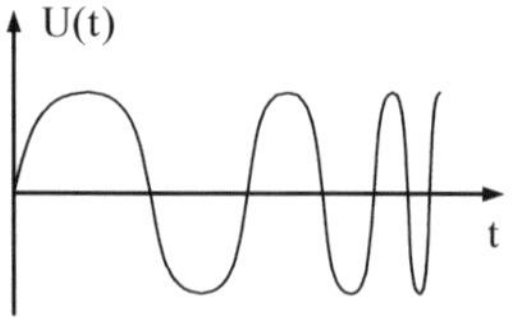

Figure 2.22 Chirp pulse

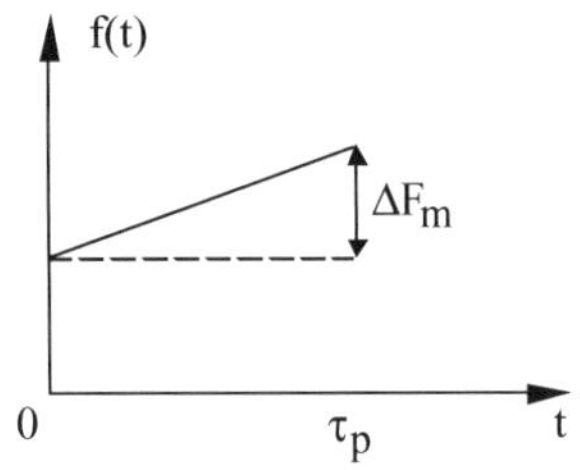

Figure 2.23 Linear FM

known as the frequency deviation and is designated the form Δf_m:

$$\Delta f_m = \frac{b}{\pi}\,\tau_p\,.$$

It is assumed hereafter that

$$\Delta f_m \tau_p \gg 1. \tag{2.69}$$

Note that the quantity b can take on negative values as well. In this case, the frequency is a linearly decreasing function of time.

The spectrum bandwidth of the chirp pulse referred to the –6 dB level is approximately equal to its frequency deviation [I.9] . It is necessary to emphasize that when condition (2.69) is satisfied, the shape of the chirp pulse spectrum may be approximated as rectangular. Therefore it can be supposed that

$$\Delta f_s \approx \Delta f_m.$$

This implies that the shape of the chirp pulse envelope at the matched filter (MF) output is given by expression (2.65).

The forming and processing of linear frequency modulation (FM) pulses is carried out by means of surface acoustic wave (SAW) devices and nonrecursive digital filters.

Let us discuss the first type of device [I.14]. A SAW device is a thin piezoelectric substrate with interdigital transducers (IDTs) of conductive material deposited on its surface by photolitography. The SAW filter is schematically shown in Figure 2.24. Voltage applied at the input IDT creates an electric field between its 'fingers' (electrodes), which, in turn, excites acoustic vibrations in the surface layer of the piezoelectric material. The acoustic waves propagate along the surface in directions perpendicular to the IDT electrodes. On reaching the output transducer, the vibrations are converted to electric voltage.

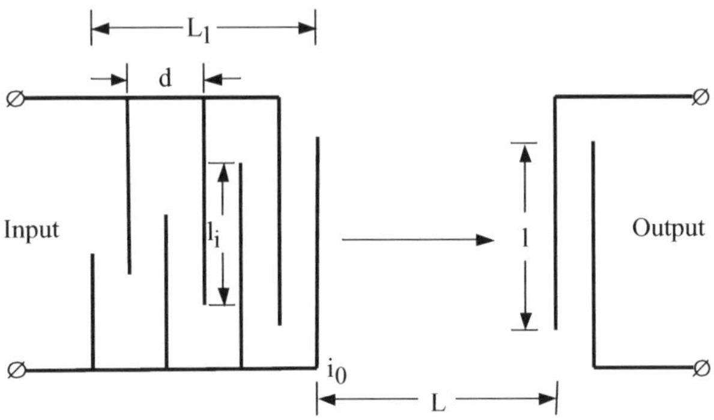

Figure 2.24 Surface acoustic wave device

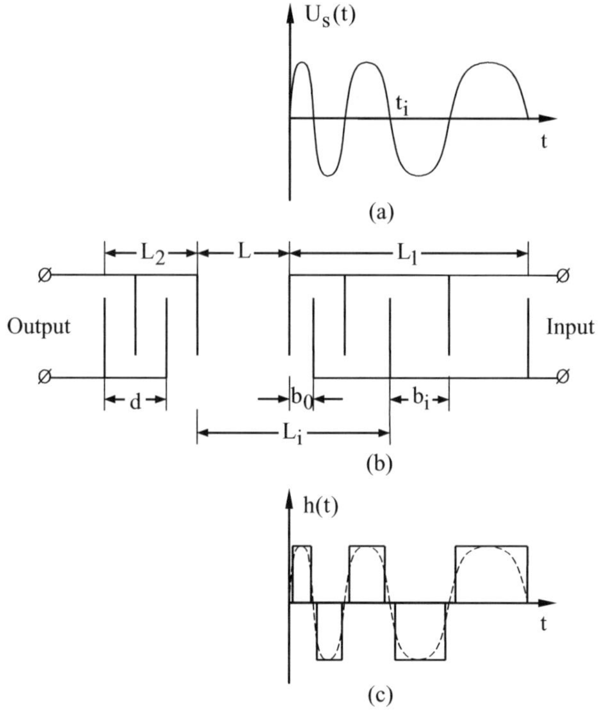

Figure 2.25 SAW filter for forming FM waveforms: (a) FM voltage to be formed; (b) arrangement of electrodes; (c) pulse responses. Graphical notations: —— $h_1(t)$; - - - - $h(t)$

It is known [I.14] that the frequency and pulse responses of an SAW filter are determined by the configuration of its input and output IDTs. For instance, a periodic IDT (i.e. an IDT with equispaced electrodes) allows a bandpass filter to be implemented with a central frequency $f_0 = V_u/d$ and a passband, $\Delta f = V_u/L_1$. Here, V_u is the speed of ultrasonic wave propagation along the surface of the piezoelectric material, d is the period of the IDT (see Figure 2.24), L_1 is the length of the IDT and $V_u \approx 3$ km/s.

Let us dwell on the principle of an SAW filter assisted forming and processing of chirp signals with any given phase (frequency) modulation law. At first, we will limit ourselves to signals whose phase is a continuous function of time. Suppose the formation of an RF pulse, $u_s(t)$, is required with a rectangular envelope of duration, τ_p, and an arbitrary phase modulation law, $u_s(t) = U_0 \cos[\omega_0 t + \varphi(t)]$. The hypothetical waveform is shown in Figure 2.25.

To form the required voltage by means of an SAW device, a nonapodized input IDT is used,[1] in which the arrangement of electrodes is similar (with a similarity coefficient V_u) to the position of nulls, $u_s(t)$, on the axis of time (see Figure 2.25). It follows from the figure that the length of the IDT should be equal to $L_1 = V_u \tau_p$. The output IDT is periodic, with a period $d = V_u/f_0$ and a length $L_2 = V_u/\Delta f_s$, where Δf_s is the signal bandwidth. Thus, the output IDT is a bandpass filter with a central frequency, f_0, and a passband, Δf_s.

Taking the pulse response of the device as the voltage to be formed, a short pulse implementing a δ-function is applied at the input IDT. To define the pulse response, an approximate model of acoustic vibrations is used, according to which the δ-function causes a pulse, $u_i(t)$, to be formed in the ith gap of

[1] A nonapodized IDT is an interdigital transducer, in which the electrodes' overlapping length (l_i in Figure 2.24) does not depend on gap number i. Otherwise, it is an apodized IDT.

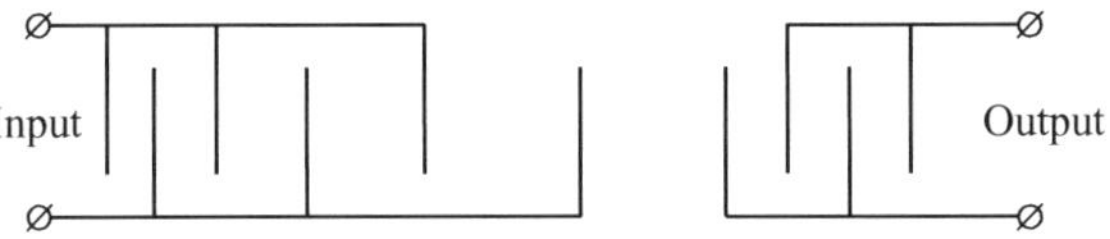

Figure 2.26 Arrangement of electrodes in an SAW filter for the processing of FM waveforms

the IDT, the duration of this pulse being

$$\tau_i = \frac{2b_i}{3V_u}.$$

Acoustic vibrations reach the output IDT with delays of $t_i = L_i/V_u$. Therefore, the pulse response of the input IDT is given by the expression

$$h_1(t) = \sum_{i=0}^{n} (-1)^i u_i(t - t_i).$$

The shape of $h_1(t)$ is shown by the solid line in Figure 2.25(c). The pulse response, $h_1(t)$, is different from the desired voltage in that it has steep fronts on certain intervals, which is indicative of the presence of higher harmonics of the carrier frequency. Since the output IDT does not pass components with frequencies outside the Δf_s band, the voltage, $h_1(t)$, smoothes out the output voltage, $h(t)$, adopting the desired waveform (the dashed curve in Figure 2.25 (c)). If the output IDT in the device shown in Figure 2.25 is placed to the right of the input IDT, as is shown in Figure 2.26, then the device will have a mirror pulse response relative to $h(t)$:

$$h_2(t) = h(t - t_0) = U_s(t_0 - t).$$

The filter in Figure 2.26 appears to be matched with the signal. Thus, the formation and matched filtering of a signal with an arbitrary phase modulation can be performed using filters with structurally identical IDTs.

Devices for the formation and matched filtering of chirp pulses are usually built in accordance with the scheme given in Figure 2.25, where the pattern of electrodes arrangement is defined by the relationship

$$\frac{\omega_0 L_i}{V_u} + \frac{\pi \Delta f_m L_i^2}{\tau_p V_u} = \frac{(2i+1)\pi}{2}.$$

Practically implemented designs of the processing device are somewhat different from the given scheme due to the necessity for sidelobe suppression in the compressed signal. One expedient way to satisfy this is to use a weighted (e.g. by a Hamming window) amplitude–frequency response of the compressing filter. Adequate realization of this technique is achieved through apodization of the input IDT by the following law:

$$l_i = l_{max}\left[0.08 + 0.92 \sin^2 \pi \left(\frac{L_i - L}{L_1}\right)\right].$$

The pulse response of the filter then takes the form

$$h(t) = \left[0.08 + 0.92 \sin^2\left(\frac{\pi t}{\tau_p}\right)\right] \sum_{i=0}^{n} u_i(t - t_i).$$

Taking a Fourier transform over $h(t)$, it can be seen that there appears to be a multiplier in the expression for the amplitude–frequency response of the filter:

$$0.08 + 0.92 \sin^2\left[\pi \frac{(f - f_{min})}{\Delta f_m}\right],$$

where f_{min} is the lower limit of the passband of the filter.

2.8.2 Phase-Coded Waveforms

Phase-coded waveforms (PCW) are obtained from the harmonic voltage by switching its phase in such a way that at certain instants it changes stepwise by a value φ. Although the shift in phase can be any chosen $\varphi = 2\pi/k$, the practical choice is often made in favour of binary PCW signals ($k = 2$), with the initial phase taking on two values: 0 and π (see Figure 2.27).

Basically, time intervals with a constant initial phase can be assigned arbitrarily. In practice, however, the said intervals (t_1 to t_4 in Figure 2.27(a) are multiples of the least interval (t_1 or t_2) referred to as the 'signal element', or 'discrete'. Phase-coded waveforms are classified into continuous (periodic) and pulsed signals. In the first case, the phase-shift pattern of the PCW signal recurs with a period τ_p. The total signal duration $\tau_s \gg \tau_p$. In the second case, the signal duration is equal to a single τ_p. The symbol τ_0 will be adopted to stand for the duration of a discrete.

It is not difficult to see that a PCW pertains to the class of complicated waveforms. Indeed, the spectrum of a PCW is the sum of spectra of several simple RF pulses, each being equal in duration to one of the intervals of the PCW. Calculations and experiments demonstrate that the spectrum bandwidth

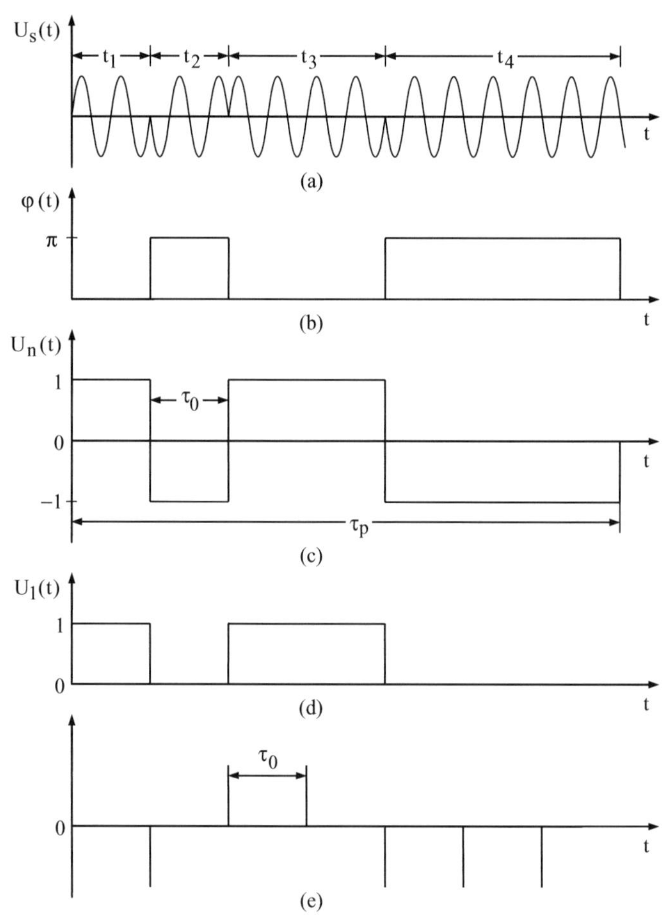

Figure 2.27 PSK signal characteristics: (a) waveform; (b) phase-shift pattern; (c) normalized complex envelope; (d) code; (e) lattice function

of a PCW signal is determined by the spectrum of its shortest interval, i.e. the discrete; $\Delta f_s \approx 1/\tau_0$. It is reasonable to assume that the base of a periodic signal is equal to the base of a pulsed signal:

$$B = \tau_p \Delta f_s = \frac{\tau_p}{\tau_0} = N \gg 1,$$

where N is the number of discretes on the length of a pulsed signal. This definition simplifies terminology without bringing about any essential changes. Note that a seven element PCW signal is shown in Figure 2.27.

The structure of a PCW is entirely determined by the pattern of the initial phase shift (Figure 2.27 (b)) or the pattern of a normalized complex amplitude shift (Figure 2.27 (c)). Since $\varphi(t) = 0, \pi$ in phase-coded signals, the signal's complex amplitude $\overline{U(t)} = U_0 \exp\left[j\varphi(t)\right]$ takes on two values, U_0 and $-U_0$, whereas the normalized complex amplitude

$$\overline{U_n(t)} = \frac{\overline{U(t)}}{U_0} = 1, -1.$$

If –1 is replaced with zero, the so-called code sequence of the PCW(often shortened to 'code') is obtained (see Figure 2.27 (d)). Just as with the normalized complex amplitude, the code sequence determines the structure and properties of the PCW completely. The PCW can also be characterized by a latticed function (LF) (Figure 2.27 (e)), which, by definition, is equal to

$$\mathrm{LF}(t) = \overline{U_n(t)} \sum_{i=0}^{N-1} \delta(t - i\tau_0).$$

At the same time, the latticed function may be regarded as a sampled complex envelope of the PCW signal with a sampling interval τ_0.

Codes are often based on linear recurrent sequences, in which each next symbol, u_i, is the weighted sum, m, of the previous symbols modulo 2:

$$u_i = a_1 u_{i-1} \oplus a_2 u_{i-2} \oplus \cdots \oplus a_m u_{i-m}. \tag{2.70}$$

Since commonly transmitted PCW are binary, let us focus on binary code sequences, in which the values of a_k and u_j are '0' or '1'. The quantity m is termed the 'sequence memory'. For example, let $m = 3$, $a_1 = a_3 = 1$ and $a_2 = 0$. Then

$$u_i = u_{i-1} \oplus u_{i-3}. \tag{2.71}$$

The three initial values of the code sequence are set as follows: $u_{i-3}^{(1)} = u_{i-2}^{(1)} = u_{i-1}^{(1)} = 1$. Table 2.1 is given to facilitate the calculation of u_i values. The three initial values $u_{i-1}^{(1)}$, $u_{i-2}^{(1)}$ and $u_{i-3}^{(1)}$ are entered into the three corresponding positions in the first row of the table. In the fourth (rightmost) position of this row, we write the first calculated value of $u_i^{(1)}$, which is $u_i^{(1)} = 1 \oplus 1 = 0$.

To obtain subsequent values of u_i, a one-step cyclic shift of the already known members of the sequence is performed according to the recurrent rule, i.e. $u_i^{(1)}$ iis entered in the first position of the second row, $u_{i-1}^{(1)}$ in the second position and $u_{i-2}^{(1)}$ in the third position; $u_i^{(2)}$ is then determined by formula (2.71), and so on.

Table analysis shows a periodic repetition of rows beginning with the eighth row, which is easy to explain. The set of characters in the three first positions of each row can be treated as a three-digit binary number. The entries of the fourth (rightmost) column are not taken into account at this time because they are functionally connected with u_{i-1} and u_{i-3}. Hence, the number of different rows is equal to the number of differing three-digit binary numbers, except for the number 0. An all-zero row is unacceptable

Table 2.1 Calculating u_i values

t/τ_0	u_{i-1}	u_{i-2}	u_{i-3}	u_i
1	1	1	1	0
2	0	1	1	1
3	1	0	1	0
4	0	1	0	0
5	0	0	1	1
6	1	0	0	1
7	1	1	0	1
8	1	1	1	0
9	0	1	1	1

because, should it emerge once, it will invariably recur in every subsequent step. Thus, the number of different rows in the table is

$$N = 2^3 - 1 = 7.$$

The recurrence of rows points to a periodic character of the sequence formed. Accordingly, the linear recurrent sequence in the case at issue has the period $N = 7$. It is not difficult to extend this result to sequences with any given value of memory, m. Generally, each row (less the entries of the rightmost column) is represented by an m-digit binary number. Therefore, the maximum number of different rows and, hence, the period of the sequence is equal to

$$N = 2^m - 1. \tag{2.72}$$

Sequences of the kind under consideration are called maximum-period sequences, or M-sequences.

Despite the regularity of their formation procedure, M-sequences are similar in their structure to purely random sequences of zeroes and unities. What is implied here are random sequences with equal probabilities of occurrence for '0' and '1'. Owing to structural similarity between purely random sequences and M-sequences, the latter are often referred to as pseudo-random sequences, and the resultant PCW signals as noise-like waveforms.

A typical PCW forming device consists of two parts: a code sequence generator and a waveform shaper. The M-sequence generator commonly used to obtain signal codes is based on a shift register with a modulo 2 adder in the feedback circuit (Figure 2.28). The clock pulse repetition period is chosen so as to equal the duration of the required code discrete.

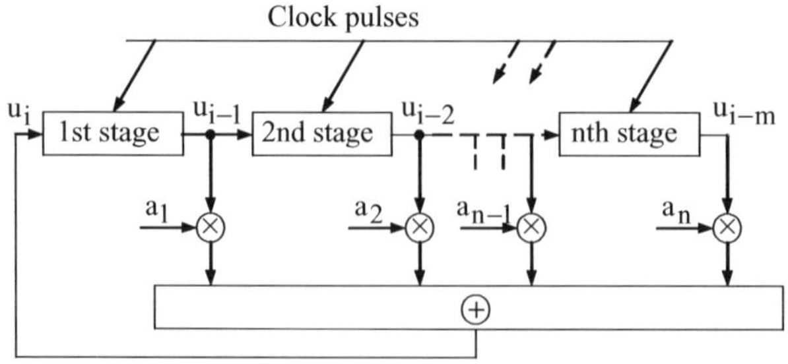

Figure 2.28 Functional diagram of the M-sequence generator

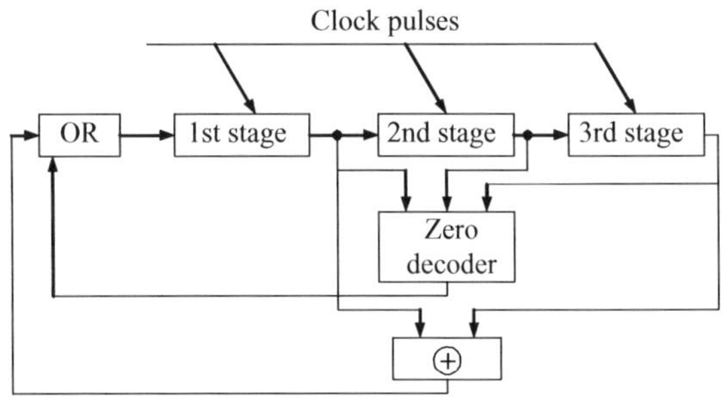

Figure 2.29 Functional diagram of a seven-element code generator

Suppose the voltage at the input of the register, u_i, represents the code sequence in question. Let us prove that it is a linear recurrent sequence. It follows from Figure 2.28 that the output voltage at the kth stage of the register is u_{i-k}. Hence,

$$u_i = a_1 u_{i-1} \oplus a_2 u_{i-2} \oplus \cdots \oplus \; a_m u_{i-m},$$

which was to be proved. Note that, since weighting coefficients $a_k = 0, 1$, the sequence former does not contain any multiplication circuits. When $a_k = 1$, the output of the kth stage of the register is connected to the adder, and when $a_k = 0$, the output of the kth stage is not connected to the adder.

The functional diagram of a seven-element M-sequence generator is shown as an illustration in Figure 2.29. Here, $a_1 = a_3 = 1$ and $a_2 = 0$. Consequently, $u_i = u_{i-1} \oplus u_{i-3}$. The operation of the circuit is clearly described in Table 2.1. When the device is activated, the states of register stages are arbitrary. It is therefore possible that all the stages yield '0' outputs. Exclusion of a zero combination can be achieved by introducing a zero decoder into the generator circuit (Figure 2.29), which yields a '1' output if there are zeroes at all its inputs. The zero decoder is also helpful in the case of soft failures in the device, which may bring about a zero combination. An alternative method to avoid a zero row consists in the periodic, forced setting of all the stages of the register to the desired state.

Figure 2.30 gives an example of a PCW forming circuit. Phase-shift coding is carried out here by means of an analogue multiplier with an output voltage given by

$$U_{\text{out}}(t) = \overline{U_{\text{n}}(t)} \cos \omega_0 t.$$

When $\overline{U_{\text{n}}(t)} = 1$, the initial phase of the output voltage is equal to zero. When $\overline{U_{\text{n}}(t)} = -1$,

$$U_{\text{out}}(t) = -\cos \omega_0 t = \cos(\omega_0 t + \pi),$$

i.e. the initial phase of $U_{\text{out}}(t)$ is equal to π.

PCW processing can be ensured through the use of correlators or matched filters in the receiving part of the radio system. This technique permits the PCW with carrier frequency or the PCW at the baseband to be processed.

First, the principle of the matched filtering of phase-coded waveforms (PCW) will be considered, regardless of its circuit realization. Let us demonstrate that a filter matched with a PCW can be built as

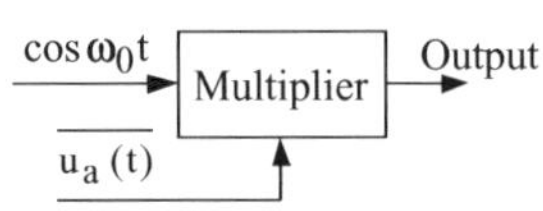

Figure 2.30 A multiplier-based keying circuit

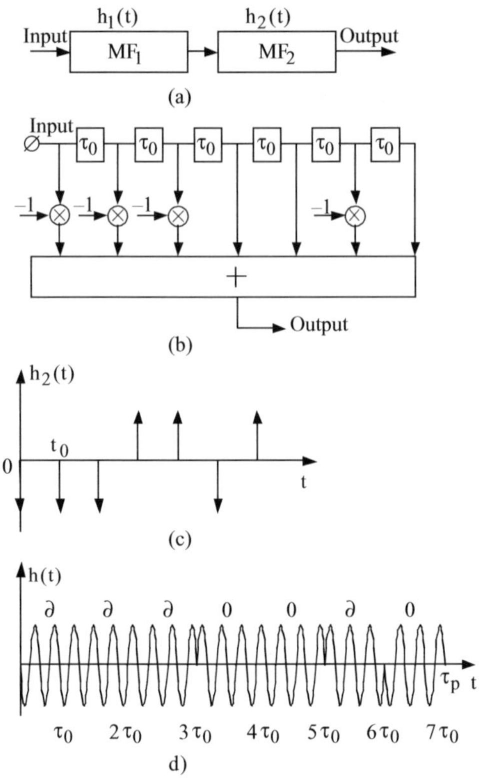

Figure 2.31 Matched filtering of the PSK waveform: (a) filter structure; (b) MF_2 structure; (c) $h_2(t)$; (d) $h(t)$

a cascade connection of two elements: MF1 and MF2 (see Figure 2.31 (a)). The filter MF1 is matched with one discrete of the PCW, i.e. with an RF pulse having a carrier frequency, f_0, and a duration, τ_0. The filter MF2 is matched with the latticed function of the PCW.

The pulse response of the filter MF1 follows the shape of a harmonic voltage pulse with a duration τ_0. Such a pulse is symmetrical relative to the middle and its mirror image coincides with that of the initial waveform. Thus,

$$h_1(t) = \begin{cases} \cos \omega_0 t, & 0 \le t \le \tau_0; \\ 0, & t < 0\,;\ t > \tau_0. \end{cases}$$

Suppose the filter being designed is to be matched with the PCW shown in Figure 2.27. Then, the pulse response, $h_2(t)$, representing the mirror image of the waveform's latticed function,

$$h_2(t) = \overline{U_n(\tau_p - t)} \sum_{i=0}^{N-1} \delta(t - i\tau_0), \tag{2.73}$$

will have a form similar to that given in Figure 2.31(c). The structure of MF2 as a transversal filter realizing the specified $h_2(t)$ is shown in the same Figure.

The resulting pulse response of the filter is shown in Figure 2.31(d). Comparing $h(t)$ with the initial PCW waveform, it can be seen (Figure 2.27) that $h(t)$ is a PCW with a phase change law that is specular relative to the initial PCW waveform.

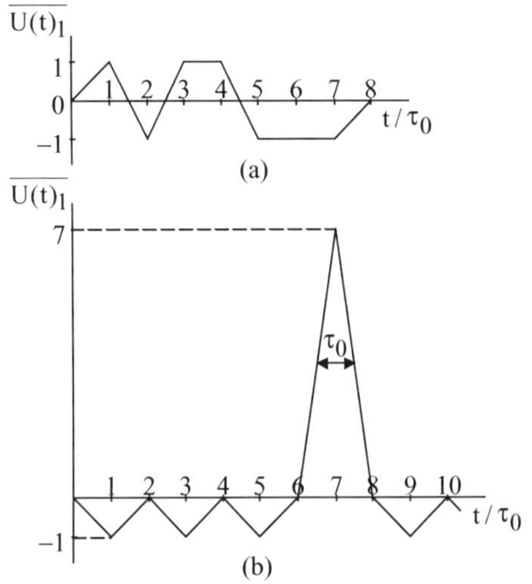

Figure 2.32 Voltage versus time diagrams for a matched filter: (a) MF1 output; (b) MF2 output

Let us examine the passage of a seven-element PCW through the matched filter shown in Figure 2.31. As stated in Section 2.2 (Figure 2.5), a rectangular RF pulse of duration τ, after passing through MF1, will be transformed into a triangular-envelope pulse of duration 2τ (at the base of the triangle). Using this result, the shape of the complex envelope of the PCW can be obtained at the output of MF1 (Figure 2.32(a)). The MF2 output voltage is a weighted sum of voltages at the taps of the delay line. For instance, the MF2 output voltage on the time interval $t = 0 - \tau_0$ will be governed by the inverted input voltage.

On the interval $t = \tau_0 - 2\tau_0$, the MF2 output voltage will be equal to the sum of inverted voltages that were present at its input on the interval $0 - \tau_0$ (after the first delay element) and on the interval $\tau_0 - 2\tau_0$ (at the input terminal). In the particular case of $t = 2\tau_0$, the MF2 output voltage is the sum of voltages opposite in phase; therefore it is equal to zero. Lastly, for the time interval $t = 6\tau_0 - 7\tau_0$, seven input voltages, each having passed a different number of delay elements, are summed in phase. Therefore, the MF2 output voltage for $t = 7\tau_0$ is seven times as great as the input voltage amplitude. The shape of the complex envelope of the voltage at the output of MF2 is given in Figure 2.32(b).

Accordingly, the duration of the main peak in the output signal of the matched filter (as referred to the level 0.5 of the maximum) is equal to the duration of the discrete τ_0. The ratio of signal durations at the MF input and output is

$$\frac{\tau_{\mathrm{p}}}{\tau_0} = B.$$

Since $\tau_0 \approx 1/\Delta f_{\mathrm{s}}$, the signal duration at the MF output is

$$\tau_{\mathrm{out}} \approx \frac{1}{\Delta f_{\mathrm{s}}},$$

which conforms to the general formula (2.66).

Matched filtering devices for the PCW with carrier frequency are easily realizable on the basis of SAW filters, just as in the case of chirp pulses. The SAW filter shown in Figure 2.33 implements the concept illustrated Figure 2.31. The periodic input IDT, which is a bandpass filter with a central f_0, and a passband $\Delta f = 1/\tau_0$, performs the function of MF1. Through this connection, the relationships $d = V_{\mathrm{u}}/f_0$ and $L_1 = V_{\mathrm{u}}\tau_0$ are satisfied.

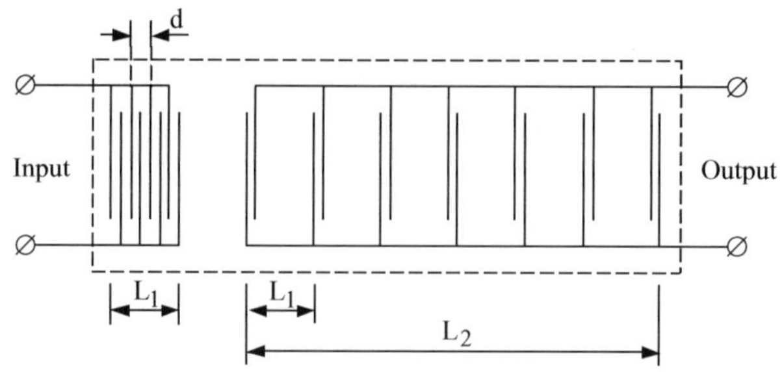

Figure 2.33 Structure of an SAW filter for a seven-element PSK waveform

The second IDT made up of seven pairs of electrodes, $L_1 = V_u\tau_0$ being the distance between adjacent pairs, has a pulse response, $h_2(t)$, shown in Figure 2.31(c). Indeed, if a single pulse (δ function) is produced in the left part of the SAW device, then, after its passage through the output IDT, the response at the output terminals of the latter will be in agreement with $h_2(t)$ (see formula (2.73)). The polarity of each individual δ-pulse within $h_2(t)$ depends on the connection order of the relevant pair of electrodes to the output terminals.

It can be seen that the SAW filter is a compact device for PCW processing. One disadvantage of this device is that it is not suitable for the processing of long signals. It was stated in Reference [I.15] that the maximum duration of a processed PCW is

$$\tau_p \approx 300\ \mu s.$$

Let us proceed to devices for the processing of PCW at baseband. Note that, again, it is practical to tune bandpass filter MF1 to the carrier frequency of the signal. Thus, for baseband signal processing, it will suffice to use the filter MF2.

Because the initial phase, φ_0, is not known when a PCW is detected, the expression for this signal has the form

$$u_s(t) = U(t)\cos\left[\omega_0 t + \varphi(t) + \varphi_0\right].$$

Here, $U(t)$ is the actual envelope having a duration τ_p. Consequently, the complex envelope of a PCW is

$$U(t) = U(t)\exp j[\varphi(t) + \varphi_0] = U_R(t) + jU_I(t).$$

The processing of a PCW at baseband can be carried out by means of a device represented schematically in Figure 2.15.

When MF2 is realized in the digital form, an analogue-to-digital converter (ADC) and a transversal filter are incorporated in each branch of the filter after the phase detector, as shown in Figure 2.34. The voltage sampling period, t_d, of the ADC must not exceed τ_0; i.e. not less than one sample per discrete of PCW signal is required for subsequent processing.

If the signal position in time is not known, as is the case in the radar detection mode, it is necessary to shorten the sampling interval in order to reduce power losses due to the mismatch of sampling pulses and peak voltages at the output of MF1. The following formula is typically recommended for estimating t_d:

$$t_d = \frac{\tau_0}{2}.$$

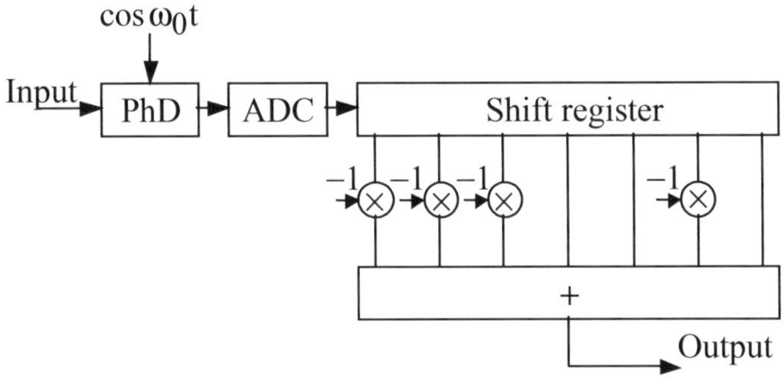

Figure 2.34 Block diagram of a digital realization of MF2

In the digital variant of MF2, a shift register assumes the role of a multitapped delay line. The clock pulse period should be equal to t_d and the number of register cells is

$$N_{reg} = \frac{\tau_p}{t_d}.$$

Thus, when $t_d = \tau_0$, $N_{reg.} = N$, and when $t_d = \tau_0/2$, $N_{reg} = 2N$. The number of taps going from the shift register to the adder is equal to N in all instances. Prior to summation, some numbers from register cells are inverted in accordance with Figure 2.31(b).

The digital correlator for the complex envelope of a PCW (Figure 2.35) consists of a phase detector (PhD), an analogue-to-digital converter (ADC), a reference voltage generator (RVG), a multiplier and an accumulator (integrator). The RVG forms a replica of the PCW waveform complex envelope $\overline{U_n(t-\tau_2)}$, which needs to be matched in time with the signal complex envelope $\overline{U_n(t-\tau_1)}$ by a search method. When $\tau_2 = \tau_1$, there is a signal maximum, NU_{in}, at the output of the correlator at the instant $t = \tau_p$.

Correlator realization requires considerably less hardware input as compared to the matched filter. However, when τ_1 is not known, it is necessary to provide for selection of this value in the correlator by the successive search of τ_2 values (see Section 2.4).

2.9 AMBIGUITY AND GENERALIZED AMBIGUITY FUNCTIONS OF RADAR SIGNALS

A common type of interference in radar is the undesirable RF signals transmitted by various radio aids or by the radar itself. Such interference comes from determinate processes and often has a structure similar, or even identical, to that of the useful signal (structural interference). Structural interference is different

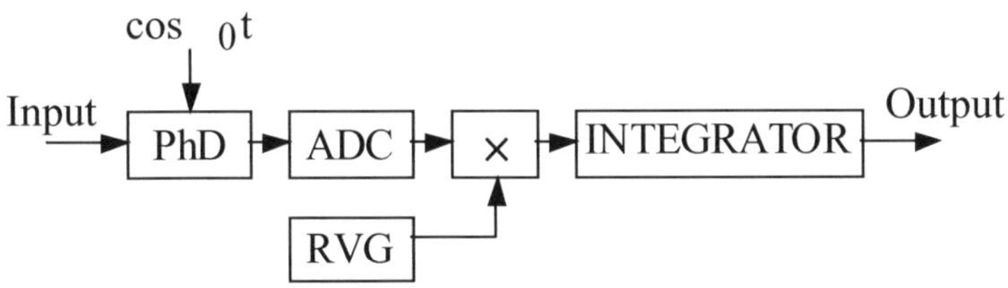

Figure 2.35 Block diagram of a digital correlator

from the useful signal only in the values of amplitude, frequency, direction and time of arrival. Radar signals reflected from unwanted objects and obscuring the useful signal represent structural interference.

In this section, we will not deal with those relatively simple cases in which frequency and/or time position of interference are known and differ considerably from the relevant parameters of the signal. In such cases, interference can be eliminated by the methods of frequency discrimination and/or time selection.

The estimation of a radar's immunity to structural interference is carried out on the assumption that the radar receiver contains a filter matched with the useful signal. The degree of immunity to interference is estimated by the ratio,U_i/U_s, of the amplitudes of interference and the useful signal at the output of the filter at the instant t_0 when the useful signal is at a maximum.

When determining the ratio U_i/U_s, it is assumed that both the useful signal and interference sources are located at the beam peak of the antenna system. If the source of interference is outside the main lobe of the antenna pattern, interference attenuation can be considered separately.

The ratio of the amplitudes of interference and the useful signal can be written as

$$\frac{U_i}{U_s} = \Psi,$$

where Ψ is the function of structural and parametric differences between the signal and the interference. To simplify calculations, the complex form for writing voltages is used.

According to Equation (2.45), the useful signal can be written in the form

$$u_s(t) = \mathrm{Re}\left[\overline{U_s(t)}\exp(\mathrm{j}\omega_s t)\right]$$

Then, using Equation (2.26), the pulse response of the matched filter can be expressed by the formula

$$h(t) = \mathrm{Re}\left\{\overline{U_s(t_0 - t)}\exp\left[\mathrm{j}\omega_s(t_0 - t)\right]\right\}.$$

Similarly, interference at the input can be represented by

$$u_i(t) = \mathrm{Re}\left[\overline{U_{i1}(t)}\exp(\mathrm{j}\omega_i t)\right],$$

Let us adopt the symbol ω to designate the difference in carrier frequencies of the signal and the interference, $\omega = \omega_s - \omega_i$. Then,

$$u_i(t) = \mathrm{Re}\left[\overline{U_{i1}(t)}\exp(-\mathrm{j}\omega t)\exp(\mathrm{j}\omega_s t)\right] = \mathrm{Re}\left[\overline{U_i(t)}\exp(\mathrm{j}\omega_s t)\right],$$

where

$$\overline{U_i(t)} = \overline{U_{i1}(t)}\exp(-\mathrm{j}\omega t).$$

Using Equation (2.29), the voltage of interference at the output of the filter can be written as

$$u_{i.out}(t) = \int \mathrm{Re}\left[\overline{U_i(x)}\exp(\mathrm{j}\omega_s t)\right]\ \mathrm{Re}\left\{\overline{U_s(t_0 + x - t)}\exp\left[\mathrm{j}\omega_s(t_0 + x - t)\right]\right\}\mathrm{d}x.$$

Taking into account Equations (2.51) and (2.53), gives

$$u_{i.out}(t) = \tfrac{1}{2}\int \mathrm{Re}\left\{\overline{U_i(x)}\ \overline{U_s(t_0 + x - t)}\exp\left[\mathrm{j}2\omega_s(t_0 + x - t)\right]\right\}\mathrm{d}x$$
$$+\tfrac{1}{2}\int \mathrm{Re}\left[\overline{U_i(x)}\ \overline{U_s(t_0 + x - t)}^{*}\exp(\mathrm{j}\omega_s t)\exp(-\mathrm{j}\omega_s t_0)\right]\mathrm{d}x. \qquad (2.74)$$

The first term in Equation (2.74), which is the integral of a fast oscillating function (multiplier $\exp(\mathrm{j}2\omega_s x)$), is much less than the second term. Therefore, the first term can be neglected in the expression obtained. Let us also take into account that

$$\int \mathrm{Re}\overline{f(x)}\,\mathrm{d}x = \mathrm{Re}\int \overline{f(x)}\,\mathrm{d}x.$$

Then,

$$u_{\mathrm{i.out}}(t) = \tfrac{1}{2}\mathrm{Re}\left[\int \overline{U_{\mathrm{i}}(x)}\ \overline{U_{\mathrm{s}}(t_0 + x - t)}^{*}\, \mathrm{d}x\ \exp(-\mathrm{j}\omega_s t_0)\ \exp(\mathrm{j}\omega_s t)\right].$$

In accordance with Equation (2.30), the amplitude of this voltage is

$$U_{\mathrm{i}}(t) = \tfrac{1}{2}\left|\int \overline{U_{\mathrm{i}}(x)}\ \overline{U_{\mathrm{s}}(t_0 + x - t)}^{*}\mathrm{d}x \exp(-\mathrm{j}\omega_s t_0)\right| = \tfrac{1}{2}\left|\int \overline{U_{\mathrm{i}}(x) U_{\mathrm{s}}(t_0 + x - t)}^{*}\mathrm{d}x\right|.$$

At the instant t_0, the amplitude of the interference voltage at the output is

$$U_{\mathrm{i}}(t_0) = \tfrac{1}{2}\left|\int \overline{U_{\mathrm{i}}(x)}\ \overline{U_{\mathrm{s}}(x)}^{*}\mathrm{d}x\right|. \tag{2.75}$$

The peak signal is obtained through substitution of $\overline{U_{\mathrm{i}}(x)} = \overline{U_{\mathrm{s}}(x)}$ into Equation (2.75):

$$U_{\mathrm{s}}(t_0) = \tfrac{1}{2}\left|\int \left|\overline{U_{\mathrm{s}}(x)}\right|^2 \mathrm{d}x\right| = E,$$

where E is the signal energy. Consequently, the interference-to-signal ratio at the output of the filter at the instant when the signal is at its maximum is equal to

$$\frac{U_{\mathrm{i}}}{U_{\mathrm{s}}}(t_0) = \frac{1}{2E}\left|\int \overline{U_{\mathrm{i}}(t)}\ \overline{U_{\mathrm{s}}(t)}^{*}\mathrm{d}t\right|.$$

The resultant ratio is a function of structural and parametrical differences of the signal and the interference. Therefore, this ratio is called the normalized cross-correlation function, or the generalized ambiguity function (GAF) of the signal. Granting equality of module of the complex conjugate quantities, the expression for the generalized ambiguity function can be written in the form

$$\mathrm{GAF} = \frac{1}{2E}\left|\int \overline{U_{\mathrm{s}}(t)}\ \overline{U_{\mathrm{i}}(t)}^{*}\mathrm{d}t\right|.$$

Let us examine the so-called similar interference, which is a special case of structural interference. In this case,

$$\overline{U_{\mathrm{i}}(t)} = \overline{U_{\mathrm{s}}(t-\tau)}\exp(-\mathrm{j}\omega t),$$

where τ is the time shift between the interference and the signal. The interference-to-signal ratio at the output of the matched filter is equal to

$$\frac{U_{\mathrm{i}}}{U_{\mathrm{s}}}(t_0) = \frac{1}{2E}\left|\int \overline{U_{\mathrm{s}}(t)}\ \overline{U_{\mathrm{s}}(t-\tau)}^{*}\exp(\mathrm{j}\omega t)\,\mathrm{d}t\right|.$$

The resultant ratio depends on the structure of the signal and is a function of the difference in frequency and the time of arrival between the signal and the interference. This ratio is called the normalized two-dimensional autocorrelation function of the signal, which is hereafter designated as

$$\Psi(\tau, F) = \frac{1}{2E}\left|\int \overline{U_{\mathrm{s}}(t)}\ \overline{U_{\mathrm{s}}(t-\tau)}^{*}\exp(\mathrm{j}2\pi F)\,\mathrm{d}t\right|. \tag{2.76}$$

In formula (2.76), $F = \omega/(2\pi)$. The squared value Ψ^2 is called the ambiguity function of the signal. Nevertheless, further on, in order to simplify terminology, this term (the ambiguity function) will be applied to function (2.76).

The geometric representation of $\Psi(\tau, F)$ takes the form of a surface over the plane τ, F (Figure 2.36). Hence, the ambiguity function is the normalized amplitude of the signal at the MF output, which

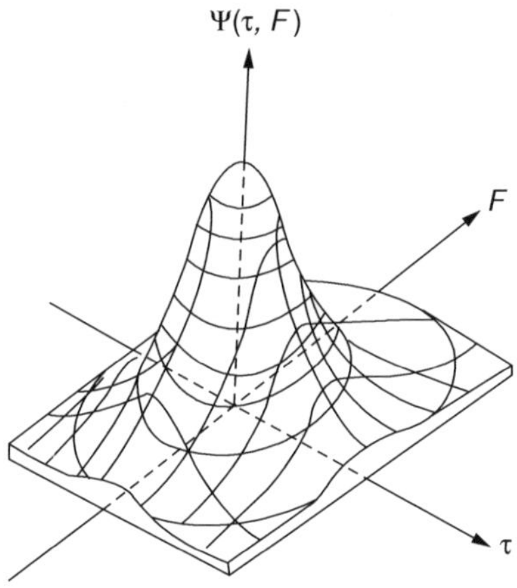

Figure 2.36 A signal ambiguity function

is different from the useful signal-in-time position and frequency; this function defines the time and frequency resolution provided by the signal of a given structure.

The matched filter and the correlator are equivalent in terms of energy. Therefore, Equation (2.76) also describes the connection between the maximum of the normalized voltage at the output of the correlator and the mismatch of input and reference voltages in time and frequency.

The most important properties of the ambiguity function are as follows:

1. $\Psi(\tau, F) \leq \Psi(0, 0)$.
2. $\Psi(-\tau, -F) = \Psi(\tau, F)$.
3. $\int_{(-\infty)}^{\infty} \int \Psi^2(\tau, F)\, d\tau dF = 1.$ (2.77)

The last relationship is of paramount importance. It states that the volume of space between the surface $\Psi^2(\tau, F)$ and the plane $\tau,\ F$ equals unity. The space enclosed by the surface $\Psi^2(\tau, F)$ is often referred to as the 'ambiguity solid' of a signal. In connection with Equation (2.77), let us examine the question of the desired shape of the ambiguity function. When the highest possible resolution of signals in time and frequency are required, it is desirable to have an ambiguity function shaped as an indefinitely narrow peak at the point $\tau = 0$, $F = 0$ (Figure 2.37(a)).

According to Equation (2.77), this is impossible since the volume of an ambiguity solid cannot equal zero. At the same time, formula (2.77) helps to find a way of obtaining a signal with an ambiguity function close to desirable. The feasible ambiguity function (that is to say, $\Psi^2(\tau, F)$) permitting a high resolution in time and frequency is shown in Figure 2.37(b). If the pedestal volume is made to approximate unity, then an extremely small peak volume can be achieved. Therefore, we assume an indefinitely small opening of the peak base along the axes τ and$F$. To make the pedestal height small, we need to have a large area of the base, i.e. time extension of the input signal and frequency extension of its spectrum should be ideally great. Thus, the signal base should be $\gg 1$.

When signals are to be resolved in time only, i.e. when $F = 0$, it is enough to know the section of the ambiguity function by the vertical plane passing through the axis τ (plane $F = 0$). The expression for this section is obtained by substituting $F = 0$ into Equation (2.76):

$$\Psi(\tau, 0) = \frac{1}{2E} \left| \int_{-\infty}^{\infty} \overline{U_s(t)}\, \overline{U_s(t-\tau)}^* \, dt \right|. \tag{2.78}$$

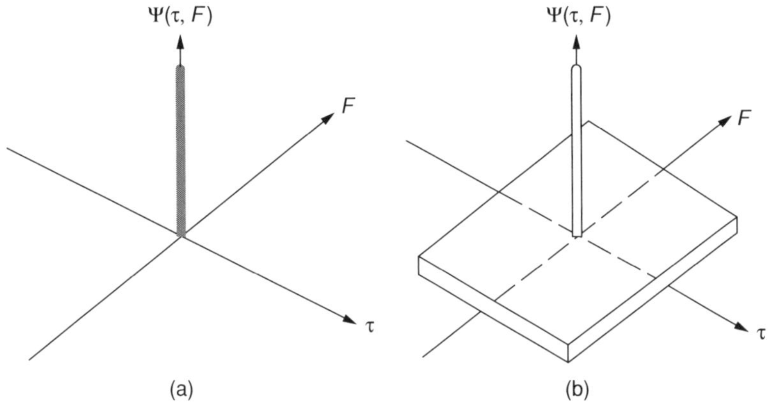

Figure 2.37 (a) Ideal and (b) practicable shapes of the ambiguity function

This section represents the modulus of the signal's normalized autocorrelation function, which, by definition, is equal to

$$r(\tau) = \frac{1}{2E} \int_{-\infty}^{\infty} \overline{U_s(t)}\, \overline{U_s(t-\tau)}^{*} \mathrm{d}t.$$

When $F = 0$, the interfering signal differs from the useful signal solely in the time shift τ. Therefore, the numerator of Equation (2.78) is the modulus of the signal amplitude at the output of the matched filter at an instant differing by τ from the instant corresponding to the maximum of the output voltage.

Consequently, $\Psi(\tau, 0)$ represents a normalized signal envelope at the output of the matched filter or, which is the same, the normalized voltage at the output of the correlator as a function of the time shift of the input signal and reference voltage.

When it is required to characterize the resolution of signals only in terms of frequency, i.e. when $\tau = 0$, it is enough to know the section of the ambiguity function by the vertical plane passing through the axis F (plane $\tau = 0$). The expression for this section is obtained by substituting $\tau = 0$ into Equation (2.76):

$$\Psi(0, F) = \frac{1}{2E} \left| \int_{-\infty}^{\infty} \left|\overline{U_s(t)}\right|^2 \exp(\mathrm{j}2\pi F t)\, \mathrm{d}t \right|. \tag{2.79}$$

It is apparent from Equation (2.79) that the given section of the ambiguity function represents the modulus of the normalized power spectrum of the signal. As the signal has a rectangular envelope, $\left|\overline{U_s(t)}\right|^2$ also has a rectangular shape. Therefore, in this case, $\Psi(0, F)$ coincides in shape with the spectrum of the signal.

Horizontal planes cutting the ambiguity function at different levels are often used to simplify the characteristic of the function. The sections thus obtained are called ambiguity contours and characterize the function in the same way as contour lines on a geographical map characterize the relief.

By way of example, we will work out ambiguity functions for selected waveforms.

2.9.1 Ambiguity Function of an RF Pulse with a Rectangular Envelope

Analytical representation of an RF pulse takes the form

$$u(t) = \begin{cases} U_{\mathrm{m}} \cos \omega_0 t & \text{if } 0 \le t \le \tau_{\mathrm{p}}, \\ 0 & \text{if } t < 0 \text{ or } t > \tau_{\mathrm{p}}. \end{cases}$$

The complex amplitude of the signal, $\overline{U_s(t)}$, coincides with its actual amplitude. The function $\overline{U_s(t-\tau)}^*$ is also a rectangle shifted by $|\tau|$ to the right or to the left along the axis of time. When $\tau > 0$, $\overline{U_s(t)}\,\overline{U_s(t-\tau)}^* = U_m^2$ for $0 \le t \le \tau_p$ and $\overline{U_s(t)}\,\overline{U_s(t-\tau)}^* = 0$ for other values of t. Besides,

$$2E = U_m^2 \tau_p.$$

Hence,

$$\Psi(\tau, F) = 1/U_m^2\tau_p \left| \int_\tau^{\tau_p} U_m^2 \exp(j2\pi F t)\,dt \right| = \left| \frac{\sin\left[\pi F \tau_p \left(1 - \tau/\tau_p\right)\right]}{\pi F \tau_p} \right|. \tag{2.80}$$

When $\tau < 0$, the limits of integration in Equation (2.80) are 0 and $\tau_p - |\tau|$. Furthermore,

$$\Psi(\tau, F) = \left| \frac{\sin\left[\pi F \tau_p \left(1 - |\tau|/\tau_p\right)\right]}{\pi F \tau_p} \right|. \tag{2.81}$$

Equation (2.81) differs from Equation (2.80) in that it contains$|\tau|$. Formula (2.80) is valid for positive values of τ only. Hence, Equation (2.81) is a general expression for the waveform in question.

Let us determine the sections $\Psi(\tau, F)$ and the ambiguity contours for an RF pulse

$$\Psi(\tau, 0) = \lim_{F\to 0} \frac{\sin \pi F \tau_p \left(1 - \tau/\tau_p\right)}{\pi F \tau_p} = 1 - |\tau|/\tau_p.$$

It is obvious from this expression that, when $|\tau| \ge \tau_p$, there is no mutual interference of the two RF pulses. Therefore, the quantity, τ_p is often adopted as a measure of signal resolution in time, $\Delta t = \tau_p$. The corresponding expression for range resolution is $\Delta R = c\tau_p/2$.

The second section is

$$\Psi(0, F) = \left|\sin \pi F \tau_p / \pi F \tau_p\right|.$$

The sections $\Psi(\tau, 0)$ and $\Psi(0, F)$ are graphically represented in Figure 2.38. The quantity $\Delta F = 1/\tau_p$, which corresponds to the main lobe boundary of the function $\Psi(0, F)$, is commonly adopted as a measure of the signal resolution in frequency. The resolution in velocity of objects is

$$\Delta V = \lambda\, \Delta F/2; = \lambda/2\tau_p.$$

Figure 2.38 also presents ambiguity contours for RF pulses of different durations. It can be seen that as τ_p increases, the time resolution degrades and the frequency resolution improves. Incidentally, the product of these resolutions is $\Delta t\ \Delta F = 1$.

2.9.2 Ambiguity Function of a Chirp Pulse

The complex envelope of a chirp pulse is

$$\overline{U_s(t)} = \begin{cases} U_m \exp\left(-jbt^2\right) & \text{if } 0 \le t \le \tau_p, \\ 0 & \text{if } t < 0,\ t > \tau_p. \end{cases}$$

Consequently, the expression for the ambiguity function of this waveform for $\tau \le 0$ takes the form

$$\begin{aligned} \Psi(\tau, F) &= \frac{1}{U_m^2 \tau_p} \left| \int_0^{\tau_p - |\tau|} U_m^2 \exp\left(-jbt^2\right) \exp\left[jb(t-\tau)^2\ \exp(j2\pi F t)\right] dt \right| \\ &= \frac{1}{\tau_p} \left| \int_0^{\tau_p - |\tau|} \exp\left[j2(\pi f - b\tau)t\right] dt \right|. \end{aligned}$$

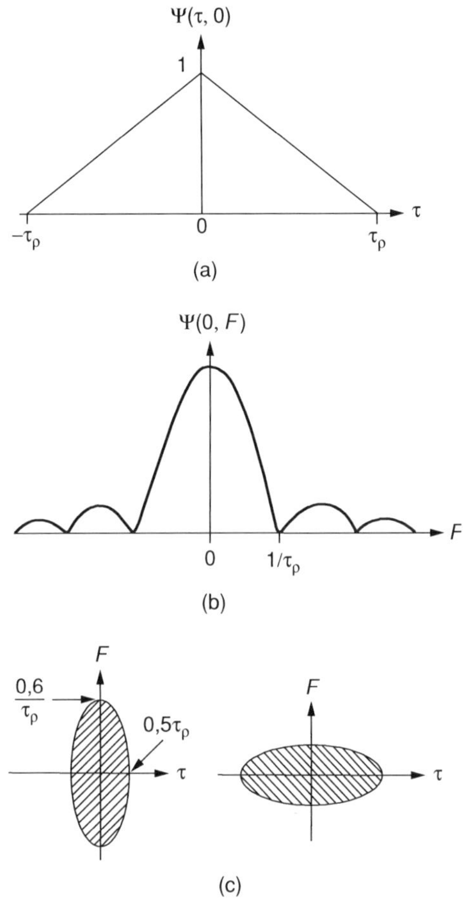

Figure 2.38 Sections (a) $\Psi(\tau, 0)$ and (b) $\Psi(0, F)$ of the ambiguity function and (c) ambiguity contours of the RF pulse

Through calculations similar to those performed in deducing Equation (2.81), granting $b\tau_p = \pi \Delta f_m$ gives

$$\Psi(\tau, F) = \left| \frac{\sin\left[\left(\pi F \tau_p - \pi \Delta f_m \tau\right)\left(1 - |/\tau|\tau_p\right)\right]}{\pi F \tau_p - \pi \Delta f_m \tau} \right|. \tag{2.82}$$

Consider the sections $\Psi(\tau, F)$ and the ambiguity contours of a chirp pulse. The section

$$\Psi(0, F) = \left| \frac{\sin \pi F \tau_p}{\pi F \tau_p} \right|$$

does not differ from its counterpart in a nonmodulated RF pulse as the shape of $\Psi(0, F)$ is only determined by the shape of the signal envelope.

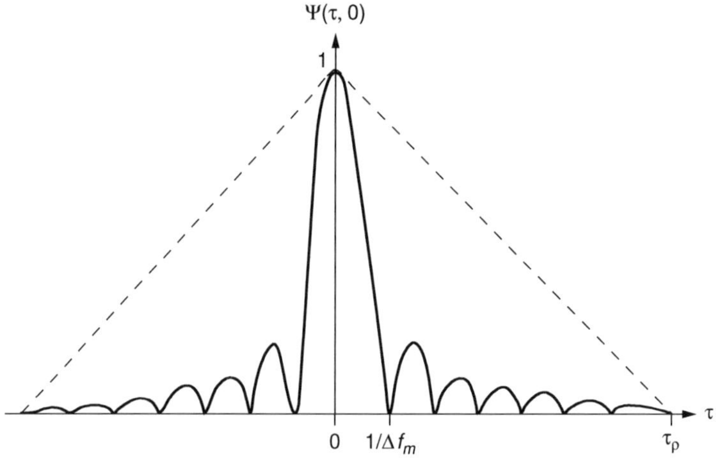

Figure 2.39 The section $\Psi(\tau, 0)$ of the chirp pulse ambiguity function

The expression for section $\Psi(\tau, 0)$ is obtained by substituting $F = 0$ into Equation (2.82):

$$\Psi(\tau, 0) = \left| \frac{\sin\left[\pi \Delta f_m \tau \left(1 - |\tau| / \tau_p\right)\right]}{\pi \Delta f_m \tau} \right| .$$

Let us first study the behaviour of $\Psi(\tau, 0)$ when $\tau/\tau_p \ll 1$. The following approximate expression for τ values within the stated interval can be written as

$$\Psi(\tau, 0) = \left| \frac{\sin \pi \Delta f_m \tau}{\pi \Delta f_m \tau} \right| . \qquad (2.83)$$

It follows from Equation (2.83) that $\Psi(\tau, 0)$ first turns to zero when $\tau = 1/\Delta f_m$. From this point onwards $\Psi(\tau, 0)$ oscillates as depicted in Figure 2.39.

Finally, the section of the ambiguity function by the vertical plane passing through a straight line,

$$F\tau_p - \Delta f_m \tau = 0 \qquad (2.84)$$

in plane τ, F will be examined. At all the points of this plane, the numerator and the denominator of Equation (2.82) turn to zero. Therefore the said section is expressed by a limit

$$\Psi\left(\tau, \frac{\tau}{\tau_p}\Delta f_m\right) = \lim_{F\tau_p - \Delta f_m \tau \to 0} \Psi(\tau, F) = 1 - |\tau|\tau_p .$$

In the section considered (Figure 2.40(a)), the ambiguity function decreases with increasing τ as slowly as the ambiguity function of a nonmodulated pulse. Moreover, with increasing F, the ambiguity function in this section decreases much more slowly than the same function of a simple RF pulse. The ambiguity contours of a chirp pulse (at level 0.5) constructed using the obtained chirp signal characteristics are given in Figure 2.40(b).

Analysis of this figure allows us to single out the basic properties of a chirp pulse as employed in radar.

1. When it is necessary to resolve signals with a small difference in frequency, the chirp waveform provides a high-range resolution, described by $\Delta R = c/(2\Delta f_m)$.
2. When the time shift between signals is insignificant, the resolution of a chirp waveform in velocity (Doppler frequency) is determined by $\Delta V = \lambda/(2\tau_p)$, just as in the case of a nonmodulated pulse.

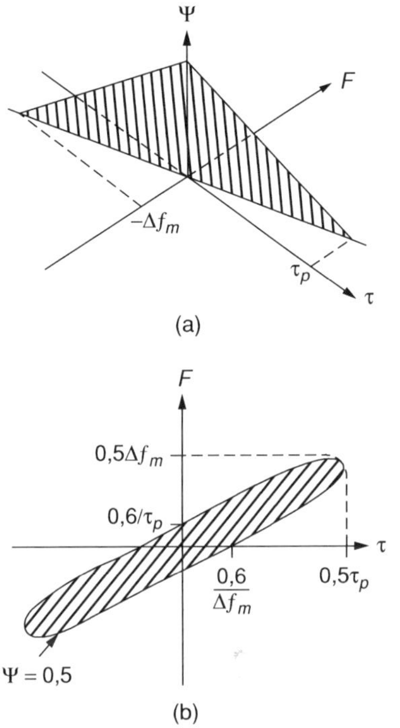

Figure 2.40 Section of (a) a chirp pulse ambiguity function and (b) ambiguity contour, if $F\tau_{\rm p} - \Delta f_{\rm m}\tau = 0$

3. With an unfavourable relationship between the time shift and frequency difference of the two signals (2.84), the range and velocity resolution of a chirp pulse may degrade sharply.
4. Owing to the peculiarity of the ambiguity function, which is shaped as a crest with its peak on the vertical plane passing through a straight line (2.84), the filter matched with the chirp pulse, $U_{\rm m}\cos\left(\omega_0 t - bt^2\right)$, appears to be nearly matched with a signal whose carrier frequency is somewhat different from ω_0. This difference should definitely be much less than the frequency deviation. If two signals with slightly differing carrier frequencies come to the input of a matched filter simultaneously, then, at the output, these signals will be shifted in time proportionally with their difference in frequency.

2.9.3 Ambiguity Function of a Phase-Coded Waveform

The ambiguity function of a phase-coded signal has the shape of a thumb-tack with its point turned upwards. The peak of the useful signal is in the centre of the time–frequency plane. It is equal to

$$\Psi(0, 0) = 1.$$

The maximum sidelobe level in the section $\Psi(\tau, 0)$ is $1/\sqrt{N}$ for a pulsed signal keyed by an M-sequence. For signals keyed by Barker codes or periodic M-sequences, this quantity equals $1/N$. On the entire time–frequency plane, the maximum sidelobe level of M-sequence coded signals seldom exceeds $3/\sqrt{N}$ [I.16]. On average, sidelobes of these signals have an order of $1/\sqrt{N}$. This observation needs to be taken into account when estimating detectability of objects on a clutter background.

3

Radar Power Budget Analysis and Radar Systems Classification

3.1 INTRODUCTION

In Section 1.3 the procedure of detecting a signal returned by a radar target was briefly discussed. Furthermore, several salient features of this procedure were singled out, namely:

- The preassigned characteristics of the radar mode under consideration are the probabilities of detection, P_d, and false alarm, P_{fa}.
- The required signal-to-noise (signal-to-interference) ratio, ρ_{min}, serves as an index of the performance of the detector; evidently it is desirable to have the least possible value of this index when implementing the preassigned P_d and P_{fa}.
- Target detection is carried out in the process of a successive survey of a predetermined plane or solid angle.
- It is assumed that the detected signal has the form of a pulse burst made up of N pulses.
- The signal detector performs the integration of a useful signal mixed with noise within a burst; the cumulative value is compared with a threshold.

To simplify the calculation of the detection procedure, an equivalent pulse burst is considered. The shape of a real pulse burst is determined by the antenna pattern of the radar (Figure 3.1(a)). In the equivalent burst (Figure 3.1(b)), the amplitudes of all the pulses are equal to the amplitude of the pulse corresponding to the beam peak. An equivalent burst duration is limited by a –3 dB beamwidth. With such a definition for the equivalent pulse burst, the total signal power is bound to be overestimated. Henceforward, the value of the hardware loss, L_a, in formula (1.26) should be increased by 1.8 dB for a single-plane scan and by 3 dB for a two-plane scan [I.17].

Signal integration is carried out differently for a coherent and a noncoherent burst. In the former case (Figure 3.2), it is expedient to use a filter matched with the burst (BMF). In the latter case, straightforward summation of pulses with a random phase is not effective. Therefore, it is necessary to use the scheme given in Figure 3.3.

The first element in this scheme is a filter matched with the coherent parts of the burst, i.e. individual pulses. To utilize the energy of the burst in full, it is necessary to go over to its envelope, which is done at

Bistatic Radar: Principles and Practice Edited by M. Cherniakov

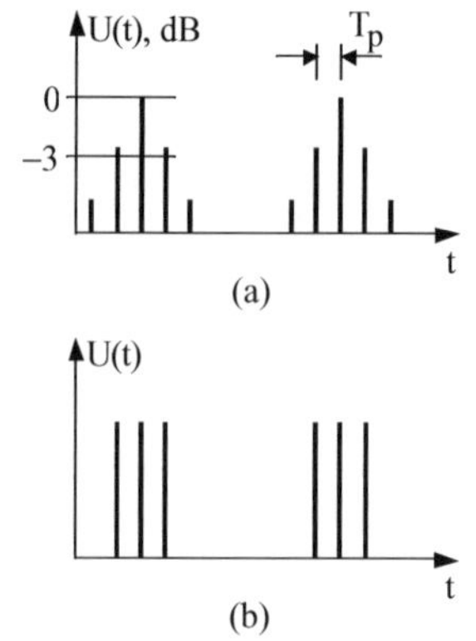

Figure 3.1 Burst signals: (a) real burst; (b) equivalent burst

the signal detection stage. Next, the envelope pulses are summed. This operation, known as noncoherent integration, provides an N-fold gain in terms of energy.

Amplitude detection is accompanied by losses that grow with a decreasing signal-to-noise ratio at the input of the detector. As the number of integrated pulses increases, a reduction in the signal-to-noise ratio at the detector input can be permissible. Thus, the loss of signal detection grows with increasing N, giving the term 'noncoherent integration loss'. It is worth mentioning here that the joint effect of signal detection and subsequent integration is always positive.

One more variety of a burst signal to be considered is the partially coherent burst. Such a burst will consist of N_n noncoherent intervals, each of them containing N/N_n coherent pulses. In this case, it is practical to use a filter matched with the coherent parts of the burst of N/N_n pulses. Signal detection is then performed and N_n samples of the envelope are summed.

Another kind of energy loss is due to peculiarities of signals returned by complex targets, which constitute the majority of radar objects (see Section 1.6.5). Figure 3.4 depicts the model of a complex target represented by a cluster of bright points, which are designated by criss-crosses.

Reflections from the vicinities of these points give the main contribution to the echo signal received by the radar. The signal, u_s, reflected by the object is a result of interference of the partial signals from n bright points:

$$u_s = \sum_{i=1}^{n} u_i \cos(\omega_0 t + \varphi_i), \qquad \varphi_i = \frac{4\pi R_i}{\lambda}. \tag{3.1}$$

Here, u_i is the amplitude, φ_i is the initial phase of the partial signal from the ith bright point and λ is the wavelength of the radar system.

When the object's aspect angle relative to the radar (angle α in Figure 3.4) varies, so do the ranges, R_i, which brings about a variation in the phases φ_i (and possibly u_i). Thus, it is easy to establish from Equation (3.1) that the parameters subject to variation are the echo signal amplitude, initial phase, Doppler shift and position of the so-called radar centre of the target. The time history of the said parameters is defined by the spectrum of fluctuations or their correlation function. The correlation window, τ_c, offers the simplest estimate of the fluctuations rate. It is usually assumed for simplicity that the values of a signal

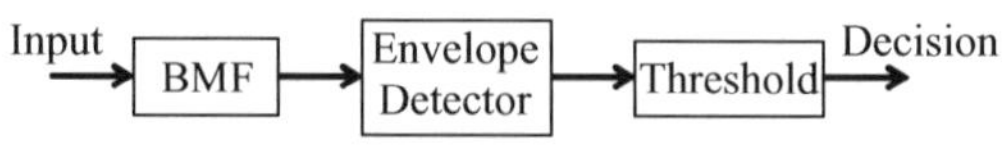

Figure 3.2 Coherent burst detection scheme

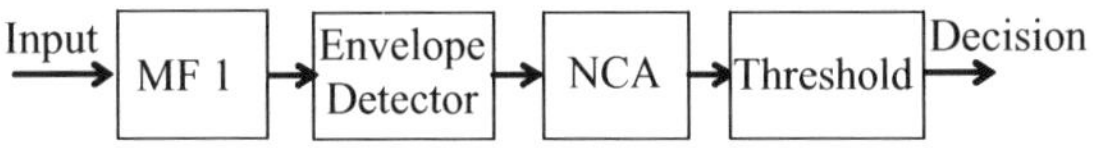

Figure 3.3 Noncoherent burst detection scheme: NCA, noncoherent accumulator

parameter are completely correlated on time interval, τ_c, while signal parameter values separated by an interval exceeding τ_c are not correlated (i.e. independent under Gaussian fluctuations). In accordance with this [I.18], the quantity N_e is introduced, which is the number of independent signal values over detection time, T_0:

$$N_e = 1 + \frac{T_0}{\tau_c}. \tag{3.2}$$

The following formula can be used for an approximate evaluation of τ_c:

$$\tau_c \approx \frac{\lambda}{4L_t\Omega_t}, \tag{3.3}$$

where L_t is the linear dimension of the target and Ω_t is the rate of its aspect angle variation. For moving targets such as aircraft, Ω_t in the first approximation is the angular rate of object fluctuations relative to its gravity centre, expressed in radians per second.

Formula (3.3) is admittedly a simple physical interpretation. The greater the values of L_t and Ω_t, the more rapid is the variation of the R_i values and, hence, that of the phase values, φ_i. This is accompanied by a reduction of τ_c. According to Equation (3.1), a growth of λ leads to a reduction in the rate of φ_i variation and, consequently, to an increase in τ_c.

It is expedient to introduce a classification of burst signal fluctuations, dividing them into four types depending on the τ_c, T_0 and T_p relationship:

- nonfluctuating signal;
- slowly fluctuating signal, which occurs when $T_0 \leq \tau_c$, where in this case, $N_e = 1$;
- fast-fluctuating signal, when $T_p \geq \tau_c$, where in this case, $N_e = N$ (N is the number of pulses in a burst);
- signal with mixed-type fluctuations, when $T_p < \tau_c < T_0$, where in this case, $1 < N_e = T_0/\tau_c < N$.

The nonfluctuating signal is the easiest to detect because its detection with the desired values of D and F requires the least signal-to-interference ratio, ρ. The greatest signal-to-interference ratio, ρ_{sf}, is required for the detection of a slowly fluctuating signal with the same values of D and F. The difference (in decibels)

$$L_{f_0} = \rho_{sf} - \rho$$

is termed the 'fluctuation loss' (for slow fluctuations).

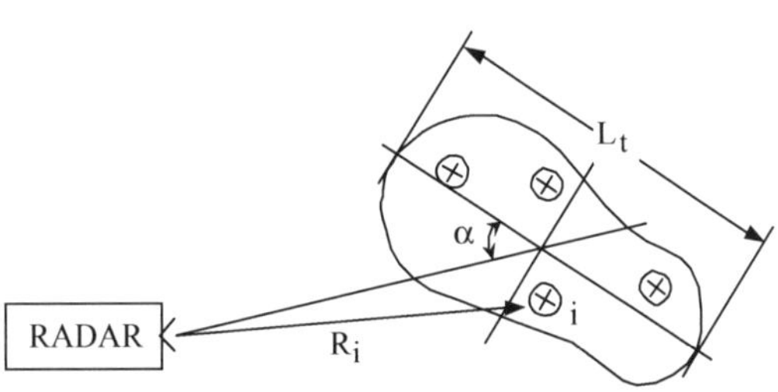

Figure 3.4 Model of a target as a cluster of bright points

3.2 BARTON'S METHOD FOR REQUIRED SIGNAL-TO-NOISE RATIO CALCULATION

Barton [I.18] proposed a simple technique for the calculation of the signal-to-noise ratio, $\rho_{\min}$, required for the detection of a noncoherent burst containing N pulses. This technique may be represented by the formula

$$\rho_{\min} = \rho_1 (P_d, P_{fa}) - 10 \lg N + L_i (\rho_1, N) + L_f (P_d, P_{fa}, N_e) \tag{3.4}$$

All the terms in Equation (3.4) are expressed in decibels. The first, third and fourth terms are evaluated using the curve families in Figures 3.5 to 3.7 [I.18]. The first term in Equation (3.4) stands for the signal-to-noise ratio necessary for the detection of a single pulse with constant amplitude. This quantity can be found using Figure 3.5.

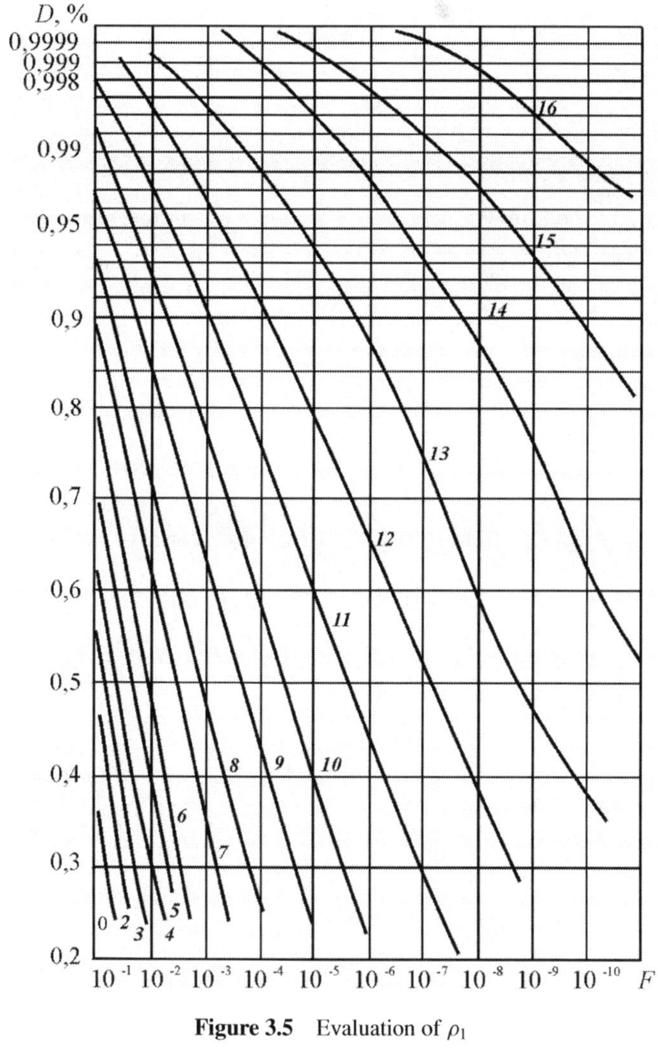

Figure 3.5 Evaluation of ρ_1

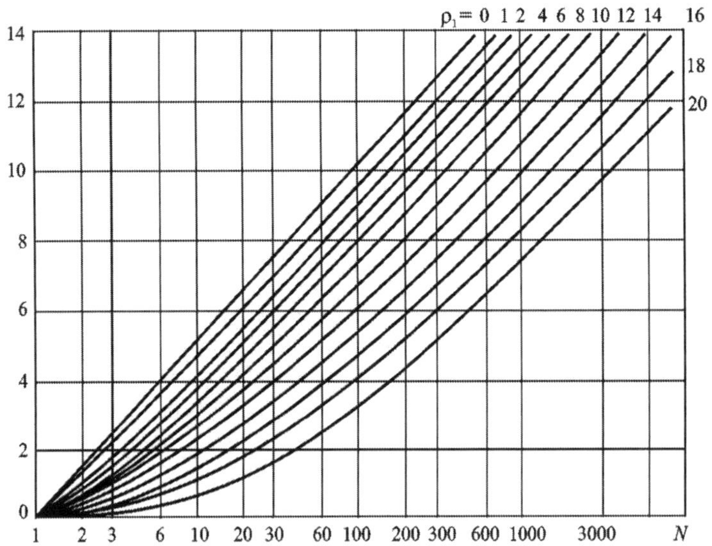

Figure 3.6 Evaluation of $L_i(\rho_1, N)$

A point with the abscissa, P_{fa}, and the ordinate, P_d, is plotted on the graph. Here ρ_1 is determined as the curve parameter closest to the plotted point. Interpolation permits ρ_1 to be determined more precisely.

The third term (noncoherent integration loss) is evaluated using Figure 3.6. Choose a curve with the ρ_1 parameter as calculated at the first step. Draw a vertical straight line with an abscissa equal to N. $L_i(\rho_1, N)$ is taken as the ordinate of the cross point of the straight line and the curve with the relevant ρ_1 parameter.

The curves plotted in Figure 3.7 are intended for determining the loss due to slow fluctuations of the echo signal, $L_{f_0}(P_d, P_{fa})$. Fluctuation loss, with N_e independent pulses in a burst, is calculated using the formula

$$L_f = \frac{L_{f_0}(\text{dB})}{N_e}.$$

Barton notes that the proposed method makes it possible to predict the additional loss arising from nonoptimal values of detector parameters. If the receiver passband, B_n, is wider than the optimal band, i.e. $B_n\tau_p > 1$, the additional loss can be estimated as L_i for $B_n\tau_p$ noncoherently integrated pulses. If the receiver is unblocked (gated) for a time interval, t_1, exceeding the echo pulse duration, τ_p, then the total loss of noncoherent integration is to be calculated for the number of pulses

$$N' = N\frac{t_1}{\tau_p}.$$

As estimated by Barton, the error resulting from his method, as a rule, does not exceed 0.5 dB.

The method considered compares favourably with other techniques because it allows calculations to be made with any values of N_e. Besides, using formula (3.4), it is possible to assess separately contributions to ρ_{min} from individual components and to set radar parameters accordingly.

Barton's method can also be used with coherent or partially coherent pulse bursts. In the former case, the term L_i is to be eliminated from formula (3.4). In the latter case, this term should be written as $L_i(\rho_1, N_n)$. Here N_n is the number of noncoherent pulses in a burst. Let us demonstrate how Barton's method can be utilized in parameters calculation for coherent continuous-wave (CW) radars. If a receiver

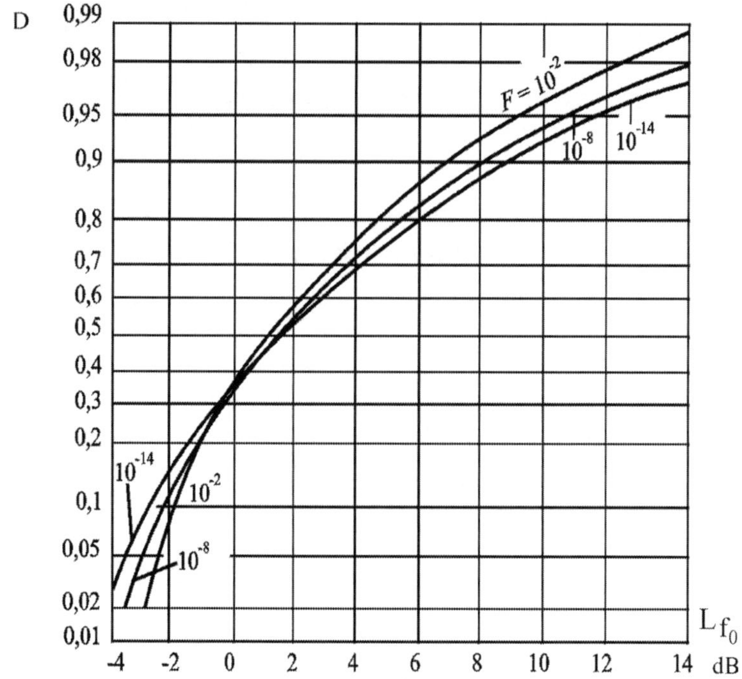

Figure 3.7 Fluctuation loss for a slowly fluctuating echo signal, L_{f_0}

bandwidth is

$$B_n = \frac{1}{T_0}, \text{ then } N = 1 \text{ and } L_i = 0.$$

If $B_n > 1/T_0$, then, after coherent processing in a filter with a passband B_n, the signal is subject to detection and noncoherent integration during the time period, T_0. In this connection, the noncoherent integration loss for $B_n T_0$ samples needs to be taken into account.

Let us illustrate Barton's method by specific calculations for a hypothetical radar. Suppose the radar is performing volume surveillance in the horizontal plane. The beamwidth is $\theta = 2°$, antenna survey rotation speed is $\Omega_{sur} = 100$ deg/s, pulse repetition frequency is $F_p = 1200$ Hz and the correlation window of echo signal amplitude fluctuations is $\tau_c = 50$ ms. Suppose the detection of the signal is required during one scan, i.e. at a rate of one burst with a detection probability $D = 0.95$ and a false alarm rate $F = 10^{-5}$.

Let us determine ρ_{min} for coherent and noncoherent bursts. To obtain comparable results, ρ_{min} will be calculated for a single pulse of echo signal in both cases. The burst duration is

$$T_0 = \theta / \Omega_{sur} = 0.02 \text{ s}.$$

Since $T_0 \le \tau_c$, we are dealing with a slowly fluctuating signal, $N_e = 1$. The number of pulses in a burst is $N = T_0 F_p = 24$. From the graphs in Figure 3.5 the ρ_1 value corresponding to the given D and F is found; $\rho_1 = 13.1$ dB and $10 \lg 24 = 13.8$ dB. For a coherent burst, $L_i = 0$. For a noncoherent burst, according to the graphs in Figure 3.6, $L_i(13; 24) = 2.8$ dB. Finally, using Figure 3.7, we determine L_f for the given $D = 0.95$, which is $L_f = L_{f_0} = 10.5$ dB. Thus, for a coherent burst, $\rho_{min} = 9.8$ dB. For a noncoherent burst, $\rho_{min} = 13.1 - 13.8 + 2.8 + 10.5 = 12.6$ dB.

The results obtained permit the following conclusions to be drawn. The use of a coherent waveform in the specified conditions gives a gain in energy of about 3 dB. The required signal-to-interference ratio is substantially dependent on the fluctuation loss.

According to Barton, there are two ways to reduce the fluctuation loss:

- transmission of radar signal at several frequencies within the ΔF range or
- use of frequency agility within the ΔF range.

Naturally, the first approach presupposes an adequate number of receiving channels and the second approach, synchronous frequency tuning of the receiver. The number of independent values of the signal is given by

$$N_e = \frac{\Delta F\, 2L_r}{c}. \tag{3.5}$$

In Equation (3.5), L_r is the projection of target length on to the radar-target line.

Radar operation within a certain range of frequencies is especially effective with respect to noncoherent bursts. In that case, the reduction in ρ_{min} (in decibels) is

$$\Delta\rho_{min} = L_{f_0} - \frac{L_{f_0}}{N_e} = L_{f_0}\left(\frac{N_e - 1}{N_e}\right).$$

If originally there is a coherent burst, realization of N_e independent pulses leads to N_e noncoherent parts and the benefit from decorrelation of slow fluctuations is reduced by the value of noncoherent integration loss, $L_i(\rho_1, N_e)$. However, when the value of L_{f_0} is great enough, decorrelation of echo signal fluctuations still offers some advantages. One exception is the situation when a high resolution in the target velocity is required, which, according to Equation (2.12), is attainable with longer coherent intervals of the echo signal.

3.3 RADAR PARALLEL AND SUCCESSIVE SURVEILLANCE

3.3.1 Introduction

When an echo signal is detected, its frequency, time and direction of arrival are unknown. In using optimum processing devices, it is necessary to survey the span of uncertainty for the above signal parameters. Along with signal detection, its parameters corresponding to the target range, velocity and angular position are approximately estimated.

Let us discuss two kinds of radar survey: parallel and successive. The parallel survey implies that signals in all the coverage elements (resolution cells) are detected simultaneously by means of a multichannel device. The number of channels is equal to the number of elements, m. The successive survey presupposes a sequential search of coverage elements by means of a single-channel device.

The number of range resolution cells (echo signals resolved by their time of arrival) is estimated by the formula

$$m_r \approx \frac{R_{max} - R_{min}}{\Delta R}. \tag{3.6}$$

In Equation (3.6), R_{max} and R_{min} are the maximum and minimum target ranges and ΔR is the range resolution. The number of angular resolution cells is determined by the formula

$$m_a \approx \frac{\psi_1\psi_2}{\theta_1\theta_2}. \tag{3.7}$$

In this formula, ψ_1 and ψ_2 are the pre-set sectors of an angular scan in two orthogonal planes; θ_1 and θ_2 denote the two-way beamwidth of the radar's antenna system in the above planes.

The number of elements in the target velocity coverage (echo signals resolved by their Doppler shift) is approximately equal to

$$m_{\mathrm{v}} = \frac{V_{R\max} - V_{R\min}}{\Delta V_R}. \tag{3.8}$$

In Equation (3.8), $V_{R\max}$ and $V_{R\min}$ are the maximum and minimum values of the target radial velocity; ΔV_R is the resolution in velocity given by formula (1.10). The total number of coverage elements (cells) is defined as the product

$$m = m_{\mathrm{r}}\, m_{\mathrm{a}}\, m_{\mathrm{v}}\,. \tag{3.9}$$

If no scanning is performed for a certain signal parameter, the value of the corresponding term in Equation (3.9) is assumed to equal unity.

Besides parallel and successive surveillance, there are varieties of combined scan. Most often, it is a combination of parallel scan for some of the signal parameters and successive scan for other parameters. Sometimes, when there are m_i cells for a target coordinate (or velocity), a detection device with m_i/n channels is used. In this case, m_i/n cells are searched in parallel and n groups of cells are searched one by one.

The present consideration of the radar survey can be reduced to an analysis of detection in a single resolution cell. To this end, it is necessary to establish the connection between the probabilities P_{ds} and $P_{\mathrm{fa.s}}$ specified for the entire scan and the probabilities P_{d} and P_{fa} concerning detection in a single cell, and to calculate the time spent on the scanning of a single cell.

Suppose there are k objects (echo signals) in the coverage and the specified probability of detection for any of them equals P_{d}. The probability of detection for any l signals out of k is given by

$$P(l) = P_{\mathrm{d}}^{l}(1 - P_{\mathrm{d}})^{k-l} C_k^l, \tag{3.10}$$

where C_k^l is the number of combinations made up of k elements each containing l signals. As shown in Reference [I.10], the mathematical expectation for the number of detected elements is

$$< l > = \sum_{l=1}^{k} l\, p(l) = \sum_{l=1}^{k} l\, P_{\mathrm{d}}^{k}(1 - P_{\mathrm{d}})^{k-l} C_k^l = k P_{\mathrm{d}}.$$

Assuming P_{ds} is the ratio of the mean number of detected targets to the total number of targets,

$$P_{\mathrm{ds}} = \frac{< l >}{k} = P_{\mathrm{d}}. \tag{3.11}$$

Thus, the required probability of detection in a single coverage cell is equal to the preassigned detection probability for one scan.

The probability of false alarm during a scan is defined as the likelihood that the interference level will cross the threshold in at least one of the cells scanned. Hence,

$$P_{\mathrm{fa.s}} = 1 - (1 - P_{\mathrm{fa}})^m.$$

Typically, $P_{\mathrm{fa}} \ll 1$. Therefore,

$$(1 - P_{\mathrm{fa}})^m \approx 1 - m P_{\mathrm{fa}}$$

and

$$P_{\mathrm{fa}} = \frac{P_{\mathrm{fa.s}}}{m}. \tag{3.12}$$

It follows from Equation (3.12) that P_{fa} demands much more stringent requirements than $P_{\mathrm{fa.s}}$.

The pre-set time of scan in the ith coordinate is limited by the value of $T_{\mathrm{s}i}$. With the parallel scanning pattern, the time of signal detection in one element is $T_{\mathrm{e}} = T_{\mathrm{s}i}$. With the ordinary successive survey, $T_{\mathrm{e}} = T_{\mathrm{s}i}/m_i$.

Let us compare the performance of the two kinds of scan by the criterion of maximum radar power budget, E_R (see formula (1.27)). The choice of this criterion was dictated by the fact that radar range coverage is substantially determined by E_R. As a basic scenario, we will consider the detection of a noncoherent burst of pulses with a pulse repetition frequency F_p. The number of pulses in a burst for the successive scan is

$$N_1 = \frac{F_p T_{si}}{m_i},$$

whereas, for the parallel scan, it is

$$N_2 = F_p T_{si} = m_i N_1.$$

According to Equation (3.5), the number of independent pulses for the successive and parallel types of scan is $N_{e1} = T_{si}/(m_i \tau_c)$ and $N_{e2} = T_{si}/\tau_c = m_i N_{e1}$ respectively.

3.3.2 Scan of Target Range Coverage

In this case, the values of all the quantities in Equation (1.27), except for ρ_{min}, do not differ subject to survey type. The minimum value of ρ_{min} corresponds to the maximum value of E_R. Therefore, the minimum of ρ_{min} is taken as a criterion for the scanning performance assessment. Using Barton's formula, expressions can be written for the required ρ_{min} values for the successive (ρ_s) and parallel (ρ_p) types of scan:

$$\rho_s = \rho_1 - 10 \lg N_1 + L_i(\rho_1, N_1) + \frac{L_s}{N_{e1}} \tag{3.13}$$

$$\rho_p = \rho_1 - 10 \lg N_1 - 10 \lg m_i + L_i(\rho_1, N_2) + \frac{L_s}{N_{e2}} \tag{3.14}$$

Using relation (3.14),

$$\begin{aligned} \rho_p &= \rho_1 - 10 \lg N_1 - 10 \lg m_i + L_i(\rho_1, N_1) + 5 \lg m_i + \frac{L_s}{N_{e2}} \\ &= \rho_1 - 10 \lg N_1 - 5 \lg m_i + L_i(\rho_1, N_1) + \frac{L_s}{N_{e2}} \end{aligned}$$

The difference between the required signal-to-noise ratios

$$\rho_s - \rho_p = 5 \lg m_i + L_s \left(\frac{1}{N_{e1}} - \frac{1}{N_{e2}} \right)$$

is always positive, as

$$\frac{1}{N_{e1}} - \frac{1}{N_{e2}} \geq 0.$$

Hence, the parallel survey of range coverage is more effective. The benefit derived from the parallel survey grows with increasing the number of scanned elements, m_i. Apart from this, in certain situations the difference of fluctuation losses may be considerable. This also adds to the advantages of the parallel surveying pattern.

The result obtained holds true for the scan of target velocity coverage. The multichannel design of the detector entails a certain hardware input. The detector belongs to the type of devices that are relatively easy to miniaturize. At the same time, the achieved gain, in terms of energy, allows the power and size of the transmitter to be reduced.

3.3.3 Angular Survey

For the successive angular survey it is practical to employ either identical transmitting and receiving antennas or a combined transmit–receive antenna. The antenna beamwidth, θ, is determined by the preassigned angular resolution. The gain of each of the antennas is denoted by G_s.

For the parallel angular survey, the beamwidth of the transmitting antenna should be equal to the preassigned angular scan sector. Therefore, the gain of the transmitting antenna is $G_p = G_s/m_i$. The receiving antenna is to have m_i beams, each beam having a width θ. This kind of antenna can be realized on the basis of an active phased array. The receiver is to have m_i independent channels as well. In this case, the power budget expressions for the successive and parallel scanning patterns will differ in two parameters: G_t and ρ_{min}. Thus, the minimum of the ρ_{min}/G_t ratio, or $\rho_{min}(\mathrm{dB}) - 10 \lg G_t$ in decibels, can be assumed as a criterion for performance evaluation. For the scanning patterns under consideration,

$$\rho_s - 10 \lg G_s = \rho_1 - 10 \lg N_1 + L_i(\rho_1, N_1) + \frac{L_{fs}}{N_{e1}} - 10 \lg G_s;$$

$$\rho_p - 10 \lg G_p = \rho_1 - 10 \lg N_1 - 10 \lg m_i + L_i(\rho_1, N_1) + 5 \lg m_i + \frac{L_s}{N_{e2}} - 10 \lg \frac{G_s}{m_i}.$$

The difference between these quantities is

$$\rho_s - \rho_p = L_{fs}\left(\frac{1}{N_{e1}} - \frac{1}{N_{e2}}\right) - 5 \lg m_i. \tag{3.15}$$

Since the terms in Equation (3.15) have opposite signs, there is no certainty as to which kind of survey is more favourable in terms of energy. If m_i is great, then $\rho_s < \rho_p$, and the successive is preferable. If m_i is small and P_d is close to unity, the first term in Equation (3.15) may turn out to be greater than the second term (in absolute value). Under such circumstances the parallel scan is more favourable.

Let us consider an example. Suppose the solid angle of scan is $\psi = 60^\circ \times 10^\circ$, the antenna beamwidth for the successive scan is $3^\circ \times 3^\circ$. The detection probability is $P_d = 0.9$, the scan duration is 2 s and the signal fluctuations correlation window is 50 ms. The number of coverage cells, in accordance with Equation (3.7), is $m_i = 60 \times 10/(3 \times 3) \approx 70$. The number of independent pulses is $N_{e1} = 1$ for the successive scanning pattern and $N_{e2} = 40$ for the parallel scanning pattern. Using Figure 3.7, it is found that $L_{fs} = 8$ dB. Hence, $L_{fs}(1/N_{e1} - 1/N_{e2}) \approx 8$ dB. From formula (3.15),

$$\rho_s - \rho_p = L_{fs}\left(\frac{1}{N_{e1}} - \frac{1}{N_{e2}}\right) - 5 \lg 70 = -0.5\,\mathrm{dB}.$$

This representative example shows that the two scan types at issue are, on the whole, equivalent. With different initial data, one of the two scanning patterns may prove to be more advantageous in terms of energy than the other. However, in this particular case, in contrast to the scan of range coverage, the parallel scan has no explicit superiority in energy. If the simplicity of hardware implementation peculiar to the successive angular scan is also mentioned, it becomes clear why this scanning pattern is commonly used.

3.4 COHERENT AND NONCOHERENT PULSED RADARS

It follows from the contents of Sections 2.1 and 3.2 that radars with a coherent and a noncoherent transmission have both their benefits and limitations. The coherent radar offers the following advantages:

- a high resolution in frequency (i.e. velocity of radar targets);
- a higher power budget owing to the absence of noncoherent integration loss.

The high resolution in frequency makes it possible to discriminate targets with a small difference in their speeds. It also gives the radar the more important capability to reject clutter (see Section 7.2).

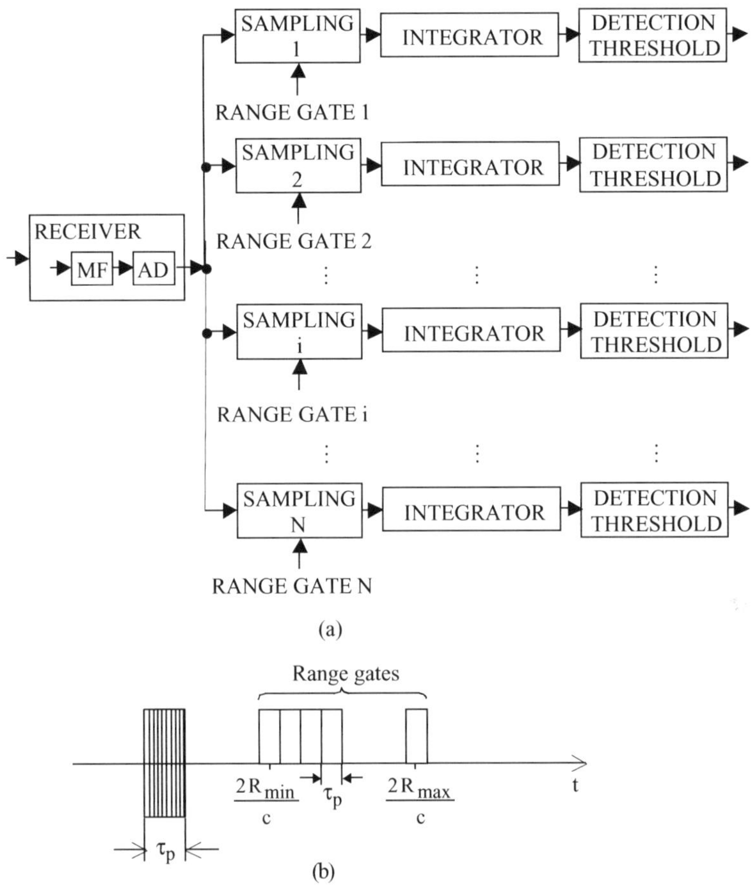

Figure 3.8 Non coherent burst detector: (a) block diagram of the detector; (b) flowcharts of pulses. MF, matched filter; AD, amplitude detector

The main benefit of a noncoherent radar is the simplicity of its signal processing circuitry. The block diagram of a noncoherent detection circuitry with parallel scanning of range coverage is given in Figure 3.8.

At the end of the receiver is a filter matched with each pulse of the burst. After passing through the amplitude detector, video pulses come to N gates. At the output of the ith gate, a burst of pulses is returned by the target located at range $i\ \Delta r$. The quantity Δr represents a range cell. Next, the N pulses of the burst are summed in the ith channel of the detector. The sum is then compared with a threshold. The range of the detected target is estimated by the number of channels, wherein the threshold has been crossed. The number of channels is

$$m_R = \frac{R_{max} - R_{min}}{\Delta r}, \tag{3.16}$$

where R_{max} and R_{min} are the limits of range coverage.

The structure of a coherent detection circuitry with parallel scanning of target range coverage and target velocity coverage is given in Figure 3.9. In this case, the receiver does not have to contain a matched filter and an amplitude detector. Signals are divided in this kind of detector into N range

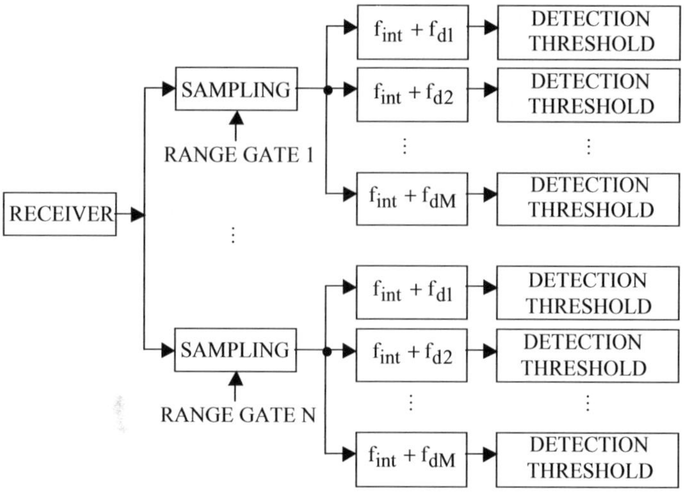

Figure 3.9 Block diagram of a coherent burst detector. f_{int}, intermediate frequency

channels (just as in Figure 3.8). The next stage is represented by combs of bandpass filters matched with signal bursts in each range channel. Here, the filter combs are supposed to span the Doppler frequency range, $f_{g\max} - f_{g\min}$. The number of filters in each comb is

$$m_f = \frac{f_{d\max} - f_{d\min}}{\Delta f}, \tag{3.17}$$

where $f_{d\max}$ and $f_{d\min}$ are the limits of the Doppler frequency coverage and $\Delta f = 1/(NT_p)$ is the passband of each filter matched with the coherent burst.

Each filter in the comb is followed by a threshold detector. The range of a target is estimated by the number of range channels wherein the threshold has been crossed. The Doppler shift, f_d, is determined by the number of the filter in the comb, wherein the threshold has been crossed. The value of f_d can be used to determine the target's radial velocity.

Sometimes, the functions of volume surveillance and accurate target position measurement are assigned to two different radars. In this case, the second radar will has target designation. Hence it has a single-channel signal processing circuitry. It will comprise a gated stage performing selection of signals in time and a filter matched with the pulse burst. The latter element is responsible for the integration of burst pulses and, simultaneously, for their selection by frequency (target velocity selection).

3.5 CW RADARS WITH NONMODULATED AND MODULATED SIGNALS

3.5.1 Introduction

Continuous-wave (CW) radars possess certain advantages that make them quite expedient in a number of cases, these advantages being:

- the possibility of obtaining a greater (as compared to pulsed radars) mean power of the transmitter with a limited peak power;
- optimal potentialities for moving target selection and velocity measurement;
- an almost zero value of minimum detection and position measurement ranges;
- simplicity of radar design unless it is required to resolve targets in range or measure target range.

The first of the listed advantages of CW radars can be explained using the radar power budget. The radar power budget (1.27), which substantially determines its range coverage, is dependent on the mean power of the radar transmitter, P_m. In a continuous-wave radar

$$P_{m.cw} = P_p, \tag{3.18}$$

where P_p is the peak power. In a pulsed radar the mean transmitter power is equal to

$$P_{m.p} = P_p \cdot \frac{\tau_p}{T_p} = \frac{P_{m.cw}}{Q}, \tag{3.19}$$

where $Q = T_p/\tau_p \gg 1$.

Comparison of Equations (3.18) and (3.19) shows that, with other conditions being equal, to obtain the same range coverage, the mean transmitter power in a pulsed radar should be equal to the power of a continuous-wave radar. This implies that the peak power of a pulsed transmitter should be Q times as great as that of a continuous-wave transmitter. The value of quantity Q may vary from several units to several hundreds.

CW radars with a nonmodulated carrier have another essential advantage: unambiguity of target velocity measurement and target resolution in velocity. This is accounted for by the fact that the spectrum of a CW echo signal contains one peak at a frequency $f_0 + f_d$, where f_0 stands for the carrier frequency and f_d for the Doppler shift in frequency. At the same time, looking at Figure 2.4, it can be seen that the spectrum of a coherent pulse waveform has several peaks of about the same level.

The main disadvantage of continuous-wave radars is the inevitable penetration of a portion of the transmitted energy into the reception path, which is known as 'feedthrough'. The said factor can affect the sensitivity of the receiver. This subject is addressed below.

The detection capability of continuous-wave radars is limited to moving objects that have a nonzero velocity component towards the radar (i.e. the nonzero range rate). This limitation is related to the necessity to separate the signal returned by the target from the transmitter's leakage within the receiver.

For range measurements and resolving targets in range, CW radars are designed to transmit frequency or phase modulated waveforms. Radar systems of the considered type can be divided into two groups:

- radars with a nonmodulated carrier wave;
- radars with frequency modulation (FM) or a phase-coded waveform.

These subclasses of continuous-wave radars are discussed in Sections 3.5.2 and 3.5.3.

3.5.2 Radar with a Nonmodulated Carrier

To start with, let us take a very simple CW radar system (Figure 3.10), which can be used for detection of targets at short ranges. The transmitter generates a harmonic signal at a frequency f_0. The signal returned by a moving target has a frequency $f_0 + f_d$, where $f_d = 2v_R/\lambda$. The voltage of the target signal at the output of the mixer has a frequency f_d, as the heterodyning voltage in this design is formed from part of the transmitted signal. The Doppler shift in frequency usually lies within a range from 100 Hz to 100 kHz; e.g. for an object moving at 600 m/s and a wavelength of 10 cm,

$$f_d = \frac{2 \times 600}{0.1} = 12 \text{ kHz}.$$

It is desirable to match the frequency response of the Doppler frequency amplifier with the spectrum of the returned signal. If the duration of the returned signal is bound by the quantity T_0 (as a consequence of antenna rotation or a specific detection time requirement), then a quasi-optimal filter such as a bandpass

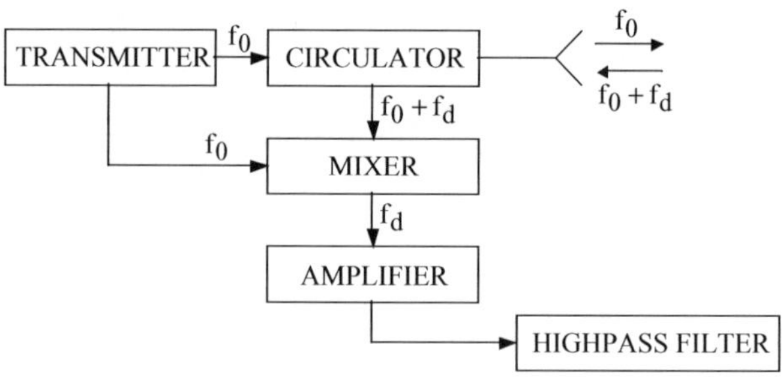

Figure 3.10 Block diagram of a simple CW radar

filter can be used with a pass-band

$$\Delta f \approx \frac{1}{T_0}.$$

More often than not, the range of Doppler frequencies

$$\Delta F_d \gg \Delta f.$$

In this case, the optimum solution would be to use a filter bank of m filters, the pass-band of each filter being equal to Δf. The bank is supposed to span the frequency range ΔF_d; therefore

$$m = \frac{\Delta F_d}{\Delta f}.$$

The output signal of each filter is detected and compared with a threshold. The number of filters, i, at whose output a target signal has been found to be present, gives an approximate evaluation of the Doppler-shifted frequency

$$f_d \approx f_{d\,min} + (i - 1)\Delta f.$$

To reduce hardware input, designers often depart from the optimal design of a frequency-selective system. Thus, sometimes a filter bank with a smaller number of filters, m_1, is used, the pass-band of each filter being equal to Δf_1:

$$\Delta f_1 \approx \frac{\Delta F_d}{m_1}$$

or a single filter with a passband, ΔF_d. Since in these cases the filter passband is wider than the optimal band, the time of coherent integration is less than T_0. Therefore, to avoid excessive energy loss it is necessary to detect the output signal of the filter and integrate it noncoherently during the time T_0. If the passband of the filter is Δf_1, the time of coherent integration in this filter is

$$\tau_c \approx \frac{1}{\Delta f_1}.$$

Consequently, there are N parts on the signal length ($N = T_0/\tau_c = T_0\,\Delta f_1$) that are subject to noncoherent integration. Apropos, Barton's method can be used in this case to calculate the required signal-to-noise ratio. The quantity $N = T_0\,\Delta f_1$ is taken as the number of noncoherently integrated pulses. When an optimum comb filter is used, noncoherent integration is ruled out and $N = 1$. If a single filter is used in the Doppler frequency amplifier having a passband ΔF_d, then

$$N = T_0\,\Delta F_d\,.$$

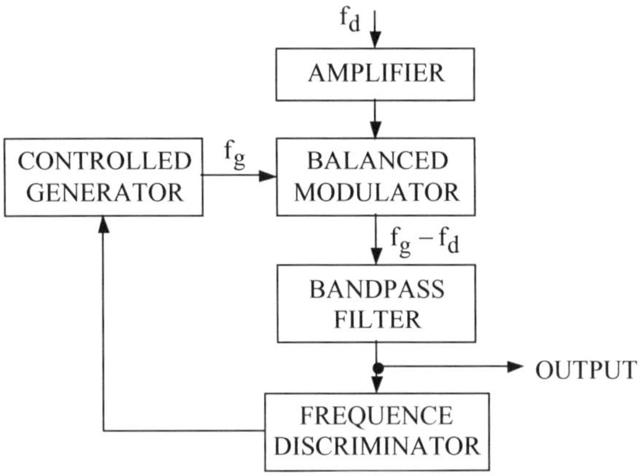

Figure 3.11 Block diagram of a frequency tracker

When it is desired to detect and track one object, a tracking filter with a passband, Δf, is sometimes employed in the amplifier. The tracking filter is based on the frequency, or phase-locked loop, of a voltage controlled oscillator (VCO). The structure of a tracking filter with a frequency-locked loop is shown in Figure 3.11. In this circuit, the filter adjustment frequency, F_f, is constant and coincides with the null in the response of the frequency discriminator. The balanced modulator converts the output signal of the amplifier to the frequency $f_d + f_g$. When $f_d + f_g \neq F_f$, the discriminator output voltage is $U_d \neq 0$, which causes the VCO frequency to be readjusted so as to satisfy the equality

$$f_d + f_g = F_f.$$

Note that the useful signal passes through the bandpass filter without loss. With the frequency f_g measured, the Doppler shift can be estimated using the formula $f_d = F_f - f_g$.

One distinctive feature of the simple CW radar (Figure 3.10) is that the amplification path of the receiver begins with a low-frequency amplifier. Such a receiver will have rather a low sensitivity. This is due to an increase of the spectral density of noise (noise factor) in the mixer and amplifier stages at low frequencies. Analysis of the noise properties of various electronic devices shows that thermal and shot noise at frequencies below 1 MHz is complemented by a noise component (flicker noise), whose spectral density is the reciprocal of frequency.

Besides the low-frequency noise, sensitivity of the receiver can be affected by the presence of vibronoise in radar deployed on moving platforms. Reduced receiver sensitivity can be tolerated when the radar is not expected to provide a great detection range, but major importance is attached to its size and weight. If receiver sensitivity cannot be sacrificed, the circuitry has to be complicated to provide for signal amplification at the intermediate frequency (Figure 3.12). In this case, the value of intermediate frequency, f_h is chosen from a range of several megahertz to several tens of megahertz, which brings the noise factor to a value typical of radar receivers. Frequency selectivity is ensured by the above-mentioned means in the intermediate frequency amplifier.

The most significant disadvantage of CW radars consists in the presence of a leakage signal at the input of the receiver, which is a small part of the transmitted signal. A similar phenomenon exists in pulsed radars as well, but, as a rule, it does not lead to undesirable consequences since the transmitted pulse and the echo pulses do not concur.

The level of the leakage in a CW radar is usually several orders higher than that of the target return. To specify the magnitude of the leakage, the notion of isolation between the transmitter and the receiver is used. This isolation is defined as the ratio of the transmitted signal power (P_t) to the power of the

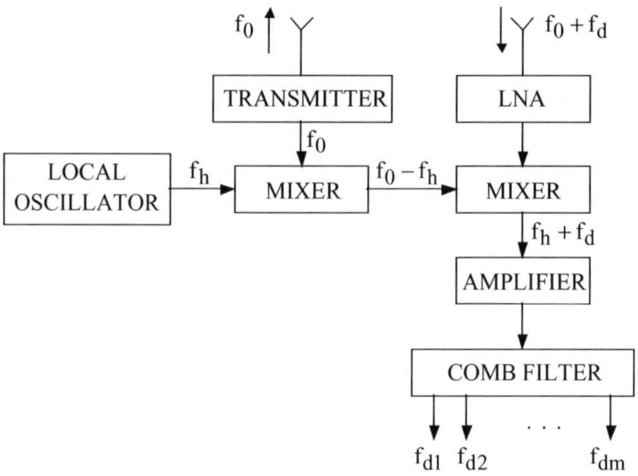

Figure 3.12 Block diagram of a modern CW radar

leakage (P_f):

$$\nu = \frac{P_t}{P_f}.$$

When the same antenna is used for both the transmission and the reception of signals, the value of the quantity, ν, does not normally exceed 30 dB. When separate transmitting and receiving antennas are employed, the isolation is dependent on their sidelobe levels, the distance between them and the presence of natural or manmade screens. The isolation may decrease due to reflections of the transmitted signal from terrain features in the neighbourhood of the radar. According to Skolnik [I.3], with separate antennas it is possible to achieve isolation values of over 80 dB.

The major part of the leakage in the receiver is not hard to eliminate using well-known methods of frequency selection. However, the situation is complicated by the fact that the transmitted signal is not entirely monochromatic. The amplitude and phase of the transmitted signal fluctuate as a consequence of the electronic noise generated by certain components of the transmitter. Amplitude and phase fluctuations, which are known as amplitude noise and phase noise respectively, bring about an expansion of the transmitted signal spectrum (Figure 3.13).

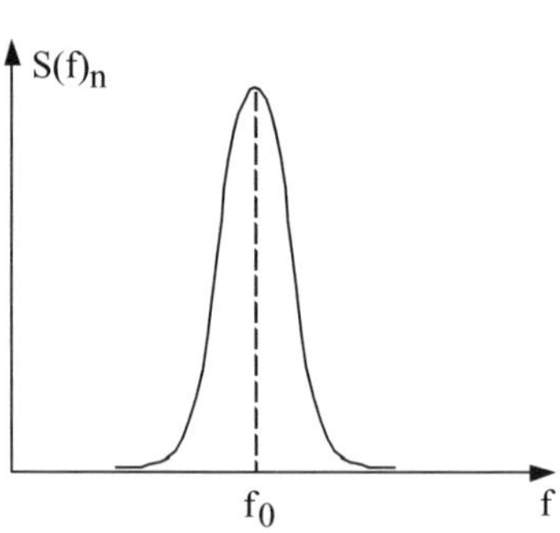

Figure 3.13 A feedthrough spectrum

The spectral components of transmitter noise, spaced by f_d from the carrier frequency, fall into the Doppler range and can create additional (to the receiver's own noise) disturbance. The same is applicable to the noise of the receiver's first local oscillator.

It is convenient to express the spectral density of noise in a per unit system. The relative spectral density of generator noise is defined as the ratio of the noise spectral density to generator power at the carrier frequency:

$$S_0(f) = \frac{S(f)}{P_g}.$$

This quantity is expressed in 1/Hz or dB/Hz. Noise spectral density of the leakage at the input of the receiver can be expressed as

$$S_R(f) = \frac{S(f)}{\nu} = \frac{S_0(f)\,P_t}{\nu}.$$

Transmitter noise will not significantly affect receiver sensitivity, provided the following inequality is satisfied:

$$S_R(f) < NkT$$

or

$$\frac{S_0(f)}{\nu}P_t < NkT. \tag{3.20}$$

Equation (3.20) determines the required isolation under the given $S_0(f)$ and P_t or the permissible value of $S_0(f)$ under the given P_t and ν. If the limit values $S_0(f)_{min}$ and ν_{max} are known, the relation given by (3.20) determines the maximum value of $P_{t.max}$, which is the reasonable upper boundary for transmitter power:

$$P_{t.max} \leq \frac{\nu_{max}}{S_0(f)_{min}} NkT.$$

Raising the transmitter power above $P_{t.max}$ leads to a proportional growth of the target signal and noise at the input of the receiver; i.e. there is no enhancement in range coverage of the radar.

Let us take an example. Suppose $P_t = 100\,\text{W}$, $S_0(f_d) = 10^{-10}$ 1/Hz and $N = 10$. Then the required isolation is given by the inequality

$$\nu \geq \frac{100\,W \times 10^{-10}\,1/\text{Hz}}{10 \times 4 \times 10^{-21}\,\text{W/Hz}} = 2.5 \times 10^{11},$$

i.e. the isolation is to exceed 114 dB.

Now, let us calculate the isolation, ν, for the case when the transmitter and the receiver of the radar are separated by a certain distance, R (Figure 3.14). The transmitter power is P_t, transmitting antenna

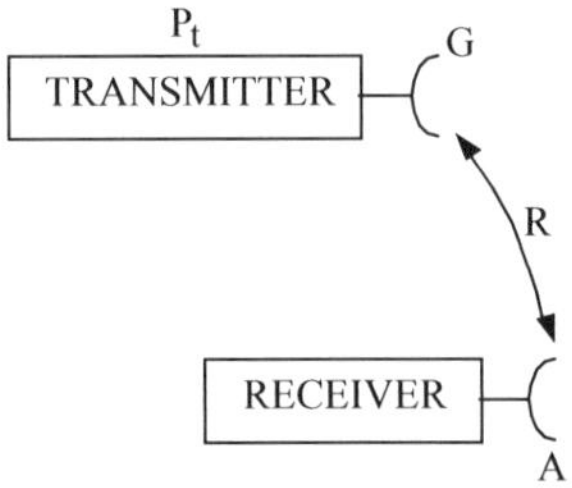

Figure 3.14 A feedthrough power calculation

gain is G, effective aperture of the receiving antenna is A, transmitting antenna sidelobe level towards the receiver is η_1 and receiving antenna sidelobe level towards the transmitter is η_2. Note that η_1 and η_2 are given on a per unit basis. It is not difficult to see that the leakage power at the output of the receiving antenna is

$$P_F = \frac{P_t G A \eta_1 \eta_2}{4\pi R^2}.$$

Hence, the isolation between the transmitter and the receiver is

$$\nu = \frac{P_t}{P_F} = \frac{4\pi R^2}{G A \eta_1 \eta_2}. \quad (3.21)$$

As an example, suppose that $R = 100$ m, $G = 10^3$, $A = 2$ m^2 and $\eta_1 = \eta_2 = 10^{-3}$. Then, according to Equation (3.21),

$$\nu = \frac{4\pi \times 10^4}{10^3 \times 2 \times 10^{-6}} \approx 6 \times 10^7$$

and the leakage level for $P_t = 100$ W is

$$P_F = \frac{P_t}{\nu} \approx 1.7 \times 10^{-6} \text{ W}.$$

A technique and a device for leakage reduction in radars with nonmodulated carrier are described in References [I.19] and [I.20]. The technique consists of channelling a small portion of transmitter power to the receiver (by means of a cable or a dedicated antenna). The voltage thus supplied to the receiver is then subtracted from the leakage signal. The amplitude and phase of the supplied voltage are adjusted so as to minimize the resultant difference and, by doing so, cancel the leakage.

3.5.3 CW Radars with a Modulated Carrier

Continuous-wave radars with a modulated carrier are employed when it is necessary to measure target range and resolve targets in range. This often involves periodic linear frequency modulation with the use of triangular or sawtooth waveforms (Figure 3.15). In some cases, phase-coded waveforms (PCW) are used. As applied to the stated problem, chirp modulation offers a significant advantage over PCW since it does not require Doppler correction. This is related to the comb-shaped ambiguity function of the chirped waveform (see Section 2.9).

3.5.3.1 Chirped Radars

When a chirp-modulated carrier is used, the two feasible and practical methods of optimum echo signal processing are matched filtering and spectral analysis. In the first case, the matched filter puts out a

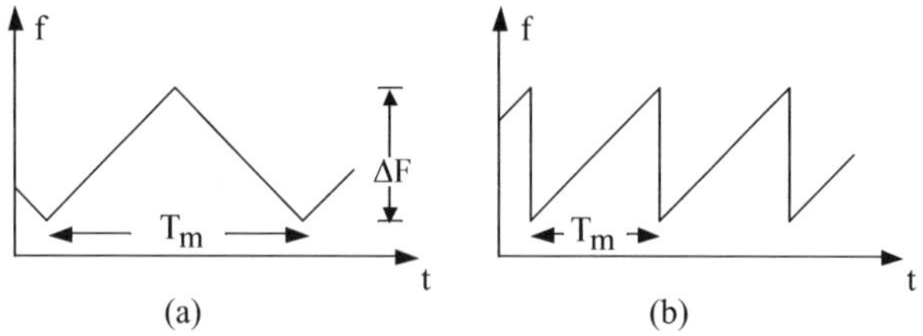

Figure 3.15 (a) Triangular and (b) sawtooth FMs

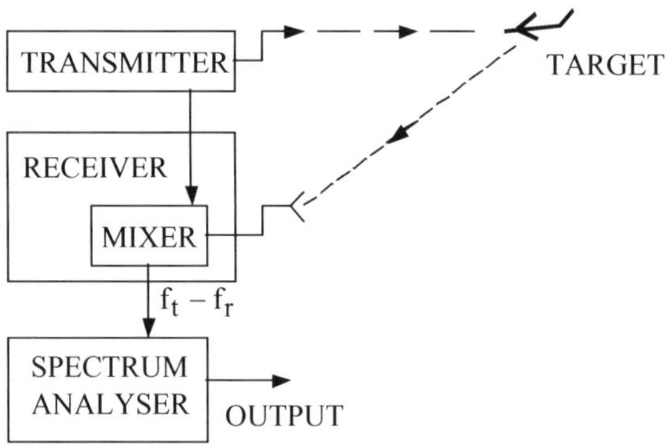

Figure 3.16 Block diagram of a radar with FM signals

compressed pulse at the end of each modulation period. The pulse delay

$$t_d = \frac{2R}{c} \tag{3.22}$$

can be used to determine the target range.

Let us examine the second alternative taking the triangular chirp waveform as an example. To this end, we will refer to Figures 3.16 and 3.17. Figure 3.16 shows a simplified block diagram of a radar. The transmitter generates a signal with a current frequency, f_t. First, consider the reception of the signal returned by a stationary target ($f_d = 0$). In this case, the signal, f_r, returned by the target follows the same frequency variation pattern as the transmitted signal, but with a delay, t_d:

$$f_r(t) = f_t(t - t_d).$$

The transformed signal at the output of the mixer with a frequency

$$f_{dif}(t) = |f_t(t) - f_r(t)|$$

comes to the spectrum analyser, which yields an estimate of the target range, R. The frequencies f_t, f_r and f_{dif} are plotted versus time in Figure 3.17. The symbols '+' and '−' on the flowchart of the transformed signal (Figure 3.17(b)) designate the adjacent halves of the modulation period. The purpose

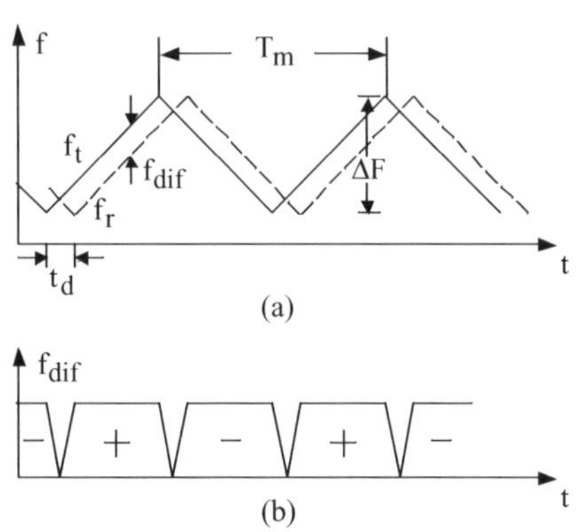

Figure 3.17 Time-frequency relations with $f_d = 0$: (a) transmitter frequency (solid curve) and reflected signal frequency (dashed curve); (b) transformed signal frequency

of this designation is to underscore the contrary variations of phase on these time intervals. The '+' sign corresponds to an increasing phase shift while the '–' sign corresponds to a decreasing phase shift.

It is worth mentioning that the phase shift is, on a certain time interval, equal to the integral of the signal frequency. The trapeziums marked by the symbols '+' and '−' in Figure 3.17 are identical in area. Hence, the phase shifts on these intervals are identical in absolute value. Since the phase shifts have opposite signs, the phases of the transformed signal in the beginning and at the end of the period, T_m, are identical. In accordance with the definition given in Section 2.1, the transformed signal is a periodic signal of the second type of coherence. Consequently, this signal has a discrete spectrum. The spectral components are located at the frequencies that are multiples of $F_m = 1/T_m$.

During most of the modulation period, T_m, the frequency, f_{dif}, is constant:

$$f_{dif} = \frac{\Delta F}{T} t_d. \tag{3.23}$$

In Equation (3.23), $T = T_m/2$. Combining formulae (3.22) and (3.23), the following expression for target range is obtained:

$$R = f_{dif} \frac{cT}{2\Delta F}. \tag{3.24}$$

Therefore, to determine target range, it suffices to know the frequency, f_{dif}, of the transformed signal. To estimate f_{dif}, the spectrum of the transformed signal is calculated. At the same time, the problem will be somewhat simplified assuming that f_{dif} is given by formula (3.23) throughout.

With $f_d = 0$, the spectra of the periodic pulse trains marked by the symbols '+' and '−' are identical. The spectrum of the first of these pulse trains is calculated. The shape of the spectrum envelope, as follows from Section 2.1, coincides with the spectrum of a single pulse, $S_1(\omega)$:

$$S_1(\omega) = UT \frac{\sin\left(\dfrac{\omega - \omega_{dif}}{2} T\right)}{\dfrac{\omega - \omega_{dif}}{2} T}. \tag{3.25}$$

In Equation (3.25), U is the amplitude of the pulse and $\omega_{dif} = 2\pi f_{dif}$. As the spectrum components lie at frequencies $i\omega_m$, where $\omega_m = 2\pi F_m$, it is necessary to substitute $i\omega_m$ for ω in Equation (3.25) to obtain the spectrum of the chirp pulse train. Then, the expression for the spectrum at issue takes the form

$$S_1(i\omega_m) = UT \frac{\sin[(\omega_{dif} - i\omega_m)T/2]}{(\omega_{dif} - i\omega_m)T/2}. \tag{3.26}$$

If ω_{dif}/ω_m is an integer, then the spectrum is at its peak when $i = \omega_{dif}/\omega_m$ and its magnitude is $S_{max} = UT$. The amplitudes of the adjacent components (relative to the peak) are obtained by substitution of

$$i = \frac{\omega_{dif}}{\omega_m} \pm 1$$

into Equation (3.26). Hence,

$$S(\omega_{dif} \pm \omega_m) = \frac{2UT}{\pi} = \frac{2}{\pi} S_{max}.$$

The next pair of spectral components numbered $i = \omega_{dif}/\omega_m \pm 2$ have zero amplitudes. The results obtained are illustrated in Figure 3.18.

Thus, the main lobe of the transformed signal spectrum contains a small number of spectral components (three in this example). It is not difficult to single out the maximum peak component, whose frequency is $i\omega_m = \omega_{dif}$ or $if_m = f_{dif}$. However, it is more typical for ω_{dif}/ω_m not to be an integer. In such cases, the number of the greatest spectral component is determined by the relationship

$$\left| i - \frac{\omega_{dif}}{\omega_m} \right| \leq 0.5.$$

This situation is illustrated in Figure 3.18(b).

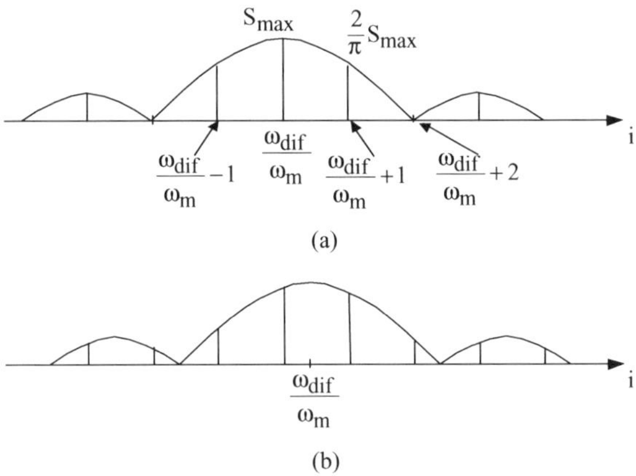

Figure 3.18 Transformed signal spectra: (a) ω_{dif}/ω_m, integer; (b) ω_{dif}/ω_m, noninteger

One characteristic feature of the spectrum under consideration is that, by varying the target range, the value of the quantity, ω_{dif}, varies continuously. The position of the spectrum envelope fluctuates with varying ω_{dif}. Furthermore, the positions of the spectral components remain the same while their amplitudes vary. The frequency, f_{dif}, characterizing the target range is measured with an error, δf_{dif}, which does not exceed $0.5F_m$.

Using Equation (3.25), the maximum target range measurement error is found to be

$$\delta R_{max} = \frac{cT}{2\Delta F}\delta f_{dif.\,max} = \frac{c}{8\Delta F}. \tag{3.27}$$

Let us proceed to the estimation of range resolution. As two targets are closing in, their returns are still perceived separately if the maxima of their spectra are spaced by $\Delta f \geq 2f_m$. Indeed, with $\Delta f = f_m$, it is impossible to ascertain whether a signal with such a spectrum belongs to one target or two. Substituting the difference between frequencies required to resolve two targets ($\Delta f = 2f_m$) for f_{dif} in Equation (3.27), the range resolution is obtained on the left side of the said equation:

$$\Delta R = \frac{c}{2}2f_m\frac{T}{\Delta F} = \frac{c}{2\Delta F}.$$

Note that the same expression has been obtained for ΔR as in the case of complicated pulse signals. Let us go on to consider the general case of processing a signal returned by a moving target. When $f_d \neq 0$, the frequencies of the transformed signals on the intervals marked by the symbols '+' and '−' will be different. Examining Figure 3.19, it can be seen that $f_{dif.1} = f_{dif} - f_d$ in the first case and $f_{dif.2} = f_{dif} + f_d$ in the second case. Here, f_{dif} is the frequency of the transformed signal for $f_d = 0$. As a consequence, the spectrum components on the '+' intervals are offset to the left by f_d, and those on the '−' intervals are offset to the right by the same value. The frequency, f_{dif}, required for the measurement of the target range can be calculated by the formula

$$f_{dif} = \frac{f_{dif.1} + f_{dif.2}}{2}.$$

When we wish to measure target velocity, the required value of f_d is calculated by the formula

$$f_d = \frac{f_{dif.1} - f_{dif.2}}{2}.$$

To have the values of $f_{dif.1}$ and $f_{dif.2}$ available, a two-channel device is necessary for separate processing of the transformed signals marked by the symbols '+' and '−'. Realization of such a device presents no difficulty.

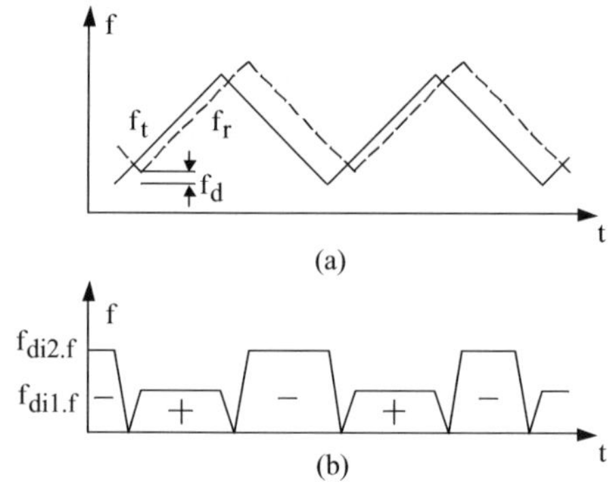

Figure 3.19 Frequency flowcharts with $f_d \neq 0$: (a) transmitted and reflected signals frequencies; (b) transformed signal frequency

3.5.3.2 Radars with a Phase-Coded Signal

Phase-coded waveforms ensure unambiguous measurement of the target range and resolution of targets in the range up to the maximum

$$R_{max} \approx \frac{cT_m}{2},$$

where T_m, as before, denotes the signal modulation period.

For echo signal processing it is practical to use a filter matched with one period of the signal (see Section 2.8). The ambiguity function of a phase-coded waveform has the shape of a thumb-tack. Therefore, the uncertainty of the Doppler frequency shift of the echo signal, f_d, results in a reduced matched filter response [I.21]. Binary phase coding is the method of phase modulation most commonly used in radar, in which the initial phase of the transmitted signal takes on either zero or π values. When there is a Doppler shift, the signal at the discrete instants of time, $n\tau_0$, can be expressed by the formula

$$U \cos\left[2\pi\,(f_0 + f_d)\,n\tau_0 + \varphi\,(n\tau_0)\right],$$

where $n = 1\text{–}N$, f_0 is the carrier frequency and $\varphi\,(n\tau_0) = 0, \pi$ is the coding pattern. Considering the complex envelope gives

$$U \cos\left[2\pi f_d n\tau_0 + \varphi\,(n\tau_0)\right].$$

Let us assume that the matched filter used is a transversal filter (see Section 2.3). The filter coefficients (from end to beginning) are $\cos[\varphi\,(n\tau_0)]$. Hence, the output voltage at the instant $N\tau_0$ is equal to

$$U_{out} = 0.5 \sum_{n=1}^{N} U \cos(2\pi f_d n\tau_0).$$

When $f_d = 0$, $U_{out,0} \approx NU/2$.

The loss factor for $f_d \neq 0$ is given by the equation

$$L = \frac{\sum_{n=1}^{N} U \cos(2\pi f_d n\tau_0)}{N}. \tag{3.28}$$

The terms of sum in the numerator of Equation (3.28) decrease with increasing 'n'. The term with $n = 1$ takes on the greatest value (≈ 1), while the term with $n = N$ has the least value. The latter equals

$$\cos(2\pi f_d N\tau_0).$$

The sum in the numerator of Equation (3.28) can be roughly estimated by

$$N\frac{1+\cos(2\pi f_{\rm d} N\tau_0)}{2} = \frac{N}{2}[1+\cos(2\pi f_{\rm d} T_{\rm m})].$$

The averaged estimate of the loss factor due to uncertainty of $f_{\rm d}$ will then be

$$L \approx \frac{1+\cos(2\pi f_{\rm d} T_{\rm m})}{2}. \tag{3.29}$$

The extreme cases are $f_{\rm d} = 0$, $L = 1$, $2\pi f_{\rm d} T_{\rm m} = \pi/2$ and $L = 0.5$.

It follows from Equation (3.29) that the requirements for the value of L, with the expected value of $f_{\rm d.\,max}$, impose a restriction on the maximum allowable value of the signal period, $T_{\rm m}$:

$$T_{\rm m} \leq \frac{\arccos(2L-1)}{2\pi f_{\rm d}} \tag{3.30}$$

For example, when $L = 0.88$ (1 dB) and $f_{\rm d.\,max} = 3$ kHz, $T_{\rm m} \leq 37\mu$s. Under the circumstances, the maximum unambiguously measured range is 5.6 km. To eliminate this limitation on $T_{\rm m}$, it may be advisable to employ a matched filter with multiple frequency channels [I.21] or to use a multiphase coding pattern approximating the chirp waveform [I.21]. Quite naturally, these measures bring about an extra complexity of hardware.

If a radar contains a frequency tracker before a matched filter, we may find that $f_{\rm d} = 0$. Therefore the restriction (3.30) can be neglected.

4

Target Tracking

4.1 INTRODUCTION

The devices for target tracking in angular position, range and velocity perform the measurement of target coordinates, elimination of unwanted background returns and partial reduction of the receiver's noise. Such devices are usually designed as tracking systems with first-, second- or-third-order astaticism. Target coordinates have a certain relationship to parameters of the echo signal. Target range is related to the time delay of the echo signal, target velocity to its Doppler shift and the angular position to the position of its phase front. The terms 'target coordinate tracker' and 'signal parameter tracker' will therefore be used synonymously.

Trackers can be constructed in either analogue or digital form. In Section 4.2 the generalized structure will be considered of a device suitable for target tracking in one or another coordinate, irrespective of its specific implementation. The design principles of trackers with simultaneous and successive comparison of signals will also be examined. Section 4.3 gives a brief description of analogue trackers. Section 4.4 provides an insight into the performance of digital trackers as the prevalent class. Section 4.5 is dedicated to the main errors in tracking systems.

Angle tracking will also be discussed in Section 4.6 and range and velocity tracking in Section 4.7. Valuable material on various aspects of radar tracking and numerous references to other sources can be found in the book by Skolnik [I.3].

4.2 TRACKING SYSTEM STRUCTURE

The input signal of a tracker is a function of time and of the parameter to be measured (α_0 will be used to denote this parameter in general form):

$$U_s(t, \alpha_0).$$

The tracker generates a reference (reference voltage) similar to the signal and contains an estimate of the parameter, which will be designate by the letter α. The tracker consists of two units: a discriminator and a controller (Figure 4.1). In the former unit, the estimate, α , is compared with the parameter value,

Bistatic Radar: Principles and Practice Edited by M. Cherniakov

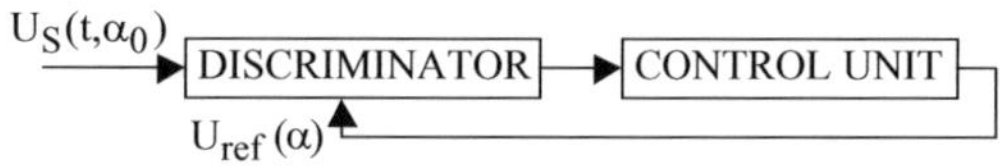

Figure 4.1 Block diagram of a tracker

α_0, and the value proportional to their difference (mismatch),

$$U_d = K_d \, \Delta\alpha = K_d(\alpha_0 - \alpha),$$

comes to the input of the controller. Based on the mismatch, the controller forms a corrective change to α. Correction of the estimate, α, goes on until (ideally) the estimate becomes equal to the true parameter value α_0.

In the simplest case, the discriminator contains a correlator to whose inputs the received signal and the reference voltage are fed. It is obvious that the correlator output voltage U_{cor} is at a maximum when $\alpha = \alpha_0$.

Measuring α_0 by the maximum method is unproductive for two reasons. First, the plot of $U_{cor}(\alpha_0 - \alpha)$ often has a nearly flat peak (Figure 4.2 (a)), which accounts for a low sensitivity to mismatch. Second, having a single value of U_{cor}, it is impossible to decide in which direction α should be changed to approach α_0. Therefore, the differential method is commonly used.

Two values of correlation integral are taken (either successively or simultaneously); the first is taken at the reference voltage $U_{ref}(\alpha + \delta\alpha)$ and the second at $U_{ref}(\alpha - \delta\alpha)$. At the output of the discriminator we have the difference $U_{cor}(\alpha_0 - \alpha - \delta\alpha) - U_{cor}(\alpha_0 - \alpha + \delta\alpha)$. The method is illustrated in Figure 4.2 (b). It follows from analysis of Figure 4.2(b) that $\alpha = \alpha_0$ when $U_{cor}(\alpha_0 - \alpha - \delta\alpha) = U_{cor}(\alpha_0 - \alpha + \delta\alpha)$.

In the case of $U_{cor}(\alpha_0 - \alpha - \delta\alpha) < U_{cor}(\alpha_0 - \alpha + \delta\alpha)$, we have $\alpha > \alpha_0$, and when $U_{cor}(\alpha_0 - \alpha - \delta\alpha) > U_{cor}(\alpha_0 - \alpha + \delta\alpha)$, $\alpha < \alpha_0$. It is not difficult to draw the inference that the dependence of the discriminator's output voltage on the mismatch $U_{discr}(\alpha_0 - \alpha)$ called the 'discriminator characteristic' conforms to the curve shown in Figure 4.3.

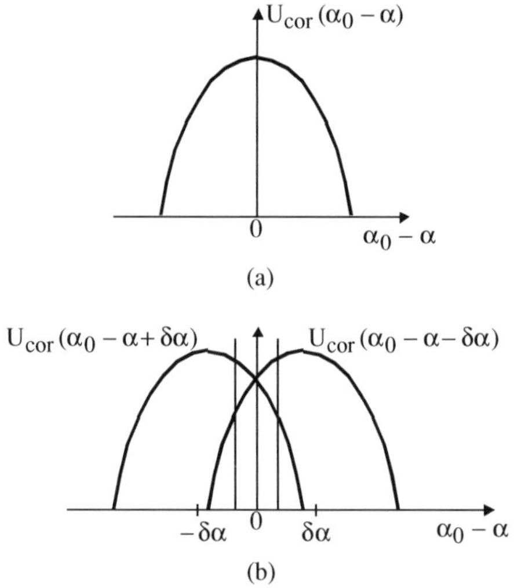

Figure 4.2 Illustration of the principle of tracker functioning

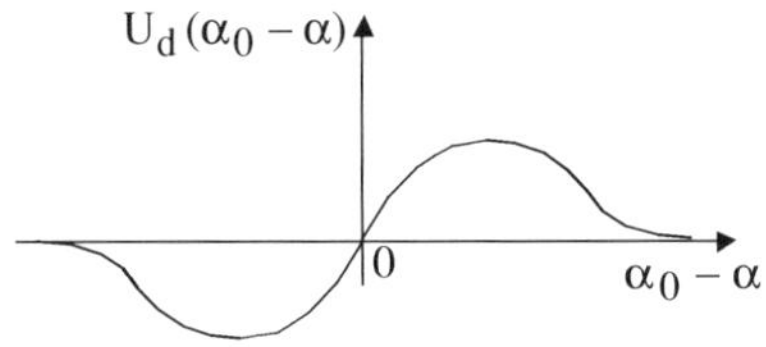

Figure 4.3 Discriminator characteristic

Because of a small time constant, the discriminator is usually regarded as an inertialess element. The derivative of a discriminator characteristic at the point $\alpha = \alpha_0$ is recognized as its gain, K_d. Its dimension is $|K_d| : V/\alpha_0$ in analogue devices or $|K_d|$: unit/α_0 in digital devices.

Note that for a simultaneous comparison of correlator output voltages $U_{cor}(\alpha_0 - \alpha - \delta\alpha)$ and $U_{cor}(\alpha_0 - \alpha + \delta\alpha)$, it is necessary to use two correlators with reference voltages $U_{ref}(\alpha + \delta\alpha)$ and $U_{ref}(\alpha - \delta\alpha)$ respectively.

The controller consists of integrators and a converter. The latter is necessary to convert the voltage (number) at the integrator output to a value of parameter estimate, α. The converter is also usually recognized as an instantaneous element with a conversion gain K_{conv}. Its dimension is $|K_{conv}| : \alpha_0$/volt in analogue devices or $|K_{conv}| : \alpha_0$/unit in digital devices.

4.3 ANALOGUE TRACKING DEVICES

The block diagram of a first-order tracker (one integrator) is given in Figure 4.4. The transfer function of an analogue integrator is equal to K_i/p, where K_i is the gain with a dimension 1/s and p is the Laplacian. The closed-loop transfer function is

$$K(p)_1 = \frac{K_1}{K_1 + p}, \quad \text{where} \quad K_1 = K_d K_i K_{conv}.$$

The –3 dB bandwidth of the tracker is

$$\Delta F_1 = \frac{K_1}{2\pi}. \tag{4.1}$$

Its error transfer function is

$$K_\Delta(p)_1 = 1 - K(p)_1 = \frac{p}{p + K_1}. \tag{4.2}$$

The block diagram of an analogue second-order tracker, presented in Figure 4.5, includes two integrators in addition to the discriminator and the converter. One of the integrators contains a correction circuit (in the form of a capacitor Ć$_c$) ensuring the desired stability of the tracker.

The closed-loop transfer function is defined by the formula

$$K(p)_2 = \frac{K_2(1 + pT_c)}{p^2 + K_2(1 + pT_c)},$$

Figure 4.4 Structure chart of a first-order tracker

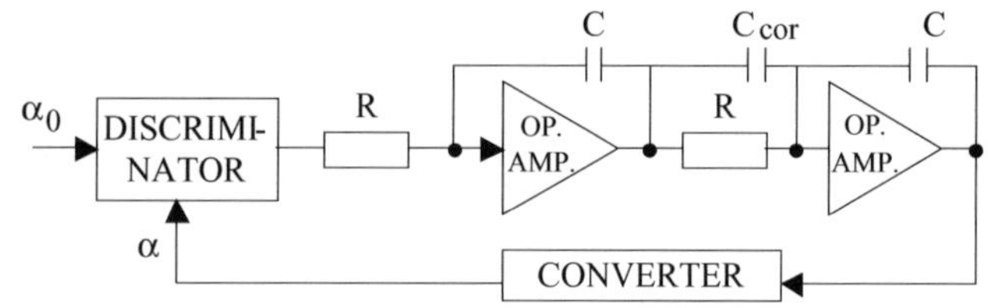

Figure 4.5 Structure chart of a second-order tracker

where $T_c = RC_{cor}$ and $K_2 = K_d K_i^2 K_{conv}$. The bandwidth of the tracker is

$$\Delta F_2 = \frac{\sqrt{K_2}}{2}\left(\xi + \frac{1}{4\xi}\right), \quad \text{where} \quad \xi = \frac{T_c}{2}\sqrt{K_2}$$

and, since typically $\xi = 1$–2, it can be assumed that

$$\Delta F_2 \approx \sqrt{K_2}. \tag{4.3}$$

The error transfer function is

$$K_\Delta(p)_2 = 1 - K(p)_2 = \frac{p^2}{p^2 + K_2(1 + pT_c)}. \tag{4.4}$$

Generally, if the tracker's astaticism order is equal to n, the error transfer function is given by

$$K_\Delta(p)_n = \frac{p^n}{p^n + K_n\varphi(p)_n}, \tag{4.5}$$

where $\varphi(p)_n$ is a function describing the behaviour of the correction circuit and $\varphi(p)_n \neq 0$ when $p = 0$.

4.4 DIGITAL TRACKING DEVICES

The discriminator is usually constructed according to one of two possible designs. The first design presupposes an analogue discriminator whose output voltage undergoes analogue-to-digital conversion. In the second alternative, the input action is digitized, and the subtraction operation

$$\Delta\tilde{\alpha} = \tilde{\alpha}_0 - \tilde{\alpha}$$

is performed digitally.

Depending on the discriminator design, the converter has an analogue or digital output. If the converter is an analogue element, a digital-to-analogue converter (DAC) is brought in between the integrator and the converter. The scaling circuit (scaler) with a gain, m, is a linear inertialess element. The digital integrator is normally constructed as an accumulator (Figure 4.6).

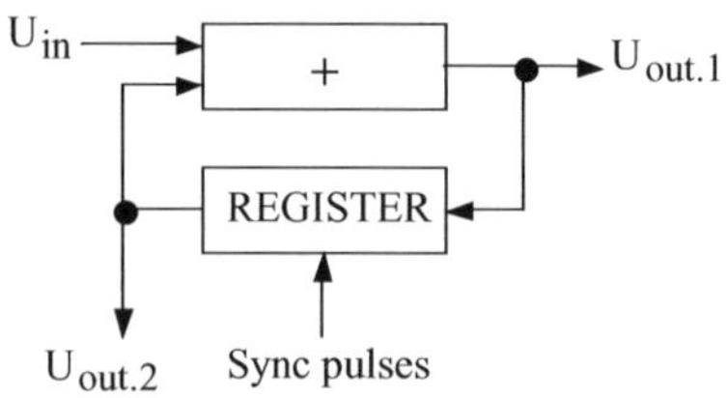

Figure 4.6 Block diagram of a digital integrator

Either the accumulator output $\tilde{U}_{\text{out.1}}$ or the register output $\tilde{U}_{\text{out.2}}$ can be used as the output number. The tilde over the symbol U implies that it is a number. The term T_i will be used for the clock cycle. Let us define the integrator transfer function for $\tilde{U}_{\text{out.1}}$ used as the output number. The difference equation for $\tilde{U}_{\text{out.1}}$ has the form

$$\tilde{U}_{\text{out.1}}(n) = U_{\text{in}}(n) + \tilde{U}_{\text{out.1}}(n - 1).$$

Taking a z-transform of each side of the equation gives

$$U_{\text{out.1}}(z) = U_{\text{in}}(z) + U_{\text{out.1}}(z)z^{-1}.$$

Hence, the transfer function of the integrator is

$$H_{\text{i1}}(z) = \frac{U_{\text{out.1}}(z)}{U_{\text{in}}(z)} = \frac{z}{z - 1}. \tag{4.6}$$

When the second integrator output is used,

$$U_{\text{out.2}}(z) = U_{\text{out.1}}(z)z^{-1}.$$

Consequently, the transfer function of the integrator in this case is

$$H_{\text{i2}}(z) = \frac{1}{z - 1}. \tag{4.7}$$

The block diagram of a first-order digital tracker is presented in Figure 4.7.

It is seen from the figure that the second integrator output is used here. In this case it is dictated by the necessity of separating the moment of feeding the next number to the input and the response of the tracker by one clock period. Otherwise, the system status (output number) would be indeterminate.

An obvious expression for the open-loop transfer function may be written as

$$H(z)_{\text{o.l.}} = K_{\text{d}} m \frac{1}{z - 1} K_{\text{conv}} = \frac{K_0}{z - 1},$$

where $K_0 = K_{\text{d}} m K_{\text{conv}}$. We will now determine the dimension of the quantity K_0. Since m is a dimensionless multiplier,

$$|K_0| = |K_{\text{d}}|\ |K_{\text{conv}}| = \frac{\text{unit}}{\alpha} \frac{\alpha}{\text{unit}},$$

i.e. K_0 is a dimensionless quantity.

Let us find an expression for the closed-loop transfer function of the tracker (referred to the converter output):

$$H(z)_{\text{l}} = \frac{H(z)_{\text{o.l.}}}{1 + H(z)_{\text{o.l.}}} = \frac{K_0/(z - 1)}{1 + K_0/(z - 1)} = \frac{K_0}{z - 1 + K_0}. \tag{4.8}$$

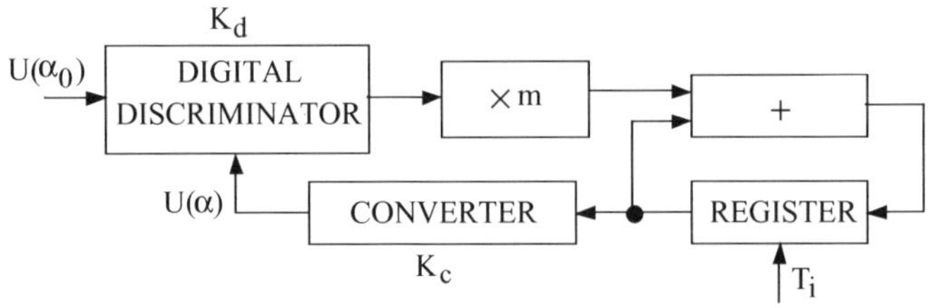

Figure 4.7 Block diagram of a first-order digital tracker

Thus, K_0 is the only parameter of the transfer function. From Equation (4.8) it is easy to determine the range of K_0 values, within which the tracking system is stable. Indeed, the only pole of $H(z)_1$ is

$$z_1 = 1 - K_0.$$

Since the necessary and sufficient condition for system stability is

$$|z_1| < 1,$$

we obtain

$$|1 - K_0| < 1.$$

It follows from the last inequality that

$$0 < K_0 < 2.$$

It will be shown further that it is often expedient to use a more stringent condition when specifying K_0:

$$0 < K_0 < 1.$$

Let us now define the 'frequency response' of a first-order system. To this end, $\exp(j\omega_n)$ $(\omega_n = \omega T_i)$ is substituted for the variable z in Equation (4.8). This yields

$$\overline{K(\omega_n)} = \frac{K_0}{\exp(j\omega_n) + K_0 - 1} = \frac{K_0}{\cos\omega_n + K_0 - 1 + j\sin\omega_n}.$$

The amplitude–frequency response of the system is

$$K(\omega_n) = \frac{K_0}{\sqrt{(\cos\omega_n + K_0 - 1)^2 + \sin^2\omega_n}} = \frac{K_0}{\sqrt{1 + (1 - K_0)^2 - 2(1 - K_0)\cos\omega_n}}. \tag{4.9}$$

Analyse the amplitude–frequency response obtained. For $\omega_n = 0$, $K(0) = 1$. For $\omega_n = \pi$,

$$K(\pi) = \frac{K_0}{2 - K_0}.$$

For $K_0 = 1$, $K(\omega_n) = 1$, i.e. the tracking system does not filter the input action, which is unacceptable in most cases.

The –3 dB bandwidth $\Delta\omega_n$ of the tracker can be found from the equation

$$K^2(\Delta\omega_n) = \frac{K_0^2}{1 + (1 - K_0)^2 - 2(1 - K_0)\cos\Delta\omega_n} = 0.5.$$

By simple manipulation,

$$\Delta\omega_n = 2\arcsin\left(\frac{0.5K_0}{\sqrt{1 - K_0}}\right).$$

If $K_0 \leq 0.7$, then

$$\arcsin\left(\frac{0.5K_0}{\sqrt{1 - K_0}}\right) \approx \frac{0.5K_0}{\sqrt{1 - K_0}}$$

and

$$\Delta\omega_n \approx \frac{K_0}{\sqrt{1 - K_0}}; \quad \Delta\omega = \frac{\Delta\omega_n}{T_i} = \frac{K_0}{T_i\sqrt{1 - K_0}}. \tag{4.10}$$

Figure 4.8 shows the amplitude–frequency responses of a digital tracker calculated by formula (4.9) for several values of K_0. It follows from the figure and from Equation (4.10) that the system bandwidth

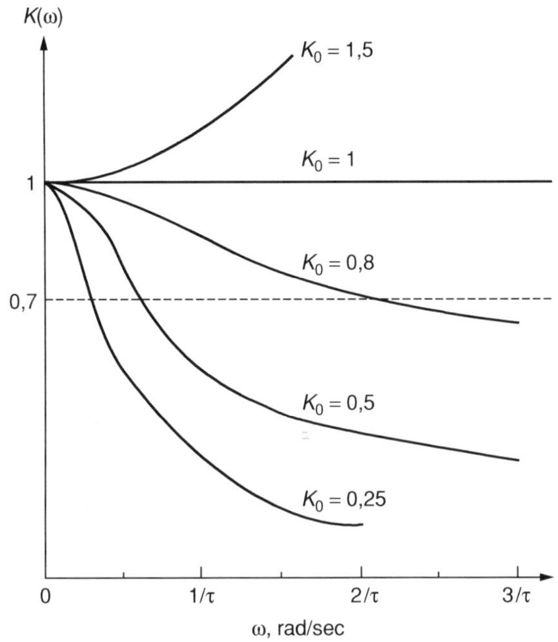

Figure 4.8 Amplitude–frequency responses of a first-order digital tracker

decreases with decreasing K_0 and/or increasing T_i. Note that the variance of tracking error due to noise accompanying the useful signal is defined by [I.22, I.23]

$$\sigma^2 = a\frac{\Delta\omega}{\mathrm{S/N}}.$$

Where a is the proportionality coefficient and S/N is the signal-to-noise ratio at the input of the tracking system.

An expression will now be found for the pulse response of a first-order tracker. The transfer function of a linear system is equal to a z-transform of its pulse response. Consequently, the pulse response of a tracking system is equal to an inverse z-transform of its transfer function. The calculation of the pulse response for a first-order system is very simple. Taking an inverse z-transform of function (4.8) gives

$$h(n) = \begin{cases} 0 & \text{for} \quad n = 0; \\ K_0(1 - K_0)^{n-1} & \text{for} \quad n \geq 1. \end{cases} \tag{4.11}$$

Using Equation (4.11), the tracker's response to one or another input action can be calculated with the help of a discrete convolution:

$$U(n)_{\mathrm{out}} = \sum_{i=1}^{n} u(n-i)_{\mathrm{in}} h(i).$$

The lower limit of summation is equal to unity because $h(0) = 0$. In particular, if $u(n - i) = 1$ is substituted into this equation the response can be found of a first-order system to a unit step (transient):

$$U_{\mathrm{out.1}} = \sum_{i=1}^{n} K_0(1 - K_0)^{i-1} = 1 - (1 - K_0)^n.$$

Table 4.1 Rounded values of n_1 as a function of open-loop gain K_0

K_0	0.2	0.3	0.4	0.5	0.6	0.7	0.8
n_1	10	6	5	3	2	2	1

The duration of the transient $n.1$ can be defined as the number of synchronizing pulses before the output value reaches $U_{\text{out.1}} = 0.9$. Then,

$$n.1 = -\frac{1}{\lg(1 - K_0)}. \tag{4.12}$$

Table 4.1 provides rounded values of $n.1$ as a function of open-loop gain K_0.

4.4.1 Digital Tracking System with Second-Order Astaticism

The block diagram of a second-order tracker is given in Figure 4.9. By definition the system comprises two integrators. Just as in an analogue tracker, one of these integrators (the first one in this case) contains a correction circuit to ensure the system has the required stability. This element is what distinguishes the structure shown in Figure 4.9 from that of a first-order tracker.

The circuit consists of a linear inertialess element with a gain M. The diagram of such an integrator with a correction circuit is shown separately in Figure 4.10. The output of this circuit is summed with the output number of the accumulator. In this case, the transfer function of the integrator as such (disregarding the correction circuit) is defined by Equation (4.6). The transfer function of the integrator including the correction circuit is

$$H_{i1}(z)_{\text{corr}} = \frac{z}{z-1} + M = \frac{z(1+M) - M}{z-1}.$$

Consequently, the open-loop transfer function is expressed as

$$H(z)_{\text{o.l.2}} = H(z)_{\text{o.l.1}} H_{i1}(z)_{\text{corr}} = \frac{K_0\,[z(1+M) - M]}{(z-1)^2}.$$

Hence, the closed-loop transfer function of a second-order tracker referred to the converter output can be written in the form

$$H(z)_2 = \frac{K_0\,[z(1+M) - M]/(z-1)^2}{1 + K_0\,[z(1+M) - M]/(z-1)^2} = K_0 \frac{z(1+M) - M}{(z-1)^2 + K_0(1+M)z - K_0 M}. \tag{4.13}$$

The restrictions imposed on tracking system parameters by the stability requirement can be cited without derivation: $K_0 > 0$, $M > 0$, $K_0(1 + 2M) < 4$. Substituting $z = \exp(j\omega_n)$ into Equation (4.13)

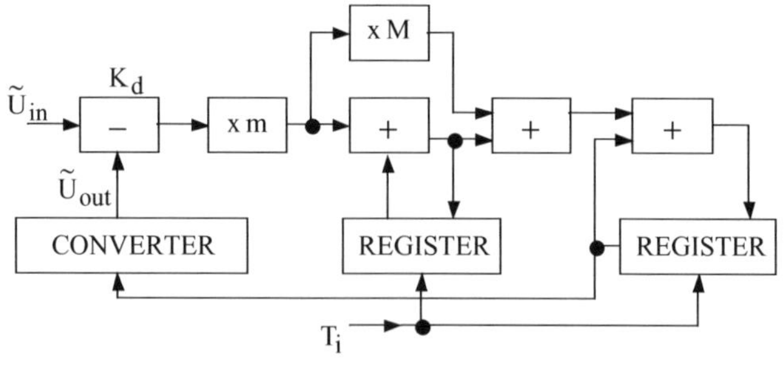

Figure 4.9 Block diagram of a second-order tracker

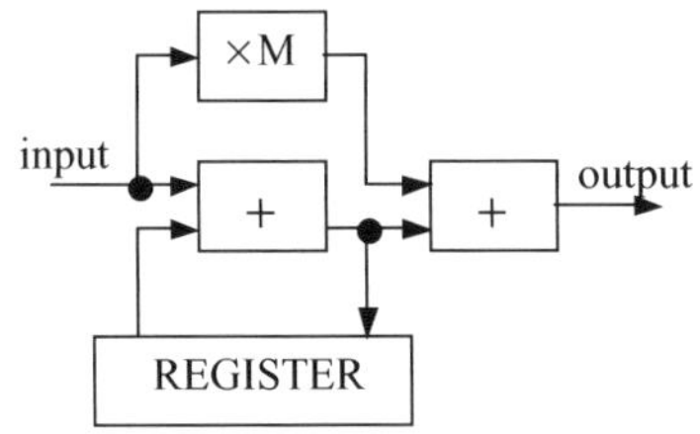

Figure 4.10 Digital integrator with correction

gives the complex frequency response of the system. Its modulus yields the amplitude–frequency response of the tracking system:

$$K(\omega_n) = K_0 \frac{\sqrt{2M^2 + 2M + 1 - 2(M^2 + M)\cos\omega_n}}{\sqrt{1 + a^2 + b^2 + 2a(1+b)\cos\omega_n + 2b\cos 2\omega_n}}. \tag{4.14}$$

Equation (4.14) contains the following notation:

$$a = K_0(1+M) - 2 \qquad b = 1 - K_0 M.$$

Analysis of relationship (4.14) shows that if $M = 0$ (no correction), there is always a certain critical value of frequency $\omega_{n.cr}$, at which the tracking system is unstable. The critical frequency $\omega_{n.cr}$, which is determinable by equating the denominator in Equation (4.14) to zero, is

$$\omega_{n.cr} = \arccos(1 - 0.5K_0). \tag{4.15}$$

Examples given in Figure 4.11 are plots of $K(\omega)$ obtained with the following values of tracking system parameters:

$$\begin{aligned} K_0 &= 1, & M &= 1(\text{curve } 1);\\ K_0 &= 0.5, & M &= 1(\text{curve } 2);\\ K_0 &= 0.2, & M &= 1(\text{curve } 3);\\ K_0 &= 0.2, & M &= 0.5(\text{curve } 4);\\ K_0 &= 0.5, & M &= 0(\text{curve } 5). \end{aligned}$$

For an approximate estimation of a –3 dB bandwidth of a second-order tracker, we can use the relationship

$$\Delta\omega \approx \frac{1.3\pi K_0}{T_i}; \qquad \Delta f \approx \frac{0.65K_0}{T_i}. \tag{4.16}$$

4.5 MAIN ERRORS IN TRACKING RADARS

The main types of errors encountered in signal parameter measurements are as follows:

- noise error due to signal distortion by the receiver's own noise or by external noise interference;
- dynamic errors relating to temporal variations of the measured parameter and dependent on structural peculiarities of the tracker;
- fluctuation errors arising from accidental variations of the measured parameter due to fluctuations of the radar centre of the target;
- specific error of a tracker with successive comparison of signals in the discriminator, which manifests itself under the conditions of signal amplitude fluctuations (amplitude noise).

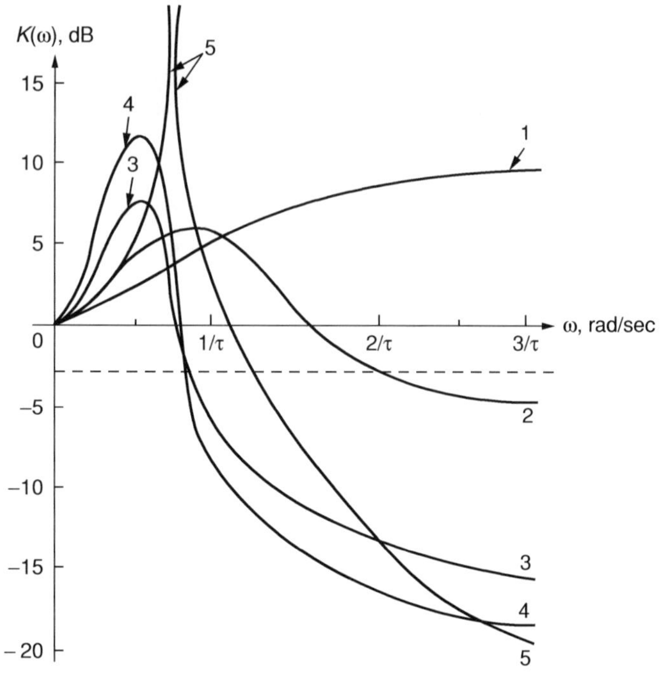

Figure 4.11 Amplitude–frequency responses of a digital second-order tracker

Let us begin with a computational technique for noise and dynamic errors which the hardware design engineer always has to estimate. Formulae will be presented for the calculation of root-mean-square values of frequency error (σ_f) and signal time-of-arrival (σ_t) and direction-of-arrival (σ_β) errors. For a *pulsed signal*,

$$\sigma_f = \frac{1}{t_{coh}}\sqrt{\frac{\Delta F}{\rho F_p}}; \tag{4.17}$$

$$\sigma_t = \frac{1}{\Delta f_s}\sqrt{\frac{\Delta F}{\rho F_p}}; \tag{4.18}$$

$$\sigma_\beta = \theta\sqrt{\frac{\Delta F}{\rho F_p}}. \tag{4.19}$$

In Equations (4.17) to (4.19), t_{coh} is the coherent interval of the signal; for example, $t_{coh} = T_0$ for a coherent burst of duration T_0, while for a noncoherent burst, $t_{coh} = \tau_p$ (duration of a single pulse). In the equations Δf_s is the spectrum bandwidth of the pulse, θ is the antenna beamwidth (as a rule, in the receive mode), ΔF is the bandwidth of the tracker, ρ is the signal-to-noise ratio at the input of the tracker and F_p is the pulse repetition frequency. For a *continuous signal*, it is necessary to substitute the receiver bandwidth for the quantity F_p.

Of all the quantities appearing in formulae (4.17) to (4.19), one (ρ) is dependent on the target range. For an active radar,

$$\rho = \frac{a}{R^4} \tag{4.20}$$

and for a semi-active radar,

$$\rho = \frac{b}{(R_1 \cdot R_2)^2}. \tag{4.21}$$

In Equations (4.20) and (4.21), a and b are quantities independent of the target range.

4.5.1 Dynamic Errors in Target Tracking

The operation of a tracking system has two distinctive conditions: transient process and steady state. A polynomial approximation for the law of measured signal parameter variations may be written as

$$\alpha_0(t) = \sum_{l=0}^{L} a_l t^l, \tag{4.22}$$

where L is the order of a nonzero senior derivative.

If all the coefficients a_l = constant on a time interval exceed the transient period (e.g. Equation (4.12) for a first-order system), we have a steady state condition. The transient process arises at the moment when the system turns on or as a result of a change in coefficients a_l in Equation (4.22). The extent of errors may be quite considerable during a transient, their duration decreasing when the open-loop gain of the tracking system, k_0, increases (see, for example, Table 4.1).

4.5.2 Dynamic Errors in Tracking Systems

Dynamic errors of an analogue tracking system in the steady state condition can be calculated by [I.22]:

$$\Delta_d = \sum_{l=0}^{L} S_l \frac{d^l \alpha_0(t)}{dt^l} \tag{4.23}$$

Thus, Δ_d is the weighted sum of the derivatives $\alpha_0(t)$. The weight coefficients (error factors) S_l are defined by the equation

$$S_l = \frac{1}{l!} \left| \frac{d^l H_\Delta(p)}{dp^l} \right|_{p=0} \tag{4.24}$$

From Equations (4.22), (4.23) and (4.24), it is not difficult to deduce a particular inference: if the astaticism order of the tracker is higher than the order of the higher derivative of the measured parameter, the dynamic error in the steady state condition is equal to zero.

Dynamic errors in a digital tracking system in the steady state condition can be estimated by a method similar to that developed for analogue systems. It can be shown that formula (4.23) holds just as well for a digital tracker if the weight coefficients are assumed to be equal to

$$S_l = \frac{T_i^l}{l!} \left| \frac{d^l H_\Delta(z)}{dz^l} \right|_{z=1}, \tag{4.25}$$

where $H_\Delta(z)$ is the error transfer function for a digital system and

$$H_\Delta(z) = 1 - H(z).$$

Let us calculate the static (for a motionless object) and dynamic error factors for tracking systems with first- and second-order astaticism. For a first-order system,

$$H_\Delta(z) = 1 - \frac{k_0}{z - 1 + k_0} = \frac{z - 1}{z - 1 + k_0}.$$

The static error factor is

$$S_0 = H_\Delta(z)|_{z=1} = 0,$$

i.e. there is no static error. The error factor S_1 is

$$S_1 = T_i \frac{d}{dz}\left(\frac{z-1}{z-1+k_0}\right)\bigg|_{z=1} = \frac{T_i}{k_0}.$$

Therefore, given only the first derivative of the function $\alpha_0(t)$, the dynamic error is

$$\Delta d = \frac{T_i}{k_0}\frac{d\alpha_0(t)}{dt}.$$

For a second-order system,

$$H_\Delta(z) = 1 - H(z) = \frac{(z-1)^2}{(z-1)^2 + k_0(1+M)z - k_0 M}.$$

The static error equals zero since $S_0 = 0$.

It is not difficult to establish that $S_1 = 0$, i.e. given only the first derivative of the signal parameter being tracked, there is no dynamic error. The factor S_2 is

$$S_2 = \frac{T_i^2}{2!}\frac{d^2 K_\Delta(z)}{dz^2}\bigg|_{z=1} = \frac{T_i^2}{k_0}.$$

Therefore, given the second derivative of the signal parameter being tracked,

$$\Delta_d = \frac{T_i^2}{k_0}\frac{d^2\alpha_0(t)}{dt^2}.$$

In general, if the order of the tracker is higher than the order of a certain derivative of the function $\alpha_0(t)$, there is no dynamic error due to this derivative. Based on this statement, it is possible to choose the order of the system by the criterion of the greater derivative of the function $\alpha_0(t)$.

4.5.3 Fluctuation Errors (Target Glint)

In Section 3.1, we touched upon the fluctuations of the radar centre of the target. Reference [I.24] contains formulae for calculating root-mean-square deviations of the target centre from its midposition:

$$\sigma_c \approx 0.35L, \tag{4.26}$$

where L is the maximum size of the target along one or another coordinate axis. The root-mean-square value of target centre fluctuations along the axis of range σ_R is expressed by (4.26), provided the target length along the said axis is substituted into this formula. To estimate the root-mean-square value of angular fluctuations (in radians), the following formula is used:

$$\sigma_\beta = 0.35\frac{L_2}{R}, \tag{4.27}$$

where L_2 is target length in the transverse direction and R is the target range. The root-mean-square value of the Doppler shift fluctuations is determined by the formula

$$\sigma_f \approx 0.35\frac{2L_2}{\lambda}\frac{d\beta}{dt}. \tag{4.28}$$

In (4.28), β is the aspect angle of the target and λ is the wavelength.

Owing to the filtering action of trackers, signal parameter measurement errors are less than σ_R, σ_β and σ_f. For the estimation of measurement errors it is necessary to integrate the product of the spectrum density of target radar centre fluctuations, $S(\omega)$, and the square of the frequency response of the tracker,

$k(\omega)$. $S(\omega)$ is a decreasing function of frequency. Using an exponential function to approximate the correlation factor of the fluctuations, we will come to an expression for $S(\omega)$ given by

$$S(\omega) = \frac{S(0)}{1 + \omega^2/\omega_a^2}. \qquad (4.29)$$

In Equation (4.29), $S(0) = 4\sigma_c^2 \tau_c$ and $\omega_a = 1/\tau_c$. A similar formula is presented in Reference [I.24].

The fluctuation error variance is defined by the formula

$$\sigma_{fl}^2 = \frac{1}{2\pi} \int_0^\infty S(\omega) k^2(\omega)\, d\omega. \qquad (4.30)$$

In particular, for a first-order tracker,

$$\sigma_{fl}^2 = \frac{1}{2\pi} \int_0^\infty \frac{S(0)}{1 + \omega^2/\omega_a^2} \frac{k^2\, d\omega}{k^2 + \omega^2} = \frac{\sigma_c^2 k}{k + \omega_a}. \qquad (4.31)$$

The quantity k in Equation (4.31) is the open-loop gain of the tracking system.

In angle tracking of a low-flying object, the complex target is represented by a true object and its antipode (see Section 1.9). The distance between the object and the antipode is considerably greater than the size of the object. Therefore, the elevation error variance may be great, which often results in mistracking. Sometimes, automatic elevation tracking collapses under such circumstances. An approximate estimate of this coordinate is then obtained from the output voltage of the relevant discriminator [I.24].

4.5.4 Angle Tracking Error in a Conical-Scan Radar Due to Amplitude Noise

The following formula can be used to calculate the variance, σ_a^2, of the error at issue [I.25]:

$$\sigma_a^2 = \frac{\theta^2}{4\pi^2 f_{sc}^2} \frac{\Delta F}{\tau_c} \qquad (4.32)$$

In Equation (4.32), θ is the beamwidth of the antenna, f_{sc} is the scanning frequency, ΔF is the bandwidth of the tracker and τ_c is the correlation window of amplitude noise. A similar result is presented in Reference [I.24].

4.6 ANGLE TRACKING DEVICES

The differential method of target angle tracking (angle measurement) in a single plane is realized with the use of either two receive beams (1 and 2) slightly offset with respect to one another (Figure 4.12(a)), or one beam set in turns to positions 1 and 2 with a scanning frequency F_{sc} (Figure 4.12(b)). In the former case, a discriminator with a simultaneous comparison of signals is employed, while in the latter a discriminator with a successive comparison is used. The measuring channel performs simultaneous or successive comparison of the amplitudes of signals received on the two beams.

The straight line passing through the intersection point of the two antenna patterns (APs) is referred to as the radar boresight or the equisignal line (ESL). If the signal source is on the ESL, the signals received on the two beams are identical. When the signal source is offset from the ESL, the signals received on the two beams exhibit differing amplitudes (or phases). With a simultaneous comparison of signals, their difference is proportional to the offset of the signal source from the ESL. With a successive comparison, the offset of the signal source from the ESL brings about amplitude modulation of the antenna output voltage with a frequency equal to the beam-switching frequency (F_{sc}).

The controller turns the AP in the required direction until the signal source is on the ESL. Such a condition is indicated by zero difference between the two received signals when they are compared

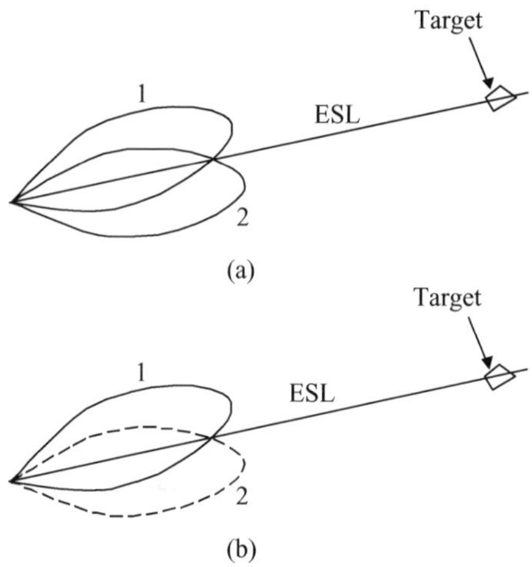

Figure 4.12 (a) Simultaneous and (b) successive comparison of signals

simultaneously. Under a successive comparison, the indication is the absence of signal amplitude modulation with the frequency F_{sc}.

Let us consider the most common kinds of angle tracking devices, which may be broadly divided into two classes. The first class includes trackers implementing a simultaneous comparison of signals. Such devices are called monopulse trackers. Other tracking devices pertain to a class with a successive comparison of signals.

4.6.1 Sum Difference Based Monopulse Tracker with a Comparison of Signal Phases

To understand the principle of operation of monopulse trackers, a device with mirror antennas may be taken as an example. The peculiarities of using phased array antennas will also be examined. Figure 4.13 gives a schematic representation of a monopulse antenna intended for the measurement of a single angular coordinate of the target, β (in the figure plane). The antenna consists of two elements and their phase centres are designated by numerals 1 and 2. In addition, the microwave part of the device includes an add–subtract circuit designated by the symbols '+' and '−', and a phase-shifter performing a phase shift by $\pi/2$.

If the target is on the axis AB of the antenna, the signal paths to phase centres 1 and 2 are of equal length. Hence, the received signals have the same phase. Signal amplitudes V_1 and V_2 are also equal, since the axes of the antenna patterns are parallel. Therefore, the difference of V_1 and V_2 (the difference signal, U_{dif}) is equal to zero.

A propos, this implies that the antenna axis AB is the equisignal line. The sum signal is

$$U_{\text{sum}} = V_1 + V_2 = 2V_1.$$

If the direction to target 2 forms an angle, β , with the ESL, the phase difference of signals V_1 and V_2 is equal to

$$\varphi = \frac{2\pi\ell\sin\beta}{\lambda}.$$

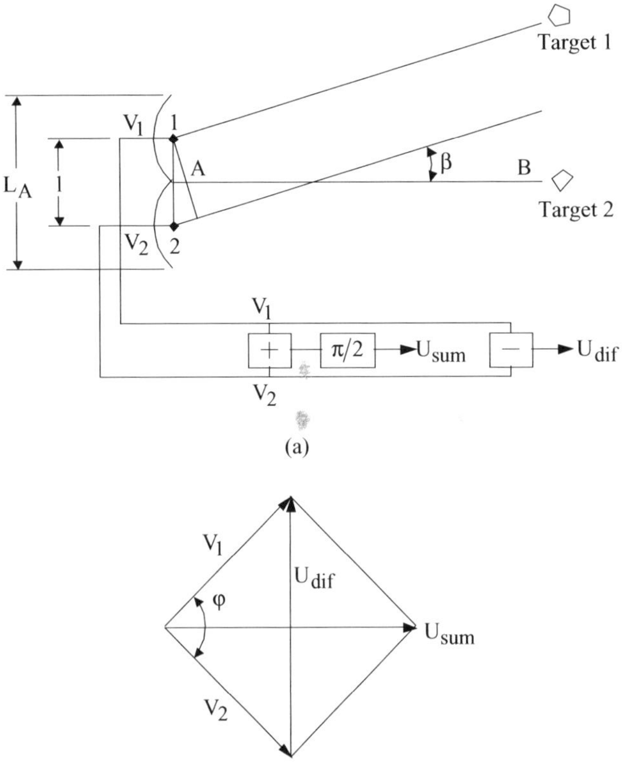

(a) (b)

Figure 4.13 Sum-difference method of angular position measurement (one chanel): (a) antenna elements and a sum-difference device; (b) vector diagram

Furthermore, the amplitudes and phases of $U_{\rm dif}$ and $U_{\rm sum}$ can be found with the help of the vector diagram in Figure 4.13(b):

$$U_{\rm sum} = 2V_1 \cos\frac{\varphi}{2}; \qquad U_{\rm dif} = 2V_1 \sin\frac{\varphi}{2};$$

$\angle(U_{\rm sum}, U_{\rm dif}) = \pi/2$ or $-\pi/2$ depending on the sign of the error β. On undergoing a phase shift by $\pi/2$ in the phase-shifter, the vectors of $U_{\rm sum}$ and $U_{\rm dif}$ have either equal or opposite phases depending on the sign of β.

Combining the formulae obtained for $U_{\rm sum}$ and $U_{\rm dif}$, we arrive at an expression connecting their amplitudes:

$$U_{\rm dif} = U_{\rm sum}\,{\rm tg}\frac{\varphi}{2} = U_{\rm sum}\,{\rm tg}\left(\frac{\pi\ell\sin\beta}{\lambda}\right).$$

For small angular errors,

$$\sin\beta \approx \beta \text{ and } {\rm tg}\frac{\pi\ell\sin\beta}{\lambda} \approx \frac{\pi\ell\beta}{\lambda}.$$

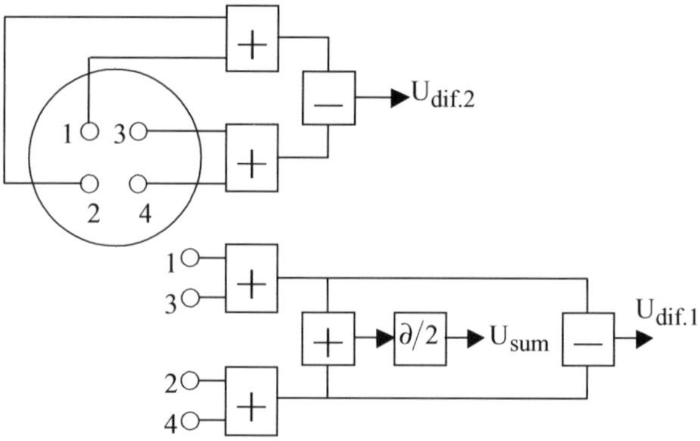

Figure 4.14 Device of a phase comparison of signals in two mutually perpendicular planes

Hence,

$$U_{\text{dif}} \approx U_{\text{sum}} \frac{\pi \ell \beta}{\lambda} = U_{\text{sum}} \frac{\pi L \beta}{2\lambda} \approx \mu U_{\text{sum}} \beta. \tag{4.33}$$

In Equation (4.33), $\mu = \pi/2\theta \approx 1.57/\theta$. Here, θ is the width of the beam formed by both antenna elements. The quantity θ is expressed in the same units as the error β (degrees or radians).

It is often necessary to measure two angular coordinates of the target in two mutually perpendicular planes; therefore the receiving antenna contains four elements separated pairwise in two orthogonal planes. The phase centres of these elements (1, 2, 3, 4) and the scheme for obtaining the sum voltage, U_{sum}, and the two difference voltages $U_{\text{dif.1}}$, $U_{\text{dif.2}}$ are shown in Figure 4.14.

The amplitudes of U_{sum}, $U_{\text{dif.1}}$ and $U_{\text{dif.2}}$ are simply expressed in terms of output voltages of antenna elements V_1 to V_4:

$$\begin{aligned} U_{\text{sum}} &= U_1 + U_2 + U_3 + U_4; \\ U_{\text{dif.1}} &= U_1 + U_3 - (U_2 + U_4) = \mu U_{\text{sum}} \beta_1 \\ U_{\text{dif.2}} &= U_1 + U_2 - (U_3 + U_4) = \mu U_{\text{sum}} \beta_2. \end{aligned} \tag{4.34}$$

Where β_1 and β_2 are the components of angular error in two mutually perpendicular planes.

Figure 4.15 is the block diagram of a monopulse radar receiver with three channels. The outputs of the phase detectors are the outputs of the two discriminators of the devices measuring two angular coordinates of the target. The value of U_{sum} in Equations (4.34) can vary by several orders because of changes in the target range and RCS. To make the slope of the discriminator characteristic independent

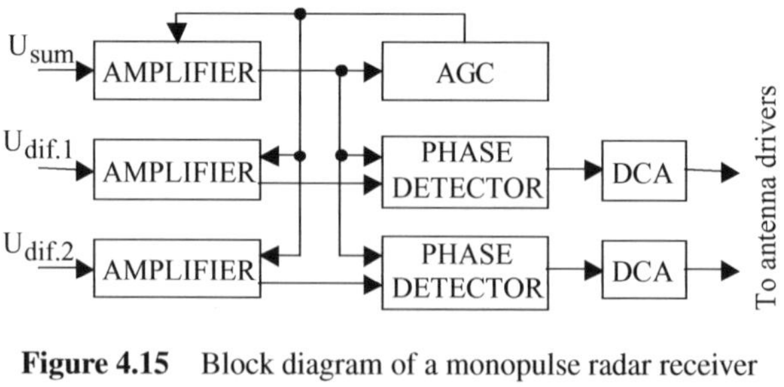

Figure 4.15 Block diagram of a monopulse radar receiver

of the sum signal amplitude, the receiver contains an automatic gain control (AGC) circuit. The AGC is common for all three channels and is controlled by the voltage of the sum signal.

As a simplification, suppose the AGC maintains the amplitude of the output voltage in the sum channel equal to U_0. Furthermore, it is assumed that all three channels are identical; i.e. the gains in all the channels are equal. Then, the amplitudes of the output voltages in the channels are

$$U_{\text{sum.out}} = U_0;$$
$$U_{\text{dif.1.out}} = \mu U_0 \beta_1; \qquad U_{\text{dif.2.out}} = \mu U_0 \beta_2.$$

Controllers in tracking systems consist of direct current amplifiers (DCA) and antenna drives, ensuring its movement in two orthogonal planes. By means of the drives the mismatches β_1 and β_2 are reduced to zero and the angular position of the target is determined by the direction of the ESL. The tracker under consideration has first-order astaticism; the antenna drives also function as integrators. The open-loop gain of the tracker is

$$K_1 = K_{\text{d}} K_{\text{DCA}} K_{\text{DRV}} = K_{\text{PD}} \mu U_0 K_{\text{DCA}} K_{\text{DRV}}, \tag{4.35}$$

where K_{DRV} is drive gain, $K_{\text{d}} = K_{\text{PD}} \mu U_0$ is the discriminator gain and K_{PD} is the phase detector gain. The remaining coefficients do not require interpretation.

If the gains of the receiving paths change in the process of the operation of the tracker, the tracker gain may change as well. Suppose the gain of the difference channel has changed by ΔK_{dif} and the gain of the sum channel has changed by ΔK_{sum} (ΔK_{sum} and ΔK_{dif} are independent and can be positive or negative). Then the expression for the gain of the tracker will take the form

$$K_1' = \frac{K_{\text{rec}} + \Delta K_{\text{dif}}}{K_{\text{rec}} + \Delta K_{\text{sum}}} K_1,$$

where K_1' is the altered gain and K_{rec} is the initial gain of the receiver. If the difference of the phase shifts produced by the sum signal and difference signal amplifiers reaches $\Delta\psi$, then it is necessary to introduce an additional multiplier, $\cos \Delta\psi$ into Equation (4.35).

4.6.2 Phased Array Based Monopulse Radar

Let us cite the main relationships in a phased array antenna, beginning with a one-dimensional (linear) array (Figure 4.16). Deflecting the beam of a phased array through an angle γ with respect to the normal

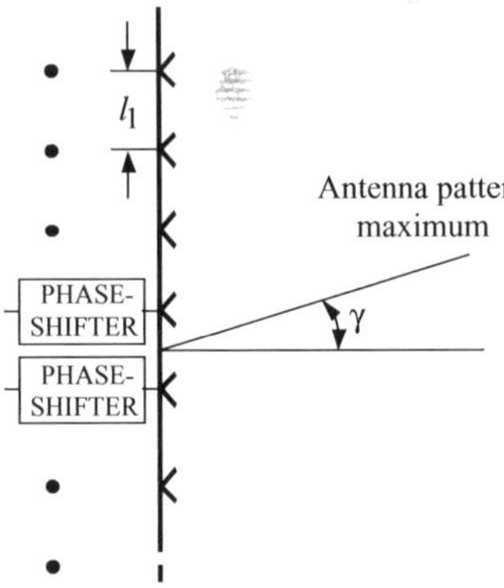

Figure 4.16 Structure of a linear phase array

direction requires the difference of phases in the phase-shifters of adjacent elements to be

$$\varphi_1 = \frac{2\pi}{\lambda}\ell_1 \sin\gamma, \tag{4.36}$$

where ℓ_1 is the distance between any neighbouring radiators of the phased array. Then

$$\sin\gamma = \varphi_1 \frac{\lambda}{2\pi\ell_1}.$$

In order to control γ, it is sufficient to set the phase gradient φ_1. The phase shift in the ith phase-shifter ($i = 0, 1, 2, \ldots, N-1$) is

$$\varphi_i = i\,\varphi_1.$$

For beam steering in two planes (two-dimensional phased array) it is necessary to set the gradients φ_1 and ψ_1 in these planes. The phase shift in the array element, residing in the ith row and in the jth column, is then equal to

$$\varphi_{ij} = i\varphi_1 + j\psi_1.$$

As a rule, phased array antennas employ discrete phase-shifters consisting of r sections arranged as a cascade, which are controlled by digital codes. The phase shift in the Kth section takes on one of the two values: 0 or φ_k. We will write φ_0 for the minimum change of phase shift (phase shift increment). In an r-bit phase-shifter,

$$\varphi_0 = \frac{2\pi}{2^r}.$$

Hence, the phase shift can be changed in increments of φ_0 from zero to $2\pi - \varphi_0$. For example, for $r = 3$, $\varphi_0 = 2\pi/8 = 45°$.

Phase-shifters are controlled by r-bit numbers, and the Kth bit of the number controls the phase shift in the Kth section of the phase-shifter. If the Kth bit of the number contains '0', the phase shift in the respective section is zero. If the Kth bit of the number contains '1', the phase shift is

$$\varphi_k = 2^{(k-1)}\varphi_0.$$

If the phase gradient φ_1 has been calculated using formula (4.36), the phase shift in the ith element of the phased array is $\varphi_i = i\varphi_1$. Next, φ_i is quantized with a quantization step φ_0, and the actual set phase shift is equal to

$$]\, i\varphi_1/\varphi_0\, [. \tag{4.37}$$

In (4.37), the inverted brackets]·[designate the operation of rounding the enclosed number to the nearest integer.

One distinctive feature of the phased array based monopulse tracker design is that instead of a mechanical drive it employs a digital phase computer (Figure 4.17) for beam steering, which begins with an analogue-to-digital converter. Figure 4.17 shows a channel for measuring a single angular coordinate. The dimensionless gain of a phased array based tracker is

$$K_0 = \frac{\mu U_0 K_{\mathrm{PD}} K_{\mathrm{ADC}} m \varphi_0 \lambda}{2\pi\ell_1 \cos\alpha \times 2^n},$$

where α is the angle between the equisignal line and the normal to the array. In fact, the value of this angle is contained in the computer and the value of $\cos\alpha$ can be compensated. Therefore, assuming that $\cos\alpha = 1$, the formula below can be used to calculate K_0 :

$$K_0 = \frac{\mu U_0 K_{\mathrm{PD}} K_{\mathrm{ADC}} m \varphi_0 \lambda}{2\pi\ell_1 \times 2^n} \tag{4.38}$$

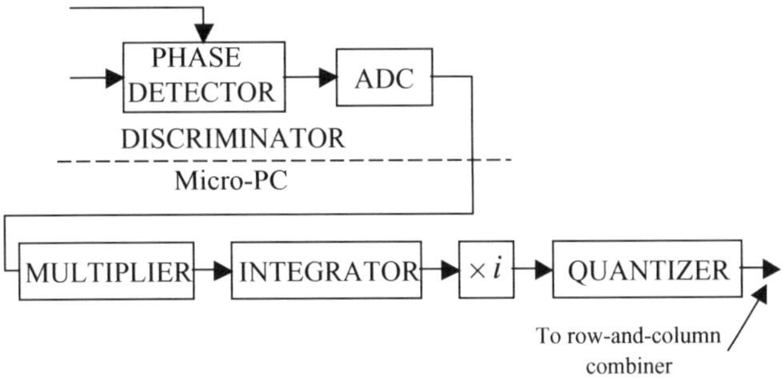

Figure 4.17 Control digit formation in a phased array

In Equation (4.38), 2^n is the reciprocal of the gain of the digital quantizer controlling the phase-shifters, m is a scale multiplier and

$$n = N_c - r,$$

where N_c is the bit capacity of the computer.

To ensure the tracker has the desired stability, it is necessary and sufficient to satisfy the inequality

$$K_0 < 2.$$

To have a limited bandwidth ΔF of the tracker, the condition to be satisfied is

$$K_0 < 1,$$

and the less the value of K_0, the narrower the bandwidth ΔF. For $K_0 \leq 0.7$,

$$\Delta F \approx \frac{K_0}{\sqrt{1 - K_0}} \frac{1}{2\pi T_i},$$

where T_i is the clock cycle of the digital integrator.

4.6.3 Conical Scan Tracker

Figure 1.20 illustrates the principle of measuring a single angular coordinate with a successive comparison of signals, which is carried out by switching the antenna beam between positions 1 and 2. However, in practice it is commonly required to measure two angular coordinates. This is achieved by switching the beam to four positions in a plane perpendicular to the ESL (this plane is also known as the 'picturesque plane') or by conical scanning of the beam with a frequency Ω_{sc}.

The first method is suitable for implementation in a phased array antenna, while the second is more suitable for a mechanically controlled mirror antenna. In the latter case, the scan axis (OA in Figure 1.20(b)) is the equisignal line.

When the signal source is offset from the ESL, the received signal carries amplitude modulation at the scanning frequency, where the modulation depth, in the first approximation, is proportional to the offset β and the initial phase of modulation conforms to the offset direction ψ. Figure 4.18 shows the relationships in the picturesque plane.

Point O corresponds to the ESL. Point A_1 indicates the peak power direction of the AP and point B the direction towards the target. The line segment OA_1 corresponds to the angle between the ESL and the peak power line, and the segment OB corresponds to the angle between the ESL and the direction towards the target, i.e. the current error in the measurement of β. The components of this error along the

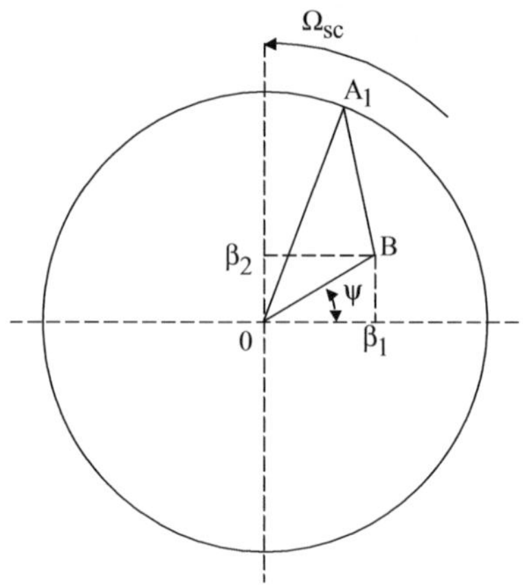

Figure 4.18 Illustration of conical scanning

two coordinate axes are

$$\beta_1 = \beta \cos \psi,$$
$$\beta_2 = \beta \sin \psi.$$

The target signal amplitude as a function of time may be expressed by an approximation formula

$$U_{\mathrm{m}}(t) = U_{\mathrm{m}} \left[1 + 0.55\mu\beta \cos(\Omega_{\mathrm{sc}}t - \psi)\right],$$

where, U_{m} denotes the signal amplitude when the signal source is on the equisignal line. To simplify this formula, we adopt the designation

$$\mu_1 = 0.55\mu.$$

In addition, a harmonic factor is introduced. The signal is then given by

$$U_{\mathrm{s}}(t) = U_{\mathrm{m}} \left[1 + \mu_1\beta \cos(\Omega_{\mathrm{sc}}t - \psi)\right] \cos \omega_0 t, \tag{4.39}$$

where ω_0 is the carrier frequency expressed in rad/s. The block diagram of a conical scan tracker is given in Figure 4.19.

The feed of a mirror antenna is rotated by the drive at an angular rate Ω_{sc}. The axle of this drive is connected with the rotor of a reference voltage generator. The outputs of the two orthogonal windings of the generator are reference voltages for the phase detectors:

$$U \cos \Omega_{\mathrm{sc}}t \qquad \text{and} \qquad U \sin \Omega_{\mathrm{sc}}t.$$

The receiver ends with an amplitude detector whose output voltage is (see formula (4.39))

$$K_{\mathrm{rec}}U_{\mathrm{m}} + K_{\mathrm{rec}}U_{\mathrm{m}}\mu_1\beta \cos(\Omega_{\mathrm{sc}}t - \psi), \tag{4.40}$$

where K_{rec} is the gain of the receiver.

The AGC circuit with a bandwidth $\Delta F_{\mathrm{AGC}} \ll F_{\mathrm{sc}}$ stabilizes the constant voltage component in formula (4.40), i.e.

$$K_{\mathrm{rec}}U_{\mathrm{m}} = U_0.$$

The variable component in formula (4.40) is filtered and supplied to the phase detectors (PhD). A phase detector is a cascade connection of a multiplier and a lowpass filter (LPF). The passband of the LPF is

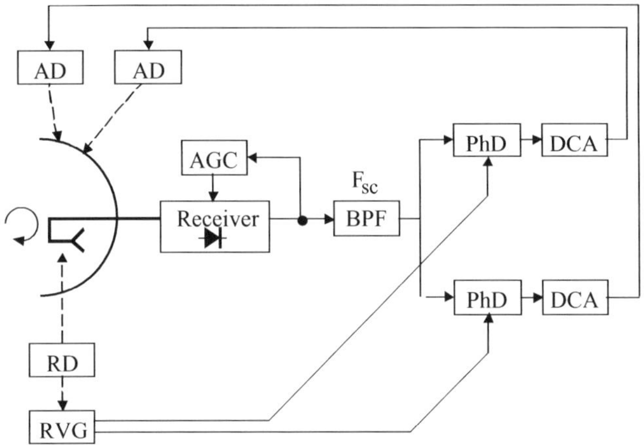

Figure 4.19 The block diagram of a conical scan tracker. AD, antenna driver; RD, radiator driver; RVG, reference voltage generator; BPF, bandpass filter; PhD, phase detector; DCA, direct current amplifier; AGC, automatic gain control

equal to $2\pi\,\Delta f \ll \Omega_{sc}$. Therefore, at the outputs of the PhDs,

$$U_1 = K_{PD} U_0 \mu_1 \beta \cos\psi \qquad \text{and} \qquad U_2 = K_{PD} U_0 \mu_1 \beta \sin\psi,$$

where K_{PD} is the gain of the phase detector.

U_1 and U_2 are the output voltages of the tracker's discriminator. The controller of the tracker incorporates two DCAs and drives rotating the antenna in two orthogonal planes until the moment when the mismatches $\beta\cos\psi$ and $\beta\sin\psi$ vanish. The direction towards the target is determined by the position of the equisignal line (in this case, the axis of the antenna).

4.6.4 Quadrant Scan (Sequential Lobing) Tracker

When conical scanning is used in a radar with a phased array, the beam is moved discretely. If the discrete value of beam movement is chosen to be 90°, what is called quadrant scanning is achieved. This implies that the beam is periodically set to four positions, one position in each quadrant. This scanning method is illustrated in Figure 4.20, where the coordinate axes X and Y are in the plane parallel to the plane of the phased array antenna. Points 1,2,3 and 4 represent intersections of the picture plane with the beam axis of the phased array at successive moments of time.

The method of quadrant scanning was briefly described in the monograph [I.3] as applied to a mirror antenna with a mechanically controlled feed periodically adopting four positions. This method is used in several aerological radars. In this section the application of the quadrant scanning method to a radar with a phased array antenna will be considered.

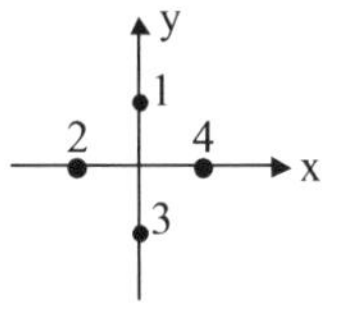

Figure 4.20 Illustration of the method of quadrant scanning

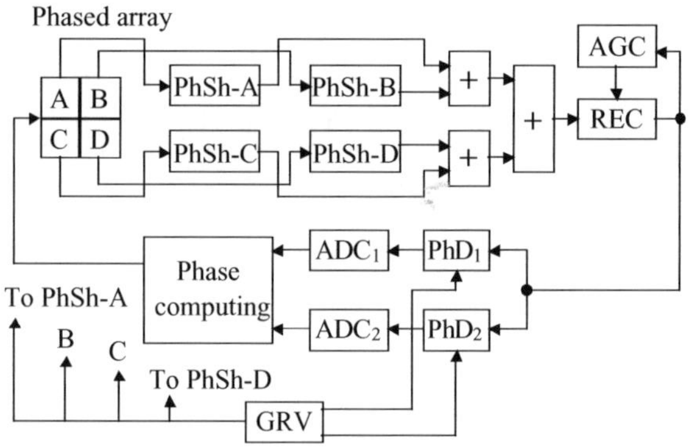

Figure 4.21 Block diagram of a quadrant scan tracker: PhSh, phase-shifter; AGC, automatic gain control; REC, receiver; PhD, phase detector; ADC, analogue-to-digital converter; GRV, reference voltages generator

Quadrant scanning can be implemented in two ways. The first method consists of using an appropriate routine of changing the phase distribution across the phased array with a view to consecutively switching the beam according to the pattern 'up–left–down–right'. The second technique presupposes the use of four controllable 0 to $\pi/2$ phase-shifters with subsequent adders installed at the outputs of the four subarrays of the antenna. We would recommend the second technique whose advantages will become clear from the text to follow.

The block diagram of a radar implementing the scanning method at issue is given in Figure 4.21. The summed voltages at the outputs of the four subarrays A, B, C and D are fed to the controllable phase-shifters. The phase shifts in the phase-shifters assume the values 0 or $\pi/2$. The control numbers are supplied by the reference voltage generator. The '0' control number causes a zero phase shift, while the '1' number causes a phase shift by $\pi/2$. When the shifts in all the phase-shifters are zero, the beam of the phased array antenna is at a conditional-zero position determined by the phase distribution in the subarrays. This direction coincides with the equisignal line of the antenna.

Time diagrams for phase shifts in the phase-shifters are presented in Figure 4.22. On the first interval of time, there are $\pi/2$ shifts in phase-shifters A and B and zero shifts in phase-shifters C and D. As a result, the beam of the antenna is moved upwards. It deflects in this particular direction because the beam peak is in such a position at which the spatial phase shifts in elements C and D are equal to the introduced phase shifts in elements A and B.

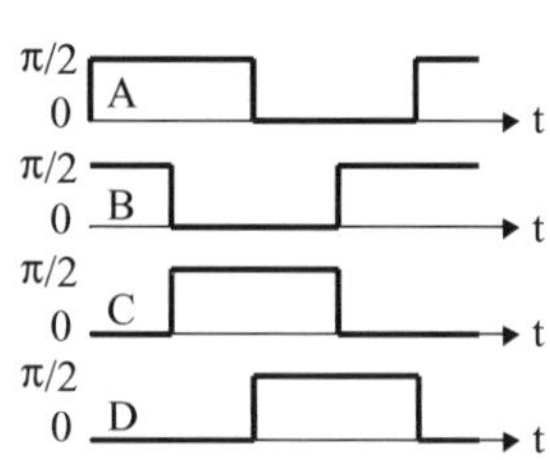

Figure 4.22 Phase shift flowcharts

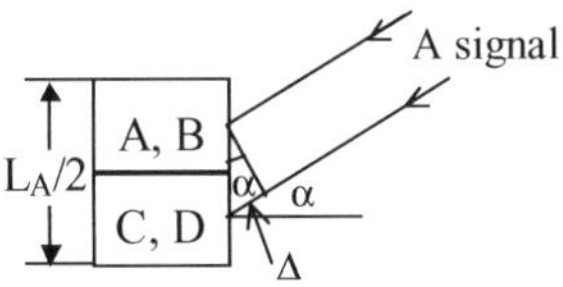

Figure 4.23 Calculation of a beam deflection angle in a phased array

Let us calculate the deflection angle of the beam of the phased array, α, with the help of Figure 4.23. The spatial phase shift in elements C and D is equal to

$$\varphi_{sp} = \frac{2\pi\Delta}{\lambda} = \frac{\pi L_A \sin\alpha}{\lambda}.$$

Assuming that $\varphi_{sp} = \pi/2$, we obtain

$$\sin\alpha = \frac{\lambda}{2L_A} \approx \tfrac{1}{2}\theta_{sum},$$

where θ_{sum} is the sum signal beamwidth of the antenna. Because in practice the value of θ_{sum} equals just a few fractions of a radian,

$$\sin\alpha \ll 1.$$

At the same time,

$$\sin\alpha \approx \alpha.$$

As a result,

$$\alpha = 0.25\theta_{sum}. \qquad (4.41)$$

If there are zero phase shifts in elements A and B, and elements C and D yield phase shifts equal to $\pi/2$, then obviously the beam of the antenna will move downwards by a value of $\alpha = -0.5\theta_{sum}$. Therefore, when the phase shifts are 'swapped' at the outputs of subarrays A + B and C + D, the beams of the phased array antenna intersect at the −3 dB level relative to the peak.

Similarly, if $\pi/2$ phase shifts are set in elements A and C while there is no phase shift in elements B and D, the beam of the phased array antenna will be deflected leftwards. Lastly, if $\pi/2$ phase shifts are set in elements B and D with zero phases preserved in elements A and C, the beam of the phased array will deflect to the right of the ESL.

The dependence will now be determined of the shape of the voltage envelope at the input of the receiver on the sign of the angular offset β of the signal source from the equisignal line. First, assume that $\beta > 0$, i.e. the signal source is offset upwards from the equisignal line. Then the received signal will be at a maximum when the beam is moved upwards with respect to the ESL, i.e. when there are $\pi/2$ phase shifts in elements A and B. When the beam peak deflects downwards (there are $\pi/2$ phase shifts in elements C and D), the received signal will be at a minimum. When $\pi/2$ phase shifts are set in elements A and C or B and D, the received signal will take on a certain average value.

The plot of the echo signal envelope against time (for $\beta > 0$) is given in Figure 4.24(a). Reference voltages of the phase detectors are shown in Figures 4.24(b) and (c). Matching the said figures, we come to the following conclusions:

- The output voltage of the phase detector PhD_1 is greater than zero.
- The output voltage of the phase detector PhD_2 is equal to zero.

These results indicate that the offset $\beta > 0$ (in the vertical plane) and that there is no offset of signal source relative to the ESL in the horizontal plane.

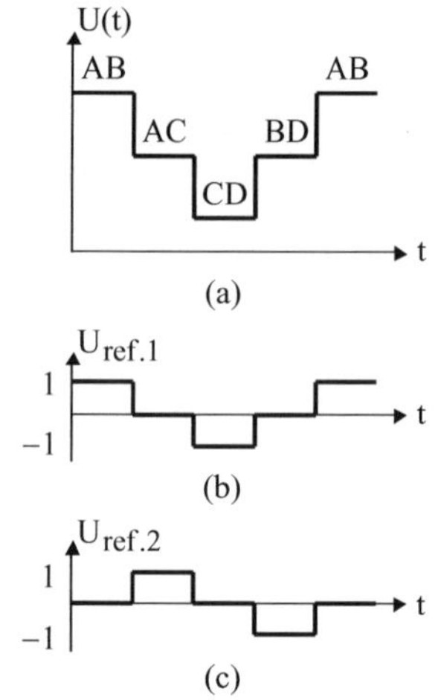

Figure 4.24 (a) Signal envelope and (b and c) reference voltages

Quadrant scanning permits loss in the signal-to-noise ratio to be reduced in the range tracking channel (in comparison with conical scanning). To achieve this, it is sufficient to make all the phase-shifters zero at the outputs of the subarrays at the moments of radar signal transmission. Moreover, since the transmitted signal is not modulated with the scanning frequency, a potential enemy has no knowledge about the value of this frequency and therefore cannot generate a jamming response.

In the device under consideration, it is possible to set a rather high scanning frequency (up to a quarter of the pulse repetition frequency). This allows, in accordance with Equation (4.32), the error due to amplitude noise to be considerably reduced. In terms of hardware input, the quadrant scanning method is equivalent to the conical scanning method, but it outperforms the latter in speed. This advantage is attributed to the fact that changing the beam position requires changing phases in just four one-bit phase-shifters, but not in all the phase-shifters of the phased array.

4.7 TARGET RANGE AND TARGET VELOCITY TRACKERS

4.7.1 Target Range Tracker

The principle of target range measurement in a continuous-wave radar was discussed in Section 3.5.3. In the present section we will focus on the structure and performance of a digital range tracker in a pulsed radar. The structure of a first-order range tracker with an ADC added at the input conforms to that shown in Figure 4.7. Shown in Figure 4.25 is the block diagram of a split-gate range tracker.

Each pulse of the echo signal is digitized in the ADC and fed to the input of the time discriminator. The number at the output of the discriminator is proportional to the difference (mismatch δt) between the

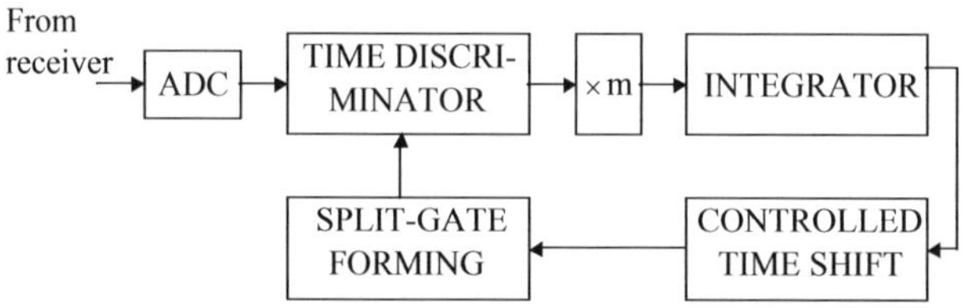

Figure 4.25 Block diagram of a range tracker

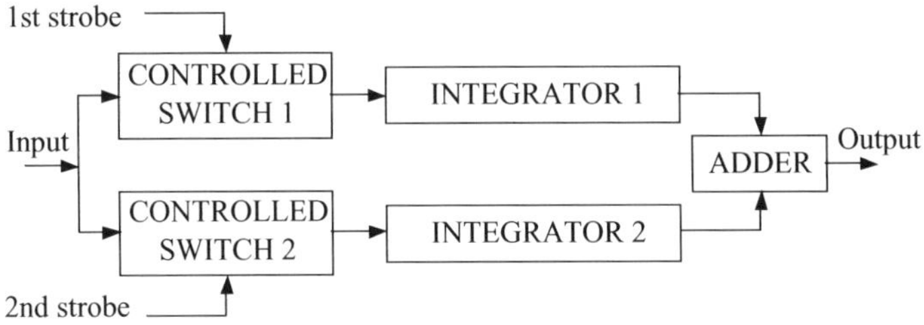

Figure 4.26 Structure chart of a time discriminator

time positions of the target signal and the split gate. The latter is made up of two half-gates, sometimes referred to as an 'early gate' and a 'late gate'. The number U_{int} at the integrator output changes at a rate proportional to the mismatch. Thus, the time delay of the half-gates (the converter in Figure 4.7) controlled by U_{int} changes in the appropriate direction. As a result, the mismatch falls off to zero if a stationary target is being tracked. The number at the integrator output is proportional to the target range.

The structure of the discriminator is shown in Figure 4.26 and its flowcharts are given Figure 4.27. The discriminator consists of two normally closed switches opened by the first half-gate and the second half-gate respectively, two accumulators (integrators) and an adder. The outputs of the accumulators are zeroed by the leading edge of the first half-gate. After that, the first accumulator adds negatively the discrete samples of the target signal within the limits of the first half-gate. We will write Σ_1 for their sum. At the end of the first half-gate the first switch closes, and once the second half-gate appears, the second switch opens. Signal samples within the limits of the second half-gate are added positively in the second accumulator. We will write Σ_2 for their sum. By the end of the second half-gate, the number $\Sigma_2 + \Sigma_1$ is formed at the output of the adder, which is proportional to the mismatch δt.

In the example illustrated by Figure 4.27, the discriminator output number $\Sigma_2 + \Sigma_1 < 0$. This causes the output number of the integrator to decrease. Consequently, the delay of the split gate also decreases, which results in a reduced mismatch. In the subsequent pulse periods the mismatch is reduced to zero, provided the target is stationary and the signal-to-noise ratio is greater than unity.

With small values of the signal-to-noise ratio, a considerable noise error $\sigma_{R.N.}$ (where R.N. is an indication of noise error of range measurement) may arise. The tracking of a moving target may be

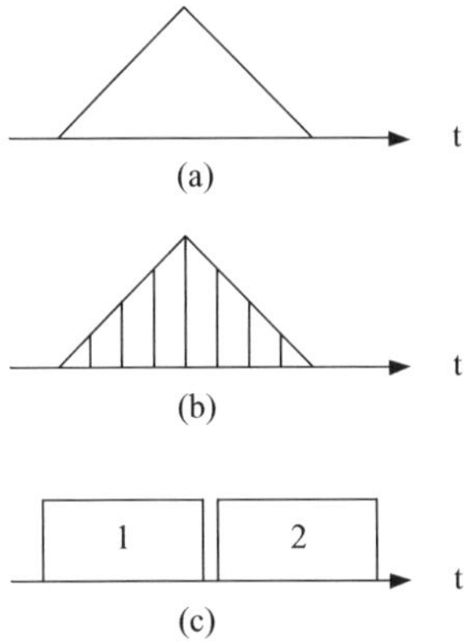

Figure 4.27 Flowcharts of a digital discriminator: (a) signal envelope; (b) digitalized signal envelope; (c) strobes

accompanied by dynamic errors (Section 4.5). The error due to the target radial velocity, V_R, often has the greatest value. There are two ways to reduce this error to an admissible value ΔR_{adm}:

- using a range tracker with a second-order astaticism;
- using a first-order tracker with a gain determined by the formula

$$K_0 = \frac{T_p V_R}{\Delta R_{adm}}. \tag{4.42}$$

In the latter case, it is necessary to make sure that the noise error $\sigma_{R.N.}$ does not exceed the admissible value σ_{adm}. For this purpose, it is necessary to calculate the tracker bandwidth ΔF by the formula (4.20):

$$\Delta F = \frac{0.25 K_0}{\pi\sqrt{1 - K_0}}.$$

Next, the noise error $\sigma_{R.N.}$ is determined using formula (4.22). If $\sigma_{R.N.} \leq \sigma_{adm}$, a first-order tracker can be used. Otherwise, it is necessary to use a second-order tracker. In this case, the admissible value of Δf is calculated from Equation (4.18), and then Equation (4.20) is used to find K_0.

4.7.2 Frequency Trackers

Target signal frequency tracking is aimed at accomplishing two tasks:

- to keep signal frequency within the limits of the passband of the matched filter;
- to measure the Doppler shift in signal frequency (radial velocity of the target).

The structure of an analogue frequency tracker is presented in Figure 4.28. The tracking system as such is enclosed in a frame. Outside the frame are the units employed in the case when a Doppler shift measurement is required.

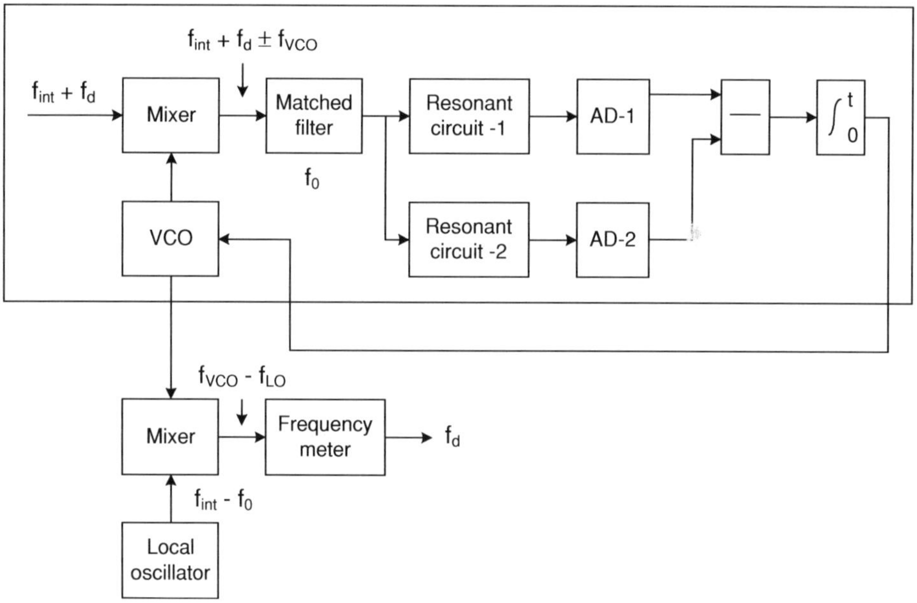

Figure 4.28 Block diagram of an analogue frequency tracker. Abbreviations: VCO, voltage controlled oscillator; AD, amplitude delector. Frequencies: f_{int}, intermediate frequency; f_d, Doppler frequency; f_{VCO}, controlled oscillator frequency; f_0, matched filter frequency tuning; $f_{LO} = f_{int} - f_0$

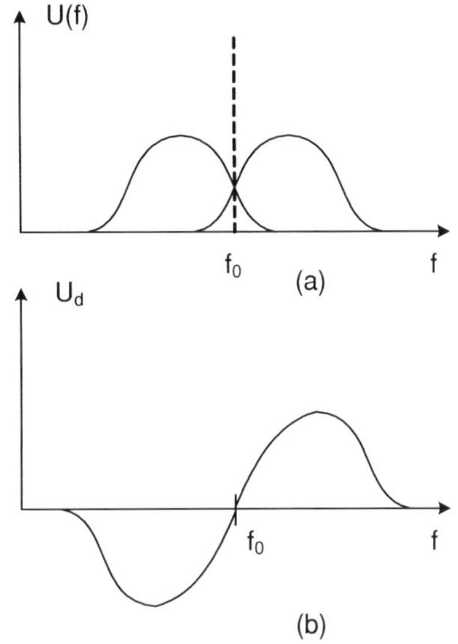

Figure 4.29 (a) Frequency responses of resonant circuits and (b) the discriminator characteristic

The discriminator of this tracking system includes a voltage controlled oscillator, a mixer, a matched filter and a frequency detector. An analogue frequency detector can be constructed as two staggered resonant circuits. Frequency responses of the circuits are shown in Figure 4.29.

If the frequency of the voltage-controlled oscillator (VCO) is

$$f_{VCO} = f_{int} - f_0 + f_d,$$

where f_{int} is the intermediate frequency, then the signal frequency after passing through the mixer is equal to f_0, and no voltage is present at the output of the discriminator. At the same time, the output voltage of the integrator remains unchanged. If the Doppler shift rises by δf, the frequency of the input voltage of the discriminator will increase by the same value. Besides, according to Figure 4.29, a voltage proportional to δf will appear at the output of the discriminator. The output voltage of the integrator will rise until the oscillator frequency f_{VCO} is increased by δf. This will bring the frequency of the voltage at the input of the matched filter back to the value f_0 and the output voltage of the discriminator will fall to zero.

Figure 4.30 shows the digital embodiment of a frequency tracker. The discriminator uses a counter of oscillation periods at the output of the matched filter and a subtraction circuit. A constant number

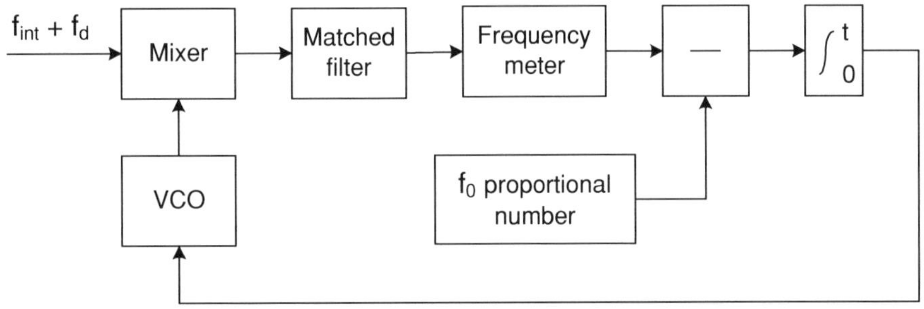

Figure 4.30 Block diagram of a digital frequency tracker

proportional to the resonant frequency of the matched filter, f_0, is fed to the second input of this circuit. The number at the output of the subtractor is

$$f_0 - f_1.$$

If $f_1 \neq f_0$, there is a nonzero voltage at the input of the integrator. The changing voltage at the output of the integrator adjusts the frequency of the controlled generator so that

$$f_1 \to f_0.$$

5

Radar Antennas

5.1 PURPOSE OF RADAR ANTENNAS AND THEIR FUNDAMENTAL PARAMETERS

Radar antennas are designed to fulfil the following functions:

- target angular position measurement;
- angular resolution of targets;
- rejection of unwanted signals (interference) whose directions of arrival differ from that of the useful signal;
- judicious utilization of transmitter power achieved by restricting the illuminated sector;
- maximization of useful received signal power.

These functional properties of radar antennas can be realized with reasonable antenna sizes only in the microwave range ($f = 1-100$ GHz). Radars make use of receiving and transmitting antennas. According to the reciprocity theorem, an antenna has identical directional patterns for receive and transmit modes of operation. Therefore, if a characteristic of the transmitting antenna is known, it can be used as the relevant characteristic for the receiving antenna and vice versa.

Let us discuss some fundamental characteristics (parameters) of radar antennas. The antenna gain $G(\alpha)$ can be defined for the transmitting antenna as

$$G(\alpha) = \frac{\mathrm{PD}_1(\alpha)}{\mathrm{PD}_2}.$$

In this formula, $\mathrm{PD}_1(\alpha)$ is the power flux density produced by the antenna at a point M (Figure 5.1). Point M is on a straight line forming an angle α with the direction of the $\mathrm{PD}_1(\alpha)$ peak. PD_2 is the power flux density created at point M by a uniform-gain (isotropic) radiator co-located with the antenna under consideration. The maximum gain will be denoted as G. Note that $G(\alpha)$ denotes the antenna gain in terms of power.

The antenna pattern (AP) in a given plane (sometimes called *unidirectional AP*) is defined as the relation

$$\mathrm{AP}(\alpha) = \frac{G(\alpha)}{G}. \tag{5.1}$$

Bistatic Radar: Principles and Practice Edited by M. Cherniakov

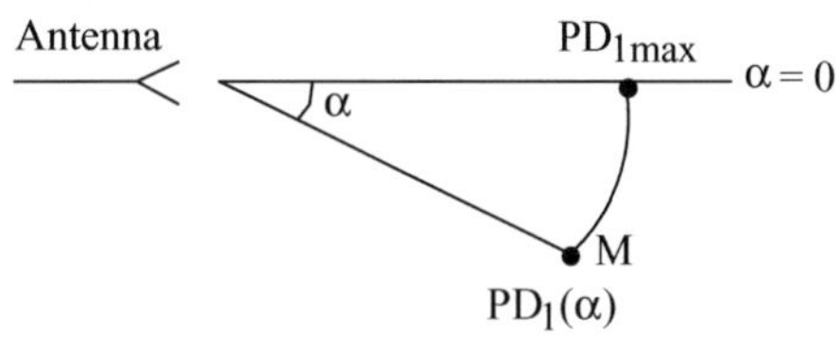

Figure 5.1 Antenna gain definition

The function AP(α), as well as $G(\alpha)$, is a power characteristic of the antenna. Along with this definition, another is used to characterize antennas, which is known as the field (strength) pattern:

$$\mathrm{AP}(\alpha)_E = \sqrt{\mathrm{AP}(\alpha)}. \tag{5.2}$$

An antenna pattern comprises a main lobe and sidelobes. The shape of AP(α) is qualitatively represented in Figure 5.2. The antenna beamwidth θ (referred to the –3 dB level) is the width of the main lobe between the points at which $G(\alpha) = 0.5$ (–3 dB).

The value of the quantity θ depends on the field distribution across the antenna aperture, or *illumination*. The level of the sidelobes (relative to the main lobe) is also dependent on illumination. Thus, with uniform illumination,

$$\theta(\mathrm{deg}) = \frac{51\lambda}{D}, \tag{5.3}$$

where λ is the wavelength and D is the linear dimension of the aperture. Furthermore, the level of the first sidelobe is [3]

$$U_{\mathrm{s.l.}} = -13.2\,\mathrm{dB}. \tag{5.4}$$

With uniform illumination, the field pattern can be expressed by the formula [I.3]

$$\mathrm{AP}(\alpha)_E = \frac{\sin\left[\pi\,(D/\lambda)\sin\alpha\right]}{\pi\,(D/\lambda)\sin\alpha}. \tag{5.5}$$

If we are only interested in AP values within ± 30°, then

$$\sin\alpha \approx \alpha,$$

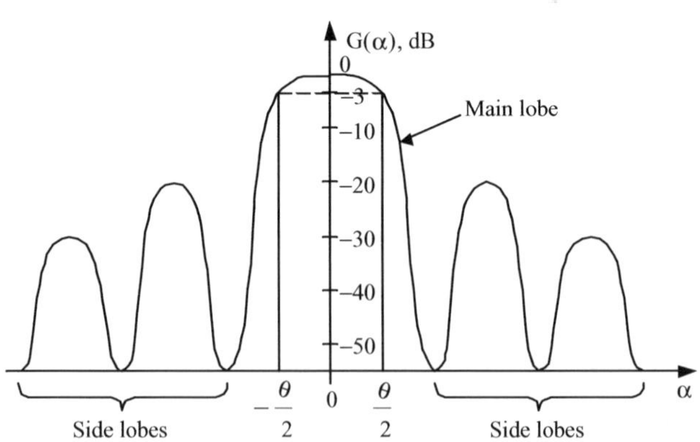

Figure 5.2 Typical antenna radiation pattern

where α is expressed in radians, and formula (5.5) may be simplified as follows:

$$\mathrm{AP}(\alpha)_E \approx \frac{\sin[\pi\,(D/\lambda)\,\alpha]}{\pi\,(D/\lambda)\,\alpha}. \tag{5.6}$$

The directivity of an antenna can be fully expressed with the help of antenna patterns in two mutually perpendicular planes (e.g. the horizontal plane and the vertical plane).

Another major characteristic of a radar antenna system is the two-way antenna pattern. This characteristic is the product of the patterns of the transmitting and the receiving antennas. The beamwidth θ_2 for the two-way antenna pattern is a measure of the angular resolution. As a rule, the angular resolution of the radar is adopted to be equal to θ_2. If the same antenna or two identical antennas are used for transmission and reception, then

$$\theta_2 \approx 0.7\theta, \tag{5.7}$$

where θ is unidirectional beamwidth or the beamwidth for each of the antenna patterns. If the transmitting antenna and the receiving antenna have differing patterns, then

$$\theta_2 < \min(\theta_{\mathrm{t}},\ \theta_{\mathrm{r}}),$$

where θ_{t} and θ_{r} are the beamwidths for the transmit and receive patterns respectively.

For an approximate estimation of the gain of an antenna (either transmitting or receiving), the following expression can be used [I.26]:

$$G = \frac{26\,000}{\theta_\beta \theta_\varepsilon} \tag{5.8}$$

In formula (5.8), θ_β and θ_ε are the beamwidths for unidirectional antenna patterns in two mutually perpendicular planes (in degrees).

One important parameter of the receiving antenna is the area of its effective aperture, A, which is defined by the relationship

$$P_{\mathrm{r}} = A\ \mathrm{PD}, \tag{5.9}$$

where PD is the returned signal power flux density at the antenna location and P_{r} is the power of the received signal. The value of A is simply expressed in terms of the geometric area of the antenna aperture, S_{A}:

$$A = \rho_{\mathrm{a}} S_{\mathrm{A}}, \tag{5.10}$$

where ρ_{a} is the efficiency of the antenna. The values of ρ_{a} lie within the range 0.5–0.8. The effective aperture and gain of the antenna are connected by the relationship

$$G = \frac{4\pi A}{\lambda^2}. \tag{5.11}$$

The gain of the transmitting antenna and the effective aperture of the receiving antenna are included in the expression for the radar power budget (1.27) and, hence, have a considerable bearing on the radar range.

Pulsed radars normally employ the same antenna for the transmission of the radar output and the reception of the target return. This concept presupposes the use of a duplexer. This element connects the antenna to the output of the transmitter during the transmission of a radar pulse. During the intervals between the pulses, the duplexer connects the antenna to the input of the receiver.

AP sidelobes are undesirable phenomena. Sidelobes in the pattern of a transmitting antenna are sources of interference for other radio systems. Sidelobes of a receiving antenna may conduct different kinds of unwanted signals into the receiver, such as clutter and jamming signals (see Section 7.5), feedthrough signal in continuous-wave radars (see Section 3.5) and unintended interference from other radio aids. A high level of sidelobes takes place when illumination, or aperture-field distribution, is uniform. This statement holds true for any pair of functions connected by Fourier transform (e.g. pulse and frequency responses of a linear filter, signal and its spectrum, illumination and antenna pattern).

Reduction of the sidelobe level is achieved with what is termed tapered illumination, which means nonuniform distribution with the field decreasing towards the periphery of the aperture. The price to pay for reduced sidelobes is a wider beam. For instance, when a Hamming function based illumination pattern is used, the sidelobe level will not exceed –40 dB, inclusive of the first sidelobe. However, this results in a 1.5-fold increase of antenna beamwidth [I.27].

Another method of countering sidelobe interference is discussed in Section 7.5.

5.2 MAIN TYPES OF ANTENNAS USED IN RADARS

The types of antennas most commonly used in modern radars are the mirror antenna and the phased array. Lens antennas and end-fire arrays have limited applications in radars [I.3] and therefore are not discussed in this chapter.

Mirror (reflector) antennas are represented by four varieties:

- parabolic antennas;
- Cassegrain antennas;
- parabolic-cylinder antennas;
- antennas with cosecant-squared vertical-plane pattern (cosecant-squared antennas).

A parabolic antenna is a conductive mirror (reflector) shaped as a segment of a paraboloid of revolution with a focal feed. An antenna section with a plane passing through its axis is given in Figure 5.3.

Sections with planes perpendicular to the axis have a circular shape. The feed may take the form of dipoles of various configuration or horns. If the middle of the feed coincides with the focus of the paraboloid, the radiation peak is on its axis. When the feed is outside the focus, the radiation peak is offset to the opposite side.

Achieving a fairly high efficiency with a parabolic antenna (efficiency factor of 0.6–0.7) is only possible through the use of a directive feed. The latter's illumination should fall solely on the reflector. Excessive width of the feed pattern leads to two undesirable phenomena. First, not all of the transmitter power falls on the antenna dish. Second, it brings about significant side- and backlobes. The backward radiation is just as unpleasant as a high sidelobe level. It is believed [I.3] that the strength of the field on the edges of the reflector should be 10 dB less than that in the centre.

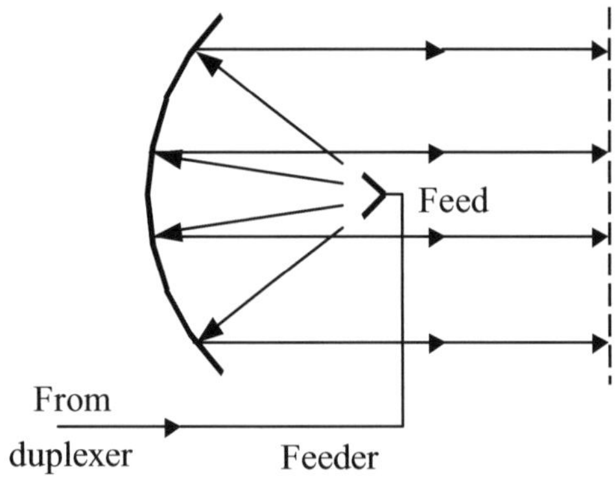

Figure 5.3 Parabolic reflector antenna illustration

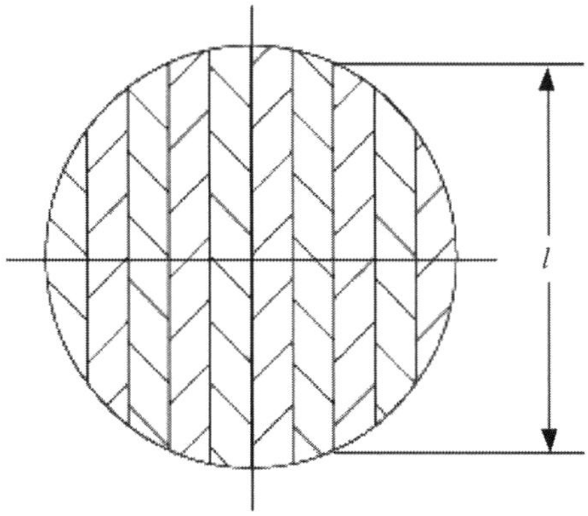

Figure 5.4 Illustration of field unevenness in a round antenna aperture

If reflector illumination power per unit area is constant throughout the circular aperture, the beamwidth is

$$\theta(\text{deg}) \approx \frac{58\lambda}{D},$$

and the first sidelobe level is–17 dB [I.3]. These figures differ from those in formulae (5.3) and (5.4). The difference is attributable to amplitude taper (the actual reduction of the field amplitude from the centre to the periphery) in the circular aperture of a paraboloid. This observation is illustrated in Figure 5.4. In this figure the antenna aperture is split into small strip-like elements. Power density is the same on all the strips, while the strip length, l, decreases with increasing separation from the centre. This leads to a respective reduction in the field strength. As a result, the antenna beam becomes somewhat wider and the level of sidelobes decreases.

It is worth mentioning here that in practice the directional pattern of a parabolic antenna is slightly distorted because the feed, the feeder and some structural elements lie in the radiating area of the reflector. The common approach to avoid this drawback is to remove the feed from the radiating area (Figure 5.5). This, however, gives rise to other unwanted effects, including complications of designing a mechanical support for the antenna [I.3].

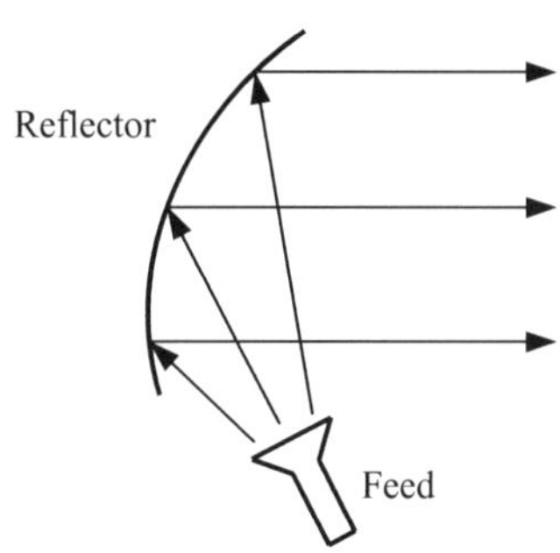

Figure 5.5 Parabolic antenna with offset feed. (After Skolnik [I.3]. Reproduced by permission)

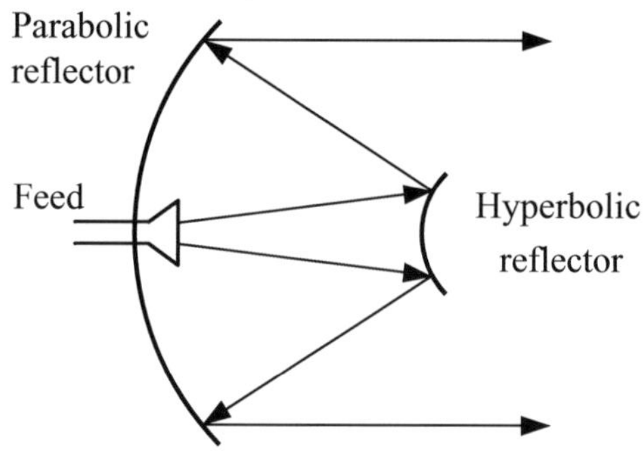

Figure 5.6 Cassegrain antenna. (After Skolnik [I.3]. Reproduced by permission)

A Cassegrain antenna comprises two reflectors: a main parabolic dish and a hyperboloidal subreflector (Figure 5.6). The feed is situated near the vertex of the paraboloid. The feed illuminates the subreflector (with the antenna operating on transmit). The latter directs the signal towards the main reflector, which forms a nearly plane wave. When the antenna is in the receive mode, the target signal falls on the main dish. Reflected by the main dish, the signal is intercepted by the subreflector and further directed to the feed.

The key advantages of the Cassegrain antenna over the single-reflector parabolic antenna are as follows:

- There is no feeder going to the feed across the radiating area of the main reflector.
- The shape and size of the subreflector can be better matched with those of the main reflector.

Figures 5.7 and 5.8 illustrate a method [I.3] that makes it possible to considerably reduce the blockage of the main reflector by the subreflector. The method is based on the use of a twist reflector and a subreflector, which reflects a signal with horizontal polarization.

Suppose we have a horizontally polarized signal at the output of the feed. In this case, the subreflector can be likened to a grating of horizontal conductors. Such a reflector redirects a horizontally polarized signal towards the main reflector. The latter converts signal polarization from the horizontal to the vertical so that the signal passes freely through the subreflector.

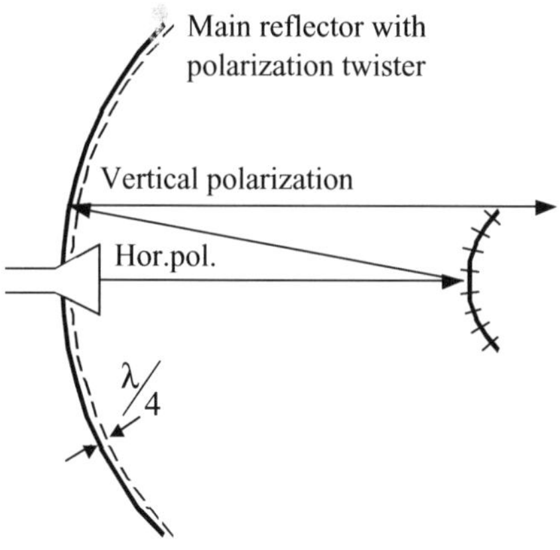

Figure 5.7 Polarization-twist Cassegrain antenna. (After Skolnik [I.3])

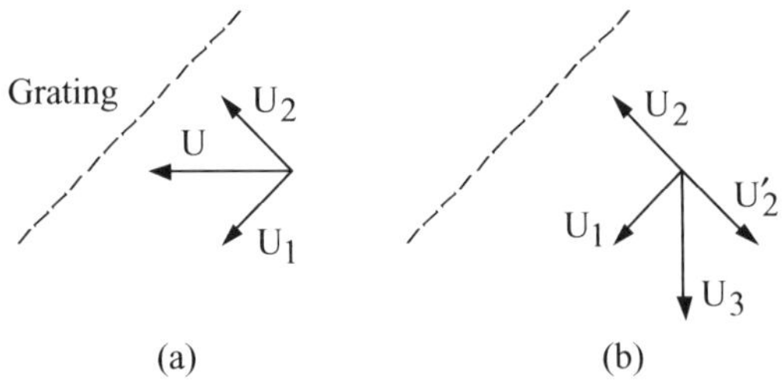

Figure 5.8 Polarization vector $\pi/2$ turn in a twist reflector

In the receive mode, the vertically polarized target signal reaches the main reflector (after passing without loss through the subreflector). When it is returned towards the subreflector, the vertical polarization is converted to the horizontal. The resultant signal is then directed by the subreflector to the feed.

Let us examine the stages the signal goes through when it is reflected from the main reflector in the transmit mode. As mentioned above, the subreflector sends the feed's signal (with horizontal polarization) into the main reflector. The main reflector is composed of two parts: a solid screen and a grating of conductors mounted in front of the screen at a distance $l = \lambda/4$. The conductors are positioned at an angle of 45° to the horizontal polarization vector (the dashed line in Figure 5.7). The incoming signal, U, can be represented as the sum of two components, namely U_1 and U_2 (Figure 5.8 (a)). The component U_1 is returned by the grating to the subreflector. The component U_2 passes through the grating and bounces from the solid reflector. While travelling an additional distance $\lambda/2$ (from the grating to the reflector and back), U_2 undergoes a phase shift by π with respect to U_1. U_2' will be used to designate the phase-shifted signal. It follows from Figure 5.8 (b) that the sum $U_1 + U_2' = U_3$ has a vertical polarization and passes freely through the subreflector towards the target.

The above principle of operation for the main reflector (Figure 5.7) drives the radar wavelength to $4l$. Consequently, the reflector is a narrowband device. However, methods of overcoming this restriction have been proposed and described in the literature on the subject [I.27–I.29].

A *parabolic-cylinder antenna* is practical when it is necessary to form a radiation pattern with differing beamwidths in two mutually perpendicular planes [I.3]. This type of antenna is schematically represented in Figure 5.9. Its sections by horizontal planes are parabolic in form. The foci of the parabolas are situated on the vertical axis AA′.

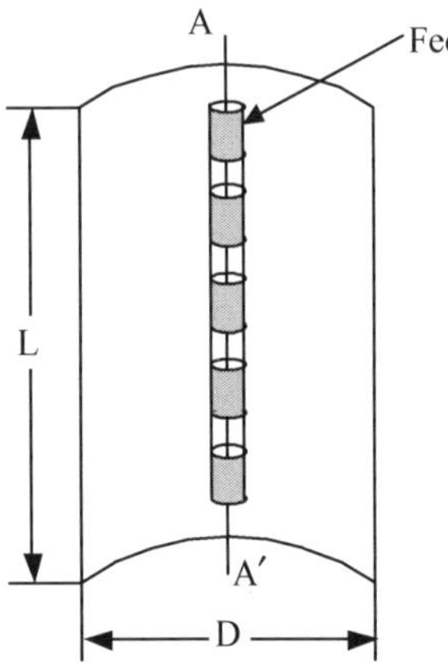

Figure 5.9 Parabolic cylinder antenna

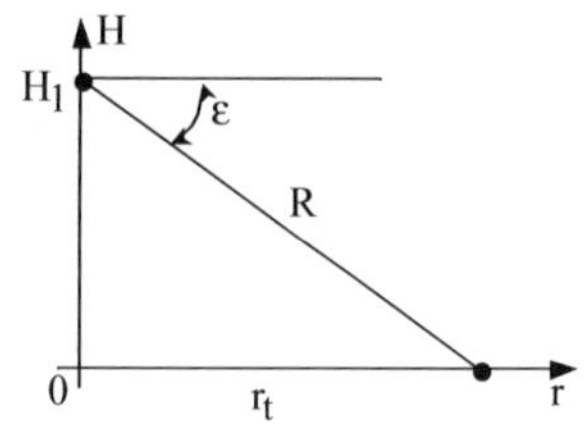

Figure 5.10 Slant range R calculation: r_t, ground range

The feeds are arranged along the vertical axis. In such a design, the beamwidth in the horizontal plane is dependent on the aperture dimension, D. The shape of the AP in the vertical plane is determined by the distribution of power of the feeds. In particular, by varying this distribution it is possible to realize an AP with a low level of sidelobes in the direction towards the Earth's surface.

Cosecant-squared antennas are most frequently employed in airborne ground-surveillance radars. Obtaining a high-quality image of the Earth's surface requires that the power of the echo signal depends on the object's RCS only and is independent of the distance to the object. This requirement can be satisfied by implementing a certain shape of antenna pattern in the vertical plane. Suppose the radar platform (e.g. an aircraft) is flying at a height H (Figure 5.10). According to Equation (1.24), the power of the echo signal from an object located at a distance R from the aircraft is expressed by the formula

$$P = \text{constant}\frac{G_t(\varepsilon)A(\varepsilon)}{R^4}.$$

However, $R = H \operatorname{cosec} \varepsilon$ and A, from Equation (5.11), is equal to

$$A = \frac{G\lambda^2}{4\pi}.$$

Hence,

$$P = \text{constant}\frac{\lambda^2 G_t^2(\varepsilon)}{4\pi H^4 \operatorname{cosec}^4 \varepsilon}. \tag{5.12}$$

If the shape of the AP in the vertical plane is

$$G_t(\varepsilon) = \operatorname{cosec}^2 \varepsilon, \tag{5.13}$$

then Equation (5.12) turns to become

$$P = \frac{\text{constant}\,\lambda^2}{4\pi H^4}. \tag{5.14}$$

The shape of AP thus calculated is qualitatively depicted in Figure 5.11.

According to Equation (5.14), the echo signal power does not depend on the target range. At the same time, formula (5.13), somewhat approximately, expresses the requirements for the shape of the antenna pattern, since Equation (5.12) contains the quantity of flying height. If the radar range does not depend on height, the shape of the AP should vary as a function of the flying height. Moreover, the

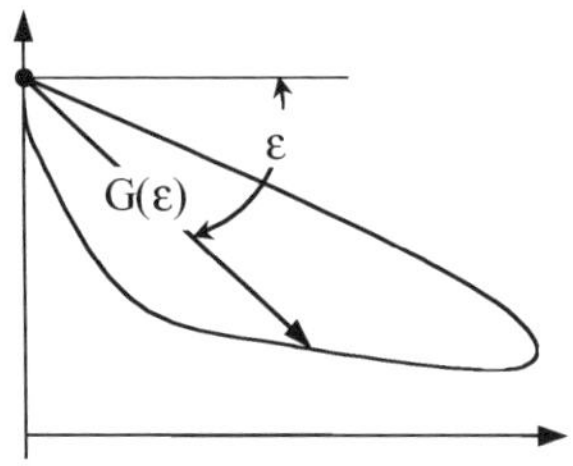

Figure 5.11 Cosecant-squared shape of the AP (qualitatively)

reflectivity of objects on the Earth's surface is a function of the angle ε. Therefore, the AP should be calculated in the vertical plane for some average value of the flying height, causing some less than ideal AP shape deviations.

5.3 ELECTRONICALLY STEERABLE ANTENNAS

5.3.1 Introduction

The phased array is a set of surface-mounted transmitting (receiving) modules. Each module incorporates radiators, controllable phase-shifters and possibly amplifiers and attenuators. Attenuators allow aperture-field illumination to be set and thus a desired antenna pattern to be formed. Adjustment of phase shifts with the help of phase-shifters ensures practically inertialess beam steering within a given angular sector.

Depending on the geometry, phased arrays are divided into linear (one-dimensional), planar (two-dimensional) and conformal types. The modules of a linear phased array are usually arranged on a straight-line segment and provide beam steering within an angular coordinate (in a single plane). In a planar phased array, the modules are arranged within the confines of a certain geometric shape on a plane. Such antennas may have apertures of rectangular, circular or other shapes. They ensure beam steering in two mutually perpendicular planes. The elements of a conformal phased array may be arranged on different types of surface, e.g. on the skin of the aircraft carrying the radar.

Phased array antennas are also classified into two large groups: passive arrays and active arrays. Elements (modules) of passive phased arrays contain radiators and phase-shifters only, with band-pass filters and attenuators added on occasion. Active phased arrays also include amplifiers. The elements of a transmitting active array constitute the output stage of the transmitter and the transmitting antenna, transmitter power being determined by the total power of the amplifiers. The voltage of the driving generator is fed to the inputs of all the amplifiers, determining the shape of the emitted output.

The set of modules of a receiving active array is equivalent to a receiving antenna with the high-frequency section of the receiver. Apart from the elements used to shape and steer the beam, the receive module in an active phased array contains a low-noise amplifier (LNA). This allows a high sensitivity receiver to be realized. The output voltages of the receive modules are summed and the resultant voltage is fed to the input of the common reception path. Radars are often equipped with a combined phased array antenna for transmission and reception of pulsed signals. The block diagram of a transmit–receive module in an active phased array is given in Figure 5.12 [I.30].

Modern phased arrays are designed with electrically controlled attenuators and digital phase-shifters. Phase shift can be changed in increments of $\Delta\varphi = 2\pi/2^n$. The phase-shifter consists of n

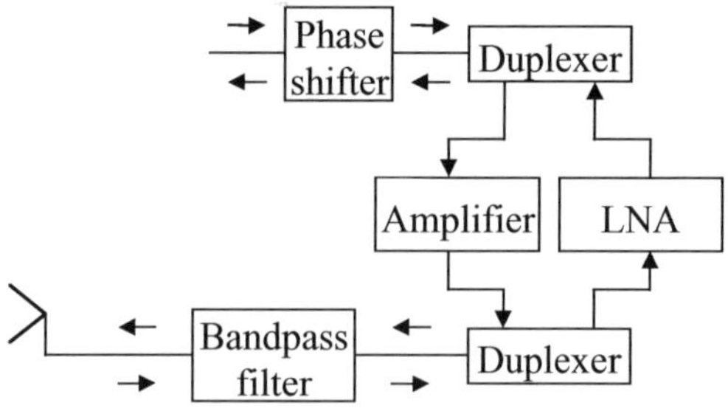

Figure 5.12 Block diagram of an active receiver–transmitter array element: LNA, low-noise amplifier

elements controlled by the digits in n bit positions of a binary number. A '1' in the ith bit position of the control number triggers the ith element of the phase-shifter giving a shift equal to $2^{i-1}2\pi/2^n$. For example, when $n = 4$, $\Delta\varphi = 2\pi/16 = 22.5°$. If the control code is represented by the number 0110, a phase shift of $\varphi = 45° + 90° = 135°$ is obtained.

The phased array possesses a number of advantages over other types of antenna. One of these is the practically inertialess control of AP shape and position, previously mentioned above. In particular, it is possible to form nulls in the receive pattern towards sources of interference, to realize multiple beams for simultaneous detection and tracking of several targets, and to employ intricate methods of angular scanning.

In a number of cases, phased array based designs feature a reduced volume of hardware as beam movement can be carried out without rotating the antenna. Conformal phased arrays can be mounted on a variety of surfaces.

Other important benefits inherent in active phased arrays are multifunctionality and high reliability. For example, an airborne radar with an active phased array is capable of fulfilling the functions of target search, target tracking, fire control and obstacle detection. Failure of some elements in a phased array (of the order of 5 %) does not result in a failure of the radar; moreover, as stated in Reference [I.30], it only slightly affects the radar's performance.

Nevertheless, the phased array is not without certain disadvantages as compared to a mechanically controlled antenna. The greatest disadvantage of phased arrays, and *active* phased array in particular, is their fairly high cost. However, the continuous technological advancement of microelectronics will drive the cost of active arrays down.

5.3.2 Directional Pattern of a Phased Array Antenna

Let us define the unidirectional pattern of a phased array antenna, i.e. transmit-only or receive-only AP. To simplify calculations and to ensure simple interpretation of results, we will limit ourselves to considering a one-dimensional (linear) equally spaced array consisting of N identical elements. Suppose that the elements of the array emit signals of equal power and that phase shifts in phase-shifters φ_i can be changed continuously. The effect of stepwise phase variation will be examined below.

The directed radiation of an array stems from interference of the oscillations excited by all its elements. For the calculation of field strength at a remote point, M, the direction to which forms an angle, β, with the axis of the array (Figure 5.13 (a)), the partial field strengths from each element are represented by vectors with the same amplitude, E_0. Angular displacement of the partial vector, E_i, with respect to the reference axis is equal to the difference of phase shift in the ith phase shifter, φ_i, and spatial phase incursion during wave propagation from point M to the ith element of the array (Figure 5.13 (b)).

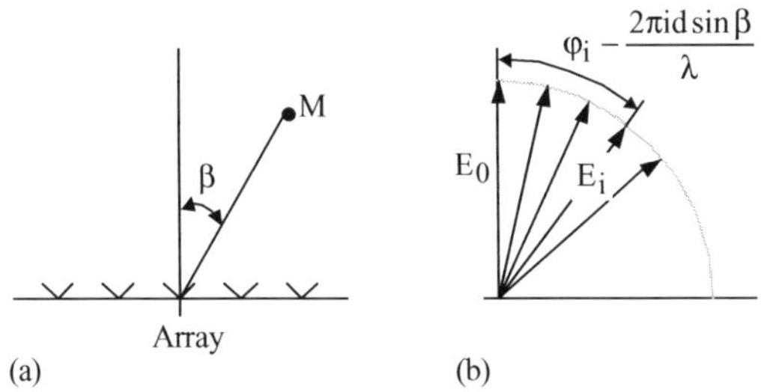

Figure 5.13 Calculation of phases in array elements: (a) point M disposition; (b) voltages phases in array elements

To begin with, suppose the array is made up of uniform-gain radiators. The amplitude of the resultant field intensity at point M is then

$$E_a(\beta) = \left| \sum_{i=0}^{N-1} E_0 \exp\left[j\left(\varphi_i - \frac{2\pi i d \sin\beta}{\lambda} \right) \right] \right|, \tag{5.15}$$

where d is the distance between two neighbouring radiators. When the radiation peak in the direction of the array axis OO′ is desired (normal to the array, $\beta = 0$), all the $\varphi_i = 0$ and formula (5.15) yields $E(0) = NE_0$. Varying the value of β, an expression for the field pattern can be obtained, provided that the AP is at its maximum when $\beta = 0$:

$$E(\beta) = \left| E_0 \sum_{i=0}^{N-1} \exp\left(j \frac{2\pi i d \sin\beta}{\lambda} \right) \right|. \tag{5.16}$$

In Equation (5.16) the sum of a geometric series can be found from the ratio $\exp\left[j\,(2\pi d \sin\beta/\lambda) \right]$. This sum equals

$$E_0 \frac{1 - \exp\left[j\,(2\pi d/\lambda)\, N \sin\beta \right]}{1 - \exp\left[j\,(2\pi d/\lambda) \sin\beta \right]}.$$

The modulus of this expression is

$$E(\beta) = E_0 \left| \frac{\sin\left(\frac{\pi d N}{\lambda} \sin\beta \right)}{\sin\left(\frac{\pi d}{\lambda} \sin\beta \right)} \right| \tag{5.17}$$

Normalizing Equation (5.17) to $E(0) = NE_0$ and squaring both sides of the equality gives the gain of the array as a function of the angle β:

$$G_a(\beta) = \frac{1}{N^2} \frac{\sin^2\left[(\pi d N/\lambda) \sin\beta \right]}{\sin^2\left[(\pi d/\lambda) \sin\beta \right]}. \tag{5.18}$$

The function $G_a(\beta)$ is frequently referred to as the 'array factor' since Equation (5.18) does not take account of directive properties of a real radiator. A full expression for phased array AP is obtained through multiplying $G_a(\beta)$ by the normalized pattern of a single radiator $G_r(\beta)$:

$$G(\beta) = G_a(\beta)\, G_r(\beta) = G_r(\beta) \frac{1}{N^2} \frac{\sin^2\left[(\pi d N/\lambda) \sin\beta \right]}{\sin^2\left[(\pi d/\lambda) \sin\beta \right]}. \tag{5.19}$$

It follows from Equation (5.19) that the maximum of the normalized gain is

$$G(0) = G_a(\beta)\, G_r(\beta) = 1.$$

To determine the width of the main lobe of the AP referred to the –3 dB level, θ, the formula (5.19) is used. Set the left side of Equation (5.19) equal to 0.5. Substitute the following values into the right side of this equation:

$$\beta = \frac{\theta}{2} \quad \text{and} \quad G_r(\beta) = 1.$$

Solving the resulting equation for θ gives

$$\theta \approx \frac{0.886\lambda}{Nd}(\text{rad}) \approx \frac{51\lambda}{Nd}(\text{deg}). \tag{5.20}$$

Suppose it is required to position the radiation peak of the phased array antenna at an angle, β_0, with the axis of the array. Then, with the appropriate values of phases φ_i, the normalized gain is

$$G(\beta) = G_r(\beta) \cos \beta_0. \tag{5.21}$$

The origin of the multiplier, $\cos \beta_0$, is explained by Figure 5.14. It follows from the figure that the aperture length of a phased array antenna is generally equal to

$$A'B' = AB \cos \beta_0,$$

where AB is the aperture size for $\beta_0 = 0$. Since the antenna gain is proportional to the aperture size (5.11), the quantity G_α is proportional to $\cos \beta_0$. Equation (5.21) furnishes an explanation for the restriction of the scan sector to $\beta_0 = \pm(45°–60°)$.

Antenna beamwidth is inversely proportional to its gain. Therefore, when the beam is pointed in the direction β_0, the width of the main lobe of the phased array is

$$\theta_{\beta_0} \approx \frac{51\lambda}{Nd \cos \beta_0}. \tag{5.22}$$

Let us calculate the array factor $G_a(\beta)$ for the AP maximum pointed at an angle β_0 with the normal. Analysis of Figure 5.15 shows that the uncompensated phase shift of the voltages in adjacent elements

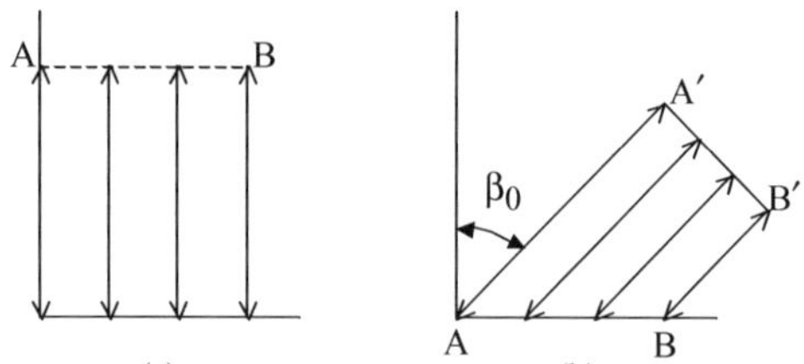

Figure 5.14 Array aperture dependence on the scanning angle β_0

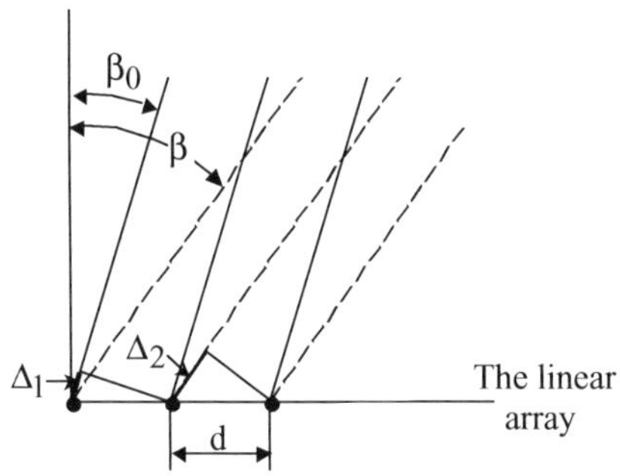

Figure 5.15 Compensated Δ_1 path difference by angle β_0 and noncompensated $\Delta_2 - \Delta_1$ path difference

of the array is

$$\varphi = \frac{2\pi}{\lambda}(\Delta_2 - \Delta_1) = \frac{2\pi d}{\lambda}(\sin\beta - \sin\beta_0).$$

Hence, the normalized AP is given by

$$G(\beta, \beta_0) = G_r(\beta)\frac{1}{N^2}\frac{\sin^2[(\pi dN/\lambda)(\sin\beta - \sin\beta_0)]}{\sin^2[(\pi d/\lambda)(\sin\beta - \sin\beta_0)]}. \quad (5.23)$$

Using Equation (5.23) the maximum allowable value of d can be determined. Note that besides the main lobe with $\beta = \beta_0$, there may appear grating lobes when

$$\frac{d}{\lambda}(\sin\beta - \sin\beta_0) = n, \quad n = 1, 2, 3, \ldots.$$

Furthermore, both the numerator and the denominator in Equation (5.23) turn to zero, and evaluation of the indeterminate form yields $G(\beta, \beta_0) = G_r(\beta)$. The necessary and sufficient condition for the absence of grating lobes is obvious:

$$\frac{d}{\lambda} < \frac{1}{|\sin\beta - \sin\beta_0|}. \quad (5.24)$$

Since $\sin\beta$ and $\sin\beta_0$ may have opposite signs and β may equal $\pi/2$, the condition given by (5.24) takes the form

$$\frac{d}{\lambda} \leq \frac{1}{1 + |\sin\beta_0|}. \quad (5.25)$$

It follows from (5.25) that the grating lobes are excluded for $d/\lambda \leq 0.5$ since β_0 is always less than $\pi/2$. Several values of the quantity d/λ as a function of the scan sector β_0 are presented in Table 5.1. In some cases, it is possible to increase the relation d/λ beyond the limit set by formula (5.25). To this end, it is necessary to calculate grating lobe suppression by the pattern of a single radiator for a definite case.

Using formulae (5.15) to (5.19) and (5.23), an expression can be obtained for the normalized pattern of a planar phased array of rectangular shape in two principal planes [I.3]. This expression is the product

Table 5.1 Connection between d/λ and β_0

β_0	25°	45°	60°
$\frac{d}{\lambda}$	0.7	0.586	0.536

of two patterns in the principal planes B and E (Figure 5.16):

$$G(\beta,\ \varepsilon) = G_{r1}(\beta)\ G_{r2}(\varepsilon)\frac{\sin^2[N\pi(d/\lambda)\sin\beta]}{N^2\sin^2[\pi(d/\lambda)\sin\beta]}\ \frac{\sin^2[M\pi(d/\lambda)\sin\varepsilon]}{M^2\sin^2[\pi(d/\lambda)\sin\varepsilon]}. \tag{5.26}$$

Where N is the number of columns in the array, M is the number of rows and β and ε are the angles with respect to the vertical passed to the array in planes B and E. $G_{r1}(\beta)$ and $G_{r2}(\varepsilon)$ are the principal-plane radiator patterns.

If the antenna beam is set in the directions β_0 and $\varepsilon_0$relative to the normal to the array plane, Equation (5.26) becomes

$$\begin{aligned} G_{\beta_0\varepsilon_0}(\beta,\varepsilon) &= G_{r1}(\beta)\ G_{r2}(\varepsilon)\cos\beta_0\cos\varepsilon_0 \\ &\times\frac{\sin^2[N\pi(d/\lambda)(\sin\beta-\sin\beta_0)]}{N^2\sin^2[\pi(d/\lambda)(\sin\beta-\sin\beta_0)]}\ \frac{\sin^2[M\pi(d/\lambda)(\sin\varepsilon-\sin\varepsilon_0)]}{M^2\sin^2[\pi(d/\lambda)(\sin\varepsilon-\sin\varepsilon_0)]}. \end{aligned} \tag{5.27}$$

5.3.3 Phased Array Bandwidth

In the previous subsections it was assumed that the transmitted and the returned signals have a narrowband spectra. A single value of wavelength appears in all the calculations. The returned signal at the input of the ith module of the phased array is delayed relative to the signal in the number zero module by

$$t_i = \frac{i\Delta}{c} \tag{5.28}$$

(see Figure 5.17). Strictly speaking, to obtain in-phase partial signals at the outputs of the modules, these should contain delay lines. However, with the current level of technology, fitting each module with such a device would entail a considerable array complication and a significant rise in cost. Therefore, phase shifters are employed in phased array modules instead of delayers.

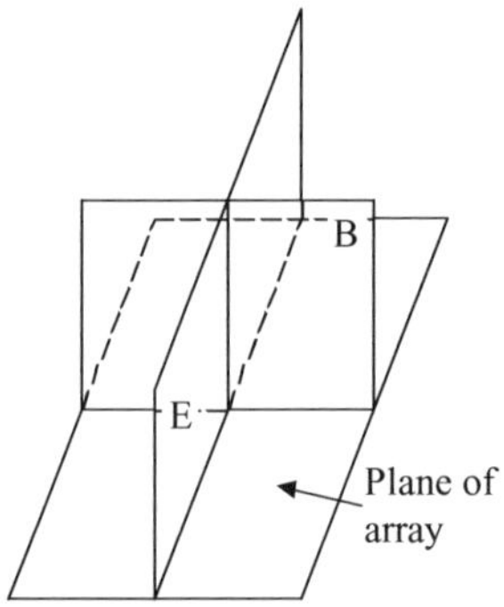

Figure 5.16 The two principal planes B and E and the plane of array

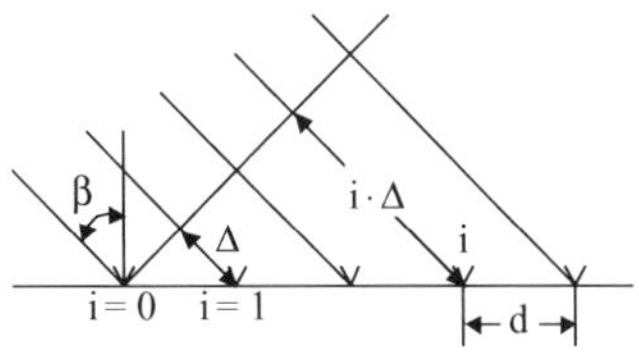

Figure 5.17 Calculation of signal time delay in the ith array element

If the radar transmits a narrowband signal (at the limit, a sinusoidal waveform with frequency f), the use of phase-shifters does not have any undesirable consequences. It follows from Equation (5.28) that the phase shift of the signal in the ith module is

$$\varphi_i = 2\pi f t_i = \frac{2\pi f i \Delta}{c}.$$

Hence, for a signal with constant frequency, its phase shift is definitively related to the time delay. However, it is quite another matter when a signal with a rather wide spectrum is used. The phase-shifters permit a phase shift variation from zero to 2π. At the same time, the spatial phase shift of the signal at the inputs of array elements may vary within a considerably greater range due to greater values of $i\Delta$. This leads to a beam-pointing error (or a widened beam).

In Reference [I.31], the connection is defined between the angular offset of the beam and the relative spectrum bandwidth of the signal, $\Delta f_{\text{rel.}}$. When the scan sector of the phased array antenna does not exceed $\pm 60^\circ$ and the allowable beam offset is equal to a quarter of its width θ, the allowable value of $\Delta f_{\text{rel.}}$ is

$$\Delta f_{\text{rel.}}(\%) = \theta(\text{deg}). \tag{5.29}$$

For instance, when $\theta = 2^\circ$, the allowable spectrum bandwidth makes 2 % of the signal's carrier frequency.

It is pointed out in Reference [I.3] that the signal spectrum can be expanded several times compared to the limitation set by Equation (5.29). This is achieved by dividing the array into k subarrays (Figure 5.18).

Suppose each subarray is composed of p modules. Subarray modules contain phase-shifters and there are controllable delayers at the outputs of all the subarrays (in a receiving antenna). No delayer is mounted at the output of the first subarray. The time delay at the output of the second subarray is set to $T_1 = (p-1)\Delta/c$. The quantity Δ is equal to $d \sin \beta$ (Figure 5.17). As the subarray number is increased by one, the delay at its output increases by T_1. Hence, the delay at the output of the kth subarray is

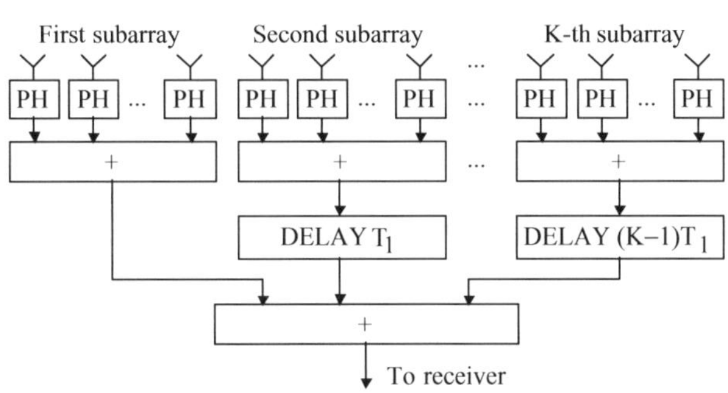

Figure 5.18 Simplified block diagram of antenna array with "K" subarrays; PH – phase shifter (after Howard D.D. [25], Skolnik M.I. [3])

$(k-1)T_1$. It is not difficult to see that the allowable signal spectrum bandwidth becomes k times greater. Thus, an array constructed according to the design illustrated in Figure 5.18 represents a compromise between signal spectrum bandwidth requirements and hardware cost.

5.3.4 Target Angular Position Errors Due to Discrete Phase Control

It is obvious that angular errors stem from discrete operation of the phase-shifters. In order to set the null of the difference pattern of a monopulse tracker in a direction towards the target, it is necessary to set a certain phase distribution across the array, $\varphi(i)$, where i is the number of the phased array element. This distribution is estimated accurately and then quantized with an adopted increment, φ_0.

The quantized distribution, $\varphi_q(i)$, differs from the required phase distribution. Because of this, the null of the difference pattern does not coincide with the desired direction. Estimates of angular errors due to phase quantization are presented in Reference [I.31], and also in Reference [I.32]. According to Reference [I.31], the angular error, $\delta\beta$, is expressed by the formula

$$\delta\beta_1 = \frac{9\theta}{N2^n}, \tag{5.30}$$

where θ is the beamwidth of the phased array for $\beta_0 = \varepsilon_0 = 0$, N is the number of elements in a row (column) of the array and n is the number of bits in the phase-shifter.

According to Reference [I.32], the maximum value of the error is

$$\delta\beta_2 = \frac{\pi\theta}{4\times 2^n}. \tag{5.31}$$

With average values of N, the estimates calculated by formulae (5.30) and (5.31) exhibit little difference. The value of $\delta\beta_2$ is always equal to $\delta\beta_2 = 0.785\theta/2^n$ irrespective of N. The value of $\delta\beta_1$ equals $\delta\beta_1 = 0.9\theta/2^n$ for $N = 10$ and $\delta\beta_1 = 0.45\theta/2^n$ for $N = 20$. For smaller values of N, formula (5.31) provides more optimistic estimates, while formula (5.30) is more suitable for greater values of N. More often than not, it is sufficient to have $n = 4$ in order to reduce *discreteness errors* to acceptable values. For example, if $\theta = 2°$, formula (5.31) yields $\delta\beta_2 = 0.785 \times 2/16 \approx 0.1°$. However, in small-sized phased arrays operating at the L-band (1–2 GHz) and having wider beams, the error in question may turn out to be beyond that which is acceptable. For instance, when $\theta = 6°$, $\delta\beta_2 = 0.3°$.

In such cases it is required either to increase the bit capacity of the phase-shifters or to resort to error compensation. The latter measure is feasible since the estimated phase distribution, $\varphi(i)$, is known. The quantized distribution, $\varphi_q(i)$, is known as well. The difference between these distributions, $\Delta\varphi(i) = \varphi(i) - \varphi_q(i)$, permits the current angular error to be estimated and, consequently, compensated. It should be noted here that this method provides no compensation for beam-steering discretion or sidelobes of the antenna pattern due to discrete operation of phase-shifter. Imperfections of phase-shifter manufacturing cannot be compensated either.

In digital arrays (Section 5.4), phase distribution quantization does not lead to degradation of antenna performance.

5.4 CONCEPT OF DIGITAL ARRAYS

Digital array antennas are arrays wherein the desired phase shifts in array elements are performed by means of mathematical operations on the received (transmitted) signals. Weighting the amplitude distribution across the aperture is also carried out with the help of multipliers.

Figure 5.19 illustrates the principle of operation of a practically used digital active, equally spaced, receiving array of N elements (channels). At first, it will be assumed that the aperture-field illumination is uniform and that all the feeds are isotropic.

The elements following the feed at the input of each channel are a band pass filter tuned to carrier frequency, a low-noise amplifier, a mixer and an intermediate-frequency filter. Then, the signals are digitized using an ADC. After that, a digital converter is used for transition to the complex envelope of the form

$$S_i(t) = U(t)\exp\left(\mathrm{j}\frac{2\pi di}{\lambda}\right)\sin\beta,$$

where β is the signal's direction of arrival and i is the channel number. Next, the signals $S_i(t)$ are multiplied by weight coefficients:

$$W_i = \exp\left(-\mathrm{j}\frac{2\pi di\sin\beta_0}{\lambda}\right),$$

where, β_0 is the desired direction of the antenna beam. Hence, the signal at the output of the ith channel can be written as

$$U(t)\exp\left(\mathrm{j}\frac{2\pi id}{\lambda}\right)(\sin\beta - \sin\beta_0).$$

The signal at the output of the adder is

$$U(t)\sum_{i=0}^{N-1}\exp\frac{\mathrm{j}\pi id}{\lambda}(\sin\beta - \sin\beta_0).$$

Proceeding in the same manner as when deducing formula (5.17), we can estimate the voltage amplitude at the output of the adder

$$U_{\mathrm{out}}(\beta) = \left|U(t)\frac{\sin\left[(\pi Nd/\lambda)(\sin\beta - \sin\beta_0)\right]}{\sin\left[(\pi/d\lambda)(\sin\beta - \sin\beta_0)\right]}\right|. \tag{5.32}$$

Normalizing Equation (5.32) to $U(t)$ and N with further squaring of the result, we arrive at an expression for the array factor:

$$G_{\mathrm{a}}(\beta) = \frac{1}{N^2}\frac{\sin^2\left[(\pi Nd/\lambda)(\sin\beta - \sin\beta_0)\right]}{\sin^2\left[(\pi/d\lambda)(\sin\beta - \sin\beta_0)\right]}. \tag{5.33}$$

It follows from Equation (5.33) that the main radiation peak of the antenna pattern occurs in the direction β_0 with respect to the normal of the array. To obtain full knowledge of the AP, $G_{\mathrm{a}}(\beta)$ should be multiplied

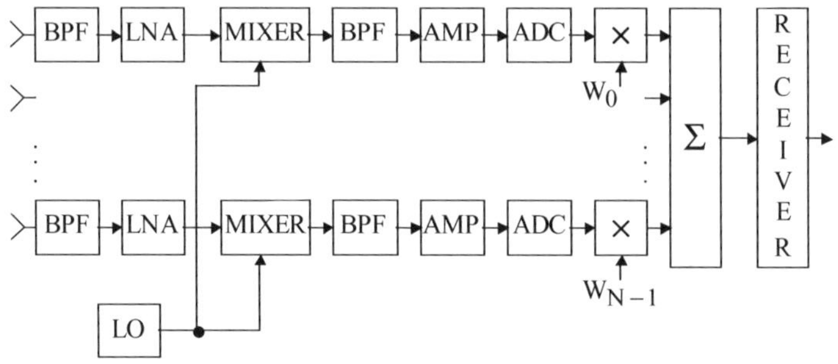

Figure 5.19 Block diagram of a digital array antenna: BPF, bandpass filter; LNA, low-noise amplifier; AMP, amplifier; ADC, analogue-to-digital converter; LO, local oscillator

Table 5.2 Illumination functions. (After Cheston and Frank [I.31])

Illumination functions	Efficiency	Peak sidelobe (dB)	Beamwidth factor, $K = \theta L_a/\lambda$
Uniform	1	−13.3	50.8
Dolph–Chebyshev 1	0.72	−50	76.2
Dolph–Chebyshev 2	0.66	−60	82.5
Hamming	0.73	−43	74.2
Taylor $n = 3$	0.9	−26	60.1
Taylor $n = 3$	0.8	−36	67.5

by the pattern for a single feed, $G_r(\beta)$:

$$G(\beta) = G_a(\beta)\, G_r(\beta) = G_r(\beta)\frac{1}{N^2}\frac{\sin^2[(\pi Nd/\lambda)(\sin\beta - \sin\beta_0)]}{\sin^2[(\pi d/\lambda)(\sin\beta - \sin\beta_0)]}. \tag{5.34}$$

Comparing formulae (5.34) and (5.23) it can be seen that they are identical.

A digital phased array primarily differs from the analogue array in that it replaces analogue phase-shifters and adders with their digital counterparts. This distinction gives the following cardinal advantages to radars with digital phased arrays:

- practically smooth beam-steering, elimination of angular errors and sidelobe growth arising from discrete phase control and manufacturing imperfections inherent in analogue phase-shifters;
- capability of simultaneously forming $M \leq N$ beams in desired directions at the cost of reasonable extra hardware input;
- capability of array adaptation to jamming and point-like clutter by forming nulls in the directions towards jamming and clutter sources (also see Section 7.4).

Summing the contents of this section, optimism should be expressed for a bright outlook for digital phased arrays.

5.5 SIDELOBES REDUCTION

As was observed in Section 5.1, the greatest sidelobe (SL) level (especially near the main lobe) occurs with a uniform illumination pattern. To reduce the SL level, it is necessary to employ tapered illumination. This is achieved by weighting the amplitude distribution with the help of an illumination function, i.e. multiplying the uniform distribution by a weighting function.

Reference [I.3] offers a comparison of phased array characteristics (efficiency, SL level, mainlobe width) obtained with the use of different illumination functions, namely Dolph–Chebyshev, Taylor and Hamming. The result of this comparison are given in Table 5.2, borrowed from Reference [I.3]. Theoretically the best results are achieved with Dolph–Chebyshev illumination functions However, the application of these types of function is rather limited. Specifically, they can only be applied to arrays with a small number of elements. There is no serious objection to the use of the two other illumination functions. As seen from Table 5.2, the results for the Taylor function with the parameter $n = 8$ are somewhat superior. It is noted in Reference [I.3] that it is this illumination function that is most frequently used in radar antennas.

6

Synthetic Aperture Radar

6.1 INTRODUCTION

Radar surveillance of the Earth's surface (or the surface of another planet) from space or from an aircraft is a challenging scientific and technological objective. To obtain high-quality imagery of the surface, the radar on a spaceborne or airborne platform should possess a high resolution in range, ΔR, and in azimuth, $\Delta\ell$ (Figure 6.1).

To avoid distortion of the image, resolution cells should be of the same order of magnitude ($\Delta\ell \approx \Delta R$) and have values within several metres. Range resolution is equal to

$$\Delta R = \frac{c}{2\Delta f_s},$$

where Δf_s is the spectrum bandwidth of the transmitted signal: ΔR is independent of the distance between the target area and the radar, and the azimuth resolution, $\Delta\ell$, is proportional to the range. As follows from Figure 6.1,

$$\Delta\ell = \theta R, \tag{6.1}$$

where θ is the width of the two-way antenna pattern (AP) of the radar expressed in radians. It is defined by the formula

$$\theta \approx \frac{0.7\lambda}{L_{\mathrm{ph}}} \tag{6.2}$$

where λ is the wavelength of the radar and L_{ph} is the antenna size in the horizontal plane.

As an example, suppose it is required that $\Delta\ell = \Delta R = 10\,\mathrm{m}$ when the objects of interest are at a distance of 500 km from the radar. Then the required signal spectrum bandwidth is

$$\Delta f_s \geq \frac{c}{2\Delta R} = \frac{3 \times 10^8\,\mathrm{m/s}}{2 \times 10\,\mathrm{m}} = 1.5 \times 10^7\,\mathrm{Hz} = 15\,\mathrm{MHz}.$$

Bistatic Radar: Principles and Practice Edited by M. Cherniakov

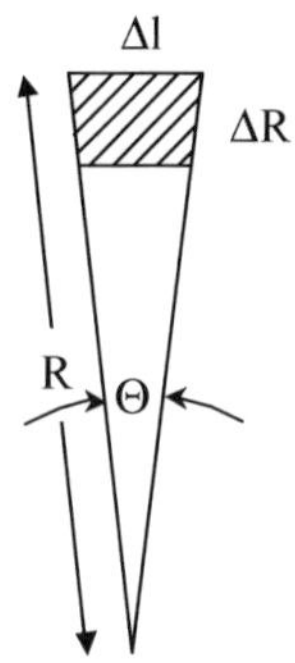

Figure 6.1 Resolution element

To form such a signal presents no difficulty. Obtaining a $\Delta\ell$ of the same order at a range of 500 km requires, in accordance with Equation (6.1), that the value of θ should not be more than

$$\theta = \frac{\Delta\ell}{R} = \frac{10\,\text{m}}{5\times 10^5\,\text{m}} = 2\times 10^{-5}\,\text{rad}.$$

We specify $\lambda = 10$ cm. According to Equation (6.2), realization of such an AP requires that the antenna size, L_{ph} (the index 'ph' stands for 'physical', denoting a real antenna), be at least

$$L_{ph} = \frac{0.7\lambda}{\theta} = 3.5\,\text{km}.$$

A dedicated aperture synthesis technique has been developed for the realization of such a large antenna aperture. Radars employing this technique are known as 'synthetic aperture radars' (SARs). Radar equipment installed on a spacecraft (satellite), or an aircraft, performs sensing of the Earth's surface in the manner of a side-looking radar (Figure 6.2).

This concept may be considered from the standpoint of analogy with either a phased array antenna or a filter matched with a linear frequency modulation (LFM) signal. In both these cases, the pulses returned by a certain object on the Earth's surface are stored for a time interval, T_s, and are then processed to give a combined image.

Two kinds of SAR operation are known: the *stripmap* mode and the *spotlight* mode. In the former case, the antenna axis is normal with respect to the carrier velocity vector. As the SAR moves along, a radar image is gradually formed of either one strip or two strips of terrain on both sides of the carrier's path.

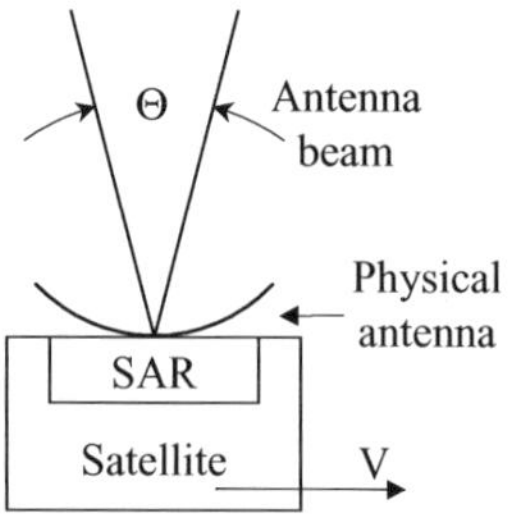

Figure 6.2 Satelliteborne SAR

In the spotlight mode, the beam of the radar antenna dwells on a small area of terrain for a given period of time. Sections 6.2–6.5 is mostly dedicated to stripmap SAR. Peculiarities of the spotlight mode are pointed out in Section 6.6.

6.2 MODEL OF AN SAR AS A PHASED ARRAY

Suppose a craft carryies an antenna of a horizontal-plane size, L_{ph}. In the vertical plane, the antenna produces a wide beam, which makes it possible to illuminate objects and to receive echo signals within a given range span, R_{min}–R_{max}. The axis of the antenna is normal to the velocity vector of the carrier (Figure 6.2). The radar transmits pulsed signals with a pulse duration, τ_p, and a pulse repetition period, T_p. During the interval between each two transmissions the craft covers a distance equal to

$$\ell_1 = VT_p, \qquad (6.3)$$

where V is the velocity of the radar carrier.

During the time interval, T_s, the craft travels a distance $L_s = VT_s$. On this part of the carrier path, the antenna of the SAR goes through $N = T_s/T_p$ positions, at which the radar signal is transmitted (Figure 6.3).

If N echo signals are phased and then summed, the same effect can be achieved as that using a linear phased array antenna of a size L_s and an element spacing ℓ_1. The 'virtual array' obtained in this way with a size many times greater than that of the real array, is termed a 'synthetic(-aperture) antenna'. All the parameters of the synthetic antenna are henceforth labelled by the index 's'.

There is a physical restriction on the maximum size, L_s, for observation of an area of terrain at a range R:

$$L_s \leq \theta_{ph} R \qquad (6.4)$$

where θ_{ph} is the beamwidth of the real antenna.

This restriction arises from the fact that any point of the Earth's surface is illuminated by the SAR and returns radar signals only as long as it stays covered by the beam of the real antenna. Another restriction on the value of L_s is discussed in Section 6.5.

There is a noteworthy distinction between an SAR and a phased array radar. In a radar with a phased array, the intensity of the signal returned by the object is proportional to the sum of field intensities due to all the elements of the transmitting array. The echo signal is common for the receiving elements of the array. Therefore, the two-way AP of a transmit–receive phased array is defined as the product of its transmit and receive patterns.

By contrast, an echo signal received by a synthetic antenna element is the result of target illumination by the transmit beam of this element only. Hence the two-way AP of a synthetic aperture antenna is somewhat narrower than that of a phased array antenna of the same size.

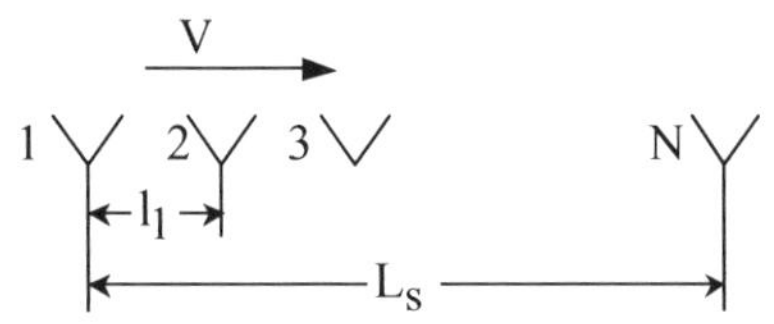

Figure 6.3 Principle of synthetic aperture forming

In view of the aforesaid, it is not difficult to write a formula for the normalized two-way field pattern of a synthetic antenna in the azimuth plane:

$$E_{\mathrm{n.sa}}(\beta) = \frac{U_{\mathrm{ph}}(\beta)}{N} \frac{\sin[(2\pi L_{\mathrm{s}}/\lambda)\sin\beta]}{\sin[(2\pi l_1/\lambda)\sin\beta]}, \tag{6.5}$$

where $U_{\mathrm{ph}}(\beta)$ is the normalized two-way field pattern of the real antenna. The function $U_{\mathrm{ph}}(\beta)$ can be approximated thus:

$$U_{\mathrm{ph}}(\beta) \approx \frac{\sin^2\left[(\pi L_{\mathrm{ph}}/\lambda)\sin\beta\right]}{(\pi L_{\mathrm{ph}}/\lambda)^2 \sin^2\beta} \tag{6.6}$$

From the relationship $E^2_{\mathrm{n.sa}}(\theta_{\mathrm{s}/2}) = 0.5$, the beamwidth of the synthetic antenna is

$$\theta_{\mathrm{s}} = 0.886\frac{\lambda}{2L_{\mathrm{s}}} \approx \frac{\lambda}{2L_{\mathrm{s}}} \tag{6.7}$$

Angular resolution of the synthetic antenna in linear units is

$$\Delta l_{\mathrm{s}} = \theta_{\mathrm{s}} R = \frac{\lambda}{2L_{\mathrm{s}}} R \tag{6.8}$$

When $L_{\mathrm{s}} = \theta_{\mathrm{ph}} R$ is at a maximum, and subject to Equations (6.2) and (6.8),

$$\Delta l_{\mathrm{s}} = \frac{\lambda R}{2\theta_{\mathrm{ph}} R} = 0.7 L_{\mathrm{ph}} \tag{6.9}$$

Just as in the case of a phased array antenna, the AP of an SAR can exhibit high-level grating lobes. When $\beta = \beta_1 = \arcsin\lambda/2l_1$, both the numerator and the denominator in Equation (6.5) become zero. Evaluating the indeterminate form yields

$$E_{\mathrm{n.sa}}(\beta_1) = U_{\mathrm{ph}}(\beta_1).$$

In order to have $E_{\mathrm{n.sa}}(\beta_1) \ll E_{\mathrm{n.sa}}(0)$, it is necessary and sufficient to satisfy the approximate inequality

$$\frac{\pi L_{\mathrm{ph}}}{\lambda}\sin\beta_1 \geq \pi \qquad \text{or} \qquad \frac{L_{\mathrm{ph}}}{\lambda}\frac{\lambda}{2l_1} \geq 1.$$

Hence another requirement follows:

$$l_1 = VT_{\mathrm{p}} \leq 0.5L_{\mathrm{ph}} \qquad \text{or} \qquad T_{\mathrm{p}} \leq \frac{0.5L_{\mathrm{ph}}}{V}. \tag{6.10}$$

6.3 SIGNAL PROCESSING IN AN SAR

One of the prerequisites for obtaining undistorted radar imagery of terrain is the absence of superposition of multirange segments of the target area. This condition is expressed by the formula

$$T_{\mathrm{p}} \geq \frac{2}{c}(R_{\max} - H), \tag{6.11}$$

where H is the height of the SAR carrier path above the Earth's surface. We shall write T_{p1} for the pulse repetition period obtained from formula (6.10) and T_{p2} for the pulse repetition period complying with condition (6.11).

Sometimes, the requirements for T_{p1} and T_{p2} may be contradictory. For instance, when $L_{\mathrm{ph}} =$ 16 m, $V = 8$ km/s, $R_{\max} = 10^3$ km and $H = 600$ km, we have $T_{\mathrm{p1}} \leq 1$ ms and $T_{\mathrm{p2}} \geq 2.67$ ms. The

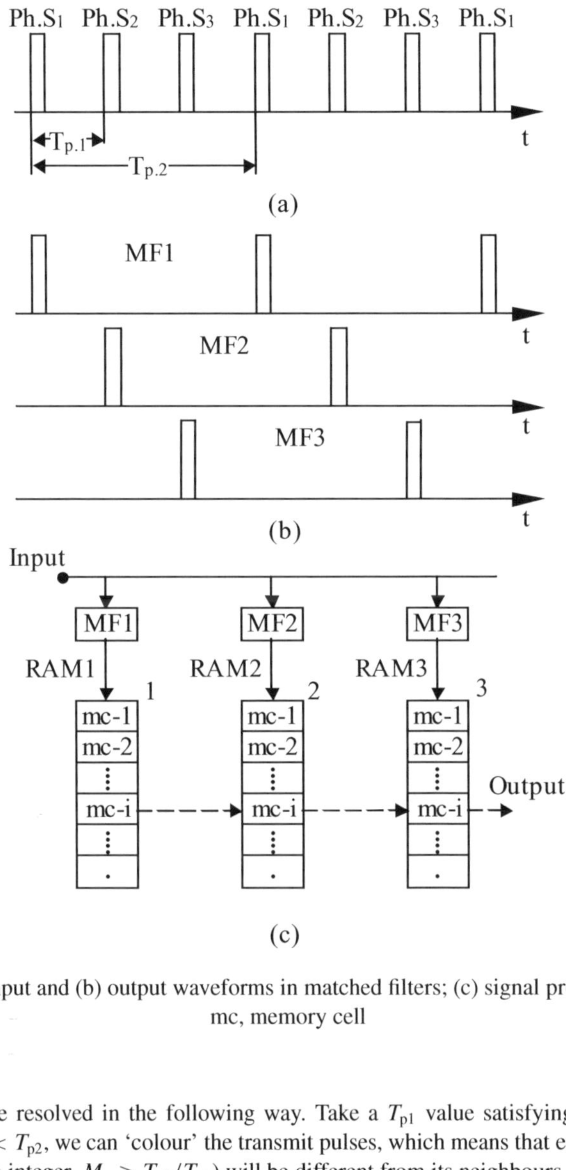

Figure 6.4 (a) Input and (b) output waveforms in matched filters; (c) signal processing schematic: mc, memory cell

contradiction can be resolved in the following way. Take a T_{p1} value satisfying condition (6.10). If it happens that $T_{p1} < T_{p2}$, we can 'colour' the transmit pulses, which means that each pulse in a train of M_n pulses (M_n is an integer, $M_n \geq T_{p2}/T_{p1}$) will be different from its neighbours. Specifically, this can be done by phase modulation of M_n successive pulses with the use of different codes, e.g. M-sequences of the same length, but different structure. In the present example, $\frac{T_{p2}}{T_{p1}} = 2.67$, i.e. $M_n = 3$. Thus, each 'triplet' of transmitted pulses will contain waveforms modulated with different codes (Figure 6.4(a)). The full period of such a train is equal to $3T_{p1} > T_{p2}$.

Echo signals are processed in three matched filters: MF1, MF2 and MF3. The output signals of these filters have a period $3T_{p1}$, i.e. the requirement (6.11) is satisfied (Figure 6.4(b)). Then, the three signals are stored in three random-access memory (RAM) units. The ith cell of each RAM stores the signal from the ith segment of the range. Numbers are fed to the azimuth processing device in the following

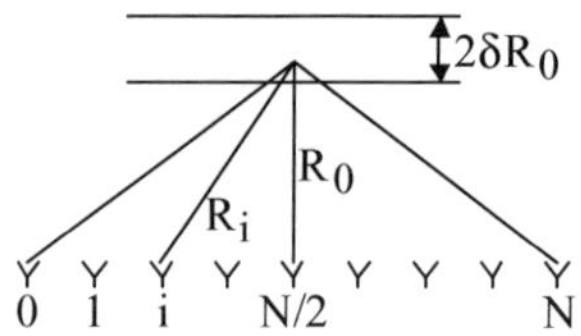

Figure 6.5 To calculation of the width of the terrain strip $2\delta R_0$

order: first, from the ith cell of the first RAM, then, after a time interval, T_{p1}, from the ith cell of the second RAM, and finally from the ith cell of the third RAM. Next, after a time interval of T_{p1}, numbers come from the $(i+1)$th cell of the first RAM, then from the $(i+1)$th cell of the second RAM, and so on. Pulses enter the azimuth processing device with a period equal to T_{p1}, i.e. the condition (6.10) is satisfied.

With the great length of the synthetic antenna, the difference of ranges from its elements to a given point on the Earth's surface can be comparable to the wavelength. Therefore, if a phased aperture is to be formed, the voltages from antenna elements need to be phased prior to summation. We shall define the dependence of the phase of a received signal on the number, i, of the antenna element and range, R_0, to the target object, the Earth's surface. The signal phase in the middle element with the number $i = N/2$ is taken as the reference point (Figure 6.5):

$$\varphi(i, R_0) = \frac{4\pi}{\lambda}\left(\sqrt{R_0^2 + \left(i - \frac{N}{2}\right)^2 \ell_1^2} - R_0\right) = \frac{4\pi}{\lambda}\left\{R_0\left[1 + \frac{(i - N/2)^2\,\ell_1^2}{R_0^2}\right]^{1/2} - R_0\right\}.$$

Since $(i - N/2)^2\,\ell_1^2/R_0^2 \ll 1$, we can write

$$\left[1 + \frac{(i - N/2)^2\,\ell_1^2}{R_0^2}\right]^{1/2} \approx 1 + \frac{(i - N/2)^2\,\ell_1^2}{2R_0^2}.$$

Consequently,

$$\varphi(i, R_0) \approx \frac{2\pi}{\lambda}\frac{(i - N/2)^2\,\ell_1^2}{R_0}.$$

Let us calculate the width of a strip of terrain, $2\delta R_0$, within which the same phasing law can be applied to the echo signal. For this purpose, φ is differentiated with respect to R_0 and the absolute value of the derivative is taken. Thus,

$$\frac{\partial\varphi}{\partial R_0} = \frac{2\pi\,(i - N/2)^2\,\ell_1^2}{\lambda R_0^2}.$$

Obviously, the derivative is a maximum when $i = 0$ or $i = N$. In both of these cases,

$$\frac{\mathrm{d}\varphi}{\mathrm{d}R_0}\max = \frac{\pi L_s^2}{2\lambda R_0^2}.$$

If we change from differentials to finite increments, this expression yields

$$\delta\varphi_{\max} = \delta R_0\frac{\pi L_s^2}{2\lambda R_0^2}.$$

It is assumed that the maximum possible value of L_s is used, which is equal to $\theta_{ph} R$. The admissible value of δR_0 is then expressed in terms of admissible out-phasing:

$$\delta R_{0.adm} = \frac{\delta\varphi_{adm}\, 2\lambda}{\pi\theta_{ph}^2} = \frac{4L_{ph}^2}{\pi\lambda}\delta\varphi_{adm}.$$

By way of example, take $\delta\varphi_{adm} = \pi/8$; then the width of the strip in question is

$$2\delta R_{0.adm} = \frac{L_{ph}^2}{\lambda}. \tag{6.12}$$

The span of object ranges is

$$R_{max} - R_{min}$$

and the number of channels in which different phasing laws are used is

$$m = \frac{(R_{max} - R_{min})\lambda}{L_{ph}^2}. \tag{6.13}$$

6.4 MODEL OF AN SAR AS A FILTER MATCHED WITH AN LFM SIGNAL

Consider an SAR carrier moving within the confines of synthetic aperture, L_s (Figure 6.6). To simplify the calculations, it will be assumed that the SAR transmits a CW signal. The Doppler shift in the frequency of the signal returned by a point-like object on the Earth's surface is equal to

$$f_d = \frac{2V\sin\alpha}{\lambda} = \frac{2V}{\lambda}\frac{L_s/2 - Vt}{\sqrt{R_0^2 + (L_s/2 - Vt)^2}} \approx \frac{2V}{\lambda R_0}\left(\frac{L_s}{2} - Vt\right),$$

as long as $L_s/2 - Vt \ll R_0$.

The moment when the SAR is at the leftmost point of the synthetic aperture is adopted as $t = 0$. Further,

$$f_d = \frac{VL_s}{\lambda R_0} - \frac{2V^2 t}{\lambda R_0}. \tag{6.14}$$

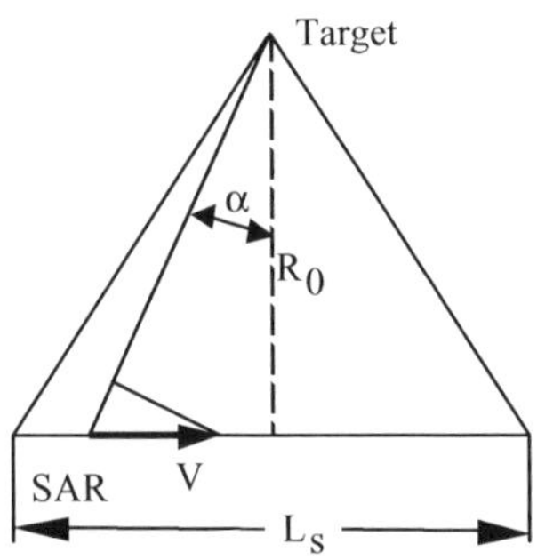

Figure 6.6 Illustration of an SAR as an LFM-matched filter

It follows from Equation (6.14) that f_d is a linear function of time, i.e. the echo signal is a frequency-modulated signal with a linear frequency variation law. When $t = 0$,

$$f_{d.0} = \frac{VL_s}{\lambda R_0}.$$

When $t = L_s/V$,

$$f_{d.1} = -\frac{VL_s}{\lambda R_0}.$$

Hence, the deviation of signal frequency is

$$\Delta f_d = f_{d.0} - f_{d.1} = \frac{2VL_s}{\lambda R_0}.$$

If a signal with a frequency deviation, Δf_d, passes through a matched filter, the output signal duration will be

$$t_s = \frac{1}{\Delta f_d} = \frac{\lambda R_0}{2VL_s}.$$

Using formulae (6.4) and (6.5) gives

$$t_s = \frac{0.7L_{ph}}{V}. \qquad (6.15)$$

The duration of a return from a point-like object is equal to the time of the object's stay within the beam of the SAR or the time it stays within an azimuth resolution cell, which is the same. The cell moves at a velocity, V. Therefore, $t_s V$ is the length of the resolution cell. However, according to Equation (6.15),

$$\Delta \ell_s = t_s V = 0.7L_{ph}.$$

This equation coincides with Equation (6.9), which was deduced by reasoning from the analogy between the SAR and the phased array.

The processing (compression) of an azimuth signal can be implemented on the basis of a digital matched filtering in accordance with the equation

$$U_{out}(t) = \int U_{in}(x - \tau)h(t - x + \tau)\mathrm{d}x. \qquad (6.16)$$

In Equation (6.16), τ depends on the azimuth of the target object. To carry out the operation described by (6.16), the values of $U_{in}(x - \tau)$ are multiplied by the values of the pulse response, $h(t - x + \tau)$, formed in the computer, with consequent integration of the products. It follows from Equation (6.16) that the processing should comprise multiple channels as there are multiple values of τ. The number of azimuth channels required is

$$N_{az} = \frac{L_s}{l_1}.$$

For example, when $L_s = 2$ km and $l_1 = L_{ph}/2 = 10$ m, $N_{az} = 200$.

The pulse response of a filter matched with the complex envelope of an azimuth signal for a certain value of τ is given by

$$U_{az}(t) = U\cos[\omega_0 t + \varphi(t - \tau)], \qquad (6.17)$$

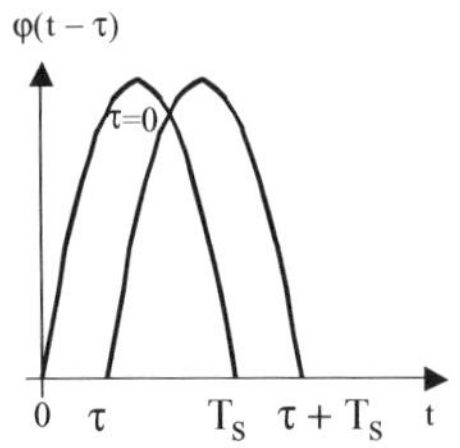

Figure 6.7 Functions $\varphi(t-\tau)$

where, ω_0 is the carrier frequency in rad/s and $\varphi(t-\tau)$ is the phase as a function of time expressing frequency modulation of the signal according to Equation (6.14). Hence,

$$\varphi(t-\tau) = 2\pi \int f_{\mathrm{d}}(t-\tau)\mathrm{d}t = 2\pi \left[\frac{VL_{\mathrm{s}}}{\lambda R_0}(t-\tau) - \frac{V^2(t-\tau)^2}{\lambda R_0} \right]. \tag{6.18}$$

The curves plotted according to Equation (6.18) for $\tau = 0$ and $\tau \neq 0$ are given in Figure 6.7.

In a pulsed radar, the time delay, τ, assumes discrete values separated by intervals, T_{pl}. Therefore, the values of the phase shifts, $\varphi(iT_{\mathrm{pl}})$, are discrete as well. The greatest difference between two adjacent values of φ, which is written as $\Delta\varphi_{\max}$, takes place with a zero argument, because at this point the derivative is a maximum:

$$\frac{\mathrm{d}\varphi_{\max}}{\mathrm{d}t} = \frac{VL_{\mathrm{s}}}{\lambda R_0}.$$

For example, with $V = 8$ km/s, $L_{\mathrm{s}} = 2$ km, $\lambda = 10$ cm and $R_0 = 800$ km,

$$\frac{\mathrm{d}\varphi_{\max}}{\mathrm{d}t} = 200\,\mathrm{rad/s}.$$

When $T_{\mathrm{pl}} = 1$ ms, $\Delta\varphi_{\max} = 0.2$ rad ≈ 11 deg. Such a value of $\Delta\varphi_{\max}$ is usually regarded as acceptable.

6.5 ADDITIONAL CONSTRAINT ON SYNTHETIC APERTURE SIZE

The synthetic aperture size, L_{s}, is also subject to a constraint differing from that given by Equation (6.4). This limitation is related to the necessity of temporal matching of the echo pulses received by different elements of the synthetic antenna. From Figure 6.5 it follows that the distances from an object on the Earth's surface to different elements of the synthetic array differ. The greatest difference in distances, δR_1, is equal to

$$\delta R_1 = R_m - R_0 = \sqrt{R_0^2 + \left(\frac{L_S}{2}\right)^2} - R_0 \approx \frac{L_S^2}{8R_0}.$$

Hence, the signals received by an extreme element and by the middle element of the synthetic antenna are shifted in time by

$$\delta t = \frac{L_{\mathrm{s}}^2}{4R_0 c}. \tag{6.19}$$

The value of δt should be much less than the signal pulse duration at the output of the matched filter, τ_{out}. Suppose the admissible time shift is

$$\delta t = K\tau_{\mathrm{out}}. \tag{6.20}$$

Table 6.1 Admissible L_s as a function of K, ΔR and R_0

	$\Delta R = 10$ m					$\Delta R = 5$ m				
R_0 (km)	50	100	200	700	1000	50	100	200	700	1000
L_s (km)	1.4	2	2.8	5.3	6.3	1	1.4	2	3.7	4.5

Combining Equations (6.19) and (6.20) gives the limitation for the size of the synthetic aperture:

$$L_s \leq 2\sqrt{R_0 K c\tau_{out}} = 2\sqrt{2}\sqrt{K R_0 \Delta R}. \tag{6.21}$$

In Equation (6.21),

$$\Delta R = \frac{c\tau_{out}}{2}$$

is range resolution of the SAR.

Table 6.1 presents estimates of admissible L_s values calculated by formula (6.21) for $K = 0.5$, $\Delta R = 5$ m and 10 m and several values of R_0.

6.6 SPOTLIGHT MODE

In the spotlight mode, the beam of the synthetic antenna tracks the position of a given object during a certain period of time. SAR beam positions at successive instants of time are shown in Figure 6.8.

Angle tracking of the object lifts the restriction imposed by Equation (6.4). At the same time, azimuth resolution is not limited by Equation (6.9) derived with the help of formula (6.4). The value of $\Delta\ell_s$ is only governed by the initial formula (6.8).

The spotlight mode can be employed for two purposes:

- to increase the aperture of the synthetic antenna and accordingly improve azimuth resolution as per Equation (6.8);
- to perform noncoherent integration of signals received from the target area on several sections of the SAR path (Figure 6.9).

Coherent echo-signal processing is carried out on each of the path sections shown in Figure 6.9. Angular resolution is determined by formula (6.8), in which L_s is substituted for L_{s1}. Noncoherent processing

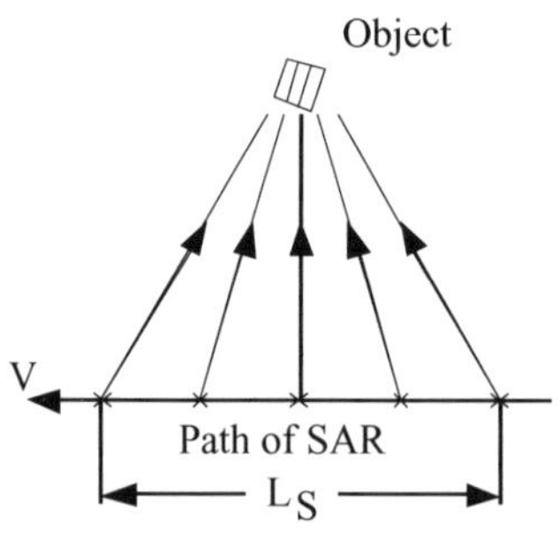

Figure 6.8 Spotlight mode and coherent integration: x, current positions of SAR; ↑; AP of SAR directions

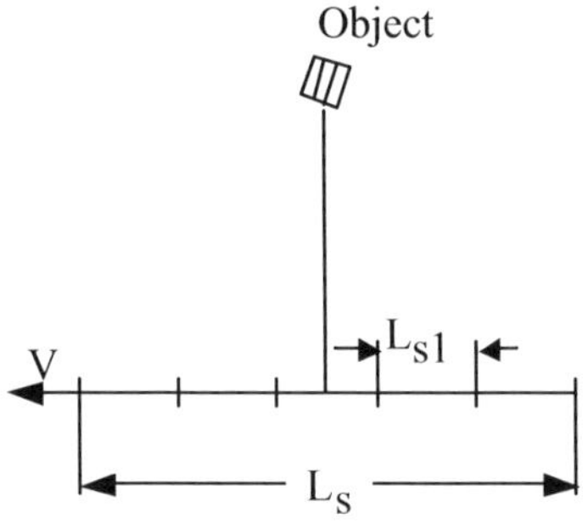

Figure 6.9 Spotlight mode and noncoherent integration

of several signals allows the signal-to-noise ratio to be improved as compared with the case when the size of the synthetic aperture is equal to L_{s1}. Besides, noncoherent integration leads to an appreciable reduction in the speckle effect in SAR imagery. The restriction on synthetic aperture size expressed by Equation (6.21) remains in force.

Beam-pointing, necessary for the realization of the spotlight mode, can be implemented by either mechanically turning a mirror antenna or by electronically controlling the phase distribution of a phased array.

7

Interference Protection

7.1 INTRODUCTION

Immunity to interference is an issue of paramount importance to radar system designers. Interference can take various forms such as natural and manmade interference, clutter and jamming, masking and deception. Apart from interference generators, electronic countermeasure (ECM) systems include signal intelligence aids to collect information about the type of radar and its key parameters (wavelength, transmitted signal bandwidth, angle tracking technique employed, direction towards the radar). Therefore, electronic counter-counter measures (ECCMs) also include means of counteracting signal intelligence.

Note that one of the possible ways of conducting an ECM is to destroy enemy radars by means of antiradiation missiles (ARMs) that use the transmitter's signal to home on the radar. This technique is not covered in this chapter.

Section 7.2 gives a brief overview of the main types of interference. Evaluation of clutter level as a function of angular and range resolution of a radar is given in Section 7.3. Section 7.4 describes clutter cancellation devices known as the moving target indicator (MTI) and the moving target detector (MTD).

Section 7.5 dwells on the method of forming nulls in the pattern of a phased array in the directions towards the sources of interference. This method is primarily used to counter jamming.

7.2 THE MAIN TYPES OF INTERFERENCE

By and large, radar interference can be classified into two groups:

- passive interference (commonly called 'clutter');
- active interference (commonly called 'jamming').

In its turn, the class of passive interference is divided in two subclasses:

- interference of natural origin (reflections from the ground and various structures, reflections caused by atmospheric effects);
- manmade interference (chaff and decoys).

Active interference generated by dedicated transmitters (jammers) is represented by two large groups:

- masking interference (usually, in the form of noise);
- deception jamming (simulated echoes).

The masking interference group can be further subdivided into barrage noise and spot jamming (continuous or discontinuous). Barrage noise has a spectrum of the order of several hundreds of megahertz and is intended for the suppression of a series of radars of the same class, working at different frequencies within the same waveband.

The spectrum bandwidth of a spot jammer is commensurat with the bandwidth of the signal transmitted by the radar being jammed. Spot jamming offers the advantage of greater spectrum density (with the same power as barrage noise). However, spot jamming is less effective against frequency-agile radars.

Jammers transmitting continuous noise are prone to attack by antiradiation missiles (ARMs). In this respect, discontinuous noise jamming is more practical as it has a considerable degrading effect on the accuracy of the guidance of such missiles.

Repeater jamming is yet another kind of active interference, intended for suppression of range, velocity and angle trackers. In the latter case, this kind of interference may cause angular tracking loss in a radar with conical scanning.

7.3 GROUND CLUTTER AND CHAFF LEVEL EVALUATION FOR PULSE AND CW MODULATED SIGNALS

Clutter sources are situated in the same radar resolution element (volume resolution cell or surface resolution cell (see Section 1.6)) as the target. Therefore, the methods of angular and range selection, as a rule, do not ensure sufficient suppression of a clutter signal. A common technique to separate the useful signal from its mix with interference consists in using the difference in velocity between the target and the source of clutter. Differing velocities cause differing Doppler shifts in the useful signal and the clutter. Frequency selection devices relying on this effect are briefly covered in Section 7.4. Nonetheless, as will be shown further, angular resolution and range resolution does permit clutter power at the receiver input to be reduced to a certain extent and facilitates clutter suppression to the desired level.

Ground clutter pertains to the class of surface-distributed objects, while chaff and precipitation clutter pertain to volume-distributed objects. According to Equation (1.16) and (1.33), the RCS of an area of terrain is defined by the equation

$$\sigma_s = \sigma_{os}\Delta S, \tag{7.1}$$

where σ_{os} is the clutter cross-section per unit area (ground reflectivity) and ΔS is the resolvable surface area of the radar equal to

$$\Delta S = \frac{\Delta R}{\cos \varepsilon} R\theta_1, \tag{7.2}$$

where ΔR and θ_1(radian) are, respectively, range resolution and angular resolution in the horizontal plane. R is the distance to the illuminated area and ε is its elevation.

According to Equation (1.15) and (1.32), the RCS of a volume-distributed object, σ_v, is expressed by the formula

$$\sigma_v = \sigma_{ov}\Delta W, \tag{7.3}$$

where σ_{ov} is the cross-section per unit volume (RCS of clutter within a cubic metre) and ΔW is the resolvable volume of the radar equal to

$$\Delta W = \frac{\pi}{4} \Delta R R^2 \theta_1 \theta_2, \tag{7.4}$$

where θ_2 (radian) is radar resolution in the vertical plane.

It follows from Equations (7.1) to (7.4) that the clutter power at the input of the radar receiver does not only depend on clutter parameters, σ_{os} and σ_{ov}, but also on the radar's resolution in range resolution and angular resolution. According to Equation (1.26), the range resolution is determined by the bandwidth of the transmitted signal

$$\Delta R = \frac{c}{2\Delta f_s}. \tag{7.5}$$

The radar range, which follows from Equation (1.26), grows with increasing duration, τ_p, of the processed coherent signal. For a continuous-wave radar with a modulated carrier, τ_p is the modulation period. Multiplying the numerator and the denominator in Equation (7.5) by τ_p gives

$$\Delta R = \frac{c\tau_p}{2\Delta f_s \tau_p} = \frac{c\tau_p}{2B}, \tag{7.6}$$

where B is the duration–bandwidth product.

It follows from Equations (1.26), (7.5) and (7.6) that to obtain a desired range coverage and a small value of ΔR, it is expedient to transmit complicated waveforms. However, when a target is detected against a background of extended objects such as clutter sources, the level of clutter is determined by the effective resolution of the radar in terms of the range, ΔR_e. Evaluation of ΔR_e is performed taking into account the clutter passage via the sidelobes of the signal's autocorrelation function. Therefore, in the case under consideration, it is practical to choose a complicated waveform with the smallest possible value of ΔR_e.

To give a practical example, let us estimate ΔR_e for a PCW waveform coded with an M-sequence and an LFM (chirp) waveform. In accordance with the results obtained in Reference [I.16], the mathematical expectation of the modulus of sidelobes in binary PSK signals is $(0.3–0.5)\sqrt{B}$, with B being equal to the signal length expressed in the number of discretes. The number of sidelobes (on both sides of the mainlobe) is equal to $2B$. Hence, the averaged power of the sidelobes is proportional to

$$\left[(0.3\text{–}0.5)\sqrt{B}\right]^2 \times 2B \approx (0.2\text{–}0.5)B^2.$$

The power of the compressed useful signal is proportional to B^2. Hence,

$$\Delta R_e = \Delta R_1 \frac{(0.2\text{–}0.5)B^2}{B^2} = (0.2\text{–}0.5)\Delta R_1, \tag{7.7}$$

where ΔR_1 is the range resolution of a simple RF pulse with a duration equal to the duration of a PCW discrete. In other words, the gain in terms of energy achieved by using a PCW signal in a clutter-polluted environment does not exceed 3–7 dB.

Better results can be achieved with an LFM signal. The sidelobes of such a signal are described by the relationship

$$y(x) = \frac{\sin x}{x}.$$

Using a Taylor function for weighting the LFM signal, it is possible to obtain the following results [I.3]: a sidelobe level of not more than −40 dB, signal loss ≈ 1 dB and a 1.4-fold widening of the mainlobe of the compressed signal. Thus, in this particular case,

$$\Delta R_e = 1.4\Delta R. \tag{7.8}$$

From a comparison of formulae (7.7) and (7.8), it can be seen that the LFM signal is preferable for the detection of targets against a background of clutter. Another advantage of this waveform is that it imposes more lenient requirements on the maximum value of τ_p (see Section 2.9).

7.4 MOVING TARGET INDICATOR AND MOVING TARGET DETECTOR

The Doppler shift of a target return is defined by the formula

$$f_d = \frac{2V_r}{\lambda},$$

where V_r is the object's velocity component towards the radar (radial velocity, or the range rate) and λ is the wavelength of the radar transmitter output. If, by way of example, a stationary radar is taken, then $V_r = V\cos\alpha$, where V is the object's velocity and α is the angle formed by the object velocity vector with the line of sight to the object from the radar.

Radial velocity of the target (in absolute value) is usually greater than the velocity of the clutter source. This brings about a difference in the Doppler shift between the useful signal and the clutter voltage, which makes it possible to attenuate the clutter voltage by filtering. The device designed for processing the mix of the useful signal and clutter, ensuring considerable reduction of the latter, is known as a 'moving target indicator' (MTI) or a 'moving target detector' (MTD).

Consider the detection of a moving object whose echo signal has a Doppler shift, f_d. Let us first make some simplifying assumptions:

- The radar signal duration and clutter voltage duration are infinitely great.
- The clutter has a zero Doppler shift.
- The pulse repetition frequency of the transmitted signal is greater than its maximum Doppler shift.

Further, these assumptions will be discarded.

Figure 7.1 shows a simplified block diagram of a pulsed radar system with MTI and Figure 7.2 shows flowcharts of voltages at key points of the transmitter. It is obvious from Figures 7.1 and 7.2 that the output voltage of the transmitter, $u_t(t)$, is a train of RF pulses with a carrier frequency f_0 and a pulse repetition period T_p. The pulse repetition frequency (PRF) is

$$F_p = \frac{1}{T_p}.$$

The voltage of the radar transmitter is expressed by the formula

$$u_t(t) = u_m(t)\cos(2\pi f_0 t), \tag{7.9}$$

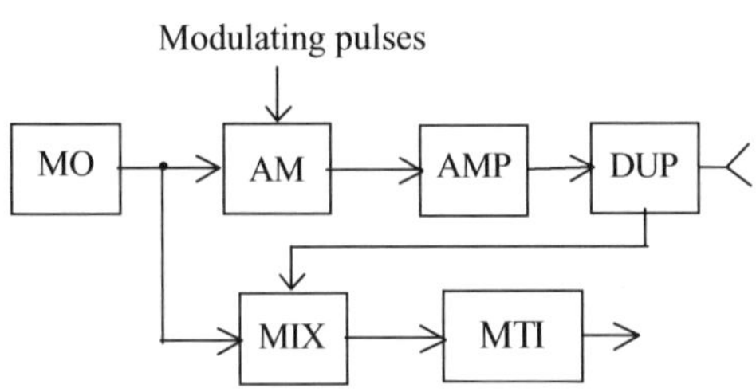

Figure 7.1 Simplified block diagram of MTI: MO, master oscillator; AM, amplitude modulator; AMP, amplifier; DUP, duplexer; MIX, mixer; MTI, moving target indicator

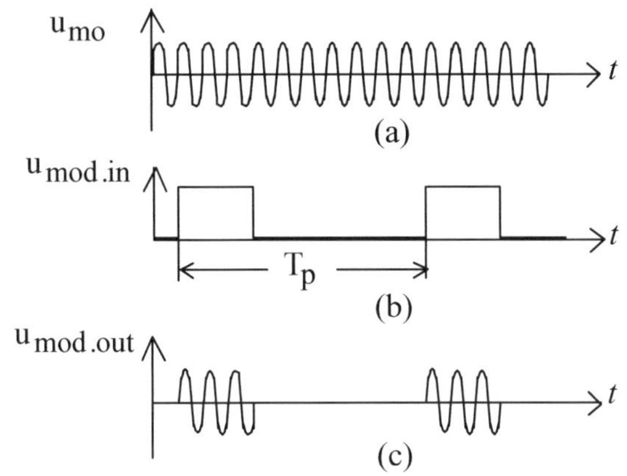

Figure 7.2 Voltage flowcharts: (a) master oscillator output; (b) modulator pulses; (c) modulator output

where $u_m(t)$ is the envelope of the transmitted waveform, i.e. a train of rectangular pulses. The voltage of clutter at the input of the receiver, $u_{cl}(t)$, in the first approximation preserves the form of the transmitter's output:

$$u_{cl}(t) = k_1 u_m(t) \cos(2\pi f_0 t + \psi),$$

where k_1 is the coefficient allowing for clutter energy reduction as compared to radiation of the transmitter and ψ is the initial phase changing slowly with time.

The voltage of the useful signal at the input of the receiver is

$$u_s(t) = k_2 u_m(t) \cos[2\pi(f_0 + f_d)t],$$

where $k_2 \ll k_1$.

The phase of the output voltage of the mixer is equal to the difference of phases of the input voltage and the voltage of the heterodyne oscillator. Hence, the phase of the output voltage of the clutter is given by

$$2\pi f_0 t + \psi - 2\pi f_0 t = \psi,$$

and the output voltage proper is

$$u_{cl.out} = k_1 u_m(t) \cos\psi. \quad (7.10)$$

The phase of the output voltage of the useful signal is

$$2\pi(f_0 + f_d)t - 2\pi f_0 = 2\pi f_d t,$$

and the output voltage proper is

$$u_{s.out} = k_2 u_m(t) \cos(2\pi f_d t). \quad (7.11)$$

As $u_m(t)$ is a train of rectangular pulses, the clutter, as per Equation (7.10), has the shape illustrated in Figure 7.3(a). According to Equation (7.11), the target signal is a train of rectangular pulses multiplied by $\cos(2\pi f_d t)$ (Figure 7.3(b)).

Thus, in the time domain, the signal of a moving target differs from the clutter in that it has amplitude modulation with Doppler frequency. In the spectral domain, the target signal components are shifted by $\pm f_d$ relative to clutter harmonics. Consequently, the target signal can be separated from the signal–clutter mix with the help of the time domain or spectral domain processing. In the former case, a delay-line

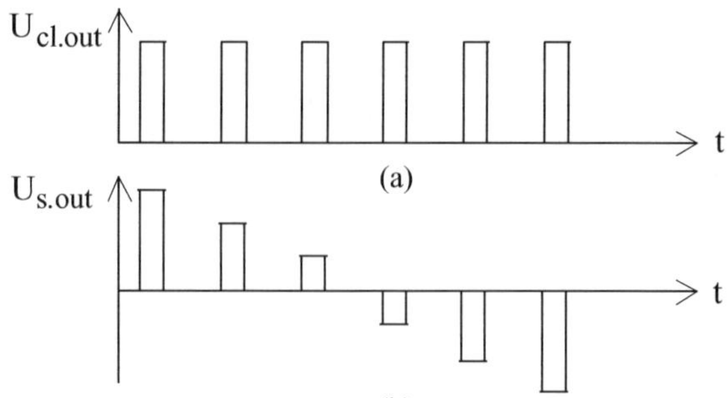

Figure 7.3 (a) Clutter and (b) target voltages (mixer output)

canceller is used (Figure 7.4). Since the pulses of the clutter are roughly identical, the voltage at the output of this device is close to zero. At the same time, a major portion of the modulated target signal passes through to the output.

Spectral domain processing relies on the difference between spectra of the clutter and the target signal. The periodic voltage of clutter, expressed by Equation (7.10), can be represented as a sum of Fourier series:

$$u_{\text{cl.out}} = \text{constant}\left[1 + 2\sum_{i=1}^{\infty} a_i \cos(i2\pi\, F_{\text{p}}\, t)\right].$$

Consequently, clutter spectrum contains only PRF harmonics $i\,F_{\text{p}}$ (the solid lines in Figure 7.5), including the constant component ($i = 0$).

According to Equations (7.10) and (7.11), the moving target signal at the output of the mixer can be written as

$$\begin{aligned} u_{\text{s.out}} &= \text{constant}\left[1 + 2\sum_{i=1}^{\infty} a_i \cos(2\pi i\, F_{\text{p}}\, t)\right]\cos(2\pi\, f_{\text{d}}\, t) \\ &= \text{constant}\left[\cos(2\pi\, f_{\text{d}}\, t) + 2\sum_{i=1}^{\infty} a_i \cos(2\pi i\, F_{\text{p}} t)\cos(2\pi\, f_{\text{d}}\, t)\right] \\ &= \text{constant}\left\{\cos(2\pi\, f_{\text{d}}\, t) + \sum_{i=1}^{\infty} a_i \cos\left[2\pi\left(i\, F_{\text{p}} - f_{\text{d}}\right) t\right] + \sum_{i=1}^{\infty} a_i \cos\left[2\pi\left(i\, F_{\text{p}} + f_{\text{d}}\right) t\right]\right\}. \end{aligned} \quad (7.12)$$

It follows from Equation (7.12) that the spectrum of the moving target signal at the output of the mixer only contains components shifted by $\pm\, f_{\text{d}}$ relative to PRF harmonics (the dashed lines in Figure 7.5).

According to Figure 7.5, to suppress clutter the amplitude–frequency response of the MTI should have rejection notches at frequency multiples of F_{p} (Figure 7.6). Specifically, such an amplitude–frequency response can be implemented in a digital highpass filter clocked by pulses with repetition frequency, F_{p}.

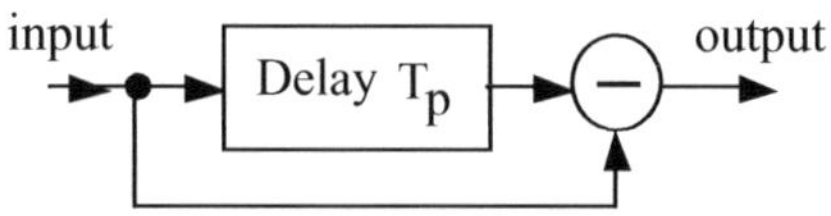

Figure 7.4 Block diagram of the delay-line canceller

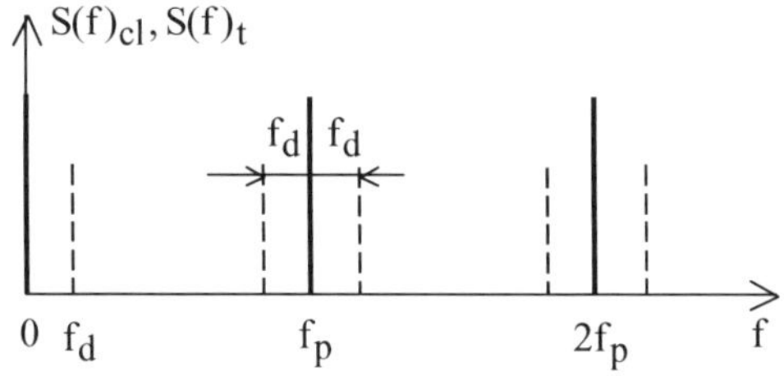

Figure 7.5 Clutter (———) and target (– – –) spectra

In practice, clutter voltage is not entirely periodic. This, as well as chaotic movements of certain parts of the clutter source, brings about a widening of its spectral components. Apart from this, the clutter source is often represented by an object moving at a small speed, V_{cl}, which results in a shift of clutter spectrum and widening of its spectral components. Hence clutter source is often represented as components by $\pm f_{d.cl} = \pm 2 V_{cl}/\lambda$. Therefore, the width of rejection notches in the highpass filter is chosen in view of the above factors, so that the spectral components of clutter lie within the limits of rejection (Figure 7.6).

Apropos, the amplitude–frequency response of the delay-line canceller given by

$$K(f)_{dlc} = |\sin(\pi f T_p)|$$

also has nulls at frequencies $f = i/T_p = i F_p$, where $i = 0, 1, 2, 3, \ldots$. It follows from Figure 7.7 that the filter with such a frequency response has a zero rejection notch and noticeable nonuniformity in its passband. The situation can be somewhat improved by employing a cascade connection of two or three delay-line cancellers. Nevertheless, preference is often given to a recursive highpass filter with a frequency response shown in Figure 7.6.

MTI performance can be improved by using a comb filter (see Figures 2.16 and 2.17 in Section 2.6.2). As noted in Reference [I.3], this solution makes it possible to discern the target from rain and birds, to estimate target velocity and to use a coherent integration of signal. Weighting the frequency response of each filter of the comb in the frequency domain is advisable for a sidelobe reduction.

To investigate blind speeds of the target, the above assumption about $F_p > f_{d.\max}$ is certainly not always justified. When $f_{d.\max} > F_p$, it is possible that

$$f_d = k\, F_p, \tag{7.13}$$

where k is an integer. Subject to the equality (7.13) being satisfied, all the spectral components of the useful signal fall into rejection notches of the digital highpass filter, and the target signal is suppressed. Target velocities corresponding to Doppler frequencies determined by formula (7.13) are termed 'blind speeds', because the radar is unable to detect the target when it moves at these speeds (the radar becomes 'blind').

In order to exclude the blind speed effect, it may be practical to use two or more values of F_p. By way of example, we shall limit ourselves to two different values of repetition frequency: $F_{p.1}$ and $F_{p.2}$.

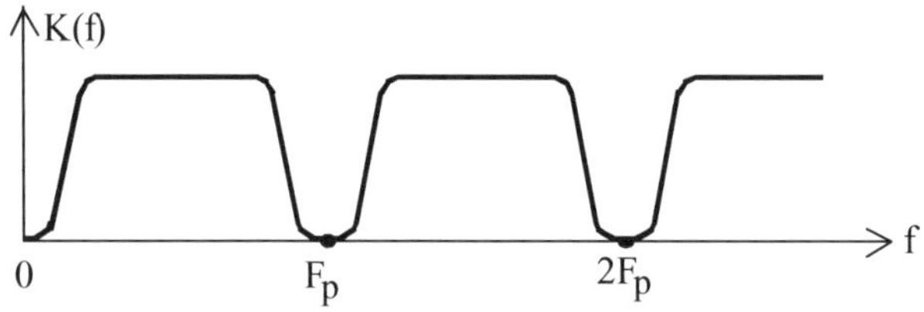

Figure 7.6 MTI amplitude–frequency characteristic

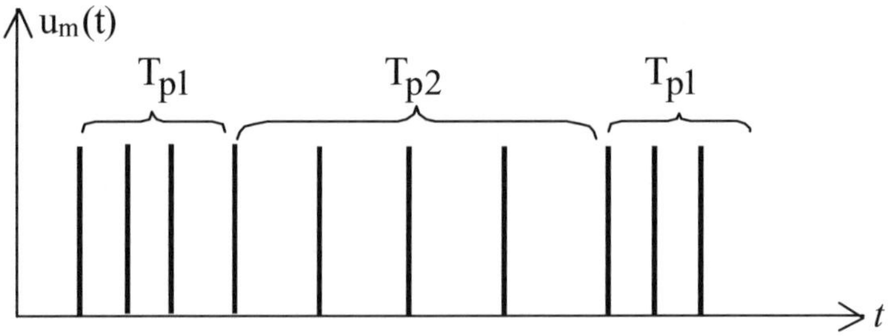

Figure 7.7 Amplitude–frequency characteristic of the delay-line canceller

In other words, suppose groups of pulses are transmitted alternately, with repetition frequencies $F_{p.1}$ and $F_{p.2}$ (Figure 7.8). When $f_d = k\ F_{p.1}$, the pulses recurring at $F_{p.1}$ vanish, while the pulses recurring at $F_{p.2}$ pass through the processing device without loss. When $f_d = k\ F_{p.2}$, contrarily, the pulses repeated at $F_{p.1}$ are allowed to pass through the processing circuitry without loss. In general, the loss does not exceed 3 dB, but there are no blind speeds in a range of frequencies from $f_{d.\min}$ to $f_d =$ LCM (least common multiple) $(F_{p.1}, F_{p.2})$. For example, for $F_{p.1} = 7$ kHz and $F_{p.2} = 5$ kHz, no blind speeds are present in the range from $f_{d.\min}$ to $f_d = 35$ kHz (Figure 7.9).

A moving target detector (MTD) differs from MTI in that it features a zero frequency filter in the comb and a clutter map. This ensures the detection of targets (primarily, aircraft) that have a zero range rate. Detection of such targets is facilitated by the effect of larger RCS occurring when the side surface of the aircraft is covered by the beam of the radar.

A simplified block diagram of an MTD is presented in Figure 7.10. The clutter map (a map of fixed objects) is formed in the RAM using the voltage from the zero-velocity channel for every range-azimuth cell. Each number of a cell on the map is a result of averaging over several tens of seconds. Therefore, the signals from fast-moving objects do not affect the map relief [I.3] that is used to form thresholds. The signals crossing the thresholds (in each range-azimuth cell), pass through to the output of the MTD.

7.5 ADAPTIVE ANTENNA ARRAYS

In counter jamming, it is practical to reduce the sidelobes of the receiving and transmitting antennas of the radar and to use complicated waveforms. Lower sidelobes of the transmit antenna make radio intelligence more difficult. Another method serving the same purpose is the expansion of the transmit signal spectrum with adequate reduction of its spectrum density. Sidelobe reduction in the receive antenna leads to a lower jamming power at the input of the receiver.

One very important method of ECCM consists in forming deep nulls in the pattern of the receive antenna in the directions towards the jamming sources. Beam nulling is most effectively implemented in phased arrays (Section 5.4).

Nulls are formed in the desired directions of the phased array pattern by introducing amplitude factors and phase shifts (weight coefficients) in each channel of the array. It is important that the weight

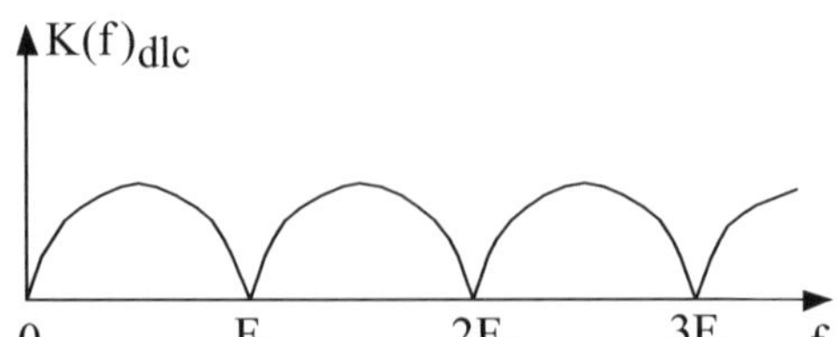

Figure 7.8 Use of two different values of F_p

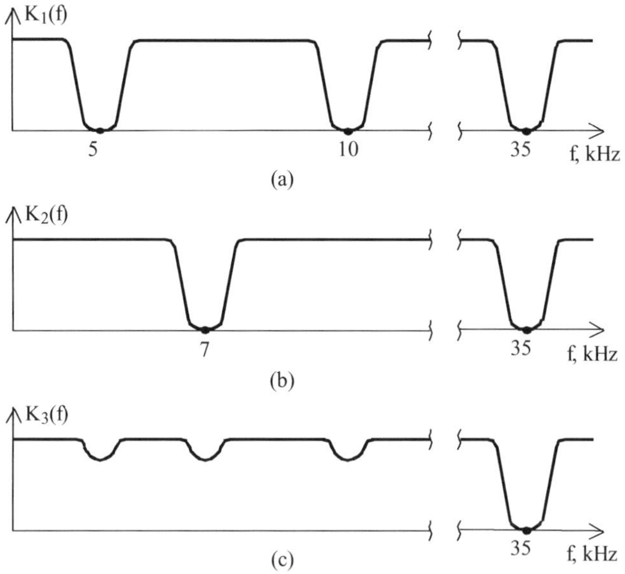

Figure 7.9 MTI frequency characteristics: (a) $F_p = 5$ kHz; (b) $F_p = 7$ kHz; (c) $F_p = 5$ kHz/7 kHz

coefficients are chosen so as to provide the required level of the useful signal along with inserting nulls in the beam pattern. This is achieved by what is called adaptation of the antenna pattern to the jamming environment.

The pattern of a phased array antenna is in fact a spatial filter having a certain space-amplitude characteristic, which is the relationship between the gain and the signal's direction of arrival. This characteristic can be altered by varying the weight coefficients of individual antenna elements. The pattern may have arbitrary values in the directions that do not present any interest and therefore the signal level in these directions is not controlled. If the directions to all the signal sources and interference sources are previously known, the problem can be solved by synthesizing the desired antenna pattern in nonreal time.

In practice, it is only the direction to the source of the useful signal that is known in advance. The number of jamming signals and the directions to their sources are usually unknown. Even so, it is possible to form a beam pattern with the required characteristics by means of adaptation. The fundamentals of the adaptive phased array theory may be found in numerous articles and books on the topic [I.34, I.35].

The adaptive phased array is a variety of the adaptive filter, wherein a certain algorithm (termed an 'adaptive algorithm') is used to perform a real-time search for optimum weight coefficients. The basics of adaptive filtering are given in a number of well-known books such as References [I.36, I.37]. Most

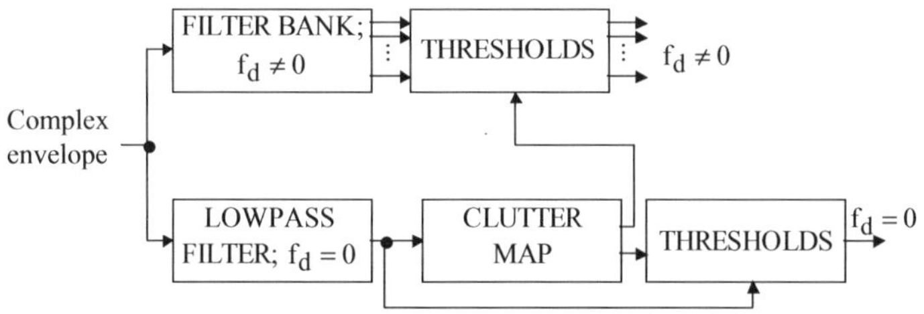

Figure 7.10 Simplified MTD block diagram

adaptive filters, including the adaptive phased array, are linear. This means that the required characteristics are achieved in such filters by linear weighting of the signals being processed, i.e. by a linear combination of these signals.

It is known that an array consisting of M elements and receiving one useful signal can attenuate $M - 1$ jamming voltages [I.34, I.35]. According to the following equation system, adaptive algorithms minimize jamming signal amplitudes A^j, with $j = 2, \ldots, M$:

$$\begin{bmatrix} h_1 A^{(1)} e^{i\omega_0 \tau_1^{(1)}} & \cdots & h_m A^{(1)} e^{i\omega_0 \tau_m^{(1)}} & \cdots & h_M A^{(1)} e^{i\omega_0 \tau_M^{(1)}} \\ \vdots & \vdots & \vdots & \vdots & \vdots \\ h_1 A^{(j)} e^{i\omega_0 \tau_1^{(j)}} & \vdots & h_m A^{(j)} e^{i\omega_0 \tau_m^{(j)}} & \vdots & h_M A^{(i)} e^{i\omega_0 \tau_M^{(j)}} \\ \vdots & \vdots & \vdots & \vdots & \vdots \\ h_1 A^{(J)} e^{i\omega_0 \tau_1^{(J)}} & \cdots & h_m A^{(J)} e^{i\omega_0 \tau_m^{(J)}} & \cdots & h_M A^{(J)} e^{i\omega_0 \tau_M^{(J)}} \end{bmatrix} = \begin{bmatrix} F(\theta^{(1)}) \\ \vdots \\ F(\theta^{(j)}) \\ \vdots \\ F(\theta^{(J)}) \end{bmatrix}, \quad (7.14)$$

where, the superscript (j) designates the number of the source of useful signal or jamming voltages received, $J = M$ is the number of useful signal and jamming voltages received, $A^{(j)}$ is the amplitude of the useful (jamming) signal received, ω_0 is the carrier frequency and $\tau_m^{(j)}$ is the time delay of the jth useful (jamming) signal in the mth antenna element relative to any other element of the array chosen as a reference. Set on the right side of the system of equations are the levels of the pattern, e.g. a unit for the useful signal ($j = 1$) and a zero for the jamming signal ($j = 2, \ldots, J$); h_m are the weight coefficients, which are used in weighting useful signals and jamming signals picked up by elements of the array.

Since the received signals are described by complex numbers, the weight coefficients, h_m , also have complex form. In this case, most methods of adaptive processing [I.34–I.40] can be applied to the stated problem, which is the suppression of signals from unknown sources of interference. Besides this, there are methods of suboptimal processing, in which weighting is carried out by means of adjustment of either phases or amplitudes. Despite the attractiveness of such methods in terms of practical realization, these types of ECCM have a number of fundamental restrictions.

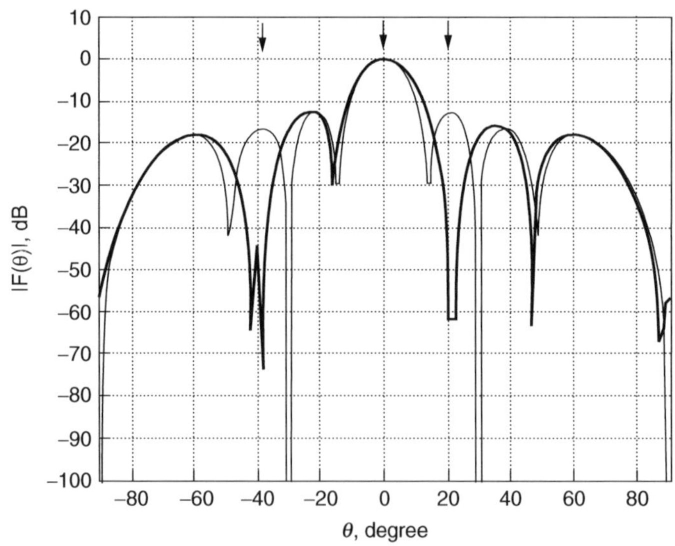

Figure 7.11 Jamming suppression by means of the LMS algorithm

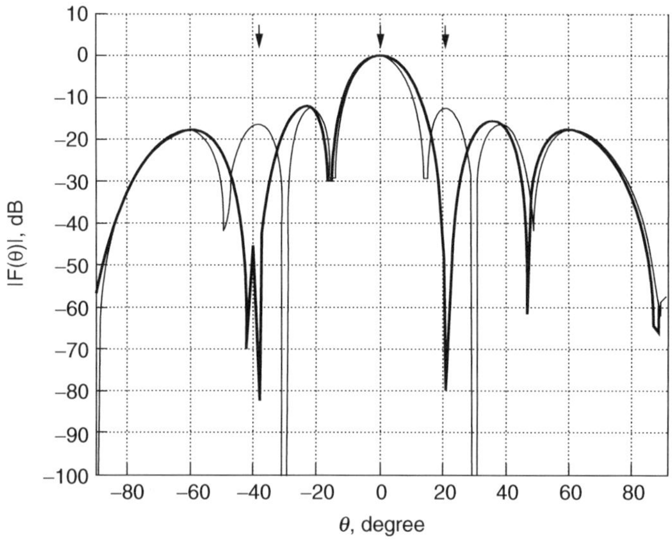

Figure 7.12 Jamming suppression by means of the RLS algorithm

When suppressing a jamming signal, it is necessary to ensure that the useful signal is unaffected. This requires some knowledge about this signal. In the case of an adaptive phased array, the angular position of the source of the useful signal can serve as such information. Jamming suppression can be performed using Frost's algorithm [I.39]. It is an algorithm of linearly constrained minimization of the output voltage of the phased array subject to the condition of constancy of the signal level from the known direction to the source of a useful signal. Thus, by minimizing the output voltage of the phased array, the level of jamming is minimized.

Given below are two examples of computer-aided simulation of the suppression of two jamming signals in an eight-element linear array ($M = 8$), with the elements spaced at half the wavelength of the carrier wave and isotropic radiators. The source of the useful signal is in the direction of the main lobe peak. Two jamming sources are situated in the directions corresponding to the peaks of the sidelobes and have equal amplitudes, which are 10 times greater than the amplitude of the useful signal. The jamming sources were positioned on opposite sides of the main lobe, in the directions ($+21°$) and ($-38°$). The required useful signal level at the output of the array is equal to unity. There is no additive noise at the input of the antenna elements. Algorithm simulation was carried out in floating-point arithmetic, which presupposes the use of a digital phased array. Processing was performed with respect to the complex envelope of the signal–interference mix. The simulation was carried out for 5000 iterations with the help of Frost's algorithm, i.e. a linearly constrained least mean squares (LMS) algorithm and a linearly constrained recursive least squares (RLS) algorithm [I.40]. The results of using these algorithms are presented in Figures 7.11 and 7.12 respectively. The vertical arrows in the figures designate the positions of the sources of signal and jammings received. The initial pattern of the phased array (with $h_m = 1/M$) is shown in thin lines, while thick lines give the shape of the pattern after the adaptation procedure performed for the stated number of iterations. It follows from the presented figures that using the RLS algorithm in this experiment ensures a 20 dB higher suppression of the first jamming source and a 10 dB higher suppression of the second jamming source than using the simpler LMS algorithm.

8

Microelectronic Aerological Radar 'MARL-A'

8.1 DESIGNATED PURPOSE OF THE RADAR

The microelectronic aerological radar 'MARL-A' (Figure 8.1) was developed in the Moscow Institute of Electronic Engineering [1.41]. It is designated to be used jointly with a radio sounder for comprehensive air sensing. The MARL-A performs the tasks of preflight checks on the radio sounder with the check results displayed on the operator's PC, automatic location and tracking of the sounder in flight and obtaining its current coordinates, reception and processing of meteorological data.

8.2 SYSTEM SPECIFICATIONS

These consist of the range of sounder automatic tracking from 100 m to 150 km and comprehensive air sensing altitude up to 40 km. The radar is fitted with an active phased array antenna. The confines of beam-steering are $\pm 210^\circ$ (mechanically turned) plus electronically scanned within a sector of $\pm 25^\circ$ in the horizontal plane and electronically scanned within a sector of -10°–100° in the vertical plane. The maximum antenna turning speed in the mechanic scan mode is 60°/s. The antenna beamwidth is:

- not more than 9° in the horizontal plane;
- not more than 6° in the vertical plane.

The peak pulse power of the transmitting part of the phased array is 115 W and the pulse duration is $0, \ldots, 5$ μs in the near field and 1–1.1 μs in the far field. The noise factor of the receiving part of the phased array (along with the subsequent signal reception path) is 3.7 dB (2.35 per unit) and the root mean square error for the sounder position measurement is:

- not more than 30 m in range;
- not more than 0.1° in angular coordinates.

The absolute error for telemetry information measurements (error in the measurement of the frequency-shift modulation period of the sounder signal) is not more than 1 μs.

Bistatic Radar: Principles and Practice Edited by M. Cherniakov

Figure 8.1 Radar MARL-A

The MARL-A ensures sounder tracking under the following conditions:

- sounder climbing speed – up to 12 m/s;
- sounder descent speed – up to 125 m/s; the horizontal speed of the sounder close to the Earth's surface is 35 m/s and at altitudes over 10 km up to 150 m/s.

The radar operator display and control system includes a Pentium-based PC (operation modes, sounder coordinates, equipment performance). A self-test mode is provided in the radar with the use of a sounder simulator.

The power supply is from 220 V, 50 Hz mains (the average power consumption does not exceed 150 W, with the peak load up to 500 W). The emergency power supply is from a storage battery.

8.3 SYSTEM STRUCTURE

The air sensing system comprises an MARL-A radar and an MR3-3 radio sounder tethered to a (free-flying) hydrogen-filled balloon. The sounder transmits a pulse-modulated signal. The carrier frequency is 1672 MHz $\leq f_0 \leq$ 1688 MHz, the pulse duration is approximately 0.6 μs and the mean value of the pulse repetition period is $T_p \approx 1.25$ μs (the mean pulse repetition frequency is $f_{p.m} \approx 800$ kHz). Receiving a signal from the sounder, it is possible to establish its current angular position. To measure the sounder range, the MARL-A incorporates a pulse interrogator. When a pulse interrogation reaches the sounder there emerges a pause of duration, T_p, in the signal of the sounder, i.e. one pulse is skipped. The current range of the sounder can be obtained by determining the delay of that pause, t_{del}, relative to the pulse transmitted by the radar:

$$R = \frac{Ct_{\mathrm{del}}}{2},$$

where C is the speed of radio propagation.

The data on air temperature and humidity at the location of the sounder is transmitted with the use of shift modulation of the sounder's pulse repetition frequency, f_{p}. The frequency, f_{p}, takes on two values in turn, namely $f_{\mathrm{p.1}}$ and $f_{\mathrm{p.2}}$ (Figure 8.2). Air temperature and humidity data are encoded by turns into the modulation period, T. Demodulation of the sounder signal in the radar permits the values T and T_1 to be obtained, which are then supplied to the PC in digital form for the calculation of weather parameters.

The height of the balloon is derived by the radar from the balloon's range and the sine of its elevation angle. This allows the atmospheric pressure to be calculated inasmuch as the temperature is known.

Another important feature of the radar should also be noted. Since the possible deviations of the sounder signal carrier frequency from the rated value (± 8 MHz) may go far beyond the radar receiver band, the receiver tunes into the frequency of the received signal automatically. Simultaneously, the radar transmitter carrier frequency is adjusted so as not to differ much from the sounder signal frequency.

The structural diagram of the radar system is shown in Figure 8.3. Let us expand on the functioning of selected components. The frequency generator (FG) shapes the voltage of the receiver's first local oscillator (carrier frequency $f_{\mathrm{LO1}} = 1480\ \mathrm{MHz} + \delta$, where δ is the deviation of the sounder signal frequency from the rated 1680 MHz) and pulses with the frequency of the transmitter (carrier frequency $f_0 = 1680\ \mathrm{MHz} + \delta$).

The FG unit also includes the input components of the signal reception path, namely a low-noise amplifier (LNA) and a mixer. The signal received from the sounder comes from the phased array antenna to the input of the LNA, and the output voltage of the mixer is supplied to the receiver at an intermediate frequency of 200 MHz. The FG unit supplies a shaped microwave pulse to the phased array antenna where it is amplified and fed via the circulator to antenna modules.

The active phased array antenna (PhAA) is composed of 64 transmit–receive modules and an equal number of coupled dipole radiators. The antenna's passive components (combiner–dividers,

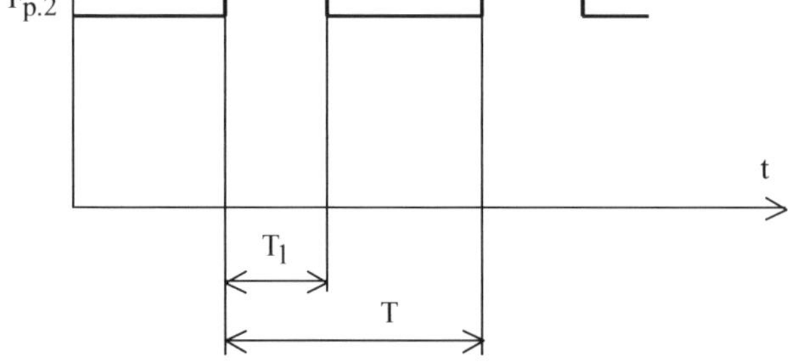

Figure 8.2 Radio sounder RPF shift modulation pattern

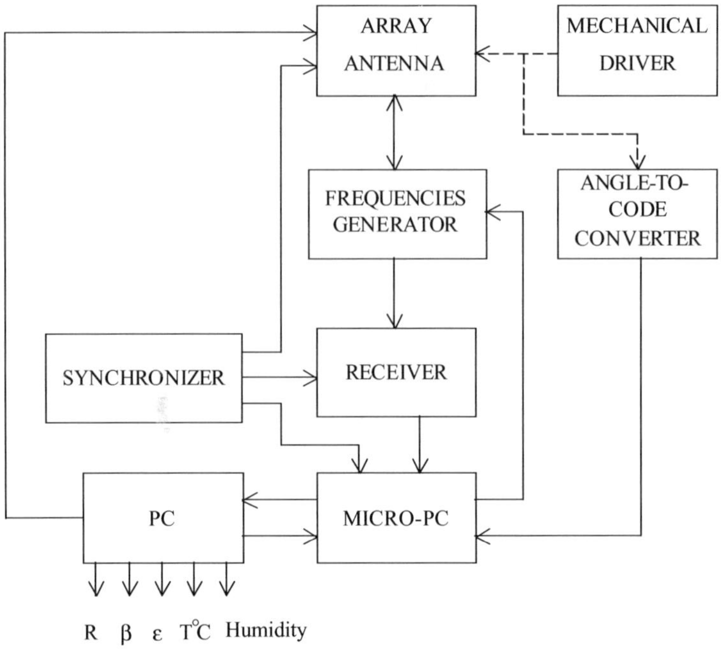

Figure 8.3 Block diagram of the radar

phase-shifters and radiators) are common to the transmitting and the receiving parts of the PhAA. Antenna excitation voltage is split by the combiner–dividers into 64 voltages, which are phased in the 64 channels to be then fed to the transmitting submodules via transmit–receive switches. In the transmitting submodules, the voltages are amplified to a power of not less than 2 W per pulse and fed to the radiators via transmit receive gates. Antenna beam steering in the vertical plane is carried out by forming a desired phase distribution of the field over the antenna aperture. The normal to the array forms an angle of 30° with the horizon. Beam deflection from the normal in the vertical plane may vary from $-40°$ to $+70°$, which makes $(-10°–+100°)$ it relative to the horizon.

Antenna beam-steering in the horizontal plane is carried out in two ways: electronic steering through adjustment of phase distribution of the field over the aperture and mechanical steering by means of an electric drive. The electronically scanned sector is not less than $\pm 25°$; the antenna can be turned mechanically through a sector of $\pm 210°$.

When measuring the azimuth of the sounder, it is necessary to take into account the mechanical turning angle of the antenna, β_{mech}. This information is furnished to the PC by an angle-to-code transducer integrated in the mechanical drive.

Phase distribution control in the PhAA is carried out by means of four-digit phase-shifters with pin-diodes being employed as controlled elements. Sounder signals picked up by the dipoles of the PhAA are gated through the protection circuits of the receiving submodules, amplified in the LNA, phased and then fed to the adding circuit. The latter puts out four signals corresponding to the four subarrays of the PhAA (Figure 8.3).

An angular position measurement is carried out by the method of quadrant scanning (see Section 4.3). This implies that the antenna pattern periodically assumes one of the four positions: the beam deflects by half the beamwidth upwards, to the left, downwards, to the right, and so on. The resultant patterns intersect along the equisignal line (ESL). Departure of the sounder from the ESL leads to amplitude modulation of the received signal with scanning frequency. The modulation depth is proportional to the current angular error, while the modulation phase corresponds to the direction of the sounder departure from the ESL. To prevent power loss (3 dB) during radar signal transmission, scanning is deactivated by zeroing the phase-shifters.

The radar receiver is of a superhet type and uses two intermediate frequencies: 200 MHz and 10.7 MHz. There are three voltages at the outputs of the receiver:

- harmonic voltage at a frequency of 10.7 MHz;
- pulses at a frequency of 800 kHz;
- voltage envelope at the scanning frequency.

The first voltage is used in the sounder carrier frequency tracker. The receiver incorporates a frequency detector and an integrator. The detector generates a number that is proportional to intermediate frequency deviation from the rated value. This number comes to the digital integrator and then to a digital-to-analogue converter (DAC). The output voltage of the DAC is fed to the FG unit to control the first local oscillator frequency and the frequency of the transmitted signal.

The second voltage (800 kHz pulses) is supplied to the range tracker and the telemetry data extractor. Distance to the sounder is calculated by the current time delay in the sounder response, which is determined by means of a split-gate range tracker (Figure 8.4). To reduce the dynamic error due to great wind speeds, a second-order astaticism range tracker is used.

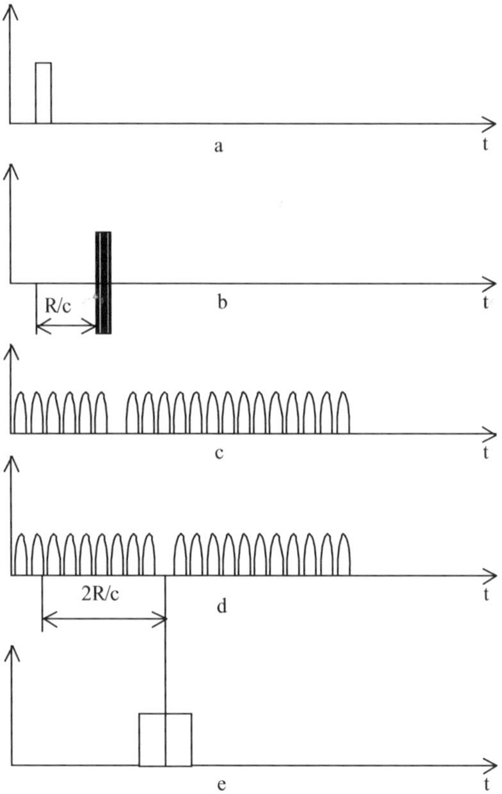

Figure 8.4 Flowcharts of voltages: (a) transmitting pulse; (b) sounder input pulse; (c) sounder reply (sounder output); (d) sounder reply (radar input); (e) split gate

The 800 kHz voltage is also supplied to the primary processing circuitry for the telemetry data, which are transmitted in the following order: reference channel, temperature, humidity, temperature, reference channel, etc. The objectives of primary processing are:

- to discriminate channel slots;
- to measure frequency-shift modulation periods carrying information about temperature and humidity;
- to measure lengths of the slots with maximum pulse repetition frequency for the identification of the reference channel (Figure 8.2).

The '800 kHz' sounder signal detected in the receiver is filtered at the input of the digital signal processor (DSP). The passband of the filter is 100 kHz. The comparator converts the signal to digital form. The low-frequency signal corresponding to the '800 kHz' frequency-shift modulation pattern is extracted by the phase lock circuit. The voltage drive generator of the DSP tunes in to the '800 kHz' signal from the sounder by the output voltage of the phase detector. The output voltage of the phase detector filtered by the LPF is the low-frequency envelope of the frequency-shift modulation of the sounder signal. The period of this voltage is determined by means of a period meter made up of digital counters clocked by a crystal oscillator with a frequency of 1 MHz. The period T contains telemetry data, while the duration of the pulse, T_1, is an attribute used for identification of the reference signal.

The measured values of the pulse repetition period are averaged in the micro-PC, with subsequent identification of the data channel that these values correspond to. Channel identification is carried out with the help of a reference channel indicator – a generator producing pulses whose duration is the boundary between the pulse durations in the reference channel and the data channel. A logic circuit compares the duration of the generator pulses with that of the signal pulses. The micro-PC interrogates the indicator, the reference channel being identified automatically, and the data channels, on the basis of the order of their succession. The durations of the pulse repetition periods in all the channels are sent to the PC for further processing.

The third voltage, that with scanning frequency, is fed to the two phase detectors, which produce error voltages (voltages proportional to the current angular errors). This procedure has been briefly described above.

8.4 RANGE COVERAGE OF THE RADAR

8.4.1 Range Coverage over the Sounder Interrogation Channel

Power flux density of the sounder signal at a distance R from the radar is

$$\mathrm{PD}_1 = \frac{P_t G_t}{4\pi R^2},$$

where P_t is the power of the transmitting PhAA and G_t is the gain of the PhAA. The radar transmitter power is

$$P_t = N p_1,$$

where N is the number of modules and p_1 is the power of a single module. The gain is $G_t = N g_1$, where g_1 is the gain of a coupled radiator.

For the PhAA used in the MARL-A radar, $N = 64$, $p_1 = 1.8$ W and $g_1 = 8$ (at the maximum). Hence, $P_t = 64 \times 1.8 = 115.2$ W and $G_{t.max} = 64 \times 8 = 512$. Consequently,

$$\mathrm{PD}_1 = \frac{115.2 \times 512}{4\pi R^2} = \frac{4696}{R^2}$$

It is shown by practice that in most cases the sounder generates a response when $\mathrm{PD}_1 \geq 10^{-7}$ W/m^2. Hence, $R_{max} = \sqrt{4969/10^{-7}} = 217$ km.

In the worst case, when the sounder falls on the Earth's surface, g_1 is reduced by approximately 30 %, and the sounder response is available at a distance of up to 177 km. This may be taken as acceptable, since the 'fall radius' of most operational sounders is by far below 200 km.

8.4.2 Range Coverage over the Sounder Signal Reception Channel

According to specifications for the MR3 sounder, the power flux density of its transmitter at a distance of 2 m is not less than 1.5×10^{-3} W/m^2. Consequently, the power flux density at a distance of 200 km will make

$$\mathrm{PD}_2 = \frac{2^2}{\left(200 \times 10^3\right)^2} 1.5 \times 10^{-3} = 1.5 \times 10^{-13}\ \mathrm{W/m^2}.$$

The effective aperture of the PhAA is

$$A = \frac{G_\mathrm{n}\lambda^2}{4\pi} = 1.3\ \mathrm{m}^2.$$

The signal processing loss in the receiver and subsequent circuitry may be assumed to equal 5 dB, which yields a loss factor $\eta \approx 0.3$. Thus, the signal received from the sounder at a distance of 200 km is

$$P_\mathrm{r} = \mathrm{PD}_2\, A\eta = 5.8 \times 10^{-14}\ \mathrm{W}.$$

Let us estimate the receiver noise power referred to its input in the band 2×10^6 Hz. The calculated receiver noise figure (taking account of input circuits, LNAs in antenna modules and subsequent circuitry) is approximately equal to 3 dB, i.e. 2 per unit. The noise power is

$$P_\mathrm{n} = (N_\mathrm{n} - 1)kT\Delta f_\mathrm{r} = \left(4 \times 10^{-21}\right)\left(2 \times 10^6\right) = 8 \times 10^{-15}.$$

The signal-to-noise ratio at a distance of 200 km is

$$\frac{P_\mathrm{r}}{P_\mathrm{n}} = \frac{5.8 \times 10^{-14}}{0.8 \times 10^{-14}} = 7.25 \rightarrow 8.5\ \mathrm{dB},$$

which, as practice shows, is quite sufficient for stable tracking of the sounder in angular position, range and frequency.

Abbreviations

AD	amplitude detector
ADC	analogue-to-digital converter
AFC	amplitude–frequency characteristic
AFR	amplitude–frequency response
AGC	automatic gain control
AP	antenna pattern
DAC	digital-to-analogue converter
DCA	direct current amplifier
DFT	discrete Fourier transform
ED	envelope detector
ESL	equisignal line
FFT	fast Fourier transform
HPF	highpass filter
IDT	interdigital transducer
LFM	linear frequency modulation
LNA	low-noise amplifier
LPF	lowpass filter
MTD	moving target detector
MTI	moving target indicator
PD	power density
PhAA	phased array antenna
PhCS	phase-coded signal
PhCW	phase-coded waveform
PhD	phase detector
RAM	random-access memory
RCS	radar cross-section

Bistatic Radar: Principles and Practice Edited by M. Cherniakov

RF	radio frequency
RVG	reference voltage generator
SAR	synthetic aperture radar
SAW	surface acoustic wave
SNR	signal-to-noise ratio

Variables

A	antenna active opening, active aperture of the antenna
B	signal base
B_n	noise band
C	radio propagation velocity
E	radar power budget
f_d	Doppler shift
f_{int}	intermediate frequency
F_n	noise factor
F_p	pulse repetition frequency
G	antenna gain
K	Boltzmann constant
K_0	transfer coefficient of the first-order tracker
$K(\omega)$	amplitude–frequency response
$\overline{K(\omega)}$	complex frequency response
L_A	antenna length
L_s	aperture of SAR
L_t	target length
$\overline{\overline{p}}$	Laplacian
P_d	detection probability
P_{fa}	false alarm probability
P_t	transmitter power
R_{max}	coverage range
T_p	pulse repetition period
T_p/τ_p	duty circle
U_{dif}	difference signal
U_{sum}	sum signal
α	general parameter of the output signal; attenuation coefficient
α_0	general parameter of the input signal

Bistatic Radar: Principles and Practice Edited by M. Cherniakov

β	angle in the horizontal plane
Δf	frequency resolution
ΔR	range resolution
Δt	time resolution
ΔV	velocity resolution
ΔW	volume resolution
$\Delta\alpha$	misadjustment $= \alpha_0 - \alpha$
$\Delta\beta$, $\Delta\varepsilon$	angular resolutions
ε	angle in the vertical plane
θ	antenna beamwidth
$\theta_{\mathrm{t.w.}}$	two-way antenna beamwidth
λ	wavelength
ν	isolation between the transmitter and the receiver
ρ	signal-to-noise ratio
$\rho_{\min}$	required signal-to-noise ratio
σ	target RCS value
τ_{c}	correlation window
τ_{p}	pulse duration
τ_0	discrete of PhCW
$\Psi(\tau, F)$	signal ambiguity function
ω	frequency (rad/s)
ω_{n}	normalized frequency (rad/s)
Ω	discrete frequency (rad/s)
Ω_{sc}	scanning frequency
Ω_{t}	rate of target aspect angle variation

Acknowledgements

Many thanks to Dr Victor N. Djigan for his contribution to Section 7.5, to Mr Alexander Maslennikov who translated Chapters 1 to 8 into English and to Ms Janna Chirkunova who performed the computer makeup.

Part II

Bistatic Radars

V.I. Kostylev

9

Different Types of Radar Systems

In the previous part the theory of monostatic radar was presented. This follows a similar pattern to many other publications (e.g. in References [II.1] and [II.2]), and is based on the fundamental assumption that the transmitter and receiver are spatially combined. However, other types of radar are also known, such as *bistatic radars*, *multistatic radars* [II.3] and *multisite radars* [II.4], for which the condition of transmitter and receiver co-location is not required. In particular, the configuration of bistatic radar assumes the presence of one transmitting station and one receiving station, separated in space (see Figure 9.1). Selected problems of bistatic radar theory will be stated in this part.

In the case of bistatic radar, the transmitter, T, receiver, R, and target, Γ, form a triangle which is shown in Figure 9.2, and is known as the *bistatic triangle*. The bistatic triangle lies in the *bistatic plane*. The sides of the triangle will be termed the *bistatic baseline* (the side between the transmitter and the receiver) TR, the *transmitter line segment* (the side between the transmitter and the target) TΓ and the *receiver line segment* (side between the receiver and the target) RΓ.

The transmitter and the receiver line segments subtend an angle, β, which is called the *bistatic angle*. The limits of this angle are from zero up to 180°: $0° \leq \beta \leq 180°$. In bistatic radar theory, the bistatic angle is a central geometrical characteristic: it appears in a number of formulae related to the bistatic radar. Sometimes a bistatic angle is also known as the cut angle or the scattering angle [II.5]. The line that bisects the bistatic angle is known as the *bistatic bisector*. The bistatic angle, as well as the bistatic bisector, are shown in Figure 9.2.

Depending on the size of the bistatic angle, the radar uses the reflection (at $\beta < 90°$) or diffraction (at $90° < \beta \leq 180°$) phenomenon. In the broadest sense, radar, using the phenomenon of the reflection of an electromagnetic wave by a target, may be termed a *back-scattering radar*, while radar, based on the use of diffraction of an electromagnetic wave on a target, may be termed *forward-scattering radar*. In more detail, forward-scattering radar is one that is specially designed for operation with bistatic angles close to 180°.

Another important parameter related to the bistatic radar is the distance, L, between the transmitter and the receiver. This is usually called the *baseline range* or simply the *baseline* [II.5, II.6]. Thus the term 'baseline' designates both a line segment and a quantity. Formally, in the case of a monostatic radar $L = 0$ and, conversely, $L \neq 0$ represents the bistatic radar in the widest meaning. A narrower standpoint

Bistatic Radar: Principles and Practice Edited by M. Cherniakov

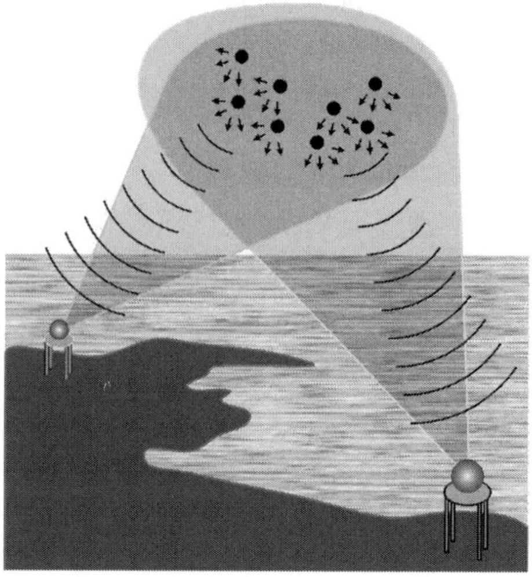

Figure 9.1 Bistatic radar

[II.7] assumes that in a case of a bistatic radar the bistatic base should considerably exceed the potential error of target range estimation by means of the radar.

From a system theory point of view, the bistatic radar may be considered as an elementary cell of a multistatic radar system [II.4]. However, it is also conclusive that the bistatic radar has an independent importance of its own. Historically, the first radars were not monostatic, but bistatic [II.7]. Only after the substantiation of a pulse method of radar and the development of antenna switches more convenient monostatic radars, where the receiving and transmitting facilities use the same antenna or adjacent antennas, were developed. Now, due to a sharp increase in remote sensing system performance requirements,

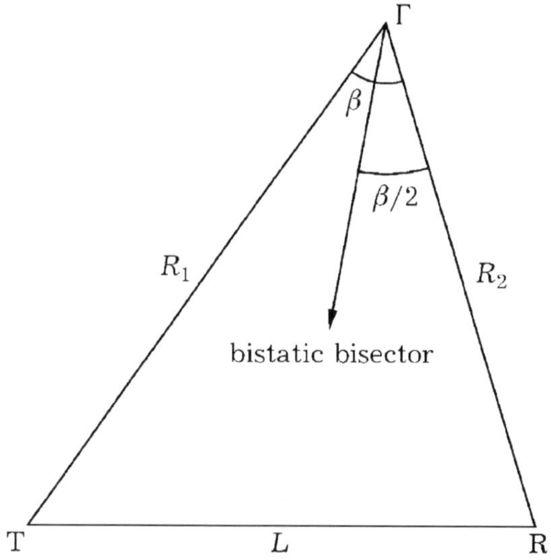

Figure 9.2 Bistatic triangle

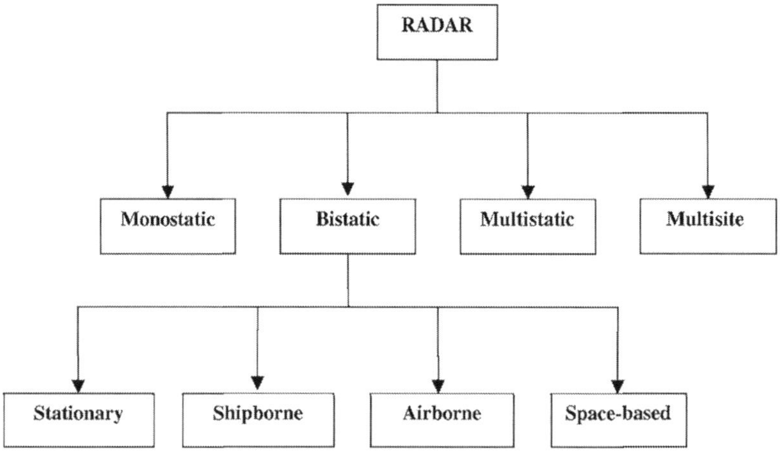

Figure 9.3 Radar classification

the radar community is revising the potential of the bistatic radar and is returning to systems where the receiving and transmitting facilities are spatially carried out at a significant distance (bistatic base).

The difference between bistatic radars, on the one hand, and both multistatic and multisite radars, on the other, will now be explained. The term 'multistatic radar' usually designates a single radar with, as a rule, one transmitting and several spatially distributed receiving stations [II.6] with joint processing of received information. Joint processing here is a paramount condition. Chernyak [II.4] introduces a wide definition of the multisite radar system. According to his work, a multisite radar system is 'a radar system including several spatially separated transmitting, receiving and (or) transmitting–receiving facilities where information of each target from all sensors are fused and jointly processed'. It follows from the above definition that a bistatic radar with spatially separated single transmitting and single receiving sites does not belong to multistatic radars or multisite radars, but requires its own class of radar system.

The bistatic radar may be stationary (sometimes stationary bistatic radars are called 'bistatic radars with a stationary baseline'), shipborne, airborne and space-based (Figure 9.3). This part is devoted to stationary bistatic radars only. It is important that the transmitter and receiver locations for a stationary bistatic radar are not changed during the operation process; hence $L = constant$.

It is clear that stationary, shipborne, airborne and space-based bistatic radars all differ in design, construction, technical characteristics, opportunities and advantages, and all have advantages and limitations. It is known that stationary bistatic radars were proposed, investigated, constructed and built earlier than other types of radar.

In certain publications (e.g. References [II.6], [II.7] and [II.8]) the baseline is considered to be one of the main distinctive features of a bistatic or multistatic radar. Of course, the bistatic baseline strongly influences the characteristics of the bistatic radar. However, the bistatic baseline is a continuous value and it is difficult to set up an exact boundary so that the transition through it will lead to abrupt changes in features and characteristics of the bistatic radar. Moreover, some important characteristics, such as the accuracy of the target angle location's measurement, depend not only upon the bistatic baseline itself but also upon the so-called effective bistatic base. According to Reference [II.4] the effective bistatic base, L_e, is the length of the bistatic baseline's projection on the plane, which is orthogonal to the bisector of the bistatic angle. It is obvious that an effective bistatic base depends not only on the configuration of the radar but also on the target location, and consequently is not characterized only by the radar. At the same time, it is always implicitly supposed that the target is located in the region of the overlapping of the main lobes of the receiving and transmitting antenna directivity diagrams. Therefore, the range of possible values of the bistatic angle can be known beforehand, and is not large. In this sense, bistatic radars may be classified by the size of their bistatic angle into three types: *quasi-monostatic* radars, actual *bistatic* radars and *shadow* radars. Quasi-monostatic radars operate with small bistatic angles and shadow radars have a bistatic angle equal or close to 180°. It is obvious, that the shadow radar is a forward-scattering radar in the narrow sense of the term.

Nevertheless, it is sometimes convenient in practical work to divide bistatic radars roughly into two classes: those with a 'short' baseline ('small' base) and those with a 'long' baseline ('big' base) [II.4]. The first class includes bistatic radars whose effective bases are essentially less than the expected transmitter line segments and receiver line segments. On the contrary, if effective bases are of the same order or greater than expected transmitter line segments and receiver line segments, such bistatic radars may be considered as bistatic radars with a long baseline [II.4]. It seems that quasi-monostatic radars always have small bases.

The hardware and software of bistatic radars with a short baseline are simpler and, as a rule, cheaper [II.4]. Geometrical relationships between signal parameters and target coordinates, as well as most processing algorithms, are also simpler. However, bistatic radars with big bases can produce higher accuracy of target coordinate and track parameter estimations.

Bistatic radars have a set of winning properties that attract interest in them. Essential in this is the fact that the receiver does not itself radiate energy. In this the aspect the bistatic radar is similar to the passive radar. Hence the receiver can be used covertly [II.9], an important feature for military applications. The specified reasons have resulted in the necessity to classify bistatic radars as cooperative and noncooperative. If transmitting and receiving stations are designed and constructed for joint functioning as a bistatic radar and also have the built-in means of synchronization, they may certainly be classed as cooperative. In the case of a noncooperative bistatic radar, a receiver is built to use another source of illumination – another radar or any other source of electromagnetic radiation. This class of bistatic radars is termed 'noncooperative' (or 'hitch-hike' [II.9]), because the illuminator is not built specifically to support bistatic radar operation. Unfortunately, the noncooperative approach, despite its initial attraction, has some obvious limitations: the waveform may be far from optimal and its parameters may not be known at the receiver.

10

Scattering Fundamentals

The widely known phenomenon of electromagnetic wave scattering is fundamental for bistatic radar. The purpose of this chapter is to review some of the basic concepts and results of electromagnetic wave scattering theory. The bases of electromagnetism are stated and some key bistatic radar concepts, such as cross-sections and scattering matrices, are presented in this chapter.

10.1 SOME BASIC CONCEPTS FROM ELECTROMAGNETIC THEORY

The electromagnetic field in any medium at any given moment of time is determined by four vectors: the intensity of the electrical field, **E**, and the electrical flux density, **D**, which characterize the electrical field; the intensity of the magnetic field, **H**, and the magnetic flux density, **B**, which characterize the magnetic field.

Sources of electromagnetic fields are charges and currents, which are commonly described by two physical values: volumetric density of a charge, ρ (charge per unit of volume), and density of a current, **J** (amount of charge per unit of time flowing across a surface normal to **J** divided by the area of the surface). Density of a current and density of a charge are connected by the equation of continuity (under the condition of the absence of macroscopic motion of matter[1]):

$$\operatorname{div}\mathbf{J}+\frac{\partial \rho}{\partial t}=0. \tag{10.1}$$

This is a mathematical formulation of the statement that the current of conductivity is caused by the motion of free electrons, which submit to the law of conservation of electrical charge.

All six values introduced above depend on three spatial coordinates, h, v and q (e.g. Cartesian coordinates x, y and z, spherical coordinates r, θ and ϕ, and others), and time t:

$$\begin{aligned} &\mathbf{E}=\mathbf{E}(\boldsymbol{r},t), \quad &&\mathbf{H}=\mathbf{H}(\boldsymbol{r},t), \quad &&\mathbf{J}=\mathbf{J}(\boldsymbol{r},t),\\ &\mathbf{D}=\mathbf{D}(\boldsymbol{r},t), \quad &&\mathbf{B}=\mathbf{B}(\boldsymbol{r},t), \quad &&\rho=\rho(\boldsymbol{r},t), \end{aligned} \tag{10.2}$$

[1]Currents of convection, caused, for example, by the motion of electrons in electrovacuum devices remain far outside the scope of this book.

Bistatic Radar: Principles and Practice Edited by M. Cherniakov

where

$$\boldsymbol{r} = h\boldsymbol{e}_h + v\boldsymbol{e}_v + q\boldsymbol{e}_q \tag{10.3}$$

is the radius vector of a given point in space and bases $\boldsymbol{e}_h$, $\boldsymbol{e}_v$ and $\boldsymbol{e}_q$ are unit vectors in the positive directions of coordinates h, v and q. Physical values given in Equations (10.2) are related by Maxwell's system of equations, the structure of which essentially depends on the used system of physical value units. In particular, in the international system SI, Maxwell's equations are as follows:

$$\operatorname{rot}\mathbf{H} = \mathbf{J} + \frac{\partial \mathbf{D}}{\partial t}, \tag{10.4}$$

$$\operatorname{div}\mathbf{D} = \rho, \tag{10.5}$$

$$\operatorname{rot}\mathbf{E} = -\frac{\partial \mathbf{B}}{\partial t}, \tag{10.6}$$

$$\operatorname{div}\mathbf{B} = 0. \tag{10.7}$$

Maxwell's equations represent a generalization of the experimental data. For example, Equation (10.6) is a mathematical formulation (in a differential form) of Faraday's law of electromagnetic induction.

In the theory of electromagnetism the uniqueness theorem is accepted. This asserts that where Maxwell's equations are applied, only one solution is possible. The importance of this theorem is clear: if a solution of an electrodynamics problem is found, by whatever means, the uniqueness theorem guarantees that it is not necessary to search for further results.

The attentive reader will notice that Maxwell's equations are not sufficient to reach a concrete solution to an electrodynamics problem, because the number of equations is less than the amount of the unknown physical quantities. In reality, even at the given sources of a field (given $\rho(\boldsymbol{r}, t)$ and $\mathbf{J}(\boldsymbol{r}, t)$), definitions of the four vector values **E**, **D**, **B** and **H** need four vector equations, while Equations (10.5) and (10.7) are scalar. Thus, it is necessary to add the system of constitutive relations to Maxwell's system, in which the influence of the physical medium on the electromagnetic phenomena would be taken into account. The elementary material equations are

$$\mathbf{D} = \varepsilon\mathbf{E}, \qquad \mathbf{B} = \mu\mathbf{H}, \tag{10.8}$$

$$\mathbf{J} = \sigma\mathbf{E}, \tag{10.9}$$

where ε, μ and σ are the permittivity, the permeability and the active conductivity of the medium. Although the constitutive relations (10.8) and (10.9) cover a large enough class of physical mediums, it is necessary to note that some of the properties of some real mediums are not taken into account by these equations.

In this part consideration will be based on Equations (10.8) and (10.9) in the assumption that constitutive parameters of the medium, ε, μ and σ, are not time varying.

In unlimited homogeneous space, permittivity, ε, and permeability, μ, represent constants. It is also acceptable to attribute permittivity and permeability for free space, namely in vacuum $\varepsilon = \varepsilon_0 = 8.85 \times 10^{-12}$ F/m and $\mu = \mu_0 = 4\pi \times 10^{-7}$ H/m. In addition, the constants ε_0 and μ_0 are named as the dielectric constant and the magnetic constant. They are related to the speed of light in vacuum, c, by a simple formula

$$c\sqrt{\varepsilon_0\mu_0} = 1. \tag{10.10}$$

Generally, space is nonuniform and the specified characteristics of the medium are functions of spatial coordinates: $\varepsilon = \varepsilon(\boldsymbol{r})$ and $\mu = \mu(\boldsymbol{r})$. Thus, on the interface between two mediums, permittivity and permeability may have discontinuities. In the latter case, differential equations for an electromagnetic field should be complemented by *boundary conditions*.

In particular, these conditions on boundaries between two lossless dielectrics are

$$\begin{aligned} \boldsymbol{n} \times \mathbf{E}_1 &= \boldsymbol{n} \times \mathbf{E}_2, & \boldsymbol{n} \cdot \mathbf{D}_1 &= \boldsymbol{n} \cdot \mathbf{D}_2, \\ \boldsymbol{n} \times \mathbf{H}_1 &= \boldsymbol{n} \times \mathbf{H}_2, & \boldsymbol{n} \cdot \mathbf{B}_1 &= \boldsymbol{n} \cdot \mathbf{B}_2, \end{aligned} \tag{10.11}$$

where the unit vector, $\boldsymbol{n}$, is normal to the border and the subscripts 1 and 2 refer to mediums on the two sides of the scattering surface. The symbol $\times$ in formulae (10.11) (and others) symbolizes the vector product of the vectors and the point symbol represents the scalar product of the vectors.

The boundary conditions at the surface of a perfect conductor are

$$\begin{aligned} \boldsymbol{n} \times \mathbf{E} &= 0, & \boldsymbol{n} \cdot \mathbf{D} &= q_S, \\ \boldsymbol{n} \times \mathbf{H} &= \mathbf{J}_S, & \boldsymbol{n} \cdot \mathbf{B} &= 0, \end{aligned} \tag{10.12}$$

where **E**, **H**, **D** and **B** designate complete vector fields on the surface of metal, q_S is the surface charge density and $\mathbf{J}_S$ is the surface current density.[2]

Taking into account Equations (10.8) and (10.9), it is possible to present Maxwell's equations as

$$\operatorname{rot} \mathbf{H} = \sigma \mathbf{E} + \varepsilon \frac{\partial \mathbf{E}}{\partial t}, \tag{10.13}$$

$$\operatorname{div} (\varepsilon \mathbf{E}) = \rho, \tag{10.14}$$

$$\operatorname{rot} \mathbf{E} = -\mu \frac{\partial \mathbf{H}}{\partial t}, \tag{10.15}$$

$$\operatorname{div} (\mu \mathbf{H}) = 0. \tag{10.16}$$

It is important to note that, unlike Maxwell's primary equations, expressions (10.13) to (10.16) do not contain a magnetic flux density, **B**, electrical flux density, **D**, or density of a current, **J**.

Power and energy are always associated with electromagnetic field. The electromagnetic energy, W, contained in some volume, V, is expressed as an integral:

$$W = \frac{1}{2} \iiint\limits_V \left(\varepsilon E^2 + \mu H^2 \right) \mathrm{d}V. \tag{10.17}$$

Thus, electromagnetic energy can change in time at the expense of two mechanisms: (a) inside the given volume, V, the electromagnetic energy can turn into other forms of energy (e.g. in heat) or arise from the nonelectromagnetic forms of energy; (b) the energy, remaining electromagnetic, can flow from (or in) volume, V, through a surface, S, enclosing the specified volume. The energy flux, Ξ, from the given volume is expressed by the formula

$$\Xi = \oiint\limits_S (\boldsymbol{n} \cdot \mathbf{S}) \, \mathrm{d}S, \tag{10.18}$$

where $\boldsymbol{n}$ is the external normal vector to a surface, S, and the vector

$$\mathbf{S} = \mathbf{E} \times \mathbf{H} \tag{10.19}$$

is known as a Poynting vector. It is useful to consider a unit vector,

$$\boldsymbol{e}_S = \mathbf{S} / \left| \mathbf{S} \right|, \tag{10.20}$$

[2] This is the amount of electricity moving in unit time on a border undressed of metal and dielectric through a line, perpendicular to the direction of moving, divided by the length of this line.

which has the same direction as the Poynting vector, as it characterizes the direction of electromagnetic wave propagation.

In applied electrodynamics two auxiliary functions of coordinates and time, vector potential, $\mathbf{A}(\boldsymbol{r}, t)$, and scalar potential, $\Phi(\boldsymbol{r}, t)$, are introduced. They are defined by

$$\mathbf{H} = \operatorname{rot} \mathbf{A}, \tag{10.21}$$

$$\mathbf{E} = -\operatorname{grad} \Phi - \mu \frac{\partial \mathbf{A}}{\partial t}. \tag{10.22}$$

Usually the vector and scalar potentials satisfy an additional restriction:

$$\operatorname{div} \mathbf{A} + \varepsilon \frac{\partial \Phi}{\partial t} = 0, \tag{10.23}$$

known as a Lorenz condition. In this case, the vector and scalar potentials can be found by solving the following nonuniform wave equations:

$$\nabla^2 \mathbf{A} - \varepsilon\mu \frac{\partial^2 \mathbf{A}}{\partial t^2} = -\mathbf{J}, \tag{10.24}$$

$$\nabla^2 \Phi - \varepsilon\mu \frac{\partial^2 \Phi}{\partial t^2} = -\frac{\rho}{\varepsilon}, \tag{10.25}$$

which in the absence of free charges and currents pass into homogeneous equations. It is known that the system of equations (10.21) to (10.25) (sometimes known as the Maxwell's equation system in terms of potentials), together with Equations (10.8), can completely replace Maxwell's equations system (10.4) to (10.7) or (10.13) to (10.16).

Alongside the vector potential, $\mathbf{A}$, and the scalar potential, Φ, the so-called Hertz vector, $\mathbf{\Pi}$, can be defined. Knowing the Hertz vector, the vector and scalar potentials can be expressed by the following simple formulae:

$$\mathbf{A} = \frac{\partial \mathbf{\Pi}}{\partial t}, \qquad \Phi = -\frac{1}{\varepsilon} \operatorname{div} \mathbf{\Pi}. \tag{10.26}$$

However, for electromagnetic problems relevant to radar, the Hertz vector is rarely used.

Many formulae presented in this chapter become essentially simpler if a considered electromagnetic wave is monochromatic (time harmonic). The time structure of the electrical and magnetic fields of a monochromatic electromagnetic wave is determined only by the multiplier $\exp(-\mathrm{j}\omega t)$, where $\omega = 2\pi f$ is a circular frequency and f is a frequency. Thus the right sides of Equations (10.2) are factorized:

$$\begin{aligned} &\mathbf{E} = \mathbf{E}_0(\boldsymbol{r}) \exp(-\mathrm{j}\omega t), && \mathbf{H} = \mathbf{H}_0(\boldsymbol{r}) \exp(-\mathrm{j}\omega t), && \mathbf{J} = \mathbf{J}_0(\boldsymbol{r}) \exp(-\mathrm{j}\omega t), \\ &\mathbf{D} = \mathbf{D}_0(\boldsymbol{r}) \exp(-\mathrm{j}\omega t), && \mathbf{B} = \mathbf{B}_0(\boldsymbol{r}) \exp(-\mathrm{j}\omega t), && \rho = \rho_0(\boldsymbol{r}) \exp(-\mathrm{j}\omega t), \end{aligned} \tag{10.27}$$

and Maxwell's equations become a little simpler:

$$\begin{aligned} &\operatorname{rot}\mathbf{E} = \mathrm{j}\omega\mu\mathbf{H}, && \operatorname{div}(\varepsilon\mathbf{E}) = \rho, \\ &\operatorname{rot}\mathbf{H} = (\sigma - \mathrm{j}\omega\varepsilon)\mathbf{E}, && \operatorname{div}(\mu\mathbf{H}) = 0. \end{aligned} \tag{10.28}$$

In the solution of electromagnetic wave scattering problems for radar targets, we will always take as true, unless otherwise stipulated, that the medium where the electromagnetic wave is propagated is homogeneous, isotropic and free from charges (where only induced charges are possible on a scattering surface). Otherwise, we will believe that the scattering body is located in free space.

For free space the following simple relationships are useful:

$$\begin{aligned} &\mathbf{B} = \mu_0\mathbf{H}, && \mathbf{D} = \varepsilon_0\mathbf{E}, \\ &\sigma = 0, && \rho = 0. \end{aligned} \tag{10.29}$$

Formulae (10.28) can then be simplified:

$$\begin{aligned} \operatorname{rot}\mathbf{E} &= \mathrm{j}\omega\mu_0\mathbf{H}, \quad \operatorname{div}\mathbf{E} = 0, \\ \operatorname{rot}\mathbf{H} &= -\mathrm{j}\omega\varepsilon_0\mathbf{E}, \ \operatorname{div}\mathbf{H} = 0. \end{aligned} \tag{10.30}$$

Besides, it is known that if an electromagnetic wave is monochromatic, its vector and scalar potentials are also time harmonic, viz.

$$\mathbf{A} = \mathbf{A}_0(\boldsymbol{r})\exp(-\mathrm{j}\omega t), \qquad \Phi = \Phi_0(\boldsymbol{r})\exp(-\mathrm{j}\omega t). \tag{10.31}$$

In this case, the wave equations (10.24) and (10.25), determining the potentials, can be transformed to the form

$$\nabla^2\mathbf{A} + k^2\mathbf{A} = -\mathbf{J}, \tag{10.32}$$

$$\nabla^2\Phi + k^2\Phi = -\rho/\varepsilon, \tag{10.33}$$

where

$$k = \omega\sqrt{\varepsilon\mu} \tag{10.34}$$

is a *wave number*. In this case, Equation (10.26) becomes

$$\mathbf{A} = -\mathrm{j}\omega\mathbf{\Pi}, \qquad \Phi = -\frac{1}{\varepsilon}\operatorname{div}\mathbf{\Pi}. \tag{10.35}$$

The Hertz vector satisfies the homogeneous wave equation

$$\nabla^2\mathbf{\Pi} + k^2\mathbf{\Pi} = 0. \tag{10.36}$$

If any solution of Equation (10.36) is found, the vectors **E** and **H** can be defined using relations (10.35), (10.21) and (10.22), and, thereby, the electromagnetic process can be determined. At the same time, it follows from Equations (10.35) that for the monochromatic fields, the Hertz vector $\mathbf{\Pi}$ and vector potential **A** differ only by a constant multiplier and there is no special necessity for a Hertz vector.

Returning to the initial Maxwell equations, it can be seen that carrying out the curl operation for both parts of Equation (10.6) and taking into account Equation (10.4), in the absence of free charges ($\rho = 0$) the electric field intensity obeys the following homogeneous differential equation with separated variables:

$$\nabla^2\mathbf{E} - \mu\sigma\frac{\partial\mathbf{E}}{\partial t} - \varepsilon\mu\frac{\partial^2\mathbf{E}}{\partial t^2} = 0. \tag{10.37}$$

A similar equation can also be written for the intensity of magnetic field, **H**.

For a monochromatic wave, Equation (10.37) reduces to

$$\nabla^2\mathbf{E} + \omega^2\mu\left(\varepsilon + \mathrm{j}\frac{\sigma}{\omega}\right)\mathbf{E} = 0. \tag{10.38}$$

Equation (10.38) demonstrates the expediency of introducing the complex dielectric permittivity, $\dot{\varepsilon}$, viz.

$$\dot{\varepsilon} = \varepsilon + \mathrm{j}\frac{\sigma}{\omega}. \tag{10.39}$$

Typical data for the values appearing in formula (10.39) are: $\varepsilon \approx 10\varepsilon_0$ and $\sigma \approx 10^{-3}$ S/m for the ground surface and $\varepsilon \approx 80\varepsilon_0$ and $\sigma \approx 4$ S/m for sea water. Therefore, according to Reference [II.10], in radar practice (when $\lambda < 1$ m) it is always possible to neglect the imaginary part of the complex dielectric permittivity of the ground; however, for sea water a similar disregard is only possible if $\lambda < 1$ cm. On the contrary, for good conductors, in formula (10.39) the real part of dielectric permittivity, ε, can be neglected for practically all frequencies used in radar practice. Thus the complex dielectric permittivity of metal becomes purely imaginary:

$$\dot{\varepsilon} \simeq \mathrm{j}\frac{\sigma}{\omega}. \tag{10.40}$$

The concept of complex dielectric permittivity allows the generalization of formula (10.34) for the wave number, namely to enter the concept of the complex wave number:

$$\dot{k} = k_{\mathrm{R}} + \mathrm{j}k_{\mathrm{I}} = \omega\sqrt{\dot{\varepsilon}\mu}. \tag{10.41}$$

For the case of good conductors, the substitution of Equation (10.40) into (10.41) gives

$$\dot{k} = \frac{1+j}{d}. \tag{10.42}$$

Thus, for a conductor the real and imaginary parts of the complex wave number are identical. In Equation (10.42), parameter d has the dimension of length and is known as *skin depth*.

A rather important characteristic of a medium is the loss tangent, representing the ratio of imaginary and real parts of complex dielectric permittivity. As is readily appreciated, the loss tangent can be evaluated by the equation

$$\tan\delta = \frac{\sigma}{\omega\varepsilon}, \tag{10.43}$$

where the angle, δ, is the loss angle. It is most likely that, the name is connected with the presence of active conductivity, σ, which entails the heating of the dielectric medium and, hence, losses of electromagnetic energy.

Taking into account Equation (10.41), Equation (10.38) will correspond to the widely known homogeneous Helmholtz equation

$$\nabla^2\mathbf{E} + \dot{k}^2\mathbf{E} = 0. \tag{10.44}$$

It is possible to obtain a similar wave equation for the intensity of a magnetic field:

$$\nabla^2\mathbf{H} + \dot{k}^2\mathbf{H} = 0. \tag{10.45}$$

Obviously in the Cartesian coordinate system, the vector wave Helmholtz equation (10.44) (or (10.45)) may be divided into three scalar differential wave equations, also termed Helmholtz equations. The scalar Helmholtz equation generally looks like

$$\nabla^2\psi + \dot{k}^2\psi = 0, \tag{10.46}$$

where ψ is a component of either the electric or the magnetic field:

$$\begin{aligned} \nabla^2 E_x + \dot{k}^2 E_x &= 0,\\ \nabla^2 E_y + \dot{k}^2 E_y &= 0,\\ \nabla^2 E_z + \dot{k}^2 E_z &= 0,\\ \nabla^2 H_x + \dot{k}^2 H_x &= 0,\\ \nabla^2 H_y + \dot{k}^2 H_y &= 0,\\ \nabla^2 H_z + \dot{k}^2 H_z &= 0. \end{aligned} \tag{10.47}$$

One of the most commonly known solutions of wave equations are so-called *plane waves*. According to the definition, the plane electromagnetic wave is an electromagnetic field whose equiphase planes are infinite and parallel. If a plane wave is travelling in an unbounded medium in the z direction, the electric and magnetic fields can be expressed as

$$\mathbf{E} = \mathbf{E}_0(z)\exp(-\mathrm{j}\omega t), \qquad \mathbf{H} = \mathbf{H}_0(z)\exp(-\mathrm{j}\omega t). \tag{10.48}$$

Substituting Equation (10.48) into Maxwell's equations, it is possible to obtain

$$E_z = 0, \qquad H_z = 0. \tag{10.49}$$

This means that the plane electromagnetic waves are transverse waves. It is possible to see also that the Poynting vector of a plane wave has only the z component:

$$\mathbf{S}_x = 0, \qquad \mathbf{S}_y = 0, \qquad \mathbf{S}_z = \frac{1}{2}\left(E_x H_y^* - E_y H_x^*\right). \tag{10.50}$$

Thus, the direction of plane wave propagation in an unbounded media is along the z axis (either positive or negative). The expressions for the other components of electromagnetic field can be easily obtained from the Helmholtz equation:

$$\begin{aligned}
E_x &= A \exp\left[\mathrm{j}\left(\dot{k}z - \omega t\right)\right] + B \exp\left[-\mathrm{j}\left(\dot{k}z + \omega t\right)\right], \\
E_y &= C \exp\left[\mathrm{j}\left(\dot{k}z - \omega t\right)\right] + D \exp\left[-\mathrm{j}\left(\dot{k}z + \omega t\right)\right], \\
H_x &= \dot{\Upsilon}\left\{-C \exp\left[\mathrm{j}\left(\dot{k}z - \omega t\right)\right] + D \exp\left[-\mathrm{j}\left(\dot{k}z + \omega t\right)\right]\right\}, \\
H_y &= \dot{\Upsilon}\left\{A \exp\left[\mathrm{j}\left(\dot{k}z - \omega t\right)\right] - B \exp\left[-\mathrm{j}\left(\dot{k}z + \omega t\right)\right]\right\},
\end{aligned} \tag{10.51}$$

where

$$\dot{\Upsilon} = \sqrt{\frac{\dot{\varepsilon}}{\mu}} = \frac{\dot{k}}{\omega\mu} \tag{10.52}$$

is the wave admittance.

The constants A, B, C and D are determined by boundary conditions. For the wave travelling in a positive direction of the z axis, the coefficients B and D have to be equal to zero and Equations (10.51) reduce to

$$\begin{aligned}
E_x &= A \exp\left[\mathrm{j}\left(\dot{k}z - \omega t\right)\right] = A \exp\left(-k_\mathrm{I} z\right) \exp\left[\mathrm{j}\left(k_\mathrm{R} z - \omega t\right)\right], \\
E_y &= C \exp\left[\mathrm{j}\left(\dot{k}z - \omega t\right)\right] = C \exp\left(-k_\mathrm{I} z\right) \exp\left[\mathrm{j}\left(k_\mathrm{R} z - \omega t\right)\right], \\
H_x &= -\dot{\Upsilon} C \exp\left[\mathrm{j}\left(\dot{k}z - \omega t\right)\right] = -\dot{\Upsilon} C \exp\left(-k_\mathrm{I} z\right) \exp\left[\mathrm{j}\left(k_\mathrm{R} z - \omega t\right)\right], \\
H_y &= \dot{\Upsilon} A \exp\left[\mathrm{j}\left(\dot{k}z - \omega t\right)\right] = \dot{\Upsilon} A \exp\left(-k_\mathrm{I} z\right) \exp\left[\mathrm{j}\left(k_\mathrm{R} z - \omega t\right)\right].
\end{aligned} \tag{10.53}$$

Equations (10.50) for such a wave become

$$\mathbf{S}_z = \frac{A^2 + C^2}{2} \dot{\Upsilon}^* \exp\left(-2k_\mathrm{I} z\right). \tag{10.54}$$

Thus, at $B = D = 0$ there is an attenuating plane wave, which propagates in a positive direction of axis OZ. On the contrary, having put $A = C = 0$ into Equations (10.51), an attenuating plane wave propagating in a negative direction of axis OZ is obtained.

In free space (in vacuum) the wave number is real: $\dot{k} \equiv k_0 = \omega\sqrt{\varepsilon_0 \mu_0} = \omega/c = 2\pi/\lambda_0$, where λ_0 is the wavelength. Hence, $k_\mathrm{I} = 0$ and the electromagnetic plane wave in vacuum does not attenuate.

Applying the variable separation method for Equation (10.46), the general solution of the Helmholz equation can be written in the following form:

$$\psi = A \exp\left\{\mathrm{j}\left[\dot{k}\left(\alpha x + \beta y + \gamma z\right) - \omega t\right]\right\}. \tag{10.55}$$

If parameters A, α, β, and γ are real, constant and also

$$\alpha^2 + \beta^2 + \gamma^2 = 1, \tag{10.56}$$

the wave in Equation (10.55) is named a 'generalized plane wave'.

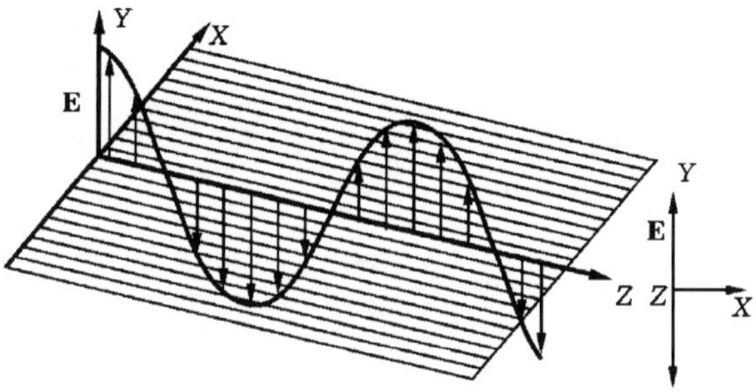

Figure 10.1 Illustration of linear polarization

In truth, it is possible to consider the parameters α, β and γ as direction cosines of a new axis OZ', along which the plane wave propagates:

$$\psi = A \exp\left[\mathrm{j}\left(\dot{k}z' - \omega t\right)\right]. \tag{10.57}$$

Simultaneously, parameters α, β and γ are the direction cosines of a Poynting vector, $\mathbf{S}$, i.e. $\boldsymbol{e}_{\mathbf{S}} = \mathbf{e}_{z'}$.

Taking into account Equation (10.20), Equations (10.55) and (10.57) can be transformed into the following formula:

$$\psi(\boldsymbol{r}, t) = A \exp\left[\mathrm{j}\left(\dot{k}\boldsymbol{e}_{\mathbf{S}} \cdot \boldsymbol{r} - \omega t\right)\right]. \tag{10.58}$$

Equation (10.58) represents the most general formula of a scalar plane wave. It describes any Cartesian component of vectors of intensity **E** or **H** in a plane wave.

As already noted, plane waves are transverse waves: they have no longitudinal components in the direction of propagation. Besides, plane waves have a plane wave front of an infinite size. Therefore, it is clear that plane waves cannot exist in a practical sense; the plane wave is an idealized model.

A key characteristic of a plane electromagnetic wave is its polarization, i.e. the orientation of components of electrical, **E**, or magnetic, **H**, fields intensity in a plane, which is perpendicular to the direction of wave propagation. For clarity, a nonattenuating plane wave will be considered:

$$\mathbf{E} = \mathbf{E}_0 \exp\left[\mathrm{j}\left(k_0 z - \omega t\right)\right], \tag{10.59}$$

propagating in vacuum along the axis *OZ*, as shown in Figure 10.1. If the vector of an electrical field constantly lies in one plane with an axis *OZ*, for example, lying in plane *YOZ*, the wave is referred to as 'linearly polarized'. This indicates that the projection of the hodograph of an electrical field intensity vector to plane *XOY* represents the segment of a direct line. In another case of 'elliptic polarization', the specified projection represents an ellipse, as shown in Figure 10.2. Thus, if the projection

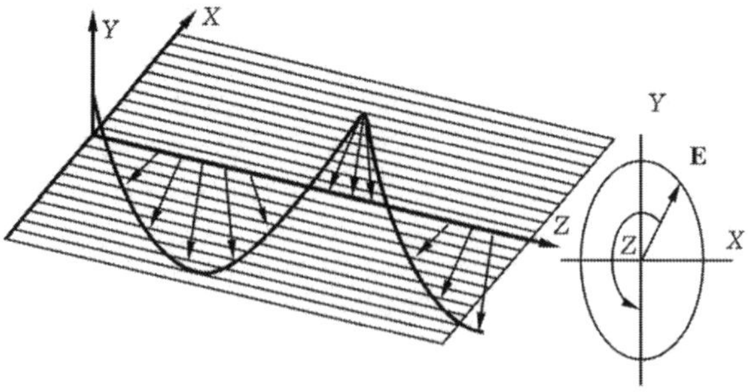

Figure 10.2 Illustration of elliptical polarization

of vector **E** on plane *XOY* moves clockwise, as observed along the direction of propagation (i.e. in a direction of axis *OZ*), it is referred to as 'right elliptic polarization'. Otherwise polarization is referred to as 'left'. Thus, as shown in Figure 10.2, the illustrative example corresponds to the case of right polarization.

In the specific case when the ellipse degenerates into a circle, polarization is termed as 'circular'. It is known that the elliptical polarized wave can be submitted as a linear combination of the two circular polarized waves.

An arbitrary linearly polarized wave with an electric field vector in the *XOY* plane can be split into two waves with orthogonal polarization:

$$\mathbf{E}_0 = \mathbf{E}_{0x} + \mathbf{E}_{0y} = E_{0x}\boldsymbol{e}_x + E_{0y}\boldsymbol{e}_y, \tag{10.60}$$

where $\boldsymbol{e}_x$ and $\boldsymbol{e}_y$ are unit orts of coordinate axes, *OX* and *OY*. Equation (10.60), written in a rectangular Cartesian system of coordinates, supposes a simple generalization in the case of the curvilinear system of coordinates:

$$\mathbf{E}_0 = \mathbf{E}_{0h} + \mathbf{E}_{0v} = E_{0h}\boldsymbol{e}_h + E_{0v}\boldsymbol{e}_v = \begin{pmatrix} \boldsymbol{e}_h & \boldsymbol{e}_v \end{pmatrix} \begin{pmatrix} E_{0h} \\ E_{0v} \end{pmatrix} = \begin{pmatrix} \boldsymbol{e}_h & \boldsymbol{e}_v \end{pmatrix} \mathsf{E}. \tag{10.61}$$

In Equation (10.61) (see also Equation (10.3)), complex orts $\boldsymbol{e}_h$ and $\boldsymbol{e}_v$ represent unit vectors of coordinates h and v. They form a polarizing basis. It is supposed that orts $\boldsymbol{e}_h$ and $\boldsymbol{e}_v$ are orthonormal and lie on a flat surface which is parallel to the front of a plane wave. The coordinates h and v can be, for example, Cartesian coordinates x and y, as in Equation (10.60), or angular coordinates of the spherical system, θ and ϕ, as in Reference [II.2].

The particular choice of vectors $\boldsymbol{e}_h$ and $\boldsymbol{e}_v$ in the horizontal (H) and vertical (V) directions is widely used. The vertical direction (direction of the ort $\boldsymbol{e}_\mathrm{V}$) is in a direction normal to the Earth's surface and the horizontal direction (direction of the ort $\boldsymbol{e}_\mathrm{H}$) is tangential to the surface of the Earth. Thus, certainly, the condition of a mutual orthogonality of vectors $\boldsymbol{e}_\mathrm{H}$, $\boldsymbol{e}_\mathrm{V}$ and $\boldsymbol{e}_\mathrm{S}$ should take place.

It should also be noted that appearing in Equation (10.61) the vector column

$$\mathsf{E} = \begin{pmatrix} E_{0h} \\ E_{0v} \end{pmatrix}$$

is known in the scientific literature as the 'Jones vector' [II.11]. It is obvious that if $E_{0h} \neq 0$, formula (10.61) can be rewritten as

$$\mathbf{E}_0 = E_{0h}\left(\boldsymbol{e}_h + q\boldsymbol{e}_v\right) = E_{0h}\begin{pmatrix} \boldsymbol{e}_h & \boldsymbol{e}_v \end{pmatrix}\begin{pmatrix} 1 \\ q \end{pmatrix}. \tag{10.62}$$

The parameter

$$q = E_{0v}/E_{0h} \tag{10.63}$$

is named as the 'polarizing relation' for the intensity of an electrical field. It is obvious that the concept of the polarizing relation can also be entered for the intensity of a magnetic field.

Sometimes, instead of q, the angle value $\gamma = \operatorname{atan} q$ is used, where atan is an inverse tangent. Another angular parameter, describing the polarization, ψ, which is the phase shift between components E_{0h} and E_{0v}, is also known. The main axes of a polarizing ellipse, large axis a and small axis b, also quantitatively characterize the polarization of a wave.

However, it is sometimes more convenient to use angular parameters instead, such as $\alpha = \operatorname{atan}(b/a)$ and inclination angle, δ, equal to the angle between the large axis of an ellipse, a, and ort, $\boldsymbol{e}_h$. Thus, it is clear that the set of parameters introduced has redundancy, since two parameters are enough to describe the polarization properties of a plane monochromatic wave illuminating the object.

In 1852 Stokes brought new parameters into scientific usage which now bear his name. Using our designations for an electrical field, the Stokes parameters can be defined as

$$I = |E_{0h}|^2 + |E_{0v}|^2, \qquad Q = |E_{0h}|^2 - |E_{0v}|^2,$$
$$U = 2\mathrm{Re}\left\{E_{0h}E_{0v}^*\right\}, \qquad V = 2\mathrm{Im}\left\{E_{0h}E_{0v}^*\right\}. \tag{10.64}$$

It is known that four Stokes parameters are satisfied by a simple formula:

$$I^2 = Q^2 + U^2 + V^2, \tag{10.65}$$

which follows from definitions (10.64). Besides

$$Q = I\cos(2\alpha)\cos(2\delta), \qquad U = I\cos(2\alpha)\sin(2\delta), \qquad V = I\sin(2\alpha). \tag{10.66}$$

From Equations (10.66) it is obvious that it is possible to enter into consideration vector **I**; the Stokes parameter, I, becomes equal to the module of a vector **I**, and the parameters Q, U and V coincide with rectangular Cartesian projections of vector **I**. Logically vector **I** is known as a 'Stokes vector'.

The polarizing conditions of an electromagnetic wave are evidently displayed in the spatial diagram shown in Figure 10.3, and known as the 'Poincare sphere'. The radius of the Poincare sphere is unity. Each wave polarization state is represented by a single point on the surface of the Poincare sphere. The Cartesian coordinates of this point are the normalized Stokes parameters, Q/I, U/I and V/I, and the spherical coordinates (longitude and latitude) coincide with parameters 2α and 2δ. Thus the longitude, 2δ, is counted from an initial diameter, hv, which lies in an equatorial plane of the Poincare sphere, and latitude, 2α, from this plane.

All points on the Poincare sphere equator are in mutual-unequivocal conformity with the linearly polarized waves. In particular, point h represents a wave with linear polarization in a horizontal direction, and the opposite point, v, displays a wave with linear polarization in a vertical direction. The points of the Poincare sphere that are not lying on the equator or at the poles correspond to waves with no degenerate elliptic polarization. Thus the points of the top hemisphere correspond to a case of right polarization and

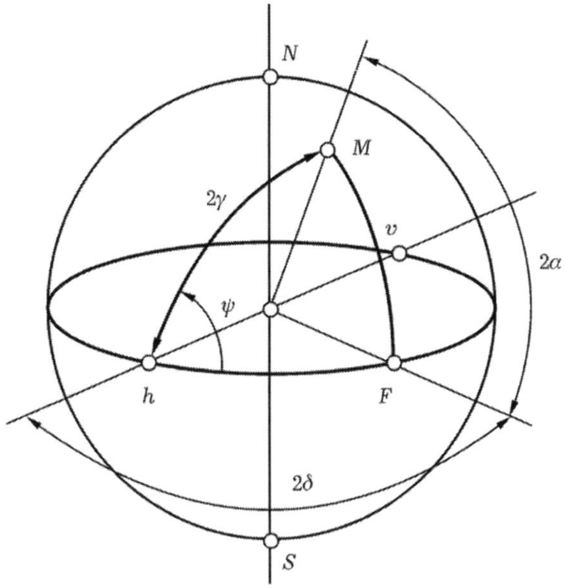

Figure 10.3 The Poincare sphere

the points of the bottom hemisphere correspond to a case of left polarization. The poles of the Poincare sphere, N and S, display the circular polarized waves with various directions of rotation of a vector of an electric field.

10.2 PLANE WAVE INCIDENCE ON A SMOOTH, FLAT INTERFACE BETWEEN TWO MEDIUMS

In the previous section we have given a sufficiently detailed description of a plane electromagnetic wave. Let us now consider the incidence of this wave on a flat interface between two infinite homogeneous mediums. Let permittivity and permeability of the first medium be ε_1 and μ_1, but for the second medium be ε_2 and μ_2 (see Figure 10.4). The direction of the axis, OZ, is specified as being along a normal to the interface into the second medium. The axis OZ' specifies the direction of the incident wave (10.57) propagation, i.e. let $\boldsymbol{e}_z = \boldsymbol{n}$ and $\boldsymbol{e}_\mathrm{S} = \boldsymbol{e}_{z'}$. The plane ZOZ' is referred to as the plane of incidence and the angle between axes OZ and OZ' is referred to as the angle of incidence, θ_1.

In the considered task it is of essential importance that there is polarization of an incident wave. The polarizing basis $\boldsymbol{e}_\parallel$ and $\boldsymbol{e}_\perp$ will be chosen so that ort $\boldsymbol{e}_\parallel$ lies in a plane of incidence. Thus, certainly, ort $\boldsymbol{e}_\perp$ is perpendicular to both the plane of incidence and to the direction of wave propagation. Then

$$\mathbf{E}_0^{\mathrm{in}} = \mathbf{E}_{0\parallel}^{\mathrm{in}} + \mathbf{E}_{0\perp}^{\mathrm{in}} = E_{0\parallel}^{\mathrm{in}}\boldsymbol{e}_\parallel + E_{0\perp}^{\mathrm{in}}\boldsymbol{e}_\perp, \qquad \mathbf{H}_0^{\mathrm{in}} = \mathbf{H}_{0\parallel}^{\mathrm{in}} + \mathbf{H}_{0\perp}^{\mathrm{in}} = H_{0\parallel}^{\mathrm{in}}\boldsymbol{e}_\parallel + H_{0\perp}^{\mathrm{in}}\boldsymbol{e}_\perp. \tag{10.67}$$

Similar equations can be written down for reflected waves as

$$\mathbf{E}_0^{(1)} = \mathbf{E}_{0\parallel}^{(1)} + \mathbf{E}_{0\perp}^{(1)} = E_{0\parallel}^{(1)}\boldsymbol{e}_\parallel + E_{0\perp}^{(1)}\boldsymbol{e}_\perp, \qquad \mathbf{H}_0^{(1)} = \mathbf{H}_{0\parallel}^{(1)} + \mathbf{H}_{0\perp}^{(1)} = H_{0\parallel}^{(1)}\boldsymbol{e}_\parallel + H_{0\perp}^{(1)}\boldsymbol{e}_\perp \tag{10.68}$$

and refracted waves as

$$\mathbf{E}_0^{(2)} = \mathbf{E}_{0\parallel}^{(2)} + \mathbf{E}_{0\perp}^{(2)} = E_{0\parallel}^{(2)}\boldsymbol{e}_\parallel + E_{0\perp}^{(2)}\boldsymbol{e}_\perp, \qquad \mathbf{H}_0^{(2)} = \mathbf{H}_{0\parallel}^{(2)} + \mathbf{H}_{0\perp}^{(2)} = H_{0\parallel}^{(2)}\boldsymbol{e}_\parallel + H_{0\perp}^{(2)}\boldsymbol{e}_\perp. \tag{10.69}$$

It is known that the appropriate components are proportional to each other and the factors of proportionality carry the name of the great French physicist Fresnel. The Fresnel coefficients can be

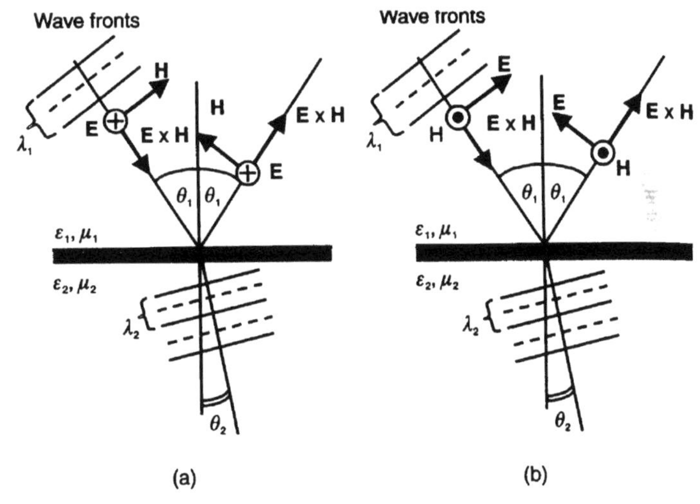

Figure 10.4 Reflection and refraction at an interface of two infinite dielectric layers for two polarizations of an electric field: (a) perpendicular to the plane of incidence and (b) parallel to it. (From Reference [II.12]. Reproduced by permission of Artech House)

expressed through trigonometric functions as follows:

$$R_\perp = \frac{E_{0\perp}^{(1)}}{E_{0\perp}^{\text{in}}} = \frac{\sin(\theta_1 - \theta_2)}{\sin(\theta_1 + \theta_2)}, \tag{10.70}$$

$$R_\parallel = \frac{E_{0\parallel}^{(1)}}{E_{0\parallel}^{\text{in}}} = \frac{\tan(\theta_1 - \theta_2)}{\tan(\theta_1 + \theta_2)}, \tag{10.71}$$

$$T_\perp = \frac{E_{0\perp}^{(2)}}{E_{0\perp}^{\text{in}}} = \frac{2\sin\theta_2\cos\theta_1}{\sin(\theta_1 + \theta_2)}, \tag{10.72}$$

$$T_\parallel = \frac{E_{0\parallel}^{(2)}}{E_{0\parallel}^{\text{in}}} = \frac{2\sin\theta_2\cos\theta_1}{\sin(\theta_1 + \theta_2)\cos(\theta_1 - \theta_2)}, \tag{10.73}$$

where θ_2 is an angle of refraction determined by Snell's law,

$$\sqrt{\varepsilon_1\mu_1}\sin\theta_1 = \sqrt{\varepsilon_2\mu_2}\sin\theta_2. \tag{10.74}$$

The Fresnel coefficient for the intensity of a magnetic field can easily be obtained from Equations (10.70) to (10.73) in view of the general relationship $\mathbf{H}_0 = \Upsilon \boldsymbol{e}_{\mathbf{S}} \times \mathbf{E}_0$, having a place in both the case of an incident wave and the case of reflected or refracted waves.

It is clear that if $\theta_1 + \theta_2 = \pi/2$, then from Equation (10.74) it follows that

$$\tan\theta_1 = \sqrt{\frac{\varepsilon_2\mu_2}{\varepsilon_1\mu_1}}. \tag{10.75}$$

The angle $\theta_{1\text{Br}}$ satisfying Equation (10.75), is known as a 'Brewster angle'. From (10.71) it is easy to show that if the angle of incidence of a plane electromagnetic wave on a flat interface between two mediums is equal to the Brewster angle, then $E_{0\parallel}^{(1)} = 0$. It is possible to show, that in this case, the condition $H_{0\perp}^{(1)} = 0$ also appears. Substituting the last two equations in (10.68) gives

$$\mathbf{E}_0^{(1)} = \mathbf{E}_{0\perp}^{(1)} = E_{0\perp}^{(1)}\boldsymbol{e}_\perp, \qquad \mathbf{H}_0^{(1)} = \mathbf{H}_{0\parallel}^{(1)} = H_{0\parallel}^{(1)}\boldsymbol{e}_\parallel. \tag{10.76}$$

This means that the linear polarization of a reflected wave occurs when the angle of incidence is equal to a Brewster angle. Thus, the plane of polarization is perpendicular to the plane of incidence.

In view of Snell's law, it is possible to rewrite Equations (10.70) and (10.71) [II.12] as

$$R_\perp = \frac{\eta_\perp\cos\theta_1 - \sqrt{1 - \nu^2\sin^2\theta_1}}{\eta_\perp\cos\theta_1 + \sqrt{1 - \nu^2\sin^2\theta_1}}, \tag{10.77}$$

$$R_\parallel = -\frac{\eta_\parallel\cos\theta_1 - \sqrt{1 - \nu^2\sin^2\theta_1}}{\eta_\parallel\cos\theta_1 + \sqrt{1 - \nu^2\sin^2\theta_1}}, \tag{10.78}$$

where the designations

$$\eta_\perp = \frac{\Upsilon_1}{\Upsilon_2} = \sqrt{\frac{\varepsilon_1\mu_2}{\mu_1\varepsilon_2}}, \qquad \eta_\parallel = \frac{1}{\eta_\perp} = \frac{\Upsilon_2}{\Upsilon_1} = \sqrt{\frac{\varepsilon_2\mu_1}{\mu_2\varepsilon_1}}, \qquad \nu = \sqrt{\frac{\mu_1\varepsilon_1}{\mu_2\varepsilon_2}}$$

are entered. From Equations (10.77) and (10.78) it is not difficult to see that in a normal incidence ($\theta_1 = 0$) the Fresnel coefficients are identical:

$$R_\perp = R_\parallel = \frac{\sqrt{\mu_2\varepsilon_1} - \sqrt{\mu_1\varepsilon_2}}{\sqrt{\mu_2\varepsilon_1} + \sqrt{\mu_1\varepsilon_2}}. \tag{10.79}$$

Certainly, $\mu_1 = \mu_0$ and $\varepsilon_1 = \varepsilon_0$ in the case of incidence of a plane electromagnetic wave from the air to a conductor. In the case of a perfect conductor, the conductivity of the second medium, σ, tends to infinity,

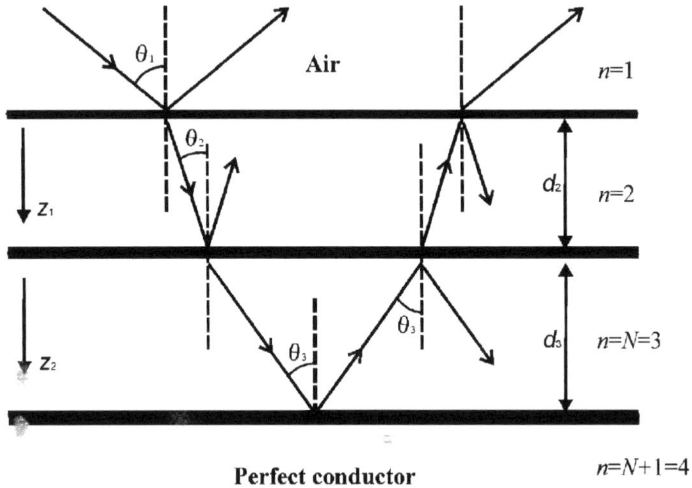

Figure 10.5 Reflection and refraction at interfaces of several infinite dielectric layers

owing to $R_\perp = R_{||} = -1$. Complete density of a surface's electrical current, $\boldsymbol{n} \times \mathbf{H} = 2\boldsymbol{n} \times \mathbf{H}_1$, is equal to the double projection (to the border between the air and the conductor) of a magnetic field intensity, $\mathbf{H}_1$. Density of a magnetic current $\boldsymbol{n} \times \mathbf{E}$ is equal to zero in an ideal case, but in reality $\boldsymbol{n} \times \mathbf{E} \neq 0$ [II.12].

The case of $N + 1$ layers was also considered in Reference [II.12]. Using Snell's law, Equation (10.74), and the geometry of rays shown in Figure 10.5, Shirman has obtained (more generally than in Reference [II.13]) electrical and magnetic fields in the nth layer as a superposition of travelling and standing waves:

$$\begin{pmatrix} E_n(z_n) \\ H_n(z_n) \end{pmatrix} = \begin{pmatrix} A_n \exp(-\mathrm{j}\alpha_n) + B_n \exp(\mathrm{j}\alpha_n) \\ \Upsilon_n \left[A_n \exp(-\mathrm{j}\alpha_n) - B_n \exp(\mathrm{j}\alpha_n)\right] \end{pmatrix}, \tag{10.80}$$

where z_n is a coordinate ($0 \leq z_n \leq d_n$), which is measured from the beginning of the nth layer in a direction perpendicular to the borders of the layers; A_n and B_n are amplitudes of travelling waves in level n (where $n = 2, 3, \ldots, N$); $\alpha_n = k_n z_n \cos\theta_n$ is a phase delay; and Υ_n is the wave admittance of the nth level [II.12]. For the first $n = 0$ and last $n = (N + 1)$ levels, amplitudes of travelling waves are considered only on surfaces $z_n = z_{N+1} = 0$, so $\alpha_n = \alpha_{N+1} = 0$. The wave admittance is $\Upsilon_n = \Upsilon_{n\perp} = \Upsilon_{n0} \cos\theta_n$ for an electrical field polarized perpendicularly to a plane of incidence and $\Upsilon_n = \Upsilon_{n||} = \Upsilon_{n0}/\cos\theta_n$ for an electrical field polarized in the plane parallel to a plane of incidence [II.12]. Here, $\Upsilon_{n0} = \sqrt{\varepsilon_n\varepsilon_0/(\mu_n\mu_0)}$ is a wave admittance in the nth layer.

Shirman also introduced a matrix

$$\mathbf{F}(\alpha, \Upsilon) = \begin{pmatrix} \cos\alpha & -\mathrm{j}\Upsilon^{-1}\sin\alpha \\ -\mathrm{j}\Upsilon\sin\alpha & \cos\alpha \end{pmatrix} \tag{10.81}$$

and received a rule of transformation of tangential components of fields in the multilayered structure:

$$\begin{pmatrix} E_N(d_N) \\ H_N(d_N) \end{pmatrix} = \mathbf{F}\begin{pmatrix} E_1(d_1) \\ H_1(d_1) \end{pmatrix} = \mathbf{F}\begin{pmatrix} A_1 + B_1 \\ Y_1(A_1 - B_1) \end{pmatrix}, \tag{10.82}$$

where $\mathbf{F} = \mathbf{F}(k_n d_n \cos\theta_n, \Upsilon_n)\,\mathbf{F}(k_{n-1}d_{n-1}\cos\theta_{n-1}, \Upsilon_{n-1}) \cdots \mathbf{F}(k_2 d_2 \cos\theta_2, \Upsilon_2)$.

In the specific case 'air – $(N - 1)$ lossy layers – perfect conductor' [II.12], the first ($n = 1$) layer is air, the last ($n = N + 1$) layer is a perfect conductor and the intermediate layers are lossy dielectric mediums characterized by their complex permeability, $\dot{\mu}$, and permittivity, $\dot{\varepsilon}$. For the described scattering problem, Shirman obtained the following equation for the complex reflection coefficient, R:

$$R = -\frac{F_{11} + F_{12}\Upsilon_1}{F_{11} - F_{12}\Upsilon}, \tag{10.83}$$

where F_{11} and F_{12} are elements of the matrix $\mathbf{F}$ (see Equation (10.81)).

In the specific case of two absorbing layers, Shirman considered a structure consisting of a lossless air layer ($n = 1$), a pair of lossy layers ($n = 2, 3$) and a perfect conductor layer ($n = N + 1 = 4$). He found that $F_{11} = \cos\alpha_2 \cos\alpha_3 - Y_2 Y_3^{-1} \sin\alpha_2 \sin\alpha_3$ and $F_{12} = -\mathrm{j}(\Upsilon_2^{-1} \sin\alpha_2 \cos\alpha_3 + \Upsilon_3^{-1} \cos\alpha_2 \sin\alpha_3)$.

According to Reference [II.12], layers having negative imaginary parts of parameters α_2 and α_3 reduce to an absolute value of the reflection coefficient, $|R|$. If $\alpha_2 = 0$ and $\alpha_3 = 0$, then $|R|$ is equal to zero.

10.3 ROUGH SCATTERING SURFACES

In Cartesian coordinates any surface can be described by the general equation

$$F(x, y, z) = 0. \tag{10.84}$$

Solving Equation (10.84) gives $z = \zeta(x, y) = \zeta(\boldsymbol{\rho})$, where $\boldsymbol{\rho} = x\mathbf{e}_x + y\mathbf{e}_y$ is a radius vector of a point on a coordinate plane, XOY. Usually $\zeta(\boldsymbol{\rho})$ is treated as the height of a rough surface above a point $\boldsymbol{\rho}$ of a plane XOY. Some radar targets, such as an air concrete rise-landing strip, a motorway, etc., can be considered as smooth flat surfaces. In this case $\zeta(\boldsymbol{\rho}) = z_0 \equiv$ *constant* and the scattering is *mirror reflection* according to Snell's law (see the previous section). However, the majority of targets have a rough surface with $\zeta(\boldsymbol{\rho}) \neq$ *constant*.

The criterion of the smooth or roughness of a surface was formulated by Rayleigh. Consider two incident beams, 1 and 2, which are falling on to a rough surface (see Figure 10.6). Let $h = \max[\zeta(\boldsymbol{\rho})] - \min[\zeta(\boldsymbol{\rho})]$. The difference of a path of beams is $\Delta r = 2h\cos\theta$, where θ is an angle between an incident beam and a normal to the surface. Therefore the difference of phases is $\Delta\varphi = k_0 \Delta r = (4\pi h/\lambda)\cos\theta$. For an absolutely smooth surface $h = z_0 - z_0 = 0$ and $\Delta\varphi = 0$. The increase in $\Delta\varphi$ entails partial destruction of beams (so-called 'diffuse reflection' occurs); under the condition $\Delta\varphi = \pi$ the specified destruction becomes complete. In the latter case, the reflected power is redistributed in other directions that do not submit to Snell's law, and is referred to as 'diffusion power'. According to Rayleigh's criterion,

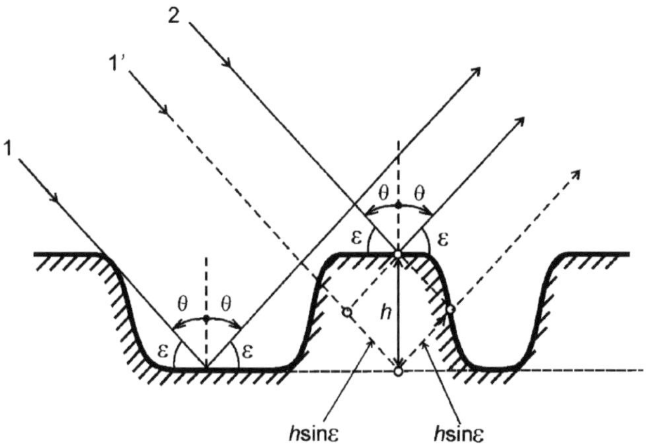

Figure 10.6 Illustration of the Rayleigh criterion

Figure 10.7 Earth surface scattering depending on the roughness level

the smooth surface is such a surface for which $\Delta\varphi < \pi/2$. Thus, the condition of a smooth surface is

$$h < \frac{\lambda}{8\cos\theta}. \tag{10.85}$$

It is necessary to emphasize that at grazing incidence angles the surface becomes practically smooth and reflects like a mirror. Thus, it is possible to consider a surface as smooth in two cases: if $h \approx 0$ or $\theta \approx 90°$. For example, let $\lambda = 3$ cm and $\theta = 30°$. In this case the surface is smooth if $h < 0.75$ cm. However, in the case of the same wavelength λ and $\theta = 89°$, a smooth condition is $h < 21.5$ cm.

Figure 10.7 shows an example of the scattering of an electromagnetic wave on the surface of the Earth depending on a level of roughness. In the literature, either deterministic or random functions $\zeta(\boldsymbol{\rho})$ have been used for rough surface modelling [II.14]. In particular, when formulating the Rayleigh criterion, it was believed, for example, that $\zeta(\boldsymbol{\rho})$ is determined.

The stochastic model assumes that the function $\zeta(\boldsymbol{\rho})$ describing the surface is a casual field. Thus, the deterministic value $\bar{\zeta} = \langle\zeta(\boldsymbol{\rho})\rangle$ is an average height, and $\sigma = \sqrt{\langle[\zeta(\rho) - \bar{\zeta}]^2\rangle}$ is the root-mean-square characterization of surface roughness, where the symbol < > designates statistical averaging. The Rayleigh criterion for casual surfaces establishes that it is necessary to consider a surface as rough if $\sigma > \lambda/(8\cos\theta)$.

The correlation properties of a surface are defined through a function of autocorrelation,

$$\mathbf{A}(\rho) = \langle\zeta(\rho + \rho')\,\zeta(\rho')\rangle_{\rho'}. \tag{10.86}$$

The homogeneous casual surface is characterized by the spatial power spectrum, $G(\mathbf{k})$, of eminences, which is the Fourier transform of a correlation function:

$$G(\boldsymbol{k}) = \frac{1}{(2\pi)^2}\iint \mathbf{A}(\rho)\exp(-\mathrm{j}\boldsymbol{k}\rho)\,\mathrm{d}\rho. \tag{10.87}$$

where $\boldsymbol{k}$ is a wave vector. If a casual field, $\zeta(\boldsymbol{\rho})$, is Gaussian, it is possible to restore $\zeta(\boldsymbol{\rho})$ from $G(\boldsymbol{k})$ by means of an integration operation:

$$\zeta(\rho) = \frac{1}{(2\pi)^2}\iint a(\boldsymbol{k})\cos[\mathbf{k}\rho + \chi(\mathbf{k})]\sqrt{2G(\boldsymbol{k})}\,\mathrm{d}\boldsymbol{k}. \tag{10.88}$$

where $a(\boldsymbol{k})$ are independent casual Rayleigh amplitudes with unit averages and $\chi(\boldsymbol{k})$ are casual phases having uniform distribution in the range $(-\pi, \pi]$.

Very often in theoretical research only the section of a field $\zeta(\boldsymbol{\rho})$ by a plane that includes an axis OZ is used. Thus, the casual field $\zeta(\boldsymbol{\rho})$ is transformed to the stochastic process $\zeta(x)$. Equations (10.86) to (10.88) will also be transformed accordingly:

$$\mathbf{A}(x) = \langle\zeta(x + x')\,\zeta(x')\rangle_{x'}, \tag{10.89}$$

$$G(k) = \frac{1}{2\pi}\int \mathbf{A}(x)\exp(-\mathrm{j}kx)\,\mathrm{d}x, \tag{10.90}$$

$$\zeta(x) = \frac{1}{2\pi}\int a(k)\cos[kx + \chi(k)]\sqrt{2G(k)}\,\mathrm{d}k, \tag{10.91}$$

where $k = 2\pi/\lambda$ is a wave number and λ is a wavelength of roughness. The width of the main maximum of the autocorrelation function (10.89) may be characterized by the quantity L, named the autocorrelation length. This specifies the spatial scale, which describes a statistical dependence of two points on a surface.

The model of dispersion described in Reference [II.14] is based on the assumption that the height of a perfectly conductive surface can be described by the stationary Gaussian field, $\zeta(\boldsymbol{\rho})$. The autocorrelation function of the specified heights was assumed to be exponential, and the root-mean-square deviation is to obey the law

$$\sigma(\alpha) = \cot(\alpha_0)\exp\left(-\frac{\tan\alpha}{\tan\alpha_0}\right), \tag{10.92}$$

where α is an angle between axis OZ and the local normal to a surface in a point of reflection and α_0 is an angle characterizing the degree of rough surface inclination; therefore $\tan\alpha_0 = 2\sigma/L$.

Another model is offered in Reference [II.15], viz:

$$G(k) = \frac{4\pi^2 L_0}{k^2} = L_0\lambda^2. \tag{10.93}$$

The value L_0 is referred to as the 'topothesy' of a surface. It has a dimension of length and characterizes both geometrical and statistical properties of an isotropic surface. The experimental normalized power spectra for 23 values of L_0 are given in Reference [II.15] as functions of wavelength of roughness. The dependence given by Equation (10.93) is also presented in the figure in Reference [II.15] as a straight line. It should be noted that this graph covers eight decades of wavelength, and over this entire range a unique convergence of results occurs for all structures of surfaces.

A more general model than Equation (10.93) was used in Reference [II.16], viz.:

$$G = L_0\lambda^\alpha = L_0\lambda^{2H+1} = L_0\lambda^{7-2D}, \tag{10.94}$$

where α is the spectral parameter, H is the Hurst parameter and D is the fractal dimension of a surface, and

$$\alpha = 2H + 1, \tag{10.95}$$

$$D = 3 - H. \tag{10.96}$$

For Brownian surfaces, when the field $\zeta(\boldsymbol{\rho})$ is Gaussian, Equation (10.93) follows from Equations (10.94) and (10.95) (or (10.96)), as in this case $H = 1/2$ and $D = 2.5$ However, generally when the field $\zeta(\boldsymbol{\rho})$ is not Gaussian, the magnitudes of parameter α, according to Reference [II.16], are in an interval [1.07, 3.03], which corresponds to the fractal dimension lying in a range from 2 up to 3.

The realization of a new approximation of the data from Reference [II.15] has allowed the construction of a histogram of estimations of the spectral parameter, α. The obtained experimental values of the spectral parameter α are grouped near the Brownian value 2, but are allocated in a range from 1 up to 3 [II.16].

In summary, it should be noted that, for example, a turbulent sea surface must be characterized by the function $\zeta(\boldsymbol{\rho}, t)$, which depends not only on spatial coordinates but also on time.

10.4 THE SCATTERING PROBLEM FOR SMALL TARGETS

The scattering of electromagnetic waves on small radar targets will now be considered. Certainly, small targets have shapes and sizes, but even their maximal sizes are small in comparison with the resolution cell of a radar. In this sense the occurrence of small targets is named in Reference [II.2] as 'long pulse illumination'.

According to Chapter 9, in the case of a bistatic radar the direction from the target to the receiver differs from the direction to the transmitter. Therefore, the scattering problem in a bistatic radar is formulated a little differently from that in a monostatic radar.

Consider a scattering body (target, object, discontinuity of medium) Γ having some given form and given physical properties. It is located in free space and is illuminated with a monochromatic electromagnetic wave. This wave is radiated by the antenna of the transmitter, and with reference to the target is an incident wave. In relation to the antenna of the transmitter, the surrounding space is usually divided into three regions: reactive near-field, radiating near-field (Fresnel) and far-field (Fraunhofer) regions. Since the scattering body can be treated as a source of secondary radiation, there is the opportunity to introduce the concepts of near and far regions of the radar target.

It is assumed that the target is located in free space in a far region of the transmitting antenna and therefore the incident wave represents a plane wave. The mathematical expression for a plane incident wave is given by formulae similar to Equation (10.58):

$$\begin{aligned}\mathbf{E}^{\text{in}}(\boldsymbol{r},t) &= \mathbf{E}_0^{\text{in}} \exp\left[\mathrm{j}\left(k_0 \boldsymbol{e}_{\mathbf{S}}^{\text{in}} \cdot \boldsymbol{r} - \omega t\right)\right],\\ \mathbf{H}^{\text{in}}(\boldsymbol{r},t) &= \mathbf{H}_0^{\text{in}} \exp\left[\mathrm{j}\left(k_0 \boldsymbol{e}_{\mathbf{S}}^{\text{in}} \cdot \boldsymbol{r} - \omega t\right)\right],\end{aligned} \tag{10.97}$$

where $\mathbf{E}^{\text{in}}$ is the incident electric field, $\boldsymbol{r}$ is a radius vector of a point of observation, $\mathbf{E}_0^{\text{in}}$ is an amplitude of the incident electric field, $\boldsymbol{e}_{\mathbf{S}}^{\text{in}}$ is the unit vector in the direction of a Poynting vector of an incident wave, $\mathbf{H}^{\text{in}}$ is the incident magnetic field and $\mathbf{H}_0^{\text{in}} = \sqrt{\varepsilon_0/\mu_0}\, \boldsymbol{e}_{\mathbf{S}}^{\text{in}} \times \mathbf{E}_0^{\text{in}}$ is the amplitude of intensity of an incident magnetic field. If the point of observation (determined by the radius vector $\boldsymbol{r}$) is located on the surface of the target, Γ, the incident wave is a function of spherical coordinates θ_{T} and ϕ_{T} of the transmitter relative the target (see Figure 10.8), i.e. $\mathbf{E}^{\text{in}} = \mathbf{E}^{\text{in}}(\theta_{\text{T}}, \phi_{\text{T}}, t)$. Similarly, $\boldsymbol{e}_{\mathbf{S}}^{\text{in}} = \boldsymbol{e}_{\mathbf{S}}^{\text{in}}(\theta_{\text{T}}, \phi_{\text{T}})$.

Analogously to Equation (10.62), the complex amplitude of the incident electric field can be expanded by components:

$$\mathbf{E}_0^{\text{in}} = \mathbf{E}_{0h}^{\text{in}} + \mathbf{E}_{0v}^{\text{in}} = E_{0h}^{\text{in}}\boldsymbol{e}_h + E_{0v}^{\text{in}}\boldsymbol{e}_v = \left(\boldsymbol{e}_h\ \boldsymbol{e}_v\right)\mathsf{E}^{\text{in}} = E_{0h}^{\text{in}}\left(\boldsymbol{e}_h\ \boldsymbol{e}_v\right)\begin{pmatrix}1\\ q^{\text{in}}\end{pmatrix}, \tag{10.98}$$

where

$$\mathsf{E}^{\text{in}} = \begin{pmatrix}E_{0h}^{\text{in}}\\ E_{0v}^{\text{in}}\end{pmatrix}$$

is an incident Jones vector and

$$q^{\text{in}} = \frac{E_{0v}^{\text{in}}}{E_{0h}^{\text{in}}} \tag{10.99}$$

is a polarization ratio (it is supposed that $E_{0h}^{\text{in}} \neq 0$), determining polarization of an incident wave.

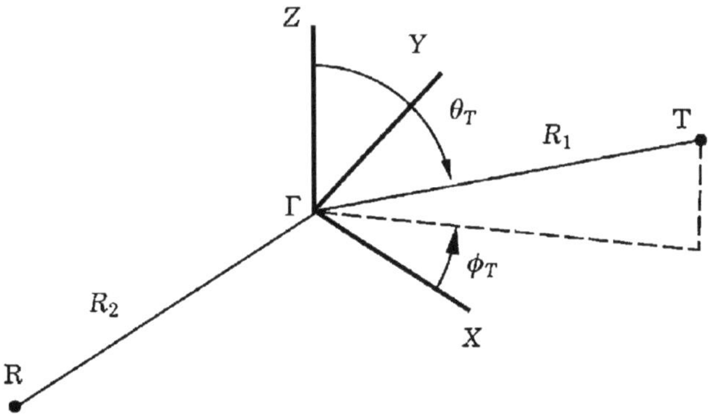

Figure 10.8 The spherical angle coordinates of the transmitter relative target

The power of the electric field is defined by the magnitude of the Poynting vector and power density is proportional to the squared amplitude of the incident electric field:

$$\text{Power per unit area in the wave incident on the target} = \kappa \left|\mathbf{E}^{\text{in}}\right|^2 = \kappa \left(\left|E_{0h}^{\text{in}}\right|^2 + \left|E_{0v}^{\text{in}}\right|^2\right). \quad (10.100)$$

In the absence of a target it would be possible to measure the magnitude of the constant incident wave amplitude, $\mathbf{E}_0^{\text{in}}$. The presence of a targets introduces perturbation to an incident wave and, therefore, in each point of space there is some field, **E**, whose amplitude and initial phases change from point to point. In the far region of the target the field **E** can be treated as a result of the interference of two waves: an incident plane wave $\mathbf{E}^{\text{in}}$ and a scattering (not plane in the general case) wave $\mathbf{E}^{\text{sc}}$. Field **E** is required to satisfy the boundary conditions, mentioned in Section 10.1, and the scattering field, $\mathbf{E}^{\text{sc}}$, is required to be outgoing from radar targets at large distances. Thus, in order to solve the scattering problem of a bistatic radar the scattered fields, $\mathbf{E}^{\text{sc}}$ and $\mathbf{H}^{\text{sc}}$, in particular, need to be defined in the far region of a target.

Further, a scattering wave that is understood as a wave that considerably exceeds the size of a scattering body over distances and on some limited sites has a spherical phase front [II.17]. Such a definition of a scattering field assumes that a scattering body is motionless and the incident and scattering waves are monochromatic waves.

Defining scattering fields as limits,

$$\mathbf{E}^{\text{sc}}(\boldsymbol{r}, t) = \lim_{r\to\infty}\left[\mathbf{E}(\boldsymbol{r}, t) - \mathbf{E}^{\text{in}}(\boldsymbol{r}, t)\right], \qquad \mathbf{H}^{\text{sc}}(\boldsymbol{r}, t) = \lim_{r\to\infty}\left[\mathbf{H}(\boldsymbol{r}, t) - \mathbf{H}^{\text{in}}(\boldsymbol{r}, t)\right], \quad (10.101)$$

it is possible to obtain

$$\mathbf{E}^{\text{sc}}(\boldsymbol{r}, t) = A(\theta, \phi)\,\frac{\exp\left[\mathrm{j}\left(k_0 \boldsymbol{e}_{\mathbf{S}}^{\text{sc}} \cdot \boldsymbol{r} - \omega t\right)\right]}{r},$$
$$\mathbf{E}^{\text{sc}}(\boldsymbol{r}, t) = \frac{A(\theta, \phi)}{\varsigma}\,\frac{\exp\left[\mathrm{j}\left(k_0 \boldsymbol{e}_{\mathbf{S}}^{\text{sc}} \cdot \boldsymbol{r} - \omega t\right)\right]}{r}, \quad (10.102)$$

where $\boldsymbol{e}_{\mathbf{S}}^{\text{sc}}$ is a unit vector in the direction of a Poynting vector of a scattering wave and $r = |\boldsymbol{r}|$. Thus, the vectors of amplitudes of a scattering field look like

$$\mathbf{E}_0^{\text{sc}}(\boldsymbol{r}) = \frac{A(\theta, \phi)}{r}, \qquad \mathbf{H}_0^{\text{sc}}(\boldsymbol{r}) = \frac{A(\theta, \phi)/\varsigma}{r}. \quad (10.103)$$

It should be remembered that in formulae (10.102) and (10.103), the spherical system of coordinates (r, θ, ϕ) is used with an origin at the target.

The characteristics of a scattered field by target radiation depend on the form, sizes and material of a scattering body, and also on its orientation (foreshortening) in relation to the transmitter and receiver lines. As a scattering wave is not, in general, a plane, the direction of vector $\boldsymbol{e}_{\mathbf{S}}^{\text{sc}}$ is various in different points of space: $\boldsymbol{e}_{\mathbf{S}}^{\text{sc}}$ is a function not only of the coordinates of the transmitter, θ_{T} and ϕ_{T}, but also a function of the coordinates (relative target) r, θ and ϕ of the observation point. Thus, irrespective of whether a receiving antenna is present or not at a point of observation with coordinates r, θ and ϕ, the unit wave vector, $\boldsymbol{e}_{\mathbf{S}}^{\text{sc}} = \boldsymbol{e}_{\mathbf{S}}^{\text{sc}}(r, \theta, \phi; \theta_{\mathrm{T}}, \phi_{\mathrm{T}})$, determines the direction of propagation of a scattered wave and in this sense it is similar to the unit vector, $\boldsymbol{e}_{\mathbf{S}}^{\text{in}}$.

The amplitude of a scattered wave can be expanded in the same polarization basis, as the amplitude of an incident wave, viz.

$$\mathbf{E}^{\text{sc}} = \mathbf{E}_{0h}^{\text{sc}} + \mathbf{E}_{0v}^{\text{sc}} = E_{0h}^{\text{sc}}\boldsymbol{e}_h + E_{0v}^{\text{sc}}\boldsymbol{e}_v = \left(\boldsymbol{e}_h \ \boldsymbol{e}_v\right)\mathsf{E}^{\text{sc}} = E_{0h}^{\text{sc}}\left(\boldsymbol{e}_h \ \boldsymbol{e}_v\right)\begin{pmatrix} 1 \\ q^{\text{sc}} \end{pmatrix}, \quad (10.104)$$

where

$$\mathsf{E}^{\text{sc}} = \begin{pmatrix} E_{0h}^{\text{sc}} \\ E_{0v}^{\text{sc}} \end{pmatrix}$$

is the Jones vector of scattering field and

$$q^{sc} = \frac{E_{0h}^{sc}}{E_{0v}^{sc}} \tag{10.105}$$

is a polarization ratio, determining polarization properties of a scattered wave (it is supposed that $E_{0v}^{sc} \neq 0$).

It is known [II.2] that if the distance between the target and receiver is large enough so that a scattered wave in the aperture of a receiving antenna may be approximated by a plane wave, we can write [II.2]

$$\begin{aligned}&\text{Power per unit area in total scattered wave at receiving antenna}\\ &= \kappa\,|\mathbf{E}^{sc}|^2 = \kappa\left(\left|E_{0h}^{sc}\right|^2 + \left|E_{0v}^{sc}\right|^2\right) = \kappa\left|E_{0h}^{sc}\right|^2\left(1 + |q^{sc}|^2\right).\end{aligned} \tag{10.106}$$

It is clear that Equation (10.104) is taken into account here.

10.5 BISTATIC CROSS-SECTIONS

An essential characteristic of the scattering properties of a target is its scattering cross-section (SCS). The back-scattering cross-section is a quantitative characteristic of the target's ability to scatter (reflect) power in the opposite direction to that of an incident wave. This is useful in the theory and practice of monostatic radars.

It is obvious that for bistatic radars a more general characteristic that takes into account the difference between the directions from a target to the transmitting and receiving antennas is necessary. The required characteristic is called the bistatic scattering cross-section (BSCS) of a target.

According to the generally accepted definition (see, for example, Reference [II.4]), the SCS of a target is equal to the surface area of a symbolic object that scatters total incident power isotropically and creates, at a distant receiving point, the same power flux density as the target. In terms of the electric field complex amplitude, the SCS of a target can be expressed as

$$\sigma^{sc} = \lim_{R_2\to\infty} 4\pi R_2^2 \frac{\begin{array}{c}\text{Power per unit area in}\\ \text{total scattered wave at}\\ \text{the receiving antenna}\end{array}}{\begin{array}{c}\text{Power per unit area in}\\ \text{the wave incident}\\ \text{on the target}\end{array}} = \lim_{R_2\to\infty} 4\pi R_2^2 \frac{|\mathbf{E}^{sc}|^2}{\left|\mathbf{E}^{in}\right|^2}, \tag{10.107}$$

where R_2 is the range of the receiver relative to the target (see Figure 10.8). The definition given above holds both for the back-scattering cross-section and the bistatic scattering cross-section.

It should also be noted that since $|\mathbf{E}^{sc}|^2$ is inversely proportional to range square, R_2^2 (see, for example, Equations (10.102)), the value R_2^2 in Equation (10.107) is cancelled out in the process of calculations [II.4]. The limit in Equation (10.107) is only necessary to guarantee that the receiving antenna is in the far region of the target so that the scattering wave has a plane wave front [II.2]. It is widely known that Equation (10.107) remains valid if the electric components of a field are replaced by magnetic components.

In the bistatic radar the intensity of the electric field of an incident wave is a function of the vector $\boldsymbol{e}_{\mathbf{S}}^{in}$, but the received electric field is a function of this vector, $\boldsymbol{e}_{\mathbf{S}}^{in}$, and an analogue vector, $\boldsymbol{e}_{\mathbf{S}}^{sc}$, in the scattering wave, i.e.

$$\mathbf{E}^{\mathrm{in}} = \mathbf{E}^{\mathrm{in}}\left(\boldsymbol{e}_{\mathbf{S}}^{\mathrm{in}}\right), \qquad \mathbf{E}^{\mathrm{sc}} = \mathbf{E}^{\mathrm{sc}}\left(\boldsymbol{e}_{\mathbf{S}}^{\mathrm{in}}, \boldsymbol{e}_{\mathbf{S}}^{\mathrm{sc}}\right). \tag{10.108}$$

Taking account of Equations (10.108), the equation for the BSCS may be obtained from Equation (10.107) as

$$\sigma^{\mathrm{sc}} = \sigma^{\mathrm{sc}}\left(\boldsymbol{e}_{\mathbf{S}}^{\mathrm{in}}, \boldsymbol{e}_{\mathbf{S}}^{\mathrm{sc}}\right) = \lim_{R_2\to\infty} 4\pi R_2^2 \frac{\left|\mathbf{E}^{\mathrm{sc}}\left(\boldsymbol{e}_{\mathbf{S}}^{\mathrm{in}}, \boldsymbol{e}_{\mathbf{S}}^{\mathrm{sc}}\right)\right|^2}{\left|\mathbf{E}^{\mathrm{in}}\left(\boldsymbol{e}_{\mathbf{S}}^{\mathrm{in}}\right)\right|^2}. \tag{10.109}$$

It is therefore clear that the back-scattering cross-section is a special case of a bistatic scattering cross-section under the condition $\boldsymbol{e}_{\mathbf{S}}^{\mathrm{sc}} = -\boldsymbol{e}_{\mathbf{S}}^{\mathrm{in}}$. Finally, substituting Equation (10.106) into Equation (10.107), it is straightforward to find that

$$\sigma^{\mathrm{sc}} = \lim_{R_2\to\infty} 4\pi R_2^2 \frac{\left|E_{0h}^{\mathrm{sc}}\right|^2}{\left|\mathbf{E}^{\mathrm{in}}\right|^2}\left(1 + |q^{\mathrm{sc}}|^2\right). \tag{10.110}$$

Another more important characteristic of this study is the radar cross-section (RCS). It is known that, in general, the SCS and RCS are not equal to each other: quantitatively the difference depends on the relation between the type of polarization of the incident electromagnetic field and the polarization properties of the receiving antenna. In Reference [II.2] the RCS, σ, was defined as the part of the SCS that corresponds to the polarization opportunity of the receiving antenna, i.e.

$$\sigma = \sigma\left(\boldsymbol{e}_{\mathbf{S}}^{\mathrm{in}}, \boldsymbol{e}_{\mathbf{S}}^{\mathrm{sc}}\right) = \lim_{R_2\to\infty} 4\pi R_2^2 \frac{\begin{array}{c}\text{Power per unit area in a scattered}\\ \text{wave at the receiving antenna,}\\ \text{which is in the polarization of}\\ \text{the receiving antenna}\end{array}}{\begin{array}{c}\text{Power per unit area in}\\ \text{the wave incident}\\ \text{on the target}\end{array}}. \tag{10.111}$$

From the definitions it is obvious that constantly $\sigma \leq \sigma^{\mathrm{sc}}$.

As an example, the dependence of the bistatic radar cross-section, σ_{b}, on the bistatic angle, β, is shown in Figure 10.9 for the model of a large aircraft [II.18]. The model has a wingspan of about 50 m and a fuselage length of about 20 m. Two frequency bands were investigated: (a) from 2.25 to 3.75 GHz and (b) from 0.15 to 0.25 GHz. Figure 10.9 evidently illustrates the dependence of the BRCS on the bistatic angle and frequency.

As before in the case of the scattering cross-section, the definition of the radar cross-section given by Equation (10.111) holds both for monostatic and bistatic radars. Clearly the monostatic radar cross-section (MRCS), $\sigma_{\mathrm{m}}\left(\boldsymbol{e}_{\mathbf{S}}^{\mathrm{in}}\right)$, is a special case of the bistatic radar cross-section (BRCS), $\sigma_{\mathrm{b}}\left(\boldsymbol{e}_{\mathbf{S}}^{\mathrm{in}}, \boldsymbol{e}_{\mathbf{S}}^{\mathrm{sc}}\right)$, viz:

$$\sigma_{\mathrm{m}}\left(\boldsymbol{e}_{\mathbf{S}}^{\mathrm{in}}\right) = \sigma_{\mathrm{b}}\left(\boldsymbol{e}_{\mathbf{S}}^{\mathrm{in}}, -\boldsymbol{e}_{\mathbf{S}}^{\mathrm{in}}\right). \tag{10.112}$$

Hence, in the particular case shown in Figure 10.9, where BRCS is a function of the bistatic angle, i.e. $\sigma_{\mathrm{b}} = \sigma_{\mathrm{b}}(\beta)$, Equation (10.112) may be transformed into $\sigma_{\mathrm{m}} = \sigma_{\mathrm{b}}(0)$.

The evident fact that in general the bistatic radar cross-section is not equal to its monostatic counterpart is confirmed, for example, by Figure 10.10 (see also Reference [II.4] and [II.18]). Here the MRCS and BRCS of the generic airborne platform shape are shown as a function of the aspect angle. The transmitting antenna radiates a wave with the horizontal polarization so that the vector $\mathbf{E}^{\mathrm{in}}$ is parallel to the wings. The receiving antenna is adjusted on the reception of a wave with the same horizontal polarization. Platform characteristic dimensions are much larger than the wavelength. The bistatic angle, β, is 160°.

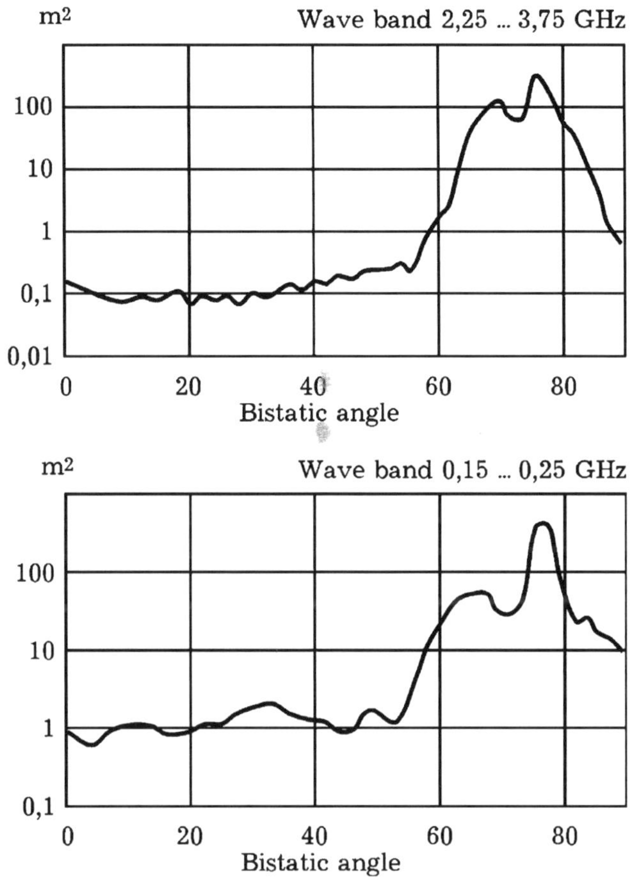

Figure 10.9 Bistatic RCS in two frequency bands calculated for the model of a large aircraft given via the bistatic angle for zero course-, pitch- and roll-aspect angles. (From Reference [II.18]. Reproduced by permission of ©1988 IEEE)

In Figure 10.11, the BRCS of a generic airborne platform shape with dependence on the aspect angle is shown for other values of the bistatic angle and two different polarization types. Thus, VV (HH) polarization means that both the transmitter antenna and receiver antenna have a vertical (horizontal) polarization. Thus, the cross-polarization is absent in Figures 10.10 and Figure 10.11. Nevertheless, Figure 10.11 can be considered as an illustration of the dependence of the RCS from the polarization of electromagnetic waves.

It follows from the definitions that the unit of all (bistatic, monostatic, radar or scattering) cross-sections is square metres (m^2) (see Figure 10.9), which can be transformed at will into dB/m^2 (as, for example, in Figure 10.10). Often, however, it is useful to define a relative bistatic radar cross-section (RBRCS), σ_r, which is measured in dB:

$$\sigma_r = \sigma_r\left(\boldsymbol{e}_{\mathbf{S}}^{\text{in}}, \boldsymbol{e}_{\mathbf{S}}^{\text{sc}}\right) = -10\log\frac{\sigma_b\left(\boldsymbol{e}_{\mathbf{S}}^{\text{in}}, \boldsymbol{e}_{\mathbf{S}}^{\text{sc}}\right)}{\sigma_m\left(\boldsymbol{e}_{\mathbf{S}}^{\text{in}}\right)}. \tag{10.113}$$

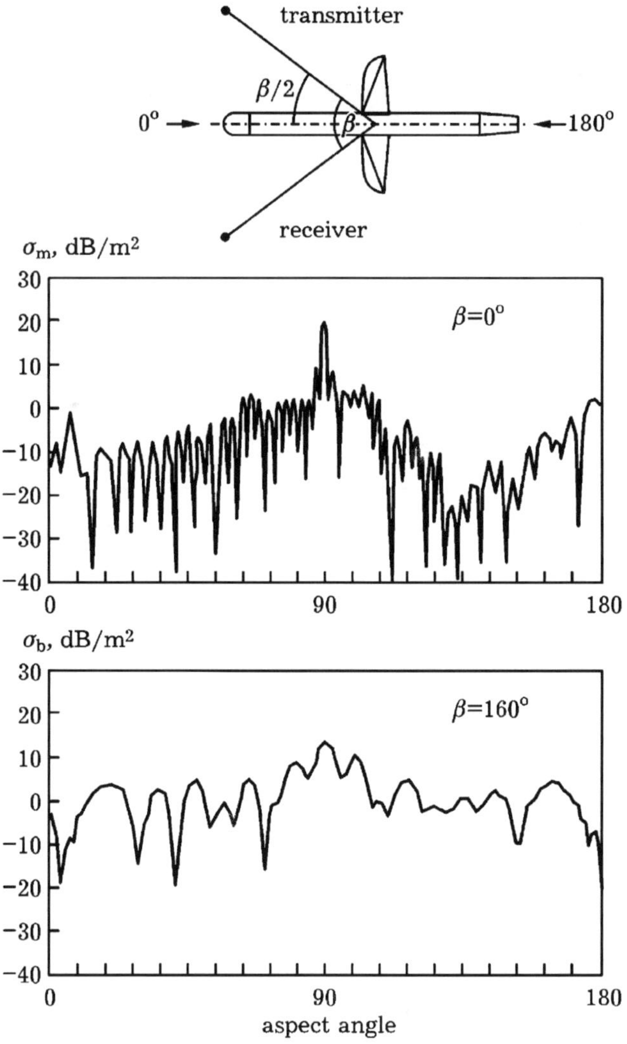

Figure 10.10 MRCS and BRCS of the generic airborne platform shape measured at the frequency 1.25 GHz. (Adapted from Reference [II.18]. Reproduced by permission of ©1988 IEEE)

Well-known measurements provided by Ewell and Zehner (see, for example, Reference [II.1] and [II.19] to [II.21]) on relatively large coastal freighters actually represent an RBRCS.[3] These results are shown in Figure 10.12. In particular, Figure 10.12 illustrates that the difference between σ_b and σ_m may be rather large, especially outside a quasi-monostatic state (e.g. under condition $\beta > 5°$). Thus, it is important to note that the general statement, 'bistatic RCS of certain targets can be much greater

[3]The ranges of a median MRCS of these ships at X-band were as follows:

The Kansas	23.5–31.4 dB/m^2
The Hellenic Challenger	33.0–37.6 dB/m^2
The Phillipean Rizal	28.5–35.7 dB/m^2
The Delaware Sun	25.3–31.9 dB/m^2

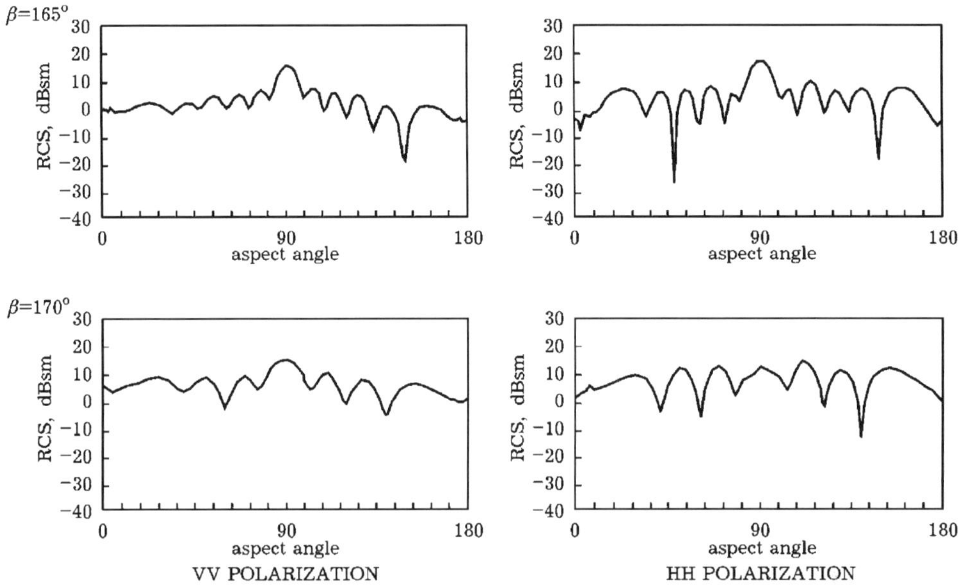

Figure 10.11 Bistatic RCS of a generic airborne platform shape measured at the frequency 1.25 GHz. (From Reference [II.18]. Reproduced by permission of ©1988 IEEE)

than monostatic RCS', is true. First of all, this refers to targets developed by special technologies like 'Stealth', which reduce the MRCS and consequently reduce the detectability of aircrafts and cruise missiles. The target can be so shaped that at certain aspect angles, incident radar signals are reflected away in a different direction. Such shaping is usually designed to give a 'cone-in-silence' about the direction of flight of an aircraft or a missile [II.4].

For the bistatic example shown in Figure 10.13, where the bistatic case is given for zero course-, pitch- and roll-aspect angles, it can be clearly seen that the BRCS increases rapidly at bistatic angles of 70° to 80°.

Thus, in general, as can be understood from the definition in Equation (10.111) and examples given in this section, the bistatic radar cross-section of a target depends on the type of polarization of the transmitting and receiving antennas and on the five parameters, four angles and the length of the incident wave, i.e. for the given polarization properties of the transmitting antenna and receiving antenna:

$$\sigma_b = \sigma_b(\theta_T, \phi_T, \theta_R, \phi_R; \lambda), \qquad (10.114)$$

where θ_R, and ϕ_R are the receiver's spherical angle coordinates measured relative to the target. Thus it is clear that any of the angles appearing in Equation (10.114) can be replaced by a bistatic angle, β. At the same time it is known [II.17] that if the scattering body is a body of revolution, the number of different angles in Equation (10.114) is reduced to three. The relative bistatic radar cross-section of a sphere of given radius, a, depends only on the wavelength, λ, and the bistatic angle, β.

The monostatic radar cross-section depends only on two angles and a wavelength, i.e.

$$\sigma_m = \sigma_m(\theta_T, \phi_T; \lambda). \qquad (10.115)$$

It is not difficult to understand that the MRCS of a sphere with a given radius depends on the wavelength, λ, only, and is independent from any angles.

The dependence of the RBRCS on the bistatic angle or spherical angle coordinates (of transmitter or receiver) will be termed a 'scattering indicatrix'. Dependence of any RCS (MRCS, BRCS or RBRCS)

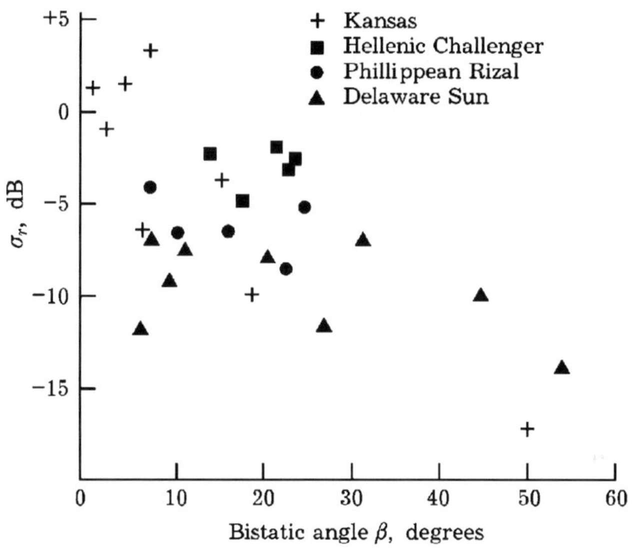

Figure 10.12 Relative RCS of ships. (Adapted from Reference [II.19]. Reproduced by permission of ©1980 IEEE)

on a wavelength may be named as the power wavelength characteristic. The scattering indicatrix has already been demonstrated in Figure 10.7. The power wavelength characteristics of radar targets, as a rule, cannot be analytically determined simultaneously for all ranges of wavelength from zero to infinity. Therefore in the study of power wavelength characteristics it is accepted that three basic wavelength regions should be distinguished: (a) the Rayleigh region, (b) the resonance region and (c) the optical region (see Figure 10.14). The borders between these areas are indistinct and scatterer dependent. However, it is usually assumed that in the limits of the Rayleigh region, the maximal size of a body does not exceed 0.2 of the wavelength, and in the optical region it is more than three[4] wavelengths [II.17, II.22].

Within the Rayleigh region, the field phase distribution over the surface feature varies little, resulting in an MRCS approximately inversely proportional to the fourth power of the wavelength. Within the resonance region, the phase begins to vary rapidly over the surface feature, enabling surface waves, which enhance the MRCS under certain conditions. In the optical region, phase variations become indistinguishable, leading to a more steady state of scattering conditions. The described properties are illustrated with an example of a perfectly conducting sphere power wavelength characteristic, shown in Figure 10.14.

Let us, however, return to Equation (10.111), which shows that only part of the scattering wave power at the aperture of the receiving antenna is used in the definition of the RCS. The receiving antenna may have some polarization, but we are able to divide an arbitrary polarized scattering wave into two parts: one, $\mathbf{E}_A^{sc}$, to which the antenna responds (which has the antenna's polarization) and another, $\mathbf{E}_N^{sc}$, to which the antenna does not respond [II.2], viz.:

$$\mathbf{E}^{sc} = \mathbf{E}_A^{sc} + \mathbf{E}_N^{sc} = E_{A0}^{sc}\boldsymbol{e}_A + E_{N0}^{sc}\boldsymbol{e}_N. \tag{10.116}$$

[4]Some experts define an electrically large object if its size is greater than 10 wavelengths.

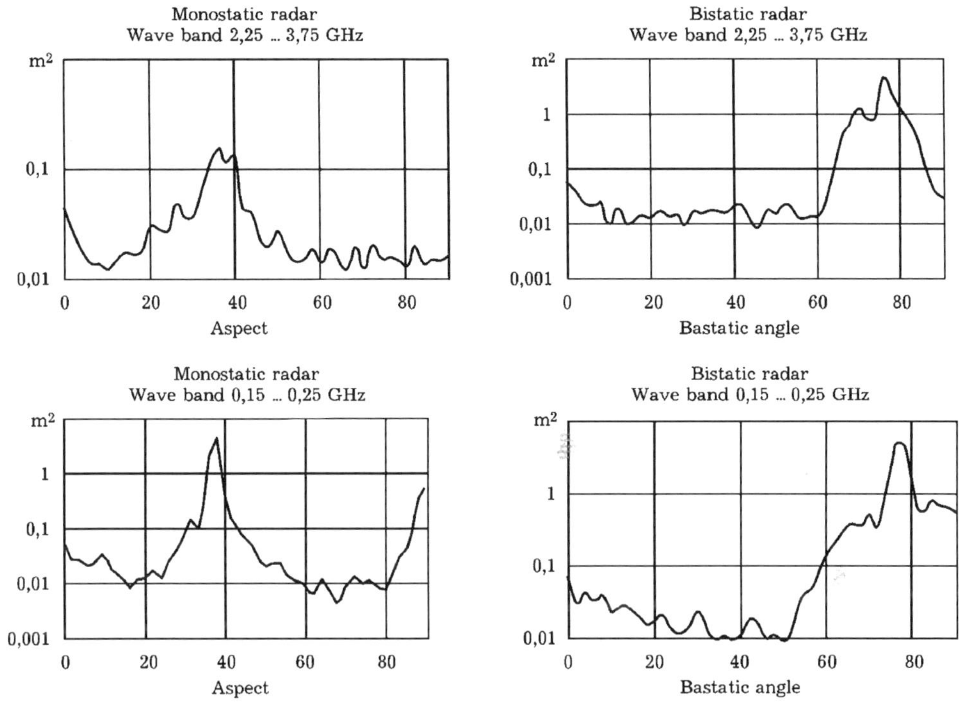

Figure 10.13 Monostatic and bistatic RCSs in two frequency bands calculated for the model of a large aircraft with a completely covered surface. (From Reference [II.18]. Reproduced by permission of ©1988 IEEE)

It is clear that formula (10.116) is a special case of the general Equation (10.104). Then [II.2]

$$\sigma = \lim_{R_2 \to \infty} 4\pi R_2^2 \frac{\left|\mathbf{E}_A^{\mathrm{sc}}\right|^2}{\left|\mathbf{E}^{\mathrm{in}}\right|^2}. \tag{10.117}$$

The definition in Equation (10.117), in comparison with Equation (10.107), shows clearly the difference between a scattering cross-section, σ^{sc}, and a radar cross-section, σ. However, we can write

$$\sigma = \sigma^{\mathrm{sc}} \rho_{\mathrm{pol}}, \tag{10.118}$$

where

$$\rho_{\mathrm{pol}} = \lim_{R_2 \to \infty} \frac{\left|\mathbf{E}_A^{\mathrm{sc}}\right|^2}{\left|\mathbf{E}^{\mathrm{sc}}\right|^2} \tag{10.119}$$

is called the polarization efficiency [II.2] or the polarization match factor [II.23]. The polarization efficiency is, as follows from Equation (10.119), the fraction of the scattered power that the receiver or receiving antenna can receive. From Equation (10.118), the RCS is smaller than the SCS by a factor ρ_{pol}.

In analogy with Equations (10.62) and (10.104) $\mathbf{E}_A^{\mathrm{sc}}$ is represented as

$$\boldsymbol{E}_A^{\mathrm{sc}} = E_{A0h}^{\mathrm{sc}} \boldsymbol{e}_h + E_{A0v}^{\mathrm{sc}} \boldsymbol{e}_v = E_{A0h}^{\mathrm{sc}} (\boldsymbol{e}_h + q_A \boldsymbol{e}_v) = E_{A0h}^{\mathrm{sc}} \begin{pmatrix} \boldsymbol{e}_h & \boldsymbol{e}_v \end{pmatrix} \begin{pmatrix} 1 \\ q_A \end{pmatrix}, \tag{10.120}$$

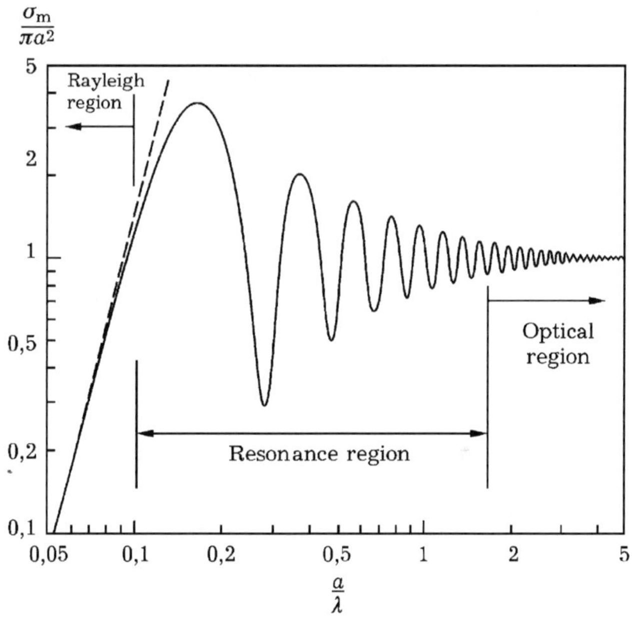

Figure 10.14 MRCS of a perfectly conducting sphere as a function of the radius-to-wavelength ratio, a/λ

where $E^{sc}_{A0h} \neq 0$ is assumed and q_A is the polarization ratio that defines the polarization of the wave component $\mathbf{E}^{sc}_A$. The component $\mathbf{E}^{sc}_N$ is represented as

$$\boldsymbol{E}^{sc}_N = E^{sc}_{N0h}\boldsymbol{e}_h + E^{sc}_{N0v}\boldsymbol{e}_v. \tag{10.121}$$

According to Reference [II.2] the parameters E^{sc}_{A0h}, E^{sc}_{N0h} and E^{sc}_{N0v} may be determined by two conditions: (a) Equation (10.116) is true and (b) the components $\mathbf{E}^{sc}_A$ and $\mathbf{E}^{sc}_N$ are orthogonal. The second condition requires that [II.2]

$$\boldsymbol{E}^{sc}_A \cdot \mathbf{E}^{sc^*}_N = E^{sc}_{A0h}\left(E^{sc^*}_{N0h} + q_A E^{sc^*}_{N0v}\right) = 0. \tag{10.122}$$

Then,

$$E^{sc}_{N0h} = -q^*_A E^{sc}_{N0v}. \tag{10.123}$$

Taking into account Equations (10.123) and (10.116) to (10.121), it is possible to obtain, following Reference [II.2], that

$$E^{sc}_{A0h} = \frac{E^{sc}_{0h}\left(1 + q^*_A q^{sc}\right)}{1 + |q_A|^2}, \tag{10.124}$$

$$E^{sc}_{N0h} = \frac{-E^{sc}_{0h} q^*_A \left(q^{sc} - q_A\right)}{1 + |q_A|^2}, \tag{10.125}$$

$$E^{sc}_{N0v} = \frac{E^{sc}_{0h}\left(q^{sc} - q_A\right)}{1 + |q_A|^2}. \tag{10.126}$$

The power associated with the wave component $\mathbf{E}_A^{\text{sc}}$ can be received by the antenna and its density is [II.2]:

$$\text{Power per unit area in } \mathbf{E}_A^{\text{sc}} = \kappa \left|\mathbf{E}_A^{\text{sc}}\right|^2 = \frac{\kappa \left|E_{0h}^{\text{sc}}\right|^2 \left|1 + q_A^* q^{\text{sc}}\right|^2}{1 + |q_A|^2}. \tag{10.127}$$

This result can be used to determine the RCS from Equation (10.117) as

$$\sigma = \lim_{R_2 \to \infty} 4\pi R_2^2 \frac{\left|E_{0h}^{\text{sc}}\right|^2}{\left|\mathbf{E}^{\text{in}}\right|^2} \frac{\left|1 + q_A^* q^{\text{sc}}\right|^2}{1 + |q_A|^2} \tag{10.128}$$

and the polarization efficiency is

$$\rho_{\text{pol}} = \frac{\left|1 + q_A^* q^{\text{sc}}\right|^2}{\left(1 + |q_A|^2\right)\left(1 + |q^{\text{sc}}|^2\right)}. \tag{10.129}$$

It should be remembered that all results regarding the SCS and RSC in this section are applied to small targets only. However, bistatic radars can also operate with spatially distributed targets (see Section 10.3). Sea and ocean surfaces, and land surfaces, including ice, snow, wood, desert, etc., are the most commonly known examples of area-extensive targets. This term means that the returned signal is obtained from electromagnetic energy reflected by the whole surface covered by the radar beam.

To generalize the concept of the scattering cross-section in the case of area-extensive targets, spatially distributed targets will be presented that consist of an infinite number of small targets with differential scattering cross-sections, $d\sigma^{\text{sc}}$ [II.24]. The latter can be written as the product of the scattering cross-section, σ_0^{sc}, of a unit area of the target, on the differential of the area, ds, viz.:

$$d\sigma^{\text{sc}} = \sigma_0^{\text{sc}}\, ds. \tag{10.130}$$

The quantity σ_0^{sc} is known as the scattering cross-section coefficient (SCSC). It is dimensionless and is usually expressed in decibels.

Thus, the scattering cross-section coefficient represents a derivative from the scattering cross-section on the area

$$\sigma_0^{\text{sc}} = \frac{d\sigma^{\text{sc}}}{ds}. \tag{10.131}$$

Similarly,

$$\sigma_0 = \frac{d\sigma}{ds} \tag{10.132}$$

is the radar cross-section coefficient (RCSC).

Sometimes the SCSC can be written in terms of incident and scattered fields, viz.:

$$\sigma_0^{\text{sc}} = \lim_{R_2 \to \infty} \frac{4\pi R_2^2}{S_A} \frac{|\mathbf{E}^{\text{sc}}|^2}{\left|\mathbf{E}^{\text{in}}\right|^2}, \tag{10.133}$$

where S_A is an area of a part of a surface located inside the resolution cell of the transmitting and receiving antennas.

In the case of a statistical model of the rough surface of the target, the scattering (or radar) cross-section coefficient should be determined in the sense of an average meaning, i.e. Equations (10.131) and (10.132) should be corrected:

$$\sigma_0^{\text{sc}} = \frac{d\langle\sigma^{\text{sc}}\rangle}{ds}, \qquad \sigma_0 = \frac{d\langle\sigma\rangle}{ds}. \tag{10.134}$$

10.6 TARGET SCATTERING MATRICES

Though RCS is a widely used parameter it does not include all the available information about the scattering properties of a target. Additional information about the radar target is available through a specification of the scattered field components. There are eight such quantities: two complex components of the incident wave and two complex components of the scattering wave (see, for example, Equations (10.98) and (10.104)). These eight characteristics are best described with the help of a two-by-two complex matrix called a scattering matrix [II.2, II.4, II.17].

Note that the possible variation in the choice of coordinate systems can bring into existence a number of different types of scattering matrices. First of all a scattering matrix called the Jones matrix is developed. The definition of this scattering matrix supposes that the coordinate system is applied to the target, as shown in Figure 10.8. The mentioned four complex field components are E_{0h}^{in}, E_{0v}^{in}, E_{0h}^{sc} and E_{0v}^{sc}. Obviously, the scattering wave components E_{0h}^{sc} and E_{0v}^{sc} depend on both incident wave components E_{0h}^{in} and E_{0v}^{in}. The component of E_{0h}^{sc} (or E_{0v}^{sc}) caused by E_{0h}^{in} (or E_{0v}^{in}) may be thought of as a co-polarized term [II.2]. The component of E_{0h}^{sc} (or E_{0v}^{sc}) caused by E_{0v}^{in} (or E_{0h}^{in}) can be considered as a cross-polarized term [II.2]. Therefore, following Reference [II.2], it is possible to write

$$E_{0h}^{\text{sc}} = \left(T_{hh}E_{0h}^{\text{in}} + T_{hv}E_{0v}^{\text{in}}\right)\frac{\exp\left(-jk_0R_2\right)}{2\sqrt{\pi}R_2}, \tag{10.135}$$

$$E_{0v}^{\text{sc}} = \left(T_{vh}E_{0h}^{\text{in}} + T_{vv}E_{0v}^{\text{in}}\right)\frac{\exp\left(-jk_0R_2\right)}{2\sqrt{\pi}R_2}. \tag{10.136}$$

By substituting Equations (10.135) and (10.136) into Equation (10.104) it is possible to obtain a formula for the amplitude of a scattering electric field in the aperture of the receiving antenna:

$$\mathbf{E}^{\text{sc}} = \frac{\left(T_{hh}E_{0h}^{\text{in}} + T_{hv}E_{0v}^{\text{in}}\right)\boldsymbol{e}_h + \left(T_{vh}E_{0h}^{\text{in}} + T_{vv}E_{0v}^{\text{in}}\right)\boldsymbol{e}_v}{2\sqrt{\pi}R_2}\exp\left(-jk_0R_2\right). \tag{10.137}$$

Parameters T_{pq} in Equations (10.135) to (10.137) are the complex constants of proportionality [II.2], where the first subscript represents the coordinate corresponding to the response while the second subscript represents the incident field's coordinate. For example, T_{vh} corresponds to the cross-polarized case and component E_{0h}^{in} gives rise to component E_{0v}^{sc}. The complex factor [II.2] $\exp\left(-jk_0R_2\right)/\left(2\sqrt{\pi}R_2\right)$ accounts for a phase shift of the scattering wave and the decreasing scattering power.

In matrix form Equations (10.135) and (10.136) may be rewritten as

$$\mathbf{E}^{\text{sc}} = \frac{\exp\left(-jk_0R_2\right)}{2\sqrt{\pi}R_2}\begin{pmatrix} T_{hh} & T_{hv} \\ T_{vh} & T_{vv} \end{pmatrix}\mathbf{E}^{\text{in}}. \tag{10.138}$$

It is necessary to note that this equation connects the Jones vectors (which were defined in Section 10.4) in the incident and the scattered waves. The 2×2 matrix

$$\mathbf{T} = \begin{pmatrix} T_{hh} & T_{hv} \\ T_{vh} & T_{vv} \end{pmatrix}$$

is known as the Jones matrix [II.2, II.23].

In the particular case of a monostatic radar, it is known that $T_{hv} = -T_{vh}$ [II.2, II.23]. However, for the bistatic radar in general, the Jones matrix includes eight different real quantities. However, the phase of one of the matrix elements may be factored from the matrix and included in the exponential term of Equation (10.138). The result is a relative matrix requiring specification of only four amplitudes and three relative phases [II.2]. The new parameters introduced above also allow Equation (10.105) to be transformed for the polarization ratio of a scattered wave, namely to present it as

$$q^{\mathrm{sc}} = \frac{T_{vh}E_h^{\mathrm{in}} + T_{vv}E_v^{\mathrm{in}}}{T_{hh}E_h^{\mathrm{in}} + T_{hv}E_v^{\mathrm{in}}} = \frac{T_{vh} + T_{hh}q^{\mathrm{in}}}{T_{hh} + T_{hv}q^{\mathrm{in}}}. \tag{10.139}$$

The alternative to a Jones matrix is a Sinclair matrix. This corresponds to the case where the origin of a coordinate system does not coincide with a target position, as in the case of the Jones matrix, but with the phase centre of a receiving antenna. By using procedures similar to those for the Jones matrix it is possible to obtain the follows formulae for scattering field components:

$$E_h^{\mathrm{sc}} = \left(S_1 E_h^{\mathrm{in}} + S_3 E_v^{\mathrm{in}}\right)\frac{\exp\left(-\mathrm{j}k_0 R_2\right)}{2\sqrt{\pi}R_2}, \tag{10.140}$$

$$E_v^{\mathrm{sc}} = \left(S_4 E_h^{\mathrm{in}} + S_2 E_v^{\mathrm{in}}\right)\frac{\exp\left(-\mathrm{j}k_0 R_2\right)}{2\sqrt{\pi}R_2}, \tag{10.141}$$

where h and v are the coordinates in the polarization basis relative to the receiver and the S_p are the new complex constants of proportionality. The 2×2 matrix

$$\mathbf{\Sigma} = \begin{pmatrix} S_1 & S_3 \\ S_4 & S_2 \end{pmatrix}$$

is called the Sinclair matrix [II.2, II.23]. According to Reference [II.2], the Jones matrix and the Sinclair matrix are related to each other by the following equation:

$$\begin{pmatrix} T_{hh} & T_{hv} \\ T_{vh} & T_{vv} \end{pmatrix} = \begin{pmatrix} S_1 & S_3 \\ -S_4 & -S_2 \end{pmatrix}. \tag{10.142}$$

In the particular case of a monostatic radar it is known that $S_3 = S_4$ and

$$\mathbf{\Sigma} = \begin{pmatrix} S_1 & S_3 \\ S_3 & S_2 \end{pmatrix}, \tag{10.143}$$

i.e. the Sinclair matrix is symmetrical.

Recently, scattering matrices have found an application in the solution of problems of classification and identification of targets. This has a simple explanation: the scattering matrix is more informative than a cross-section. However, in spite of the fact that the RCS can be determined from a known scattering matrix, in practice the radar cross-section is more commonly measured.

In addition to scattering matrixes and cross-sections, it is sometimes useful to define the target vector, $\tau = 1/2\,\mathrm{Tr}\,\{\Sigma\psi\}$, which can be treated as a vectorization of the scattering matrix, $\mathbf{\Sigma}$ [II.25]. Here ψ is a set of 2×2 complex basis matrices, which are known in quantum mechanics as the Pauli matrices, viz.:

$$\psi = \sqrt{2}\left[\begin{pmatrix} 1 & 0 \\ 0 & 1 \end{pmatrix}\begin{pmatrix} 1 & 0 \\ 0 & -1 \end{pmatrix}\begin{pmatrix} 0 & 1 \\ 1 & 0 \end{pmatrix}\begin{pmatrix} 0 & -j \\ j & 0 \end{pmatrix}\right]. \tag{10.144}$$

Alongside the target scattering matrices, the so-called Kennaugh matrix is also used in the theory and practice of radar scattering. Like the scattering matrix, which links the two Jones vectors, incident and scattered (see, for example, Equation (10.138)), the Kennaugh matrix, **K**, connects the incident and the scattered Stokes vectors, viz.:

$$\begin{pmatrix} I^{\mathrm{sc}} \\ Q^{\mathrm{sc}} \\ U^{\mathrm{sc}} \\ V^{\mathrm{sc}} \end{pmatrix} = \mathbf{K}\begin{pmatrix} I^{in} \\ Q^{\mathrm{in}} \\ U^{\mathrm{in}} \\ V^{\mathrm{in}} \end{pmatrix}. \tag{10.145}$$

One more approach to the solution of a scattering problem is connected to the use of so-called phenomenological parameters. During the 1970s, Huynen published [II.26] his 'phenomenological theory of radar targets' in which he defines nine physical parameters – viz. A_0, B_0, B, C, D, E, F, G, H – for the case of monostatic radars. These parameters may be expressed by elements of a monostatic Sinclair matrix as

$$
\begin{aligned}
A_0 &= \tfrac{1}{4}\left(|S_1 + S_2|^2\right), \qquad B_0 = \tfrac{1}{4}\left(|S_1 - S_2|^2\right) + |S_3|^2, \\
B &= \tfrac{1}{4}\left(|S_1 - S_2|^2\right) - |S_3|^2, \\
C &= \tfrac{1}{2}\left(|S_1 - S_2|^2\right), \qquad D = \operatorname{Im}\left(S_1 S_2^*\right), \\
E &= \operatorname{Re}\left[S_3^*\left(S_1 - S_2\right)\right], \qquad F = \operatorname{Im}\left[S_3^*\left(S_1 - S_2\right)\right], \\
G &= \operatorname{Im}\left[S_3^*\left(S_1 + S_2\right)\right], \qquad H = \operatorname{Re}\left[S_3^*\left(S_1 + S_2\right)\right].
\end{aligned}
\tag{10.146}
$$

These parameters draw out the information about the geometrical properties and physical structure of a target. The parameters A_0, $B_0 + B$ and $B_0 - B$ are called generators of the target structure: A_0 is related to the symmetry of the target, $B_0 - B$ to the nonsymmetry of the target and $B_0 + B$ measures the target's irregularity. Huynen's parameters are not independent: they are tied together by four well-known equations [II.11]

$$
\begin{aligned}
2A_0\left(B_0 + B\right) &= C^2 + D^2, \qquad 2A_0E = CH - DG, \\
2A_0\left(B_0 - B\right) &= G^2 + H^2, \qquad 2A_0F = CG + DH.
\end{aligned}
\tag{10.147}
$$

The generalization of Huynen's phenomenological theory for the case of a bistatic radar is not inconsequential. The difficulties are connected with the fact that in the bistatic case the form of the scattering matrix, $\mathbf{\Sigma}$, is generally nonsymmetrical. In Reference [II.11] one successful idea is offered: to break down this nonsymmetrical scattering matrix into a sum of two matrices, a symmetric matrix, $\mathbf{\Sigma}_S$, and a skew-symmetric matrix, $\mathbf{\Sigma}_{SS}$:

$$
\mathbf{\Sigma} = \begin{pmatrix} S_1 & S_3 \\ S_4 & S_2 \end{pmatrix} = \begin{pmatrix} S_1 & S_5 \\ S_5 & S_2 \end{pmatrix} + \begin{pmatrix} 0 & S_6 \\ -S_6 & 0 \end{pmatrix} = \mathbf{\Sigma}_S + \mathbf{\Sigma}_{SS},
\tag{10.148}
$$

where

$$
S_5 = \frac{S_3 + S_4}{2}, \qquad S_6 = \frac{S_3 - S_4}{2}.
\tag{10.149}
$$

It is possible to assume that the skew-symmetric matrix depends only on the additional information due to the bistatic configuration.

Furthermore, Germond *et al.* show in Reference [II.11] that in the bistatic case the Kennaugh matrix can be broken down as the sum of three matrices $\mathbf{K}_S$, $\mathbf{K}_{SS}$ and $\mathbf{K}_C$, i.e.

$$
\mathbf{K} = \mathbf{K}_S + \mathbf{K}_{SS} + \mathbf{K}_C,
\tag{10.150}
$$

where $\mathbf{K}_S$ is already known. Because the Kennaugh matrix corresponds to the monostatic case, matrix $\mathbf{K}_S$ is consequently symmetric, $\mathbf{K}_{SS}$ is a 4×4 diagonal matrix and $\mathbf{K}_C$ is an antisymmetric matrix with zero elements in its diagonal. Thus, the expressions for the matrix elements are obtained in Reference [II.11] as

$$
\begin{aligned}
K_{11} &= \tfrac{1}{4}\left(|S_1 + S_2|^2 + |S_1 - S_2|^2\right) + |S_5|^2 + |S_6|^2, \\
K_{21} &= \tfrac{1}{2}\left(|S_1|^2 - |S_2|^2 - |S_4|^2 + |S_3|^2\right), \\
K_{31} &= \operatorname{Re}\left[S_5^*\left(S_1 + S_2\right) - S_6^*\left(S_1 - S_2\right)\right], \\
K_{41} &= \operatorname{Im}\left[S_5^*\left(S_1 - S_2\right) - S_6^*\left(S_1 + S_2\right)\right], \\
K_{12} &= \tfrac{1}{2}\left(|S_1|^2 - |S_2|^2 + |S_4|^2 - |S_3|^2\right),
\end{aligned}
$$

$$
\begin{aligned}
K_{22} &= \tfrac{1}{4}\left(|S_1+S_2|^2+|S_1-S_2|^2\right)-|S_5|^2-|S_6|^2,\\
K_{32} &= \mathrm{Re}\left[S_5^*(S_1-S_2)-S_6^*(S_1+S_2)\right],\\
K_{42} &= \mathrm{Im}\left[S_5^*(S_1+S_2)-S_6^*(S_1-S_2)\right],\\
K_{13} &= \mathrm{Re}\left[S_5^*(S_1+S_2)+S_6^*(S_1-S_2)\right],\\
K_{23} &= \mathrm{Re}\left[S_5^*(S_1-S_2)+S_6^*(S_1+S_2)\right],\\
K_{33} &= \tfrac{1}{4}\left(|S_1+S_2|^2-|S_1-S_2|^2\right)+|S_5|^2-|S_6|^2,\\
K_{43} &= \mathrm{Im}\left(S_1S_2^*-S_4S_3^*\right),\\
K_{14} &= \mathrm{Im}\left[S_5^*(S_1-S_2)+S_6^*(S_1+S_2)\right],\\
K_{24} &= \mathrm{Im}\left[S_5^*(S_1+S_2)+S_6^*(S_1-S_2)\right],\\
K_{34} &= \mathrm{Im}\left[S_1S_2^*+S_4S_3^*\right],\\
K_{44} &= -\tfrac{1}{4}\left(|S_1+S_2|^2-|S_1-S_2|^2\right)+|S_5|^2-|S_6|^2.
\end{aligned}
\tag{10.151}
$$

Reference [II.11] also offers new bistatic physical parameters A, I, J, K, L, M and N, through which the Kennaugh matrix can is expressed as

$$
\boldsymbol{K}=\begin{pmatrix}
A_0+B_0+A & C+I & H+N & F+L\\
C-I & A_0+B-A & E+K & G+M\\
H-N & E-K & A_0-B-A & D+J\\
F-L & G-M & D-J & -A_0+B_0-A
\end{pmatrix}.
\tag{10.152}
$$

Here,

$$
\begin{aligned}
A &= |S_6|^2, & G &= \mathrm{Im}\left[S_5^*(S_1+S_2)\right],\\
A_0 &= \tfrac{1}{4}\left(|S_1+S_2|^2\right), & H &= \mathrm{Re}\left[S_5^*(S_1+S_2)\right],\\
B &= \tfrac{1}{4}\left(|S_1-S_2|^2\right)-|S_5|^2, & I &= 1/2\left(|S_4|^2-|S_3|^2\right),\\
B_0 &= \tfrac{1}{4}\left(|S_1-S_2|^2\right)+|S_5|^2, & J &= \mathrm{Im}\left(S_4S_3^*\right),\\
C &= \tfrac{1}{2}\left(|S_1|^2-|S_2|^2\right), & K &= \mathrm{Re}\left[S_6^*(S_1+S_2)\right],\\
D &= \mathrm{Im}\left(S_1S_2^*\right), & L &= \mathrm{Im}\left[S_6^*(S_1+S_2)\right],\\
E &= \mathrm{Re}\left[S_5^*(S_1-S_2)\right], & M &= \mathrm{Im}\left[S_6^*(S_1-S_2)\right],\\
F &= \mathrm{Im}\left[S_5^*(S_1-S_2)\right], & N &= \mathrm{Re}\left[S_6^*(S_1-S_2)\right].
\end{aligned}
\tag{10.153}
$$

The definitions (10.153) should be complemented with equations similar to (10.147). In a bistatic case they are [II.11]

$$
\begin{aligned}
2A_0(B_0+B) &= C^2+D^2, & 2A_0E &= CH-DG,\\
2A_0(B_0-B) &= G^2+H^2, & 2A_0F &= CG+DH,\\
2A(B_0-B) &= I^2+J^2, & 2AE &= JM-IN,\\
L(B_0-B) &= JH-IG, & 2AF &= -(JN+IM),\\
K(B_0-B) &= -(HI+JG).
\end{aligned}
\tag{10.154}
$$

According to Reference [II.11], Equations (10.154) reveal interdependence between the 16 bistatic characteristic parameters of the target.

11

Geometry of Bistatic Radars

The most obvious and evident difference of a bistatic radar from a monostatic radar is the separate location of the transmitter and the receiver, i.e. the difference in their geometries. Both cases, the three-dimensional (3D) and two-dimensional (2D) geometries, have been well documented in the existing scientific literature.

11.1 3D GEOMETRY OF BISTATIC RADARS

A central question within the theory of bistatic radar concerns the choice of the system of coordinates allowing the specification of the location of the target in space. In order to characterize the position of a target, not only in space but also in time, the system of coordinates is supplemented by a clock (in the broadest sense, any physical process allowing in principle time to be determined), and therefore the system of coordinates becomes a system of reference.

11.1.1 Systems of Coordinates

Many publications are dedicated to the problem of the choice of a coordinate system in a bistatic radar. The most sophisticated method of this problem's solution is presented in Reference [II.2].

Peebles [II.2] considers the 3D bistatic geometry shown in Figure 11.1. It can be seen from Figure 11.1 that Peebles takes into consideration three Cartesian right-hand coordinate systems and three spherical systems of coordinates, so that the transmitter, T, the target, Γ, and the receiver, R, are located at the origins of one Cartesian and one spherical coordinate system. This represents the most general but complicated method.

From the present point of view, the axes of Cartesian coordinates X_2, Y_2 and Z_2 are expedient for choosing one of a number of key directions, e.g. axes of symmetry of a target, normal to the surface of a target, and so on. Sometimes a certain benefit may be gained by choosing the Cartesian coordinates (X_2, Y_2, Z_2) as the main axes of inertia of a target. If the target is assumed to be an airplane, then the X_2 axis is usually directed along the fuselage (station line), the Y_2 axis is directed across the fuselage (butt line) and the Z_2 axis is chosen as vertical (water line) [II.27].

Bistatic Radar: Principles and Practice Edited by M. Cherniakov

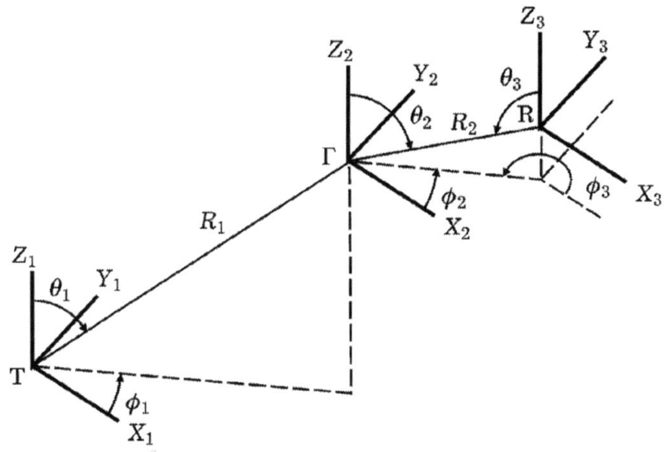

Figure 11.1 Three-dimensional coordinate systems

For simplicity, Peebles assumes that axes X_1, X_2 and X_3 are parallel among themselves, and, moreover, axes Y_1, Y_2 and Y_3 are also parallel among themselves. From there it follows that three Cartesian coordinate systems can be completely merged by translation (without rotation of the coordinate axes). Thus, the distinction between the three Cartesian systems of coordinates actually disappears. It is therefore possible to be limited only to one Cartesian system of coordinates, as shown in Reference [II.3] by Caspers.

Caspers proposed a single Cartesian coordinate system, with its origin at the midpoint of the baseline and the Xaxis directed along the baseline from the transmitter to the receiver. It is obvious that the coordinates of the transmitter (x_T, y_T, z_T) and receiver (x_R, y_R, z_R) in this case are

$$\begin{aligned} x_T &= -L/2, & x_R &= L/2, \\ y_T &= 0, & y_R &= 0, \\ z_T &= 0, & z_R &= 0. \end{aligned} \tag{11.1}$$

Moreover, Caspers did not attach any coordinate system to the target. However, he used two spherical systems of coordinates connected to the transmitter and the receiver. Thus, Caspers used two spherical and one Cartesian system of coordinates.

Kremer [II.6] was satisfied with one Cartesian and one spherical system of coordinates, whose common origin coincided with the transmitter. It seems probable, however, that for the analysis of noncooperative systems this is not the best choice. It is more logical to co-locate the origin of the coordinate system with the receiver. However, in fairness it should be noted that in the monograph [II.6] only the cooperative systems were analysed.

In this section, the 3D coordinate system is introduced, as shown in Figure 11.2. Here x_A, y_A and z_A are the Cartesian coordinates of any point A in space and ${}_Tr_A$, ${}_T\theta_A$ and ${}_T\phi_A$ are the spherical coordinates of the same point A in space, determined relative to the transmitter; ${}_Rr_A$, ${}_R\theta_A$ and ${}_R\phi_A$ are the spherical coordinates of the same point A, but are now determined relative to the receiver. According to radar theory and practice the coordinates ${}_Tr_A$ and ${}_Rr_A$ are known as 'ranges', whereas the coordinates ${}_T\phi_A$ and ${}_R\phi_A$ are termed 'azimuths'.

The two spherical coordinate systems attached to the transmitter and receiver and the Cartesian coordinate system with its origin in the middle point of the baseline one now introduced. The left-hand subscripts, T and R, represent the spherical coordinates determined relative to the transmitter and receiver respectively. Note that Cartesian coordinates have no left-hand subscript indexes. A right-hand

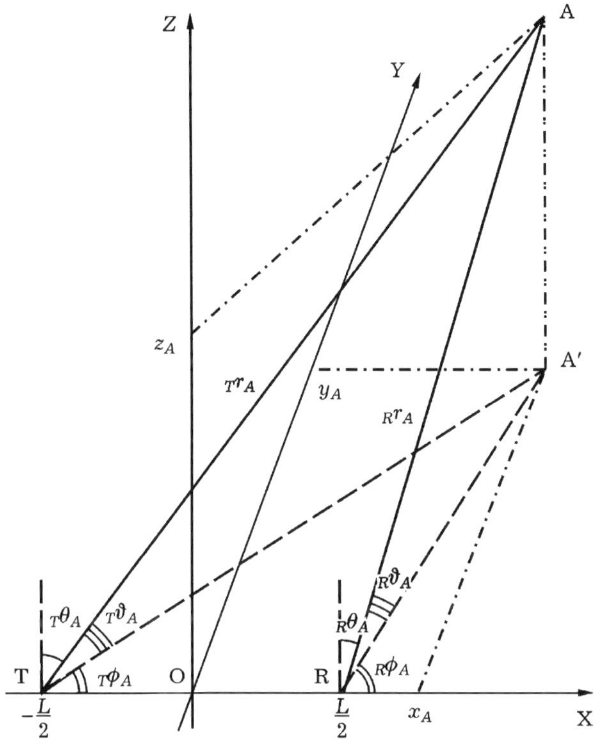

Figure 11.2 3D coordinate systems

subscript, where present, carries the information about the point in space, whose location is determined by its coordinate. For example, $_Tr_R$ means the range of the receiver from the transmitter, but $_Rr_T$ means the range of the transmitter from the receiver. Certainly, $_Tr_R = {}_Rr_T = L$, where L is the base of the bistatic radar. For the target, Γ, the coordinate $_Tr$ represents the range of the target from the transmitter, whereas $_Rr$ represents the range from the receiver (for simplicity, the right-hand subscript Γ is not used). Conventionally $_Tr$ and $_Rr$ are denoted as R_1 and R_2 respectively (see Figure 9.2). The sum $R_1 + R_2$ is known as the 'total range', or the 'range sum'.

In the Cartesian coordinate system, introduced above, there is a certain ambiguity in the directions of the coordinate axes *OY* and *OZ*. However, it is also obvious that, in the case of ground-based stationary bistatic systems presented here, axis *OZ* is naturally directed to the zenith. Thus, the coordinate plane *XOY* becomes horizontal.

It is important to bear in mind that in radar theory and practice another angular coordinate, namely elevation, $_T\vartheta_A = 90^\circ - {}_T\theta_A$ (such as elevation $_R\vartheta_A = 90^\circ - {}_R\theta_A$), is often used. The coordinates $_T\vartheta_A$ and $_R\vartheta_A$ are also shown in Figure 11.2.

Through $_T\mathbf{r}_A$, a radius vector of a point, A, will be designated in space relative to the transmitter. It is apparent that $_T\theta_A$ is the angle between this radius vector and the *OZ* axis and may vary from 0° to 180°. Elevation, $_T\vartheta_A$, however, is the angle between the same radius vector $_T\mathbf{r}_A$ and the *XOY* plane, and its region of variation is from –90° to 90°. It is possible to make similar statements for $_R\theta_A$ and $_R\vartheta_A$.

Lastly, the angles $_\beta\phi = {}_R\phi - {}_T\phi$ and $_\beta\vartheta = {}_R\vartheta - {}_T\vartheta$, which are projections of a bistatic angle, β, on the horizontal and vertical planes, are frequently used instead of angles $_R\phi$ and $_R\vartheta$. It is straightforward to demonstrate (see, for example, Reference [II.17]) that $\cos\beta = \cos_\beta\phi\cos_\beta\vartheta$.

Cartesian and spherical coordinates of any space point, A, are connected with each other by known formulae:

$$\begin{aligned} x_A &= {}_Tr_A \sin {}_T\theta_A \cos {}_T\phi_A - L/2, & x_A &= {}_Rr_A \sin {}_R\theta_A \cos {}_R\phi_A + L/2, \\ y_A &= {}_Tr_A \sin {}_T\theta_A \sin {}_T\phi_A, & y_A &= {}_Rr_A \sin {}_R\theta_A \sin {}_R\phi_A, \\ z_A &= {}_Tr_A \cos {}_T\theta_A, & z_A &= {}_Rr_A \cos {}_R\theta_A. \end{aligned} \tag{11.2}$$

It is easy to check that Equations (11.1) are a special case of Equations (11.2).

Another set of equations, which are analogous with Equations (11.2), are given in Reference [II.3]. They can be used after some minor updating, for example:

$$\begin{aligned} x_A &= -\frac{L \sin({}_T\phi_A + {}_R\phi_A)}{2 \sin({}_R\phi_A - {}_T\phi_A)}, \\ y_A &= \frac{L \sin {}_T\phi_A \sin {}_R\phi_A}{\sin({}_R\phi_A - {}_T\phi_A)}, \\ z_A &= \frac{L \sin {}_R\phi_A}{\sin({}_R\phi_A - {}_T\phi_A)} \tan {}_T\vartheta_A. \end{aligned} \tag{11.3}$$

According to Reference [II.3], formulae similar to Equations (11.3) are used in bistatic radar practice for the calculation of the coordinates of the targets.

11.1.2 Systems of Reference

To pass from a system of coordinates to a system of reference, it is necessary to first of all understand the Doppler effect. Imagine that during one interval of observation the target's movement relative to the bistatic radar is rectilinear. Its motion is uniform with speed $\mathbf{V}$. Let us consider two systems of reference. The first is a moving system, Σ' ($X'Y'Z't'$ or $\boldsymbol{r}'t'$), which is linked to the target. (the target, Γ, is constantly placed at the origin of coordinates, O', of the system Σ'. This is because the target is motionless relative to Σ'). The second is a motionless system, Σ ($XYZt$ or $\boldsymbol{r}t$), relative to which the phase centre of the receiving antenna is motionless. It should be remembered that, according to the special theory of relativity [II.28], time in the systems of reference, Σ and Σ', proceeds unequally; therefore $t' \neq t$. However, the initial moment of time is common for both systems of reference, i.e. $t = 0$ and $t' = 0$ simultaneously. The origin, O, of the motionless system is chosen in such a way that the time origin of the systems is also common, i.e. O and O' coincide at $t' = t = 0$.

The relationship between the spatial and time coordinates in the two systems, Σ and Σ', is given by the Lorentz transforms:

$$\boldsymbol{r}' = \boldsymbol{r} + \mathbf{V}\left[\frac{(\aleph - 1)\,\mathbf{V}\cdot\boldsymbol{r}}{V^2 - \aleph t}\right], \qquad t' = \aleph\left(t - \mathbf{V}\cdot\frac{\boldsymbol{r}}{c^2}\right), \tag{11.4}$$

where $V = |\mathbf{V}|$ is the module of the velocity vector, $\aleph = 1/\sqrt{1 - (V/c)^2}$ is a relative coefficient and c is the speed of light.

According to Reference [II.6], the range of possible velocities of the real radar target is $V < 10^{-4}c$. Therefore, $\aleph - 1 < 5 \times 10^{-9}$. Hence Equations (11.4) reduce to

$$\boldsymbol{r}' = \boldsymbol{r} - \mathbf{V}t, \qquad t' = t - \frac{\mathbf{V}\cdot\boldsymbol{r}}{c^2}. \tag{11.5}$$

Equations (11.4) and (11.5) are the direct Lorentz transforms. The *inverse* Lorentz transforms can be obtained by replacing all primed coordinates with similar coordinates but without primes, and vice versa, and also by sign reversal of the velocity vector, $\mathbf{V}$. The inverse Lorentz transformations are represented as

$$\boldsymbol{r} = \boldsymbol{r}' + \mathbf{V}t', \qquad t = t' + \frac{\mathbf{V}\cdot\boldsymbol{r}'}{c^2}. \tag{11.6}$$

Let the target reradiate (in the direction of the receiving antenna) a monochrome wave of frequency ω', whose wave front is the plane on the inside of the receiving aperture. The phase (measured relative to the system of reference, Σ', connected with the target) of the wave at the moment of time, t', in some point of space, $\boldsymbol{r}'$, is $\vartheta = \omega' t' - \boldsymbol{k}' \cdot \boldsymbol{r}'$, where $\boldsymbol{k}' = \omega' \boldsymbol{e}'_S / c = 2\pi \boldsymbol{e}'_S / \lambda$ is a wave vector; λ' is a wavelength, $\boldsymbol{e}'_S$ is the unit Poynting vector (see Section 11.1) and $\boldsymbol{k}'$, λ' and $\boldsymbol{e}'_S$ are all brought into being within the moving system of reference, Σ'.

It is known that the phase of a wave is an invariant of Lorentz transforms, i.e. it does not change during the transition from the system of reference, Σ', to system Σ. Therefore in the system Σ, linked with the receiving antenna, the same phase can be written as $\vartheta = \omega t - \boldsymbol{k} \cdot \boldsymbol{r}$, where ω and $\boldsymbol{k} = \omega\, \boldsymbol{e}_S / c = 2\pi\, \boldsymbol{e}_S / \lambda$ are measured relative to the system of reference, Σ, a cyclic frequency and a wave vector of an electromagnetic field.

Equating among themselves, the last two formulae for a phase of a wave, and using the Lorentz transforms (11.4), it is not difficult to obtain the following relationships between the parameters of a field in two systems of reference, Σ' and Σ, i.e. the parameters of the reradiated (by target) field and the received field:

$$\boldsymbol{k}' = \boldsymbol{k} + \mathbf{V}\left[\frac{(\aleph - 1)\mathbf{V} \cdot \boldsymbol{k}}{V^2} - \frac{\aleph \omega}{c^2}\right], \tag{11.7}$$

$$\omega' = \aleph(\omega - \mathbf{V} \cdot \boldsymbol{k}) = (\omega - \mathbf{V} \cdot \boldsymbol{k})\left[1 + \frac{1}{2}\left(\frac{V}{c}\right)^2 + \frac{3}{8}\left(\frac{V}{c}\right)^4 + \cdots\right]. \tag{11.8}$$

The formulae (11.7) and (11.8) define the Doppler effect. Neglecting in Equations (11.7) and (11.8) the terms containing relative speed V/c in second and higher degrees, the approximated formulae describing the first-order Doppler effect can be obtained:

$$\boldsymbol{k}' = \boldsymbol{k} - \frac{\omega \mathbf{V}}{c^2}, \qquad \frac{2\pi\, \boldsymbol{e}'_S}{\lambda'} = \frac{2\pi}{\lambda}\left(\boldsymbol{e}_S - \frac{\mathbf{V}}{c}\right), \tag{11.9}$$

$$\omega' = \omega - \mathbf{V} \cdot \boldsymbol{k} = \omega\left(1 - \frac{\mathbf{V} \cdot \boldsymbol{e}_S}{c}\right). \tag{11.10}$$

Where, as before, $\boldsymbol{e}_S$ and $\boldsymbol{e}'_S$ are unit Poynting vectors, measured in the systems of reference, Σ and Σ', respectively; λ and λ' are wavelengths in these systems.

Equations (11.8) and (11.10) describe the Doppler effect's well-known change of frequency: the frequency of a field in the aperture of the receiving antenna differs from the frequency of the radiation of a moving source of an electromagnetic field (in this case, from the frequency of the wave, reradiated by the target).

Equations (11.7) and (11.9) describe two effects that have an essential importance for bistatic radar:

1. The wavelength, λ, of a field in the aperture of the receiving antenna differs from the wavelength, λ', of a field, reradiated by the target.
2. The direction of the Poynting vector (or normal, $\boldsymbol{n}$, to a wave front) of the received field differs from the direction of the Poynting vector (or normal, $\boldsymbol{n}'$, to a wave front) of the radiated field.

The latter results in an error (known as an 'aberration') in the estimation of a target's direction (in practice the direction of a normal to the front of the received wave is used as the target direction). For small sizes of receiving aperture it is usually possible to overlook this aberration, but for larger sizes, or in multisite systems, it can be critical.

It can be seen from formulae (11.7) to (11.10) that the influence of a target's motion on the parameters of a useful signal substantially depends on the relationship between the direction of a target's velocity vector and the direction of a Poynting vector of an electromagnetic wave. In order to discover this dependence, a vector of speed, $\mathbf{V}$, will be spread out on two components: radial, $\mathbf{V}_r$, directed opposite to Poynting's vector, and tangential, $\mathbf{V}_t$, directed normally to a Poynting vector. Substituting $\mathbf{V} = \mathbf{V}_r + \mathbf{V}_t$

in Equation (11.9), gives

$$\boldsymbol{k}' = \left(1 + \frac{V_r}{c}\right)\boldsymbol{k} + \frac{\omega \mathbf{V}_t}{c^2}, \tag{11.11}$$

where $V_r = |\mathbf{V}_r| = -\mathbf{V}\cdot\boldsymbol{e}_S$. From Equation (11.11) the relationships between the modules of a wave vector (i.e. wave numbers), measured in moving, Σ', and motionless, Σ, systems of reference, can be obtained:

$$|\boldsymbol{k}'| \approx \left(1 + \frac{V_r}{c}\right)|\boldsymbol{k}|. \tag{11.12}$$

Relationships between wavelengths in these two systems of reference are

$$\lambda' \approx \frac{\lambda}{1 + V_r/c} \tag{11.13}$$

and relationships between normalized (unit) Poynting's vectors are

$$\boldsymbol{e}_S' = \boldsymbol{e}_S - \frac{\mathbf{V}_t}{(c + V_r)}. \tag{11.14}$$

From the obtained equations it can be seen that the angle between the directions of wave propagation in the motionless and moving systems of reference (aberration angle, ι) is

$$\iota = \sin^{-1}\left(\frac{V_t}{c + V_r}\right) \approx \frac{V_t}{c}, \tag{11.15}$$

where $\sin^{-1}$ means inverse sine. Thus, the radial component of a target's velocity causes a change in wave number and wavelength, and the tangent component causes a change in the visible direction of wave propagation. The angular coordinates of a target are estimated by the distribution of the phase of a field on the aperture of the receiving antenna. Therefore, a turn of the wave vector can cause an error of measurement in the target angular coordinates [II.6].

It is possible to determine the conditions in which the performance of the Doppler turn of a wave vector does not have an essential influence on the distribution of a phase of a field on the aperture of the receiving antenna. Reference [II.6] leads us to believe that the Doppler turn can be dismissed, if the maximal phase shift, $\Delta\vartheta$, caused by Doppler effect on the aperture of the receiving antenna does not exceed $\pi/8$. For this purpose we should be satisfied with the condition $\Delta\vartheta \approx D|\boldsymbol{k} - \boldsymbol{k}'| \approx D[2\pi V_t/(\lambda' c)] < \pi/8$, i.e.

$$\frac{DV_t}{\lambda' c} < 1/16, \tag{11.16}$$

where D is the overall dimension of the receiving aperture.

It is not difficult to see that the left part of an inequality (11.16) represents the ratio (see Equation (11.15)) of a target direction estimation error to the antenna beamwidth. It follows from condition (11.16) that for a median size antenna and an average wavelength the bearing angle estimation error, which occurs due to the Doppler wave front rotation, is negligibly small. However, for large antenna systems the condition (11.16) is not always true. In particular in Reference [II.6], the Russian radio telescope RATAN-600 was discussed with parameters of $D = 600$ m, $\lambda = 3$ cm, $V = 8$ km/s, where this error reaches a half of the antenna beamwidth.

The Doppler change of frequency, as follows from Equations (11.8) and (11.10), is caused, primarily, by the radial component of velocity:

$$\omega = \omega'\left(1 - \frac{V_r}{c}\right)\left[1 + \frac{1}{2}\left(\frac{V}{c}\right)^2 + \cdots\right] \approx \omega'\left(1 - \frac{V_r}{c}\right). \tag{11.17}$$

From Equation (11.17) it follows that in the absence ($V_r = 0$ and $V = V_t$) of a radial component of a target's velocity, the motion of the target causes a change in frequencies known as the 'Doppler cross effect':

$$\omega = \omega' \left[1 + \frac{1}{2} \left(\frac{V_t}{c} \right)^2 + \cdots \right]. \tag{11.18}$$

However, at the target velocities, which take place in practice in a bistatic radar, a change of frequency caused by the Doppler cross effect is usually extremely small.

11.2 2D GEOMETRY OF BISTATIC RADARS

It is evident, in Figure 11.2, that a target, Γ, is absent and therefore the bistatic plane is not determined. We can establish the bistatic plane and move from 3D to 2D geometry if a target is brought into consideration. In this case it is convenient to co-locate a coordinate plane, *XOZ*, with the bistatic plane.

11.2.1 Systems of Coordinates

In this section the widely known north-referenced coordinate system, shown in Figure 11.3, is described to represent bistatic radar 2D geometries [II.29]. The north-referenced coordinate system has now become the standard for bistatic radars, and it will be seen that this coordinate system is quite useful for specifying a target's position relative to the transmitter or the receiver.

The orientation of the north-referenced coordinate system is similar to those used to create topographical maps. The look angle, ${}_{\mathrm{T}}\theta$, of the transmitter and the look angle, ${}_{\mathrm{R}}\theta$, of the receiver are measured clockwise from the north of the coordinate system, and are restricted to the interval from $-\pi/2$ to $3\pi/2$. Thus, the bistatic angle, β, is connected to these two angles by a simple relationship, viz. $\beta = {}_{\mathrm{T}}\theta - {}_{\mathrm{R}}\theta$. The bistatic angle, β, lies in the interval $[0, \pi]$, where the target lies above (north) the baseline and in the interval $[-\pi, 0]$, where the target is lying below (south) the baseline.

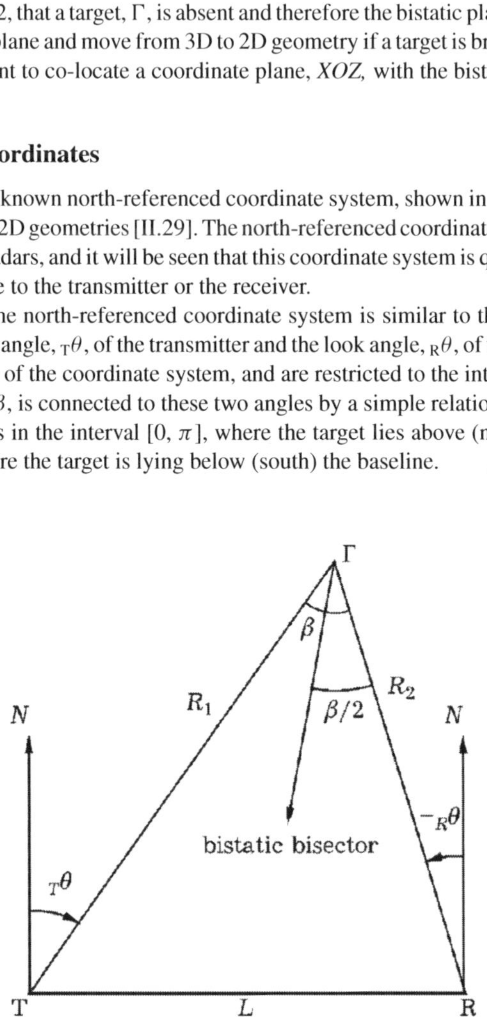

Figure 11.3 2D north-referenced coordinate system

The bistatic radar geometry can be completely specified in terms of any three of the four parameters, ${}_{\mathrm{T}}\theta$, ${}_{\mathrm{R}}\theta$, L and R. It is first assumed that R, ${}_{\mathrm{R}}\theta$ and L are known, which corresponds to the case of ground-based bistatic radars [II.7]. The transmitter-to-target range, R_1, can be calculated by solving the bistatic triangle of Figure 11.3:

$$R_1^2 = R_2^2 + L^2 - 2R_2L\cos\left(\frac{\pi}{2} + {}_{\mathrm{R}}\theta\right). \qquad (11.19)$$

Using the simple relation, $R_1 = R - R_2$, the following equations can be derived:

$$R_2 = \frac{R^2 - L^2}{2\,(R + L\sin{}_{\mathrm{R}}\theta)} \qquad (11.20)$$

and

$$R_1 = \frac{R^2 + L^2 + 2RL\sin{}_{\mathrm{R}}\theta}{2\,(R + L\sin{}_{\mathrm{R}}\theta)}. \qquad (11.21)$$

However, the cosines theorem also allows an alternative to Equation (11.19) to be written, viz.:

$$R_2^2 = R_1^2 + L^2 - 2R_1L\cos\left(\frac{\pi}{2} - {}_{\mathrm{T}}\theta\right). \qquad (11.22)$$

Using the equations written above, Tsao *et al*. [II.30] obtained

$${}_{\mathrm{T}}\theta = \begin{cases} \sin^{-1}\dfrac{(R^2 + L^2)\sin{}_{\mathrm{R}}\theta + 2RL}{(R^2 + L^2) + 2RL\sin{}_{\mathrm{R}}\theta}, & -\pi/2 \le {}_{\mathrm{R}}\theta \le \pi/2 \\[2ex] \pi - \sin^{-1}\dfrac{(R^2 + L^2)\sin{}_{\mathrm{R}}\theta + 2RL}{(R^2 + L^2) + 2RL\sin{}_{\mathrm{R}}\theta}, & \pi/2 \le {}_{\mathrm{R}}\theta < 3\pi/2 \end{cases} \qquad (11.23)$$

under the condition that $-\pi/2 < \sin^{-1}(\cdot) < \pi/2$. Thus, if the parameters R, ${}_{\mathrm{R}}\theta$ and L are known, other parameters of the bistatic geometry can be determined using Equations (11.20), (11.21) and (11.23). A similar analysis can be made for the case when parameters R, ${}_{\mathrm{T}}\theta$ and L are known.

Before the conclusions of Tsao *et al.*, Aver'yanov [II.7] assumed that the known parameters are ρ, ${}_{\mathrm{R}}\theta$ and L, where $\rho = R_1 + R_2 - L$ is the difference between the path of the useful signal (passing through the 'transmitter–target–receiver') and the path of the direct wave from the transmitter to the receiver. Thus,

$$R_2 = \frac{\rho\,(2L + \rho)}{2\,[\rho + L\,(1 + \sin{}_{\mathrm{R}}\theta)]}. \qquad (11.24)$$

At ${}_{\mathrm{R}}\theta = \pi/2$, Equation (11.24) becomes the general formula [II.7] of the target range for the case of a monostatic radar:

$$R_2 = \frac{\rho}{2}. \qquad (11.25)$$

According to Reference [II.7], bistatic systems acting with angles ${}_{\mathrm{R}}\theta$ near to $\pi/2$ are used in military applications.

In the use of bistatic radars in astronomy, an asymptotic case of Equation (11.24), appropriate to the condition $L \gg \rho$, is of interest, viz.:

$$R_2 = \frac{\rho}{1 + \sin{}_{\mathrm{R}}\theta}\left\{1 - \frac{\rho\,(1 - \sin{}_{\mathrm{R}}\theta)}{2\,[\rho + L\,(1 + \sin{}_{\mathrm{R}}\theta)]}\right\}. \qquad (11.26)$$

According to Equation (11.26), the magnitude of range, R_2, poorly depends on the base L. This condition has significant importance, as it allows the determination of the target range in the absence of information about the size of the base.

In a quasi-monostatic case, when $\rho \gg L$, Equation (11.24) easily transforms into the following equation (see Reference [II.7]):

$$R_2 = \frac{\rho}{2}\left\{1 + \frac{L\,(1 - \sin{}_{\mathrm{R}}\theta)}{\rho\,[1 + L\,(1 - \sin{}_{\mathrm{R}}\theta)\,/\rho]}\right\}. \tag{11.27}$$

It follows from Equation (11.27) that a quasi-monostatic radar system is similar to a monostatic radar, which determines the target range according to Equation (11.25).

In determining the target range with Equation (11.24), it is necessary to measure the values of the parameters ρ, ${}_{\mathrm{R}}\theta$ and L. The base, L, for ground stations is measured by various methods with sufficiently high accuracy, e.g. by topographical binding. The method of determining parameter ρ depends on the type of transmitted signal used; e.g. in a monopulse radar it is possible to measure this parameter directly through a difference of arrival times of the reflected signal and the direct signal. (According to Reference [II.7], a monopulse method is the most logical one to use for a bistatic radar.) The angle ${}_{\mathrm{R}}\theta$ can be measured by one of the established methods: phase, peak-phase or peak [II.7].

It is also important to note that having measured parameters ρ and L, it is therefore possible to receive information about the range sum:

$$R = R_1 + R_2 = \rho + L. \tag{11.28}$$

Knowing the total range, an ellipse in the bistatic plane on which the target is located can be determined. Focuses of this ellipse coincide with the positions of the receiver and the transmitter, and the major axis is equal to the range sum, R. For a concrete definition of a target's location on an ellipse of constant total range, it is necessary to measure angle ${}_{\mathrm{R}}\theta$ in addition to parameters ρ and L.

At the same time, it is obvious that for a concrete definition of a target location on the ellipse it is possible to measure not only the angle ${}_{\mathrm{R}}\theta$ but also the angle ${}_{\mathrm{T}}\theta$. Such variance is realized in onboard bistatic radars, in which the direction of the transmitted beam, and also the size and location of the base, i.e. parameters ${}_{\mathrm{T}}\theta$ and L, are known. In this case, in the receiver station, the necessity disappears for an exact definition of a target's direction: it is sufficient to have a weakly directed receiving antenna ensuring the necessary capacity of an echo signal [II.7]. The target range in this case is determined by [II.7]

$$R_2 = \frac{(L + \rho)^2 + \left[L^2 - 2L\,(L + \rho)\sin{}_{\mathrm{T}}\theta\right]}{2\,[\rho + L\,(1 - \sin{}_{\mathrm{T}}\theta)]}. \tag{11.29}$$

At ${}_{\mathrm{T}}\theta \to 0$, Equation (11.29) is converted into Equation (11.25), which is known for the monostatic radar. The system is physically feasible for $L \gg \rho$ and small angles, ${}_{\mathrm{T}}\theta \approx 0$ [II.7]. In this case the beam will pass through an area that is close enough to the receiver and allows radar observation to be carried out. It is therefore not difficult to show that Equation (11.29) will be reduced to Equation (11.25). At $\rho \gg L$, Equation (11.29) accepts a kind of

$$R_2 = \frac{\rho}{2}\left[1 + \frac{L}{\rho}\,(1 - \sin{}_{\mathrm{T}}\theta)\,\frac{1 + 2L/\rho}{1 + L\,(1 - \sin{}_{\mathrm{T}}\theta)\,/\rho}\right], \tag{11.30}$$

which is also close to Equation (11.25).

It can be seen that, in the considered case, knowing ${}_{\mathrm{T}}\theta$ and L, and having measured ρ, it is possible to determine a target's direction relative to the receiver, ${}_{\mathrm{R}}\theta$:

$$\sin{}_{\mathrm{R}}\theta = \frac{\rho\,(2L + \rho)\cos{}_{\mathrm{T}}\theta}{(L + \rho)^2 + \left[L^2 - 2L\,(L + \rho)\sin{}_{\mathrm{T}}\theta\right]}. \tag{11.31}$$

If the three parameters, ${}_{\mathrm{T}}\theta$, L and ρ, and the angle, ${}_{\mathrm{R}}\theta$, are measured in the receiver, the superfluous information generated can be used for the specification of target coordinates [II.7].

The third variant is also considered in Reference [II.7] when the measured parameters are ${}_{\mathrm{T}}\theta$, ${}_{\mathrm{R}}\theta$ and ρ. In this case the range, R_2, is determined by means of the following equation:

$$R_2 = \frac{\rho \cos {}_{\mathrm{T}}\theta}{\cos {}_{\mathrm{T}}\theta - \cos {}_{\mathrm{R}}\theta + \sin ({}_{\mathrm{T}}\theta + {}_{\mathrm{R}}\theta)}. \tag{11.32}$$

Practically, such a scenario takes place in radar astronomy, when L is not known exactly, though is known to be rather large. It is possible to believe, then, that ${}_{\mathrm{T}}\theta \to 0$ and in Equation (11.32) it is necessary to overcome uncertainty:

$$R_2 = \frac{\rho}{1 + \sin ({}_{\mathrm{R}}\theta)}. \tag{11.33}$$

Equation (11.33) has similar limitations to Equation (11.26) at $\rho \ll L$. Both Aver'yanov [II.7] and Tsao *et al*. [II.30] have also considered the case when two angular coordinates ${}_{\mathrm{R}}\theta$ and ${}_{\mathrm{T}}\theta$ and range R are known. Aver'yanov considers this case to be typical for bistatic radars with a noncooperative transmitter. The next simple dependence is true for the bistatic triangle:

$$R_1 \cos {}_{\mathrm{T}}\theta = R_2 \cos {}_{\mathrm{R}}\theta, \tag{11.34}$$

Then

$$R_2 = \frac{R \cos {}_{\mathrm{T}}\theta}{\cos {}_{\mathrm{T}}\theta + \cos {}_{\mathrm{R}}\theta} \tag{11.35}$$

and

$$R_1 = \frac{R \cos {}_{\mathrm{R}}\theta}{\cos {}_{\mathrm{T}}\theta + \cos {}_{\mathrm{R}}\theta}. \tag{11.36}$$

From a bistatic triangle it is also not difficult to determine that

$$L = R_1 \sin {}_{\mathrm{T}}\theta - R_2 \sin {}_{\mathrm{R}}\theta. \tag{11.37}$$

Uniting (11.35) and (11.37), it is possible to obtain

$$L = \frac{R \sin ({}_{\mathrm{T}}\theta - {}_{\mathrm{R}}\theta)}{\cos {}_{\mathrm{T}}\theta + \cos {}_{\mathrm{R}}\theta}. \tag{11.38}$$

Thus, the formulae (11.35), (11.36) and (11.38) again express unknown parameters through known parameters.

In Reference [II.30] it is also shown that it is possible to find the simple relationship between the range rate and target velocity. Thus, only the assumption that the transmitter and the receiver are both stationary and focus on the target motion is upheld. It is also supposed with certainty that the target is located north of the baseline; by symmetry similar conclusions can be drawn for the case of a target located to the south of the baseline.

From Figure 11.4 it is obvious that

$$\frac{\mathrm{d}}{\mathrm{d}t} R_1 = V \cos \left(\phi - \frac{\beta}{2} \right) \tag{11.39}$$

and

$$\frac{\mathrm{d}}{\mathrm{d}t} R_2 = V \cos \left(\phi + \frac{\beta}{2} \right), \tag{11.40}$$

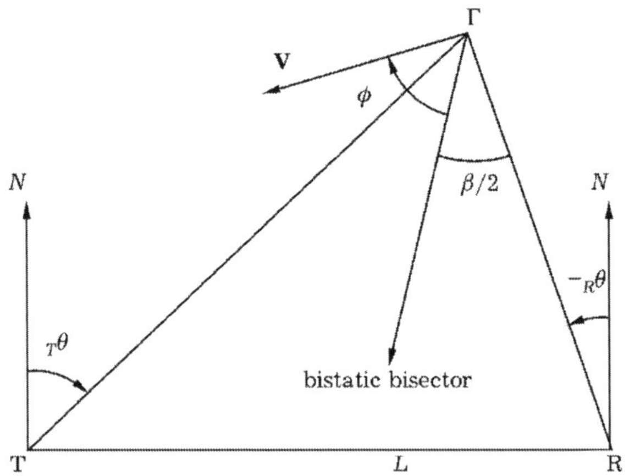

Figure 11.4 Velocity relationships in the bistatic radar geometry

where ϕ is measured clockwise from the bistatic bisector of the angle between this bisector and the target velocity vector, **V**. Hence,

$$\frac{\mathrm{d}}{\mathrm{d}t}R = \frac{\mathrm{d}}{\mathrm{d}t}R_1 + \frac{\mathrm{d}}{\mathrm{d}t}R_2 = 2V\cos\phi\cos\left(\frac{\beta}{2}\right) = 2V_\mathrm{b}\cos\left(\frac{\beta}{2}\right), \tag{11.41}$$

where $V_\mathrm{b} = V\cos\phi$ is equal to a projection of a speed vector to a bistatic bisector direction. From the factor $\cos(\beta/2)$ it can be seen that the resulting range rate is smaller in the bistatic radar than in the monostatic case, where $\beta = 0$, for the same target motion relative to the ground [II.30].

Using the results of References [II.30] and [II.31], it is possible to rewrite the previous equation as

$$\frac{\mathrm{d}}{\mathrm{d}t}R = V\cos\phi\sqrt{2\left(1 + \frac{R_2 + L\sin{}_\mathrm{R}\theta}{\sqrt{R_2^2 + L^2 + 2R_2L\sin{}_\mathrm{R}\theta}}\right)} \tag{11.42}$$

or

$$\frac{\mathrm{d}}{\mathrm{d}t}R = V\cos\phi\sqrt{2\left(1 + \frac{R_1 - L\sin{}_\mathrm{T}\theta}{2\sqrt{R_1^2 + L^2 - 2R_1L\sin{}_\mathrm{T}\theta}}\right)}. \tag{11.43}$$

11.2.2 Resolution

The resolution of a radar system is defined as its ability to separate two or more targets on the basis of range, velocity (Doppler) or angle.

11.2.2.1 Range Cell

The bistatic range cell is defined as the separation, ΔR, between two confocal concentric isorange contours (Figure 11.5). It is seen from Figure 11.5 that the separation between the two contours varies

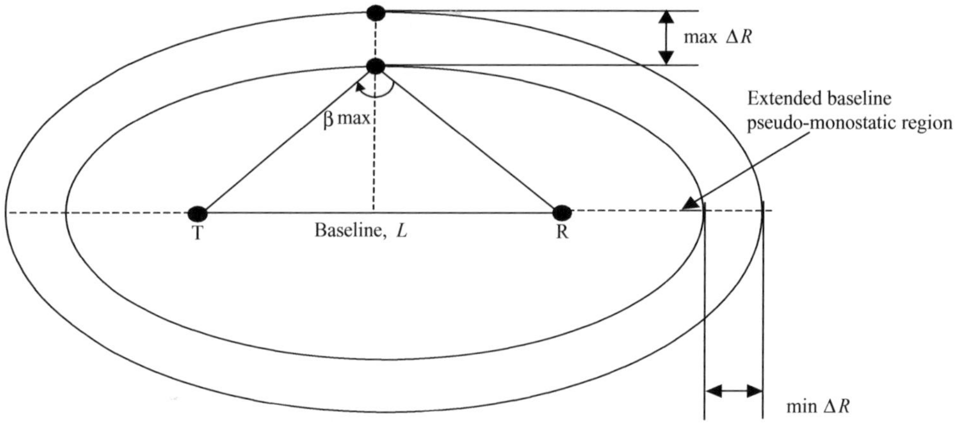

Figure 11.5 Bistatic range cell

depending upon the target position. The separation is minimum at the extended baseline and maximum at the bisector of the baseline. The bistatic range cell, ΔR (measured along the bisector of the bistatic angle of the inner ellipse), can be expressed as

$$\Delta R = \frac{c}{2\Delta f \cos(\beta/2)}, \tag{11.44}$$

where c is the velocity of light in vacuum and Δf is the bandwidth of the transmitted signal.

From Equation (11.44) it is observed that the bistatic range cell width degrades by cos $(\beta/2)$ compared to the equivalent monostatic radar. At the extended baseline $\beta = 0$ the range cell width of the bistatic radar is equal to the range cell width of monostatic radar. In this region, the bistatic radar is known to have a pseudo-monostatic geometry.

For monostatic and bistatic range resolution, an adequate degree of separation between two target echoes at the receiver is conventionally taken to be $c/(2\Delta f)$, where Δf is the bandwidth of the transmitted signal. In the bistatic case, the two targets lying on the isorange contours (Figure 11.5) must be at least physically separated by a distance given by Equation (11.44), to generate $c/(2\Delta f)$ echo separation at the receiver. The bistatic range cell is also referred to as the 'physical range resolution'.

11.2.2.2 Doppler Resolution

For monostatic and bistatic Doppler resolution, an adequate degree of Doppler separation between two target echoes at the receiver, ν_{b1} and ν_{b2} respectively, is conventionally taken to be $1/T_c$, where T_c is the receiver's coherent processing interval. Thus, the requirement for Doppler resolution is

$$|\nu_{b1} - \nu_{b2}| = \frac{1}{T_c}, \tag{11.45}$$

where, again, the equality represents a minimum requirement for Doppler separation. From Equation (11.41). The bistatic Doppler can be obtained as

$$\nu_{b1} = \frac{2V_1}{\lambda} \cos\phi_1 \cos\left(\frac{\beta}{2}\right), \qquad \nu_{b2} = \frac{2V_2}{\lambda} \cos\phi_2 \cos\left(\frac{\beta}{2}\right). \tag{11.46}$$

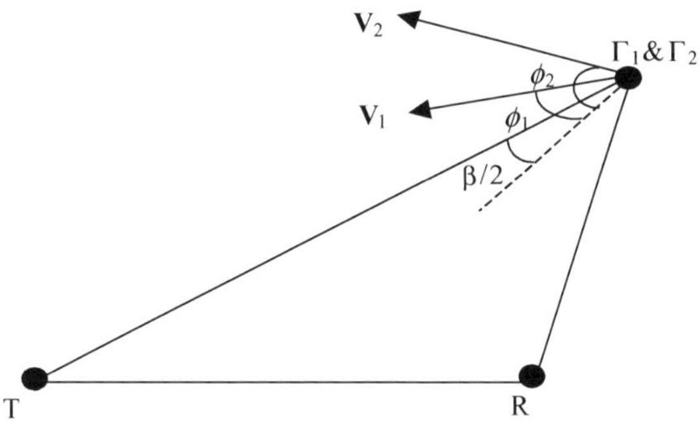

Figure 11.6 Geometry for bistatic Doppler resolution

The geometry for V_1, V_2, ϕ_1 and ϕ_2 is shown in Figure 11.6. The two targets are assumed to be collocated so that they share a common bistatic bisector. Combining Equations (11.45) and (11.46) yields

$$\Delta V_b = |V_1 \cos\phi_1 - V_2 \cos\phi_2| = \frac{\lambda}{2T_c \cos(\beta/2)}, \tag{11.47}$$

where ΔV_b is the required difference between the two target velocity vectors, projected on to the bistatic bisectors, for sufficient bistatic Doppler resolution. It is seen from Equation (11.47) that the velocity resolution degrades as the target approaches the baseline, $\beta \to 180°$; in fact there is a complete loss of resolution when the target is on the baseline. For $\beta = 0$, Equation (11.47) represents the monostatic case.

11.2.2.3 Angle Resolution

In a monostatic radar, the angle resolution is conventionally taken as $\geq \Delta\theta$, the 3 dB (one-way) antenna beamwidth. This means that the amplitude returns of two targets can be observed to separate when the targets are separated in angle by at least the 3 dB (one-way) beamwidth. The angle resolution for a monostatic radar can be defined as

$$\Delta\theta_m = \frac{\lambda}{D}, \tag{11.48}$$

where D is the dimension of the antenna in the plane of the angle θ. The above angular separation is equivalent to a physical target separation of $R\Delta\theta$, the cross-range dimensions of the monostatic radar, where R is the monostatic target range.

For a bistatic radar, the two targets are considered to lie on the same isorange contour (see Figure 11.7) and $R_1\Delta_T\theta > R_2\Delta_R\theta$; i.e. the cross-range dimension of the 3 dB transmitting beam is larger than the receiver's beam cross-range dimensions. Thus, it is too large to contribute to angle resolution. To generate a $R_2\Delta_R\theta$ cross-range separation, the two targets must be physically separated by Δ, where

$$\Delta \approx \frac{R_2\Delta_R\theta}{\sin[\pi - \beta/2]} = \frac{R_2\Delta_R\theta}{\cos(\beta/2)}. \tag{11.49}$$

11.2.3 Ambiguity Function

In a bistatic radar, the shape of the ambiguity function (AF) is dependent on both the waveform properties and the system geometry. Tsao *et al.* in their publication [II.30] have comprehensively discussed the AF for a bistatic radar. Here we summarise their work and with the aid of simple examples illustrate the nature of the bistatic AF.

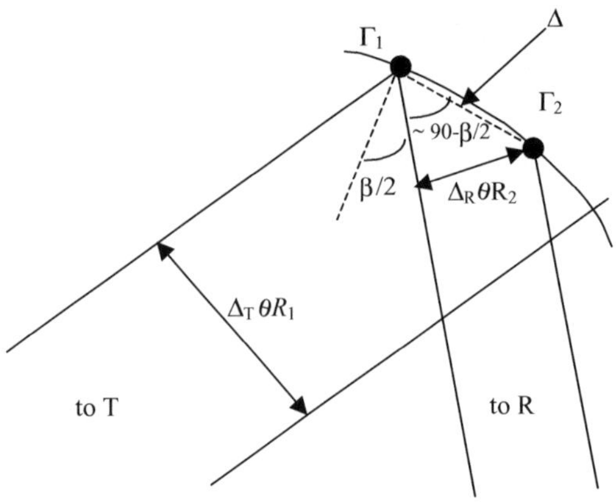

Figure 11.7 Geometry for angle resolution with only the receiver antenna beam contributing to the resolution

In Part I, it is seen that the monostatic AF is plotted as a function of delay and Doppler. The quantities delay and Doppler have a linear relationship with the target range and radial velocity. Therefore, the pair delay and Doppler or the pair range and velocity may be used in an interchangeable manner as shown by the arguments for the AF.

For a bistatic radar, the situation is quite different due to the geometry of the bistatic configuration. Here, the transmitter and the receiver are not at the same site, and the relative position and velocity of the target needs to be determined with respect to either the transmitter or the receiver or some other suitable reference point. As a result, it is of interest to evaluate how the receiver response varies with respect to the relative position and velocity in a corresponding manner. In the following discussion it is assumed that the reference point is at the receiver site; i.e. the range R_2 and the look angle ${}_{\mathrm{R}}\theta$ (see Figure 11.3) are used as the position parameters.

Let us first calculate the bistatic range sum and Doppler shift in terms of the receiver position parameters, R_2 and ${}_{\mathrm{R}}\theta$. It follows from Equation (11.19) that the range sum, R, can be expressed as

$$R = R_1 + R_2 = \sqrt{R_2^2 + L^2 + 2R_2 L \sin {}_{\mathrm{R}}\theta} + R_2. \tag{11.50}$$

The total time delay is given by

$$\tau_{\mathrm{b}}(R_2, {}_R\theta, L) = \frac{R_1 + R_2}{c} = \frac{\sqrt{R_2^2 + L^2 + 2R_2 L \sin {}_{\mathrm{R}}\theta} + R_2}{c}. \tag{11.51}$$

Recalling Equation (11.41), the bistatic Doppler shift can be expressed as

$$\nu_{\mathrm{b}} = \frac{2V}{\lambda} \cos\phi \cos\left(\frac{\beta}{2}\right) = \frac{2V_{\mathrm{b}}}{\lambda} \cos\left(\frac{\beta}{2}\right). \tag{11.52}$$

Using the cosine theorem the bistatic angle can be expressed in terms of R_2 and ${}_{\mathrm{R}}\theta$ as

$$\cos\beta = \frac{R_2 + L \sin {}_{\mathrm{R}}\theta}{\sqrt{R_2^2 + L^2 + 2R_2 L \sin {}_{\mathrm{R}}\theta}} \tag{11.53}$$

and

$$\cos\frac{\beta}{2} = \sqrt{\frac{1}{2} + \frac{R_2 + L \sin {}_{\mathrm{R}}\theta}{2\sqrt{R_2^2 + L^2 + 2R_2 L \sin {}_{\mathrm{R}}\theta}}}. \tag{11.54}$$

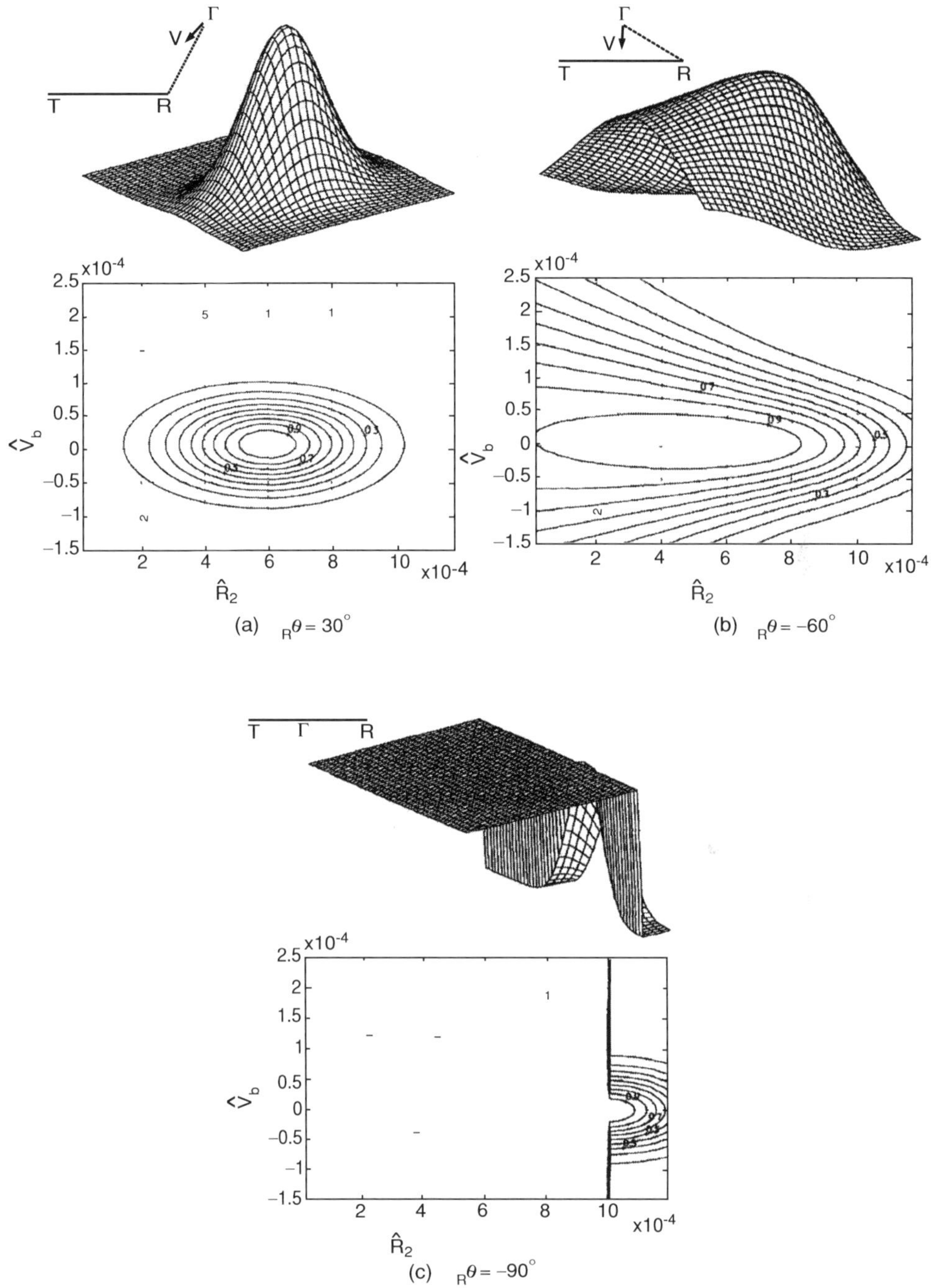

Figure 11.8 Ambiguity function of a simple Gaussian pulse: $L = 100$ km; $R_2 = 60$ km; $V_b = 600$ m/s. (From Tsao *et al.* [II.30]. Reproduced by permission of IEEE © 1997)

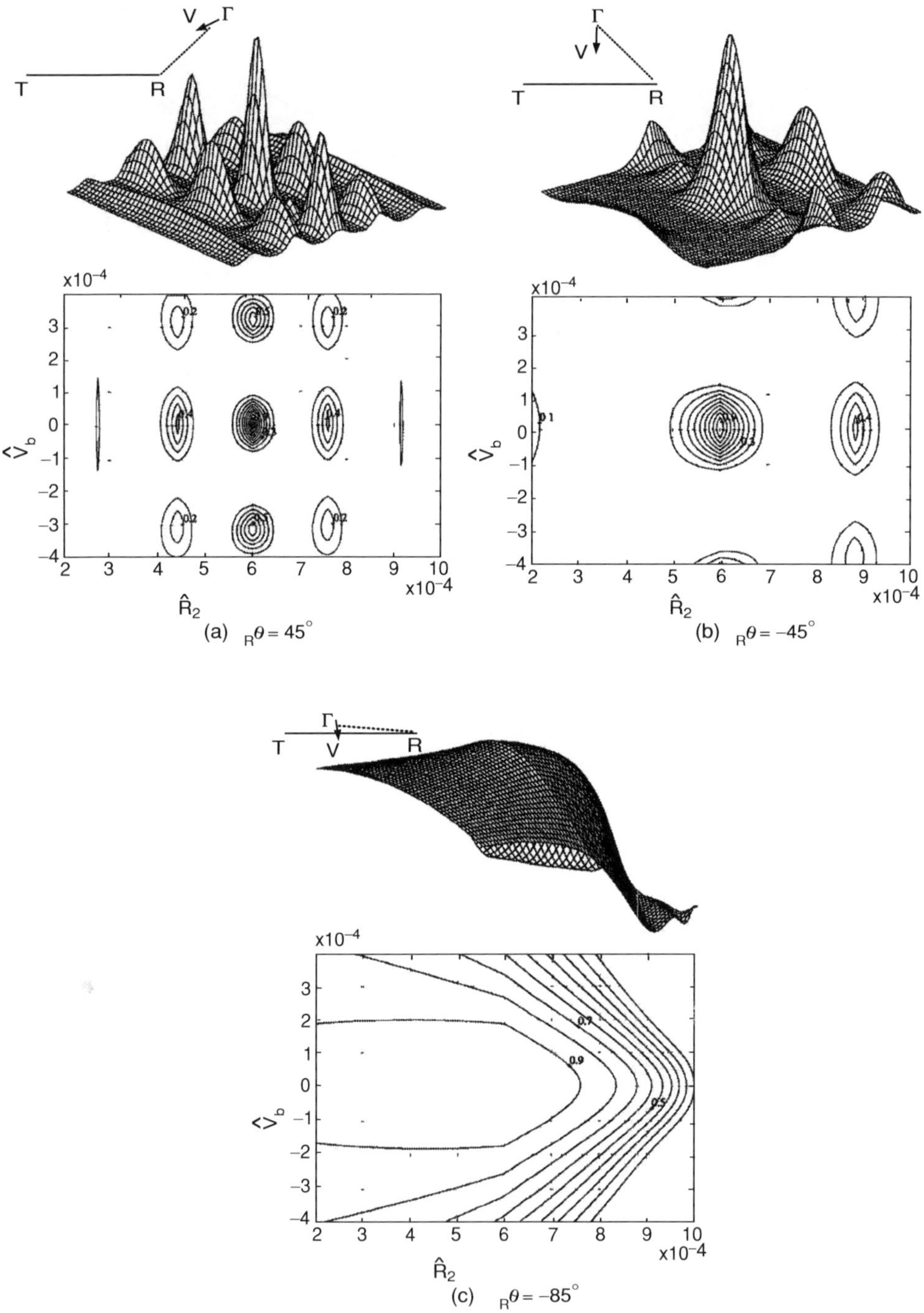

Figure 11.9 Ambiguity function of a pulse train: $L = 100$ km; $R_2 = 60$ km; $V_b = 600$ m/s. (From Tsao *et al.* [II.30]. Reproduced by permission of IEEE © 1997)

Substituting Equation (11.54) in Equation (11.52) gives

$$\nu_{\mathrm{b}} = \frac{2V_{\mathrm{b}}}{\lambda}\sqrt{\frac{1}{2} + \frac{R_2 + L\sin {}_{\mathrm{R}}\theta}{2\sqrt{R_2^2 + L^2 + 2R_2 L\sin {}_{\mathrm{R}}\theta}}}. \tag{11.55}$$

For $L = 0$, Equation (11.55) is reduced to the monostatic case.

From Equations (11.51) and (11.55) it is observed that unlike the monostatic case a nonlinear relationship exists between $(\tau_{\mathrm{b}}, \nu_{\mathrm{b}})$ and $(R_2, {}_{\mathrm{R}}\theta, L, V_{\mathrm{b}})$. Due to these nonlinear relationships, if the AF is plotted in the τ–ν plane, the behaviour of the waveform cannot immediately be infered as a function of R_2 and V_{b}, for the given values of ${}_{\mathrm{R}}\theta$ and L. Therefore, Tsao *et al*. [II.30] proposed expressing the AF for a bistatic radar as

$$\left|\chi\left(R_2, \hat{R}_2, V_{\mathrm{b}}, \hat{V}_{\mathrm{b}}, {}_{\mathrm{R}}\theta, L\right)\right| = \left|\int_{-\infty}^{\infty} s(\eta)s^*\left[\eta + \tau\left(R_2, \hat{R}_2, {}_{\mathrm{R}}\theta, L\right)\right]\exp\left[-\mathrm{j}2\pi v\left(R_2, \hat{R}_2, V_{\mathrm{b}}, \hat{V}_{\mathrm{b}}, {}_{\mathrm{R}}\theta, L\right)\eta\right]\mathrm{d}\eta\right|. \tag{11.56}$$

where $\tau = \tau_{\mathrm{b}}(R_2, {}_{\mathrm{R}}\theta, L) - \tau_{\mathrm{b}}\left(\hat{R}_2, {}_{\mathrm{R}}\theta, L\right)$ and $\nu = \nu_{\mathrm{b}}(R_2, V_{\mathrm{b}}, {}_{\mathrm{R}}\theta, L) - \nu_{\mathrm{b}}\left(\hat{R}_2, \hat{V}_{\mathrm{b}}, {}_{\mathrm{R}}\theta, L\right)$. The symbol ^ denotes the estimated values.

The basic concept of AF remains unchanged: it is a result of correlating the received signal with the delayed and Doppler shifted replica of the transmitted signal. As an example, Tsao *et al*. [II.30] calculated and plotted (see Figure 11.8) the ambiguity function of a simple Gaussian pulse,

$$s(t) = \frac{\exp\left[-t^2/(2T^2)\right]}{\sqrt[4]{\pi}\sqrt{T}}, \quad -\infty \le t \le \infty, \tag{11.57}$$

for several bistatic geometries and $T = 1.2 \times 10^{-4}$ s.

In the second example in Reference [II.30], the complex envelope of the signal is assumed to be a pulse train consisting of three rectangular pulses with a subpulse duration $T = 4 \times 10^{-5}$ s, period $T_{\mathrm{p}} = 10^{-4}$ s and carrier frequency $\omega = 3 \times 10^8$ rad/s. The other parameters remain the same as those in the previous example.

Figure 11.9 is the bistatic ambiguity function plotted as a function of R_2 and V_{b} at given values of ${}_{\mathrm{R}}\theta$ and L. According to Tsao *et al*. [II.30], the effects of geometry are clearly shown. The geometry considered represents the receiver-centred operating region. In a similar manner, the effects of the geometry can be examined for the transmitter-centred and the co-site regions. However, it was expected by the authors of Reference [II.30] that those effects would be less noticeable when the distance to the target increases, in which case the bistatic system behaves more and more like a monostatic system.

12

Maximum Range and Effective Area

Bistatic radar target observation, i.e. their detection, measurement of their parameters, etc., is possible if the signal power, P_R, of the target at the input of the receiver is not less than a specific threshold level, $P_{R\,min}$ [II.32, II.33]. The value of $P_{R\,min}$, characterizing the sensitivity of the radar receiver, depends on a number of aspects. These include: the type and intensity of interferences at the input of the receiver; the required efficiency of target detection and the accuracy in measuring their parameters; the duration of processing the received signal; as well as a number of other factors [II.33].

The space region where the condition of $P_R \geq P_{R\,min}$ is executed is called the 'effective area' of the bistatic radar. The effective area of the bistatic radar is quite often determined by characteristics of attainable accuracy of measurements [II.32] or by target detection characteristics [II.4, II.7]. In this case, a certain level of the signal-to-noise ratio is specified. This is necessary for target detection with a required detection probability at a given probability of false alarms; or for measurement of target parameters with a specified variance of measurement error. In this case, the space region is assumed to be an effective area of the bistatic radar if a target with a specified RCS remains in this area and the signal-to-noise ratio at the receiver is not less than a certain level [II.4].

The form and size of the effective area of a bistatic radar depends both on receiver sensitivity (i.e. on $P_{R\,min}$), as well as on other aspects such as: the power of the emitted signal; the RCS of the target; the transmitting and receiving antennas' directional pattern; atmospheric attenuation of the electromagnetic waves; reflection of electromagnetic waves from the ground surface, as well as other factors [II.10, II.32, II.33].

In order to find the main criteria determining the effective area of a bistatic radar, let us consider an elementary case; namely let us assume that the target is situated in a uniform free space where attenuation of electromagnetic waves is absent. Let us imagine that at some point within the space located at distance R_1 from the transmitter there is a target. Electromagnetic waves emitted by the transmitter have a spherical wave front, limited by the antenna directional pattern of the transmitting antenna at a sufficiently large distance from the transmitter. Therefore, the power flux density of the direct electromagnetic wave in the vicinity of a target is

$$\Pi_\Gamma = \frac{P_T g_T}{4\pi R_1^2}, \tag{12.1}$$

Bistatic Radar: Principles and Practice Edited by M. Cherniakov

where P_T is an output power and g_T is a directivity factor of the transmitting antenna. It is known [II.10, II.33] that the directivity factor, g, of any antenna and its gain coefficient, G, are related by the simple formula

$$G = \eta_A g, \tag{12.2}$$

where η_A is the efficiency factor of the antenna. As a rule, the efficiency factor for radar antennas is rather high and therefore it can be assumed that $G = g$.

Thus, instead of Equation (12.1) one can write down

$$\Pi_\Gamma = \frac{P_T G_T}{4\pi R_1^2}, \tag{12.3}$$

where G_T is the gain coefficient for the transmitting antenna. In the vicinity of the phase centre for the receiving antenna of a bistatic radar, the power flux density of the wave reradiated by the target is expressed as

$$\Pi_R = \frac{P_T G_T \sigma_b}{(4\pi)^2 R_1^2 R_2^2}, \tag{12.4}$$

where just as previously σ_b is the bistatic RCS of the target and R_2 is the range of the target relative to the receiver. Multiplying the value of Π_R by the effective area, A_R, of the receiving antenna, the power can be found of the reflected signal incoming to the input of the matched receiver:

$$P_R = \frac{P_T G_T A_R \sigma_b}{(4\pi)^2 R_1^2 R_2^2}. \tag{12.5}$$

From antenna theory it is known that

$$A = \frac{G\lambda^2}{4\pi}, \tag{12.6}$$

where λ is a wavelength. Taking into account Equation (12.6), the power of the received signal is

$$P_R = \frac{P_T G_T G_R \lambda^2 \sigma_b}{(4\pi)^3 R_1^2 R_2^2}, \tag{12.7}$$

where G_R is the receiving antenna gain. Formula (12.7) is known as the equation of the bistatic radar. Obviously, assuming in Equation (12.7) that $\sigma_b = \sigma_m$ and $R_1 = R_2 = R$, the widely known (see, for example, Reference [II.33]) equation of the monostatic radar can be obtained from the equation of the bistatic radar:

$$P_R = \frac{P_T G_T G_R \lambda^2 \sigma_m}{(4\pi)^3 R^4}. \tag{12.8}$$

Formula (12.8) presumes that, in most cases of the monostatic radar, both the transmitting and receiving antennas can be different. If a pulse monostatic radar receiver and transmitter can be connected to one antenna with the help of an antenna switch then $G_T = G_R = G$, and the monostatic radar's equation can be further simplified as

$$P_R = \frac{P_T G^2 \lambda^2 \sigma_m}{(4\pi)^3 R^4}. \tag{12.9}$$

Let us note once again that the equation of the bistatic radar (12.7) was obtained in the assumption of the absence of losses. It is commonly accepted in the scientific literature that any losses can be accounted for using the following coefficients [II.5, II.34]: L_{P1} = losses that appear during wave propagation in the area of the transmitter–target; L_{P2} = losses that appear during wave propagation in the area of the target–receiver; L_s = system losses. Thus, the equation of the bistatic radar can be written in the form [II.34]:

$$P_R = \frac{P_T G_T G_R \lambda^2 \sigma_b}{(4\pi)^3 R_1^2 R_2^2 L_{P1} L_{P2} L_s}, \tag{12.10}$$

which is transformed into the monostatic case as

$$P_R = \frac{P_T G^2 \lambda^2 \sigma_m}{(4\pi)^3 R^4 L_P L_s}. \tag{12.11}$$

Thus, Equations (12.7) and (12.10) generalize the equations of the monostatic radar.

Let us introduce the notation

$$\Omega = \frac{P_T G_T G_R \lambda^2}{(4\pi)^3}, \tag{12.12}$$

which allows the equation of the bistatic radar (12.7) to be rewritten in the form of

$$P_R = \frac{\Omega \sigma_b}{R_1^2 R_2^2}. \tag{12.13}$$

The expediency of introducing the parameter Ω is due to its independence from the target coordinates. However, the bistatic RCS of a target, σ_b, considerably depends on its location, even in the simplest case when the target is an absolutely conductive sphere with an ideal shape.

Nevertheless, in the literature [II.4, II.5, II.7, II.34] concerning bistatic radars, the effective area is usually determined with the assumption that $\sigma_b = \sigma_0$ = constant. Following this convention, let us obtain the dependency of the power of the signal received, P_R, from the target location. To perform this procedure let us locate the coordinate origin, O, in the middle point of the baseline between the receiver and transmitter. In this case, the distance ρ_0 of the target relative to the coordinate origin is equal to the median length of the bistatic angle (Figure 12.1). The OX axis is directed in such a way

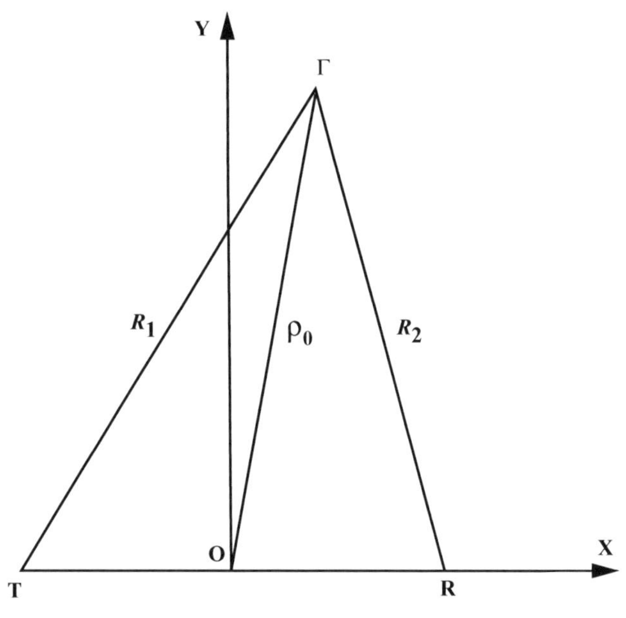

Figure 12.1 The median length of the bistatic angle

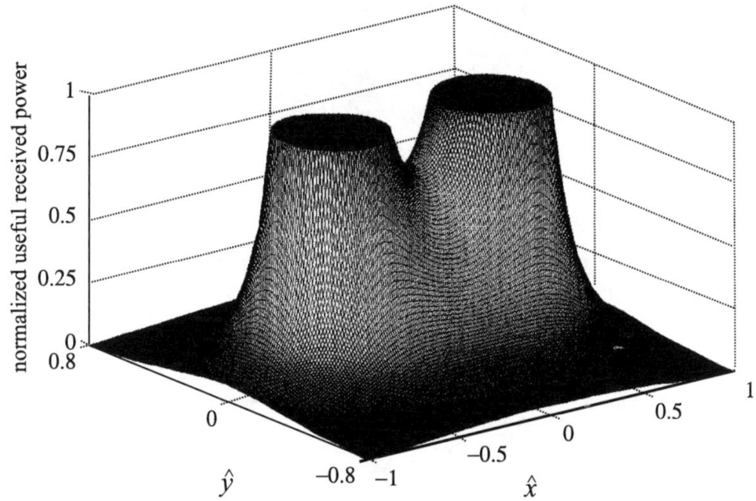

Figure 12.2 Dependence of normalized useful received power on the target location

that it passes through the phase centres of the transmitter and receiver antennas, while the *OY* axis is chosen to be arranged in the bistatic plane perpendicular to the *OX* axis. To continue the generalization, the normalized coordinates of the target, $\hat{x} = x/L$ and $\hat{y} = y/L$, are introduced and, as previously in this part, L is the base of the bistatic radar. Thus, according to Equation (12.1), the normalized coordinates of the transmitter are $\hat{x}_T = -0.5$, $\hat{y}_T = 0$, while those of a receiver are, $\hat{x}_R = 0.5$, $\hat{y}_R = 0$. Figure 12.2 demonstrates the dependence of normalized useful received power on the target location.

Note that formula (12.13) was deduced for the case of spherical waves. The wave is not spherical in the near-field area of the transmitting antenna. Therefore, if a target is located in this area Equations (12.7) and (12.13) are not applicable. Similarly, if a target is located quite close to the phase centre of the receiving antenna, then a secondary wave re-emitted by the target is also a nonspherical one; hence, formulae (12.7) and (12.13) again become incorrect.

It follows from Figure 12.2 that the possibility of operating with the required characteristics in the case of a bistatic radar depends not only on the target range but also on the angle position (azimuth) of a target relative to the baseline of the radar. Therefore, the use of the idea of the radar range [II.4] in the bistatic case is not as fruitful as in the monostatic scenario.

Substituting $P_R = P_{R\,min}$ in Equation (12.7), where just as earlier $P_{R\,min}$ is the sensitivity of the receiving system (the minimum possible value of received power), the following relation is obtained:

$$(4\pi)^3 R_1^2 R_2^2 P_{R\,min} = P_T G_T G_R \lambda^2 \sigma_b, \tag{12.14}$$

describing the closed line arranged in the bistatic plane. This closed line is the external boundary of the effective area of the bistatic radar. Assuming independence of the bistatic RCS, σ_b, on the parameters of the bistatic triangle, the external boundary of the effective area of the bistatic radar is a locus of points where a product of distances to two specified points is a constant value. In the present case, these are the points where the transmitter and receiver are situated. In mathematics these curves are known as Cassini curves.

Sometimes the ratio of power of the probing signal to receiver sensitivity, $\varsigma = P_T/P_{R\,min}$, is chosen as the independent radar characteristic and is called the 'energy potential' [II.10]. By using the idea of an energy potential, the formula describing the external boundary of an effective area of the bistatic radar

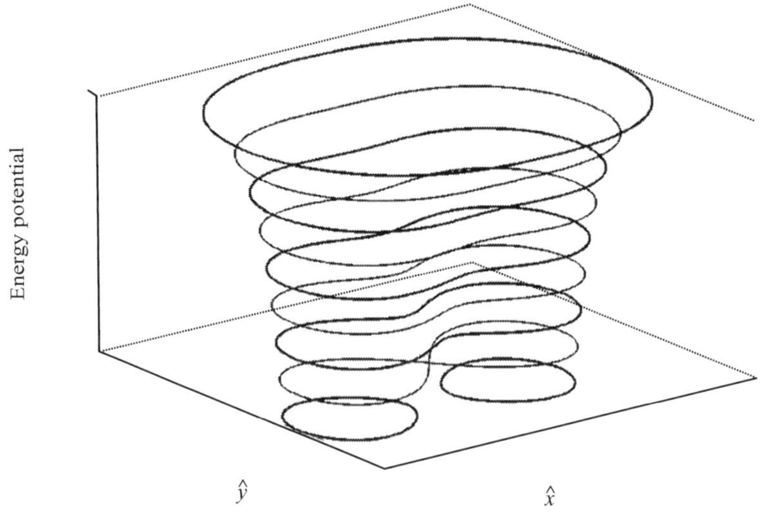

Figure 12.3 External boundaries of the effective area for different values of the energy potential

takes the form

$$(4\pi)^3 R_1^2 R_2^2 = \varsigma G_T G_R \lambda^2 \sigma_b. \tag{12.15}$$

In Figure 12.3, external boundaries of the effective area of a bistatic radar are presented for different values of the energy potential. For an analysis of effective areas of the bistatic radar it seems useful to introduce the geometric mean value,

$$\rho = \sqrt{(R_1 R_2)_{max}} = \sqrt[4]{\frac{\varsigma G_T G_R \lambda^2 \sigma_b}{(4\pi)^3}}, \tag{12.16}$$

for the range of a target situated on the Cassini oval. Parameter ρ is then called the 'effective range' of the bistatic radar.

Through analysis of Equation (12.16), it is possible to make the following conclusions:

1. The effective range of a bistatic radar increases quite slowly with an increase in the energy potential. For example, in order to raise the effective area of a bistatic radar by 2 times, its energy potential should be increased by 16 times, i.e. by 12 dB.
2. The effective range of a bistatic radar also slightly depends on the gain coefficients of the transmitting and receiving antennas. In order to double the effective range, it is necessary to increase both of these coefficients by 6 dB.
3. Dependence of the effective range of a bistatic radar on the wavelength is more complicated than it seems from Equation (12.16), since the parameters G_T, G_R and σ_b considerably depend on the wavelength (see, for example, Figure 10.14).
4. Dependence of the effective range of a bistatic radar on the RCS is also rather weak. An increase or decrease of the RCS by several decibels is equal, by implication, to the corresponding change in energy potential by the same value in decibels. Therefore, in order to increase (decrease) the effective area of the bistatic radar by 2 times, the RCS should be increased (reduced) by 12 dB.

It is seen from Figure 12.3 that where a high energy potential is present, the shape of the external boundary for the effective area of a bistatic radar is similar to an elliptic one. It can be shown [II.7] that the effective area of a bistatic radar has a similar convex form under the condition $\rho \geq L/\sqrt{2}$, where L is the bistatic base of the radar. With a decrease in the energy potential, the effective area of a bistatic radar is extended: for $L/2 < \rho < L/\sqrt{2}$, this area of bistatic radar remains a simple connected domain, but it ceases to be convex. For $\rho = L/2$, the boundary of the effective area of a bistatic radar is a closed curve known as the Bernoulli lemniscate. This is the case when the effective area of a bistatic radar consists of two symmetrical regions with the same shape and size contacting each other in one and the same point, the middle point of the baseline. For further decreases of the energy potential (for $\rho < L/2$), the effective area of a bistatic radar ceases to be simply connected; it is divided into two separate areas surrounding the transmitter and the receiver. In this case, the external boundary of the effective range of the radar is a so-called [II.7] degenerated Cassini oval. A more detailed discussion of the properties of effective areas of the bistatic radar bound with Cassini curves is given in Reference [II.7].

It should be remembered that for a monostatic radar, the effective region is a circle and its area is equal to

$$S = \pi R_{\max}^2, \tag{12.17}$$

where $R_{\max}$ is the maximum range of a monostatic radar. The area of the effective region for a bistatic radar can be calculated using the formula

$$S = 4 \int_0^{\pi/2} \int_0^{\rho_0} r\,\mathrm{d}r\,\mathrm{d}\varphi, \tag{12.18}$$

where ρ_0 is a target range relative to the coordinate origin point, O, situated in the middle point of the baseline of the radar (see Figure 12.1); r and φ are polar coordinates of the target relative to the same coordinate origin, O. The value of parameter ρ_0 can be calculated by [II.7]

$$\rho_0^2 = \frac{1}{4} L^2 \cos 2\varphi \pm \sqrt{\rho^4 - \frac{1}{16} L^4 \sin^2 \varphi}. \tag{12.19}$$

Formula (12.19) represents some other settings of the boundary for the effective area of a bistatic radar.

Aver'yanov [II.7] derived that, for $\rho \gg L/2$, the area of the effective region of a bistatic radar can be calculated using the approximate formula

$$S \simeq \pi \rho^2 \left(1 - \frac{1}{64} \frac{L^4}{\rho^4}\right). \tag{12.20}$$

According to Reference [II.7], for $\rho > L$ an error or the formula (12.20) does not exceed 10%. For the case of $\rho = L/2$, the area of the effective region of a bistatic radar can be calculated exactly by [II.7]

$$S = \rho^2. \tag{12.21}$$

In the case of $\rho \ll L/2$, the area of the effective region for a bistatic radar is [II.7]

$$S \simeq \pi \frac{\rho^4}{L^2}. \tag{12.22}$$

It can be seen from the formulae (12.20) to (12.22) that the influence of radar parameters, such as the energy potential, gain coefficients of transmitting and receiving antennas, wavelength, as well as the target RCS, on the value of the area for the effective region of a bistatic radar is determined by the effect of these parameters on the effective range length of the bistatic radar. From a comparison of Equations (12.17) to (12.22), it follows that for equal radar parameters and target RCSs, the effective range of a bistatic radar is less in area than the effective region of a monostatic (equivalent) radar [II.7].

One of the assumptions made in this chapter was that the derivation of the equation of the bistatic radar did not take into consideration the Earth's influence on the process of radio-wave propagation. This assumption allows us to consider both electromagnetic wave spherical forms, one emitted by the transmitting antenna and the other re-emitted by the target. If the bistatic plane is near the ground surface then the formulae represented in this chapter should be corrected. For example, instead of Equation (12.13),

$$P_{\mathrm{R}} = \frac{\Omega\sigma_{\mathrm{b}}}{R_1^{\alpha} R_2^{\alpha}} \tag{12.23}$$

should be written, where α is a real parameter that does not exceed 2. Certain values of this parameter depend on a number of factors: a degree of smoothness or roughness of the ground surface; electrophysical properties of the ground; the height of the transmitter, the receiver and target over the ground surface; polarization of the probing and received electromagnetic waves, and so on. By simple participation of the author of this chapter in experimental study (see, for example [II.35] and [II.36]), it was determined that in a certain situation parameter α can be close to 1.5.

According to Equation (12.23), the main equation of a bistatic radar must also be changed:

$$R_1^{\alpha} R_2^{\alpha} = \frac{\varsigma G_{\mathrm{T}} G_{\mathrm{R}} \lambda^2 \sigma_{\mathrm{b}}}{(4\pi)^3}. \tag{12.24}$$

However, from Equation (12.24) one can simply obtain

$$R_1^2 R_2^2 = \left[\frac{\varsigma G_{\mathrm{T}} G_{\mathrm{R}} \lambda^2 \sigma_{\mathrm{b}}}{(4\pi)^3}\right]^{2/\alpha}, \tag{12.25}$$

and from here it directly follows that, just as before, the external boundary of the effective area of a bistatic radar has the shape of a Cassini oval. Thus, taking into account the Earth's influence results only in a scaling change for Cassini curves.

As for the effective range of a bistatic radar, taking into consideration the Earth's influence results in

$$\rho = \left[\frac{\varsigma G_{\mathrm{T}} G_{\mathrm{R}} \lambda^2 \sigma_{\mathrm{b}}}{(4\pi)^3}\right]^{1/(2\alpha)}. \tag{12.26}$$

In particular, for $\alpha = 1.5$, expression (12.26) takes the form

$$\rho = \frac{\sqrt[3]{\varsigma G_{\mathrm{T}} G_{\mathrm{R}} \lambda^2 \sigma_{\mathrm{b}}}}{4\pi}. \tag{12.27}$$

Thus, if the radar bistatic plane is arranged near the ground surface, then in order to increase the effective range of the radar by 2 times it is sufficient to increase the energy potential of the radar by only 9 dB, not by 12 dB as was previously noted.

Until now, it has been assumed that the antennas of the transmitter and the receiver are omnidirectional and that the gain coefficients of transmitting, G_{T}, and receiving, G_{R}, antennas are constant (independent from the target direction). The use of a phased array as the receiving and the transmitting antennas of a bistatic radar can require a consideration of the dependence of parameters G_{T} and G_{R} on a target location while deriving the equation of the bistatic radar. A detailed discussion of this issue can be found in Reference [II.4].

13

Signal Models

Within radar, a target's related information can be obtained by analysing an electromagnetic (EM) field reradiated by the target. Only a small portion of this reradiated field is recovered by the aperture of the receiving antenna. The radar signal, in the form of current or voltage, is formed as the result of EM wave interaction with this antenna, and necessary information about the object is obtained via signal processing. It goes without saying that a part of the information is lost during the conversion of the electromagnetic wave to the electric signal. For example, information relevant to target distance is contained in the space structure of a spherical continuous wave formed by a point target. However, if the size of the receiving antenna aperture is rather small compared to the target distance, and this is a very common situation, the wave front is considered to be plane in the limits of the receiving antenna. Thus, information about the target distance is lost in the formation process of the electric signal. For an understanding of the bistatic radar system, it is important to obtain the relationships between EM waves reradiated by a target and the radar signal parameters.

13.1 SIGNALS FORMED BY A MOTIONLESS TARGET

If the moving speed (relative to the bistatic radar) of a target is not large, models that are correct for a stationary target can be used [II.6]. Consequently, we begin by considering the field formed by a stationary target.

As the main target model, we accept a point isotropic reradiative target that creates a spherical wave in a uniform environment. Usage of such a model allows the main laws to be acertained and the results to be found in a visual form [II.6]. Furthermore, it is known that a real target is often well described by the model of a bright point aggregate. Finally, Reference [II.6] maintains that equations which are correct for point targets are applicable for low-sized targets with small errors, if the sizes are significantly less than the proper radar resolution cell sizes.

The radius vector of the target, Γ, relative transmitter, T, is designated as **R**. It is directed from the phase centre of the transmitting antenna to the point of the target. The radius vector, $\boldsymbol{r}$, is the vector directed from the phase centre, R, of the receiving antenna to the point of the target, Γ. The vector $\boldsymbol{\rho}$

Bistatic Radar: Principles and Practice Edited by M. Cherniakov

is directed from the phase centre, T, of the transmitting antenna to the phase centre, R, of the receiving antenna. Obviously, these vectors form the bistatic triangle: $|\mathbf{R}| = R_1$, $|\boldsymbol{r}| = R_2$ and $|\boldsymbol{\rho}| = L$.

Let us consider the probing signal which is radiated from the phase centre of the transmitting antenna, T:

$$s_{\mathrm{T}}(t) = \mathrm{Re}\left[\dot{s}_{\mathrm{T}}(t)\right] = A\mathrm{Re}\left[\dot{s}_0(t)\right] = A\mathrm{Re}\left[\dot{U}(t)\exp(\mathrm{j}\omega_0 t)\right], \tag{13.1}$$

where A is a coefficient, which is constant in the observing area at parallel radar surveillance and is proportionate to the amplifying coefficient of the transmitting antenna at serial radar surveillance; $\dot{s}_0(t)$ is the complex signal of unit power; $\dot{U}(t) = U(t)\exp\left[\mathrm{j}\upsilon(t)\right]$ is its complex envelope; $U(t)$ is the law of the amplitude modulation; $\upsilon(t)$ is the law of the angle modulation; and $\omega_0 = 2\pi f_0$ is the cyclic carrier frequency.

The complex field intensity of a wave that reaches the target is defined by

$$\dot{u}_\Gamma(t) = \frac{C_1}{R_1}\dot{s}_0\left(t - \frac{R_1}{c}\right) = \frac{C_1}{R_1}\dot{U}\left(t - \frac{R_1}{c}\right)\exp\left[\mathrm{j}\omega_0\left(t - \frac{R_1}{c}\right)\right], \tag{13.2}$$

where c is the speed of propagation of the electromagnetic oscillations and C_1 is a constant factor. The current, which is induced in the target by the field of Equation (13.2), according to Reference [II.6] is defined by

$$i_\Gamma(t) = C_2 u_\Gamma(t) = \frac{C_1 C_2}{R_1}\mathrm{Re}\left[\dot{s}_0\left(t - \frac{R_1}{c}\right)\right] = \frac{C_1 C_2}{R_1}\mathrm{Re}\left\{\dot{U}\left(t - \frac{R_1}{c}\right)\exp\left[\mathrm{j}\omega_0\left(t - \frac{R_1}{c}\right)\right]\right\}, \tag{13.3}$$

where C_2 is a constant.

The complex envelope of the field in the point, R, of the phase centre of the receiving antenna is [II.6]

$$\dot{u}_{\mathrm{R}}(t) = \frac{C_1 C_2 C_3}{R_1 R_2}\dot{s}_0\left(t - \frac{R}{c}\right)\exp(\mathrm{j}\upsilon_\Gamma), \tag{13.4}$$

where $C_3 = \sqrt{\sigma}$ is a constant factor, σ is a constant RCS of the point isotropic scattering target, υ_Γ is the change in the wave phase reflecting from the target and $R = R_1 + R_2$ is the total range (see Section 11.1). In accordance with Reference [II.6], $C_3\exp(\mathrm{j}\upsilon_\Gamma)$ can be interpreted as a complex coefficient of target reflection. It is efficient to rewrite Equation (13.4) as

$$\dot{u}_{\mathrm{R}}(t) = \frac{R_1}{R_2}\dot{\alpha}\dot{s}_0\left(t - \frac{R}{c}\right) = \frac{R_1}{R_2}\dot{\alpha}\dot{U}\left(t - \frac{R}{c}\right)\exp\left[\mathrm{j}\omega_0\left(t - \frac{R}{c}\right)\right], \tag{13.5}$$

where $\dot{\alpha} = (C_1 C_2 C_3 / R_1^2)\exp(\mathrm{j}\upsilon_\Gamma)$ is a complex normalizing coefficient.

It is evident that Equation (13.5) is an assumption about the constancy of the amplifier coefficient of the transmitting antenna in the work sector of angles. However, directed properties of the transmitting antenna, if they exist, are easy to consider. In Equation (13.2) and below, the factor of attenuation, $1/R_1$, should be replaced by $\Phi(\mathbf{R})/R_1$, where $\Phi(\mathbf{R})$ is the amplifier coefficient of the transmitting antenna (throughout the field intensity) in the direction $\mathbf{R}$ [II.6]. This is possible if, as is usually the case in practice, radar targets are in the far-field region of the transmitting antenna. In this case, the antenna pattern or the directivity factor, which does not exist if the target position is in the near-field region of the antenna [II.6], can be used.

With the help of a simple equation,

$$s(t) = \mathrm{Re}\left[\dot{I}\,\dot{u}_{\mathrm{R}}(t)\right], \tag{13.6}$$

the expression for finding the signal model, which is created by the stationary target in a bistatic radar, can be easily obtained from Equation (13.5):

$$s(t) = \frac{R_1}{R_2}\mathrm{Re}\left[\dot{\alpha}\dot{I}\dot{s}_0\left(t - \frac{R}{c}\right)\right] = \frac{R_1}{R_2}\mathrm{Re}\left\{\dot{\alpha}\dot{I}\dot{U}\left(t - \frac{R}{c}\right)\exp\left[\mathrm{j}\omega_0\left(t - \frac{R}{c}\right)\right]\right\}. \tag{13.7}$$

Here $\dot{I} = I\exp(\mathrm{j}\Delta\upsilon)$ and factor I makes the connection between the intensity of the amplitude of the incident electromagnetic wave to the receiving antenna and the useful electric signal, where $\Delta\upsilon$ describes the fluctuation of the initial phase, which appears in the process of the conversion of the electromagnetic wave to an electric signal.

13.2 SIGNAL MODEL OF THE MOVING TARGET

The Doppler effect in bistatic radar theory has important features that need to be taken into consideration in comparison with monostatic radar theory. This means, equally, that with the variation of frequency it is necessary to take into consideration the variation of the wave vector of the field. Therefore, the frequency, the wavelength and the direction of Poynting's vector of the receiving electromagnetic wave differ from the same parameters of the wave which are reradiated by the moving target. The connection between these parameters can be found in terms of the Lorentz transform, i.e. using relativist methods (see Section 11.1.2).

In Chapter 11, the Doppler effect is considered using the simplest example of a plane monochromatic wave. However, the point target radiates a spherical wave. Moreover, a radar signal is not usually monochromatic. Therefore the field created by the moving target will be considered in this case.

It is assumed that at the interval of observation, the target moves linear by and evenly relative to the bistatic radar with velocity **V**. As in Section 11.1.2, two systems of reference will be considered: the moving system, Σ', connected with the target (the space–time coordinates are $\boldsymbol{r}'t'$) and the motionless (relative to the radar) system, Σ (the space–time coordinates are $\boldsymbol{r}t$).

13.2.1 Passive Radar

If the form of the signal, radiated by the target, is defined by the function $\dot{s}_0(t') = \dot{U}(t')\exp(\mathrm{j}\omega_0 t')$, the field of the spherical wave in the system, Σ', connected with the target is

$$\dot{u}(t', \boldsymbol{r}') = \frac{C_0}{|\boldsymbol{r}'|}\dot{s}_0\left(t' + t_\mathrm{p} - \frac{|\boldsymbol{r}'|}{c}\right) \approx \frac{C_0}{|\boldsymbol{r}'|}\dot{s}_0\left(t' + \frac{R_\Gamma}{c + {}_\Gamma V_\mathrm{r}} - \frac{|\boldsymbol{r}'|}{c}\right), \tag{13.8}$$

where C_0 is a constant, $t_\mathrm{p} \approx R_\Gamma/(c + {}_\Gamma V_\mathrm{r})$ is the propagation time of the signal, which is acted in point Γ at the moment $t = 0$, and R_Γ and ${}_\Gamma V_\mathrm{r}$ are the range and radial velocity of the target relative to point Γ at the moment $t = 0$.

This field can be written in the system, Σ, if the Lorentz transform is used:

$$\dot{u}(t, \boldsymbol{r}) = \frac{C_0}{|\boldsymbol{r} - \mathbf{V}t|}\dot{s}_0\left(t - \frac{\mathbf{V}\cdot\boldsymbol{r}}{c^2} + \frac{R_\Gamma}{c + {}_\Gamma V_\mathrm{r}} - \frac{|\boldsymbol{r} - \mathbf{V}t|}{c}\right). \tag{13.9}$$

We will now consider part $|\boldsymbol{r} - \mathbf{V}t|$ in Equation (13.9). Suppose that the range on which the target moves during the signal processing time is usually smaller than the range to target; then condition $r \gg Vt$ is true for any point of the aperture of the receiving antenna and we can use the expansion

$$|\boldsymbol{r} - \mathbf{V}t| = \sqrt{r^2 + V^2t^2 - 2\mathbf{V}.\boldsymbol{r}t} \approx r\left(1 - \frac{V_\mathrm{r}t}{r} + \frac{V_\mathrm{t}^2t^2}{2r^2}\right), \tag{13.10}$$

where $V_\mathrm{r}t$ and $V_\mathrm{t}t$ are radial and tangential target transferences by the time interval $(0, t)$.

If the processing time, T, does not exceed the time of signal propagation on the target–antenna line, it is not possible to go beyond the first two terms of expansion (13.10). If the processing time, T, is essentially bigger than the time of signal propagation on the target–antenna line, it is necessary to suppose terms in the order of $(Vt/r)^2$. The value of Vt, under which it is necessary to account such terms, will be found below (see Equations (13.15)). It is not difficult to see that these terms describe a linear frequency modulation of the receiving signal, which is formed by the target moving.

Substituting Equation (13.10) into formula (13.9), the equation for the field of aperture of the receiving antenna can be taken as

$$\dot{u}(t, \boldsymbol{r}) \approx \frac{C_0}{r} \dot{s}_0 \left[\left(1 - \frac{V_r}{c}\right)\left(t - \frac{r}{c}\right) + \frac{R_\Gamma}{c + {}_\Gamma V_r} - \frac{V_t^2 t^2}{2cr} \right], \tag{13.11}$$

where r is the distance from the target to a point of the antenna at the moment the signal reaches point Γ.

As formula (13.11) shows, the carrier frequency, the initial phase and the complex envelope of the field are changed as a consequence of the Doppler effect: at $V_r > 0$ the signal temporal increases, but the spectrum width decreases. However, as calculations show in Reference [II.6], aberrations of the complex envelope can be disregarded in a typical situation. At the same time, the field intensity, which is created by the moving target on the receiving antenna aperture, is described by

$$\dot{u}(t, \boldsymbol{r}) \approx \frac{C_0}{r} \dot{U}\left(t - \frac{r}{c} + \frac{R_\Gamma}{c + {}_\Gamma V_r}\right) \exp\left\{ j\omega_0 \left[\left(1 - \frac{V_r}{c}\right)\left(t - \frac{r}{c}\right) + \frac{R_\Gamma}{c + {}_\Gamma V_r} - \frac{V_t^2 t^2}{2cr} \right] \right\}. \tag{13.12}$$

To get a dependence (in an explicit form) of the total phase, $\vartheta(t, \boldsymbol{\rho})$, from a coordinate, $\boldsymbol{\rho}$, of the point in the aperture of the receiving antenna, we put $\boldsymbol{r} = \boldsymbol{r}_1 + \boldsymbol{\rho} = \boldsymbol{\rho} - \mathbf{R}_\Gamma$ in formula (13.9), where $\boldsymbol{r}_1$ is the position of the receiving antenna centre in the system, Σ, and factorize $|\boldsymbol{\rho} - \mathbf{R}_\Gamma - \mathbf{V}t|$ to power series, confining small terms by the order of Vt/R_Γ, ρ/R_Γ and $(\rho/R_\Gamma)^2$. As a result,

$$\vartheta(t, \boldsymbol{\rho}) \approx \omega_0 \left[1 - \frac{{}_\Gamma V_r}{c} + \frac{\boldsymbol{\rho} \cdot {}_\Gamma \mathbf{V}_t}{cR_\Gamma}\right] t - \frac{\omega_0 V_t^2 t^2}{2cR_\Gamma} + \frac{\omega_0}{c} \boldsymbol{\rho} \left(\frac{\mathbf{R}_\Gamma}{R_\Gamma} - \frac{\mathbf{V}}{c}\right) + \frac{\omega_0 \rho^2}{2cR_\Gamma} - \frac{\omega_0 (\boldsymbol{\rho} \cdot \mathbf{R}_\Gamma)^2}{2cR_\Gamma^3}, \tag{13.13}$$

where ${}_\Gamma V_r$ and ${}_\Gamma V_t$ are modules of radial and tangential constituents of target velocity relative to the phase centre of the receiving antenna.

Equation (13.13) shows that the Doppler frequency shift of the signal in the receiving antenna aperture changes according to the law

$$\Omega_D(\boldsymbol{\rho}) \approx \omega_0 \left(\frac{{}_\Gamma V_r}{c} - \frac{\boldsymbol{\rho} \cdot {}_\Gamma \mathbf{V}_t}{cR_\Gamma} + \frac{V_t^2 t}{cR_\Gamma} \right). \tag{13.14}$$

Using Equation (13.13), it is possible to define the conditions in which a spherical wave (created by a moving target) can be approximated by a plane wave. Supposing that this approximation is possible, if the terms in Equation (13.13), which characterize the distinction of the spherical wave from a plane wave, give phase incursion that do not exceed $\pi/8$ it is found that

$$V_t T < \sqrt{\frac{\lambda R_\Gamma}{8}} \tag{13.15}$$

under the condition that the target is in the far region of the receiving antenna.

Therefore, if the target is in the far region and the conditions (13.15) are true, a spherical wave can be approximated by the plane wave with a frequency

$$\omega \approx \omega_0 \left(1 - \frac{{}_\Gamma V_r}{c}\right) \tag{13.16}$$

and wave vector

$$\boldsymbol{k} \approx \frac{\omega_0}{c}\left(\frac{\boldsymbol{r}}{r} + \frac{\mathbf{V}}{c}\right). \tag{13.17}$$

These parameters of the wave are a demonstration of the Doppler effect considered earlier in Section 11.1.2: a change of frequency, wavelength and the direction of the wave propagation.

13.2.2 Active Radar

When an active radar is used, the probing signal is radiated from point R (the phase centre of the transmitting antenna) and its form can be determined by function $\dot{s}_0(t) = \dot{U}(t)\exp(\mathrm{j}\omega_0 t)$. In this case, in the system of reference, Σ, the complex intensity of a field that is incident to the target is (see Equation (13.2))

$$\dot{u}_\Gamma(t, R_1) = \frac{C_1}{R_1}\dot{s}_0\left(t + t_\mathrm{p} - \frac{R_1}{c}\right). \tag{13.18}$$

With the help of the Lorentz transforms it is found that, by referring to the system of reference, Σ',

$$\dot{u}_\Gamma\left(t', \mathbf{R}'\right) = \frac{C_1}{|\mathbf{R}' + \mathbf{V}t'|}\dot{s}_0\left(t' + \frac{\mathbf{V}\mathbf{R}'}{c^2} + t_\mathrm{p} - \frac{|\mathbf{R}' + \mathbf{V}t'|}{c}\right). \tag{13.19}$$

The current, which is created in the target by field (13.19), is

$$\dot{i}_\Gamma\left(t', \mathbf{R}'\right) = \frac{C_1 C_2}{|\mathbf{R}' + \mathbf{V}t'|}\dot{s}_0\left(t' + \frac{\mathbf{V}\mathbf{R}'}{c^2} + t_\mathrm{p} - \frac{|\mathbf{R}' + \mathbf{V}t'|}{c}\right)\exp(\mathrm{j}\upsilon_{ui}), \tag{13.20}$$

where υ_{ui} is a shift of phase between the field and current. This current forms the field of the second radiation:

$$\begin{aligned}\dot{u}_\Gamma(t', \boldsymbol{r}') &= \frac{C_3}{R_2'}\dot{i}_\Gamma\left(t' + t_\mathrm{p} - \frac{R_2'}{c}, \boldsymbol{r}'\right)\\ &= \frac{C\exp(\mathrm{j}\upsilon_{ui})}{R_2'\,|\mathbf{R}' + \mathbf{V}t'|}\dot{s}_0\left(t' + \frac{\mathbf{V}\mathbf{R}'}{c^2} + t_\mathrm{p} - \frac{R_2' + |\mathbf{R}' + \mathbf{V}t'|}{c}\right).\end{aligned} \tag{13.21}$$

With the help of inverse Lorentz transforms and by referring again to the system Σ, an equation can be found for the field in the receiving antenna:

$$\begin{aligned}\dot{u}_\mathrm{R}(t) = {}&\frac{C\exp(\mathrm{j}\upsilon_{ui})}{R_1 R_2}\\ &\times \dot{s}_0\left(t + \frac{\mathbf{V}r}{c^2} + t_\mathrm{p} + \frac{2R_1 - |\boldsymbol{r} + \mathbf{V}(t + R_1/c)| - |\mathbf{R} + \mathbf{V}(t + R_1/c - |\boldsymbol{r} + \mathbf{V}(t + R_1/c)|)|}{c}\right).\end{aligned} \tag{13.22}$$

As was done earlier, modules of sums of vectors can be factorized in power series. By purposely choosing the start time, Equation (13.22) can be transformed to

$$\begin{aligned}\dot{u}_\mathrm{R}(t) = {}&\frac{\dot{a}R_1}{R_2}U\left(t - \frac{R_1 + R_2}{c}\right)\\ &\exp\left\{\mathrm{j}\omega_0\left[\left(1 - \frac{{}_\Gamma V_\mathrm{r} + V_\mathrm{r}}{c}\right)\left(t - \frac{R_1 + R_2}{c}\right) - \frac{{}_\Gamma V_\mathrm{t}^2\left(t - 2R_1/c^2\right)}{cR_2}\right]\right\},\end{aligned} \tag{13.23}$$

where ${}_{\Gamma}V_r = |d\mathbf{R}/dt|$ and $V_r = |d\mathbf{r}/dt|$. From formula (13.23) it follows that if condition (13.15) is true, the receiving signal frequency of a bistatic radar is approximately determined by

$$\omega \approx \omega_0 \left(1 - \frac{{}_{\Gamma}V_r + V_r}{c}\right). \tag{13.24}$$

The wavelength change, which corresponds to the change in Doppler frequency, is

$$\lambda \approx \lambda_0 \left(1 - \frac{{}_{\Gamma}V_r + V_r}{c}\right). \tag{13.25}$$

The condition under which it is necessary to account for an aberration and the condition under which a spherical wave can be approximated by a plane wave will remain the same. Equations for the aberration angle, which were found earlier in Section 11.1.2, also do not require modification.

Finally, putting Equation (13.23) into Equation (13.6), we can determine the formula (or analytic model) of a useful signal of a bistatic radar for the occurrence of moving targets.

13.3 SIGNAL MODEL IN A FORWARD SCATTERING RADAR

As already noted, the use of the phenomenon of electromagnetic wave (EMW) diffraction for radar targets is a fundamental aspect of the operation of a forward scattering radar (FSR). The effect of EMW diffraction is essentially different in comparison with the back-scattering effect, which is the basis of traditional monostatic and bistatic radar operations. For this reason, the signal model in an FSR will be the subject of this section.

13.3.1 Diffraction Foundations

Diffraction of an EMW is the bending and spreading of waves when they meet an obstruction where this cannot be interpreted by wave reflections or refractions. Diffraction phenomenon does not require any new principles for its explanation and quantitative analysis: any diffraction problem may be described by Maxwell's equations and solved by the appropriate equation analysis with given boundary conditions. Unfortunately, Maxwell's equations have analytical solutions in a closed and reasonably simple form for only elementary or idealized cases. In RF and microwave (MW) electrodynamics, approximate methods for the solution of diffraction problems have a dominant role. In many cases, these approximations are based on the Huygens principle in a generalized form formulated by Fresnel or Kirchhoff [II.37].

The Fresnel zone method is highly useful. Assume that a spherical electromagnetic wave is radiated from point O, shown in Figure 13.1, and the receiver is located at point R. Concentric spheres with the centre at point R and radii. r, $r + \lambda/2$, $r + 2(\lambda/2)$, $r + 3(\lambda/2), \ldots$ are described in Figure 13.1. These divide the wave front, F, into ring areas, which are termed Fresnel regions or zones.

This may be illustrated by a simple practical experiment. An opaque screen with a circular hole is placed between the point source, T, of the electromagnetic wave and point R. The screen plane is normal to the TR axis and its centre, O, is located at the same axis (Figure 13.2). According to Fresnel, this obstacle shield eliminates an appropriate part of the wave front. The unshielded part of the wave front remains unchanged. This assumption can only be approximately true when the radius of the hole is large in comparison with the wavelength. This assumption will be used later in this section.

Let us call the plane, which is parallel to the screen (i.e. perpendicular to the TR axis) and passes through point R, the image plane. Let us also suppose that the size of the hole in Figure 13.2 can be changed. This presents the opportunity to open any number of Fresnel regions by varying the hole's diameter.

Denote a_0 and P_0 as the amplitude and power of the electromagnetic field at point R for the case of free-space propagation, i.e. the screen is absent. If only the first Fresnel region in the opaque screen is

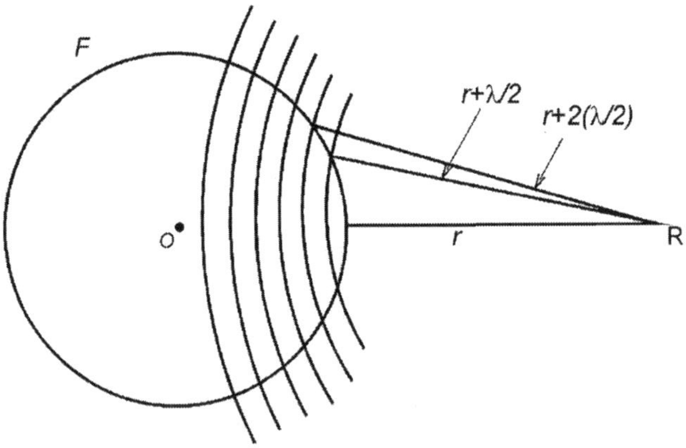

Figure 13.1 For a definition of Fresnel regions

opened, the amplitude and power in the same point R will be $a_1 = 2a_0$ and $P_1 = 4P_0$. Therefore, if the first Fresnel region is opened and the other Fresnel regions are closed, the power in the receiving point, R, will be four times bigger in comparison with free-space propagation. If we move away from point R, but remain on the image plane, a monotone decrease in power is observed.

By making the hole in the screen bigger, a wave from the second Fresnel zone will partly appear at point R. Interference of these waves causes a decrease in the electromagnetic field power at this point. For a specific hole diameter, the maximum electromagnetic field power in the image plane moves away from point R: power maximum will surround point R. When the hole is big enough to open the first and the second Fresnel regions, the signal power at point R will be nearly zero, as these two signals completely compensate each other and there will be a power minimum at R. For the optical range of electromagnetic waves this phenomenon can be visually observed on the image plane as a dark point with a surrounding bright ring.

Further enlargement of the hole's size does not influence either of the first two Fresnel regions at point R, and they remain compensated. The total field at point R is only created by opening part of the third Fresnel region. In the central point, R, of the image plane a bright spot appears which is surrounded by a dark ring. This dark ring in turn is surrounded by a bright ring. When the amount of opened Fresnel

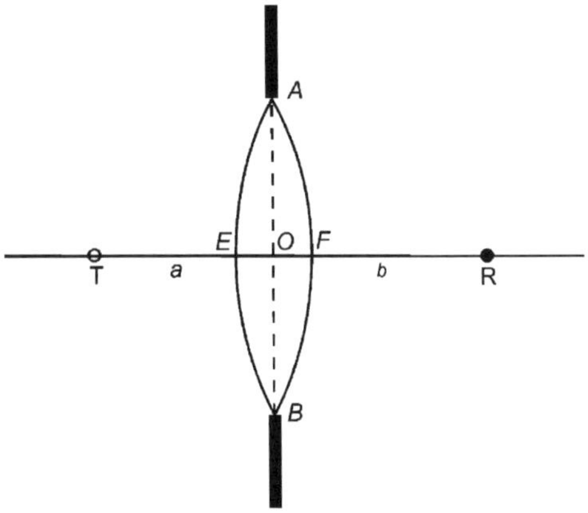

Figure 13.2 Fresnel diffraction on an opaque screen with a circular hole

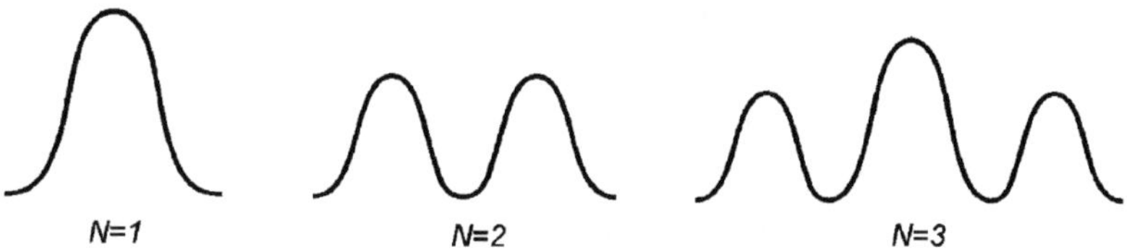

Figure 13.3 Distribution of electromagnetic wave power in the image plane

regions, N, is exactly equal to three, the power of the electromagnetic field will be equal to the case when only the third Fresnel region is opened and all others, including the first two Fresnel regions, are closed. At point R the brightness will correspond to the case of only the third opened Fresnel region. At $N = 4$, the bright spot in the centre is replaced by a dark spot. In fact, maximum electromagnetic wave power at point R takes place for odd N and the minimum power for even N. In Figure 13.3, a distribution of electromagnetic wave power in the image plane is sketched as a function of the distance between the point of observation and point R.

The sizes and quantity of Fresnel regions that remain within hole AB will now be determined (Figure 13.2). Let D be the diameter of the hole; a and b are the distances from the hole's centre, O, to points T and R, creating two spheres. They have centres at T and R and cross the edge of the hole AB. Neglecting the squares of small segments, OE and OF,

$$\left(\frac{D}{2}\right)^2 \approx \mathrm{OF}\,2a, \qquad \left(\frac{D}{2}\right)^2 \approx \mathrm{OE}\,2b. \tag{13.26}$$

From these equations,

$$\mathrm{EF} = \mathrm{EO} + \mathrm{OF} = \frac{D^2}{8}\left(\frac{1}{a} + \frac{1}{b}\right). \tag{13.27}$$

The quantity of Fresnel regions, m, may be found by dividing this segment on $\lambda/2$:

$$m = \frac{D^2}{4\lambda}\left(\frac{1}{a} + \frac{1}{b}\right). \tag{13.28}$$

If m is integer, D will be the diameter of the mth Fresnel region edge. Consequently, the radius of the mth Fresnel region is defined by

$$\rho_m = \sqrt{\frac{ab}{a+b}m\lambda}. \tag{13.29}$$

Let us envisage one more experiment. An opaque disc (circular screen), AB, is now placed between the source of the EMW (transmitter), T, and the observing point (receiver), R (Figure 13.4). The disc plane is to be normal to the TR axis and parallel to the image plane. Let DA and BE be the opened parts of the spherical front of the electromagnetic wave, which is emitted from source T. They are divided into ring Fresnel regions starting from the edge of the screen. It can be shown [II.37] that the field intensity at point R is equal to half of the field intensity generated by the first ring of the Fresnel region. Consequently, whatever the diameter of the disc, a local maximum of electromagnetic wave power is observed in centre R of the image plane. In the optical range of electromagnetic waves this maximum is a bright spot in the centre of a geometric shadow. Theoretical conclusions about the existence of this bright spot at the centre of the shadow was made by Poisson, a prominent physicist and mathematician, who believed in the theoretical possibility of its existence to be the objection for Fresnel's wave theory of light. However, Arago conducted the experiment and found the bright spot. This experiment confirmed the validity of the theoretical conclusion made by Poisson, as well as the wave theory of light proposed by Fresnel. Since then the described spot is known in science as the 'Arago–Poisson spot'.

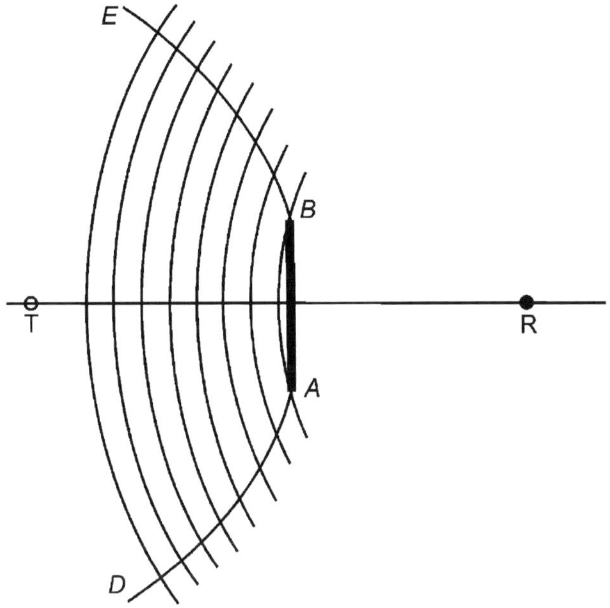

Figure 13.4 Fresnel diffraction on an opaque disc

It should be remembered that the Fresnel region method only works in the case of small wavelengths in comparison with the screen diameter. Therefore an application of this method is not restricted by optical frequencies. The Arago–Poisson spot could be observed at any frequencies where a proper relationship occurs between an obstacle and the wavelength.

If the disc centre is not located on the TR axis, more than one Fresnel region will be overlapped by the disc. In this case, or when the obstacle has a more sophisticated shape, an analytical study of the diffraction phenomena will be essentially more complex. However, an increased complexity is a source of interest for the forward-scattering radar.

In conclusion, it should be remembered that in wave physics and, in particular, in wave electrodynamics there are two accepted types of diffraction: Fresnel diffraction and Fraunhofer diffraction. In optics, Fraunhofer diffraction is known as diffraction in parallel rays and Fresnel diffraction is diffraction in skew rays. Let us describe, then, the difference between these types of diffraction in terms of application to the FSR, shown in Figure 13.5. Here, as in the earlier chapters, T is the phase centre of a transmitting antenna, R is the phase centre of a receiving antenna; Γ is a small target and **V** is the velocity vector of the target.

Assume that the target is moving along a linear trajectory that is perpendicular to the baseline, TR, of the radar with a constant speed. Let the point coordinate axis, *OX*, along the baseline of the radar be in the direction from the transmitter to the receiver and the coordinate axis, *OY*, be along the target trajectory in the direction of **V**. Therefore, the point of the coordinate origin, O, is the point where the target crosses the radar baseline.

Denote $d_\mathrm{T} = O\mathrm{T}$ as the distance between T and the coordinate origin and $d_\mathrm{R} = O\mathrm{R}$ as the distance between R and O; $r_\mathrm{T}(t)$ represents the range of the target relative to the transmitter and $r_\mathrm{R}(t)$ represents the range of the target relative to the receiver. Thus, $L = d_\mathrm{T} + d_\mathrm{R}$ is the base of the radar and Equation (13.29) for the radius of the mth Fresnel region will be

$$\rho_m = \sqrt{\frac{m\lambda d_\mathrm{T} d_\mathrm{R}}{L}}. \tag{13.30}$$

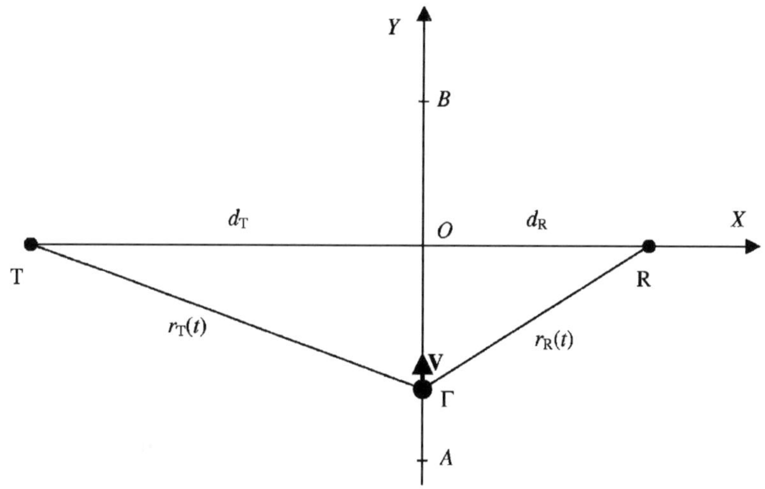

Figure 13.5 Geometry and system of coordinates for the FSR

Designate through t_0 the moment the target crosses the baseline of the FSR. In general, t_0 is not previously known, and is impractical to use as the start point for time. At the same time it is obvious that $r_T(t_0) = d_T$ and $r_R(t_0) = d_R$. Using the Pythagoras theorem it is possible to find that

$$r_T(t) = \sqrt{d_T^2 + V^2 (t - t_0)^2}, \qquad r_R(t) = \sqrt{d_R^2 + V^2 (t - t_0)^2}. \tag{13.31}$$

The time interval of the signal receiving will be denoted as T; i.e. it will be supposed that $t \in [0, T]$. It is obvious that $t_0 \in [0, T]$.

Points A and B on the target trajectory correspond to the target position at the time instant $t = 0$ and $t = T$. Therefore, $AB = VT$, $OA = Vt_0$ and $OB = V(T - t_0)$. Assume for the sake of certainty that $OA > OB$. Let us draw the circular arc with the centre in point T through point A. This arc will cross the baseline of the FSR at point O' and the coordinate axis OY at point A' (see Figure 13.6(a)). From the figure it is obvious that $\mathrm{T}A = \mathrm{T}A'$ and $OA = OA'$. Let us also draw the circle arc at the centre at point R (see Figure 13.6(b)) and through point A in the same way. This arc will cross the baseline of the FSR at point O'' and the coordinate axis OY at point A''.

Fraunhofer diffraction will take place if the biggest distance from both OO' and OO'' is negligible in comparison with the smallest from parameters d_T and d_R. In this case Equations (13.31) are transformed to

$$r_T(t) \simeq d_T, \qquad r_R(t) \simeq d_R. \tag{13.32}$$

Thus, the Fraunhofer approximation assumes independence of the target ranges, r_T and r_R, from a target location (and hence from time) during a time interval $[0, T]$. Therefore, the Fraunhofer approximation assumes a parallelism of segments TO, TA and TA', and also RO, RA and RA''. This fact is explained by the other name for Fraunhofer's diffraction – diffraction in parallel rays.

To determine Fresnel diffraction we will return to formulae (13.31). As long as the possible distance of the target, VT, is much smaller than d_T and d_R, the right sides of Equations (13.31) can be factorized to a series for small arguments $V(t - t_0)/d_T$ or $V(t - t_0)/d_R$. Fresnel diffraction will occur if the series can be limited by only two terms, namely

$$r_T(t) = d_T + \frac{V^2 (t - t_0)^2}{2d_T}, \qquad r_R(t) = d_R + \frac{V^2 (t - t_0)^2}{2d_R}. \tag{13.33}$$

Note that, in contrast to Equations (13.32), formulae (13.33) considers the curvature of EMW fronts.

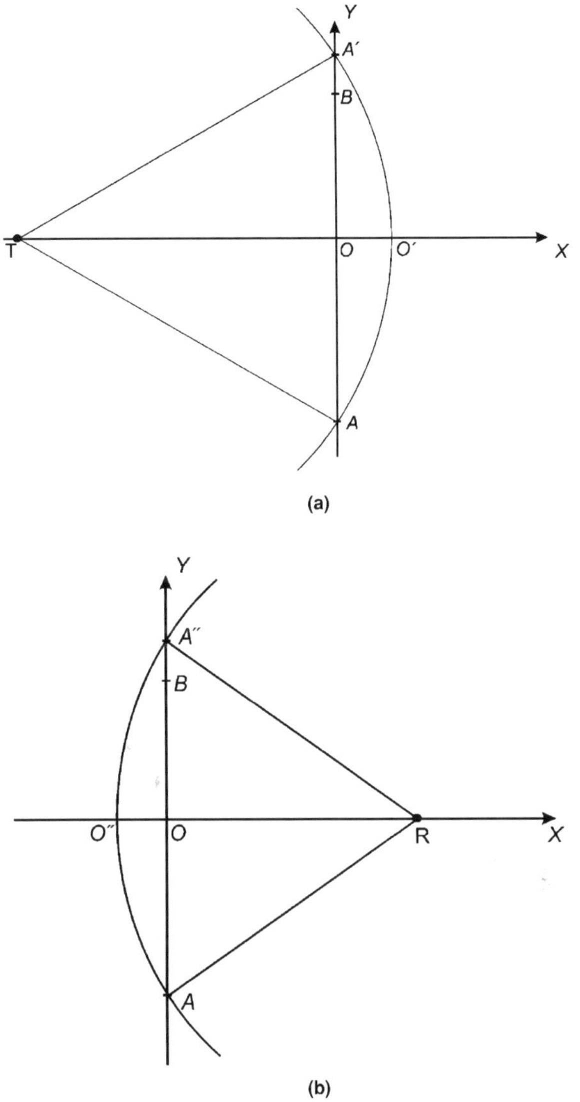

Figure 13.6 Fresnel assumption

13.3.2 Signal Model in the Case of a Small Target

Let us first consider an analytic signal model in the FSR for the case of small targets. For this case all designations in Figure 13.5 are correct. A small target differs from a point target by having a finite (nonzero) area, S. Therefore, a small target creates a shadow where a point target does not.

Using Kirchhoff's method and Babinet's principle for solution of diffraction problems [II.4, II.17], the analytical expression can be found for the receiving signal:

$$s_{\mathrm{T}}(t) = A\cos\left[2\pi\left(f_0 t - \frac{L}{\lambda}\right)\right] + \frac{2AS}{\rho_1^2}\sin\left\{2\pi\left[f_0 t - \frac{L}{\lambda} - \frac{V^2(t-t_0)^2}{2\rho_1^2}\right]\right\}, \tag{13.34}$$

where A is the amplitude factor, $f_0 = \omega_0/(2\pi)$ is the carrier frequency of the probing wave, L is the base of the radar, $\lambda = c/f_0$ is the signal wavelength (c is the speed of light) and S is the area of the small target. In addition,

$$\rho_1 = \sqrt{\frac{\lambda d_T d_R}{L}} \tag{13.35}$$

is the radius of the first Fresnel region and V is the module of target velocity. The radius of the first Fresnel region presents in expression (13.34) as a secondary (auxiliary) parameter, which quantitatively characterizes the geometry of the radar, shown in Figure 13.5.

It can be seen in Equation (13.34) that the receiving signal $s_T(t)$ is a sum of noninformative oscillations, $e(t)$, and the signal, $s_{sh}(t)$, diffracted by the target, namely

$$s_T(t) = e(t) + s_{sh}(t), \tag{13.36}$$

where

$$e(t) = A\cos\left[2\pi\left(f_0 t - \frac{L}{\lambda}\right)\right], \qquad s_{sh}(t) = \frac{2AS}{\rho_1^2}\sin\left\{2\pi\left[f_0 t - \frac{L}{\lambda} - \frac{V^2(t-t_0)^2}{2\rho_1^2}\right]\right\}. \tag{13.37}$$

Without the target, the receiver would register the oscillation $e(t)$ only (i.e. a nondistorted transmitted signal or direct signal). The signal, $s_{sh}(t)$, occurs as a result of the transmitting waveform shadowed by the target. It contains the information about target velocity, V, and moment, t_0, when the target, Γ, crosses the baseline of the FSR.

If new designations are entered,

$$f_V = \frac{V}{\rho_1}; \qquad \phi_0 = f_V t_0 = \frac{V t_0}{\rho_1}, \tag{13.38}$$

then the right side of Equations (13.37) can be presented as

$$s_{sh}(t) = \frac{2AS}{\rho_1^2}\sin\left\{2\pi\left[f_0 t - \frac{L}{\lambda} - \frac{(f_V t - \phi_0)^2}{2}\right]\right\}, \tag{13.39}$$

where parameter f_V has a frequency dimension and the parameter ϕ_0 is dimensionless.

From Equations (13.37) to (13.39), it follows that the direct measurement of the target velocity, V, and time moment, t_0, when the target crosses the radar's baseline are not possible. These parameters can be measured only indirectly. Direct measurement can be made for the auxiliary parameters, f_V and ϕ_0. If these are measured, we will then be able to calculate velocity and the moment the baseline is crossed. However, the noted computation needs prior exact knowledge of the radius of the first Fresnel region, ρ_1. Inaccurate knowledge of the last parameter decreases the accuracy of the measurement of the target's velocity and the moment the target crosses the radar's baseline.

According to Equation (13.39), the full phase,

$$\Phi(t) = 2\pi\left(f_0 t - \frac{L}{\lambda}\right) - \pi(f_V t - \phi_0)^2 - \frac{\pi}{2}, \tag{13.40}$$

of the informative (shadow) signal is given by a parabola with down-directed branches. The law of frequency modulation of the signal, Equation (13.39), can be defined as a time derivative from the full phase, Equation (13.40), namely

$$\omega(t) = \frac{d\Phi(t)}{dt} = 2\pi\left[f_0 - f_V(f_V t - \phi_0)\right] = \omega(0) - 2\pi f_V^2 t = \omega(0) - \frac{2\pi V^2 t}{\rho_1^2}, \tag{13.41}$$

where $\omega(0) = 2\pi(f_0 + f_V\phi_0)$ is the significance of frequency in the start moment. The second term in the right part of formula (13.41) is the result of Fresnel diffraction on the small target. It is not present in the case of a stationary target and, consequently, can be interpreted as a demonstration of the Doppler effect.

Therefore, the frequency of the shadow signal, $s_{sh}(t)$, decreases in time as a linear function. The maximum deviation of frequency during the observing time is

$$\Delta\omega = 2\pi T f_V^2 = \frac{2\pi T V^2}{\rho_1^2}. \tag{13.42}$$

Therefore, to measure a target velocity, the radius of the first Fresnel region must be known and the maximum derivation of the frequency of the receiving signal during the time, T, must also be measured.

Again note that necessary and sufficient conditions of *linear* frequency modulation of the shadow signal are:

1. The target has a small size.
2. Diffraction is the Fresnel diffraction.
3. There is the opportunity to use approximations of physical optics (the Kirchhoff method).

If even one of the noted conditions is not present, then the statement about linear frequency modulation of a shadow signal, strictly speaking, will not be correct.

The instant power of the signal, $s_T(t)$, may be presented as

$$P_T(t) = P_{DC} + P_{AC}(t) + P_{HF}(t), \tag{13.43}$$

where P_{DC} is a constant component of the instant power, $P_{AC}(t)$ is a comparatively slow-changing variable constituent of the power and $P_{HF}(t)$ is a high-frequency constituent of the power. The last has a frequency variation two times bigger than the frequency of the transmitting signal. It can be shown that

$$P_{DC} = \frac{A^2}{2}\left[1 + \left(\frac{2S}{\rho_1^2}\right)^2\right]; \qquad P_{AC}(t) = -\frac{2A^2 S}{\rho_1^2}\sin\left[\pi\frac{V^2(t-t_0)^2}{2\rho_1^2}\right]. \tag{13.44}$$

For a small target the assumption,

$$2S \ll \rho_1^2, \tag{13.45}$$

takes place and the first term in Equation (13.44) can be simplified as $P_{DC} \approx A^2/2$.

It is obvious that the signal $s_T(t)$ at the receiver input in the FSR is a high-frequency bandpass signal. Signal processing at high frequencies has known problems, so it is useful to convert the RF signal, $s_T(t)$, into a low frequency or to the baseband signal. Towards this end, a down-conversion receiver can be used. However, another method could be used in the FSR. If the signal $s_T(t)$ is crossed through the nonlinear element (although in practice a diode can give satisfactory results [II.35, II.36]), with, for example, a quadratic characteristic, a lowpass filter and a highpass filter (Figure 13.7), then the signal $s(t)$ on the

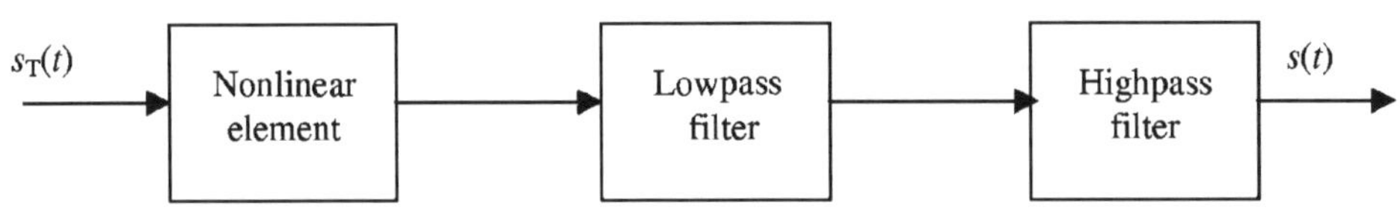

Figure 13.7 Blockdiagram of equipment for transformation of the RF signal to the baseband signal

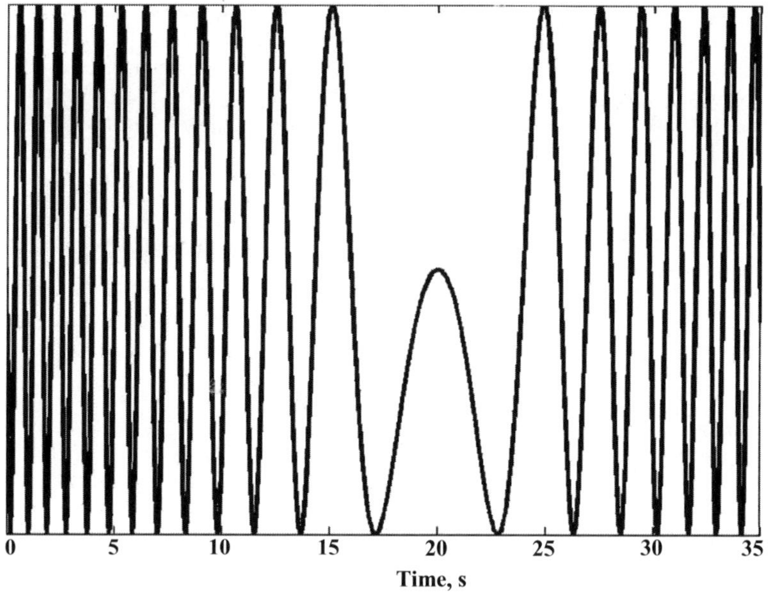

Figure 13.8 Signal from a small target

output of such a three-cascade system will concur with the component of instant power, $P_{AC}(t)$:

$$s(t) = -\frac{2A^2 S}{\rho_1^2} \sin\left[\pi \frac{V^2 (t - t_0)^2}{2\rho_1^2}\right]. \tag{13.46}$$

It can be seen from Equation (13.46) that, the signal $s(t)$ is informative. Similar to $s_T(t)$, it contains information about target velocity and the moment the target crosses the radar's baseline.

As an example, in Figure 13.8, calculated from formula (13.46), the signal $s(t)$ is shown for the following sets of values of parameters: $V = 1$ m/s, $t_0 = 20$ s, $\rho_1 = 4$ m and $T = 35$ s. It can be seen that the signal shown in the diagram has frequency modulation. At $t = t_0$ a local maximum of signal exists, which is possible to interpret as a demonstration of the Arago–Poisson spot in radio physics. Indeed, we consider the screen's motion, which is opaque for the electromagnetic wave, between a fixed transmitter and receiver. However, according to Galileo's principle of relativity, a target can be considered as motionless, but the transmitter and receiver can be considered as moving at parallel linear trajectories. Suppose that condition $VT \ll d_T$ is true; i.e. the path passed by the transmitter during the registration time is negligibly small in comparison with the transmitter–target distance. Thus, the target and the transmitter can be considered to be approximately motionless, but moving in the image plane receiver. It is clear that the samples of the signal s(t), which are proportional to the power of the incident electromagnetic field, are obtained at different points of a diffraction picture. In this sense, the curve shown in Figure 13.8 is similar to the curves shown in Figure 13.3. The local maximum of the curve in Figure 13.8 is the same phenomenon of nature as the Arago–Poisson spot.

The frequency modulation law of the signal $s(t)$, plotted in Figure 13.8, is shown in Figure 13.9, where Figure 13.9(a) represents the instant frequency calculated with the help of Equation (13.47). Figure 13.9(b) represents the spectrogram obtained and imaged by built-in means of Matlab. It can be seen from this picture that the instant frequency of the signal equals zero at the moment the target crosses

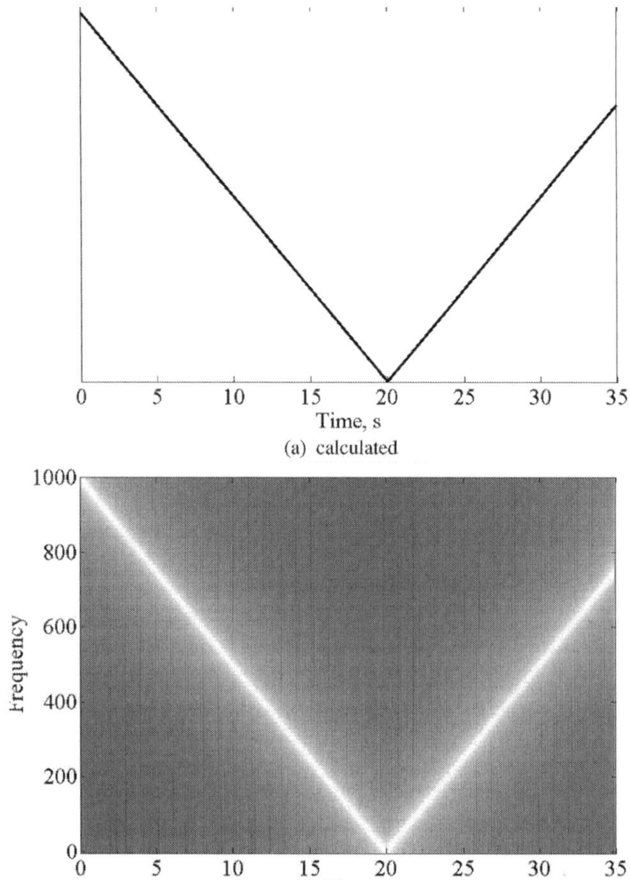

(a) calculated

(b) spectrogram

Figure 13.9 Frequency modulation law

the radar's baseline, and the graph of the dependence of frequency from time is a broken line. Strictly speaking, the law

$$\omega(t) = \frac{2\pi V^2 |t - t_0|}{\rho_1^2} \tag{13.47}$$

of frequency modulation of the signal $s(t)$ is linear only in two special cases: $t_0 = 0$ and $t_0 = T$. In general it is a piecewise law. The minimum value of the signal frequency is equal to zero at the moment the target crosses the baseline. When its maximum value is achieved, either at the start ($t = 0$) or at the end ($t = T$) of the target observation, its values are between $\pi T(V/\rho_1)^2$ and $2\pi T(V/\rho_1)^2$.

The spectrum of signal (13.46) can be found by Fourier transform, namely

$$\dot{G}(\omega) = -\frac{2A^2 S}{\rho_1^2} \int_0^T \sin\left[\pi \frac{V^2 (t - t_0)^2}{2\rho_1^2}\right] \exp(-j\omega t)\, dt. \tag{13.48}$$

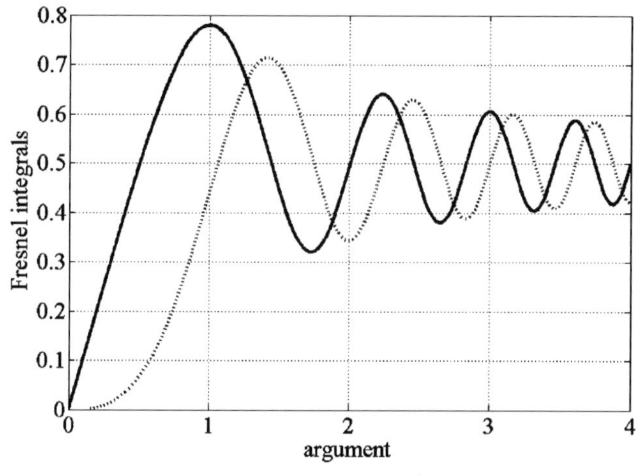

Figure 13.10 Fresnel integrals

After the difficult and awkward transformations of Equation (13.48) it is possible to obtain

$$
\begin{aligned}
\dot{G}(\omega) &= \mathrm{j}\frac{A^2 S}{\sqrt{2}\,\rho_1 V}\exp\left(-_{\mathrm{j}}\omega t_0\right) \\
&\times\left(\exp\left[-\frac{\mathrm{j}}{\pi}\left(\frac{\rho_1\omega}{2V}\right)^2\right]\{[C(X_+(\omega)) - C(_+Y_-(\omega))] + \mathrm{j}\,[S(X_+(\omega)) - S(_+Y_-(\omega))]\}\right. \\
&\left. + \exp\left[\frac{\mathrm{j}}{\pi}\left(\frac{\rho_1\omega}{2V}\right)^2\right]\{[C(X_-(\omega)) - C(_-Y_+(\omega))] - \mathrm{j}\,[S(X_-(\omega)) - S(_-Y_+(\omega))]\}\right),
\end{aligned}
\tag{13.49}
$$

where the notation

$$
\begin{aligned}
X_{\{\pm\}}(\omega) &= \frac{\rho_1\omega}{\sqrt{2\pi}\,V}\,\{\pm\}\,\frac{\sqrt{2}\,V t_0}{\rho_1}, \\
_+Y_-(\omega) &= X_+(\omega) - \frac{\sqrt{2}\,VT}{\rho_1}, \\
_-Y_+(\omega) &= X_-(\omega) + \frac{\sqrt{2}\,VT}{\rho_1}
\end{aligned}
\tag{13.50}
$$

is used. Special functions in Equation (13.49) are known as Fresnel integrals:

$$
C(z) = \int_0^z \cos\left(\frac{\pi}{2}\alpha^2\right)\mathrm{d}\alpha \qquad \text{and} \qquad S(z) = \int_0^z \sin\left(\frac{\pi}{2}\alpha^2\right)\mathrm{d}\alpha. \tag{13.51}
$$

Handbook [II.38] provides sufficiently detailed tables of Fresnel integrals. The dependence of Fresnel integrals, $C(z)$ and $S(z)$, from the argument is shown in Figure 13.10. The curve drawn by a solid line corresponds to $C(z)$. The graph drawn by a dotted line represents $S(z)$.

The Matlab programme for calculation of Fresnel integrals, based on the decomposition of Fresnel integrals to a series by the Tchebyshev polynomial (see, for example, Reference [II.39]), is:

```
function [C,S] = Fresnel(x)
% Copyright 2005 Vladimir Ivanovich Kostylev
a=[0.76435138664186000189; -0.43135547547660179313; 0.43288199979726653054; -
0.26973310338387111029; 0.08416045320876935378; -0.01546524484461381958; 0.00187855423439822018;
```

```
-0.00016264977618887547; 0.00001057397656383260; -0.00000053609339889243;
0.00000002181658454933; -0.00000000072901621186; 0.00000000002037332548; -
0.00000000000048344033; 0.00000000000000986533; -0.00000000000000017502; 0.00000000000000000272;
-0.00000000000000000009]';

b=[0.63041404314570539241; -0.42344511405705333544; 0.37617172643343656625; -
0.16249489154509567415; 0.03822255778633008694; -0.00564563477132190899; 0.00057454951976897367;
-0.00004287071532102004; 0.00000245120749923299; -0.00000011098841840868;
0.00000000408249731696; -0.00000000012449830219; 0.00000000000320048425; -
0.00000000000007032416; 0.00000000000000133638; -0.00000000000000002219; 0.00000000000000000032;
-0.00000000000000000000]';

e=[0.97462779093296822410; -0.02424701873969321371; 0.00103400906842977317; -
0.00008052450246908016; 0.00000905962481966582; -0.00000131016996757743; 0.00000022770820391497;
-0.00000004558623552026; 0.00000001021567537083; -0.00000000251114508133;
0.00000000066704761275; -0.00000000018931512852; 0.00000000005689898935; -
0.00000000001798219359; 0.00000000000594162963; -0.00000000000204285065; 0.00000000000072797580;
-0.00000000000026797428; 0.00000000000010160694; -0.00000000000003958559;
0.00000000000001581262; -0.00000000000000646411; 0.00000000000000269981; -
0.00000000000000115038; 0.00000000000000049942; -0.00000000000000022064; 0.00000000000000009910;
-0.00000000000000004520; 0.00000000000000002042; -0.00000000000000000982;
0.00000000000000000467; -0.00000000000000000225; 0.00000000000000000110; -
0.00000000000000000054; 0.00000000000000000027; -0.00000000000000000014; 0.00000000000000000007;
-0.00000000000000000004; 0.00000000000000000002; -0.00000000000000000001;
0.00000000000000000001]';

f=[0.99461545179407928910; -0.00524276766084297210; 0.00013325864229883909; -
0.00000770856452642713; 0.00000070848077032045; -0.00000008812517411602; 0.00000001359784717148;
-0.00000000246858295747; 0.00000000050925789921; -0.00000000011653400634;
0.00000000002906578309; -0.00000000000779847361; 0.00000000000222802542; -
0.00000000000067239338; 0.00000000000021296411; -0.00000000000007041482; 0.00000000000002419805;
-0.00000000000000861080; 0.00000000000000316287; -0.00000000000000119596;
0.00000000000000046444; -0.00000000000000018485; 0.00000000000000007527; -
0.00000000000000003131; 0.00000000000000001328; -0.00000000000000000574; 0.00000000000000000252;
-0.00000000000000000113; 0.00000000000000000051; -0.00000000000000000024;
0.00000000000000000011; -0.00000000000000000005; 0.00000000000000000002; -
0.00000000000000000001; 0.00000000000000000001; 0; 0; 0; 0; 0; 0]';

C=zeros(size(x)); S=zeros(size(x));
z=0.5*pi*x.*x;
inds = find(z>8);
w(inds)=exp(complex(0,-z(inds)))./sqrt(z(inds));
o(inds)=8./z(inds);
se(inds)=0; sf(inds)=0;
  for k=1:41
    n=42-k;
    m=2*(n-1);
    s1(inds)=Cheb(o(inds),m);
    se(inds)=se(inds)+e(n)*s1(inds); sf(inds)=sf(inds)+f(n)*s1(inds);
  end
hz=sqrt(0.5*pi); ww=complex(hz,-hz);
E(inds)=ww-w(inds).*complex(0.5*se(inds)./z(inds), -sf(inds));
zh=sqrt(2*pi);
C(inds)=sign(x(inds)).*real(E(inds))/zh;
S(inds)=-sign(x(inds)).*imag(E(inds))/zh;
inds = find((z<=8)&(z>0));
w(inds)=exp(complex(0,-z(inds)))./sqrt(z(inds));
```

```
sa(inds)=0; sb(inds)=0;
for k=1:18
  n=19−k;
  m=2*(n−1);
  sa(inds)=sa(inds)+a(n)*Cheb(z(inds)/8,m); sb(inds)=sb(inds)+b(n)*Cheb(z(inds)/8,m+1);
end
E(inds)=sign(x(inds)).*sqrt(z(inds)).*complex(sa(inds),sb(inds));
zh=sqrt(2*pi);
C(inds)=real(E(inds))/zh;
S(inds)=imag(E(inds))/zh;
end

function res = Cheb(x,n)
if n==0
res=ones(size(x));
else
m=n/2;
x2=2*x;
s=zeros(size(x));
for k=0:m
  nm2k=n−2*k;
  a=((−1) k)*factorial(n−k−1)/factorial(k)/factorial(nm2k);
  b=x2.^nm2k;
  s=s+a*b;
end
res=0.5*n*s;
end
```

As discussed previously, under the condition $t_0 = 0$ the signal (13.46) has linear frequency modulation. For this case Equation (13.49) for the spectrum can be modified to

$$
\begin{aligned}
\dot{G}(\omega) = {} & \mathrm{j}\frac{A^2 S}{\sqrt{2}\rho_1 V} \\
& \times \left(\exp\left[-\frac{\mathrm{j}}{\pi}\left(\frac{\rho_1\omega}{2V}\right)^2\right] \left\{ \left[C\left(\frac{\rho_1\omega}{\sqrt{2\pi}V}\right) - C\left(\frac{\rho_1\omega}{\sqrt{2\pi}V} - \frac{\sqrt{2}VT}{\rho_1}\right)\right] \right.\right. \\
& \left. + \mathrm{j}\left[S\left(\frac{\rho_1\omega}{\sqrt{2\pi}V}\right) - S\left(\frac{\rho_1\omega}{\sqrt{2\pi}V} - \frac{\sqrt{2}VT}{\rho_1}\right)\right]\right\} \\
& + \exp\left[\frac{\mathrm{j}}{\pi}\left(\frac{\rho_1\omega}{2V}\right)^2\right] \left\{ \left[C\left(\frac{\rho_1\omega}{\sqrt{2\pi}V}\right) - C\left(\frac{\rho_1\omega}{\sqrt{2\pi}V} + \frac{\sqrt{2}VT}{\rho_1}\right)\right] \right. \\
& \left.\left. - \mathrm{j}\left[S\left(\frac{\rho_1\omega}{\sqrt{2\pi}V}\right) - S\left(\frac{\rho_1\omega}{\sqrt{2\pi}V} + \frac{\sqrt{2}VT}{\rho_1}\right)\right]\right\}\right).
\end{aligned}
\tag{13.52}
$$

Another interesting case takes place at $t_0 = T/2$, when piecewise modulation of the signal (13.46) becomes symmetrical (V-type). The spectrum of the signal (13.46) at $t_0 = T/2$ can be represented in the form

$$
\begin{aligned}
\dot{G}(\omega) &= \frac{\sqrt{2}A^2 S}{\rho_1 V} \exp\left(-\mathrm{j}\frac{\omega T}{2}\right) W(\omega) \cos\left[\frac{1}{\pi}\left(\frac{\rho_1\omega}{2V}\right)^2 + \Psi(\omega)\right], \\
W(\omega) &= \sqrt{[C(X_+(\omega)) - C(X_-(\omega))]^2 + [S(X_+(\omega)) - S(X_-(\omega))]^2}, \\
\tan\Psi(\omega) &= \frac{C(X_+(\omega)) - C(X_-(\omega))}{S(X_+(\omega)) - S(X_-(\omega))},
\end{aligned}
\tag{13.53}
$$

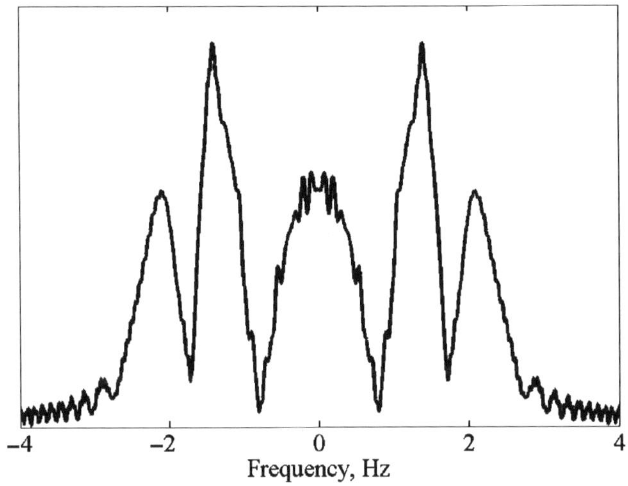

Figure 13.11 Amplitude–frequency spectrum of a signal

where now

$$X_{\{\pm\}}(\omega) = \frac{1}{\sqrt{2}}\left(\frac{\rho_1\omega}{\pi V}\{\pm\}\frac{VT}{\rho_1}\right). \tag{13.54}$$

The amplitude–frequency spectrum of the signal $s(t)$ calculated from Equation (13.49) for the same parameters, occurring in the previous two diagrams, is shown in Figure 13.11. It is not difficult to calculate the signal spectrums for other parameters using these equations.

13.3.3 Signal Model in the Case of a Target with a Rectangular Shape

Suppose now that the target is a rectangular shield, which is not a small target according to our definition. The height of the target (the size in the direction of the coordinate axis *OZ*) is denoted by h and the length of the target in the direction of its motion (the size in the direction of the coordinate axis *OY*) by l. The third size (the size in the direction of the coordinate axis *OX*) is not important here, as only the silhouette of the target is considered in the FSR [II.4, II.17]. Similar to the case of a small target, the receiving signal $s_T(t)$ is the sum of the spurious oscillation $e(t)$ and the informative shadow signal $s_{sh}(t)$ (see formula (13.36)). Using the Babinet principle, with the help of the Kirchhoff method, an analytical expression for the shadow signal can be found, making Equation (13.37) more general, namely

$$\begin{aligned} s_{sh}(t) = -A\,\mathrm{Re}\,\Big(&\mathrm{j}\,\{[C(\xi) - C(\zeta)] - \mathrm{j}\,[S(\xi) - S(\zeta)]\} \\ &\times \{[C(\Xi_+(t)) - C(\Xi_-(t))] - \mathrm{j}\,[S(\Xi_+(t)) - S(\Xi_-(t))]\}\exp\left[\mathrm{j}2\pi\,(f_0 t - L/\lambda)\right]\Big), \end{aligned} \tag{13.55}$$

where

$$\begin{aligned} \xi &= \sqrt{2}\frac{h+b}{\rho_1}, \qquad \zeta = \sqrt{2}\frac{b}{\rho_1}, \\ \Xi_{\{\pm\}}(t) &= \frac{\sqrt{2}V(t-t_0)}{\rho_1}\{\pm\}\frac{l}{\sqrt{2}\rho_1}, \end{aligned} \tag{13.56}$$

and b is the edge of shield elevation relevant to the horizontal plane which crosses the radar baseline. If the shield's rise is small, then $b \cong 0$ and Equation (13.55) can be rewritten in a much easier form:

$$s_{sh}(t) = -A\,\mathrm{Re}\big(\mathrm{j}\left[C(\chi) - \mathrm{j}S(\chi)\right] \times \{[C(\Xi_+(t)) - C(\Xi_-(t))] - \mathrm{j}[S(\Xi_+(t)) - S(\Xi_-(t))]\}\exp\left[\mathrm{j}2\pi\left(f_0 t - L/\lambda\right)\right]\big), \tag{13.57}$$

where $\chi = \sqrt{2}h/\rho_1$ is a constant, which characterizes the proportion between the target height and the radius of the first Fresnel region. If the target height is exactly equal to the radius of one of the Fresnel regions, e.g. the mth region, i.e. $h = \rho_m$, then $\chi = \sqrt{2m}$.

Expression (13.57) for the shadow signal can be rewritten in another way:

$$s_{sh}(t) = \left|\dot{U}(t)\right| \sin\left\{2\pi\left(f_0 t - L/\lambda\right) + \arg\left[\dot{U}(t)\right]\right\}, \tag{13.58}$$

where

$$\dot{U}(t) = A\left[C(\chi) - \mathrm{j}S(\chi)\right]\{[C(\Xi_+(t)) - C(\Xi_-(t))] - \mathrm{j}[S(\Xi_+(t)) - S(\Xi_-(t))]\} \tag{13.59}$$

is a complex envelope of the informational shadow signal, $s_{sh}(t)$. It is interesting to remark that the complex envelope, $\dot{U}(t)$, is not an analytical signal (the real and imaginary parts of $\dot{U}(t)$ are not connected by the Hilbert transform).

Comparing Equations (13.58) and (13.39), it is clear that in the case of diffraction in a rectangular screen, the informational signal has not only angle modulation, as in the case of a small target, but also amplitude modulation. The law of angle modulation in the case of a rectangular target of sufficiently large size is much more complex than in the case of a small target.

In Figure 13.12, the laws of amplitude and phase modulation of the shadow signal are shown for the case when the shield dimensions are $h = 3$ m and $l = 6$ m. Other parameters are: $V = 10$ m/s; $t_0 = 0$ s; $\rho_1 = 15$ m and $T = 10$ s. The linear part of the phase in Figure 13.12 is omitted.

In the case of a small target, the signal $s_T(t)$ is passed through a nonlinear device and then filters out the constant and high-frequency components (see Figure 13.7) to obtain an LF signal $s(t)$ from the RF signal $s_T(t)$. After lengthy transformations it is possible to achieve

$$s(t) = \frac{A^2}{2}\left|\dot{\Upsilon}(t)\right|^2, \tag{13.60}$$

where

$$\dot{\Upsilon}(t) = \mathrm{j} + \left[C(\chi) - \mathrm{j}S(\chi)\right]\{[C(\Xi_+(t)) - C(\Xi_-(t))] - \mathrm{j}[S(\Xi_+(t)) - S(\Xi_-(t))]\}. \tag{13.61}$$

Figure 13.13 shows signal $s(t)$ calculated with formulae (13.60) and (13.61) for the same example, for which curves in Figure 13.12 were plotted. At the moment when, $t = t_0 = 0$ s, the target crosses the radar's baseline and the signal has a local maximum, which verges on two minimums. This maximum is a demonstration in radio-physical terms of the natural phenomenon, the Arago–Poisson spot in optics, discussed above.

Suppose that the shield is motionless ($V = 0$ m/s) and its central point is located exactly above the baseline of the FSR. Then

$$s_{sh}(t) = -2A\,\mathrm{Re}\left\{\mathrm{j}\left[C(\chi) - \mathrm{j}S(\chi)\right]\left[C(\upsilon) - \mathrm{j}S(\upsilon)\right]\exp\left[\mathrm{j}\,2\pi\left(f_0 t - L/\lambda\right)\right]\right\}, \tag{13.62}$$

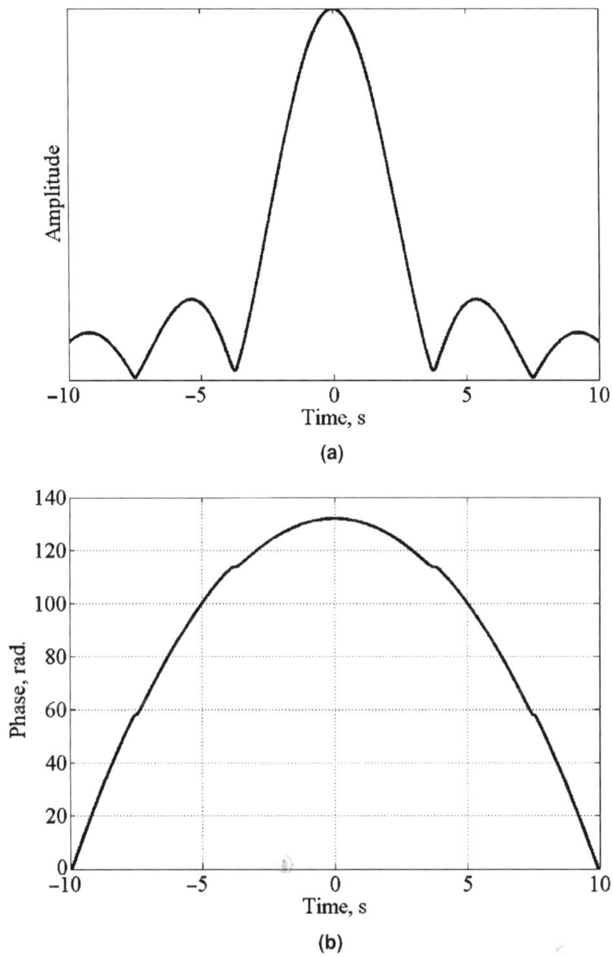

Figure 13.12 Laws of (a) amplitude and (b) phase modulation of the shadow signal from a truck

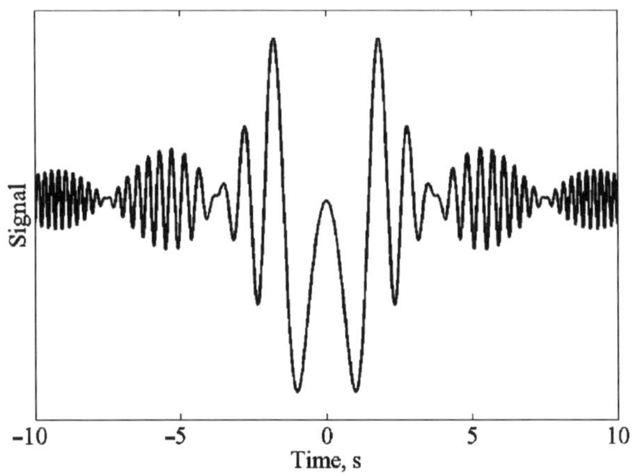

Figure 13.13 Shadow signal from a truck

where $\upsilon = \sqrt{2}l/\rho_1$. If the width of the shield is equal to the radius of a Fresnel region, e.g. the nth Fresnel region, then parameter υ can be evaluated as $\upsilon = \sqrt{2n}$.

Formally, the phase centre of the receiving antenna is located in the area of a geometric shadow from the rectangular shield. However, substituting Equations (13.62) to (13.36), it can be seen that the receiving signal does not exactly equal zero. This is a demonstration of the diffraction phenomenon.

The power of the receiving signal in the occurrence of a motionless shield is determined by the formula

$$P(t) = P_0 \left|1 - 2\mathrm{j}\left[C(\chi) - \mathrm{j}S(\chi)\right]\left[C(\upsilon) - \mathrm{j}S(\upsilon)\right]\right|, \tag{13.63}$$

where P_0 is the power of a nondisturbed signal, which would be received without a shield. If the sizes of the shield concur with radii of the mth and nth Fresnel regions then

$$P(t) = P_0 \left|1 - 2\mathrm{j}\left[C\left(\sqrt{2m}\right) - \mathrm{j}S\left(\sqrt{2m}\right)\right]\left[C\left(\sqrt{2n}\right) - \mathrm{j}S\left(\sqrt{2n}\right)\right]\right|. \tag{13.64}$$

Let $m \gg 1$; and $n \gg 1$, i.e. the target is large enough to shield a significant number of Fresnel regions. In this case, the Arago–Poisson spot is practically absent. Taking into account the properties of Fresnel integrals [II.38], it is not difficult to show that

$$\lim_{m\to\infty} C\left(\sqrt{2m}\right) = \tfrac{1}{2} \quad \lim_{m\to\infty} S\left(\sqrt{2m}\right) = \tfrac{1}{2} \quad \lim_{n\to\infty} C\left(\sqrt{2n}\right) = \tfrac{1}{2}, \quad \lim_{n\to\infty} S\left(\sqrt{2n}\right) = \tfrac{1}{2}. \tag{13.65}$$

Taking into account Equations (13.65) and (13.64) gives

$$\lim_{\substack{m\to\infty \\ n\to\infty}} P(t) = P_0 \left|1 - 2\mathrm{j}\left(\tfrac{1}{2} - \mathrm{j}\tfrac{1}{2}\right)\left(\tfrac{1}{2} - \mathrm{j}\tfrac{1}{2}\right)\right| = \left|1 - \mathrm{j}(1-\mathrm{j})^2/2\right| = 0. \tag{13.66}$$

In contrast, when the size of the first Fresnel region is much larger than the shield sizes, i.e. at $\chi \ll 1$ and $\upsilon \ll 1$, the shield degenerates to a point target. As the shadow is absent, consideration of the Arago–Poisson spot is unnecessary. At the same time,

$$\lim_{\chi\to 0} C(\chi) = 0, \qquad \lim_{\chi\to 0} S(\chi) = 0, \qquad \lim_{\upsilon\to 0} C(\upsilon) = 0, \qquad \lim_{\upsilon\to\infty} S(\upsilon) = 0 \tag{13.67}$$

and, from Equation (13.64),

$$\lim_{\substack{h\to 0 \\ l\to 0}} P(t) = P_0 \left|1 - 2\,\mathrm{j}0\right| = P_0. \tag{13.68}$$

As expected for the case of the point target (the EMWs are not disturbed), the receiving signal corresponds to the case of free-space propagation.

The relative power, $P(t)/P_0$, of the RF receiving signal, $s_T(t)$, calculated using Equation (13.63), is shown in Figure 13.14 as a function of the radius of the first Fresnel region, ρ_1. A graph drawn with a solid line is constructed for the shield with sizes $h = 1.7$ m and $l = 0.3$ m. Such a shield can be a human model. A graph is drawn with a dotted line according to the sizes of the shield where $h = 3$ m and $l = 6$ m. Such a shield has already been considered as the truck model. From the behaviour of the curves in Figure 13.14 it can be seen that the dependence of the signal power, which is received in the centre of Arago–Poisson spot from the radius of the first Fresnel region, has an appreciably nonlinear and nonmonotone character.

For simplification of the graphical representation and without loss of generality it can be supposed that in the quantitative calculations, amplitude A, included in the expressions for the shadow signal, is equal to one. For this occurrence, in Figure 13.15 a dependence of power of the informational shadow signal (13.62) from the radius of the first Fresnel region is shown. The curve in Figure 13.15(a) is built for the human model and in Figure 13.15(b) for the truck model. From the behaviour of the curves, for every shield an optimal value of the radius of the first Fresnel region can be defined (in terms of the criterion of maximum power of the information part of the Arago–Poisson spot).

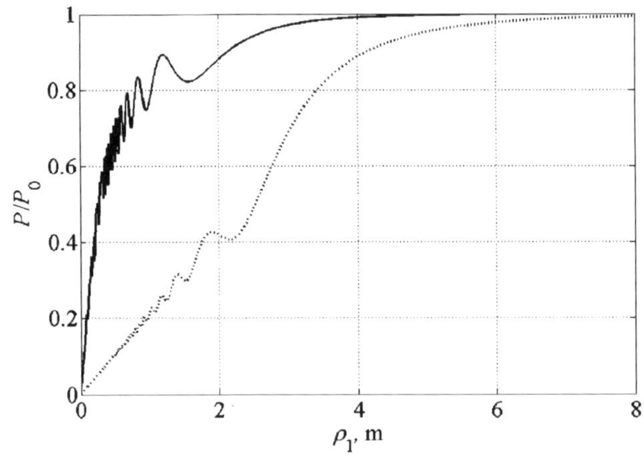

Figure 13.14 Relative power of an RF receiving signal as a function of the radius of the first Fresenel region

In connection with this, it is interesting to analyse in detail the dependence of the radius of the first Fresnel region on the radar parameters. Suppose that the base of radar, L, is fixed and its value is considered as a scale for other linear sizes. Denote by

$$\hat{\rho}_1 = \rho_1/L, \qquad \hat{\lambda} = \lambda/L, \qquad \hat{d}_T = d_T/L \tag{13.69}$$

the dimensionless normalized values of the radius of the first Fresnel region, of wavelength and of distances from the transmitter to the target. Applying new notation of Equations (13.69) to Equation (13.35), it is straightforward to obtain

$$\hat{\rho}_1 = \sqrt{\hat{\lambda}\hat{d}_T\left(1 - \hat{d}_T\right)}. \tag{13.70}$$

The results of the calculations using Equation (13.70) are shown in Figure 13.16, where the normalized radius of the first Fresnel region is shown in dependence on the normalized wavelength, $\hat{\lambda}$, and the normalized distance, $\hat{d}_T$. It can be seen from the diagram that at the adjusted metric parameters of the radar, λ and L, the maximum size of the radius of the first Fresnel region is achieved by the condition $d_T = d_R = L/2$, i.e. in the case when the radar trajectory divides the baseline of the radar into two equal halves.

For the case $d_T = d_R$, the dependence of the radius of the first Fresnel region, ρ_1, on the parameter d_T is represented in Figure 13.17. Lines are constructed for different carrier frequencies of the probing signal, namely data 1 for $f_0 = 70$ MHz data 2 for $f_0 = 169$ MHz data 3 for $f_0 = 439$ MHz and data 4 for $f_0 = 869$ MHz.

Obviously, the diameter of any Fresnel region is two times bigger than its radius. Therefore, the shield with sizes 3 m $\times$ 6 m (the truck) wholly closes the first Fresnel region if $\rho_1 \leq 3$ m. For the carrier frequency $f_0 = 70$ MHz this means that $d_T = d_R \leq 4.2$ m; for frequency $f_0 = 169$ MHz this means that $d_T = d_R \leq 10.1$ m; for frequency $f_0 = 453$ MHz this means that $d_T = d_R \leq 26$ m; and for frequency $f_0 = 869$ MHz this means that $d_T = d_R \leq 52.2$ m. The shield with sizes 1.7 m $\times$ 0.3 m (the human model) almost never wholly closes the first Fresnel region. In order for a human to close the first Fresnel region at the frequency $f_0 = 869$ MHz, $d_T = d_R$ should be ≤ 2 m. This is practically impossible.

In conclusion, it is remarked that the orthogonality of the baseline of the FSR and the linear trajectory of the target is supposed. If a nonzero projection of the target's velocity to the baseline of the radar exists, the radii of the Fresnel regions are not constants, but change with the moving process of the target; i.e. they are functions of time, $\rho_1 = \rho_1(t)$.

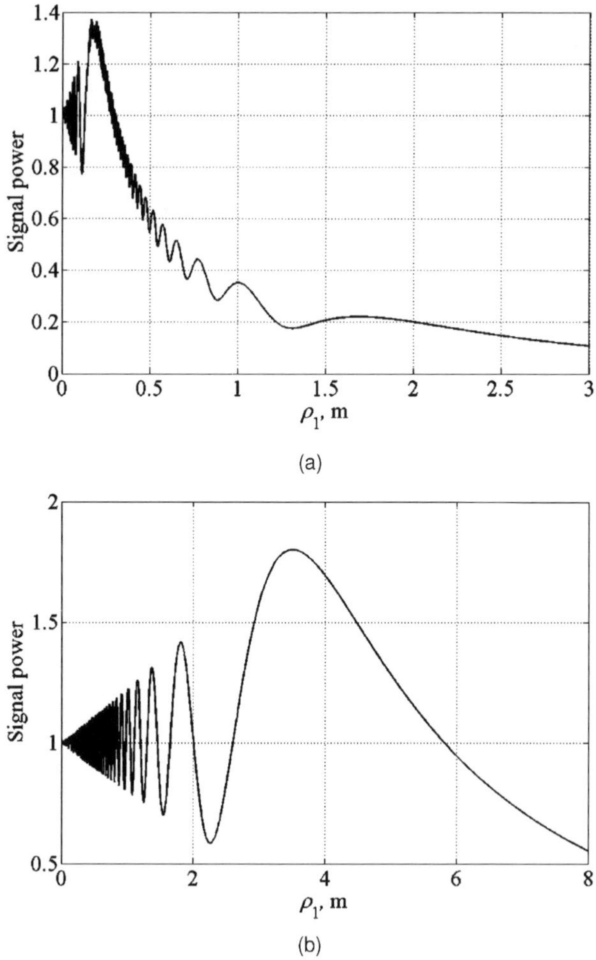

Figure 13.15 Dependence of the power of the informational shadow signal (13.62) on the radius of the first Fresnel region

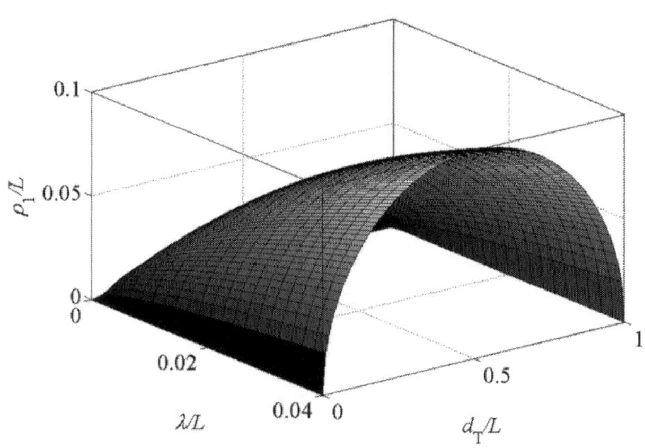

Figure 13.16 Dependence of the normalized radius of the first fresnel region on the normalized wavelength and the normalized distance

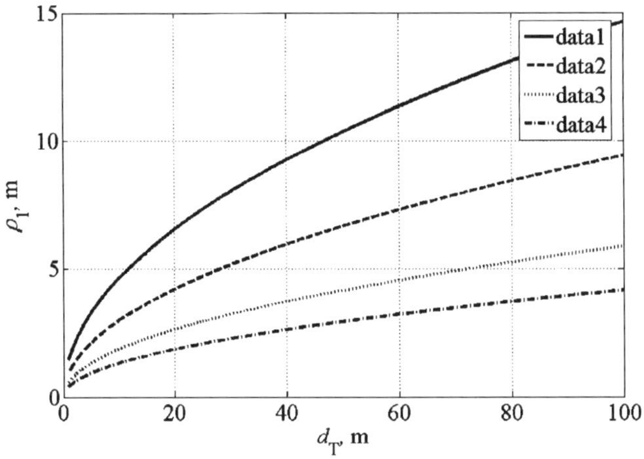

Figure 13.17 Dependence of the radius of the first Fresnel region on the parameter d_T

13.3.4 Signal Model of a Target with an Arbitrary Form

A target with an arbitrary form has a contour that can be described with the help of two functions, $h(x)$ and $b(x)$ (see Figure 13.18). The coordinate axis, *OX*, is chosen in the direction of the movement of the target. The central point, C, of the target intersects a vertical plane crossing the baseline of the FSR at the moment t_0. The receiving signal, which is formed by the receiving antenna from the electromagnetic field incident on the receiving aperture, can be represented, as before, in view of the sum of the spurious oscillation, $e(t)$, and the shadow signal, $s_{sh}(t) = \text{Re}[\dot{s}_{sh}(t)]$ (see formula (13.36). The expression for the complex shadow signal is obtained as

$$\begin{aligned} \dot{s}_{sh}(t) &= \dot{S}_{sh}(t)\exp\left[\text{j}\,2\pi\,(f_0 t - L/\lambda)\right] \\ &= -\text{j}\frac{\sqrt{2}A}{\rho_1}\int_{-l/2}^{l/2}\left\{C\left(\sqrt{2}\,h(x)/\rho_1\right) - C\left(\sqrt{2}\,b(x)/\rho_1\right) - \text{j}\left[S\left(\sqrt{2}\,h(x)/\rho_1\right)\right.\right. \\ &\quad \left.\left. - S\left(\sqrt{2}\,b(x)/\rho_1\right)\right]\right\} \times \exp\left[-\text{j}\pi\left(\frac{Vt - x_0 + x}{\rho_1}\right)^2\right]\text{d}x\,\exp\left[\text{j}\,2\pi\,(f_0 t - L/\lambda)\right]. \qquad (13.71) \end{aligned}$$

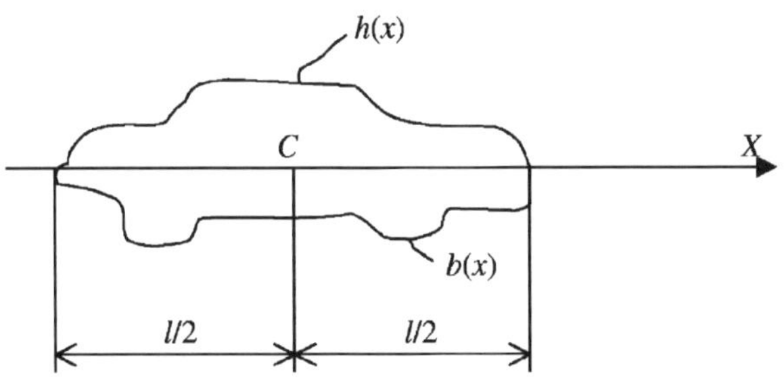

Figure 13.18 Target with an arbitrary form

Following Reference [II.40], the operation of the direct Fresnel transform is defined as

$$\boldsymbol{F}_{\Omega}(x) = \sqrt{2}\Omega \int_{-\infty}^{\infty} \boldsymbol{f}(z) \exp\left[\mathrm{j}\Omega^2 (x - z)^2\right] \mathrm{d}z \tag{13.72}$$

and the inverse Fresnel transform as

$$\boldsymbol{f}(z) = \sqrt{2}\Omega \int_{-\infty}^{\infty} \boldsymbol{F}_{\Omega}(x) \exp\left[-\mathrm{j}\Omega^2 (x - z)^2\right] \mathrm{d}x, \tag{13.73}$$

where Ω is the parameter of the Fresnel transform.

Obviously, the Fresnel transform is a convolution of the signal, $f(z)$, and the Fresnel function [II.40],

$$\boldsymbol{W}_{\Omega}(x) = \exp\left[\mathrm{j}\Omega^2 x^2\right] \tag{13.74}$$

and the inverse Fresnel transform is a convolution of the signal with the complex conjugate Fresnel function.

If we take into consideration an auxiliary function,

$$\Phi_{\sqrt{\pi}/\rho_1}(x) = \begin{cases} C\left(\frac{\sqrt{2}h(x)}{\rho_1}\right) - C\left(\frac{\sqrt{2}b(x)}{\rho_1}\right) - \mathrm{j}\left[S\left(\frac{\sqrt{2}h(x)}{\rho_1}\right) - S\left(\frac{\sqrt{2}b(x)}{\rho_1}\right)\right], & -\frac{l}{2} \le x \le \frac{l}{2}, \\ 0, & \text{else} \end{cases} \tag{13.75}$$

and determine the inverse Fresnel transform from it,

$$\phi(z) = \frac{\sqrt{2\pi}}{\rho_1} \int_{-\infty}^{\infty} \Phi_{\sqrt{\pi}/\rho_1}(x) \exp\left[-\frac{\mathrm{j}\pi}{\rho_1^2}(x - z)^2\right] \mathrm{d}x, \tag{13.76}$$

then Equation (13.71) for the complex envelope of the shadow signal can be represented as

$$\dot{S}_{\mathrm{sh}}(t) = -\mathrm{j}\frac{A}{\sqrt{\pi}}\phi(x_0 - Vt); \tag{13.77}$$

i.e. the complex envelope, $\dot{S}_{\mathrm{sh}}(t)$, of the RF shadow signal is proportional to the shifted and inverted in time inverse Fresnel transform of function $\Phi_{\sqrt{\pi}/\rho_1}(x)$, which is defined by Equation (13.75).

Choosing a concrete value,

$$\Omega = \sqrt{\pi}\frac{V}{\rho_1} = \sqrt{\frac{\pi L}{\lambda d_{\mathrm{T}} d_{\mathrm{R}}}}\, V, \tag{13.78}$$

for the parameter of the transform (13.72) and calculating the Fresnel transform $\boldsymbol{F}_{\sqrt{\pi}V/\rho_1}(\tau)$ from the complex envelope $\dot{S}_{\mathrm{sh}}(t)$ of the shadow signal, it is easy to ensure that

$$\boldsymbol{F}_{\sqrt{\pi}V/\rho_1}(t) = -\mathrm{j}\frac{A}{\sqrt{\pi}}\Phi_{\sqrt{\pi}/\rho_1}(x_0 - Vt). \tag{13.79}$$

Therefore, by knowing the ratio of the target velocity to the radius of the first Fresnel region, extracting the complex envelope of the shadow signal and calculating the Fresnel transform from it for the absolute value of parameter defined by Equation (13.78), it is possible to find an estimation of the useful function, $\Phi_{\sqrt{\pi}/\rho_1}(x)$, which is determined by Equation (13.75). The estimation is shifted along the coordinate axis and inverted relative to this axis.

If the target velocity is not known, and thus we are unable to use Equation (13.78), Fresnel transforms for different reference values of the parameter of the transform Ω should be found repeatedly. Thus, a set of estimations of the function$\Phi_{\Omega}(x)$ can be found.

It is significant that the estimated function (13.75) contains information about the target's contour. This may be illustrated by a simple example. Let us consider the target as a rectangular shield, which is moving with velocity $V = 5$ m/s. Let the sizes of the shield be $l = 6$ m and $h = 3$ m. Such a target has already been documented in this section, and is well known as the truck model. Let the signal be registered by the receiver of the radar during a time interval from 0 to 40 s, and the target centre be located under the baseline of the radar at the moment $t_0 = 19$ s. We will also consider that $d_T = d_R = 100$ m and that the carrier frequency is $f_0 = 69$ MHz. According to these conditions, a shadow signal is shown in Figure 13.19(a).

Initial data and formula (13.78) are sufficient for the calculation of the necessary value of the parameter of the Fresnel transform. The results of the calculation of the Fresnel transform from the complex envelope of the signal shown in Figure 13.19(a) can be seen in Figure 13.19(b). If we recollect that the target is a rectangular shield, it will be possible to establish that the line on the last figure gives an overview of the target's contour. The difference in the curve in Figure 13.19(b) from the ideal rectangle is perhaps due not only to the nonlinearity of the Fresnel integrals in Equation (13.75) but also to the restriction of the time interval ($T = 40$ s) and to the time discrimination in the calculation procedure.

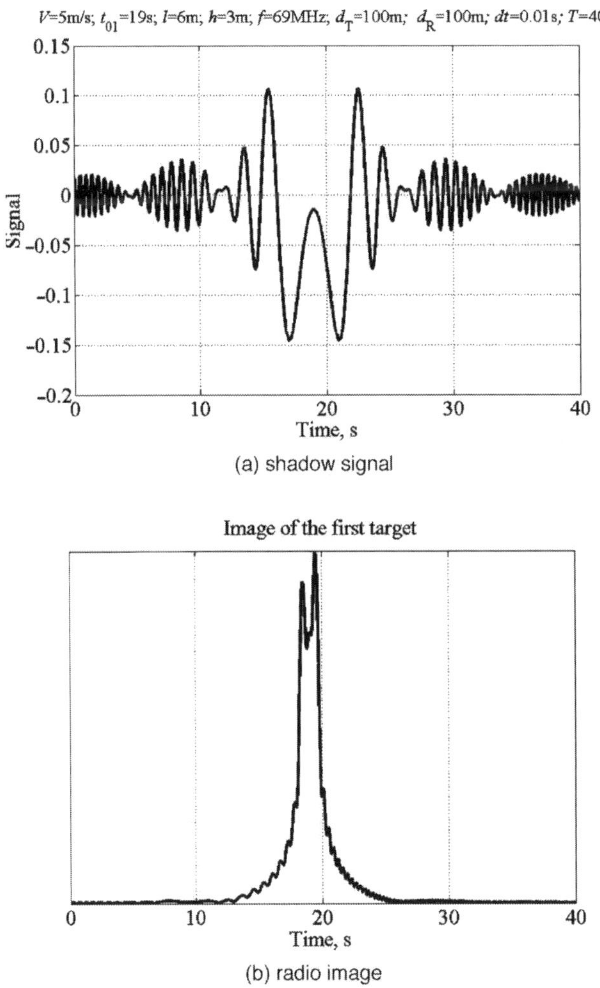

(a) shadow signal

(b) radio image

Figure 13.19 Shadow signal and radio image

Consequently, the curve shown in Figure 13.19(b) can be interpreted as a radio image of the target's contour. Therefore, the algorithm forming the shadow contour's radio image is made of the following steps:

1. Extract the complex envelope, $\dot{S}_{sh}(t)$, of the shadow signal from the receiving EMW.
2. Calculate the parameter of the Fresnel transform using formula (13.78).
3. Find the Fresnel transform from the complex envelope, $\dot{S}_{sh}(t)$, of the shadow signal.

The contour radio image mentioned above can be very useful in solving the task of target resolution. In Figure 13.20, the signal and its shadow radio image are represented for the case when two targets in series cross the baseline of the FSR. These are moving with the same velocity, $V = 3$ m/s, at a distance

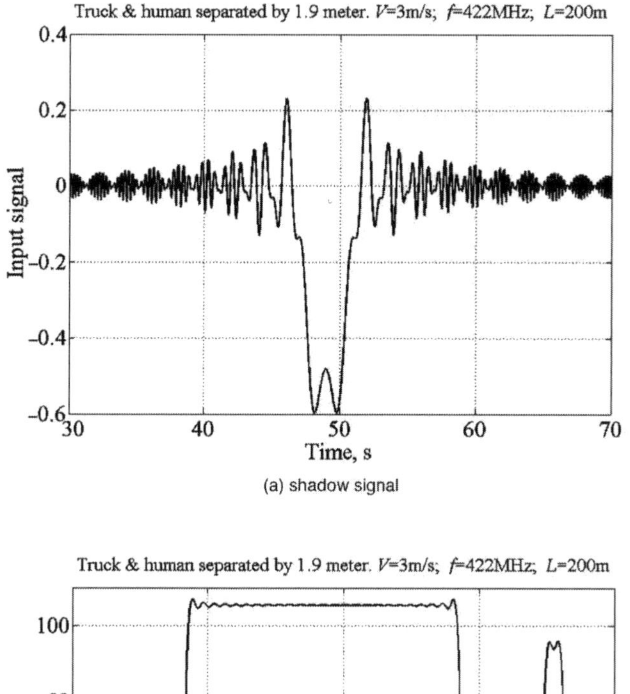

(a) shadow signal

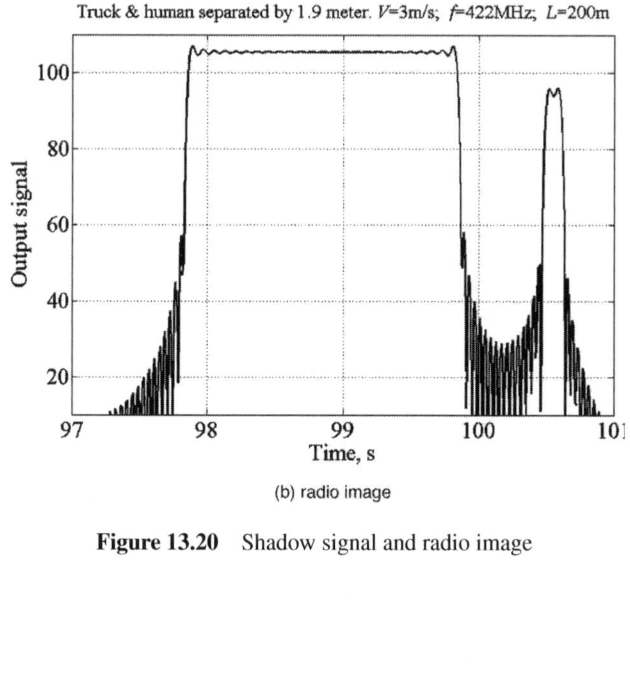

(b) radio image

Figure 13.20 Shadow signal and radio image

of 1.9 m from each other. The targets have the same square shape and sizes of $l = 6$ m and $h = 3$ m for the first and $l = 0.4$ m and $h = 1.7$ m for the second. Such rectangular targets can be models of both the truck and the pedestrian. Parameters of the radar are $d_T = d_R = 100$ m and $f_0 = 422$ MHz. As seen from the behaviour of the curves in Figure 13.20, making a decision about the existence of two targets (resolution) from the graph of the signal in the time domain is very problematic. However, by analysing the shadow radio image, the presence of two separated targets can clearly be seen.

14

Advanced Scattering

The bases of electromagnetic theory have been set out in Chapter 10. This Chapter is dedicated to reviewing the methods of analysis of bistatic scattering problems.

14.1 ELECTROMAGNETIC THEORY PRINCIPLES

A system of Maxwell's differential equations (10.4) to (10.7) and Helmholtz wave equations (10.47) form the theoretical basis of electromagnetism. However, only a few simple problems of scattering can be solved in the closed form.

14.1.1 The Kirchhoff Integral and the Stratton–Chu Equations

When solving a task involving electromagnetic wave scattering on a surface S of target Γ, it is also possible to use the so-called Green's function method for the solution of specific boundary value problems. Green's function methods are not a replacement for Maxwell's equations, but simply a method by which to solve the differential equations in partial variables or integral equations.

Green's function is the fundamental solution of the differential equation; i.e. it has to obey the original equation to be solved, but with a Dirac delta function on the right side. Thus, for the Helmholtz equation (10.46), the Green function must obey

$$\left(\nabla^2 + \dot{k}^2\right) G\left(\boldsymbol{r}, \boldsymbol{r}'\right) = -\delta\left(\boldsymbol{r} - \boldsymbol{r}'\right). \tag{14.1}$$

Here $G(\boldsymbol{r}, \boldsymbol{r}')$ is the Green function, $\boldsymbol{r}$ is the radius vector of observation point, $\boldsymbol{r}'$ is a radius vector of the point on the scattering surface of a body and $\delta(\boldsymbol{x})$ is the Dirac delta function. It is known that the general solution of Equation (14.1) has the form

$$G\left(\boldsymbol{r}, \boldsymbol{r}'\right) = G_0\left(\boldsymbol{r}, \boldsymbol{r}'\right) + H\left(\boldsymbol{r}, \boldsymbol{r}'\right) \equiv \frac{\exp\left(\mathrm{j}\dot{k}\left|\boldsymbol{r} - \boldsymbol{r}'\right|\right)}{\left|\boldsymbol{r} - \boldsymbol{r}'\right|} + H\left(\boldsymbol{r}, \boldsymbol{r}'\right), \tag{14.2}$$

Bistatic Radar: Principles and Practice Edited by M. Cherniakov

where $H(\boldsymbol{r}, \boldsymbol{r}')$ is the solution of the Helmholtz homogeneous equation. Now, using the Green theorem (refer to Reference [II.37] or [II.41]), it is possible to obtain a highly important equation for the theory of scattering electromagnetic waves, namely the Kirchhoff integral:

$$\psi(\boldsymbol{r}, t) = \frac{1}{4\pi} \oiint_S \left[\psi(\boldsymbol{r}', t) \frac{\partial}{\partial n'} G(\boldsymbol{r}, \boldsymbol{r}') - G(\boldsymbol{r}, \boldsymbol{r}') \frac{\partial}{\partial n'} \psi(\boldsymbol{r}', t) \right] \mathrm{d}\boldsymbol{r}'. \tag{14.3}$$

Equation (14.3) connects a value of the Cartesian component, ψ, of vector **E** (or **H**) in some point of space with the values of the same component and its derivative on a bounding surface. It is not difficult to rewrite formula (14.3) for vectors **E** and **H** themselves.

The Stratton–Chu equations [II.42],[1]

$$\mathbf{E} = \mathbf{E}^{\text{in}} + \frac{1}{4\pi} \oiint_S \left[(\boldsymbol{n}' \times \mathbf{H}) G_0 + (\boldsymbol{n}' \times \mathbf{E}) \times \nabla' G_0 + (\boldsymbol{n}' \cdot \mathbf{E}) \nabla' G_0 \right] \mathrm{d}S',$$

$$\mathbf{H} = \mathbf{H}^{\text{in}} + \frac{1}{4\pi} \oiint_S \left[(\boldsymbol{n}' \times \mathbf{H}) \times \nabla' G_0 + (\boldsymbol{n}' \cdot \mathbf{H}) \nabla' G_0 + (\boldsymbol{n}' \times \mathbf{E}) G_0 \right] \mathrm{d}S', \tag{14.4}$$

may appear more convenient in a number of cases. The Integrals in Equations (14.4) are taken on closed surfaces, S, for which $\boldsymbol{n}'$ is a unit normal at the point $\boldsymbol{r}'$. Either two Stratton–Chu vector equations are an integro-differential equation.

Taking into account formulae (10.101) and (14.4), the scattered fields can be represented in the form:

$$\mathbf{E}^{\text{sc}} = \frac{1}{4\pi} \oiint_S \left[(\boldsymbol{n}' \times \mathbf{H}) G_0 + (\boldsymbol{n}' \times \mathbf{E}) \times \nabla' G_0 + (\boldsymbol{n}' \cdot \mathbf{E}) \nabla' G_0 \right] \mathrm{d}S',$$

$$\mathbf{H}^{\text{sc}} = \frac{1}{4\pi} \oiint_S \left[(\boldsymbol{n}' \times \mathbf{H}) \times \nabla' G_0 + (\boldsymbol{n}' \cdot \mathbf{H}) \nabla' G_0 + (\boldsymbol{n}' \times \mathbf{E}) G_0 \right] \mathrm{d}S'. \tag{14.5}$$

The expressions (14.5) are valid both for conducting objects and for dielectrics. However, it is possible to simplify them in the case of a perfectly conducting target when $\mathbf{n}' \times \mathbf{E} = 0$ and $\mathbf{n}' \cdot \mathbf{H} = 0$; namely they can be rewritten as

$$\mathbf{E}^{\text{sc}} = \frac{1}{4\pi} \oiint_S \left[(\boldsymbol{n}' \times \mathbf{H}) G_0 + (\boldsymbol{n}' \cdot \mathbf{E}) \nabla' G_0 \right] \mathrm{d}S',$$

$$\mathbf{H}^{\text{sc}} = \frac{1}{4\pi} \oiint_S \left[(\boldsymbol{n}' \times \mathbf{H}) \times \nabla' G_0 \right] \mathrm{d}S'. \tag{14.6}$$

It is also known that the separate parts in formulae (14.4) to (14.6) can be interpreted as surface currents and charges [II.43]. For instance,

$$\mathbf{J} = \boldsymbol{n}' \times \mathbf{H} \tag{14.7}$$

is an electrical surface current [II.44], $\boldsymbol{n}' \times \mathbf{E}$ is a magnetic surface current [II.44], $\mathbf{n}' \cdot \mathbf{E}$ is an electrical surface charge and $\boldsymbol{n}' \cdot \mathbf{H}$ is a magnetic surface charge. Naturally, the appearance of a magnetic surface current and a magnetic surface charge in Equations (14.4) is essential for a dielectric material with finite conductivity.

---

[1] Some symbols in formulae in this chapter differ from similar ones in Reference [II.42] because Stratton directed a vector of normal to the scattering surface in the opposite direction.

On the assumption of perfect conductivity, and taking into account Equation (14.7), the second of Equations (14.6) can be written more briefly as

$$\mathbf{H}^{\mathrm{sc}} = \frac{1}{4\pi} \oiint_S \mathbf{J} \times \nabla' G_0 \, \mathrm{d}S'. \tag{14.8}$$

This very elegant formula connects the scattering field in a point of space with the current induced on the surface of the object. The formula for the vector potential has an even more refined form (refer to Reference [II.41]):

$$\mathbf{A} = \frac{1}{4\pi} \oiint_S \mathbf{J} G_0 \, \mathrm{d}S'. \tag{14.9}$$

It is necessary to point out that formulae (14.8) and (14.9), as well as the Stratton–Chu equations themselves, are exact. Exact equations can be useful as background equations for different models and numerical methods. Some of these models and methods will be described later in the chapter.

The solution methods of scattering problems based on exact equations are given the term 'exact methods'. However, it should be noted that this term does not imply that results obtained via this method are necessarily exact. The main advantage of the exact method is that the error of the solution can be mathematically evaluated at any given stage – from the problem statement to obtaining numerical results.

A key feature of the exact method is that the given results may be compared to those obtained via the heuristic method, where physical simplification allows the solution of mathematical difficulties. In this way, the heuristic method itself may be evaluated.

14.1.2 Electromagnetic Models

In contrast to the exact method, the heuristic method allows an approximate solution to be obtained in closed form by using a relatively simple procedure. However, a distinctive characteristic of such a method is an uncontrolled inaccuracy and a dependence on concrete conditions that varies from case to case. Thus, some heuristic methods are only applicable for a narrow range of tasks. If a problem exists that also allows the parallel solution by exact methods, then, as described above, the inaccuracy of heuristic methods is evaluated by the comparison of exact and heuristic solutions. Otherwise, the inaccuracy of heuristic methods can only be evaluated experimentally.

A physical model (hypothesis) lies at the base of each heuristic method. This model completely or partly eliminates (or greatly softens) the mathematical difficulties that seemed unsolvable under the strict statement of the task.

Let us consider some of the questions concerning the creation of the physical models with reference to electromagnetic wave scattering on extended surfaces or simple form bodies. Several types of physical models are known, from which the most interesting are phenomenological and analogue models. A phenomenological model is based on the direct empirical observation of a process, and herewith the main characteristics are formulated as attributes of the model and the unessential characteristics are rejected [II.45]. These discarded 'unessential' characteristics generate uncontrolled inaccuracy within phenomenological methods, mentioned above. Practically, this unpredictable error defines a main difference between the study of wave scattering on the basis of phenomenological models and by the means of solving the exact equations.

Most heuristic methods are used in a high-frequency (HF) range, where the distinctive dimensions of scattering bodies are large in contrast with the wavelength of the electromagnetic field. These particular methods present the greatest interest for radar applications. Among the approximate methods are: the geometrical optics method, also called 'ray approximation'; the physical optics method, also termed the 'wave optics method' or 'Kirchhoff approximation'; the aperture method; the edge waves method; the Keller geometric theory of diffraction, also termed the 'diffraction rays method'; and many others.

Let us consider briefly some of the most commonly known heuristic methods and physical models.

14.1.2.1 Geometrical Optics (or Ray Optics) [II.17, II.37, II.46]

The simplest of all heuristic methods is the geometrical optics method, ignoring interference phenomena. The following physical assumptions apply to its basis:

- The wavelength of an electromagnetic field is significantly small in comparison with the distinctive dimensions of bodies, with which the field interacts.
- In the case of a nonuniform environment the wavelength of an electromagnetic field in this environment is significantly small in comparison with the distance, on which the parameters of the environment noticeably change.
- The interfaces where the electromagnetic field interacts with bodies and environments are ideally smooth, and surface curvature is unnoticeable. Therefore, within small platforms refraction and reflection of electromagnetic waves are in accordance with Fresnel formulae for flat division border surfaces (refer to Chapter 10).
- The energy of an electromagnetic field is carried along the rays.

These assumptions, for the most part, correspond to light scattering and diffraction on macroscopic objects. The use of the geometric optics model in the microwave range is on the basis that only specific local areas of radar targets, which are mirror reflection points (spots),[2] edges, borders, shades, etc., bring a main contribution to the radar cross-section [II.46]. The geometric optics model intends that electromagnetic waves correspond to Fermat's principle.

This method allows the calculation of the intensity of the scattered electromagnetic field and, as a result, the value of the RCS. Essentially, the geometric optics method, with reference to large perfect conducting or lossless dielectric bodies, allows the calculation of the RCS without using other electromagnetism laws except Snell's law (10.74) for reflection and refraction.

Let us consider, as an example, a perfect conducting convex body of double curvature. The expression for a monostatic RCS was obtained in References [II.17] and [II.46] by means of the geometric optics model: $\sigma_m = \pi r_1 r_2$, where r_1 and r_2 are radii of curvature. Thus, the RCS of a conducting convex body of double curvature does not depend on wavelengths.

According to Refereence [II.17], the results that are obtained by the geometric optics method have an error of not more than 20 % under the condition that both r_1 and r_2 are bigger than two wavelengths. However, the radii of curvature cannot be infinitely big; it is impossible to calculate the RCS of flat plates or cylindrical and conical surfaces by the geometric optics method.

14.1.2.2 Physical Optics [II.4, II.17, II.37, II.46]

The physical optics approach, known also as the Kirchhoff method, the Huygens principle (which had been formulated for the first time long before the Maxwell equations) or the tangent plane model, unlike the geometrical optics method, is mostly applicable for bodies with endless curvature radii and a sharp bend in the surface, such as flat plates and convex cylinders of finite length. The physical optics method gives a certain precise definition in comparison with the geometric optics method for smooth convex bodies, such as a sphere or ellipsoid.

The following assumptions apply to the physical optics method basis:

- A wavelength of an electromagnetic field is of the same order as the distinctive dimensions apply scattering body.
- An incident wave causes currents on the surfaces of a scattering body, which are a source of the secondary (scattered) field.
- Currents are distributed only on a 'lighted' surface of a body, whose edge is defined by the geometric optics rules.

---

[2] A mirror reflection point is a point on target surface, in which the normal to surfaces divides the bistatic angle in half.

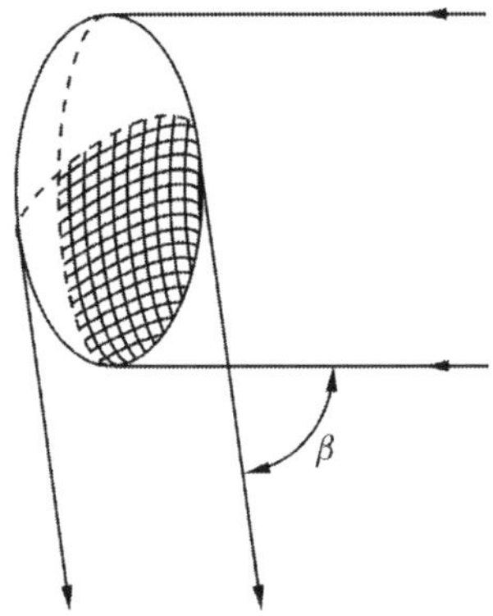

Figure 14.1 Definition of a site of a surface contributing into a scattering field at bistatic scattering in an approach of physical optics

- The current appears on each point of a 'lighted' surface, and this current is equal in magnitude and direction to the one that would be induced under other similar conditions on an infinite plane that is tangential to the surface at this point.
- The full field scattered in the given direction is always calculated by integration over only the area of illuminated surface of the body mutual for both, transmitting and receiving, stations (refer to Figure 14.1). The common part of the surface reaches the maximum and is equal to the whole 'lighted' surface in a monostatic case; the common part reaches the minimum and degenerates into a shadow contour in the case of forward scattering [II.17].

Under these assumptions, the physical optics method is applicable to perfect conducting bodies and bodies with finite, but sufficiently large, conductivity. Kobak supposes in Reference [II.17] that with some change in the original assumptions, the physical optics method is also suitable for absolutely black bodies.

A fundamental admission of the physical optics method is that incident and scattered waves are linearly connected by means of the Fresnel coefficient, R, depending on polarizations and the local angle of incidence. In this case, integral equations propose only surface integration of the incident wave. Let us suppose that a plane incident wave can be represented as

$$\mathbf{E}^{\mathrm{in}}(\boldsymbol{r},t)=\mathbf{E}_0^{\mathrm{in}}\exp\left[\mathrm{j}\left(k\boldsymbol{e}_{\mathrm{S}}^{\mathrm{in}}\cdot\boldsymbol{r}-\omega t\right)\right]=E_0^{\mathrm{in}}\boldsymbol{e}^{\mathrm{in}}\exp\left[\mathrm{j}\left(k\boldsymbol{e}_{\mathrm{S}}^{\mathrm{in}}\cdot\boldsymbol{r}-\omega t\right)\right], \tag{14.10}$$

where $\boldsymbol{e}^{\mathrm{in}}$ is the unit vector along the direction of linear polarizations of the incident wave. The result given by the physical optics method is

$$\mathbf{E}_0^{\mathrm{sc}}(\boldsymbol{r})=E_0^{\mathrm{in}}k\boldsymbol{e}_{\mathrm{S}}^{\mathrm{sc}}\times\oint\!\!\!\oint\limits_S\mathbf{P}\left(\boldsymbol{e}^{\mathrm{in}},\boldsymbol{e}_{\mathrm{S}}^{\mathrm{sc}},\boldsymbol{e}_{\mathrm{S}}^{\mathrm{in}},\boldsymbol{n}',R\right)\exp\left(\mathrm{j}k\boldsymbol{r}\cdot\boldsymbol{r}'\right)\mathrm{d}S', \tag{14.11}$$

where the vector function $\mathbf{P}$ depends on the polarization of the incident wave, directions of the incident and scattered wave propagation, direction of the normal to the scattering surface and the Fresnel coefficient. Readers may refer to Reference [II.47] or [II.48] for more in-depth detail. It should be noted that under the geometrical optics approximation the last formula appeared in a simpler form:

$$\mathbf{E}_0^{\mathrm{sc}}(\boldsymbol{r})=E_0^{\mathrm{in}}k\boldsymbol{e}_{\mathrm{S}}^{\mathrm{sc}}\times\mathbf{P}_0\left(\boldsymbol{e}^{\mathrm{in}},\boldsymbol{e}_{\mathrm{S}}^{\mathrm{sc}},\boldsymbol{e}_{\mathrm{S}}^{\mathrm{in}},\boldsymbol{n}',R\right)\oint\!\!\!\oint\limits_S\exp\left(\mathrm{j}k\boldsymbol{r}\cdot\boldsymbol{r}'\right)\mathrm{d}S'. \tag{14.12}$$

It is well known that waves with a short wavelength (in comparison with target dimensions) are scattered in the given direction almost completely by the mirror oriented spots on a target surface [II.4,

II.46]. During a moonlit night or solar daytime human beings have the opportunity to observe brilliant spots on the surface of an open pond.

It has been shown analytically in Reference [II.46] that the physical optics model is applicable for objects with limited dimensions, when the main scattering is mirror-like and accomplished by reflecting points and their nearest neighbours. It is necessary to emphasize that the physical optics method ensures a correct solution only in the HF limit [II.46]. In this way it is similar to, and no more precise than, the geometrical optics method. The advantage of the physical optics method over geometrical optics is that it can be adapted to a much broader range of objects including flat plates, wedge-shaped bodies, conical bodies, etc., for which the geometrical optics method is inapplicable [II.46].

In a physical sense, this model assumes that the target surface is sufficiently flat to be presented by planes. This assumption is not true for the case of the big angles of incidence (refer to References [II.14] and [II.49]). It has been noted in Reference [II.50] that under a slip incident wave some more complex mechanisms, which are not taken into account by the physical optics model, are engaged: shadowing, diffraction, multiple scattering, atmospheric channels and waves capture. Listed mechanisms begin to play a significant role. In such conditions all phenomenological models are not a success.

The physical optics method does not take into account multiple reflecting and thus, in the strictest sense, is applicable for convex bodies only. However, the combination of the physical and geometric optics methods in a number of cases also succeeds in resolving scattering problems for nonconvex bodies as well. The essence of this combined method consists of the following: instead of a current distribution on a body surface, a field distribution on some auxiliary surface (aperture) surrounding a body is defined and, then, a scattered field is calculated in the far region by integration on the chosen aperture. Field distribution on the aperture is found by geometric optics rules. The combined method, as far as it does not require the determination of surface currents, is applicable not only to conducting but also to dielectric bodies. Usually, the name 'physical optics method' is given to the combined method. However, Kobak [II.17] considers the latter with reference to flat metallic surfaces as a separate model, which he terms an 'aperture model'.

14.1.2.3 Aperture Model [II.17]

This method occupies an intermediate position between the geometrical and physical optics methods. It is mainly used for the calculation of the scattered field, which arises from multiple wave reflection between flat conducting surfaces. In particular, the aperture method is highly suitable for the calculation of the scattering characteristics of the corner reflectors. This method assumes that flat surfaces are large in comparison with the wavelength and that the secondary reflection happens under geometric optics laws.

Under the aperture method, a scattering problem is reduced to a radiation problem, whose solution is known in many cases. The reduction procedure is founded on geometric optics laws, but the radiation problem solution is based on the laws of physical optics.

In the case of dihedral and trihedral corner reflectors with right angles between the edges, it appears that the main contribution to the scattered field is caused by apertures, which correspond to double or triple reflection. Moreover, these apertures are always located in the perpendicular direction to the incident wave.

Calculations transpire to be much more complex for the corner constructions with no right angles between edges.

14.1.2.4 The Edge Wave Model [II.51]

This model renders a physical optics approximation more precise and allows the diffraction by the sharp curves of the scattering body surface to be taken into account. Similarly to the physical optics method, the edge wave method is applicable, first of all, for perfect conducting convex bodies, for instance disc, rectangular plate and finite cylinder.

The edge wave method is based on the following physical assumptions:

- The wavelength of an electromagnetic field is small in comparison with the distinctive dimensions of a scattering body and longitudinal radii of sharp bends curvature of its surface (edges).
- An incident wave causes currents on the surfaces of a scattering body, which are a source of the secondary scattered field.
- Surface currents consist of two parts: even, which is defined by physical optics rules, and uneven, which originates from the influence of sharp bends in the surface.
- The uneven part of the current near some point, which lies on the sharp bend edge, is taken as equal to the current on the infinite wedge surface. The edge of the wedge is tangent to the sharp bend edge. Faces of the wedge are tangent to the body's surface in the given point.
- The uneven part of the current has the nature of an edge wave propagating from the sharp bend edge and fading when retreating from it.
- An edge wave that appears on some edges can reach other edges of that body and be reflected from them (secondary diffraction); an edge wave can be reflected not only once but can also multiply.

The scattering characteristics of a disc, finite cylinder, finite cone, finite paraboloid of revolution and spherical segments during the incidence of a wave along the axis of rotation [II.51], as well as rectangular plates [II.52], are determined by means of the edge wave method. The results of the monostatic RCS calculation differ from the measured values by no more than 0.5–1 dB in the broad sector (up to the sliding angle) of angles of incidence at the condition $k_0a > 3$, where a is a characteristic dimension of the scattering body (diameter, length) [II.17].

14.1.2.5 Geometrical diffraction theory (GDT) [II.17, II.53]

This method represents the development and generalization of the geometrical optics method and is based on the following physical assumptions:

- The wavelength of an electromagnetic field is small in comparison with the distinctive dimensions of a scattering body, radii of its surface curvature and longitudinal radii of sharp bends curvature of its surface (edges).
- Electromagnetic wave energy distributes along rays.
- Diffracted rays also exist, other than incident, reflected and refracted rays.
- Rays that fall normally or oblique to fluently curled (smooth) areas of a scattering body surface are reflected and refracted according to the geometrical optic rules; no diffracted rays appear herewith.
- Rays that thatare incident at a tangent to the smooth areas of a surface (i.e. under the right angle of incidence), at any angle on edges, as well as at any angle on the apexes (cone and pyramid edges), generate diffracted ray beams.
- An amplitude of a field corresponding to any diffracted ray is proportional to the incident field amplitude in the diffraction point (in the point of incidence).
- A structure of diffracted ray beams and coefficients of proportionality in diffraction points (coefficients of diffraction) differ for rays that fall tangential on smooth surface areas, on edges and on tops.
- A diffracted ray beam has a form of the fan dispersing from the geodesic line (Figure 14.2) if the ray falls tangential on a smooth surface area.
- A diffracted ray beam has the form of a crater (the surface of a circular cone) with the apex in the diffraction point and with the axis tangential to the edge (Figure 14.3).
- A diffracted ray beam is taken as isotropic, as for the point-like source, for the ray falling on the apex.
- A diffracted ray, which appears in any point of the diffraction surface of a scattering body, can reach other diffraction points and generate secondary diffracted ray beams.
- The full scattered field in some direction is defined by the summation of fields of all reflected, refracted and diffracted rays that spread in the given direction.

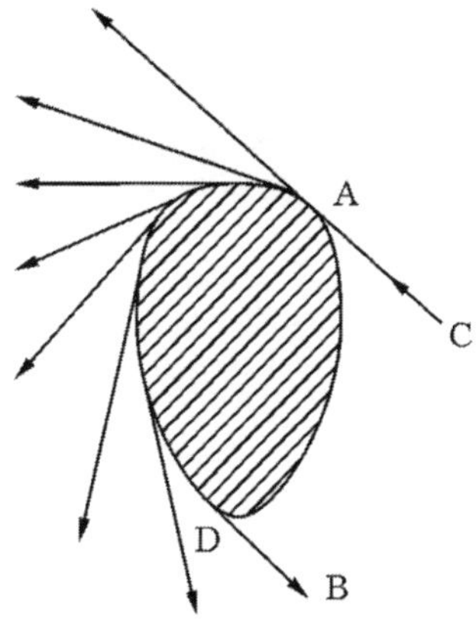

Figure 14.2 Diffractional rays arising on a smooth site of a surface of a scattering body: CA is an incident ray that is tangent to a smooth surface, A is a point of diffraction and DB is a diffractional ray that is extended in the opposite direction

The geometrical diffraction theory has been designed mainly for perfect conducting bodies. In particular, the MRCS of the perfectly conducting flat rectangular plate derived from geometrical diffraction theory, as well as by means of physical optics, is shown in Figure 14.4. Experimental results are also presented in Figure 14.4. The plate's centre coincides with the origin of coordinates O; the plate's plane coincides with the *XOY* plane; and the plate sides with length at 20.32 cm for each were oriented along the horizontal, O*X*, and vertical, O*Y*, axes. The incident wave is a plane wave with a frequency of 9.23 GHz ($\lambda = 3.25$ cm) and a linear vertical or horizontal polarization. Theoretical values, which have been obtained by the method of geometrical diffraction theory, shown in Figure 14.4, are sufficiently exact under values of angle lying in the range $0 \leq \phi \leq 80^\circ$. Singularity has a place for $\phi > 80^\circ$, where the geometrical diffraction theory gives unacceptable results.

Recently, a great number of tasks has been solved by means of this theory. In particular, the geometrical diffraction theory has allowed the definition of scattering characteristics not only for a rectangular plate, disc and finite cylinder but also for the finite cone, finite truncated cone, as well as for a number of other bodies of a more complex shape [II.17].

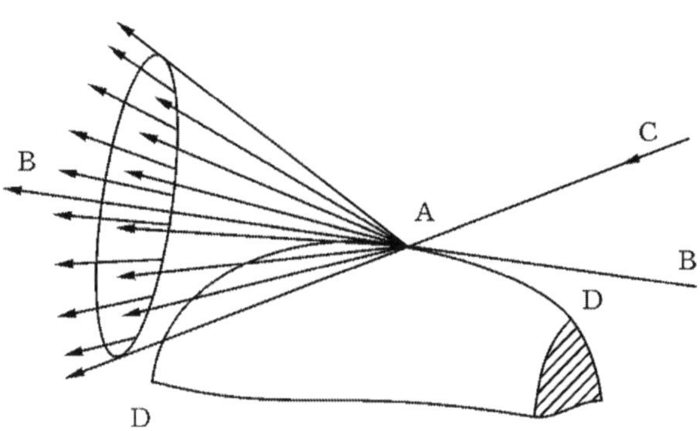

Figure 14.3 Diffractional rays arising on an edge: CA is an incident ray, A is a point of diffraction; BB is a tangent to the edge line in a point of diffraction and DD is an edge of a scattering body

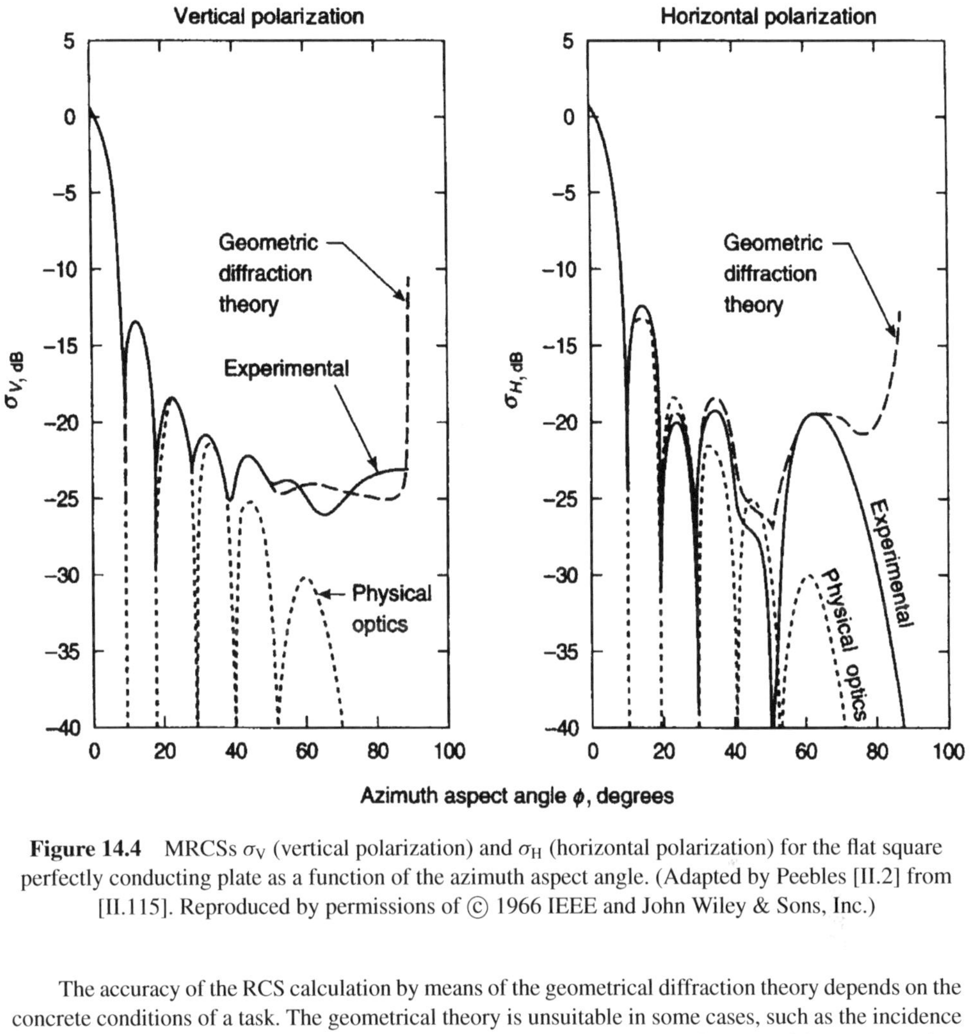

Figure 14.4 MRCSs σ_V (vertical polarization) and σ_H (horizontal polarization) for the flat square perfectly conducting plate as a function of the azimuth aspect angle. (Adapted by Peebles [II.2] from [II.115]. Reproduced by permissions of © 1966 IEEE and John Wiley & Sons, Inc.)

The accuracy of the RCS calculation by means of the geometrical diffraction theory depends on the concrete conditions of a task. The geometrical theory is unsuitable in some cases, such as the incidence of a slide wave on to a rectangular plate. However, there are only a few such cases. If they are not taken into account, then the error can be evaluated according to Reference [II.17] as 0.5–1 dB when $k_0a > 3$; this is the same as the case using the edge wave method.

14.1.2.6 Fock Model [II.46, II.54]

Nearly 60 years ago, the Soviet physicist V.A. Fock created a theory of electromagnetic field, inducted on or next to the surfaces of the convex conducting bodies with greater (in comparison with the wavelength) curvature radii. The first purpose of Fock's work was to demonstrate that surface waves in the penumbra area, which surrounds the geometrical shade edge, are local. Surface waves are dependent on the radius of curvature and physical characteristics of the surface only in the direct vicinity of the shade edge. Fock used several methods to obtain equivalent and consistent results. This method defines the field components in the vicinity of the shade edge as the scalar solutions of a parabolic differential equation of the diffusion type, and most clearly illustrates physical assumptions. Fock developed essential physical arguments to justify the use of a parabolic differential equation instead of an elliptical one, such as the Helmholtz equation.

It is necessary to note that Fock obtained solutions that describe the field on or near a body surface and only in the direct vicinity of the shade edge. However, the Fock method can be extended by means of a small modification so that it is possible to obtain the approximate formula for surface fields, which are valid in irradiated areas and beyond, in the shade area or on the shade edge [II.46].

14.1.2.7 The Small Perturbation Method (SPM)

This method belongs to a larger family of methods of wave equation solutions by means of the disturbance of a general solution and is presented in References [II.55] to [II.57]. It is widely used in the analysis of electromagnetic wave scattering on rough surfaces. The main idea of the SPM consists in finding a solution that satisfies boundary conditions in terms of plane waves. A series expansion for a field on surfaces is carried out as follows:

$$\mathbf{E} = \mathbf{E}^{(0)} + \mathbf{E}^{(1)} + \mathbf{E}^{(2)} + \cdots, \tag{14.13}$$

where $\mathbf{E}^{(n)}$ is a field depending on the n-power of deflection of the rough surface height from the average value (see Section 10.3). In particular, the field $\mathbf{E}^{(0)}$ corresponds to the field scattering on an ideally flat surface. The content of the considered approach consists in the small surface efficient currents on average planes substituting the rough edges itself. It is clear from Reference [II.50] and [II.47] that the method is applicable only for surfaces with low rough edges under small inclines of the edges and does not depend on the curvature radius.

The Fourier series expansion for small-scale rough edges is carried out and, consequently, the influence of rough edges on the scattered wave is analysed in terms of wave harmonics. The result of a similar analysis is that the frequencies that satisfy the Bragg condition play a predominant role in so far as they give a coherent contribution to the field.

The microwave Bragg resonance is similar to the Bragg resonance, and the latter is known in X-ray spectroscopy, where X-rays are scattered by atoms of a crystal lattice [II.58]. In both cases, the Bragg condition for constructive interference has the form $2d \sin\theta = n\lambda$, where λ is the wavelength of the radiating wave, d is the characteristic spatial dimension (scale) of the scatterer and θ is an incident angle [II.28]. In the case of X-ray spectrography, d is the interatomic spacing in the crystal lattice, but in the case of microwave remote sensing of the sea d is the wavelength of the spectrum component of capillary waves [II.47].

The Bragg scattering competes with the dominant diffuse scattering under a small rise in the angles of the incident wave [II.59] in the frequency range of the global positioning system (GPS).

14.1.2.8 Two-Scale (Composite or Hybrid) Model

When there are two relevant scales in the surface roughness (one large and one small), the results from the rough scale model can be used by incoherently averaging the effective incidence and scattering angles with the large scale slopes. Consider the two-scale model with reference to ocean remote sensing. This is carried out in four steps:

1. The surface height fluctuation, ζ, is divided into a large-scale and a small-scale fluctuation, $\zeta = \zeta_l + \zeta_s$, each with its own spectra, W_l and W_s, and it is assumed that the total spectrum (which is the sum of the two spectra) is effectively one or the other depending on whether the ocean wavelength is larger or smaller than a given wavelength [II.60].
2. The small-scale spectrum is used to compute the SPM scattering coefficient, $\sigma_{\pi}^{0}\left(\theta^{\text{in}}, \theta^{\text{sc}}, \phi^{\text{sc}}\right)$, (see Chapter 10), in the Bragg mechanism.

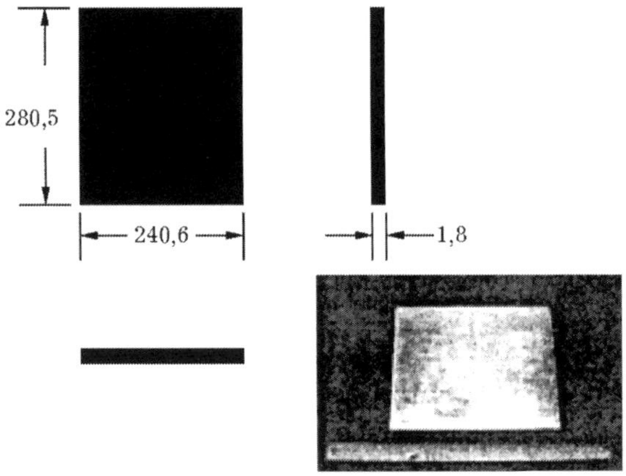

Figure 14.5 Object A. (From Eigel *et al.* [II.61]. Reproduced by permission of © 2000 IEEE)

3. Using the large-scale statistics, and the resulting tilting angle, Ω, statistics, $P(\Omega)$, the SPM scattering coefficient is averaged,

$$\sigma_{\mathrm{rt}}^{\prime 0}\left(\theta^{\mathrm{in}}, \theta^{\mathrm{sc}}, \varphi^{\mathrm{sc}}\right)=\int \mathrm{d}\Omega\ P(\Omega) \sigma_{\mathrm{rt}}^{0}\left(R^{\Omega}\left[\theta^{\mathrm{in}}, \theta^{\mathrm{sc}}\right], \phi^{\mathrm{sc}}\right). \tag{14.14}$$

   In this expression, $R^{\Omega}\left[\theta^{\mathrm{in}}, \theta^{\mathrm{sc}}\right]$ stands for the local incident and scattering angles after rotation of the locally tilted reference frame.
4. Finally, a Kirchhoff large-scale model cross-section coefficient is added to $\sigma_{\mathrm{rt}}^{\prime 0}$, $\sigma_{\mathrm{T}}^{0}=\sigma_{\mathrm{rt}}^{\prime 0}+\sigma_{\mathrm{K}}^{0}$, to account for the specula missing in the SPM. Adding this last step, which is not crucial for back-scattering but is important for forward-scattering, creates the composite model.

One of the problems with this approach is that the real surface is not correctly described by two scales, and some formulae must be used to divide the spectrum.

The comparative analysis of two models – physical optics (PO) and the physical theory of diffraction (PTD) – was performed in Reference [II.61], where the scattering properties of the three different conducting objects – A, B and C – were studied. Object A was an aluminium plate, shown in Figure 14.5 (dimensions in mm).

The transmitter illumination angle[3] was broadside to the flat plate ($\alpha=0$). This produces a distinct specula scattering signature characterized by a narrow main lobe at $\delta=0^{\circ}$, side lobe activity through 60–70° and diffraction effects through edge-on incidence ($\delta=90^{\circ}$). Figure 14.6 shows (with dependence upon the receiver look angle) the pattern for 8 GHz VV polarization and HH polarization for the measured, the PO and the PO/PTD solutions. Eigel *et al.* [II.61] noticed, as expected (at the EC Joint Research Centre (JRC), Ispra, Italy), a similarity in the lobing structure, amplitude and null placement for the first 30°–40°. The PO solution begins to diverge, especially at HH polarization, from the measured data due to a lack of diffraction inclusion. The PTD solution correlates with the measured data to a greater extent, but on average predictions are higher than those measured through $\delta=90^{\circ}$ and the lobes appear to be skewed towards higher bistatic angles. At $\delta=90^{\circ}$, a large discontinuity appears in all of the predicted solutions. This may be explained, in the opinion of Eigel *et al.*, by limitations of the PO/PTD approaches. At low grazing angles, the orthogonal component of each facet face becomes exceedingly small, prompting the software to estimate the solution via a different process whose exact formulation is

[3] The bistatic arrangement has been defined by the transmitter illumination angle α and the receiver look angle δ. Therefore, the bistatic angle becomes $\beta=\delta-\alpha$.

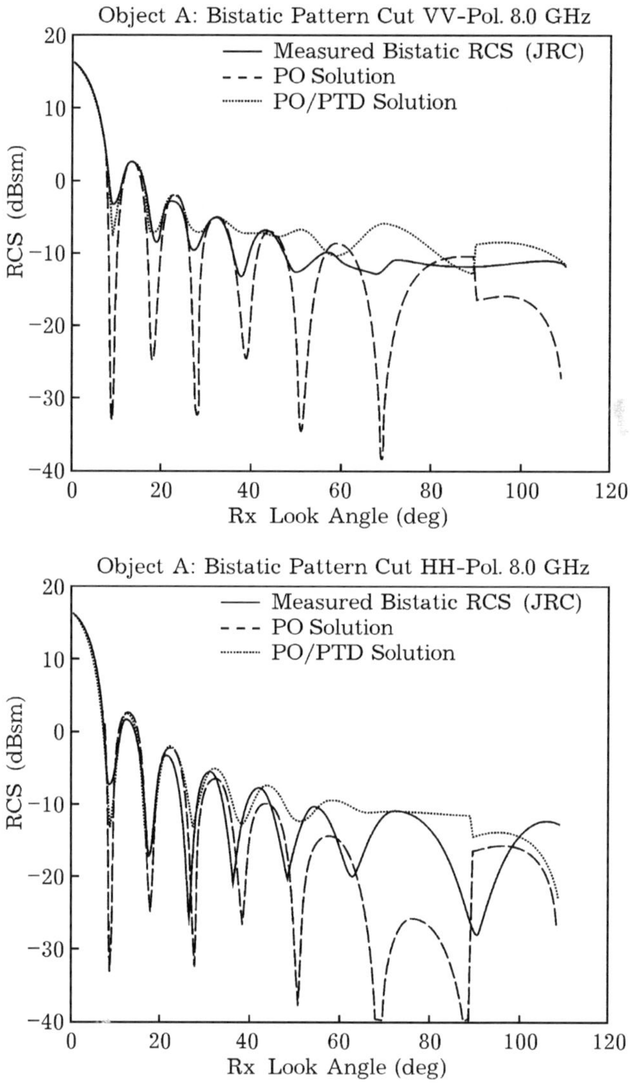

Figure 14.6 Object A: bistatic RCS physical optics/physical theory of diffraction (PO/PTD) predictions versus measured (JRC) data, 8 GHz. (From Eigel *et al.* [II.61]. Reproduced by permission of © 2000 IEEE)

unknown at the time of writing [II.61]. The process is demonstrably inaccurate where diffraction effects dominate. According to Eigel *et al.*, the discrepancy in the null/lobe positions could indicate that either the measured data is faulty (e.g. the incident field is not truly planar) or the models for smaller bistatic angles are in error.

Objects B and C have more complex geometries (see Figures 14.7 and 14.8). The canonical features of objects B and C are specifically designed by the authors of Reference [II.61] to be more complex in nature, yet support relatively simple scattering mechanisms. The short cylinder in objects B and C is solid

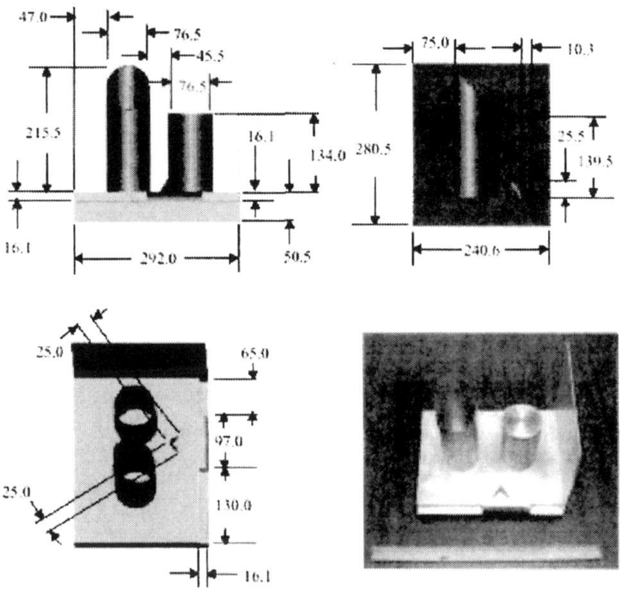

Figure 14.7 Object B: Aluminium cylinders mounted in a base shadowed by flat plate (dimensions in mm). (From Eigel *et al.* [II.61]. Reproduced by permission of © 2000 IEEE)

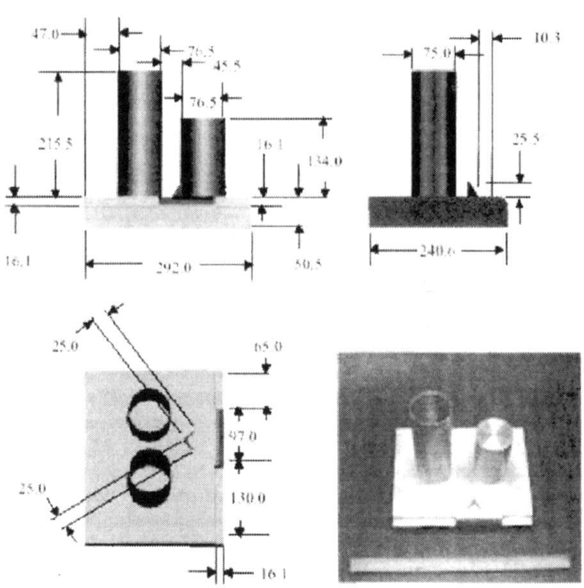

Figure 14.8 Object C: Aluminium cylinders mounted in a base without a shadowing plate (dimensions in mm). (From Eigel *et al.* [II.61]. Reproduced by permission of © 2000 IEEE)

and the larger cylinder is hollow. A small triangular dihedral is placed just forward of the cylinders in both objects. Object B was considered more complex than object C because of the large shadowing plate and the canted hollow cylinder. None of the object features are larger than 300 mm in any dimension. Note that the flat plate attached to one end of object B was used by itself as object A.

The analysis for both objects was conducted in Reference [II.61] simultaneously to highlight the model's weaknesses and unanticipated strengths and to discuss the various scattering mechanisms that form the overall scattered signature. It was noted in Reference [II.61] that object features are between 0.67 and 14.03 wavelengths depending on the frequency of interest.

Eigel *et al.* believe that object C is simpler in the sense that no large shadowing geometry is present, nor is there a canted cylinder into which directly incident energy can couple and be scattered. In evaluating the measured data's lobing structure in Figures 14.9 and 14.10, what appears to be two main reflection sources beating in and out of phase can be seen over most of the covered region. These are most likely the specula returns from the cylinder bodies as evidenced through the similarity between polarizations. An inspection of the bistatic image plots in Figures 14.11 and 14.12 supports this hypothesis.[4]

It has been argued in Reference [II.61] that a small amount of multibounce interaction between cylinders is present, but is probably small enough so as not to compete with the main cylinder specula effects. The dihedral is not a significant scatterer at these frequencies or a larger return would be present near the transmitter illumination angle. The PTD predictions are surprisingly accurate, even at the lower frequency. The higher-frequency lobes in Figure 14.10 appear to be slightly skewed towards higher bistatic angles than the true data, but the amplitudes are reasonably correct. The discrepancies may be due to a slight misalignment of the object during measurement or small, second-order surface wave and diffraction effects.

The discussion in Reference [II.61] has demonstrated that the PO/PTD model can predict reasonably accurate bistatic signatures from more complex geometries, even if they do not meet the standard electrically large criteria. Such objects need to be comprised of surfaces that support several specula reflections with a wide angular distribution (e.g. cylinders). This analysis leads Eigel *et al.* [II.61] to another conclusion regarding the models. If it can be inferred, through inspection, that the dominant bistatic reflection mechanisms from an object are specular in nature, the simulations should be fairly accurate. Conversely, if this is not possible, then the software will be less accurate. The lobing structure of the patterns may become skewed towards larger bistatic angles at higher frequencies, but lobe amplitude and null presence should be close to the measured data. One clearly distinctive characteristic of this type of object is a lack of significant shadowing features that could mask specula effects.

To reinforce the conclusion, inspection in Reference [II.61] should be accomplished for an object for which large specula reflections cannot be guaranteed for all α and δ angles. Object B provides just such a geometry, because of the large flat plate on one end. The first 110° of receiver coverage are shown in Figures 14.13 and 14.14 (8 and 15 GHz respectively). Clearly, the PTD solution is more dissimilar than before. Discrepancies are explained as follows. For the first $\delta = 45°$, the receiver is essentially looking at the flat plate, which shadows the remaining object geometry for much of that region. At larger bistatic angles (δ near 0° in the figure) mostly sidelobe activity associated with the plate's specula reflection is expected, which occurs near $\delta = -45°$ (not shown). The lobe structure should be uniform with steadily decreasing amplitude as smaller bistatic angles (δ near 30°) are approached, and this is exactly what is present. The PTD predictions are consistent with the measured data (albeit shifted again) until close to the transmitter illumination angle, at which point they diverge. Between $\delta = 30°$ and $\delta = 40°$ the software predicts higher than expected results for the lower RF. The exact cause of this is unknown, but inaccuracy of the PTD implementation is suspected. Clearly the PTD solution is more dissimilar than before. Discrepancies in Reference [II.61] are explained as follows.

[4] An efficient algorithm for near-field synthetic aperture radar (SAR) imaging developed at the Joint Research Centre (JRC) was used in Reference [II.61] to form both monostatic and bistatic images. The dynamic range of all images is 30 dB.

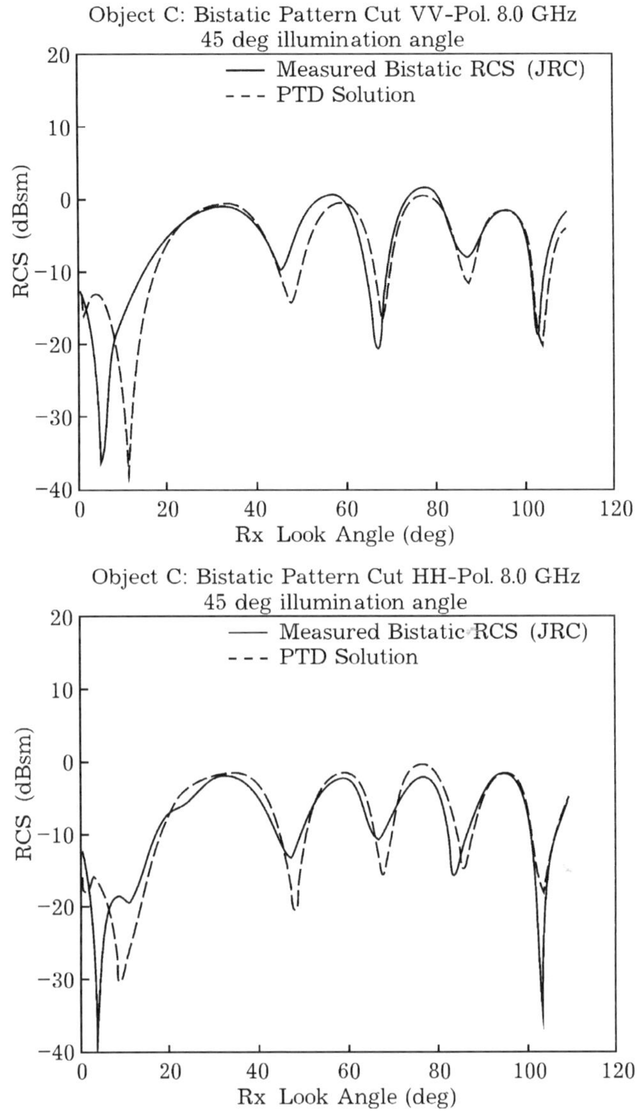

Figure 14.9 Object C: bistatic RCS PTD predictions versus measured (JRC) data, 8 GHz. (From Eigel *et al.* [II.61]. Reproduced by permission of © 2000 IEEE)

Beyond $\delta = 45^\circ$, the canted cylinder is completely visible to both the transmitter and the receiver, but the shorter cylinder remains unilluminated. Therefore, only a single direct reflection, analogous to one of object C's speculars (from the canted cylinder), is now present. The nature of the remaining scattering pattern is changed accordingly. The lobing structure of Figure 14.14 suggests that, once again, two scatterers are beating against each other at VV polarization, but probably not at HH polarization. This implies the presence of additional scattering mechanisms whose amplitudes are roughly the same as those of the specula in this region. Some nonspecular effects will be more noticeable at certain polarizations.

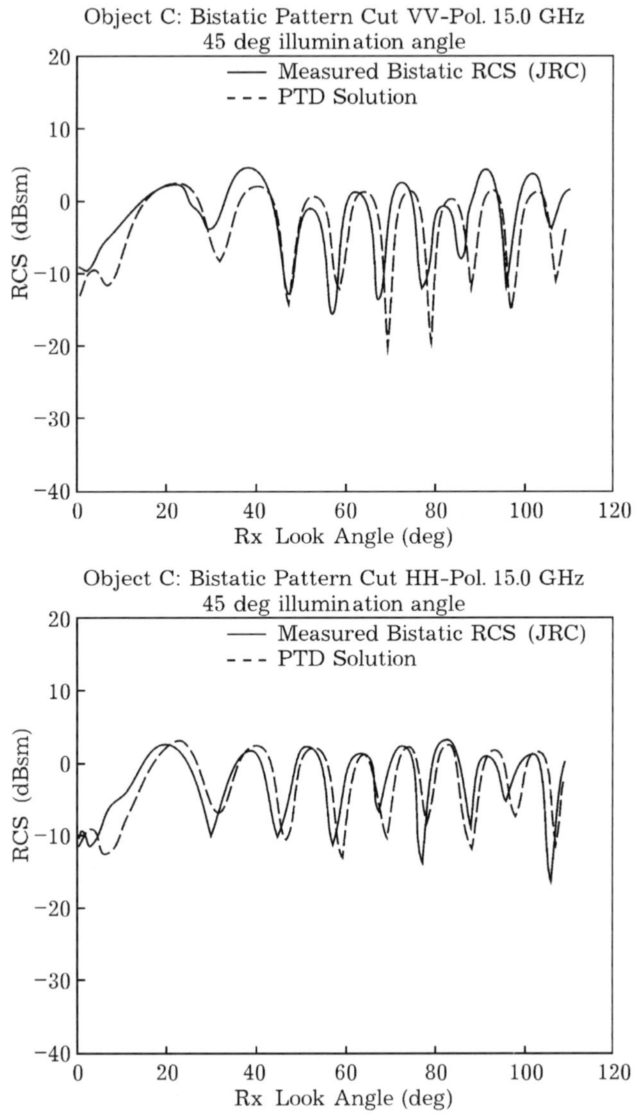

Figure 14.10 Object C: bistatic RCS PTD predictions versus measured (JRC) data, 15 GHz. (From Eigel *et al.* [II.61]. Reproduced by permission of © 2000 IEEE)

The anticipated trend is that they will contribute to the measured signature more at lower frequencies, especially at HH polarization, and cause the simulated and measured data to be more dissimilar than for object C. A single specular reflection from the canted cylinder, diffraction from the plate edge, multibounce between cylinders, some specular or cavity effects from within the canted cylinder and nominal surface wave effects are all suspected to exist throughout much of this region. The imaging plots shown in Figures 14.15 and 14.16 reveal a multitude of scattering sources. At VV polarization, the two dominant scatterers are most likely an edge diffraction from the flat plate and the specula from the

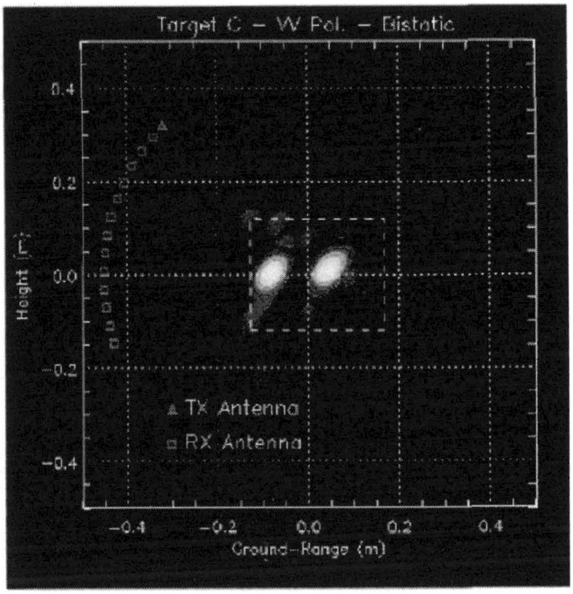

Figure 14.11 Object C: bistatic image, VV polarization, 45° transmitter angle, −20°–41° receiver angle. (From Eigel *et al.* [II.61]. Reproduced by permission of © 2000 IEEE)

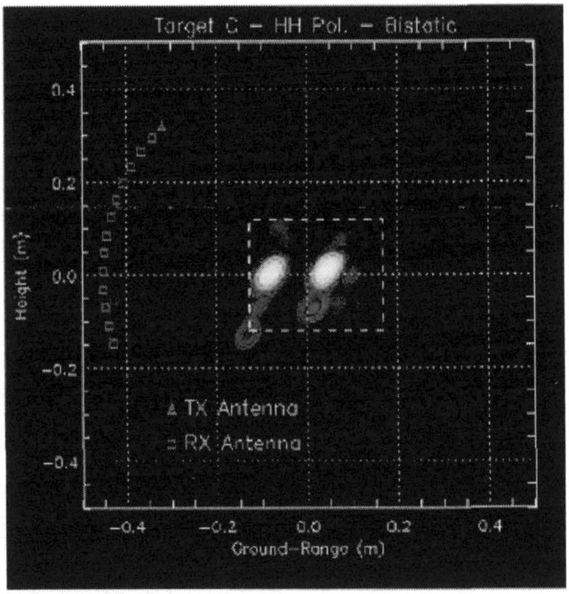

Figure 14.12 Object C: bistatic image, HH polarization, 45° transmitter angle, −20°–41° receiver angle. (From Eigel *et al.* [II.61]. Reproduced by permission of © 2000 IEEE)

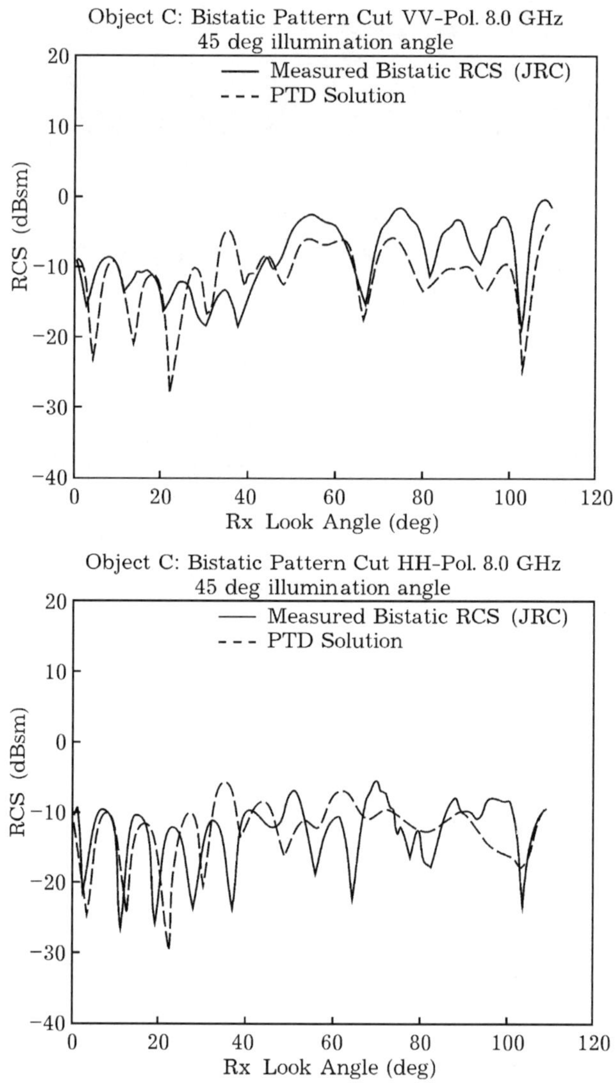

Figure 14.13 Object B: bistatic RCS PTD predictions versus measured (JRC) data, 8 GHz. (From Eigel *et al.* [II.61]. Reproduced by permission of © 2000 IEEE)

canted cylinder. Because neither is as strong as the cylinder specula noted on object C, other nonspecular effects may contribute more or less to the overall measured pattern and may not be predicted by the PTD model. The effect is observed in lower simulation amplitudes than those measured at larger bistatic angles, as shown in Figure 14.13 and 14.14. At HH polarization, the plate edge diffraction does not exist, leaving other sources free to compete with the cylinder specula. Surface wave effects on the cylinders may be more noticeable at this polarization, but because their effect diminishes with increasing frequency one expects better correlation of the HH polarization data at 15 GHz. This is indeed observed in the Figures Overall, the presence of nonspecular components, whose amplitudes are of the same order as one speculum, yield a greater dissociation between the measured and PTD data.

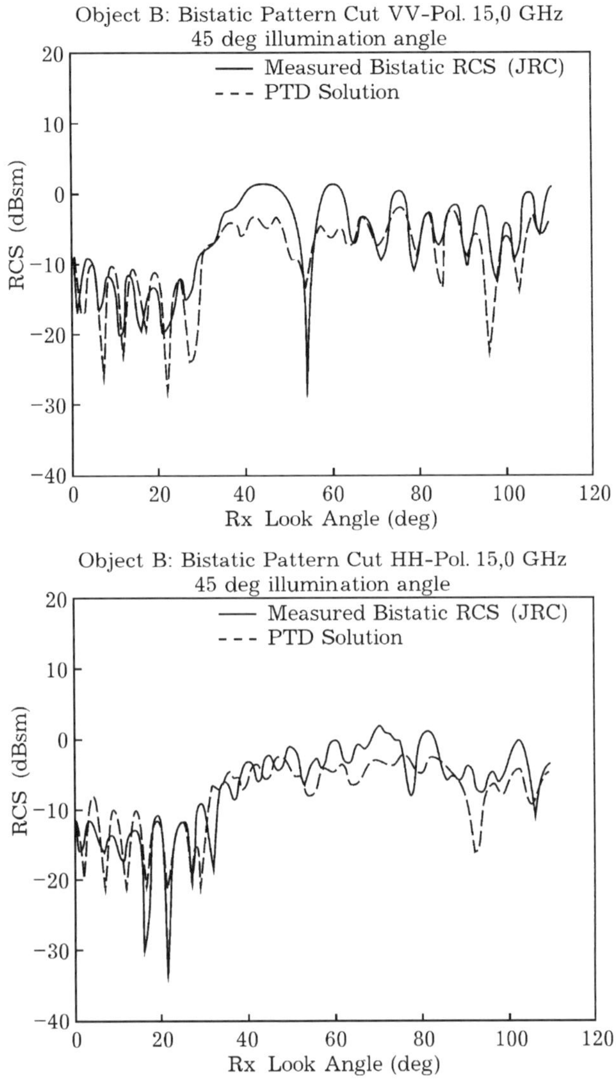

Figure 14.14 Object B: bistatic RCS PTD predictions versus measured (JRC) data, 15 GHz. (From Eigel *et al.* [II.61]. Reproduced by permission of © 2000 IEEE)

Eigel *et al.* [II.61] noted that the observation can be summarized as another conclusion for the PTD model as a whole. Therefore, if upon inspection of an object it can be determined that a large shadowing feature or cavity exists, then bistatic PTD predictions may be lower than the true signature. If only a single direct specular reflection is likely to be present, especially those with a low, broad angular distribution (i.e. from curved surfaces), additional nonspecular or multibounce scattering mechanisms can arise and be of a similar amplitude. These could compete with the specula in such a way as to make the PTD prediction even less accurate, especially when electrically sizes of objects are small. At worst, the model will be incapable of accurately predicting the bistatic RCS with any degree of certainty. Imaging can lend significant insight into determining the presence and origin of the specula to assist in

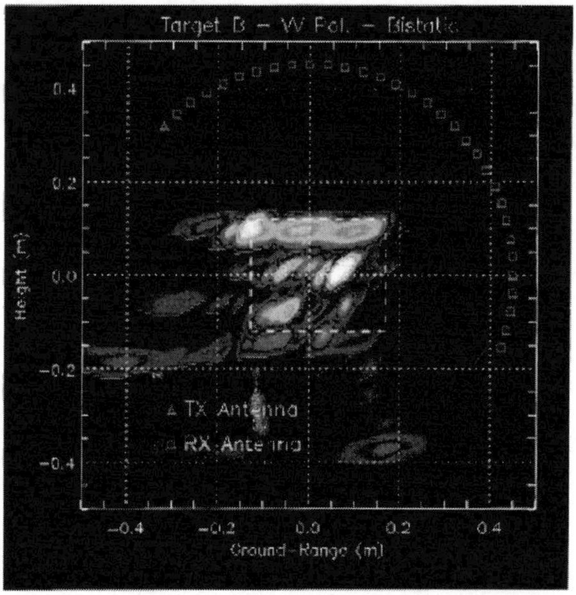

Figure 14.15 Object B: bistatic image, VV polarization, 45° transmitter angle, 49°–200° receiver angle. (From Eigel *et al.* [II.61]. Reproduced by permission of © 2000 IEEE)

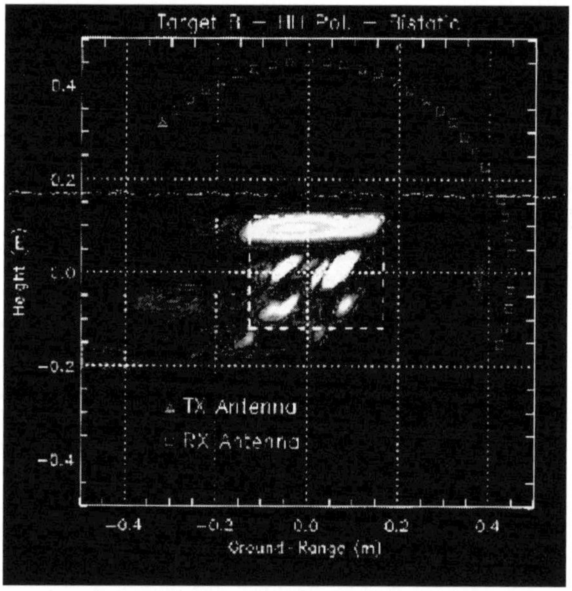

Figure 14.16 Object B: bistatic image, HH polarization, 45° transmitter angle, 49°–200° receiver angle. (From Eigel *et al.* [II.61]. Reproduced by permission of © 2000 IEEE)

the evaluation. This conclusion is not entirely unexpected. By its very nature, a PTD solution would not be sufficient for smaller objects and lower frequencies. Nonasymptotic techniques (e.g. the Method of Moments (MOM), which will be considered in the next part) are better poised to handle predictions in this area.

As a conclusion, Eigel *et al.* [II.61] proposed that the PTD software's bistatic predictive capabilities are less than adequate in some situations, but surprisingly good in others. It computes edge diffraction incorrectly near receiver edge-on incidence, but predicts a reasonably accurate bistatic RCS when wide-angle specula dominate the pattern, even if the object does not meet the electrically large criteria. It tends to predict low RCS when second-order scattering contributions are prevalent, but this is primarily a limitation of the PO formulation and not its implementation within the code.

14.1.3 Numerical Solutions

There are many areas of electromagnetic scattering theory where computer and numerical techniques can be used. For instance, they were used in Reference [II.46] on spheres and cylinders to sum the exact series solution. In addition, they were used to solve the static problem of a body immersed in static fields in order to obtain the low-frequency electromagnetic scattered fields from an object [II.46]. As a third example, the computer could be used to evaluate a physical optics integral for scattering from a complex body in the high-frequency limit. In all of these cases, the computer is used to solve the scattering problem. However, in the first example given the exact series solution is already known, and the computer is solely used as an aid in performing the summation. In the second and third examples, the scattering problem has already been reduced and restricted to the low- or high-frequency region. Hence, the computer is being used to solve an approximation in the first place, and the numerical solution is only as valid as the original approximation. Therefore, in these examples the computer is used as a secondary tool in the solution, since either the exact solution is already known or else we are starting with an approximate solution in the first place. Nonetheless, the computer is an invaluable and indispensable tool in the evaluation of these types of problems [II.46].

The general usage of the term 'numerical solution' of scattering problems, however, does not include the preceding examples in which the computer is used as a secondary tool in the problem solution. Originally, the term applies to the direct solution of the exact Maxwell equations or their integral forms by numerical techniques. The bodies or targets for which the scattered fields are sought are of sufficient complexity in shape that an exact analytical solution cannot be derived as it can, for instance, for the sphere. In beginning with an exact formulation upon which the numerical solution is to be based, rather than an approximation, the solution for the scattered field is not restricted to certain frequency regions or body sizes. In fact, the main value of a numerical solution proceeding from an exact formulation is its validity in the resonance region where neither high- nor low-frequency approximations are generally applicable.

14.1.3.1 First Numerical Solutions

The numeral techniques and computer programs of the first investigators (at the end of the 1950s and the early 1960s) differ on some points, but possess many similar characteristics. The general similarities between the properties of these techniques are as follows (see Reference [II.46]):

- The size and shape of the scattering target, the frequency, the incidence and scattering angles of interest and the incident wave polarization direction are used as input parameters.
- The output of such programs can be the far-region scattered field strength, phase angle and polarization directions. The scattering cross-section is also available. Most of the programs can also give the surface currents on the scatterer (assuming the scatterer is perfectly conducting).

- All of the solutions reduce to an infinite set of linear equations in an infinite number of unknowns by enforcing boundary or constraint conditions at an infinite number of points on the surface of the body. This infinite set is truncated in a suitable fashion, resulting in a finite set of equations. The number of equations is chosen to be large enough so that the spacing between points is generally smaller than a wavelength.
- Once the unknowns of the simultaneous equations have been calculated, they must be either summed in some manner or numerically integrated over the surface of the body to obtain the scattered far-region field.

The main advantage of numerical solutions lies in the ability to obtain an immediate estimation of the cross-section of a given scatterer and the cross-section variation as a function of aspect angle, frequency and polarization.

The main disadvantage of the first numerical solutions is the lack of insight they provide into the nature of the scattering process. The geometrical theory of diffraction, for example, while only a high-frequency approximation, not only provides solutions but relates scattered field behaviour to certain areas on the scatterer surface. Hence, an understanding of the scattering process can be obtained and the knowledge can be used to shape the surface better in order to either enhance or conceal the object at certain frequencies and aspects. Conversely, it is hoped that the shape of an unknown target can be related to its cross-section signature through the knowledge gained from many of the analytical theories and approximations. On the other hand, an exact numerical solution provides a simple number as an answer; this number cannot be easily related to different parts of the scatterer, nor can the numerical solution for one given shape be readily extrapolated to other shapes.

Thus, the first numerical solutions are invaluable in situations where there is a need for a rapid, relatively accurate estimate of the cross-section. They do not, however, replace the analytical theories and models more recently developed. In most situations, both approaches to scattering problems complement each other and, therefore, hybrid approaches combining the two are currently more prevalent.

14.1.3.2 Matrix Methods

High-speed digital computers have facilitated the use of matrix methods to solve scattering problems. The basic mathematical concept is the method of moments (MOM) by which the functional equations of field theory are reduced to a set of simultaneous linear matrix equations.

Harrington [II.62, II.63] first treated the problem of a plane wave incident on wire scatters. The voltage excitation matrix, $\mathbf{V}$, contains the incident electric field tangent to each wire segment. The coupling between the different segments is obtained by an impedance matrix $\mathbf{Z}$. Equations relating to the cancellation of the incident electric field by the field scattered by the induced currents $\mathbf{I}$ yield the equation

$$\mathbf{0} = \mathbf{V} + \mathbf{Z}\,\mathbf{I}, \tag{14.15}$$

which can be solved as

$$\mathbf{I} = -\mathbf{Z}^{-1}\mathbf{V}. \tag{14.16}$$

Once the impedance matrix has been inverted, the solution may be obtained for any excitation by multiplying the inverse of the impedance matrix by the excitation column vector.

The MOM led to the development of many different computer codes. One well-known and widely used user-oriented computer code is the numerical electromagnetic code (NEC) [II.64]. The NEC has been developed at the Lawrence Livermore Laboratory, Livermore, California, under the sponsorship of the Naval Ocean Systems Center and the Air Force Weapons Laboratory, for analysing the electromagnetic response of an arbitrary structure consisting of wires and surfaces in free space or over a ground plane. It is an advanced version of the Antenna Modeling Program (AMP) developed in the early 1970s by MBAssociates. The NEC is built around the numerical solution of integral equations for currents induced on the structure by sources of incident fields. The source of excitations may be either an

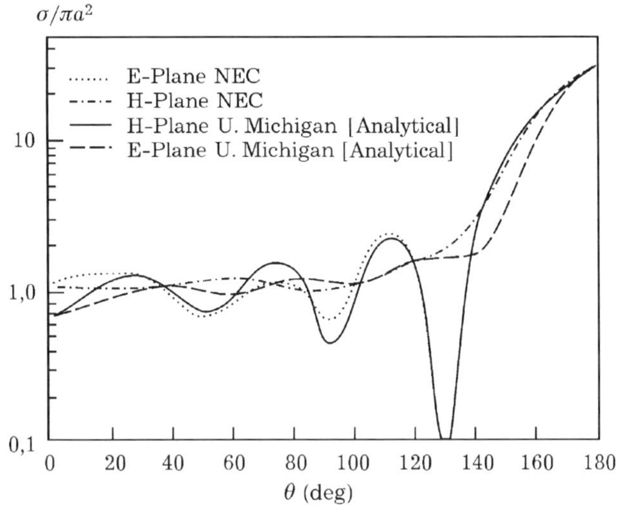

Figure 14.17 Bistatic RCS of a sphere with $ka = 5.3$. Uniform segmentation

impressed voltage source or an incident plane wave of linear or elliptical polarization, while the output may include current and charge density, an electric or magnetic field in the vicinity of the structure and radiated fields. The NEC uses both an electrical field integral equation (EFIE) and a magnetic field integral equation (MFIE) to model the electromagnetic response of general structures. The EFIE is well suited for thin-wire structures of small or vanishing conductor volume, while the MFIE works better for voluminous structures, especially those having large smooth surfaces.

The results of the NEC calculations with two types of segmentations are shown in Figures 14.17 and 14.18 for scattering by a sphere of ka (2π radius/wavelength) equal to 5.3. The uniform segmentation used 14 increments in azimuth and 14 equal mesh lengths along the vertical axis. The variable segmentation used 13 equal increments in arc length along the vertical axis, with each cell from top to bottom divided

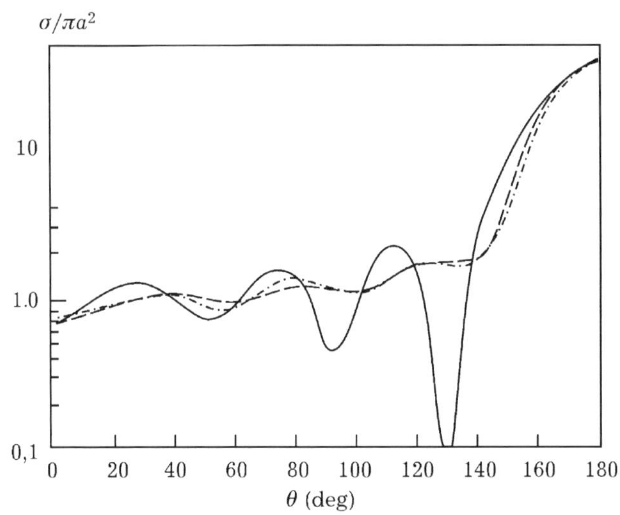

Figure 14.18 Bistatic RCS of a sphere with $ka = 5.3$. Variable segmentation

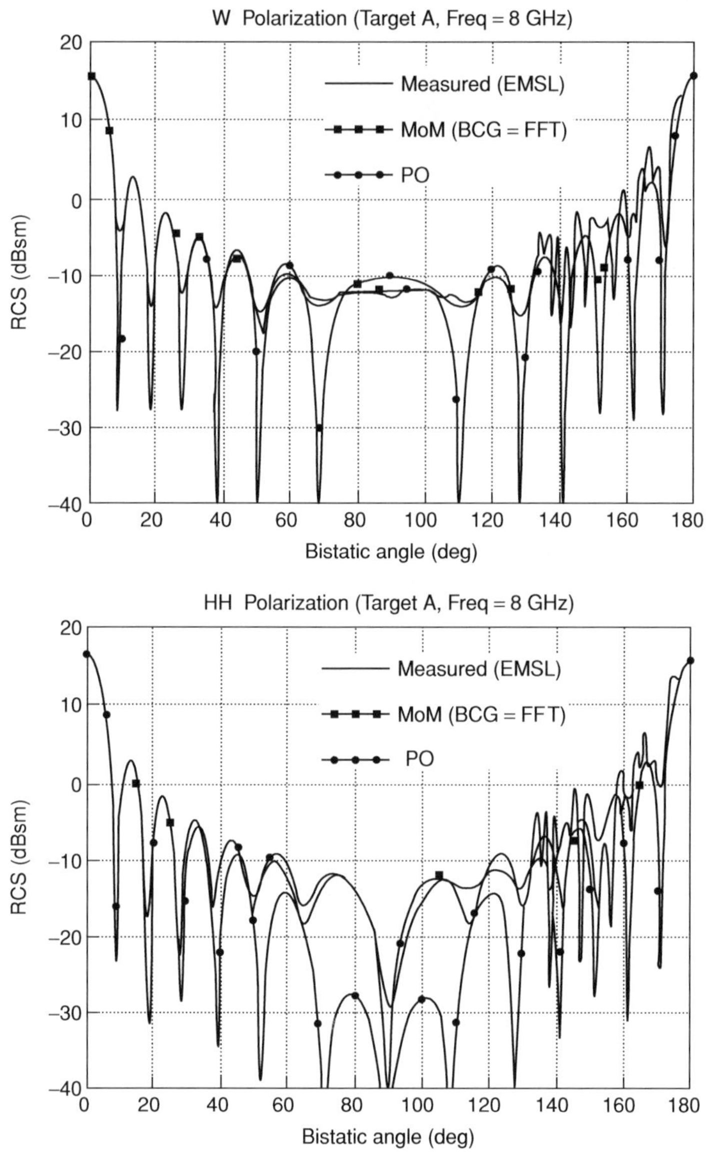

Figure 14.19 Bistatic RCS method of moments predictions versus measured (JRC) data, 8 GHz. (From Eigel *et al.* [II.61]. Reproduced by permission of © 2000 IEEE)

into the following number of patches in azimuth: 4, 8, 12, 16, 20, 24, 24, 24, 20, 16, 12, 8, 4. A greater concurrence with the experiment is obtained with variable segmentation.

Figure 14.19 from Reference [II.61] contains a comparison of the measurement results with results of the calculation by the MOM for a body, A, shown in Figure 14.5. The sound agreement indicates that the measurements are, in fact, valid and the PO/PTD data, which was considered in a previous point, is inaccurate. Therefore, all measured data are regarded as more exact in this situation.

A different technique known as the 'unimoment method' was developed by Mei to compute the scattering of a plane wave by a lossy dielectric body [II.65]. The method divides the problem into a centre core, a numerical region and an exterior region. In the numerical region, Maxwell's equations are solved directly by a finite element method to yield a linearly independent set of solutions. This set is then to be matched with analytic spherical vector waves at each region interface. The solution of the resulting set of linear simultaneous equations yields the complete scattered electromagnetic field in all regions. The method can handle bodies of arbitrary size and shape as long as the computing speed and memory are available.

To minimize storage and speed, a version of the unimoment method that applies to symmetrical bodies of revolution has been developed. This software was used successfully to analyse high-frequency distortion of a plane wave incident on a radome composed of three layers, each with its own dielectric constant and loss tangent [II.66].

14.1.3.3 Iterative Methods

Extensive work shows that a procedure based on the iterative conjugate-gradient method can be incorporated into conventional numerical methods to extend the range of application to larger geometries [II.67]. The conjugate-gradient method has both advantages and limitations when compared with conventional approaches. This method can provide a convenient means of treating scattering problems at optical wavelengths. Problems treated by this method can involve two-dimensional metal and dielectric cylinders, periodic capacitive and inductive frequency selective surfaces and rough surfaces.

14.1.3.4 Time Domain Approach

There are many ways that can be used to solve the time-domain Maxwell equations numerically. Bennett and Weeks [II.68] presented an approach to solve the time-domain integral equation rather than the frequency-domain equations. For solid conducting bodies, the equation is

$$\boldsymbol{j}(\boldsymbol{r},t) = 2\boldsymbol{e}_n \times \mathbf{H}^{\text{in}}(\boldsymbol{r},t) + \frac{1}{2\pi}\int_S \boldsymbol{e}_n \left\{ \left[\frac{1}{R^2} + \frac{1}{R}\frac{\delta}{\delta\tau} \right] \boldsymbol{j}(\boldsymbol{r}',t) \times \boldsymbol{e}_R \right\}_{\tau=t-R} \mathrm{d}S', \tag{14.17}$$

where $\boldsymbol{r}$ is a position vector to the observation point; $\boldsymbol{r}'$ is a position vector to the integration point; $R = |\boldsymbol{r} - \boldsymbol{r}'|$; $\boldsymbol{e}_R = \boldsymbol{r} - \boldsymbol{r}'/R$; t is a time unit of lightmeters, where one lightmeter is the time it takes an electromagnetic wave, moving at the speed of light c, to travel one meter; $\boldsymbol{e}_n$ is a unit normal at $\boldsymbol{r}$; S is a surface of the target; $\mathbf{H}^{\text{in}}$ is an incident magnetic field; and $\boldsymbol{j}$ is a surface current density.

The first term in Equation (14.17) yields the familiar physical optics approximation for the surface current. The integral term represents the influence of currents at other surface points on the current at $(\boldsymbol{r}, t)$. The crucial observation here is that the influence of other currents on the current at $(\boldsymbol{r}, t)$ is delayed by R, which makes the 'marching on in time' numerical solution feasible.

Once current density has been obtained, the field at any point in space, and therefore the bistatic RCS, can be computed. For example, the far scattered field is given by

$$\mathbf{H}^{\text{sc}}(\boldsymbol{r},t) = \frac{1}{4\pi}\int_S \left\{ \left[\frac{\delta \boldsymbol{j}(\boldsymbol{r}',t)}{\delta\tau} \right] \times \boldsymbol{e}_R \right\}_{\tau=t-R} \mathbf{e}_n \, \mathrm{d}S'. \tag{14.18}$$

Taflove and Umashankar [II.69] developed a finite difference time-domain (FDTD) approach for determining the RCS of general three-dimensional scatters. The FDTD method is a direct solution of Maxwell's time-dependent partial differential (curl) equations. By time stepping (i.e. repeatedly implementing a finite difference analogue of the curl equation at each cell of a corresponding space lattice), the incident wave is tracked as it first propagates to the structure and then interacts with it via surface current excitation, diffusion, penetration and diffraction. This procedure simplifies the solution by analysing the interaction of the wave front with portions of the structure surface at a given instant of time, rather than by attempting a simultaneous solution of the entire problem.

Far-field data can be obtained using the FDTD method by applying a near-field-to-far-field transformation. Equivalent electric and magnetic currents are derived from the FDTD fields on a closed surface; calculating the far fields from these currents is straightforward.

The primary advantage of the FDTD method over the MOM method is that it can handle objects with 'volumetric complexity', which refers principally to missile seekers and radars that may have apertures and cavities containing metal and dielectric structures and electronic circuitry.

Shankar *et al.* [II.70] extended computational fluid dynamics (CFD) based on upwind schemes to numerically solve Maxwell's equations for scattering from objects with layered nonmetallic sections. They utilized the numerical algorithms that had been proven to be most successful in integrating the time-dependent equations of fluid dynamics, namely the Euler and the Navier–Stokes equations. After a discussion on the character of Maxwell's equations in Reference [II.70], it was shown that they represent a linearly degenerate set of hyperbolic equations. To show the feasibility for applying CFD-based algorithms, the transverse magnetic (TM) and the transverse electric (TE) waveforms of Maxwell's equations were considered in Reference [II.70]. A finite volume scheme was developed with appropriate representations for the electric and magnetic fluxes at a cell interface, accounting for variations in material properties in both space and time. The computational solution extending from the object to a far-field boundary located a few wavelengths away constituted the near-field solution. A Green function based on a near-field-to-far-field transformation was employed to obtain the bistatic radar cross-section information.

Figure 14.20 from Reference [II.70] shows the effect of a dielectric shield over a perfectly conducting cylinder. According to Shankar *et al.*, the present method in Reference [II.70] is able to predict the detailed BRCS structure with various nulls quite accurately. Figure 14.21 shows a similar case, but with a very large dielectric value of $\varepsilon = 100$.

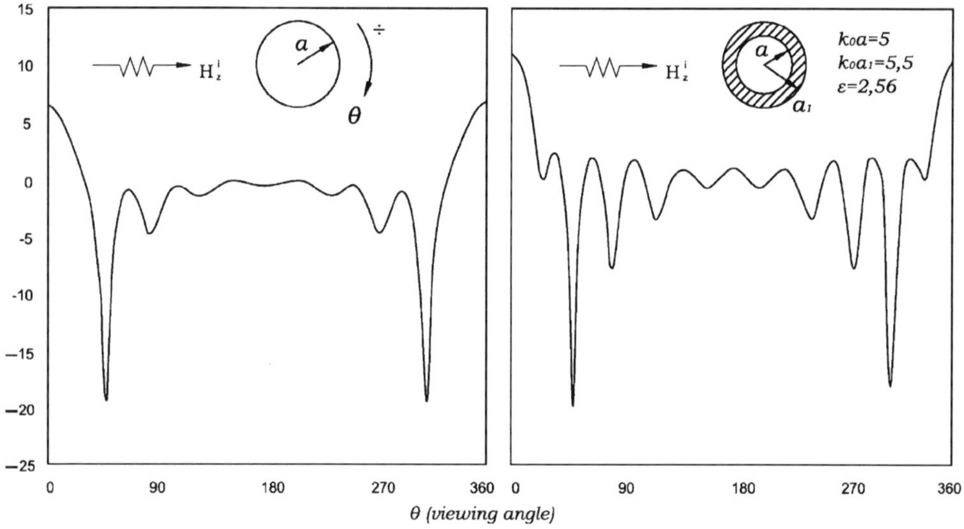

Figure 14.20 BRCS for a perfectly conducting cylinder with (right) and without (left) a dielectric shield. (From Shankar *et al.* [II.70]. Reproduced by permission of © 1989 IEEE)

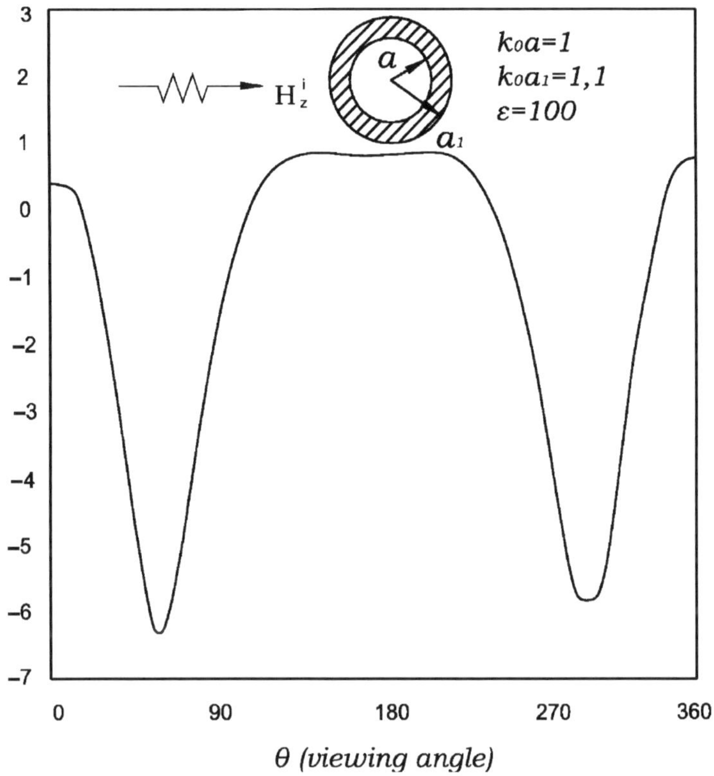

Figure 14.21 Application of the time-domain solver for a dielectric shield of large ε. (From Shankar *et al.* [II.70]. Reproduced by permission of © 1989 IEEE)

14.1.3.5 Vector Parabolic Equation Method

Parabolic equation (PE) techniques have been used for some time to solve long-range forward propagation problems for radar and sonar applications [II.71]. More recently, they have been applied to scattering problems and, in particular, to RCS calculations [II.72, II.73]. The approach of References [II.72] and [II.73] gave very encouraging results, but could not provide the full bistatic scattering pattern of a target because of intrinsic paraxial limitations.

In a subsequent publication [II.74] on acoustic scattering, it was shown that these paraxial limitations of the PE method can be extended to a large extent by decoupling the paraxial direction from that of the incident wave and solving for the scattered field rather than the total field. In Reference [II.71], Zaporozhets and Levy developed a paraxial framework for Maxwell's equations in order to treat polarization effects fully for electromagnetic scattering. They then applied the rotating PE method of Reference [II.74] to obtain bistatic scattering results at all scattering angles. The resulting vector PE algorithms provide an efficient numerical technique, which is somehow intermediate between rigorous solutions such as the MOM or the finite difference time-domain (FDTD) methods and facet-based approximations involving physical optics and the physical theory of diffraction. The field scattered by three-dimensional (3D) objects ranging in size from one wavelength or so to several tens or even hundreds of wavelengths can be calculated on a desktop computer.

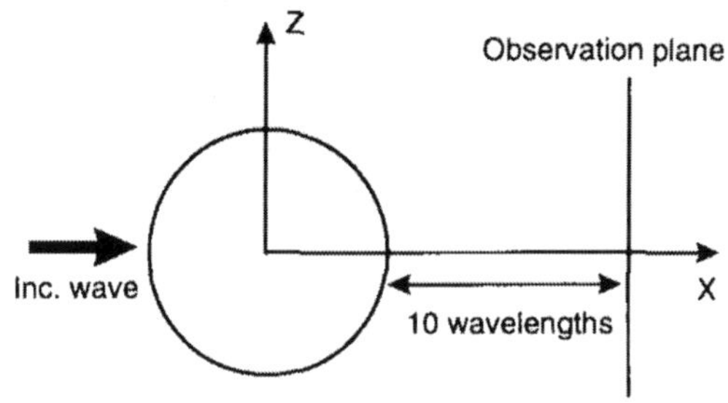

Figure 14.22 Geometry for scattering by a circular cylinder. (From Zaporozhets and Levy [II.71]. Reproduced by permission of © 1999 IEEE)

Zaporozhets and Levy [II.71] begin with a simple validation of the vector PE formulation on a two-dimensional (2D) example. A vertically polarised plane wave of wavelength λ is incident on a perfectly conducting circular cylinder of radius 5λ, as shown in Figure 14.22. Zaporozhets and Levy surmised that they could either solve the magnetic field H_y with the scalar PE or the electric field components E_x and E_z with the vector PE. Figure 14.23 shows the forward scatter results in a transverse plane located 10λ beyond the cylinder. For this 2D test, Zaporozhets and Levy used a standard narrow-angle code for the scalar PE and a single-pass full matrix inversion code for the vector PE. The E_x and E_z scalar PE results have been calculated by numerical differentiation of H_y. The scalar and vector PE results are almost identical. For comparison, Zaporozhets and Levy also show analytical results computed from a Hankel and cosine functions expansion [II.75]. These differ substantially from the PE results. The reason for this is, according to Reference [II.71], that the near-field narrow-angle PE results do not represent large-angle diffraction accurately; in particular, interference between the fields diffracted from the top and bottom of the cylinder is incorrectly modelled. Figure 14.24 shows the bistatic RCS results obtained

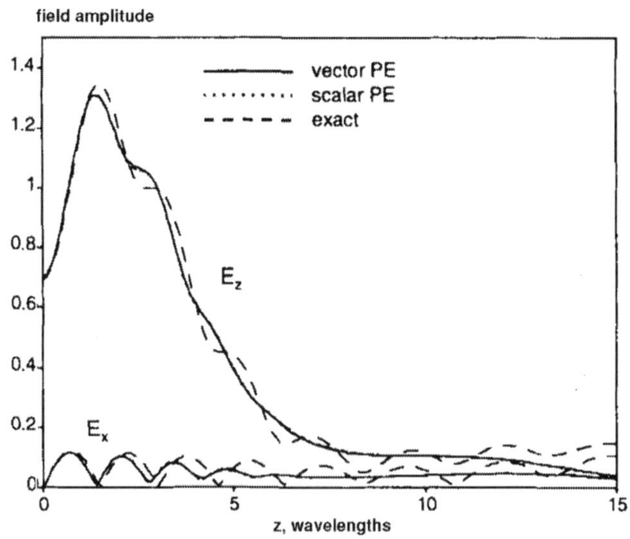

Figure 14.23 Near-field results for a perfectly conducting circular cylinder for a vertically polarized incident field. The dotted line (scalar PE result) is almost identical to the solid line (vector PE result). (From Zaporozhets and Levy [II.71]. Reproduced by permission of © 1999 IEEE)

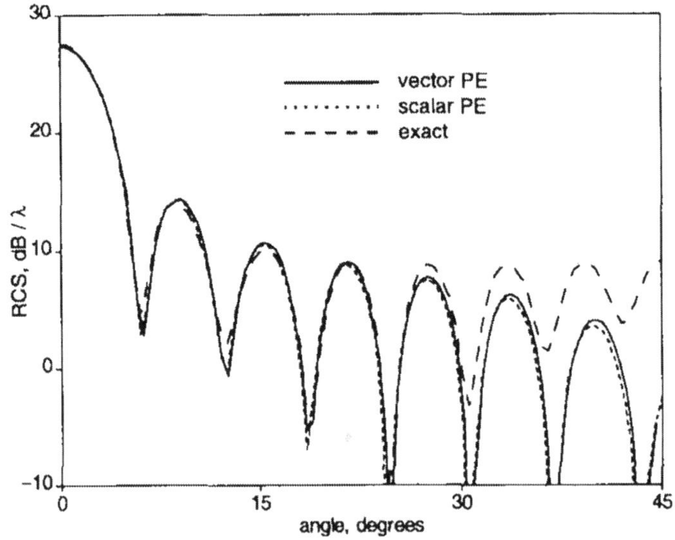

Figure 14.24 Bistatic RCS of a circular cylinder of radius 5λ from a single PE run. (From Zaporozhets and Levy [II.71]. Reproduced by permission of © 1999 IEEE)

by Zaporozhets and Levy from the three methods for angles up to 45°, assuming a wavelength of 1 m. Here, the agreement is very good for scattering angles up to 25° or so, showing that the imperfect near-field results contain the correct small angle propagation components. In order to obtain the full bistatic RCS pattern, the paraxial direction has to be rotated in 30° steps. Figure 14.25 shows the combined RCS pattern obtained with the vector PE together with the analytical solution. According to Zaporozhets and

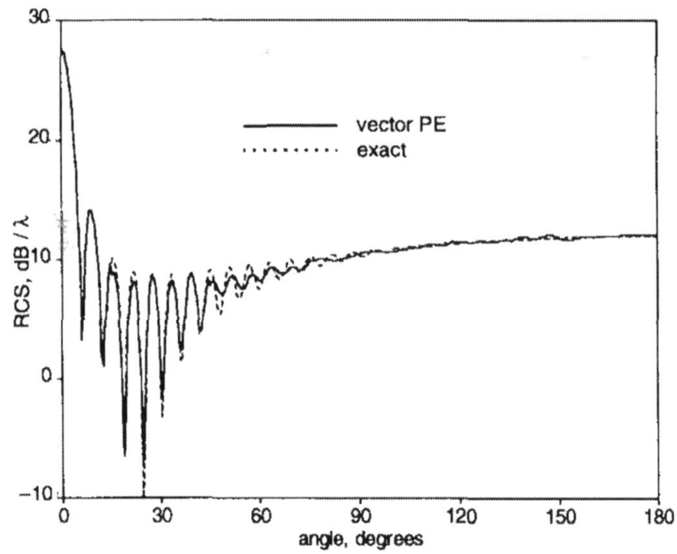

Figure 14.25 Bistatic RCS of a circular cylinder of radius 5λ from seven rotated vector PE runs. (From Zaporozhets and Levy [II.71]. Reproduced by permission of © 1999 IEEE)

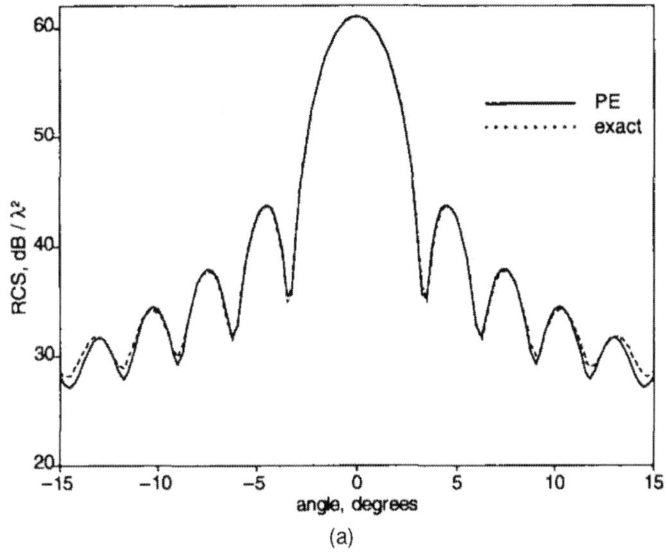

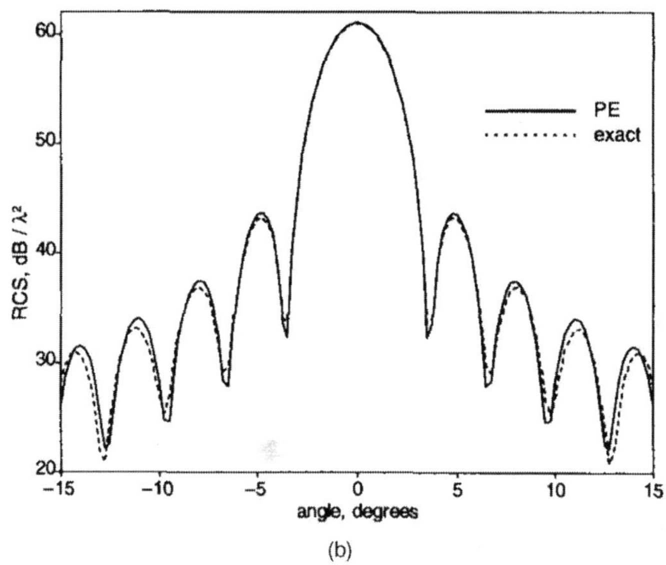

Figure 14.26 Total bistatic RCS for a perfectly conducting sphere of radius 10λ from a single forward-scatter vector PE run: (a) horizontal plane pattern and (b) vertical plane pattern. (From Zaporozhets and Levy [II.71]. Reproduced by permission of © 1999 IEEE)

Levy, the agreement is excellent over the whole angular range. In Figures 14.22 to 14.25, the angle is measured from the direction of the incident wave.

Zaporozhets and Levy also examine scattering by perfectly conducting spheres. The vector PE solution is compared to the theoretical solution given by Mie expansions [II.71]. Figure 14.26 shows forward scatter results for a single PE run for a sphere of radius 10λ for both horizontal and vertical plane patterns, assuming a wavelength of 0.1 m (frequency 3 GHz). Agreement with the theory is very

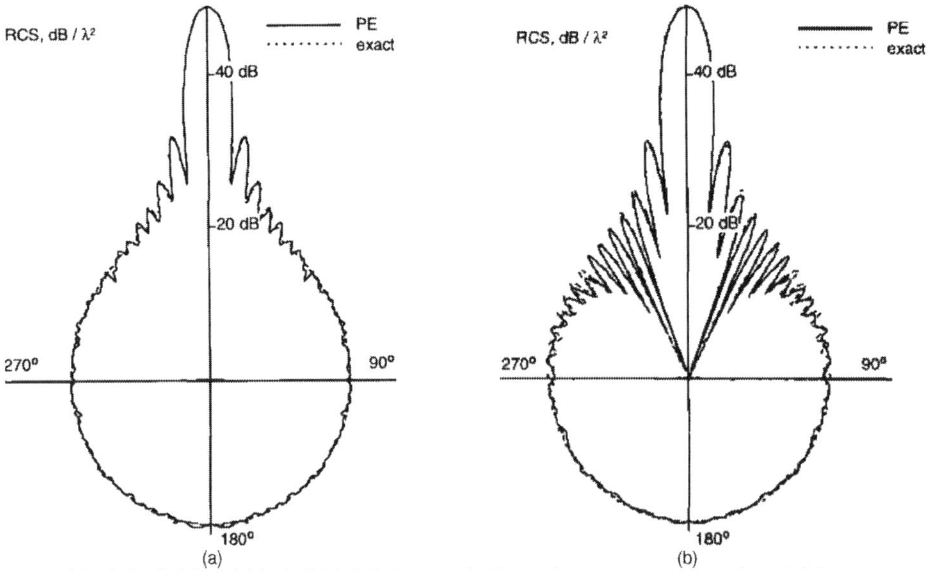

Figure 14.27 Total bistatic RCS for a perfectly conducting sphere of radius 5λ from seven rotated vector PE runs: (a) horizontal plane pattern and (b) vertical plane pattern. (From Zaporozhets and Levy [II.71]. Reproduced by permission of © 1999 IEEE)

consistent, particularly for the horizontal plane pattern, where oscillations are less marked. The error increases towards the larger angles, as the accuracy of the narrow-angle PE approximation decreases. It should be noted here that this case would be quite severe for a full-wave method like the MOM in view of the large size of the object.

Figure 14.27 shows the full bistatic scattering pattern for a sphere of radius 5λ. Again, both the vector PE and the theoretical results are shown. For each plane, the full pattern requires seven PE runs covering angular sectors of 30° each. Since the larger scattering angles require finer grids, computation times increase as the paraxial direction moves towards 180°. According to Zaporozhets and Levy, agreement with the theoretical results is excellent for this case, which would be computer-intensive for MOM codes.

When the size of the scatterer becomes comparable to the wavelength λ, the PE approximation becomes less accurate, as it neglects the effects of creeping waves (introduced in Section 14.1.2, where analytical methods were discussed), which can propagate all around the object. Zaporozhets and Levy [II.71] illustrated this in Figure 14.28, which shows bistatic scattering results for a sphere of radius 0.25λ. The PE method breaks down except for forward scatter and 90° side scatter in the vertical plane, where the creeping wave is cancelled out for reasons of symmetry. Parametric simulations show that results are reasonably accurate down to a radius of one wavelength.

Finally, Zaporozhets and Levy [II.71] use the vector PE with an idealized representation of a missile. The object is 12 m long and was made from a 2 m diameter cylinder with hemispherical caps and attached fins (Figure 14.29). It should be noted here that the PE method does not require a decomposition of the target into canonically shaped elements and can deal directly with an arbitrary shaped target. Figure 14.29 shows the amplitude of the vertical component of the scattered field computed for a vertically polarized (along the z axis) incident wave propagated along the x axis at a frequency of 300 MHz. The forward scattered field is shown just beyond the missile. Figures 14.30 and 14.31 show bistatic RCS results in horizontal and vertical planes, where the angles are measured from the direction of the incident wave.

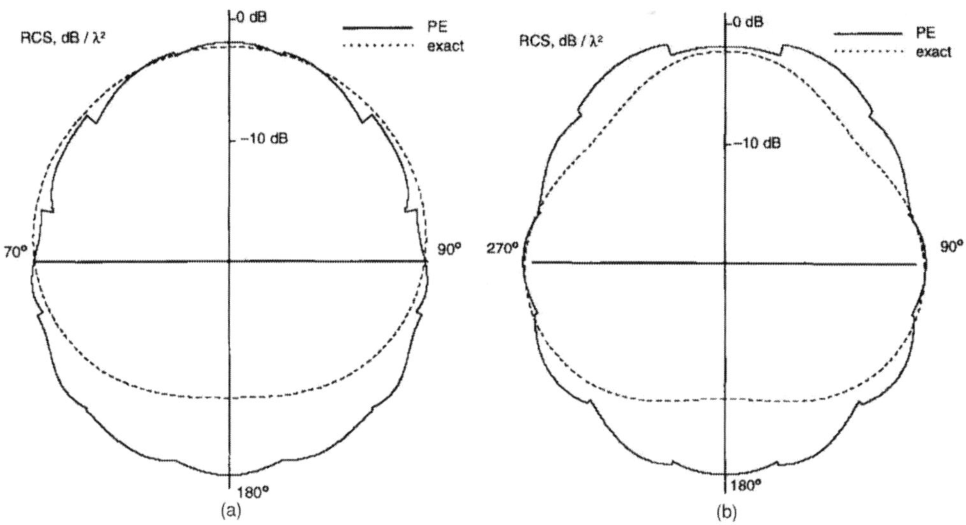

Figure 14.28 Total bistatic RCS for a perfectly conducting sphere of radius 0.25λ from seven rotated vector PE runs: (a) horizontal plane pattern and (b) vertical plane pattern. (From Zaporozhets and Levy [II.71]. Reproduced by permission of © 1999 IEEE)

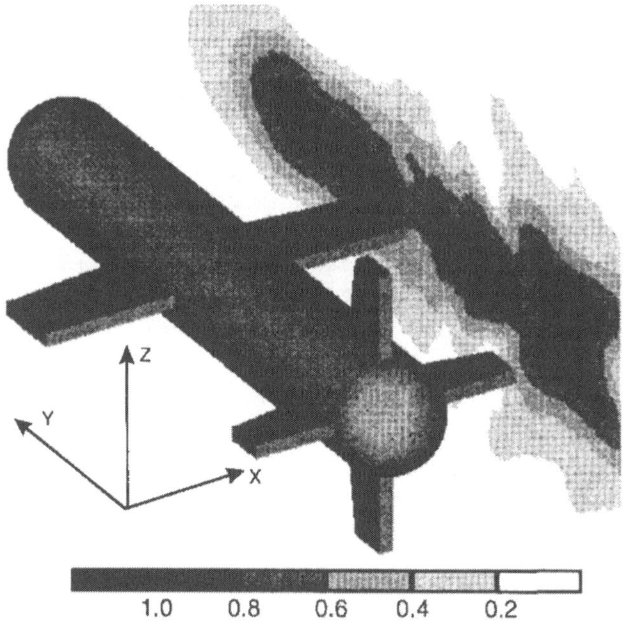

Figure 14.29 Scattering by a 12 m-long idealized missile at 300 MHz. Amplitude of a vertical (z) component of the scattered electric field for the forward PE run. The incident wave has vertical polarization. (From Zaporozhets and Levy [II.71]. Reproduced by permission of © 1999 IEEE)

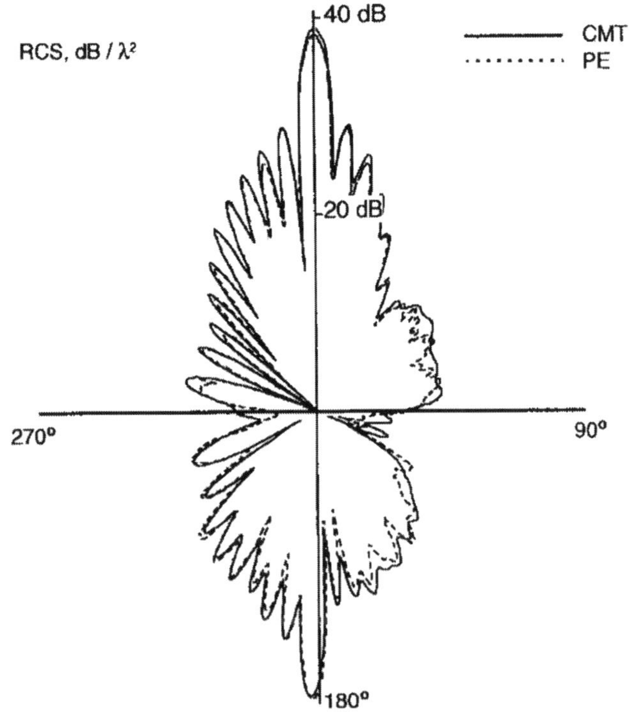

Figure 14.30 Bistatic. RCS in the horizontal plane for the 12 m-long idealized missile at 300 MHz. (From Zaporozhets and Levy [II.71]). Reproduced by permission of © 1999 IEEE)

Zaporozhets and Levy used the current marching technique (CMT) [II.71] to validate the PE results. The CMT is based on the magnetic field integral equation, which is solved with an efficient backward/forward iterative solver. The missile object was divided into 13 618 triangles and the authors used the simplest pulse basis functions and delta functions as testing functions. The PE and CMT results are in good agreement. The vertical plane PE curve (Figure 14.31) has larger errors compared to the horizontal plane data (Figure 14.30). Zaporozhets and Levy believed that the main reason for the larger error in the vertical plane data is that the missile is reasonably thin (two wavelengths in diameter).

Thus, the combination of the vector PE formulation with the rotated PE method provides a powerful tool for electromagnetic scattering calculations. The main limitation is that the paraxial framework cannot handle creeping waves accurately for objects with electrically small features. The bistatic RCS is computed from near-field results using appropriate near-field-to-far-field transformations. The examples given in Reference [II.71] demonstrate that the parabolic equation method can be applied to objects of a variety of sizes and shapes.

14.1.3.6 Trigonometric Approximations

New methods for the trigonometric approximation of the BRCS and MRCS have been presented by Giuseppe Borzì [II.76]. This method computes trigonometric approximations for the expansion coefficients of the scattered field. These approximations make it possible to trace the MRCS easily and to compute the BRCS when the incidence angle is one for which the trigonometric approximation of the

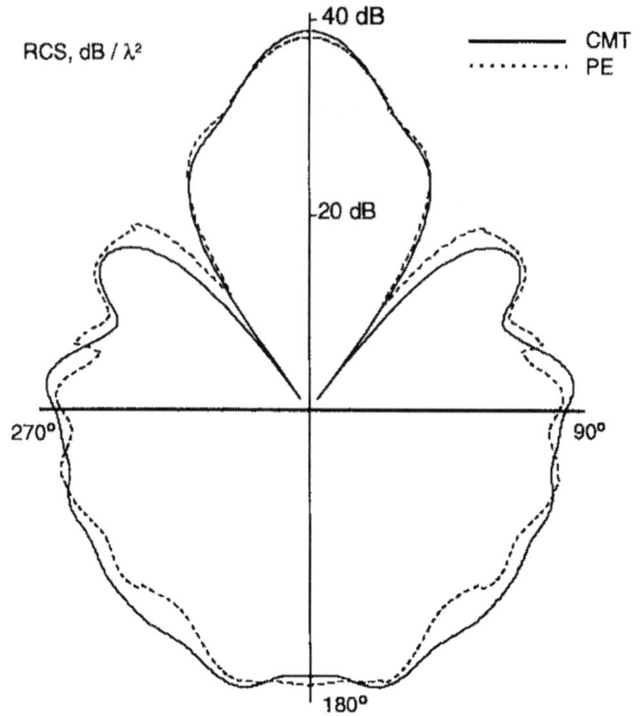

Figure 14.31 Bistatic RCS in the vertical plane for the 12 m-long idealized missile at 300 MHz. (From Zaporozhets and Levy [II.71]. Reproduced by permission of © 1999 IEEE)

scattering coefficients holds. According to Borzì, this latter feature is especially interesting for bistatic radars. The proposed method in Reference [II.76] proved to be superior to rational function approximation, since from a physical point of view the most suitable approximating functions for scattered far fields are trigonometric functions.

Scattering by a PEC square plate with three circular holes has been considered in Reference [II.76] as a demonstration of the suggested method. The plate, whose sides are 2λ long, lies on the *XZ* plane, and the circular holes have radius of 0.15λ, 0.25λ and 0.7λ, as shown in Figure 14.32, which also shows the finite element (FE) mesh used. The BRCS for coordinate $\phi = 0$, computed in Reference [II.76], is shown in Figure 14.33.

14.1.3.7 Method for Computation of the Bistatic Radar Cross-Section

After the scattering problem has been solved, the BRCS is very often the parameter of most interest [II.77]. The BRCS is generally computed in the post-processing phase of the numerical solution of the scattering problem by direct integration, which has to be performed for each observation angle. If the required number of observation points is great, the total amount of computing time is not negligible [II.77].

In Reference [II.77] it has been shown that the computation effort can be substantially reduced by taking advantage of the fact that the BRCS can be constructed using a limited set of sample points for the scattered field, which are computed in the usual way. Moreover, a very useful indicator is simply derived from this approach to check the accuracy of the numerical solution of the scattering problem.

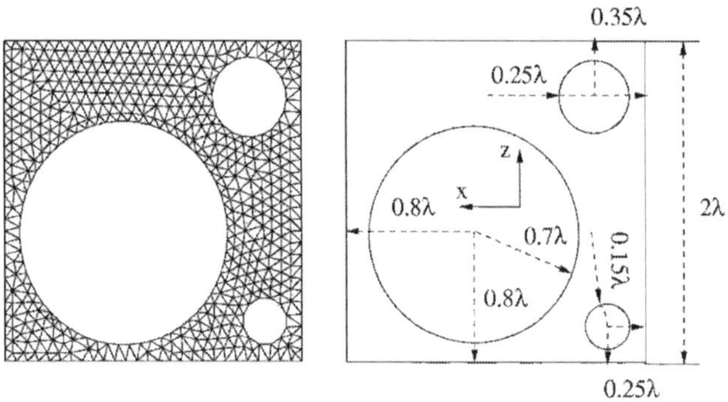

Figure 14.32 FE discretization and geometry of the PEC plate. (From Borzi [II.76]. Reproduced by permission of © 2004 IEEE)

Alfonzetti and Borzì used a well-known fact that both the incident and the scattered fields may be expanded in series of cylindrical harmonics. In their paper [II.77] two examples have been given in order to highlight the application of the proposed algorithm to the computation of the BRCS and to check the accuracy of the solution.

The first example is a homogeneous lossy dielectric cylinder having a square cross-section, each side of which is 2λ long, where λ is the wavelength. The cylinder is lit up by an E-polarized plane wave whose direction vector forms an angle of 30° with the X axis. The relative permittivity and permeability are $\varepsilon_r = 2.0 - j0.8$ and $\mu_r = 2.5 - j1.3$ respectively. The BRCS, computed by means of the suggested [II.77] fast method, is plotted in Figure 14.34.

The second example concerns a cavity in a ground plane which is filled in its deepest part with a lossless material having $\varepsilon_r = 2.0$ and $\mu_r = 1.5$. The aperture of the cavity is 2λ, whereas its depth is λ. The cross-section of the nonhomogeneous material part is a rectangle 0.8λ wide and 0.2λ high. The truncation boundary is 2.4λ wide and 0.2λ high. The cavity is lit up by an E-polarized plane wave with an incidence angle of 45°. The BRCS was computed as in Figure 14.35.

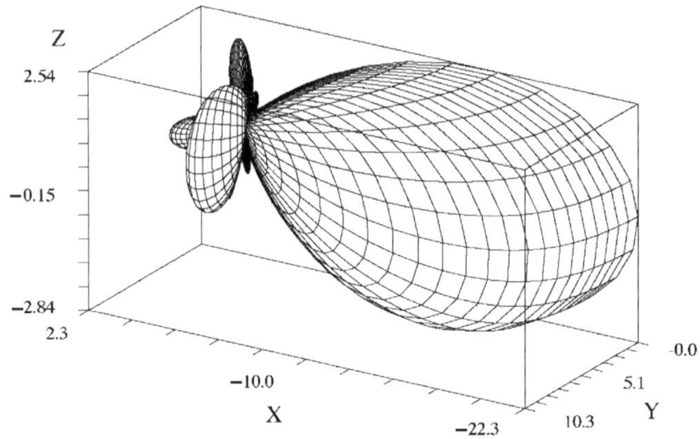

Figure 14.33 BRCS of the PEC square plate for $\phi = 0$ with 2° resolution. (From Borzi [II.76]. Reproduced by permission of © 2004 IEEE)

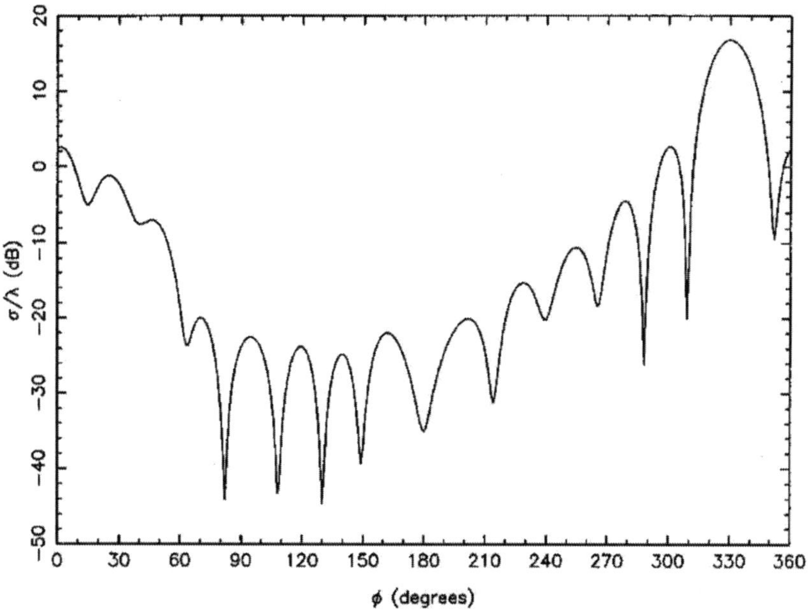

Figure 14.34 BRCS for the first example. (From Alfonzetti and Borzi [II.77]. Reproduced by permission of © 2000 IEEE)

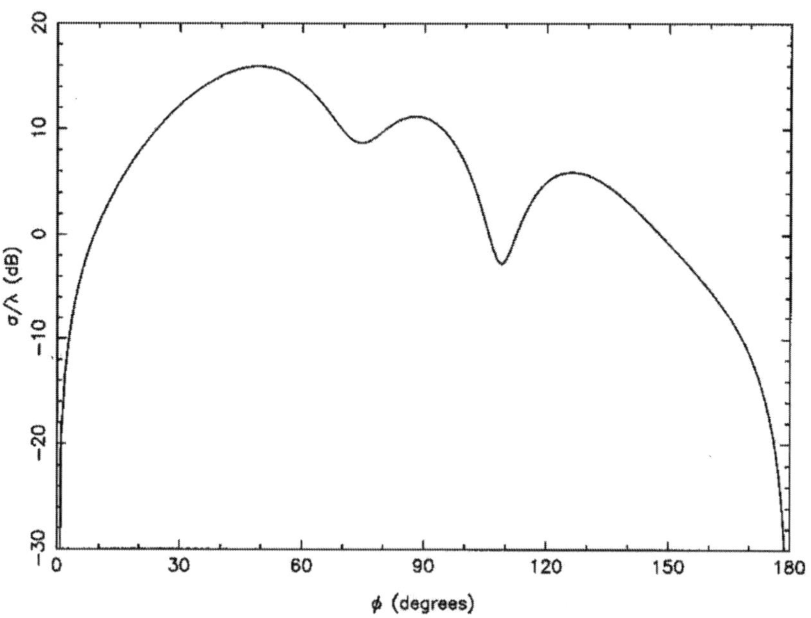

Figure 14.35 BRCS for the second example. (From Alfonzetti and Borzi [II.77]. Reproduced by permission of © 2000 IEEE)

14.1.4 Hybrid Methods

The electrical size of a body limits the tractability of numerical methods such as the MOM, and the geometric complexity of an object circumscribes the applicability of electromagnetic models. Hybrid methods incorporating both numerical and high-frequency asymptotic techniques have the potential to enlarge the class of scattering problems that can be treated substantially. In Reference [II.78] the current-based hybrid formulation was summarized for classes of two- and three-dimensional scatters. The use of Ansatz [II.78] solutions derived from physical optics (PO) and the physical theory of diffraction (PTD), and the Fock theory was illustrated for perfectly conducting, partially penetrable and totally coated bodies.

In Reference [II.78] it was shown that a hybrid approach to electromagnetic scattering problems, incorporating the high-frequency optics-derived models and the MOM techniques, has the potential to build on the strengths of these approaches, while mitigating their inherent limitations. Properly formulated, the hybrid approach offers a unified approach for EM scattering for classes of complex, electrically large scatters.

The hybrid formulation can be designed as a field-based analysis or as a current-based one (and there is also the option of a mixed field – current-based approach). In the field-based version, the geometrical theory of diffraction (GTD) solution for the fields associated with edge or surface diffraction are used as the starting point. These solutions serve as the Ansatz to the MOM formulation and representation for parts of a scatterer not conforming to a canonic geometry and, hence, not amenable to an *ab initio* GTD solution itself. Examples of such problems include rounded wedges and totally or partially penetrable surfaces (see Reference [II.79] for details). The field-based approach is particularly attractive for radiation problems where the radiator is in the proximity of large convex surfaces for which canonical GTD solutions are known.

In the current-based formulation, the analysis proceeds from Ansatz solutions for the currents, obtained from physical optics (PO), the physical theory of diffraction (PTD) and the Fock theory. Edges can be treated by equivalent edge current (EEC) representations complementing the results of the PTD. In the current-based hybrid formulation, integration of the Ansatz currents occurs in the integral operators. This is both an advantage and a drawback. Surface currents exhibit slower spatial fluctuations than fields in the vicinity of a scatterer and the integration in the operators provides additional smoothing. Thus, the approximate Ansatz currents often suffice to yield satisfactory answers. On the other hand, the complexity of the surface geometry generally precludes analytical evaluation of the operators. Numerical integration, particularly in two dimensions, is computationally unattractive. Investigations have shown that, for a number of practical cases, a combined analytical–numerical integration scheme is tractable by decomposing the scatterer surface into curvilinear strips.

By use of the Ansatz solutions, hybrid methods can be used to treat several classes of scatters that are perfectly conducting or coated with highly lossy materials with and without perfectly conducting edges. With the PO/Fock/EEC Ansatz, the hybrid technique introduced here extends the ability of the PTD to allow treatment of forward scattering as well as oblique illumination, where the tangent plane approximation is tenuous or where nonspecula phenomena such as creeping waves predominate.

The previous methods were applied in Reference [II.78] to a variety of 2D and 3D scatters. The simplest 2D case is the flat conducting strip. In Figure 14.36, the TM (transverse magnetic) polarized back-scatter cross-sections are depicted for a 2.84λ-wide strip using the coupled and uncoupled MOM hybrid formulations. The interior region of the strip away from the edges is treated with PO and the edges over a $\frac{1}{2}\lambda$-wide interval are treated using the Galerkin representation.

For comparison, an entirely MOM-based ('exact') solution is also shown as well as one in which only PO is used throughout. As expected by Medgyesi-Mitschanc and Wang [II.78], the latter progressively fails as grazing angles are approached because PO provides no contributions from the strip edges, while the hybrid solutions do. (The uncoupled MOM hybrid formulation leads to erroneous results because the strong interactions between the MOM region with the remaining surface are not included.) The results given by the coupled MOM hybrid for a small ogival cylinder for TM-polarized illumination are depicted in Figure 14.37(a). The Galerkin representation spans a $\frac{1}{2}\lambda$-wide region near the edge.

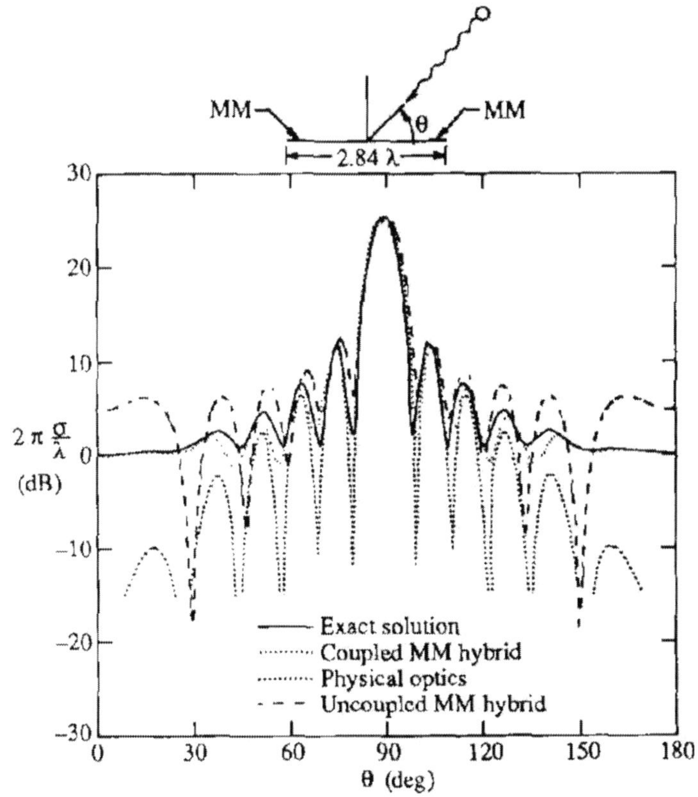

Figure 14.36 Back-scatter cross-sections of a flat conducting a 2.84 λ-wide strip. TM polarized illumination. (From Medgyesi-Mitschanc and Wang [II.78]. Reproduced by permission of © 1989 IEEE)

The computed results closely correlate with measured data by Knott *et al.*, given in Reference [II.80]. Both of the foregoing cases are in the resonance range where the hybrid approach is known to be least accurate because the underlying Ansatz PO solutions fail and the interactions of the various parts of the scatterer are pronounced. An even more limiting configureuration is a relatively small concave surface, such as the dihedral in Figure 14.37(b). The cross-sections are normalized by l, which denotes the length perpendicular to the plane of the figure. The Galerkin representation spans a 1λ-wide region near the edges and at the corner; PO spans the rest of the surface. The coupled MOM hybrid results are compared to the calculations by Knott [II.81]. The coupled MOM hybrid solution is most accurate near the specula angles; again PO by itself fails. Calculations have shown that when the dihedral angle is acute, i.e. $\beta < 45°$, the hybrid solution becomes inaccurate since the interaction effects (i.e. multiple bounces) between the walls dominate the scattering process.

This observation illustrates an important point about the hybrid approach: the Ansatz solutions must be at least approximately valid in the region where they are used. In the case of the dihedral, subtending acute angles, PO is not a good Ansatz because multiple reflections predominate. Consequently, the results are erroneous.

Application of the hybrid method to 3D scatters was illustrated by Medgyesi-Mitschanc and Wang in the following cases. In Figure 14.38, the back-scatter cross-sections of a circular conducting cone

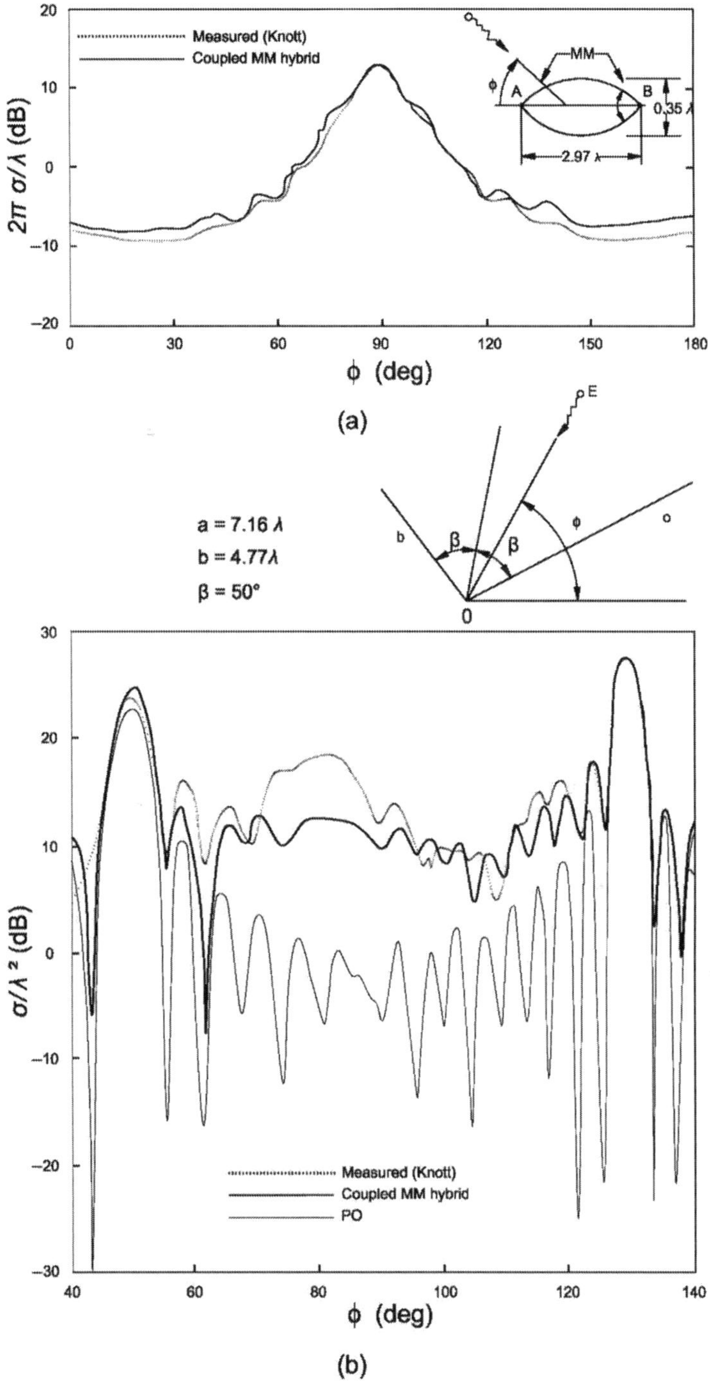

Figure 14.37 Back-scatter cross sections for TM polarized illumination: (a) ogival cylinder and (b) dihedral. (From Medgyesi-Mitschane and Wang [II.78]. Reproduced by permission of © 1989 IEEE)

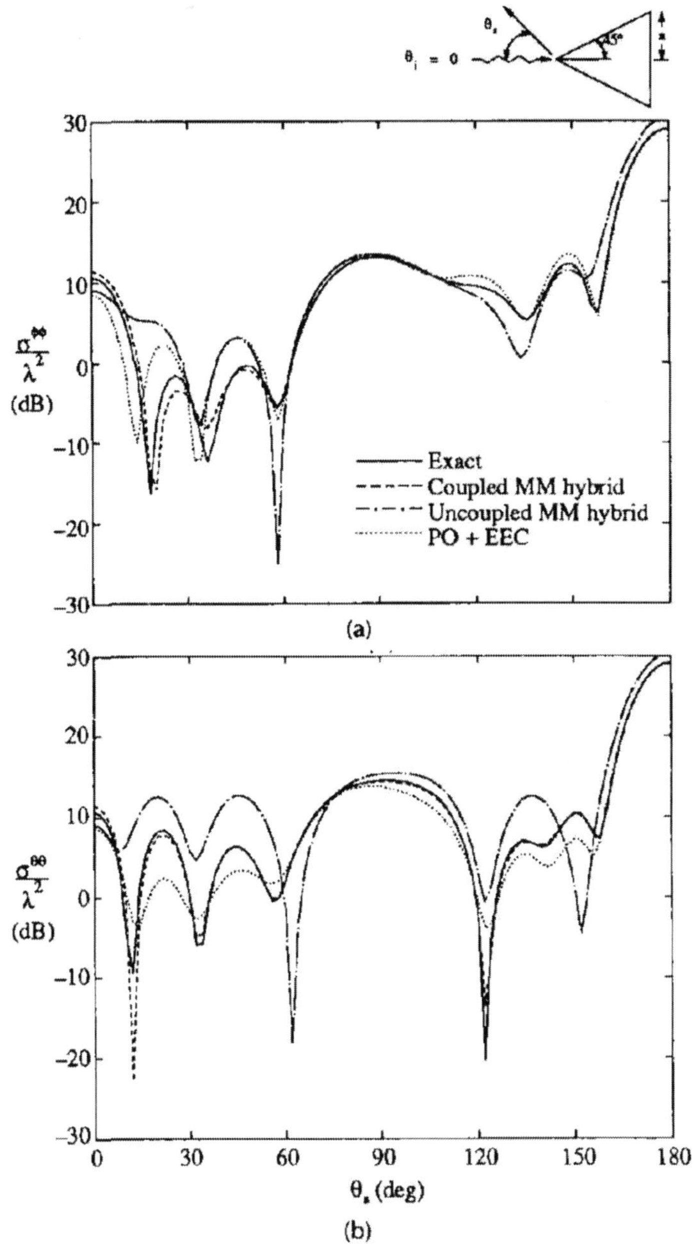

Figure 14.38 Back-scatter cross sections of a perfectly conducting circular cone. Cone half-angle = 45°, $ka = 10$. Axial illumination ($\theta_{\text{in}} = 0°$). (From Medgyesi-Mitschanc and Wang [II.78]. Reproduced by permission of © 1989 IEEE)

are given. The cone length is 1.5λ, with a cone half-angle of 45° and $ka = 10$. In the coupled and uncoupled MOM hybrid solutions, the Galerkin representation is used near the tip and the base region spanning a region $\frac{1}{2}\lambda$ wide. On the rest of the surface PO is used. For comparison, Medgyesi-Mitschanc and Wang also showed results that were obtained without the MOM term, where the EEC representation is used for the currents at the cone rim and PO is used elsewhere. These results are compared with an 'exact' (MOM) solution as a benchmark. The computations were performed for both polarizations.

In Figure 14.39, the asymptotic-based hybrid approach is applied to elliptic cones. As in the circular case above, equivalent-edge currents were used for the rim effects and PO on the rest of the surface. The depicted data are given for scattering angles $|\Theta| \leq 20°$ off the nose of the cone, where the nonspecular effects predominate, particularly in the $\theta\theta$ polarization.

The hybrid method can also be applied to certain penetrable and coated scatters. Two cases are considered in Reference [II.78]: a coated body satisfying the impedance boundary conditions (IBC) requirements [II.82] and a partially coated body. These cases are illustrated for cone spheres and ogive spheres. The hybrid solutions are compared with an entirely MOM-based solution validated earlier [II.83]. In Figure 14.40, the bistatic scattering (nose-on illumination, $\theta_{in} = 0°$) from a coated cone sphere is depicted. The cone half-angle is 45° and $ka = 10$ for the outer surface. The coating thickness has a $kd = 2.0$ and the material properties ($\varepsilon_r = 99.75 - j10.0$, $\mu_r = 1.0$) are such that the IBC criterion in Reference [II.82] is satisfied. A subdomain representation was used, spanning 1λ in the neighbourhood of the discontinuity joining the conical and spherical sections. The PO and Fock Ansatz solutions were used on the remaining conical and spherical surfaces respectively. For both polarizations, the coupled MOM hybrid approach is in good agreement with the 'exact' (entirely MOM-based) solution. For comparison, Medgyesi-Mitschanc and Wang [II.78] depict the PO results modified by the Fresnel coefficients for coatings. Figure 14.41 depicts the bistatic scattering ($\theta_{in} = 0°$) by a cone sphere partially coated with a lossless dielectric ($\varepsilon_r = 14.87$, $\mu_r = 1.0$). The coating thickness increases monotonically from $kd = 0$ at the join to $kd = 2.0$ at the centre of the spherical cap. The cone half-angle is 45° and $ka = 10$ for the outer surface. In the coupled MOM hybrid calculation, the Galerkin representation is used on both the outer and inner surfaces in the coated region. The PO Ansatz is used on the conical (uncoated) surface. The coupled MOM hybrid approach correlates well with the exact solution; however, the other two approaches produce erroneous results, especially for $\theta_{sc} \leq 60°$. Note that the lossless coating on the spherical section actually enhances tenfold the back-scatter cross-section at $\theta_{sc} = 0°$ over that of an uncoated cone–sphere.

Scattering calculations for an ogive sphere are given in Figure 14.42. The scatterer is covered with a uniform coating with $kd = 2.0$ and $\varepsilon_r = 99.75 - j10.0$, $\mu_r = 1.0$. The ogival section has $kl = 24.142$. In the coupled MOM hybrid calculation, the Galerkin representation spans the entire ogival section. The coupled MOM hybrid results using the IBC approximation correlate well with the exact solution, while the PO results using the Fresnel coefficients are in error.

Thus, the paper [II.78] discussed the application of hybrid formulations to classes of complex two- and three-dimensional scatterers. The examples chosen represented severely limiting cases where the hybrid method can be used, namely electrically small scatters, coated convex surfaces and conducting concave geometries. The calculations show that even in these limiting cases the coupled MOM hybrid approach yields results closely correlating with benchmark results.

While it is difficult to assert the conditions necessary for the hybrid method's efficacy to hold universally for complex scatters, some general principles have been stated in Reference [II.78]. These are:

(a) The high-frequency Ansatz must be valid in the region where it is applied for it to work in the hybrid construct.
(b) The low-frequency (MOM) region must extend approximately $\frac{1}{2}\lambda$ from the edge of a surface or a region of material discontinuity.

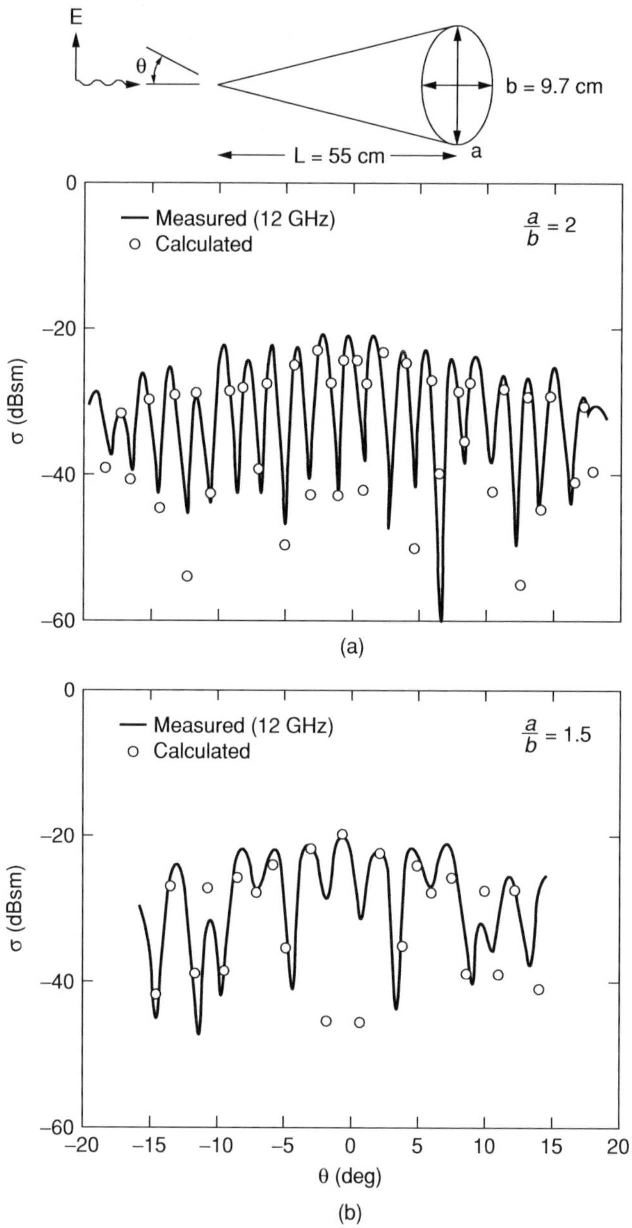

Figure 14.39 Back-scatter cross-sections of an elliptic cone: (a) $a/b = 2.0$ and (b) $a/b = 1.5$. (From Medgyesi-Mitschanc and Wang [II.78]. Reproduced by permission of © 1989 IEEE)

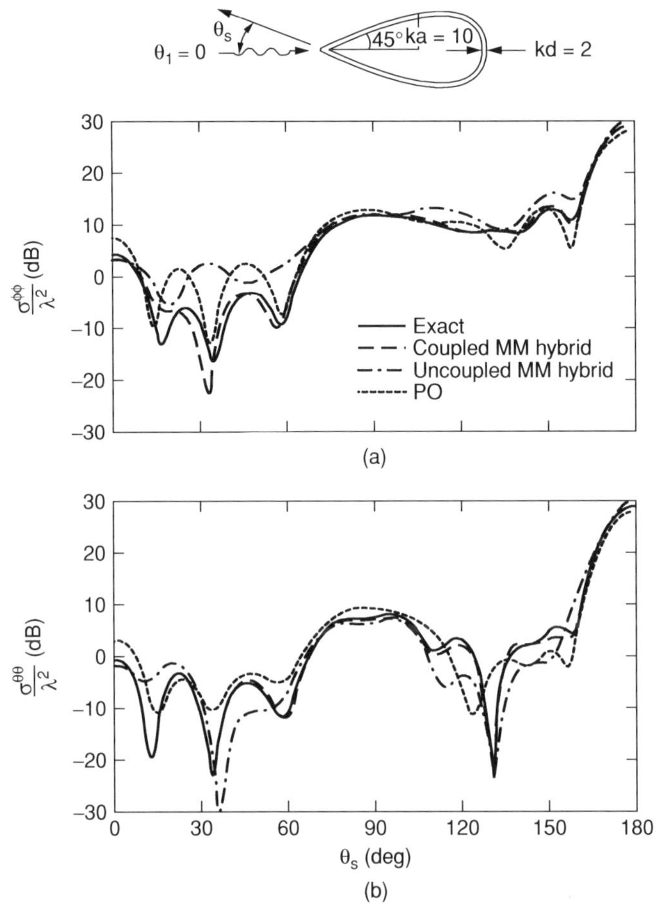

Figure 14.40 Bistatic scattering cross-sections of a coated conesphere: (a) cone half-angle = 45°, $ka = 10$, $kd = 2$, $\varepsilon_r = 99.75 - j10.0$, $\mu_r = 1$. (b) axial illumination ($\theta_{\text{in}} = 0°$). (From Medgyesi-Mitschanc and Wang [II.78]. Reproduced by permission of © 1989 IEEE)

(c) For scatters where the travelling wave contributions are significant in comparison to the edge wave and specula scattering at a given aspect angle, the coupled MOM hybrid results can be in error (a special modification of the present methods exists for correcting this deficiency on electrically large bodies of revolution (BORs), where the length-to-diameter ratio is large.
(d) Creeping wave effects can be incorporated effectively in the hybrid analysis for conducting surfaces. Creeping waves on coated surfaces can be approximately represented using modified Fock functions obtained from the IBC formulation.
(e) Varying the extent of the MOM region in the hybrid calculation provides a measure of solution convergence (reducing the MOM region generally degrades the solution accuracy and vice versa).
(f) Finally, the hybrid method works best for electrically large bodies.

The hybrid finite element method (FEM)/MOM formulation including the IBC was presented in Reference [II.84]. According to Wang *et al.* incorporating IBC in the hybrid FEM/MOM allows the analysis of geometries with thin dielectric sheets or imperfect conductors more efficiently. Three

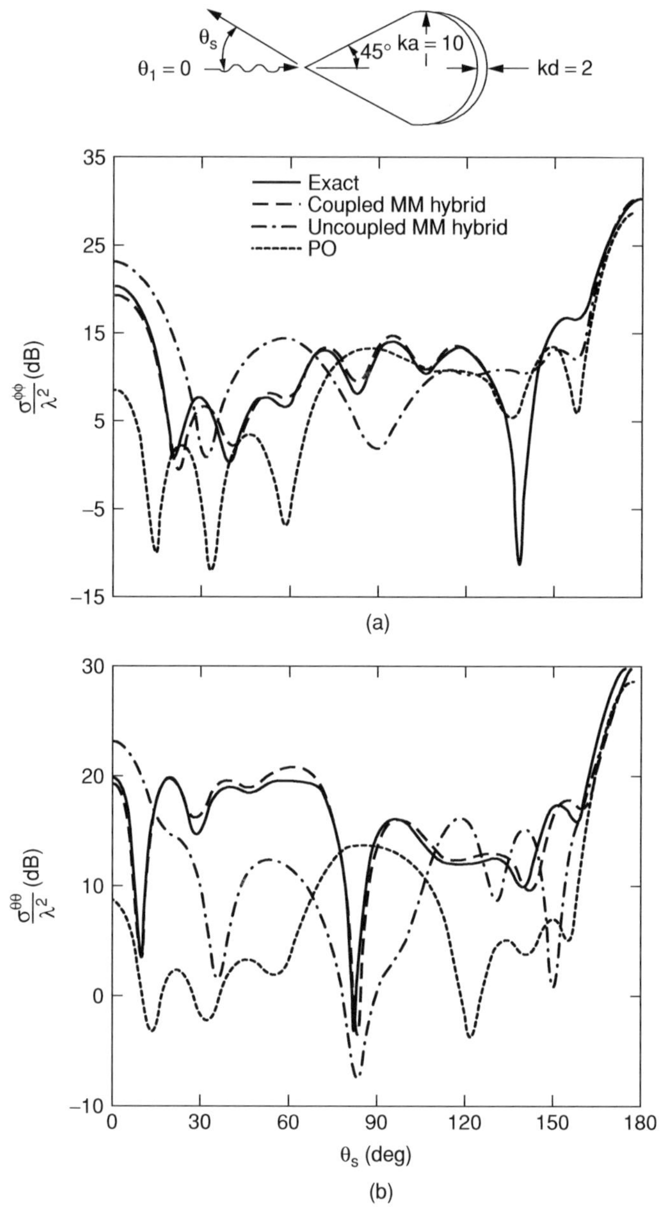

Figure 14.41 Bistatic scattering cross-sections of a partially coated conesphere. (a) cone half-angle = 45°, $ka = 10$, $kd = 2$, $\varepsilon_r = 14.87$, $\mu_r = 1$; (b) axial illumination ($\theta_{in} = 0°$). (From Medgyesi-Mitschanc and Wang [II.78]). Reproduced by permission of © 1989 IEEE)

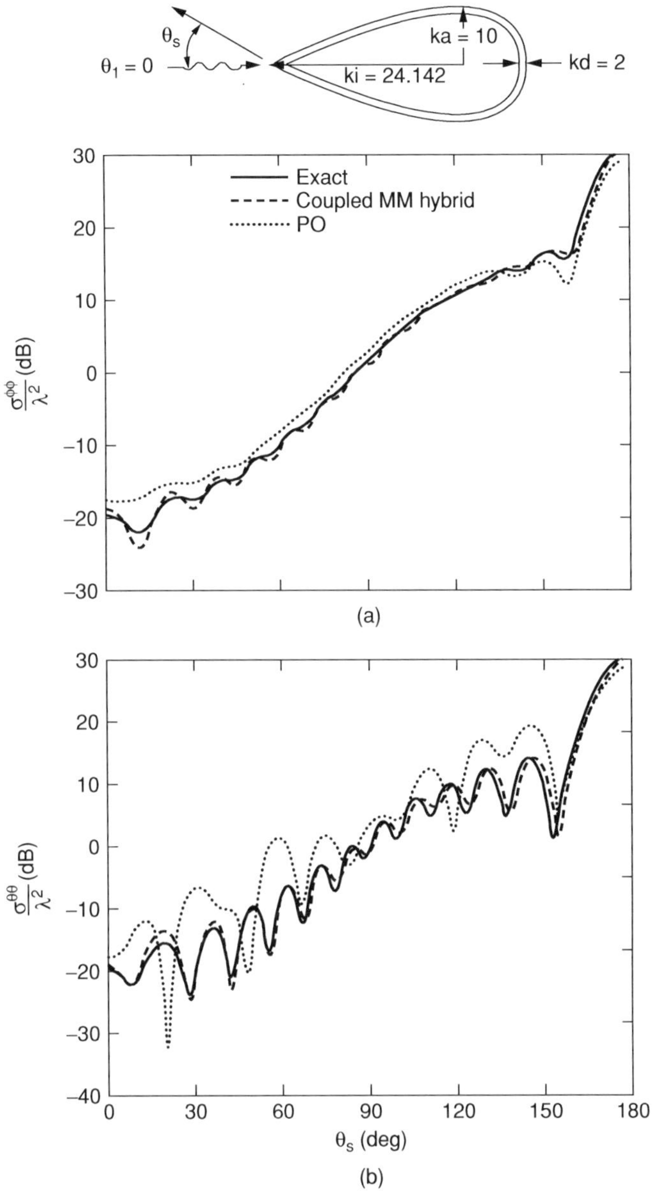

Figure 14.42 Bistatic scattering cross-sections of an ogive–sphere configuration with a homogeneous coating: (a) $ka = 10$; $kl = 24.142$, $kd = 2$, $\varepsilon_r = 99.75 - j10.0$, $\mu_r = 10$; (b) axial illumination ($\theta_{in} = 0°$). (From Medgyesi-Mitschanc and Wang [II.78]. Reproduced by permission of © 1989 IEEE)

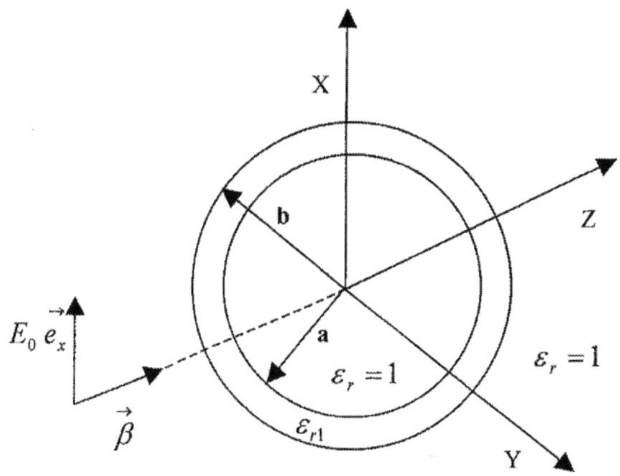

Figure 14.43 Scattering of a plane wave by a hollow coated sphere. (From Wang *et al.* [II.84]. Reproduced by permission of © 2003 IEEE)

practical problems were modelled in Reference [II.84] to validate the formulation and to demonstrate methods for applying the IBC to different geometries. The first problem was the scattered field from a thin, dielectric, spherical shell. The second problem is the analysis of a printed circuit board power bus structure whose top and bottom planes are not perfect electric conductors (PECs). In this case, the loss of the imperfect conductors must be considered to obtain accurate results. The third geometry analysed is an asymmetric stripline structure. Results were compared to either measured data or other numerical results. Good agreement was achieved between the FEM/MOM formulation and other well-established codes, analytical results or measurements. The effects of the IBC terms on the condition numbers of the final FEM/MOM matrices were also discussed.

Let us consider, following Reference [II.84], the scattered field from a thin, dielectric, spherical shell. As shown in Figure 14.43, a hollow sphere with inner and outer radii a and b, and $b/a = 1.001$ cm, was centred at the origin of the coordinate system. The shell was made of a dielectric material with relative permittivity $\varepsilon_{r1} = 1000$. The hollow sphere was located in free space.

Figure 14.44 shows the bistatic scattering cross section at 583 MHz obtained by Wang *et al.* [II.84] using FEM/MOM with an impedance boundary and Mie series. The Mie series formulation models spheres like this without difficulty and the Mie series result can be considered accurate. The difference between these two results is within 1 dB.

14.1.5 Monostatic–Bistatic Equivalence Theorem

It follows from Equation (10.112) that the BRCS contains more information than the MRCS, which is simply one projection of the BRCS. Reference [II.85] described a method of using bistatic computed data to generate multiple monostatic predictions for each incident angle. The monostatic/bistatic approximation has been used in Reference [II.85] to reduce the number of runs required by finite volume time-domain (FVTD) codes for the creation of RCS versus azimuth plots. This approximation was generated by the observation of a range test, where the transmitting and receiving antennas were separated by a few degrees in order to minimize cross-talk between the antennas. The measurements from this range were presented as a monostatic RCS, rather than bistatic RCS. The procedure of reporting the experimental bistatic RCS as the monostatic RCS, at the angle bisecting the transmitting and receiving antennas, was extended to FVTD codes. This produces excellent results, which show that the accuracy is initially high and then degrades as the bistatic angle becomes larger. Figure 14.45 shows the results from the monostatic/bistatic

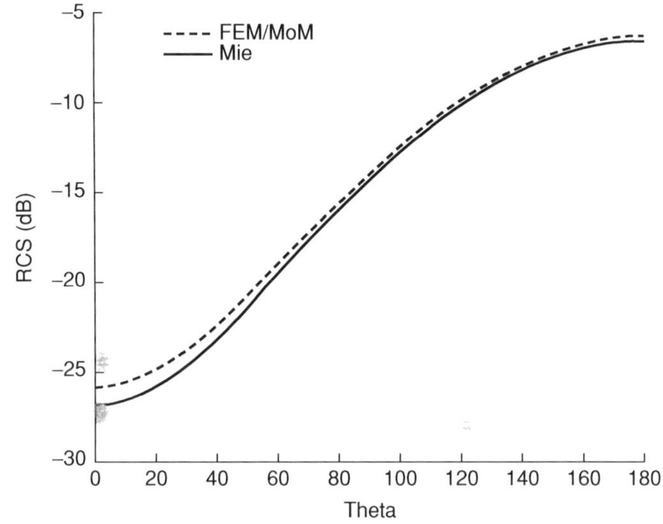

Figure 14.44 Bistatic scattering cross section of a hollow sphere. (From Wang *et al.* [II.84]. Reproduced by permission of © 2003 IEEE)

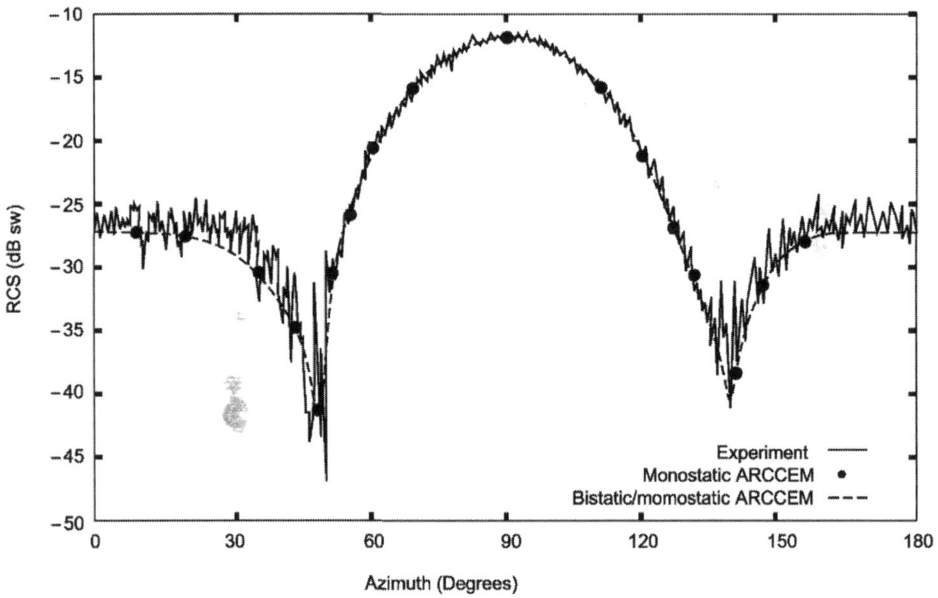

Figure 14.45 RCS for the NASA Almond, one λ long, VV polarization, from ARCCEM runs every 10° in azimuth. (From Schuh *et al.* [II.85]. Reproduced by permission of © 1994 IEEE)

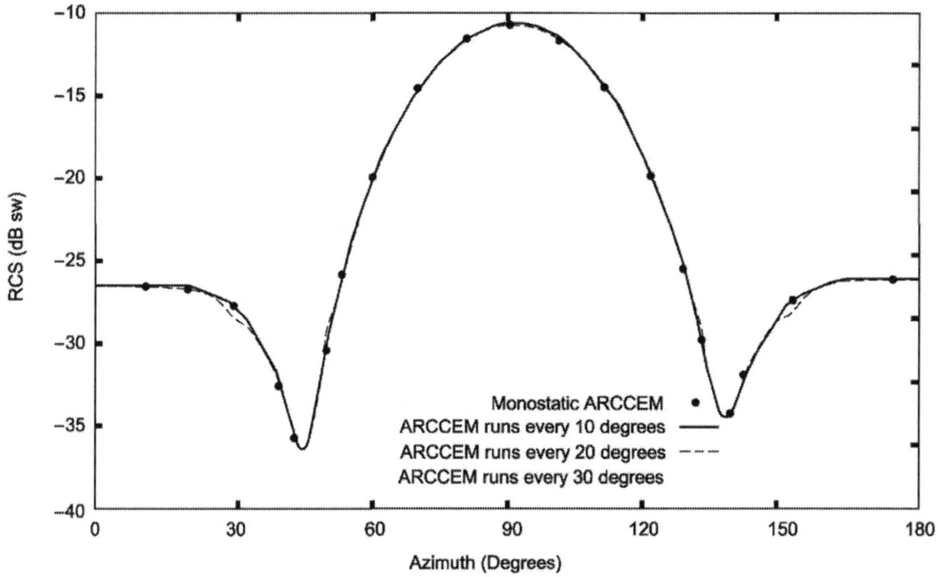

Figure 14.46 RCS for the NASA Almond, one λ long, VV polarization. A comparison of ARCCEM runs every 10°, 20° and 30° in azimuth is shown. (From Schuh *et al.* [II.85]. Reproduced by permission of © 1994 IEEE)

approximation, used with monostatic runs of ARCCEM at 10° azimuthal increments. ARCCEM is a NASA Ames FVTD code, based on the Rockwell Science Center Algorithm [II.74] and implemented on a Thinking Machines Corporation CM-5 parallel computer. This first demonstration is for the VV polarized RCS of a one-wavelength-long NASA almond, described in Reference [II.86]. This figure shows that the return from the almond is well described by using the monostatic/bistatic approximation with 19 ARCCEM runs. It also shows that there are no significant discrepancies at the double-valued points (5°, 15°, 25°, ..., 175° azimuths).

Figure 14.46 shows that expanding the angle increment between ARCCEM runs to the 20° and 30° azimuths results in larger differences between the monostatic return and the approximated monostatic return, at and near the double-valued points. This Figure also shows that the RCS, through the null between the 40° and 50° azimuths, is accurately predicted by the monostatic/bistatic approximation, from an ARCCEM run with an incident-field angle of 40° azimuth (the dashed line in Figure 14.46). This is a typical example of how well the RCS is approximated through nulls by this method.

The ARCCEM RCS results and experimental results [II.86] for a $1.25\lambda \times 1.25\lambda$ rectangular aperture by 1.875λ long cavity are presented in Figure 14.47. These show another interesting aspect of the method. The two sets of ARCCEM results are for runs where the time-domain calculation was run until the cosine incident field propagated 10λ and 20λ respectively. They were run with a cosine wave incident field, with 5° azimuthal increments between the ARCCEM runs. There are discrepancies at the double-valued points located (52.5°, 57.5°, ..., 77.5° azimuths) between the 50° and 80° azimuths for the 10λ run. The discrepancies take the form of discontinuities, and give the impression that the approximation has broken down and that more ARCCEM runs should be made at a closer azimuthal spacing. Many times this is true, but for this case the discrepancies are significantly reduced by running the code for 20λ, which shows that a longer run time was needed, rather than more ARCCEM runs. The additional run time significantly improves the agreement of the calculated ARCCEM RCS with the experimental results.

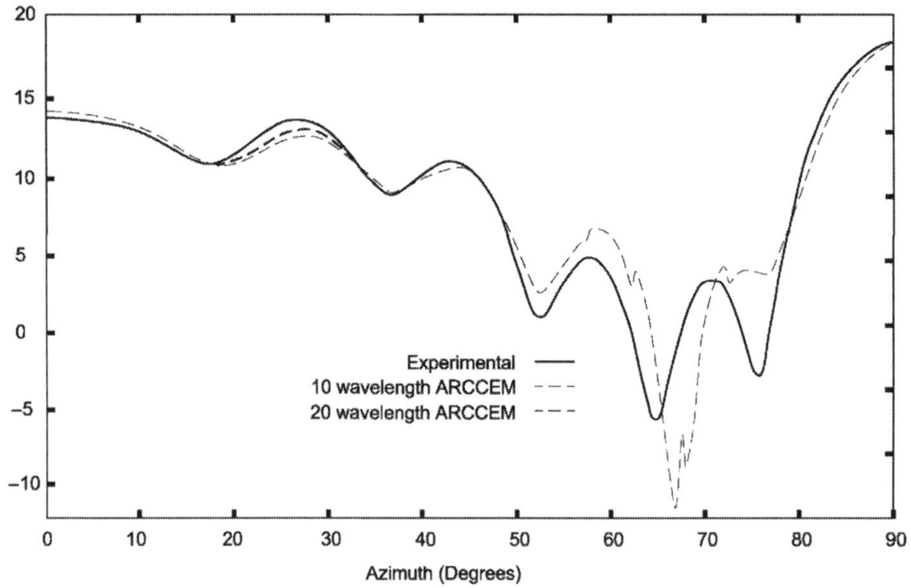

Figure 14.47 The RCS of the 1.25λ × 1.25λ rectangular aperture by the 1.875λ NASA rectangular cavity. (From Schuh *et al.* [II.85]. Reproduced by permission of © 1994 IEEE)

Thus, it is possible to obtain an MRCS from a BRCS, but the opposite is generally impossible. However, for special cases the attempts to obtain a BRCS from a known MRCS were undertaken: several monostatic–bistatic equivalence theorems have been proposed to establish such a relationship, allowing an approximate bistatic data set from monostatic information.

Crispin and his co-authors in References [II.87] and [II.88] presented a similar monostatic–bistatic equivalence theorem (MBET), which states: 'For perfectly conducting bodies which are sufficiently smooth in the limit of vanishing wavelength, the bistatic cross-section is equal to the monostatic cross-section at the bisector of the bistatic angle between the direction to the transmitter and receiver.' This theorem was derived from a physical optics model. The approached formula following from the MBET at finite (not vanishing) meanings of the wavelength is

$$\sigma_{\mathrm{b}}\left(\boldsymbol{k}_{\mathrm{in}}, \boldsymbol{k}_{\mathrm{sc}}\right) \simeq \sigma_{\mathrm{m}}\left(\frac{\boldsymbol{k}_{\mathrm{in}}-\boldsymbol{k}_{\mathrm{sc}}}{2}\right)=\sigma_{\mathrm{m}}\left(k_0 \frac{\hat{\boldsymbol{k}}_{\mathrm{in}}-\hat{\boldsymbol{k}}_{\mathrm{sc}}}{2}\right). \tag{14.19}$$

The formula (14.19) is important for BRCS evaluation of real targets whose characteristic dimensions (feature sizes) are much larger than the bistatic radar wavelength. Unfortunately, it is impossible to determine a single boundary of application of the formula (14.19) for different targets.

Again, MBET accuracy should be linked to that of the originating theory (i.e. the physical optics model). Crispin's MBET is expected to perform better for smaller bistatic angles and specular reflections. It is not expected to be particularly adept at predicting nonspecular effects (surface waves in particular) and therefore bistatic scenarios in which these effects compete should demonstrate the MBET's limitations.

As mentioned in Chapter 10, the scattering field from a target of a complex shape can, in many cases, be represented as the sum of fields from several discrete scattering centres ('flare spots'). It is usually valid if characteristic dimensions of a target are much larger than the wavelength. This representation is one of the simplest and widely practised target models. The scattering centres are usually either specular centres on singly or doubly curved surfaces, or are actual geometric discontinuities of the target (e.g. boundaries, points of inflection and so on). In the first case their positions on the target surface depend on the illumination and reception directions, whereas in the second case they are rigidly connected with those discontinuities. It is assumed here that the directivity patterns of discrete scatters are sufficiently wide. The MRCS of such a target can be written as [II.4]

$$\sigma_{\mathrm{m}}(\boldsymbol{k}_{\mathrm{in}}) = \left| \sum_{n=1}^{N} \sqrt{\sigma_n} \exp\left[\mathrm{j}\left(2k_0 \boldsymbol{r}_n \hat{\boldsymbol{k}}_{\mathrm{in}} + \varphi_n \right) \right] \right|^2, \tag{14.20}$$

where σ_n is the RCS of the nth scattering centre for the polarization of the receiving part of a radar, $\boldsymbol{r}_n$ is the radius vector of the nth scattering centre and φ_n is the initial phase shift introduced by the nth scattering centre into the reflected field. The BRCS of the target can be expressed in the form

$$\begin{aligned}\sigma_{\mathrm{b}}(\boldsymbol{k}_{\mathrm{in}}, \boldsymbol{k}_{\mathrm{sc}}) &= \left| \sum_{n=1}^{N} \sqrt{\sigma_n} \exp\left\{ \mathrm{j}\left[2k_0 \boldsymbol{r}_n \hat{\boldsymbol{e}}_{\beta/2} \cos\left(\frac{\beta}{2}\right) + \varphi_n \right] \right\} \right|^2 \\ &= \left| \sum_{n=1}^{N} \sqrt{\sigma_n} \exp\left\{ \mathrm{j}\left[\frac{4\pi \boldsymbol{r}_n \hat{\boldsymbol{e}}_{\beta/2}}{\lambda \csc(\beta/2)} + \varphi_n \right] \right\} \right|^2,\end{aligned} \tag{14.21}$$

where $\hat{\boldsymbol{e}}_{\beta/2}$ is the unit vector in the direction of the bistatic bisector. In the general case, N, σ_n, $\boldsymbol{r}_n$ and φ_n in Equation (14.21) may be not the same as in Equation (14.20).

However, at the constant target aspect in the direction of $\hat{\boldsymbol{e}}_{\beta/2}$, the values of N, σ_n, $\boldsymbol{r}_n$ and φ_n often do not change when β increases from zero up to a certain value, $\beta_{\max}$. It follows from the comparison of Equation (14.20) with (14.21) that in these cases the BRCS of a target, σ_{b}, at the wavelength λ is equal to its MRCS, σ_{m}, for a radar located at the bisector of the bistatic angle, β, and working at the wavelength λ csc(β/2).

Thus, for a target that can be represented as a set of point scatters and under the conditions considered, the MBET is also valid. In this case the MBET is known as the Kell theorem: Kell [II.89] proposed a general scattering centre-based monostatic–bistatic equivalence theorem in 1965. It relates the BRCS to the monostatic measured along the bisector at a frequency proportional to the cosine of half the relevant bistatic angle. The relationship between the MRCS and the BRCS according to this theorem is given by

$$\sigma_{\mathrm{b}}(\boldsymbol{k}_{\mathrm{in}}, \boldsymbol{k}_{\mathrm{sc}}) \simeq \sigma_{\mathrm{m}}\left(\hat{\boldsymbol{e}}_{\beta/2} k_0 \cos\left(\frac{\beta}{2}\right) \right) = \sigma_{\mathrm{m}}\left(\frac{\pi}{c} \hat{\boldsymbol{e}}_{\beta/2} f_0 \cos\left(\frac{\beta}{2}\right) \right). \tag{14.22}$$

This form of the MBET is valid for small bistatic angles. According to Reference [II.4] it was verified for cones, rectangular flat plates, cylinders and spheres using the geometrical theory of diffraction.

The nature of scattering centres gives some indication of the applicability of this approximation. Differences between the MRCS and the BRCS can arise from one of three sources: (a) phase differences among scattering centres, (b) radiative changes from scattering centres and/or (c) changes in the nature or existence of scattering centres themselves [II.4, II.61, II.90]. If changes in any of these areas should be as small as a function of the changing bistatic angle, the monostatic and bistatic returns should be similar. In essence, the MBET should prove more accurate for regions in which an equipotential phase distribution is common, namely where direct specular-type reflections dominate. In general, these will occur for smaller bistatic angles. Because diffraction effects extend for a relatively large angular region, the MBET is also expected to predict the RCS fairly well for regions dominated by monostatic diffraction effects. At larger bistatic angles, the frequency-dependent nature of individual scattering centres is of greater significance than the equivalent frequency shift provided by the cos(β/2) term, and thus Kell's MBET should begin to fail. Certain nonspecular-type effects may begin to dominate the field scattering in the bistatic region. Shadowing features and those geometries that support multibounce effects could change the nature of individual scattering centres, thus leading to a disassociation between an actual and an MBET-computed RCS.

The maximum bistatic angle, β_{max}, to which the monostatic–bistatic transformation may be usefully extended, is limited by the angular width of individual scatterer patterns, which, in their turn, depend on the type and size of scatters involved. The angular width of the scatterer patterns is governed to a great extent by the size (in wavelengths) of the associated co-phase area. If a scatterer results from direct illumination only, the size of the associated co-phase area (in wavelengths) is usually small, and the corresponding angular pattern is broad. This is a so-called 'simple scatterer'. When a scatterer is a result of multiple reflections from adjacent parts of the target surface, as in a corner reflector, then the size of the associated co-phase area can be large, so that the corresponding angular pattern is narrow. This is a so-called 'reflex scatterer'. For small β the conditions of the 'statement of the monostatic–bistatic equivalence' are usually satisfied, e.g. if $\beta < 10^\circ$. For greater β, a careful examination of both the target geometry and the MRCS, σ_m, is required to judge whether the limits posed by the restricted angular reradiation from individual scattering centres are being exceeded. The MRCS envelope is useful in this examination, because narrow reradiation lobes are generally associated with a large RCS, and RCS envelope values in excess of a few square wavelengths suggest immediately that narrow lobes from reflex scatterers may be present [II.89]. The MBET cannot be applied to depolarizing targets.

It was shown by experiments that the MBET is useful for bistatic RCS evaluation of complex targets such as aircrafts and missiles if β is as large as 60–90° and in some cases even up to 130–150° [II.4]. However, it concerns not the 'fine structure' of a bistatic RCS as a function of the target aspect but the averaged values (e.g. medians), as well as maxima and minima over certain intervals of aspect angles [II.91]. For bistatic radars with short baselines (compared with the target range) when β is small, the bistatic RCS of a target may be considered as practically equal to its MRCS.

According to Reference [II.4], experimental data concerning the bistatic RCS of real targets are rather scanty in the literature. It was noted in Reference [II.90] that the bistatic RCS of a target can be both lower and higher than the monostatic RCS. For example, the measured bistatic RCS of an aircraft turned out to be lower on average by 2–5 dB [II.92] and in other cases even by 6–8 dB [II.93, II.94] than the MRCS. This can be explained by decreasing the contribution of specular reflections from scattering centres with narrow angular patterns (engine manifolds, wing joints, etc.). At the same time the bistatic target reradiation pattern is more smoothed and the lobes are broader. This result is in accordance with the 'equivalent' wavelength increase by the factor $\csc(\beta/2)$.

As was previously mentioned, both monostatic-to-bistatic equivalence theorems are expected to perform better for specular reflections, especially near the transmitter illumination angle. Like asymptotic solutions, they are not expected to perform well in situations where shadowing effects, surface waves or multibounce dominates, because the theory upon which each is based does not predict these phenomena well. They may predict diffraction effects fairly well due to their large angular distribution. However, the treatment of an object's signature as an average response over a limited angular or time extent may have been among the author's original intentions [II.88, II.89]. This suggests another way of viewing the MBET data. Specifically, if the data were averaged in some way to reflect an average signature better, the MBET performance may improve. With this in mind, each signature is averaged with a 5° sliding window and the ensuing analysis accompanies that of the untreated returns [II.61].

14.1.5.1 Object A

Object A, which has been shown in Figure 14.5, is reviewed first for its dominant specula and distinguishable diffraction component. The measured and MBET bistatic signatures at 8 and 14 GHz are

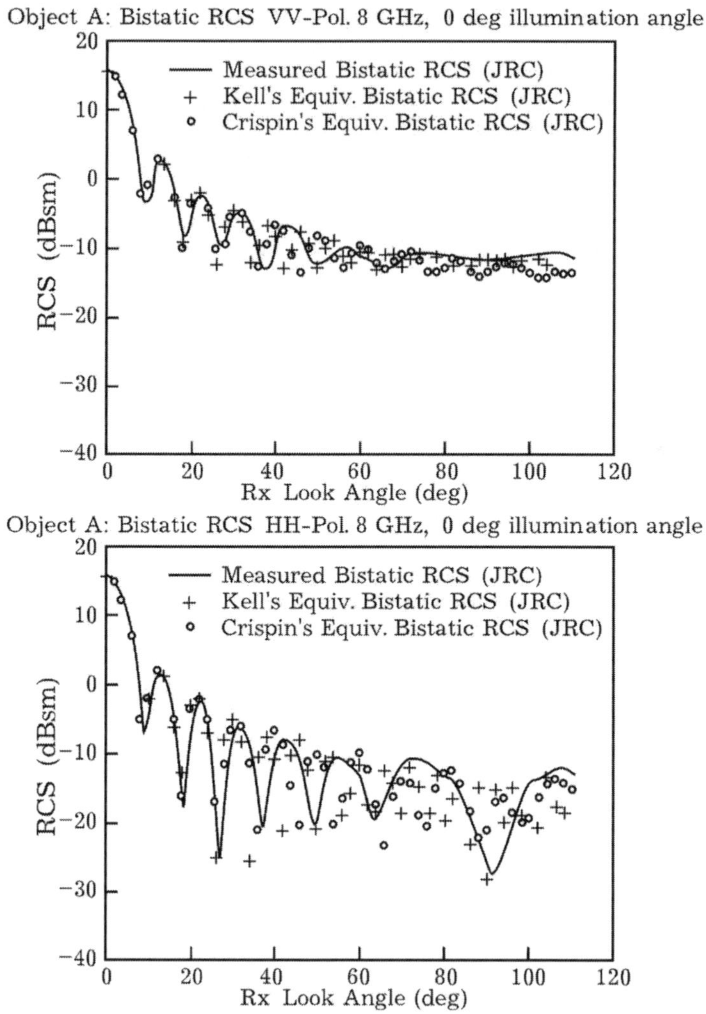

Figure 14.48 Object A: bistatic RCS MBET predictions versus measured (JRC) data, 8 GHz. (From Eigel *et al.* [II.61]. Reproduced by permission of © 2000 IEEE)

shown in Figures 14.48 and 14.49. Figures 14.50 and 14.51 show the same data averaged with a 5° sliding window. One immediately notes one of the primary limitations of Kell's theory in the 14 GHz plots of Figure 14.49; namely that due to the frequency shift accompanying the increase in bistatic angle, the workable data set for use with Kell's approximation must be larger than that used for Crispin's. Kell's MBET for an equivalent bistatic frequency of 14 GHz has a limited angular extent of 41° because the next higher angle's data corresponds to a measurement in the quasi-monostatic data set above 15 GHz. According to Reference [II.61], inspection of object A's measured and MBET signatures reveals good correlation through approximately 30–40° receiver look angle for both polarizations. The specular peaks predicted by both MBETs at 0° are close to the PO prediction (16.1 and 21.0 dB cm) and the lobing structure through approximately 30° azimuth is a close match, although the MBET predicted

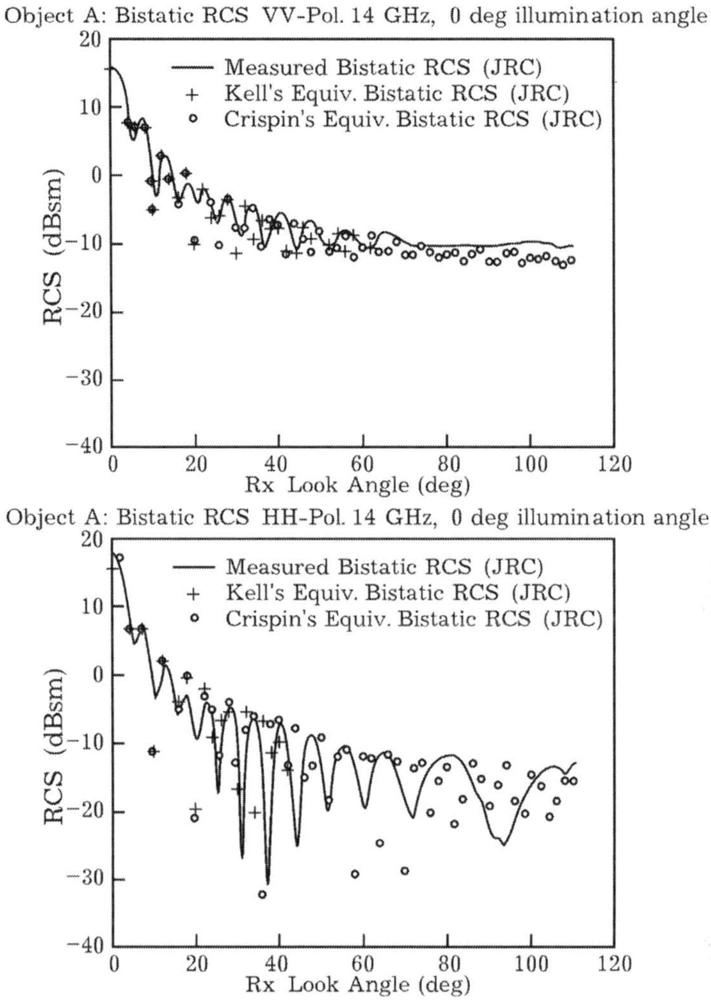

Figure 14.49 Object A: bistatic RCS MBET predictions versus measured (JRC) data, 14 GHz. (From Eigel *et al.* [II.61]. Reproduced by permission of © 2000 IEEE)

sidelobes seem to be slightly skewed towards the peak specula. The quasi-monostatic nature of the monostatic data set may account for some of the shift. Limitations of the MBETs probably become more significant at larger β angles. Beyond 40° both approximations appear equally poor at predicting the position or amplitude of the sidelobes. As the receiver nears 90° (edge-on), diffraction effects should be visible for VV polarization but not HH polarization. Both MBETs predict that the edge diffraction at VV polarization is lower than measured, with Crispin's being slightly worse. For HH polarization both MBETs predict amplitudes too high (by approximately 10 dB) at 90°, missing the null entirely. This is understandable considering where the MBET data originates. The MBET data near 90° is actually being generated from monostatic data near 45° (monostatic data along the bisector), where specular sidelobe data at HH polarization dominates.

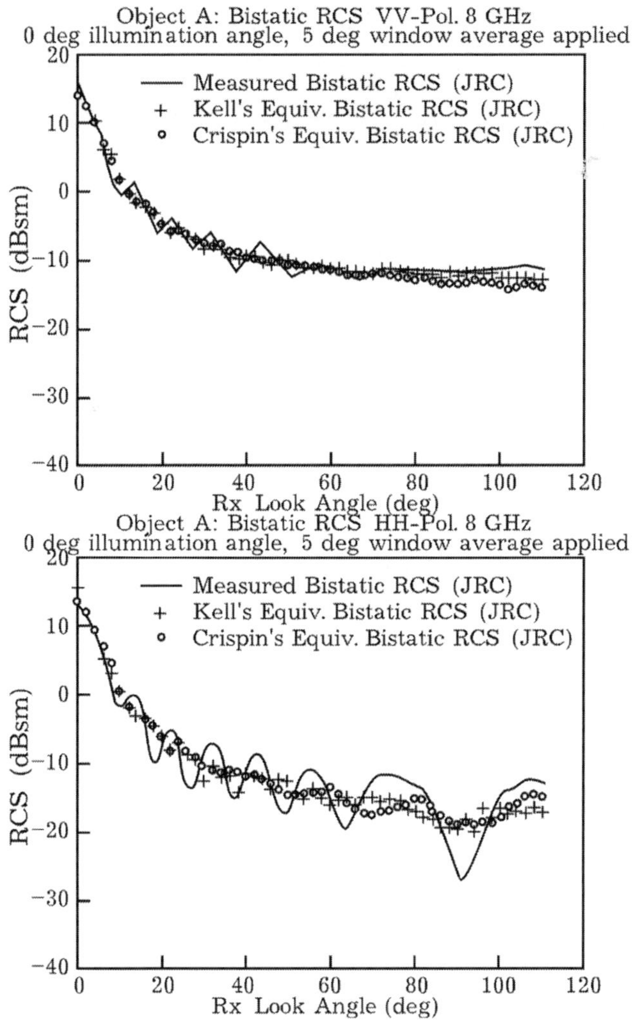

Figure 14.50 Object A: bistatic RCS MBET predictions versus measured (JRC) data, 8 GHz averaged with a 5° sliding window. (From Eigel *et al.* [II.61]. Reproduced by permission of © 2000 IEEE)

Table 14.1 from Reference [II.61] suggests that the MBET predictions tend to be lower overall than actual levels and that averaging tends to increase correlation when the window size increases for the entire region of interest. Caution should be exercised when interpreting these results [II.61]. The shifting of the MBET lobing structure and the averaging process itself could influence the eventual mean and standard deviation significantly, even though a subjective evaluation of the data reveals an improved correlation. For this reason the absolute values in Table 14.1 are deemed unimportant, but the trend between analogous values for different averaging conditions is worthy of discussion. For the entire 110° region in which data exist, the mean difference and standard deviation decrease for most situations as more data are averaged by increasing the window size. Overall, this trend appears

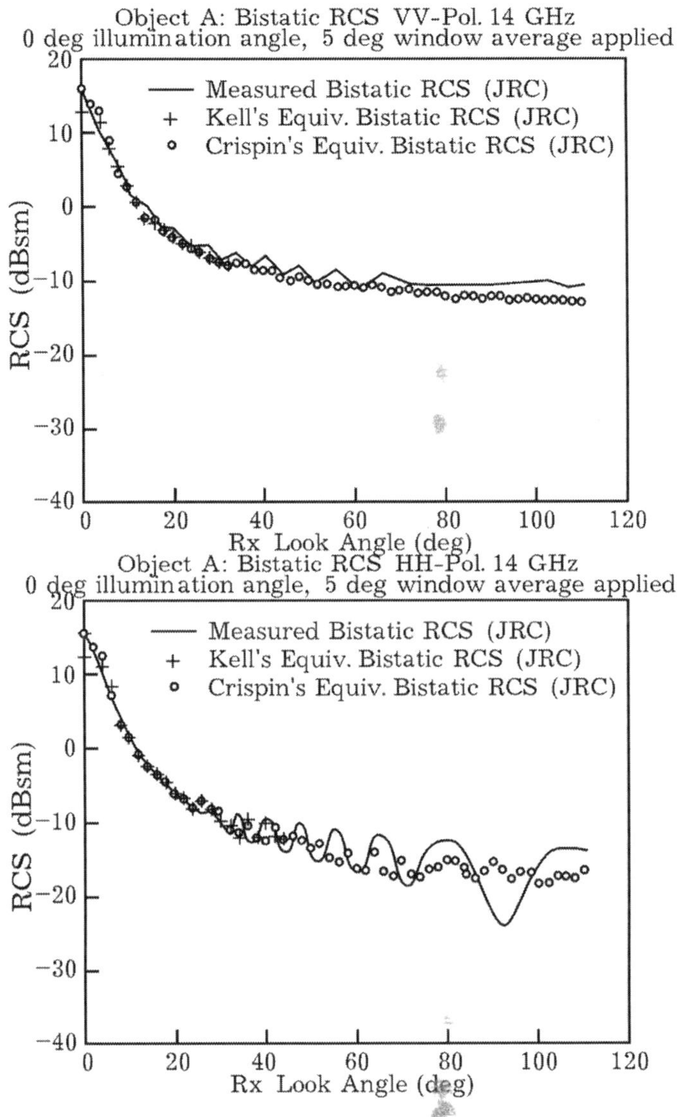

Figure 14.51 Object A: bistatic RCS MBET predictions versus measured (JRC) data, 14 GHz averaged with a 5° sliding window. (From Eigel *et al.* [II.61]. Reproduced by permission of © 2000 IEEE)

to be correct, but more insight may be gathered by looking at the amplitude behaviour within certain regions.

Averaging the data improves correlation for certain regions of the plot. The net effect of averaging this type of data is to reduce peak amplitudes of each lobe and to fill in the nulls. This increases the correlation for the first 30° of azimuth, but does not near 90° for either frequency or polarization. Notice that as the data are averaged, Kell's and Crispin's MBET solutions become more similar for comparable polarizations and frequencies.

Table 14.1 Object A: mean and standard deviation (SD) of the difference between measured and MBET amplitudes, $0° < \beta < 110°$. (From Eigel *et al.* [II.61]. Reproduced by permission of © 2000 IEEE)

		8 GHz		14 GHz	
MBET	Polarization	Mean	SD	Mean	SD
Kell's	VV	−0.4520	2.1038	−1.2523	2.8487
	HH	−0.7228	5.2145	−1.2550	5.2975
Crispin's	VV	−0.7384	1.9067	−1.2810	2.0229
	HH	−0.3987	4.2186	−1.1062	6.2325
5° window average					
Kell's	VV	−0.2887	1.3539	−0.4245	1.2473
	HH	−0.2082	3.3084	0.3504	1.3983
Crispin's	VV	−0.5713	1.4931	−1.0174	1.1400
	HH	−0.840	3.1404	−0.2718	2.7822
9° window average					
Kell's	VV	−0.2590	0.8255	−0.1740	0.7957
	HH	−0.2584	2.4867	0.6378	0.6378
Crispin's	VV	−0.4452	1.2189	−0.8707	1.0487
	HH	−0.0219	2.4815	−0.1355	2.1716

14.1.5.2 Objects B and C

Objects B and C (see Figures 14.7 and 14.8) are analysed by Eigel *et al.* [II.61] in tandem to differentiate between the specular and nonspecular interactions dominating each scattered signature. The analysis often refers to the preceding software investigation to help explain and quantify MBET performance. The same scattering mechanisms covered there apply here.

Figures 14.52 and 14.53 show object C's VV polarization MBET predicted patterns at 8 and 15 GHz for $0° < \beta < 110°$ for the raw data and the data with a 5° sliding average applied. Analogous data for object B are shown in Figures 14.54 to 14.57 at 8 and 12 GHz for both polarizations. An RF of 12 GHz for object B is chosen to allow Kell's MBET to predict a larger angular region than was accomplished for object A. Object C's data are generated from the MRC monostatic data set, so a higher RF can be accommodated at the cost of the second polarization. Tables 14.2 and 14.3 list the mean and standard deviation of the entire 110° region for objects C and B respectively.

Because the MBETs are derived from asymptotic principles similar to those incorporated into the PTD model, there should be some similarity in the data analysis. Similarities should be confined to smaller bistatic angles, since the MBET formulae are only appropriate here. If the logic follows, the MBET predictions for this region will be slightly higher than measured, and the correlation should improve if the data are averaged. It is also expected that the correlation between the MBET and measured data is greater for object C than for object B, as was previously shown. Again, object C is reviewed first in Reference [II.61].

In Figures 14.52 and 14.53, the receiver collects scattering from two primary specula sources, the cylinder bodies for the $0° < \beta < 110°$ region. The MBET data exhibits good correlation to the measured data, especially near the transmitter illumination angle, as expected. Higher RF predictions are slightly lower here than was observed in the software predictions, and this lower trend characterizes both MBETs at larger β angles. In Table 14.3, again the absolute values of the mean and standard deviation are ignored, but the averaging trend is studied. Averaging the data seems to have less of an effect than it did for object A. It seems that for complex objects whose bistatic signature is dominated

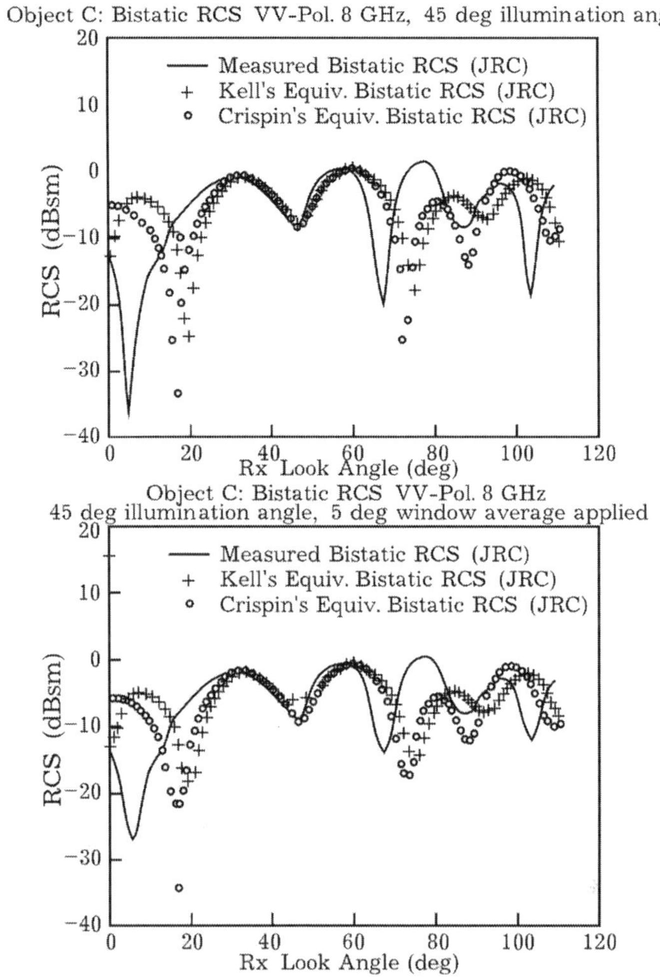

Figure 14.52 Object C: bistatic RCS MBET predictions versus measured (JRC) data, 8 GHz without/with a 5° sliding average applied. (From Eigel *et al.* [II.61]. Reproduced by permission of © 2000 IEEE)

by two wide-angle specular interactions, both MBETs can predict a reasonably accurate RCS at lower frequencies where the specular lobe widths are wider. The extent of this capability is limited to bistatic angles of less than 15–20°. As the frequency increases, the lobe widths become narrower and the MBETs begin to fail, tending to predict lower RCS than should be expected. The bistatic angles for which the approximations work is correspondingly narrowed. The nature of the scattering centres as perceived from a bistatic perspective also changes as a function of the bistatic angle. As the bistatic angle increases, the disassociation between monostatic and bistatic scattering centres becomes more apparent, leading to greater discontinuity between measured data and MBET predictions. Averaging the data provides minimal improvement to the correlation between the measured and MBET predicted data sets.

In Figures 14.54 to 14.57, the receiver looks at the flat plate of object B for the first 45° and at the cylinder bodies for larger β angles. Correlation between the MBET predictions and the measured data is

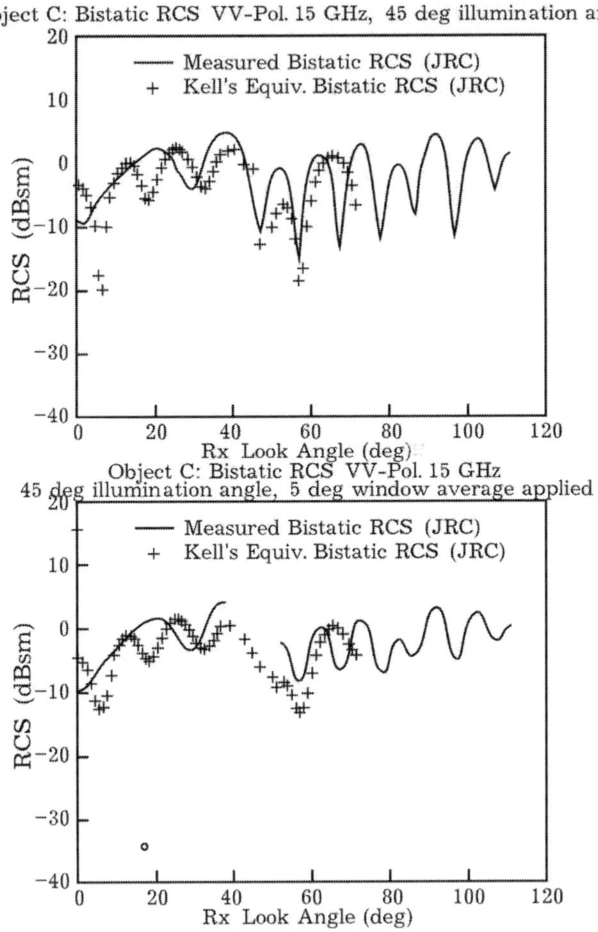

Figure 14.53 Object C: bistatic RCS MBET predictions versus measured (JRC) data, 15 GHz without/with a 5° sliding average applied. (From Eigel et al. [II.61]. Reproduced by permission of © 2000 IEEE)

expected to be less than that witnessed for object C, for many of the same reasons given in the software analysis. This is in fact the case. Discrepancies between the MBET and measured data increase as the β angle (starting at 0°) approaches the transmitter illumination angle for VV polarization. Both MBET-predicted lobing structures near 0° seem shifted, just as they did for object A's sidelobe structure. As β approaches α, the MBETs predict high, just as expected. The HH polarization patterns correlate much better than PTD predictions, but for small bistatic angles. This is probably due to the fact that the MBET plots are extrapolated from measured monostatic data, which captures the true scattering phenomena. Lower-order effects, which could not be computed by the software, are present in the measured data and are incorporated into the MBET bistatic prediction. Logically, the higher RF plots should demonstrate greater correlation as some of the nonspecular effects become smaller.

However, the useful range of both MBETs is limited to bistatic angles of approximately 10° for VV polarization and 15° at HH polarization for the higher RF. The large shadowing geometry produces

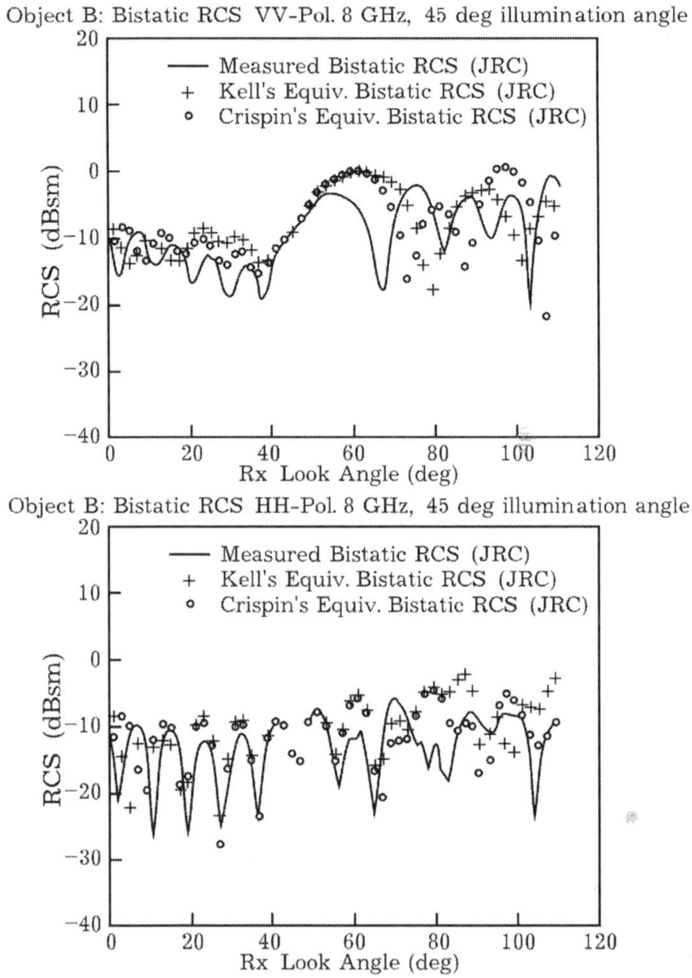

Figure 14.54 Object B: bistatic RCS MBET predictions versus measured (JRC) data, 8 GHz. (From Eigel *et al.* [II.61]. Reproduced by permission of © 2000 IEEE)

even greater changes in the nature of each scattering centre from monostatic and bistatic perspectives than is evident for object C. In other words, the monostatic data (from which the MBETs are computed) arise from scattering centres with even less similarity to the true bistatic scattering centres, which are produced by the single specular and nonspecular interactions. Table 14.3 again suggests that averaging the data does not improve the MBET correlation. In fact, it may diminish the correlation in certain situations. It can be concluded that whenever nonspecular effects contribute markedly to the overall signature, averaging techniques should not be used to improve the MBET performance.

Thus, the MBET performance is related to the complexity of the object studied. Simple objects are those whose scattered signature is dominated by a single specular mechanism from any particular vantage point. Complex objects fall into one of two categories: those whose RCS is dominated by a combination of specular interactions and those whose RCS is derived from specular and nonspecular components

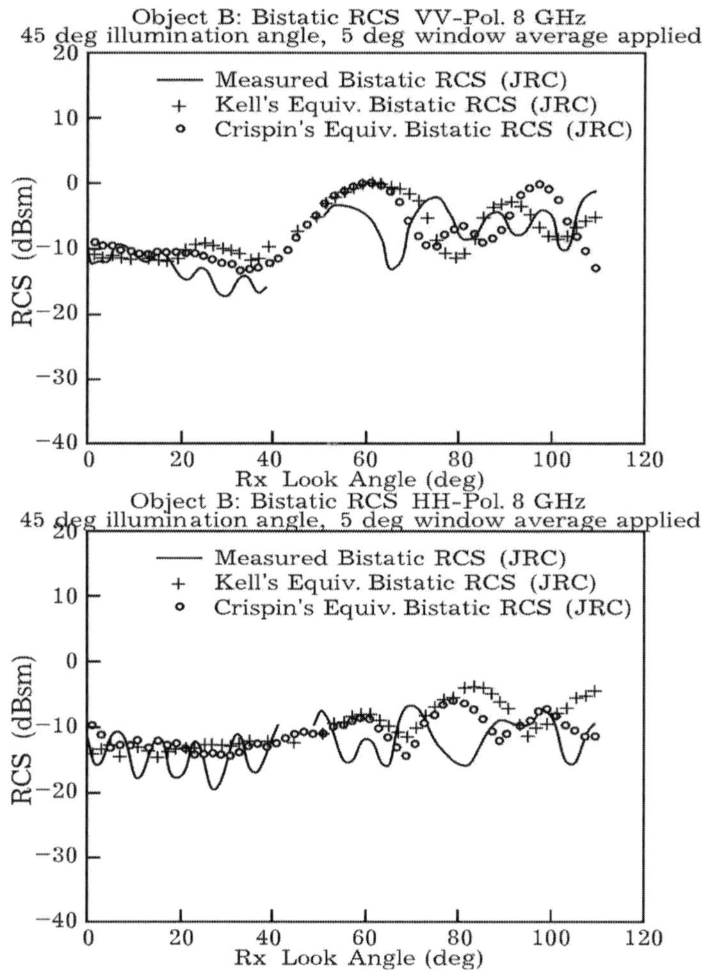

Figure 14.55 Object B: bistatic RCS MBET predictions versus measured (JRC) data, 8 GHz averaged with a 5° sliding window. (From Eigel *et al.* [II.61]. Reproduced by permission of © 2000 IEEE)

of similar amplitude. The former is referred to as 'minimally complex' and can be characterized by a geometry that generally incorporates canonical structures, producing large specular reflections. The latter is referred to as 'rigorously complex' and may incorporate large shadowing features, cavities or smoothly sloped surfaces. These features may produce multibounce, diffraction, surface waves, etc., with amplitudes analogous to any specula.

Kell's and Crispin's MBETs work well for simple geometries for at least bistatic angles of 30° (sidelobe structure and amplitudes); Crispin's has a slight edge over Kell's in general, but Kell's has a slight advantage when diffraction effects are present. Both predict edge diffraction effects lower on average than measured (2–3 dB), and averaging the data improves correlation with measured data.

For minimally complex objects, MBET accuracy diminishes to only bistatic angles less than 15–20°. Both tend to predict a bistatic RCS lower than measured for bistatic angles >15° and at higher RFs. This is primarily due to a narrowing of monostatic specular spikes at higher RFs and the changing

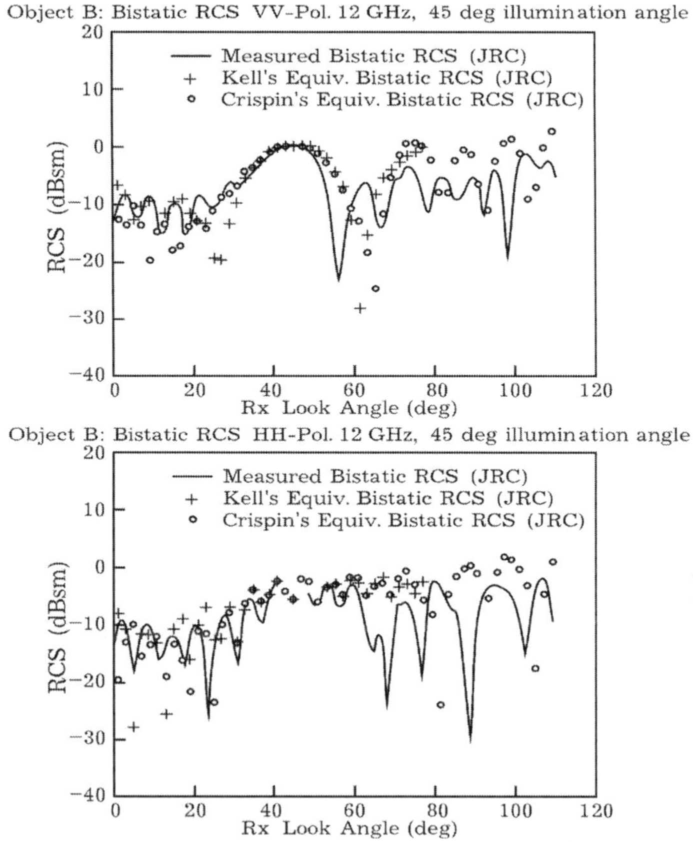

Figure 14.56 Object B: bistatic RCS MBET predictions versus measured (JRC) data, 12 GHz. (From Eigel *et al.* [II.61]. Reproduced by permission of © 2000 IEEE)

nature of the scattering centres as the bistatic angle increases. Averaging does not improve correlation with measured data sets nearly as much as it did for simple shapes, and may even decrease it in some instances.

This trend continues for rigorously complex objects. Both MBETs demonstrate reasonable accuracy for bistatic angles of no more than 5–10° and both tend to predict the RCS higher than measured for bistatic angles >10° and at higher RFs. Again, the primary reason for poor correlation is the changing nature of the scattering centres. From these types of objects, the scattering centres change more rapidly as a function of the bistatic angle because they are derived from roughly equivalent specular and nonspecular components. Averaging data does not improve data correlation with measured data.

The overall performance of Kell's and Crispin's MBETs are quite similar for all the tested objects, but there are some limitations associated with using Kell's formula. Kell's MBET requires a much larger data set than Crispin's to predict an equivalent bistatic RCS matrix. Where as a single monostatic pattern cut (i.e. single RF) can be used to predict the bistatic RCS through Crispin's formula, Kell's requires a large RF bandwidth monostatic measurement (at a very fine frequency resolution) to accomplish the same goal. The frequency shift also contributes to Kell's poor angular resolution near the transmitter illumination angle. Finally, extracting Kell's bistatic RCS from monostatic data sets proves to be much more computationally intensive than Crispin's.

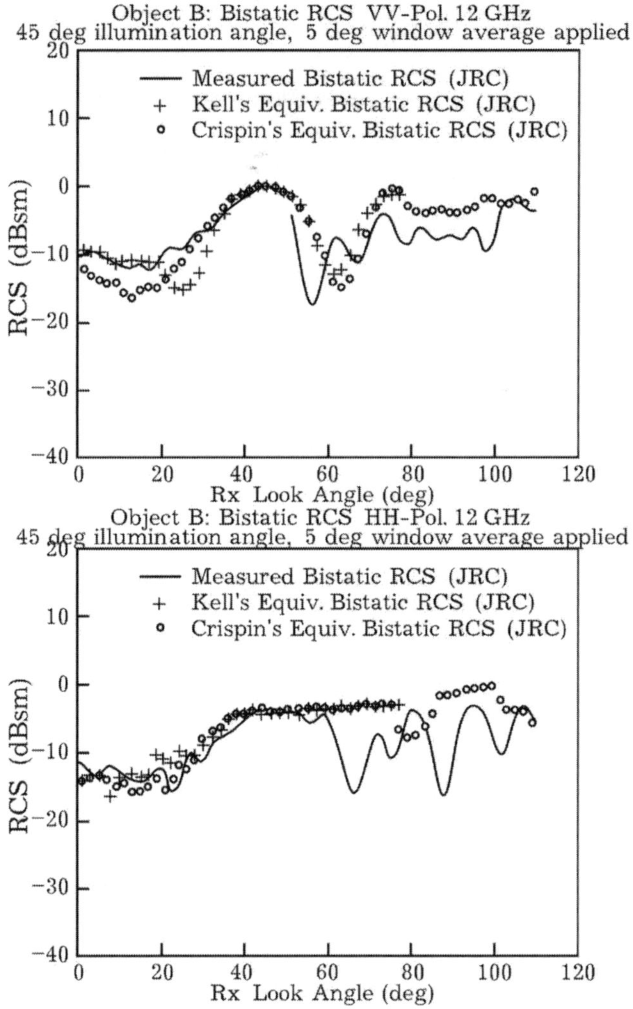

Figure 14.57 Object B: bistatic RCS MBET predictions versus measured (JRC) data, 12 GHz averaged with a 5° sliding window. (From Eigel *et al.* [II.61]. Reproduced by permission of © 2000 IEEE)

14.2 EXAMPLES OF BISTATIC CROSS-SECTIONS

Many examples of the BRSC have already been given in this part. Nevertheless, some new examples will be added in the following section.

14.2.1 Simple Shapes

In Reference [II.95] the radar cross-section of an arbitrarily shaped, homogeneous dielectric BOR was evaluated by surface integral equation (SIE) formulation and by the MOM. The method's accuracy is verified by good agreement with the exact solutions for the RCS of a dielectric sphere. Wu's solution

Table 14.2 Object C: mean and standard deviation (SD) of the difference between measured and MBET amplitudes, $0^\circ < \beta < 110^\circ$. (From Eigel *et al.* [II.61]. Reproduced by permission of © 2000 IEEE)

MBET	Polarization	8 GHz Mean	8 GHz SD	15 GHz Mean	15 GHz SD
Kell's	VV	−0.8874	9.2136	−1.5727	5.0074
Crispin's	VV	−0.5670	8.7628	−2.0441	5.8282
5° window average					
Kell's	VV	0.8183	7.9199	−1.5165	3.6837
Crispin's	VV	−0.0359	7.4203	−2.1352	5.4003
9° window average					
Kell's	VV	0.8855	6.7135	−1.3567	2.5098
Crispin's	VV	0.0604	6.3712	−1.7037	3.4829

is plotted in Figure 14.58 along with the Mautz–Harrington solution [II.96] and the exact Mie series solution.

To demonstrate the versatility of the SIE technique to analyse arbitrarily shaped BORs, the RCS for a human torso model illuminated by a broadside incident plane wave was computed in Reference [II.95] and plotted in Figure 14.59. This BOR model (Figure 14.59 inset) consists of a muscle hemisphere head, cylinder neck and conical torso. Consider the composite dielectric and conducting BOR, as depicted in the inset of Figure 14.60. It consists of a 1λ-long, fixed-radius ($a = 0.021\lambda$) conducting cylinder and a 1λ-long, varied-radius lossy dielectric BOR ($\sigma = 0.1$ and 1 S/m). Figures 14.60 and 14.61 shows the bistatic RCS of the composite BOR, which is pictured in Figure 14.62. Wu [II.95] noted that a slightly higher RCS is obtained from $\sigma = 1$ S/m than for $\sigma = 0.1$ S/m near the back-scattering direction ($\theta = 90^\circ$).

Table 14.3 Object B: mean and standard deviation (SD) of the difference between measured and MBET amplitudes, $0^\circ < \beta < 110^\circ$. (From Eigel *et al.* [II.61]. Reproduced by permission of © 2000 IEEE)

MBET	Polarization	8 GHz Mean	8 GHz SD	12 GHz Mean	12 GHz SD
Kell's	VV	1.8343	5.4581	0.3325	6.8510
	HH	3.0669	5.4992	0.2441	5.6675
Crispin's	VV	1.3639	6.4258	0.5953	6.2132
	HH	1.9024	5.3804	2.3513	7.7804
5° window average					
Kell's	VV	1.7648	4.1543	0.6208	4.1443
	HH	2.9753	4.0958	2.3818	3.7144
Crispin's	VV	1.4370	4.2663	0.4485	5.5401
	HH	1.9894	3.5521	1.9383	6.4762
9° window average					
Kell's	VV	1.6850	3.2285	0.7658	2.7382
	HH	2.9874	2.6570	2.4854	3.0734
Crispin's	VV	1.4909	2.6556	0.9368	3.0818
	HH	1.9617	1.8444	2.5387	3.2832

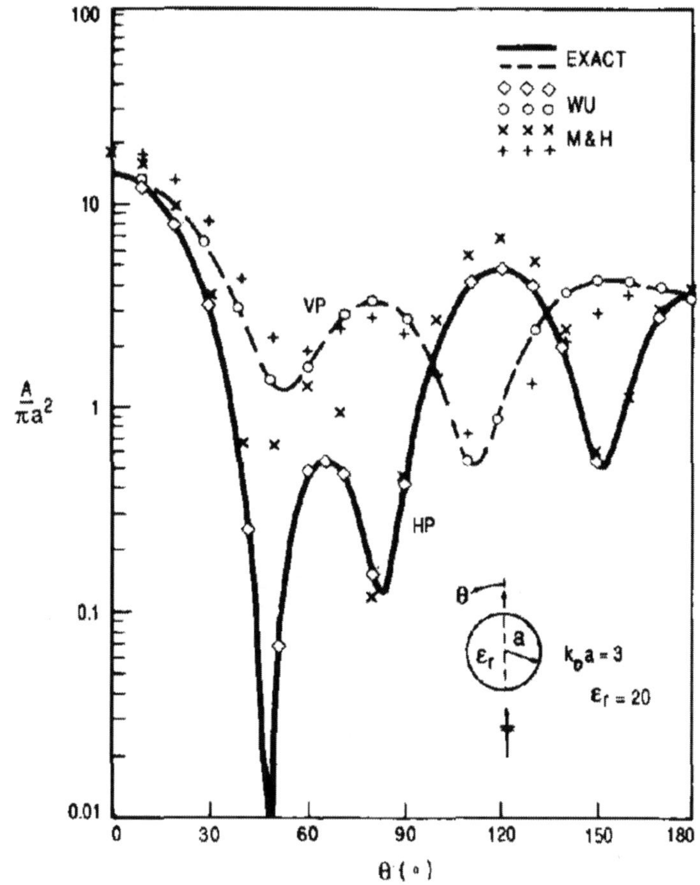

Figure 14.58 Bistatic RCS for a dielectric sphere ($k_0a = 3$, $\varepsilon_r = 20$) with an axially incident plane wave. (From Wu [II.95]. Reproduced by permission of © 1989 IEEE)

Figure 14.61 shows the bistatic RCS of the Figure 14.62 composite BOR with a 30° illumination. More resonances and a higher RCS can be observed for a higher conductivity case ($\sigma = 1$ S/m).

In Reference [II.97], the scattering from a perfectly conducting cube was examined. The cube epitomizes the complex, three-dimensional scatterer, with its multiple interactions and vertex diffraction playing a critical part in the far-field patterns of some bistatic planes. The results presented in Reference [II.97] were for a cube of the order of 7.5–3 wavelengths, the edge of which is illuminated by a plane wave at broadside incidence. The method employed was the hybrid iterative method (HIM), which utilizes an initial approximation of the surface currents on the cube faces. These currents are inserted into the magnetic-field integral equation (MFIE) to produce improved or updated approximations to these surface currents. This process is repeated to convergence by the method of successive approximations. The currents were then used to find the bistatic radar cross-section for an arbitrary plane of measurement. As the iteration process progressed, it was seen that all physical scattering processes present in this body are introduced by enforcing the MFIE.

The bistatic RCS was calculated in Reference [II.97] for both initial and first-order currents. These results (both E-plane and H-plane as described in Figure 14.63) were then compared to the contemporary

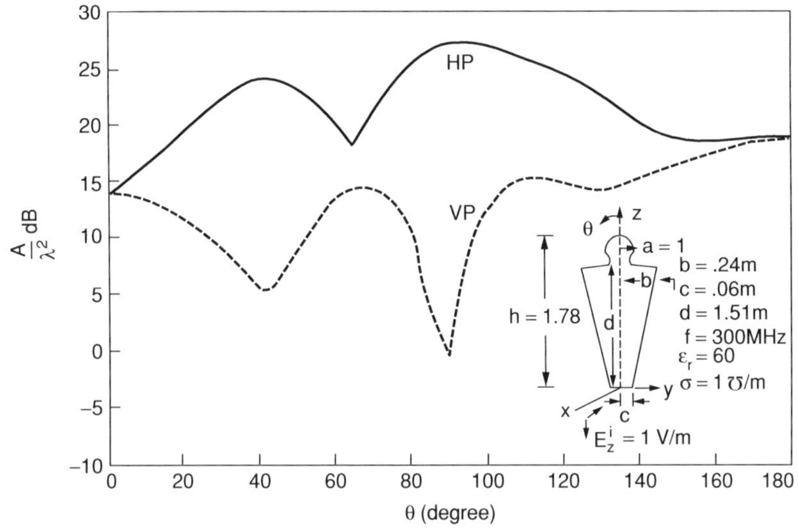

Figure 14.59 Bistatic RCS for a human torso model with a broadside incident plane wave. (From Wu [II.95]. Reproduced by permission of © 1989 IEEE)

analytical results of Yaghjian [II.98] and the measured data compiled by McCahan [II.99] for three sizes of cubes: $w = 0.755\lambda$, 1.5075λ and 3.015λ. Some of these results are shown in Figures 14.64 to 14.67. In all of these results, the back-scattered RCS corresponds to the angle $\theta = 180°$. It can be seen that the initial currents alone provide excellent results. The higher-order currents were seen to provide equally reliable results. By this is meant that the HIM results corresponded very closely, while the general shape

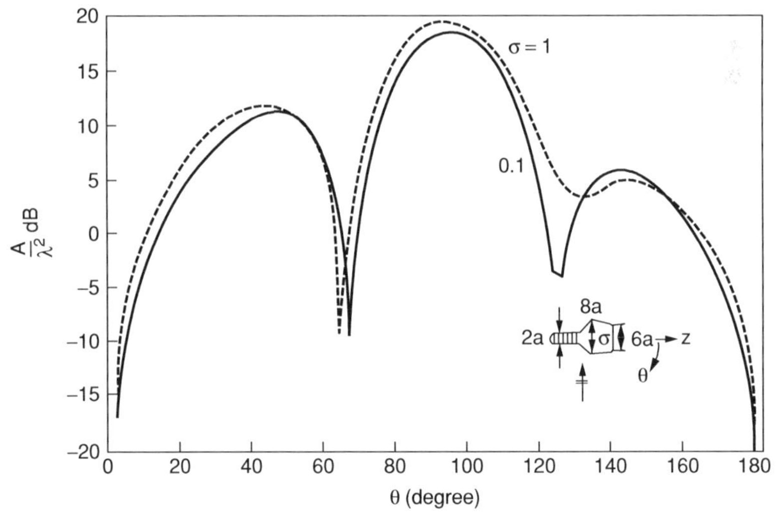

Figure 14.60 Bistatic RCS for the composite metal and dielectric body of revolution with broadside incidence. (From Wu [II.95]. Reproduced by permission of © 1989 IEEE)

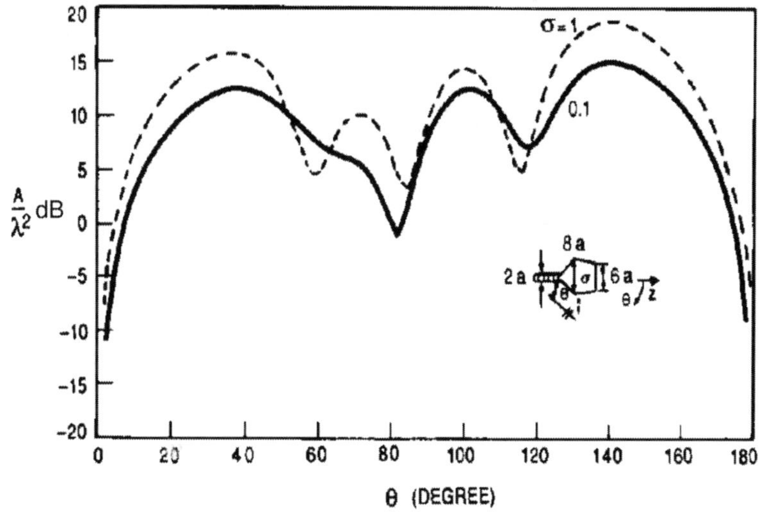

Figure 14.61 Bistatic RCS for the composite metal and dielectric body of revolution with an incident angle equal to 30°. (From Wu [II.95]. Reproduced by permission of © 1989 IEEE)

(relative maximums and minimums) was retained with respect to both Yaghjian's analytic results and McCahan's measured data. The single case of concern was the H-cut for $w = 0.755\lambda$ where the second-order results tended to deviate significantly from the first-order results. However, third-order data were calculated and were seen to corroborate with the first-order results. The fact that these higher-order results were nondiverging supports the conclusion that the first-order results were accurate to within the experimental error of the measured data, as well as the numerical and analytic accuracy of the HIM.

Figure 14.68 from Reference [II.100] plots the RCS of a conducting cylinder with a circular base, 2.60 m long and 0.40 m in diameter, at 14 GHz, for the following bistatic near-field configuration:

- cylinder axis parallel to the horizontal plane;
- analysis performed on ±5° around the direction normal to the cylinder generator (i.e. 90°);
- distance between the receiver and the target equal to 54.64 m;
- distance between the transmitter and the target equal to 24.82 m;
- angle between the transmission direction and the horizontal plane equal to +12°;
- angle between the reception direction and the horizontal plane equal to −28°.

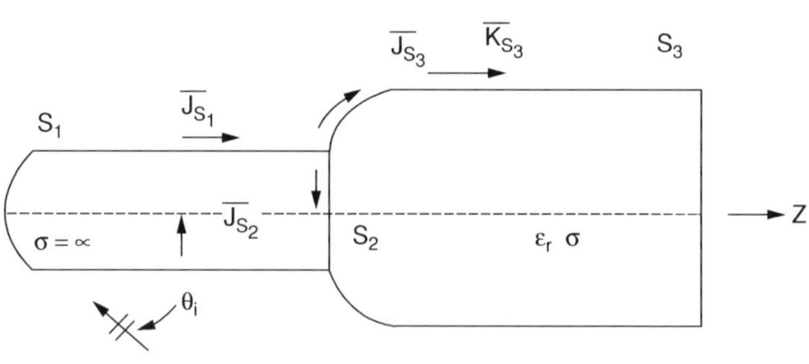

Figure 14.62 Geometry of a composite metal and dielectric body of revolution. (From Wu [II.95]. Reproduced by permission of © 1989 IEEE)

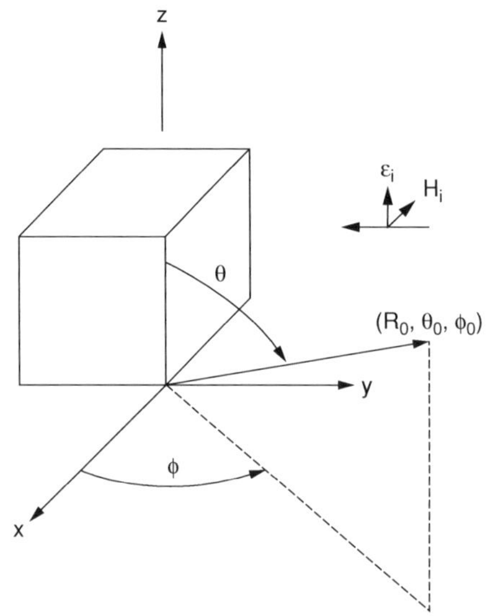

Figure 14.63 Geometry used in the evaluation of the RCS of the cube. (From Penno. *et al.* [II.97]. Reproduced by permission of © 1989 IEEE)

The bistatic angle is equal to 40° in the vertical plane. The grey curve corresponds to the measured RCS and the black curve gives the theoretical RCS. Good agreement is seen between the two results with differences below 0.5 dB on the maxima. The dashed curve plots the RCS of the same target in the far field.

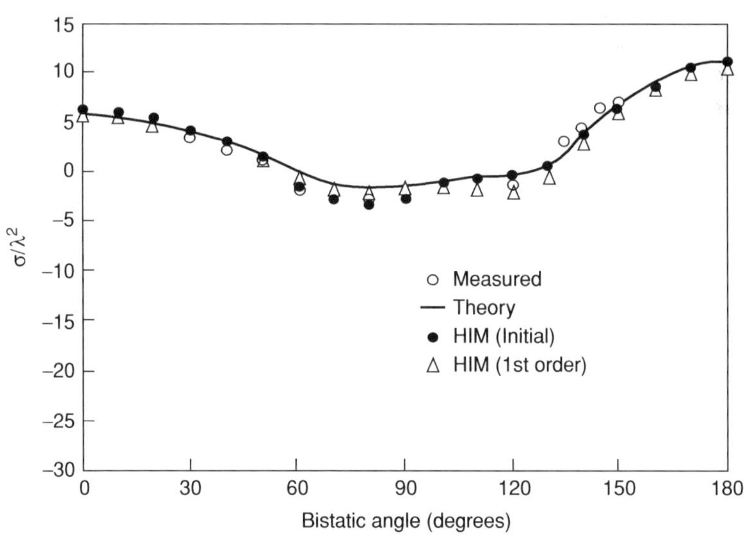

Figure 14.64 H-plane bistatic scattering from a metal cube of width 0.755λ. (From Penno *et al.* [II.97]. Reproduced by permission of © 1989 IEEE)

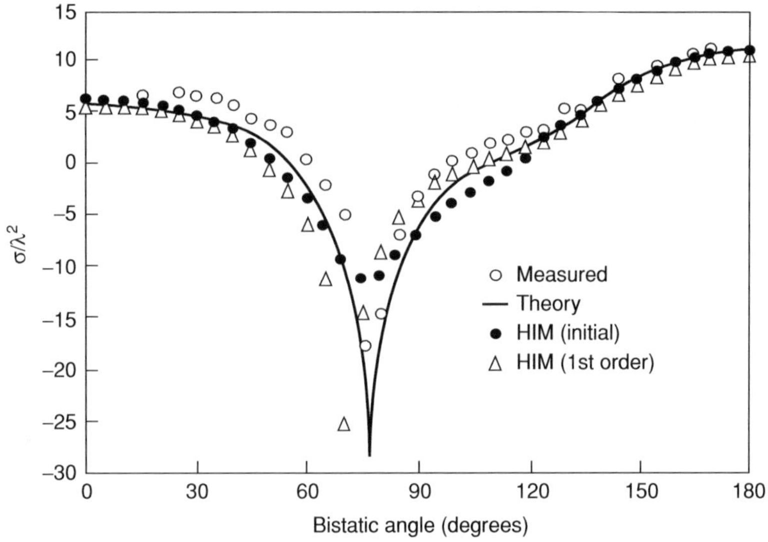

Figure 14.65 E-plane bistatic scattering from a metal cube of width 0.755λ. (From Penno *et al.* [II.97]. Reproduced by permission of © 1989 IEEE)

Robillard and Herrett [II.101] of CM-Delco Electronics presented data on the bistatic RCS of re-entry vehicles for principal and opposite circular polarized scattering from 8° half-angle cones, for bistatic angles from 30 to 160°. Their research covered four cone sizes ($D/\lambda = 0.22, 0.42, 0.84$ and 1.95) and nose bluntness variations. Figures 14.69 and 14.70 show the BRCS versus turntable angle at bistatic angles of 180° and 70° for an 8° cone, base diameter 35 mm and $\lambda = 17.9$ mm. The RCS variation with

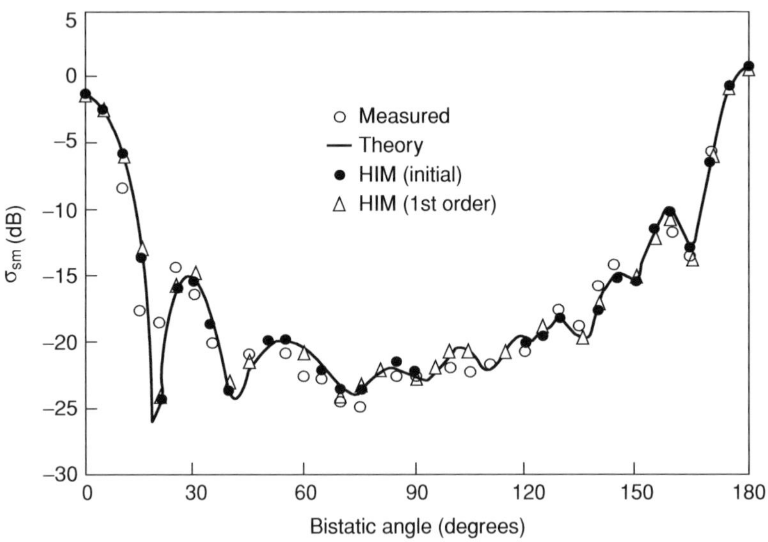

Figure 14.66 H-plane bistatic scattering from a metal cube of width 3.015λ. (From Penno *et al.* [II.97]. Reproduced by permission of © 1989 IEEE)

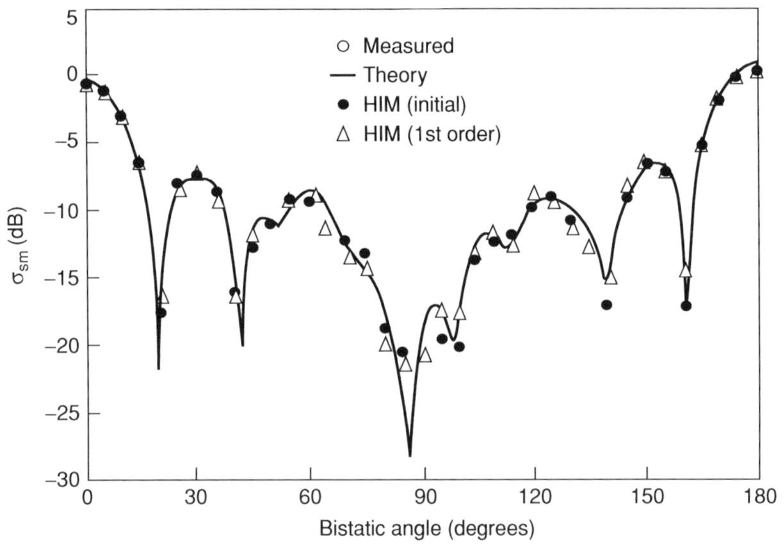

Figure 14.67 E-plane bistatic scattering from a metal cube of width 3.015λ. (From Penno *et al.* [II.97]. Reproduced by permission of © 1989 IEEE)

aspect angle in the forward-scatter case is only 7 dB, while the variation in the 70° case is more than 30 dB. Figure 14.71 shows these characteristic variations as a plot of the bistatic RCS versus the bistatic angle [II.102].

14.2.2 Random Rough Surfaces

The scattering of electromagnetic waves by rough surfaces may be described using the models presented in Section 10.3. In Figure 14.72 from Reference [II.103] the validity conditions of the stationary phase (geometrical optics), scalar approximation (physical optics) and small perturbation models are shown. Khenchaf *et al.* noticed that the scalar approximation and the small perturbation model can be simultaneously applied for $k_0L > 6$ and $k_0a < 0.3$. The stationary phase and scalar approximation can be applied, according to Khenchaf *et al.*, concurrently for $k_0L > 17$ and $ka > 3$.

In the case of processing with the Gaussian autocorrelation function, the validity conditions for the Kirchhoff model can be written as [II.103, II.104]

$$k_0L > 6 \quad \text{and} \quad k_0L > 2\sqrt{k_0\sigma\sqrt{6\pi}} \cong 4.17\sqrt{k_0\sigma}. \tag{14.23}$$

When the surface height is Gaussian distributed with zero mean, the validity conditions for the stationary phase model are [II.103]

$$k_0L > 2\sqrt{k_0\sigma\sqrt{6\pi}}, \quad k_0L > 6 \quad \text{and} \quad k_0\sigma > \frac{\sqrt{10}}{|\cos\theta_{sc} + \cos\theta_{in}|}. \tag{14.24}$$

According to Khenchaf *et al.*, after using the stationary phase approximation to simplify the scattered field expression, the scattering coefficient can be written as [II.103]

$$\sigma_{sc} = \frac{\pi\left(k_0q\left|U_{pq}\right|\right)^2}{q_z^4}P\left(h_x, h_y\right), \tag{14.25}$$

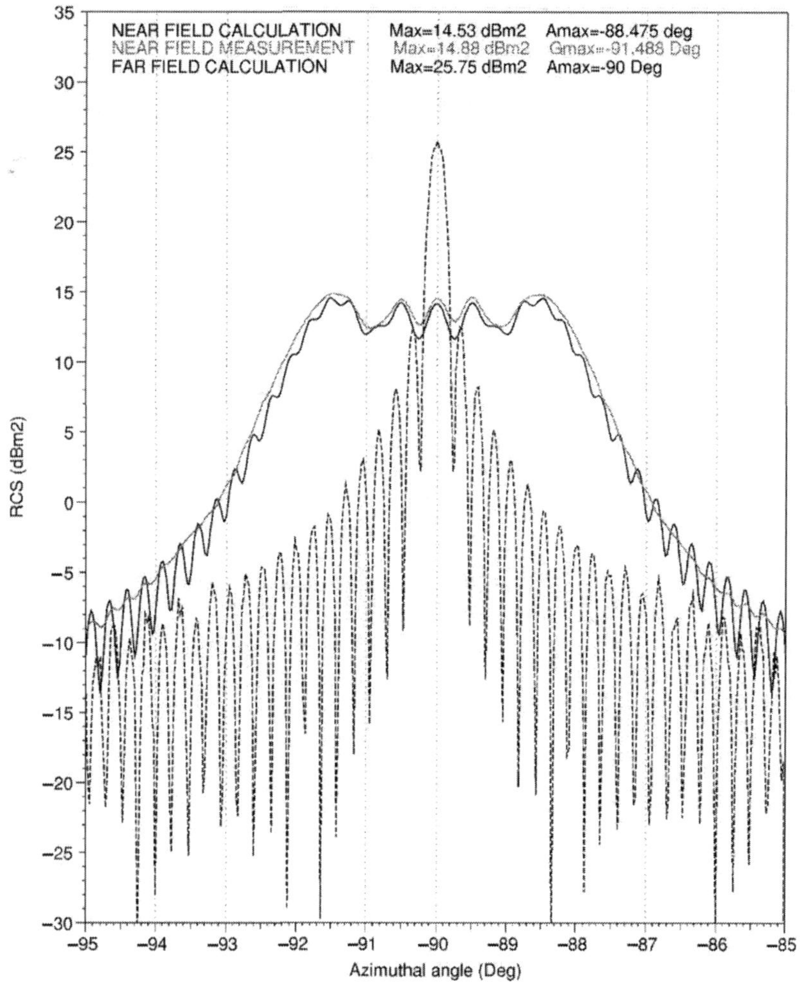

Figure 14.68 Near-field bistatic RCS of a conducting cylinder. (From Pouliguen *et al.* [II.100]. Reproduced by permission of © 2002 IEEE)

where $P(h_x, h_y)$ is the probability density function of the surface slopes,

$$q_x = k_0 [\sin\theta_{sc}\cos\phi_{sc} - \sin\theta_{in}\cos\phi_{in}],$$
$$q_y = k_0 [\sin\theta_{sc}\sin\phi_{sc} - \sin\theta_{in}\sin\phi_{in}],$$
$$q_z = k_0 [\cos\theta_{sc} + \cos\theta_{in}],$$
$$q^2 = q_x^2 + q_y^2 + q_z^2,$$

and U_{pq} are polarization-dependent coefficients.

In the geometrical optics model, it is assumed that the variance of the surface height was large compared to the incident wavelength. This restriction is not required with the physical optics

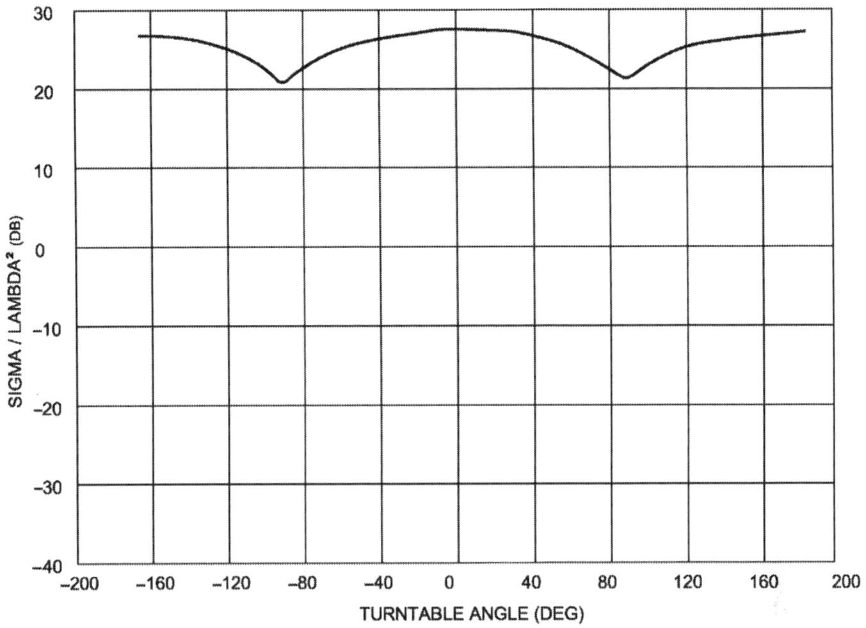

Figure 14.69 Bistatic RCS versus turntable angle for $\beta = 180°$, 8° cone, diameter = 35 mm, $\lambda = 17.9$ mm, $d/\lambda = 1.95$. (From Glaser [II.102]. Reproduced by permission of © 1989 IEEE)

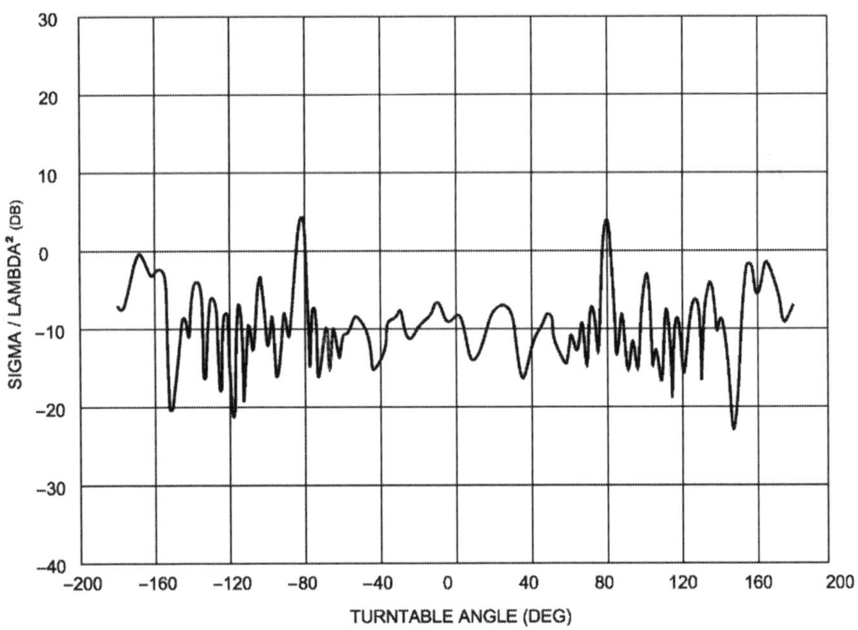

Figure 14.70 Bistatic RCS versus turntable angle for $\beta = 70°$, 8° cone, diameter = 35 mm, $\lambda = 17.9$ mm, $d/\lambda = 1.95$. (From Glaser [II.102]. Reproduced by permission of © 1989 IEEE)

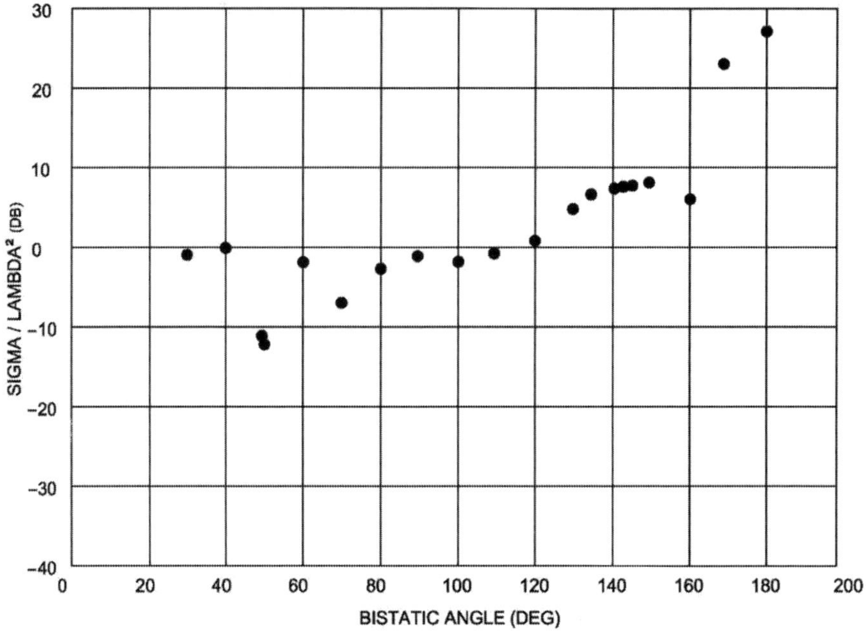

Figure 14.71 Bistatic RCS versus bistatic angle for turntable angle = 180°, 8° cone, diameter = 35 mm, $\lambda = 17.9$ mm, $d/\lambda = 1.95$. (From Glaser [II.102]. Reproduced by permission of © 1989 IEEE)

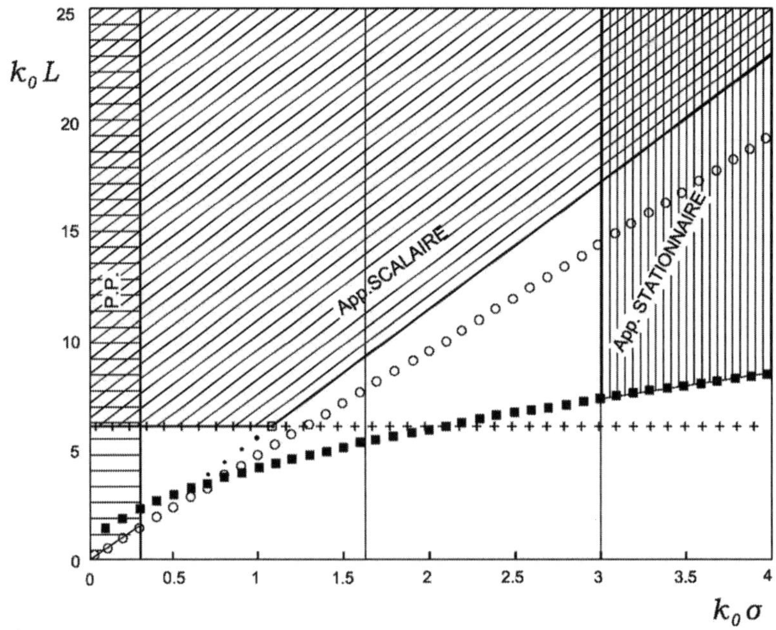

Figure 14.72 Validity conditions of Kirchhoff and small perturbation models. (From Khenchaf *et al.* [II.103]. Reproduced by permission of © 1996 IEEE)

model [II.107]. When the surface height has a Gaussian distribution, the conditions of validity are [II.103, II.104]

$$k_0 L > 6 \qquad \text{and} \qquad L > 4\sqrt{2}\,\sigma. \tag{14.26}$$

The Kirchhoff method has been applied to a surface with a horizontal roughness scale and average radius of curvature larger than the electromagnetic wavelength. When both the surface standard deviation and correlation length are smaller than the wavelength, one standard approach considered in Reference [II.103] is the small perturbation method.

The bistatic scattering coefficient for either a horizontally or vertically polarized incident wave from a slightly irregular surface within the incident medium is described by Reference [II.103]:

$$\sigma_{\text{sc}} = 8\left|k_0^2\sigma\cos\theta_{\text{in}}\cos\theta_{\text{sc}}\right|^2\left|\alpha_{pq}\right|^2 W\left(k_{0x} + k_0\sin\theta_{\text{in}}, k_{0y}\right), \tag{14.27}$$

where $k_{0x} = -k_0\cos\theta_{\text{sc}}\cos\phi_{\text{sc}}$, $k_{0y} = -k_0\sin\theta_{\text{sc}}\sin\phi_{\text{sc}}$, $k_0^2 = k_{0x}^2 + k_{0y}^2 + k_{0z}^2$, and α_{pq} are polarization-dependent coefficients. For the Gaussian case in Reference [II.103],

$$\sigma_{\text{sc}} = 4\left|k_0^2 L\sigma\left(\cos\theta_{\text{in}}\cos\theta_{\text{sc}}\alpha_{pq}\right)\right|^2\exp\left[-\frac{L^2}{4}\left(k_{0x} + k_0\sin\theta_{\text{in}}\right)^2 + k_{0y}\right]. \tag{14.28}$$

In the latter case the validity conditions are given by [II.103, II.104]

$$k_0\sigma < 0.3 \qquad \text{and} \qquad L > \frac{\sqrt{2}}{0.3}\sigma. \tag{14.29}$$

Thus, two special types of rough surfaces have been considered. The surface roughness has to be either large or small compared with the incident wavelength.

However, naturally occurring surfaces may include both types of roughness in various proportions. Some surfaces may have one continuous distribution of roughness instead of two significantly different average sizes. In this case the surfaces can be modelled as having only two average sizes of roughness with one large and the other small compared with the incident wavelength.

Surfaces that are rough only in one dimension were considered in detail by Bahar and Lee [II.105] in order to compare the fullwave results with published experimental and numerical results based on Monte Carlo simulations of rough surfaces. The full-wave solution for the vertically polarized incoherent diffuse bistatic scattering cross-sections (widths) $\langle\sigma_{\text{IN}}^V\rangle$ were evaluated for perfectly conducting one-dimensionally rough surfaces that are characterized by four-dimensional Gaussian joint probability density functions $p(h_1, h_2, h_{x1}, h_{x2})$ for the surface heights, h_1 and h_2, and surface slopes, h_{x1} and h_{x2}, at two points. The full-wave solution that accounts for all the rough surface height–slope correlations was compared with earlier solutions that were based on the *a priori* assumption that the surface heights and slopes are uncorrelated. They were also compared with the corresponding small perturbation solutions and the physical optics solutions for surfaces with small Rayleigh roughness parameters and small slopes. The impact of including the effects of self-shadow was also considered by Bahar and Lee in detail.

In Figures 14.73 and 14.74, the bistatic scattering cross-sections are plotted for the case $\delta = 0.1$ and $k_0 L = 1$ corresponding to $< h_x^2 >= 0.05$. Here $\delta = 4k_0^2 < h^2 >$ is the Rayleigh roughness parameter [II.105], k_0 is the free-space wave number, $< h^2 >$ and $< h_x^2 >$ are the mean square height and slope, and L is a correlation length, defined as the distance at which the autocorrelation function of the surface height $< h_1 h_2 >$ becomes e^{-1} times its mean square height $< h^2 >$. In Figure 14.73, $\theta_0^{\text{i}} = 45°$ and in Figure 14.74, $\theta_0^{\text{i}} = 30°$, where θ_0^{i} is an incident angle.

According to Bahar and Lee, the difference between the small perturbation and the polarization-independent physical optics solutions and the full-wave solutions increases as the slope increases ($< h_x^2 >> 0.05$), even for normal incidence. The effects of self-shadow are negligible for these cases considered, except for near grazing angles, as the mean square slope increases. Thus for the near specular direction, all five solutions are in relatively good agreement. Note that for perfectly conducting one-dimensionally rough surfaces the full-wave solutions are polarization dependent, even for scatter in

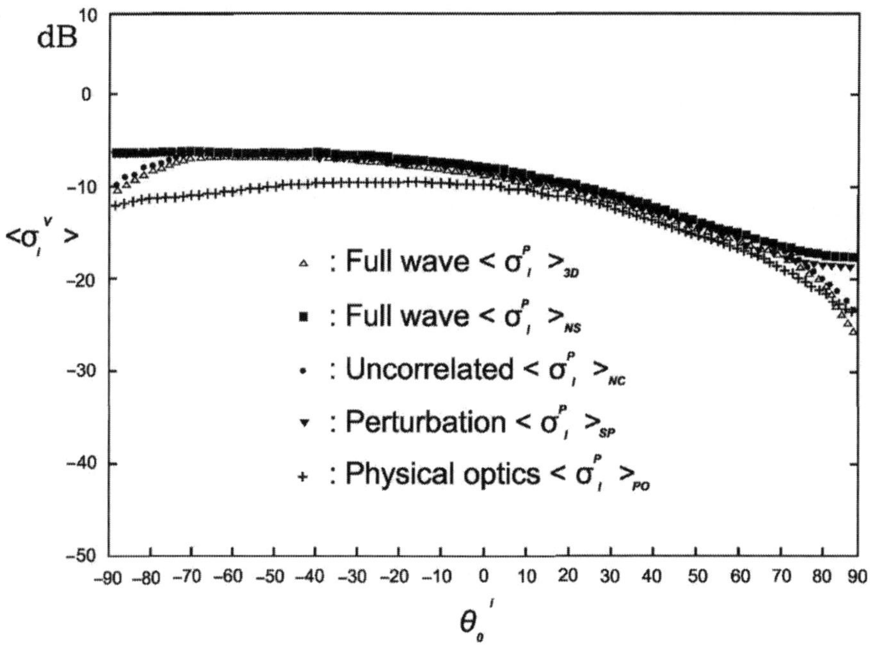

Figure 14.73 Plots of the bistatic cross sections as functions of the scatter angle $\theta_0^{\rm f}$. Incident angle $\theta_0^{\rm i} = 45°$, $\delta = 0.1$, $k_0L = 1$, $< h_x^2 > = 0.05$. (From Bahan and Lee [II.105]. Reproduced by permission of © 1995 IEEE)

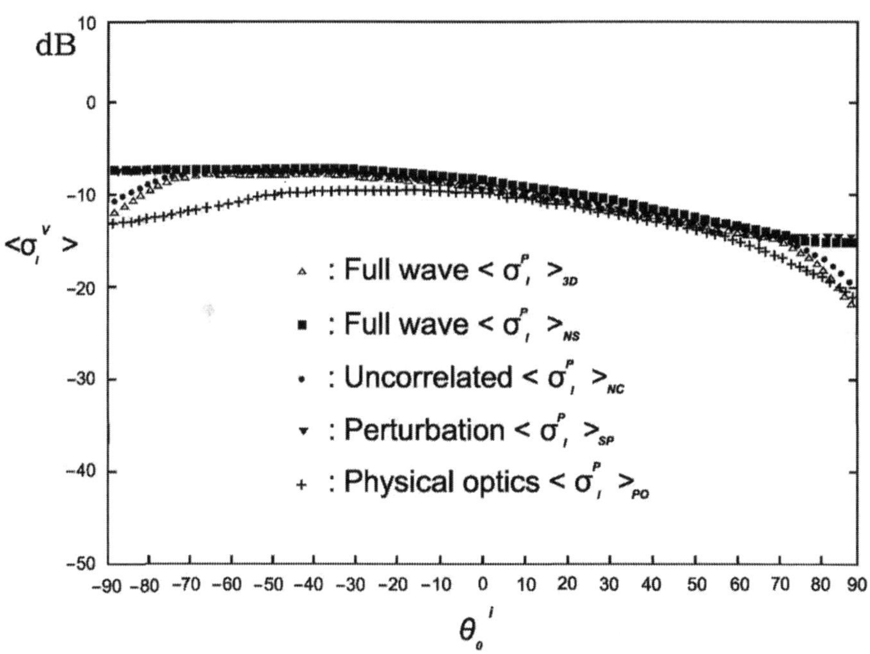

Figure 14.74 Plots of the bistatic cross sections as functions of the scatter angle $\theta_0^{\rm f}$. Incident angle $\theta_0^{\rm i} = 30°$, $\delta = 0.1$, $k_0L = 1$, $< h_x^2 > = 0.05$. (From Bahan and Lee [II.105]). Reproduced by permission of © 1995 IEEE)

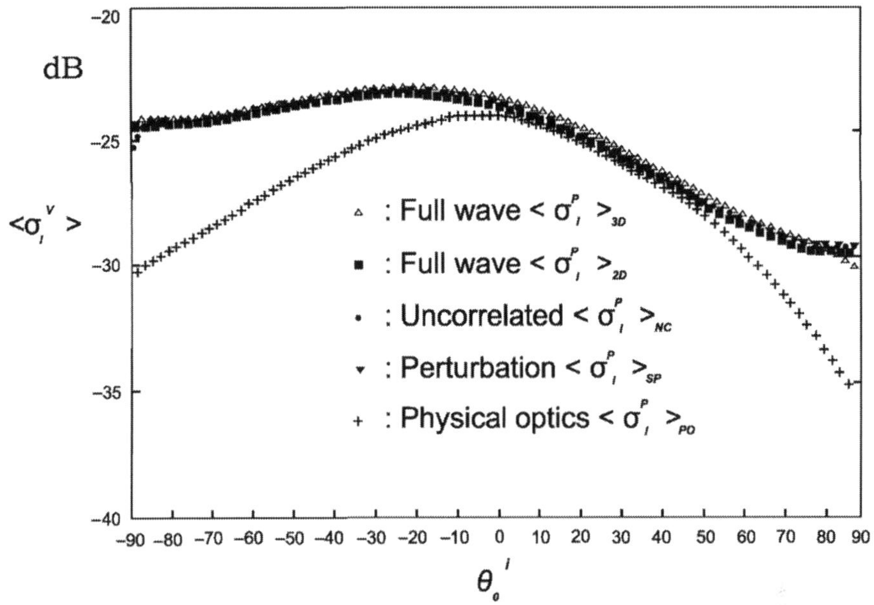

Figure 14.75 Plots of the bistatic cross sections as functions of the scatter angle $\theta_0^{\rm f}$. Incident angle $\theta_0^{\rm i} = 30°$, $k_0 h = 0.02236$, $k_0 L = 1.414214$, $< h_x^2 > = 0.0005$. (From Bahan and Lee [II.105]. Reproduced by permission of © 1995 IEEE)

the specular direction. The difference between the full-wave solution and the small perturbation solution (which is polarization independent in the specular direction) increases with increasing slope [II.105].

What is very significant in all the results of Reference [II.105] is the fact that the assumption made *a priori* that the correlation between the surface heights and slopes may be neglected is in good agreement with the results obtained when this correlation is not ignored. Although these full-wave solutions were shown to be in good agreement with the experimental results, the simplifying assumption that surface heights and slopes are uncorrelated is analytically justified in Reference [II.105]. This conclusion has a major impact on the computation of the scattering cross-section.

In Figure 14.75 the bistatic scatter cross-section for vertically polarized waves is plotted as a function of the scatter angle $\theta_0^{\rm f}$. The incident angle was $\theta_0^{\rm i} = 30°$. The rough surface parameters are $k_0 h = \sqrt{k_0^2 \langle h^2 \rangle} = \sqrt{\langle h_x^2 \rangle} = 0.02236$ and $k_0 L = \sqrt{2}$. For this case, the height and slopes are chosen to be of the same order of smallness. The first- and second-order small perturbation solution of Rice is based on the assumption that the heights and slopes are of the same order of smallness. For this case there is excellent agreement between the full-wave solution and the small perturbation solution. No higher-order terms are needed in the small perturbation analysis for this case. The physical optics approximations are very poor except for the near specular direction. Notice that in this case the correlation between the heights and slopes can be neglected [II.105].

In Figures 14.76 to 14.78 the full-wave solutions for the bistatic cross-sections are plotted as functions of scatter angle for incident angles $\theta_0^{\rm i} = 0°$, $20°$ and $40°$. The rough surface parameters were $k_0 h = 11.687$, $k_0 L = 29.468$, $< h_x^2 > = 0.315$ and $\varepsilon_{\rm r} = 1.991$.

In Figures 14.79–14.84 the bistatic radar cross-sections are also plotted as functions of the scatter angles. The rough surface parameters in Figs. 14.79 and 14.80 are $k_0 h = 11.938$, $k_0 L = 19.855$, $< h_x^2 > = 0.723$ and $\varepsilon_{\rm r} = 2.04$. In Figs. 14.81–14.84, they are $k_0 h = 3.142$, $k_0 L = 19.855$,

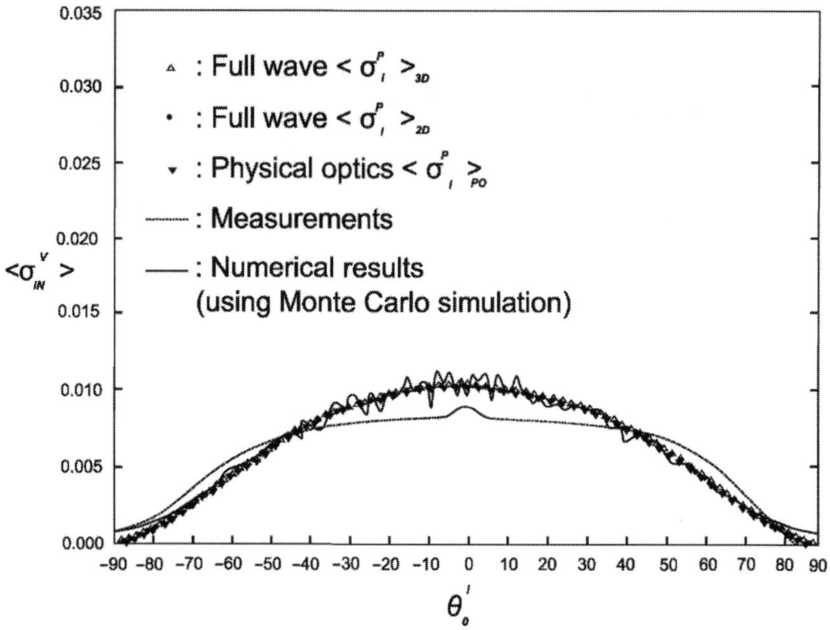

Figure 14.76 Plots of the bistatic cross sections as functions of the scatter angle $\theta_0^{\rm f}$. Incident angle $\theta_0^{\rm i} = 0°$, $k_0h = 11.687$, $k_0L = 29.468$, $< h_x^2 > = 0.315$, $\varepsilon_{\rm r} = 1.991$. (From Bahan and Lee [II.105]. Reproduced by permission of © 1995 IEEE)

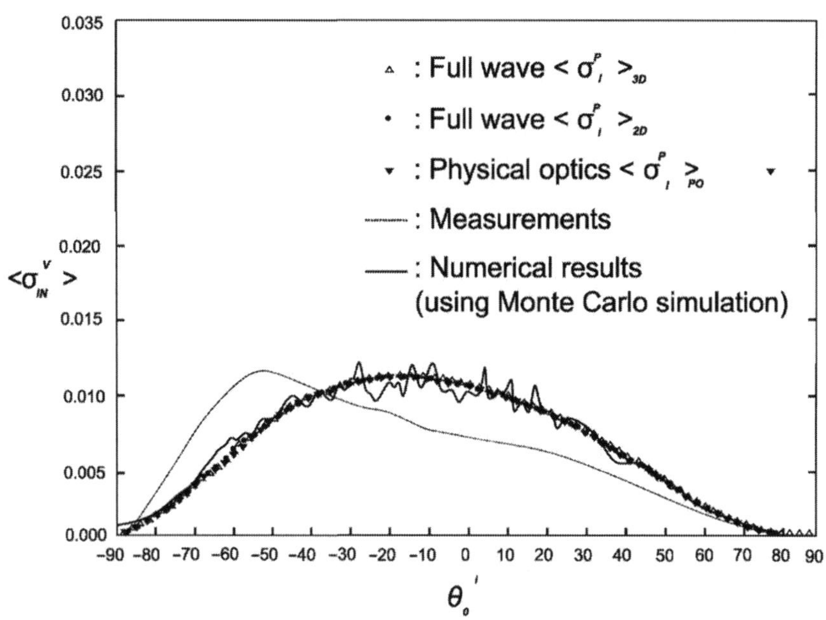

Figure 14.77 Plots of the bistatic cross sections as functions of the scatter angle $\theta_0^{\rm f}$. Incident angle $\theta_0^{\rm i} = 20°$, $k_0h = 11.687$, $k_0L = 29.468$, $< h_x^2 > = 0.315$, $\varepsilon_{\rm r} = 1.991$. (From Bahan and Lee [II.105]. Reproduced by permission of © 1995 IEEE)

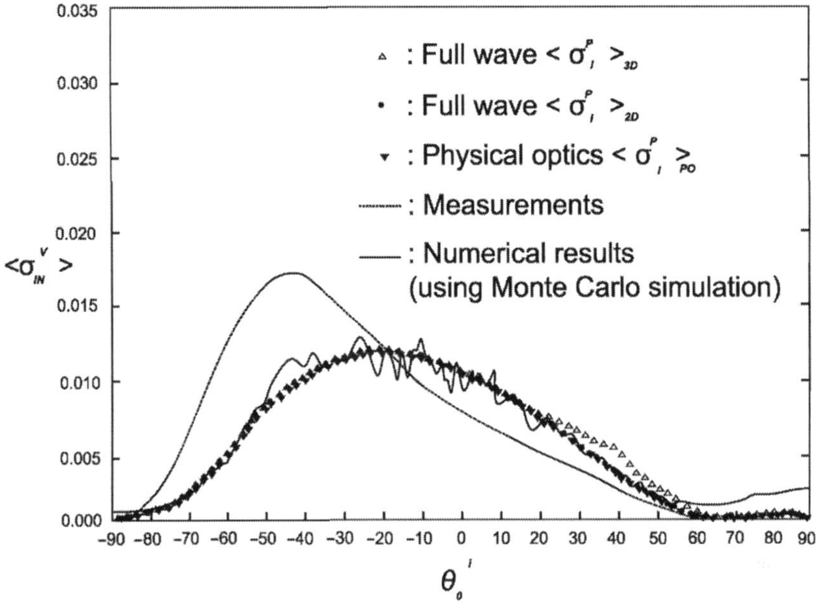

Figure 14.78 Plots of the bistatic cross-sections as functions of the scatter angle θ_0^{f}. Incident angle $\theta_0^{\mathrm{i}} = 40^\circ$, $k_0 h = 11.687$, $k_0 L = 29.468$, $< h_x^2 > = 0.315$, $\varepsilon_{\mathrm{r}} = 1.991$. (From Bahan and Lee [II.105]. Reproduced by permission of © 1995 IEEE)

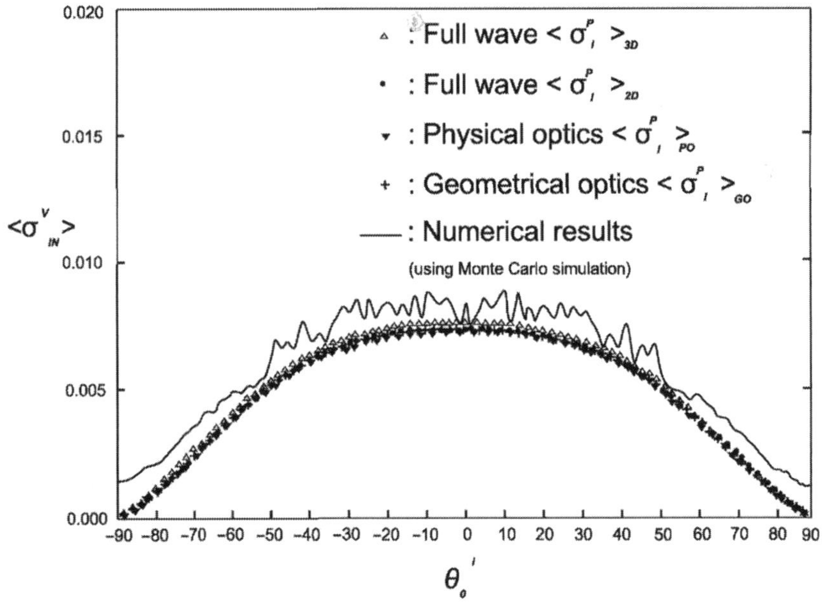

Figure 14.79 Plots of the bistatic cross-sections as functions of the scatter angle θ_0^{f}. Incident angle $\theta_0^{\mathrm{i}} = 0^\circ$, $k_0 h = 11.938$, $k_0 L = 19.855$, $< h_x^2 > = 0.723$, $\varepsilon_{\mathrm{r}} = 2.04$. (From Bahan and Lee [II.105]. Reproduced by permission of © 1995 IEEE)

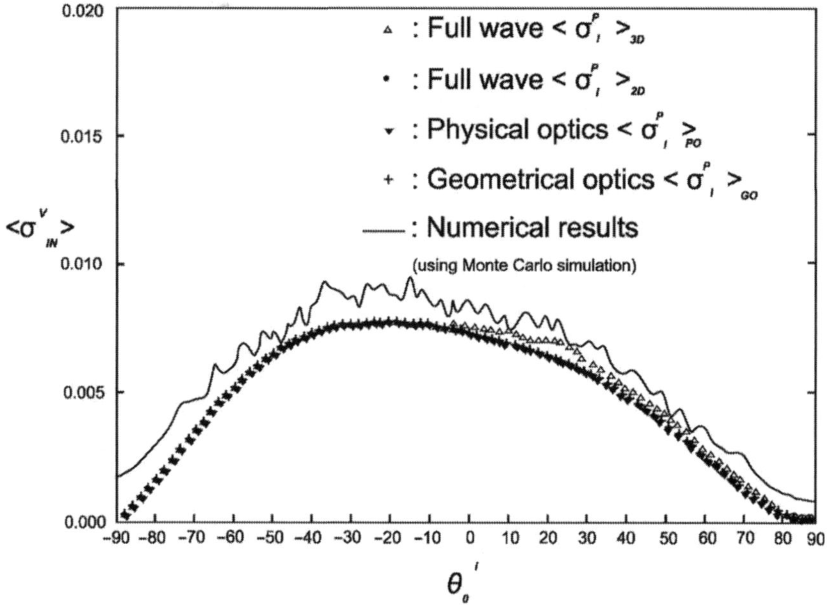

Figure 14.80 Plots of the bistatic cross-sections as functions of the scatter angle θ_0^{f}. Incident angle $\theta_0^{\mathrm{i}} = 10°$, $k_0 h = 11.938$, $k_0 L = 19.855$, $< h_x^2 > = 0.723$, $\varepsilon_{\mathrm{r}} = 2.04$. (From Bahan and Lee [II.105]. Reproduced by permission of © 1995 IEEE)

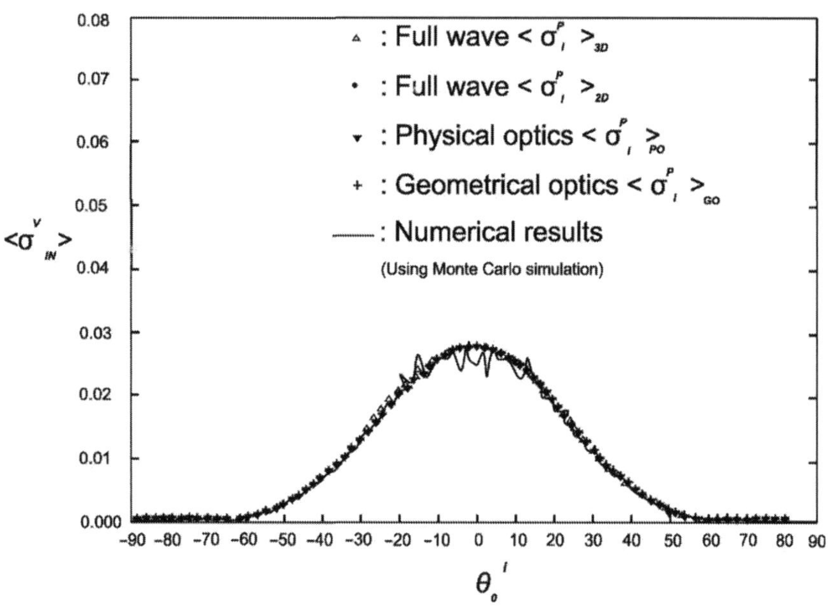

Figure 14.81 Plots of the bistatic cross-sections as functions of the scatter angle θ_0^{f}. Incident angle $\theta_0^{\mathrm{i}} = 0°$, $k_0 h = 3.142$, $k_0 L = 19.855$, $< h_x^2 > = 0.05$, $\varepsilon_{\mathrm{r}} = 2.04$. (From Bahan and Lee [II.105]. Reproduced by permission of © 1995 IEEE)

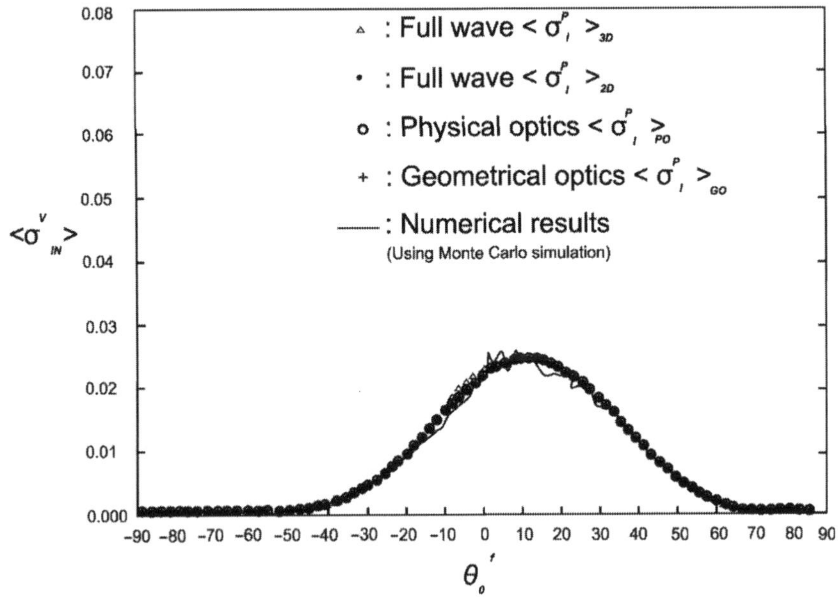

Figure 14.82 Plots of the bistatic cross-sections as functions of the scatter angle θ_0^{f}. Incident angle $\theta_0^{\mathrm{i}} = 20°$, $k_0 h = 3.142$, $k_0 L = 19.855$, $< h_x^2 > = 0.05$, $\varepsilon_{\mathrm{r}} = 2.04$. (From Bahan and Lee [II.105]. Reproduced by permission of © 1995 IEEE)

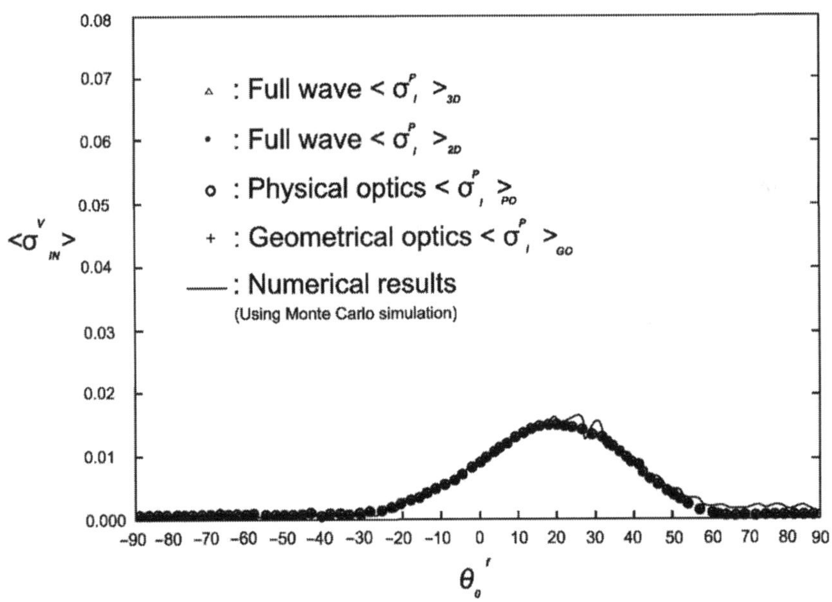

Figure 14.83 Plots of the bistatic cross-sections as functions of the scatter angle θ_0^{f}. Incident angle $\theta_0^{\mathrm{i}} = 40°$, $k_0 h = 3.142$, $k_0 L = 19.855$, $< h_x^2 > = 0.05$, $\varepsilon_{\mathrm{r}} = 2.04$. (From Bahan and Lee [II.105]. Reproduced by permission of © 1995 IEEE)

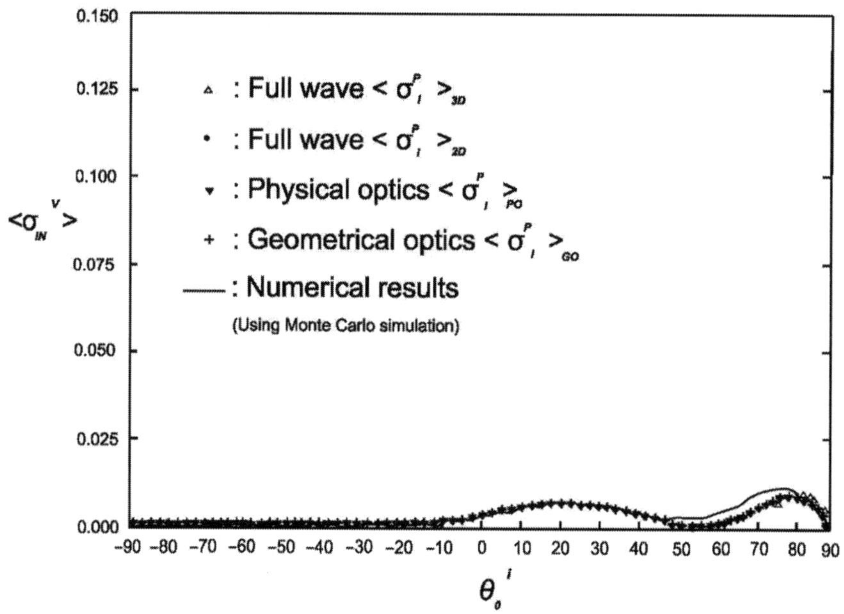

Figure 14.84 Plots of the bistatic cross-sections as functions of the scatter angle θ_0^f. Incident angle $\theta_0^i = 55°$, $k_0 h = 3.142$, $k_0 L = 19.855$, $< h_x^2 > = 0.05$, $\varepsilon_r = 2.04$. (From Bahan and Lee [II.105]. Reproduced by permission of © 1995 IEEE)

$< h_x^2 >= 0.05$ and $\varepsilon_r = 2.04$. The corresponding numerical results based on the Monte Carlo simulation [II.106] were superimposed on the full-wave results. According to Reference [II.105], there is excellent agreement between the full-wave results and the corresponding high-frequency physical optics and geometrical optics results. They are also in good agreement with the numerical results except at very near grazing angles. Bahar and Lee note the good agreement in Figure 14.82 between the full wave results, and the numerical results, even around the Brewster angle $\theta_0^f = 55°$.

The impact of the large radii of curvature assumption is also illustrated in Figures 14.79 to 14.84, where the effects of curvature are indeed shown to be negligible. However, in Figures 14.79 and 14.80 Bahar and Lee noticed a small deviation between the full-wave solutions based on the three- and two-dimensional integrals in the quasi-specular direction. The small humps in the results appear about the quasi-specular direction. They may be related to the small humps observed in the experimental results (see Reference [II.22]).

Thus, while there is good agreement between the full-wave solutions and the small perturbation solutions for surfaces with small heights and slopes, it is shown in Reference [II.105] that for the one-dimensionally rough surfaces considered, as the mean square slope increases the difference between the small perturbation solution and the full-wave solution increases. The difference between the polarization-dependent full-wave solutions for the bistatic scattering cross-sections and the corresponding physical optics solutions is very significant in the backward direction. There is excellent agreement between the first- and second-order small perturbation solution and the full-wave solution when the rough surface height and slopes are of the same order of smallness.

The impact of the self-shadow on the results for the vertically polarized waves at near grazing angles is more significant than on the corresponding results for the horizontally polarized waves. According to Reference [II.105], this is because the surface element scattering coefficient for the horizontally polarized waves vanishes for grazing angles. For perfectly conducting surfaces, the vertically polarized surface element scattering coefficient does not vanish at grazing angles.

It was shown in Reference [II.105] that for surfaces with small slopes, the full-wave results can be significantly simplified by assuming, *a priori*, that the surface heights and slopes are uncorrelated. Thus it is possible to significantly reduce the time needed to compute the scattering cross-sections without significantly reducing the accuracy of the results.

The full-wave solutions were also compared in Reference [II.105] with experimental results and numerical results based on Monte Carlo simulations. There was good agreement between the full-wave solutions and the experimental results. They were also in good agreement with the numerical results except for near grazing angles. The impact of the commonly used large radii of curvature assumption is also considered in detail.

To understand the scattering behaviour and the energy distribution of scattering strength from randomly rough surfaces, information about scattering must be known unidirectionally, not just in the specular or back-scatter direction. The model developed in Reference [II.107] was used to predict the like- and cross-polarized scattering, and to study the distribution of scatter energy along the azimuth angle from randomly rough surfaces. As expected, the single like-polarized scattering coefficient vanishes in the specular plane and the single like-polarizations reach their minima in the orthogonal plane. The cross-polarized multiple scattering shows a significant contribution in the specular plane while the like-polarized multiple scattering shows significant contributions in the orthogonal plane.

First, Hsieh considered [II.107] the comparisons of surface scattering based on the integral equation method with and without the phase terms of Green's function and its derivatives (Figures 14.85 to 14.89). The phase terms in Green's function and its derivative represent the propagation directions. According to Reference [II.107] with the consideration of the phase terms of Green's function and its derivatives, the upward and downward multiple scattering strength can be derived to study the relative contribution of single and multiple scattering from different rough surfaces along different azimuth angles.

The prediction was developed for an electromagnetic wave scattering pattern from the statistically rough surface with a normalized surface height of 0.2 (Figures 14.85 and 14.86) and 0.4 (Figures 14.87 to 14.89), respectively. The normalized surface correlation length was 1.0. The surface material was a

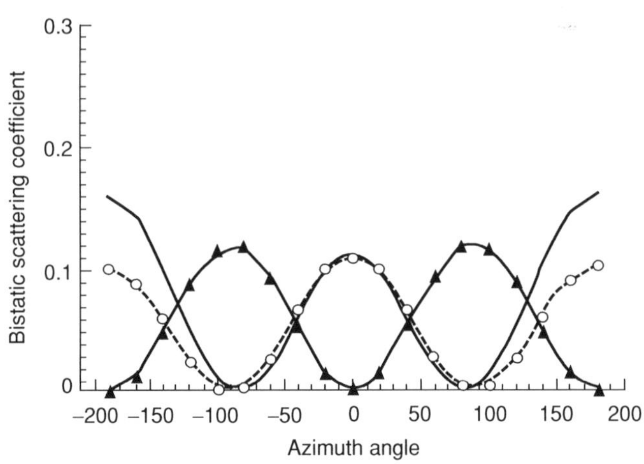

Figure 14.85 Comparisons of like- with cross-polarized single scattering strength. The normalized surface RMS height is 0.2, surface correlation length 1.0 and surface dielectric constant 63.2 along the azimuth direction: —, VV (0.2, single); –○–, HH (0.2, single); —, VH (0.2, single). —▲—, (From Hsieh [II.107]. Reproduced by permission of © 2001 IEE)

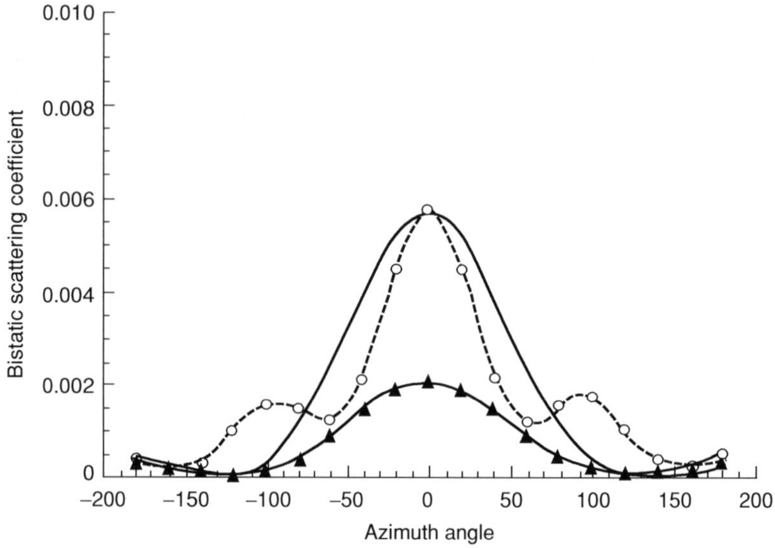

Figure 14.86 Comparisons of like- with cross-polarized multiple scattering strength. The normalized surface RMS height is 0.2, surface correlation length 1.0 and surface dielectric constant 63.2 along the azimuth direction: —, VV (0.2, single); –○–, HH (0.2, single); –▲–, VH (0.2, single). (From Hsieh [II.107]. Reproduced by permission of © 2001 IEE)

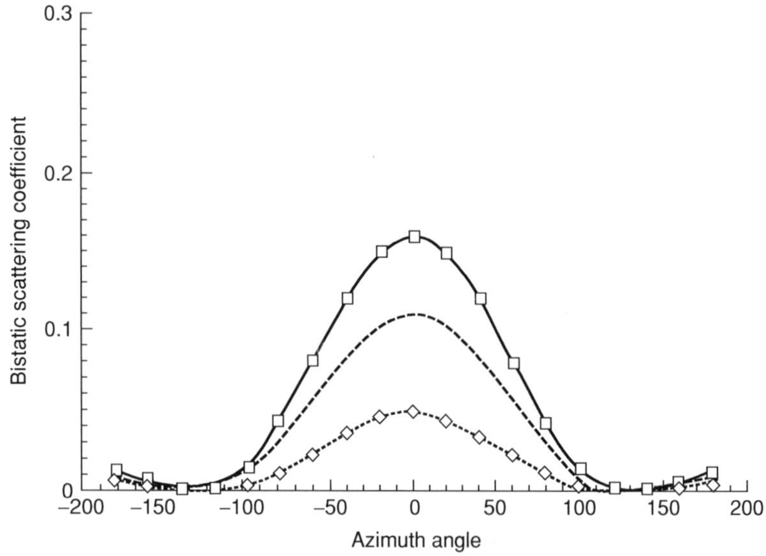

Figure 14.87 Comparisons of VV polarized single with multiple scattering strength. The normalized surface RMS height is 0.4, surface correlation length 1.0 and surface dielectric constant 63.2 along the azimuth direction: –◇–, VV (0.4, multiple); —, VV (0.4, single) —; —□—, VV (0.4, +×). (From Hsieh [II.107]. Reproduced by permission of © 2001 IEE)

water-soaked form brick with an estimated dielectric constant 63.2. The incident angle was 45° and the scatter polar angle was also 45°. A zero value of the azimuth angle corresponds to forward-scattering and an azimuth value of 180° represents the back-scatter. The azimuth angle of 90° is in the orthogonal plane.

Hsieh [II.107] compared the like- and cross-polarised scattering strength along the azimuth angle (Figures 14.85 and 14.86). Figure 14.85 compares the like- with cross-polarised single-scattering strength; the VV polarization reaches the maximum scatter power strength in the forward and backward specular plane, but reaches its null scatter on the plane orthogonal to the specular plane. On the other hand, the cross-polarized scatter strength reaches the maximum scatter power on the orthogonal plane and reaches its null point in the specular plane. In the forward-scatter direction on the plane of incidence, the contribution of the VV polarized scatter strength is similar to that of the HH polarized scatter strength, but the contribution of the VV polarized scatter strength is larger than that of the HH polarized scatter strength around the backward-scatter directions.

Figure 14.86 shows the comparisons of like- with cross-polarized multiple scattering strength on the surface. In the specular plane, the multiple like-polarized scattering powers reach their maximum contribution, while the multiple cross-polarization has the same scattering behaviour but less scatter strength. According to Hsieh, around the orthogonal plane, the like-polarized scatter strength disappears and makes no contribution to the bistatic scattering strength, while cross-polarization makes some contribution on the orthogonal plane.

To understand the contribution of multiple scattering, Hsieh examines the comparisons of the single, multiple and total scattering coefficients along the azimuth angle from random rough surfaces. For multiple VV and HH polarized scattering, the strongest scatter power occurs in the specular direction, but there is a minimum multiple scatter power in the direction orthogonal to the plane of incidence in Figures 14.87 and 14.88. The single like-polarized scattering strength has a similar scattering trend to

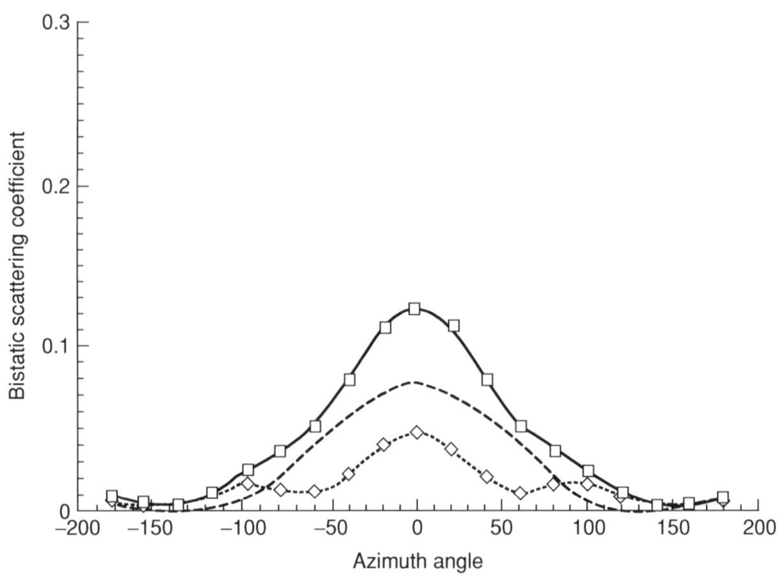

Figure 14.88 Comparisons of HH polarized single with multiple scattering strength. The normalized surface RMS height is 0.4, surface correlation length 1.0 and surface dielectric constant 63.2 along the azimuth direction: –◇–, HH (0.4, multiple); —, HH (0.4, single) —□—, HH (0.4, +×). (From Hsieh [II.107]. Reproduced by permission of © 2001 IEE)

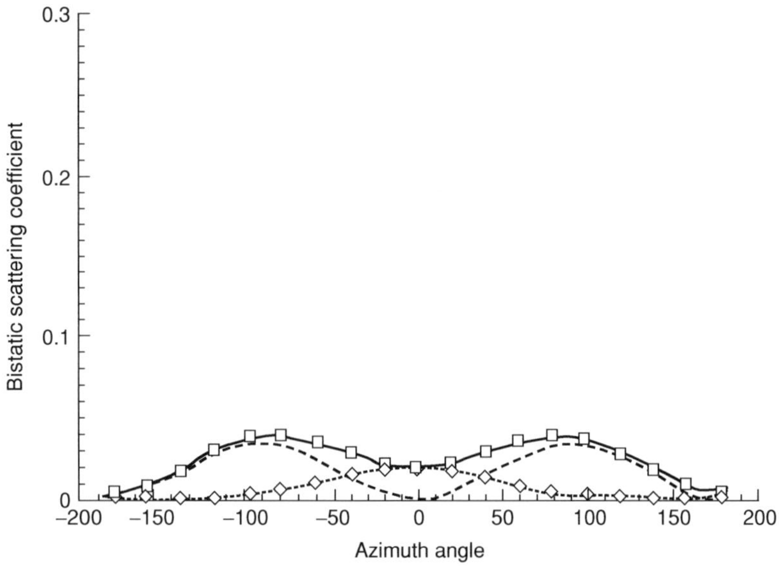

Figure 14.89 Comparisons of VH polarized single with multiple scattering strength. The normalized surface RMS height is 0.4, surface correlation length 1.0 and surface dielectric constant 63.2 along the azimuth direction: –◇–, VH (0.4, multiple); —, VH (0.4, single) —□—, VH (0.4, $+\times$). (From Hsieh [II.107]. Reproduced by permission of © 2001 IEE)

that of multiple scattering. Figure 14.89 shows the comparisons of cross-polarized single, multiple and total scattering coefficients. The expectation for the cross-polarized scattering is the strongest multiple scatter power in the orthogonal direction, but there is a minimum multiple scatter power in the direction of incidence. The cross-polarized scatter strength is less than the like-polarized scatter strength due to the small RMS surface slope and small surface parameters (normalized surface height, normalized surface correlation length) [II.107].

Finally, Hsieh compared the model prediction with the types of bistatic scattering measurements made by Hauck *et al.* [II.108]. After removing the effect of antenna gain to the bistatic scattering strength he compared the model predictions of bistatic scattering coefficients with the measured data from the artificial random rough surfaces. The surface material was water-soaked form bricks with an estimated dielectric constant of 63.2. For the isotropic surface being assumed, the scattering pattern is symmetric to the plane of incidence. Figures 14.90 and 14.91 show the comparisons of like-polarized measured data and model prediction along the azimuth angle. Because the scattering trend of cross-HV-polarized bistatic scattering is very similar to that of VH polarized bistatic scattering, Hsieh only showed the comparisons of VH polarized bistatic scattering for the cross-polarized bistatic scattering in Figure 14.92. The like- and cross-polarized incoherent scattering coefficients were measured from rough surfaces with a surface RMS height of 0.2 and correlation length 1.0 normalized to the incident wavelength respectively [II.107].

Figures 14.90 to 14.92 show the comparisons of bistatic scattering strength with and without the consideration of phase terms of Green's function and its derivative. According to Hsieh, without the contribution of multiple scattering, the excellent agreement cannot be reached in VV polarization in the orthogonal plane. For VH polarized scattering the multiple contributions cannot be ignored in the plane of incidence where the single cross-polarized scattering vanishes. Because of the contribution of

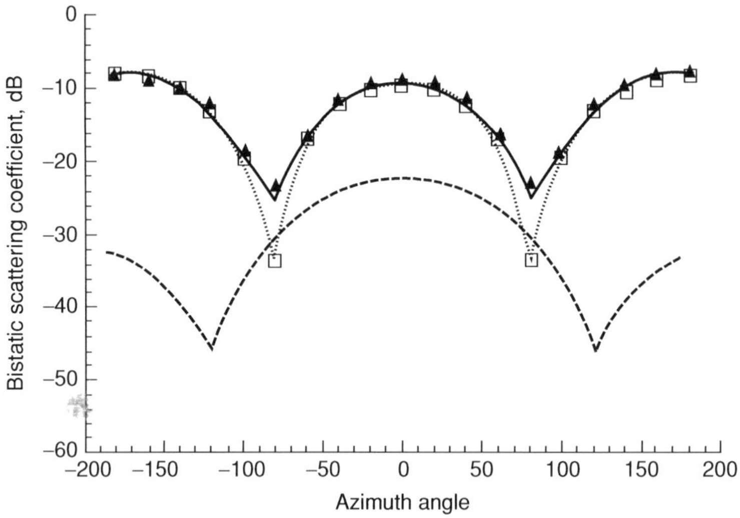

Figure 14.90 Comparisons of VV polarized single and multiple scattering. The normalized surface RMS height is 0.2 and surface correlation length 1.0 along the azimuth direction: —, VV (0.2, multipule); …□…, VV (0.2, Single); —, VV (0.2, single plus multiple); ▲, VV (0.2, data). (From Hsieh [II.107]. Reproduced by permission of © 2001 IEE)

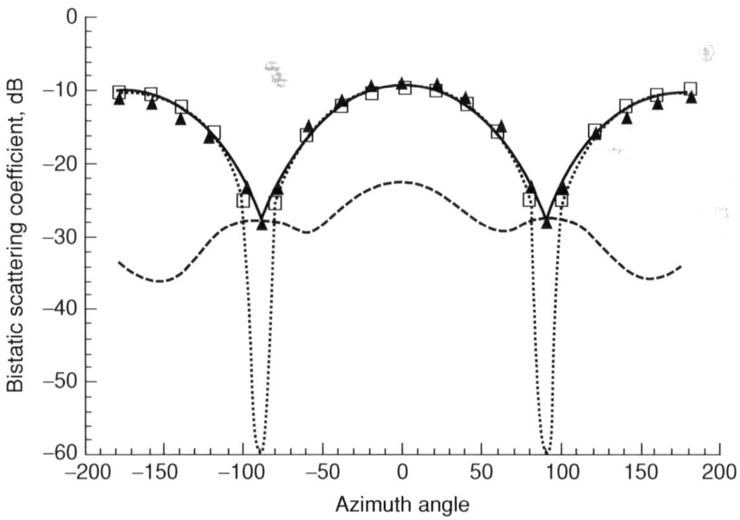

Figure 14.91 Comparisons of HH polarized single and multiple scattering. The normalized surface RMS height is 0.2 and surface correlation length 1.0 along the azimuth direction: —, HH (0.2, multiple); …□…, HH (0.2, single); —, HH (0.2, single plus multiple); ▲, HH (0.2, data). (From Hsieh [II.107]. Reproduced by permission of © 2001 IEE)

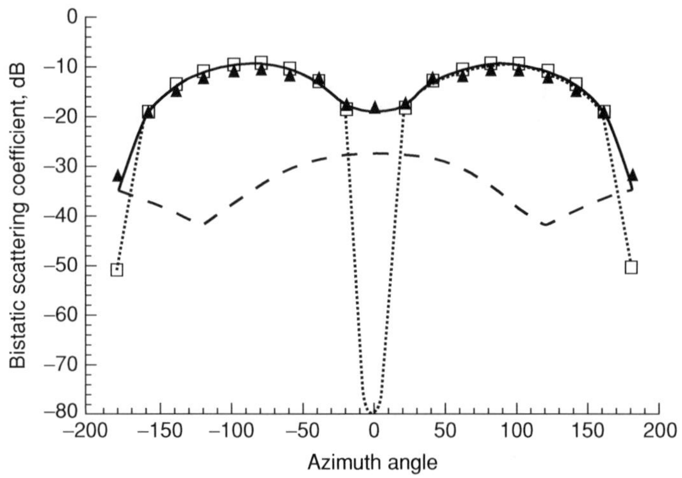

Figure 14.92 Comparisons of VH polarized single and multiple scattering. The normalized surface RMS height is 0.2 and surface correlation length 1.0 along the azimuth direction: —, VH (0.2, multiple); ...□..., VH (0.2, single); — VH (0.2, single plus multiple), ▲; VH (0.2, data). (From Hsieh [II.107]. Reproduced by permission of © 2001 IEE)

multiple scattering, the excellent agreements between the measured data and model prediction can be reached, especially for VV polarization along the orthogonal plane and cross-polarization along the specular plane [II.107].

Examination of Figures 14.90 to 14.92 indicates that there is excellent agreement between the model prediction and the measured data. This is true except for the weak bistatic scattering strength less than 30 dB for like-polarization in the orthogonal plane and cross-polarization in the plane of incidence. The difference between the model prediction and measurements is primarily caused by the measurement precision of the installed equipment. Basically, the agreement between the model prediction and measurements is good according to Reference [II.107].

The bistatic RCS behaviour of a Gaussian beam from a hemispherical boss (see Figure 14.93) on the conducting plane has been investigated by Sakurai *et al.* in [II.109]. The difference between the bistatic RCS of a Gaussian beam and that of a plane wave was compared. Figures 14.94(a) and 14.94(b) show the bistatic RCS $\sigma_s(\theta, \phi = 0)$ in the *XZ* plane and $\sigma_s(\theta, \phi = \pi/2)$ in the *YZ* plane at the incident angle $\theta_i = 90°$. The solid line shows the behaviour of the plane wave scattered field. Sakurai *et al.* noticed that the difference between the main lobe and the sidelobe is over 50 dB, while the difference between them for the 2D beam is less than 10 dB in the literature. The dotted line shows the behaviour of the Gaussian beam scattered field. It is found that the bistatic RCS behaviour of the Gaussian beam shows a much broader scattering pattern as well as a lower bistatic RCS at nadir ($\theta = \pi/2$) than the plane wave. Comparing Figure 14.94(b) with Figure 14.94(a), Sakurai *et al.* noticed that the scattering pattern of the *YZ* plane is much wider than that of the *XZ* plane.

Figures 14.94(c) and 14.94(d) show the bistatic RCS $\sigma_s(\theta, \phi = 0)$ in the *XZ* plane and $\sigma_s(\theta, \phi = \pi/2)$ in the *YZ* plane at the incidence angle $\theta_i = 60°$. It is reconfirmed that a significant scattering pattern difference exists between the Gaussian beam (dotted) and the plane wave (solid) incidences, in particular near the forward direction. The bistatic RCS pattern for the Gaussian beam is seen to be much broader than that for the plane wave. In addition, while the mainlobe beamwidth in the *YZ* plane and the *XZ* plane is seen to be equal for the plane wave, the mainlobe beamwidth of the *YZ* plane is much wider than that of the *XZ* plane for the Gaussian beam [II.109].

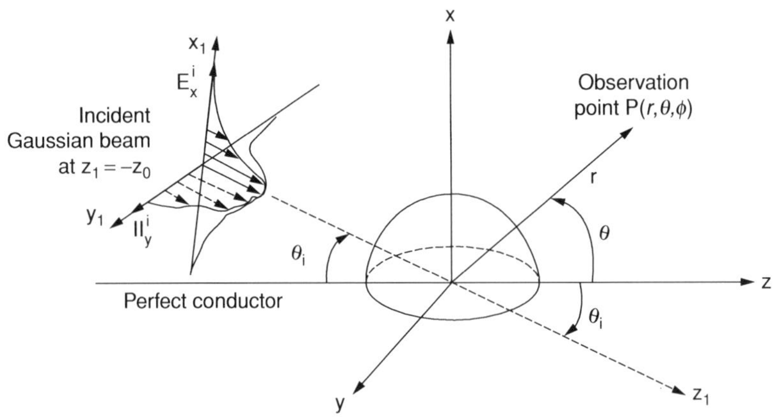

Figure 14.93 Geometry of the hemispherical boss scattering problem. (From Sakurai *et al.* [II.109]. Reproduced by permission of © 2004 IEEE)

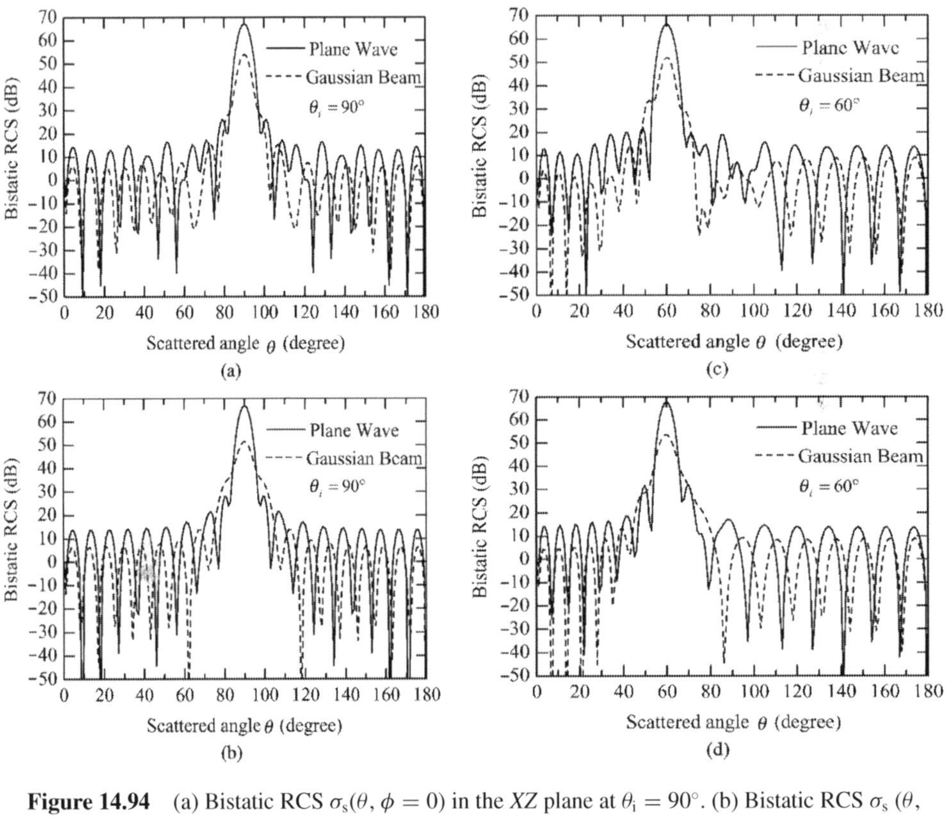

Figure 14.94 (a) Bistatic RCS $\sigma_s(\theta, \phi = 0)$ in the *XZ* plane at $\theta_i = 90°$. (b) Bistatic RCS σ_s $(\theta, \phi = \pi/2)$ in the *YZ* plane at $\theta_i = 90°$. (c) Bistatic RCS $\sigma_s(\theta, \phi = 0)$ in the *XZ* plane at $\theta_i = 60°$. (d) Bistatic RCS σ_s $(\theta, \phi = \pi/2)$ in the *YZ* plane at $\theta_i = 60°$. (From Sakurai *et al.* [II.109]. Reproduced by permission of © 2004 IEEE)

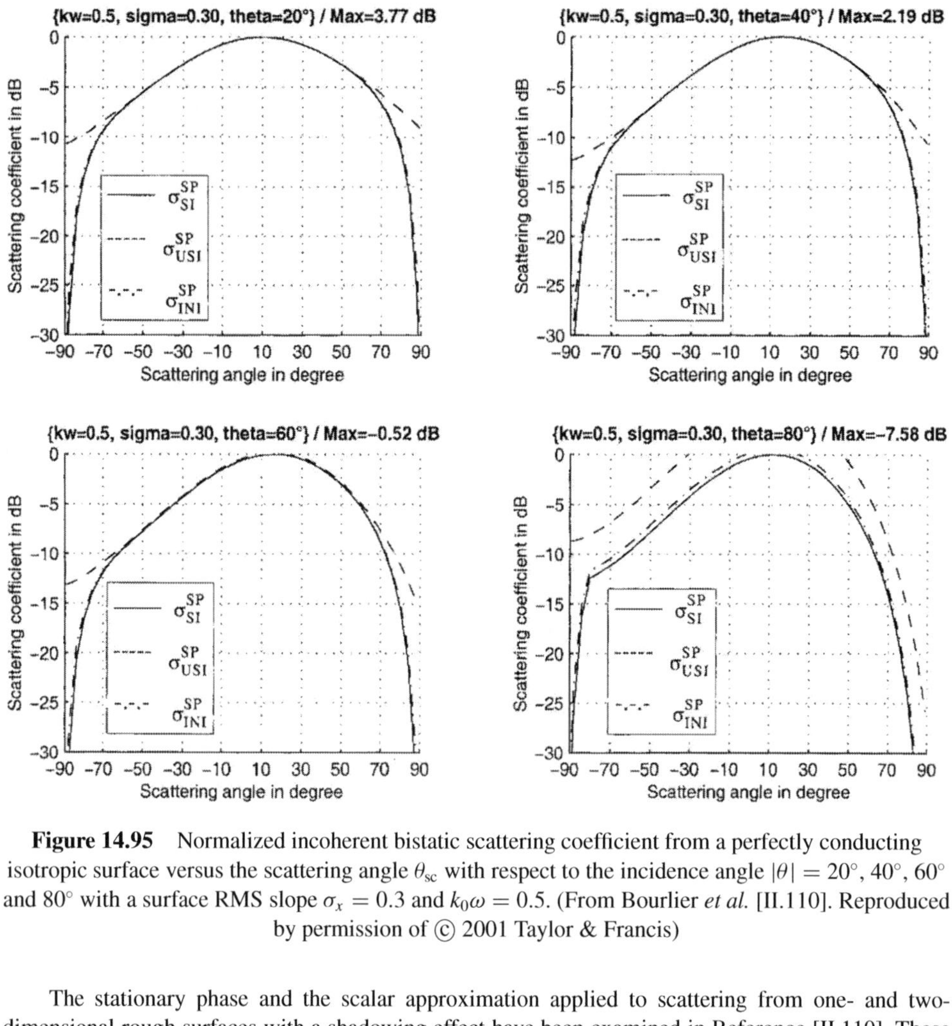

Figure 14.95 Normalized incoherent bistatic scattering coefficient from a perfectly conducting isotropic surface versus the scattering angle θ_{sc} with respect to the incidence angle $|\theta| = 20°, 40°, 60°$ and $80°$ with a surface RMS slope $\sigma_x = 0.3$ and $k_0\omega = 0.5$. (From Bourlier *et al.* [II.110]. Reproduced by permission of © 2001 Taylor & Francis)

The stationary phase and the scalar approximation applied to scattering from one- and two-dimensional rough surfaces with a shadowing effect have been examined in Reference [II.110]. These solutions were obtained from the Kirchhoff integral which assumed that both the surface RMS radius of curvature R_C and the surface length correlation L are much greater than the wavelength λ. For the first assumption, Bourlier *et al.* [II.110] also quoted

$$k_0 R_C \cos^3\theta > 1, \tag{14.30}$$

where k_0 is the wave number and θ is the incidence angle. The surface RMS (root mean square) radius of curvature was evaluated from the joint probability density function (PDF). From the stationary phase method, the scattering coefficient was derived in Reference [II.110] as a function of the geometrical problem, the Fresnel coefficients and the surface height joint characteristic function, giving the surface random behaviour. With the scalar approximation, the radar bistatic cross-section depends, according to Bourlier *et al.*, on the previous characteristic function and on the expected values, which characterize the integration over the slopes.

The statistical dependence of the shadowing function over the characteristic function was also studied in Reference [II.110]. This means that the surface height density probability, assumed to be Gaussian,

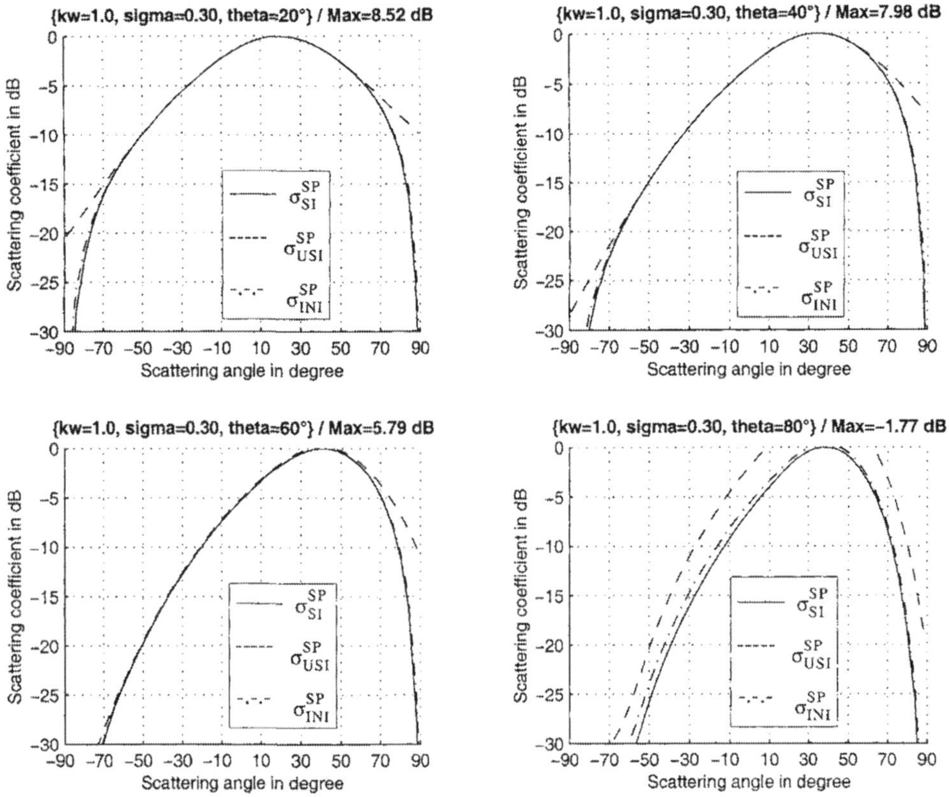

Figure 14.96 Same variation as Figure 14.95 with $k_0\omega = 1$. (From Bourlier *et al.* [II.110]. Reproduced by permission of © 2001 Taylor & Francis)

is modified by the shadow and loses its Gaussian behaviour. With the Smith shadowing function, Bourlier *et al.* showed that the characteristic function obtained from the shadowing function independence is greater than that computed with the statistical dependence; consequently, the characteristic function is overestimated.

Figures 14.95 to 14.99 represent the normalized incoherent bistatic scattering coefficient in dB from a perfectly conducting isotropic surface versus the scattering angle θ_{sc}, according to the incidence angle $|\theta| = \{20°, 40°, 60°, 80°\}$ with the surface RMS slope $\sigma_x = \{0.1, 0.3\}$ and $k_0\omega = \{0.5, 1, 4\}$, where ω is a surface RMS height. The scattering coefficient is normalized by the maximum of σ_{SI}^{SP}. The full curve is σ_{SI}^{SP} (stationary phase with shadow), the broken curve is σ_{USI}^{SP} (stationary phase without shadow) and the chain curve is σ_{INI}^{SP} (stationary phase with the shadow assumed to be independent). As depicted in Figure 14.95, for a slightly rough surface ($k_0\omega = 0.5$) the unshadowed scattering coefficient, σ_{USI}^{SP}, is in good agreement [II.110] with σ_{SI}^{SP} and σ_{INI}^{SP} around the specular direction given by $\theta_{sc} = |\theta|$. For grazing scattering angles the unshadowed scattering coefficient is overestimated, according to Bourlier *et al.*, since the shadowing effect is not negligible. From Figures 14.96 and 14.97, this behaviour is significant with respect to the surface roughness $k_0\omega = \{1, 4\}$. Comparing Figures 14.95 and 14.96 ($\sigma_x = 0.3$) with Figures 14.98 and 14.99 ($\sigma_x = 0.1$) respectively, Bourlier *et al.* noted that this overestimation decreases when the surface RMS slope, σ_x, decreases because the percentage of the hidden surface also decreases.

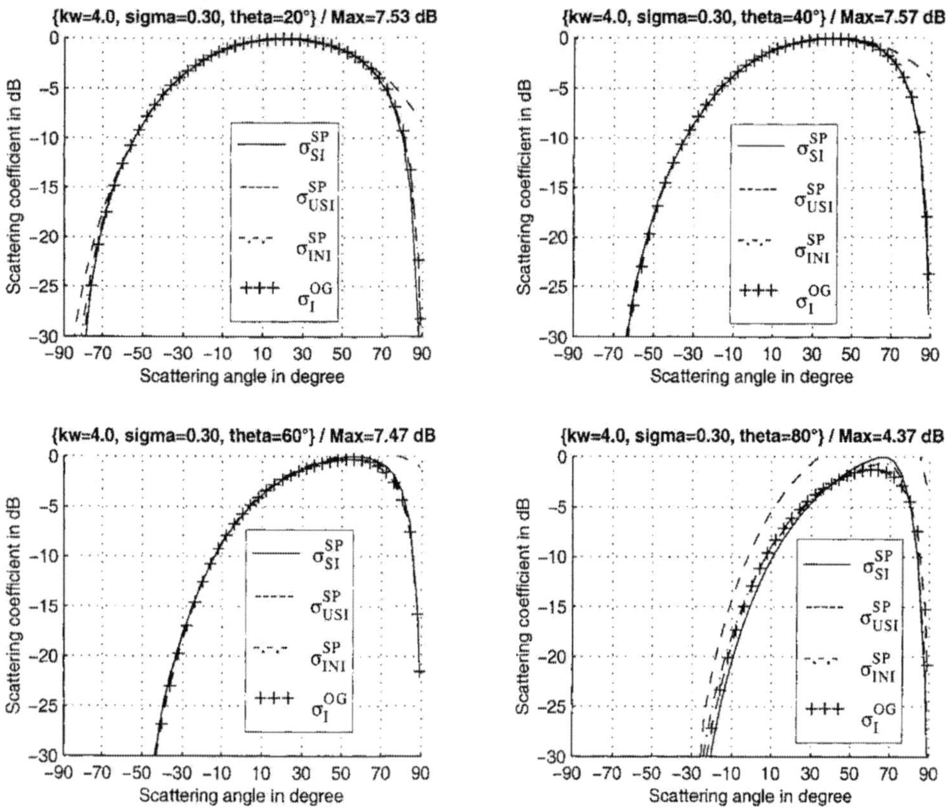

Figure 14.97 Same variation as Figure 14.95 with $k_0\omega = 4$. Moreover, $\sigma_{\mathrm{I}}^{\mathrm{OG}}$ values are plotted as crosses. (From Bourlier *et al.* [II.110]. Reproduced by permission of © 2001 Taylor & Francis)

As expected, the bistatic cross-section drops off more slowly with an increasing scattering angle as the surface RMS slope increases [II.110].

From Figures 14.95 to 14.99 and comparing $\sigma_{\mathrm{INI}}^{\mathrm{SP}}$ with $\sigma_{\mathrm{SI}}^{\mathrm{SP}}$, Bourlier *et al.* observed that $\sigma_{\mathrm{INI}}^{\mathrm{SP}}$ obtained from averaging the statistical shadowing function independently of the unshadowed bistatic cross-section is slightly overestimated and this overestimation increases with the surface roughness $k_0\omega$ and the surface slope variance. As depicted in Figure 14.97 with $k_0\omega = 4$, the geometrical optics solution with the shadow, $\sigma_{\mathrm{I}}^{\mathrm{OG}}$, plotted in crosses is correct since this example falls in the high-frequency region [II.110]. In general, since the shadowing effect is included, the bistatic cross-section is not symmetrical according to the specular direction [II.110].

In Figure 14.100, the bistatic incoherent scattering coefficient from the dielectric isotropic surface versus the scattering angle, θ_{sc}, with respect to the surface permittivity $\varepsilon_{\mathrm{r}} =$ {infinity (perfectly conducting surface case), $60 - 40\mathrm{j}$ (sea case in band X), $1.5 - 0.5\mathrm{j}$ (optical case)} and the polarization (VV or HH) with $|\theta| = 80°$ is plotted. The full curve is $\sigma_{\mathrm{SI}}^{\mathrm{SP}}$ (stationary phase method with a shadow) and the cross curve is $\sigma_{\mathrm{SI}}^{\mathrm{SA}}$ (scalar approximation with a shadow). The surface parameters are similar to those in Figure 14.99. It can be seen that when the permittivity increases, the incoherent component proportional to the squared modulus of the Fresnel coefficients decreases because the Fresnel coefficients are smaller [II.110]. Their moduli are maxima for a perfectly conducting

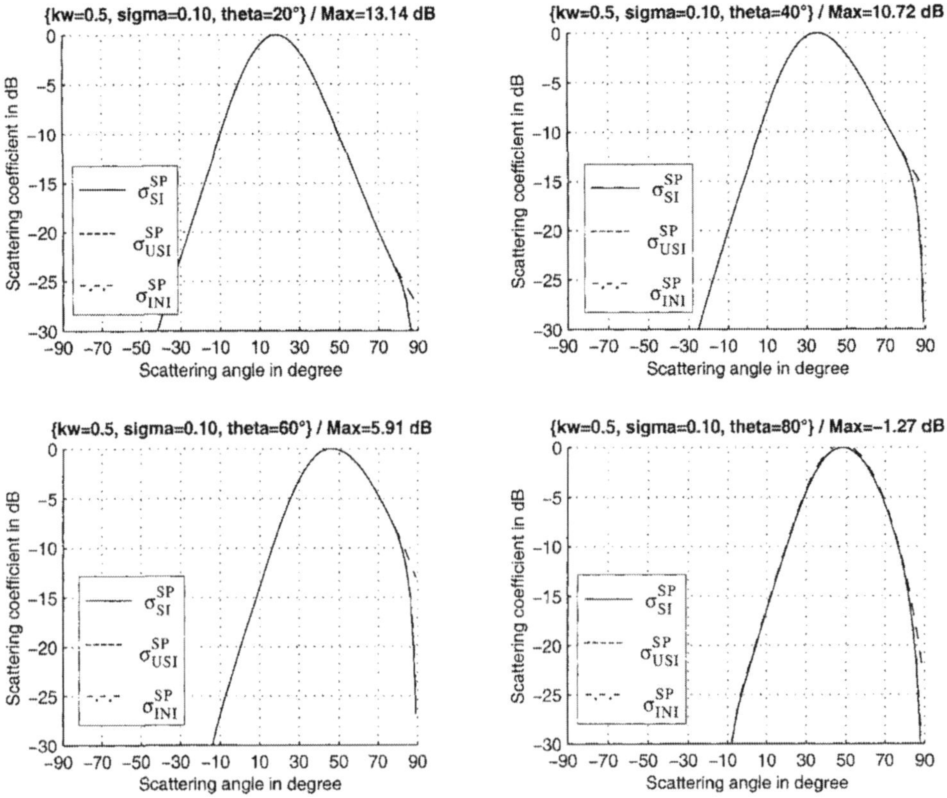

Figure 14.98 Same variation as Figure 14.95 with $\sigma_x = 0.1$. (From Bourlier *et al.* [II.110]. Reproduced by permission of © 2001 Taylor & Francis)

surface. It should also be noted that there is no deviation between the two scattering models. In vertical polarization, the Brewster phenomenon is not observed, because Brewster's angle is close to 90° where the scattering coefficient is very small [II.110].

In Figure 14.101, σ_{SI}^{SP}, σ_{INI}^{SP}, σ_{SI}^{SA} and σ_{INI}^{SA} are compared with the experimental data σ_I^{EX} ([II.111] with $\sigma_x = 0.154$). In Figure 14.101(a), $|\theta| = 20°$, $k_0\omega = 22.532$, $k_0L = 207.424$ and $k_0R_C = 679.81$. In Figure 14.101(b), $|\theta| = 20°$, $k_0\omega = 1.346$, $k_0L = 12.387$ and $k_0R_C = 40.698$. Both of these cases are in the high-frequency region and k_0R_C is much larger than unity. The experimental data were not plotted in Reference [II.110] because they are in good agreement with σ_{SI}^{SA} and slightly smaller than σ_{SI}^{SP}. It isn't difficult to see that $\sigma_{INI}^{SP} = \sigma_{SI}^{SP}$ and $\sigma_{INI}^{SA} = \sigma_{INI}^{SA}$, because the shadowing effect can be neglected [II.110].

The surface parameters of Figure 14.101(c) are exactly the same as in Figure 14.101(a). The only difference is in the incident angle. Although Bourlier *et al.* have $\{k_0L, k_0R_C\}$ much greater than unity, for scattering angles smaller than the specular direction the models deviate from the experimental data. Although the shadowing effect is included, at near grazing scattering angles, an overestimation according to σ_I^{EX} was observed in Reference [II.110]. Bahar and Lee [II.112] have also studied a full-wave solution without a shadow. From Figure 14.102 of Reference [II.112], it is observed that at near grazing angles, their model underestimates the bistatic cross-section. Bahar and Lee explain this deviation as being due

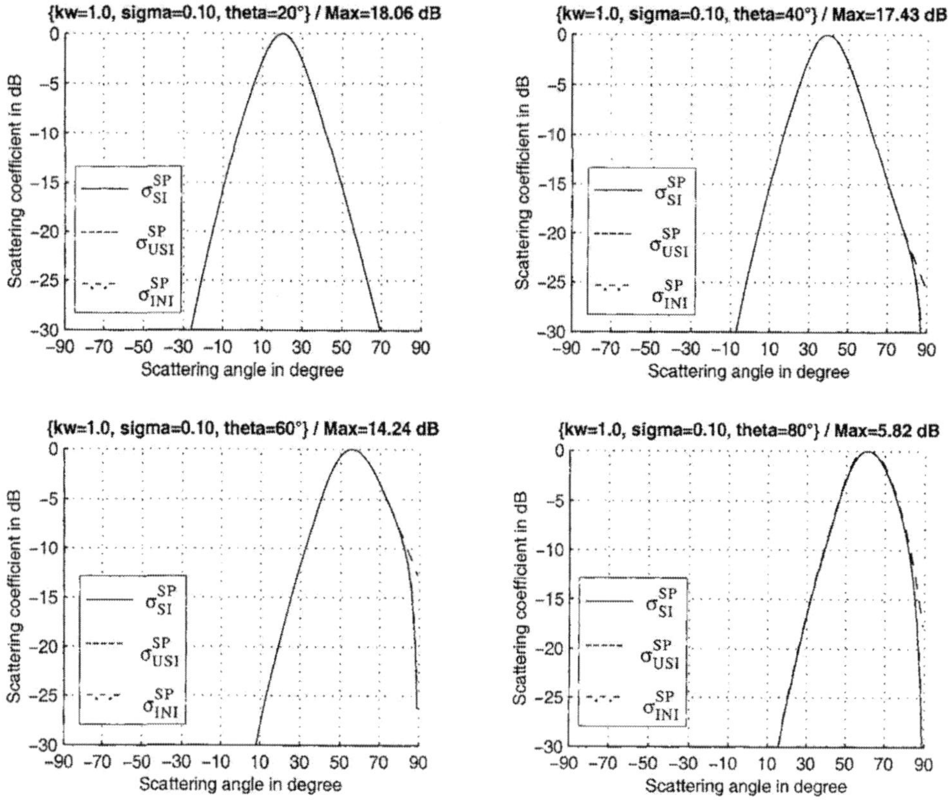

Figure 14.99 Same variation as Figure 14.96 with $\sigma_x = 0.1$. (From Bourlier *et al.* [II.110]. Reproduced by permission of © 2001 Taylor & Francis)

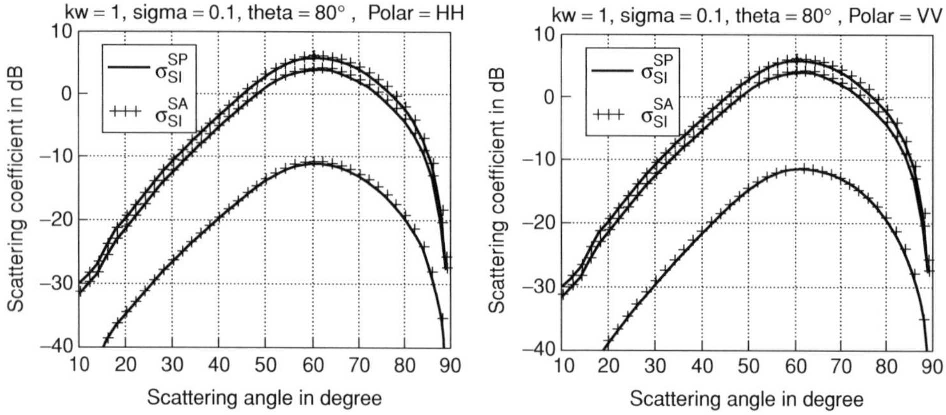

Figure 14.100 Bistatic incoherent scattering coefficient from a dielectric isotropic surface versus the scattering angle. (From Bourlier *et al.* [II.110]. Reproduced by permission of © 2001 Taylor & Francis)

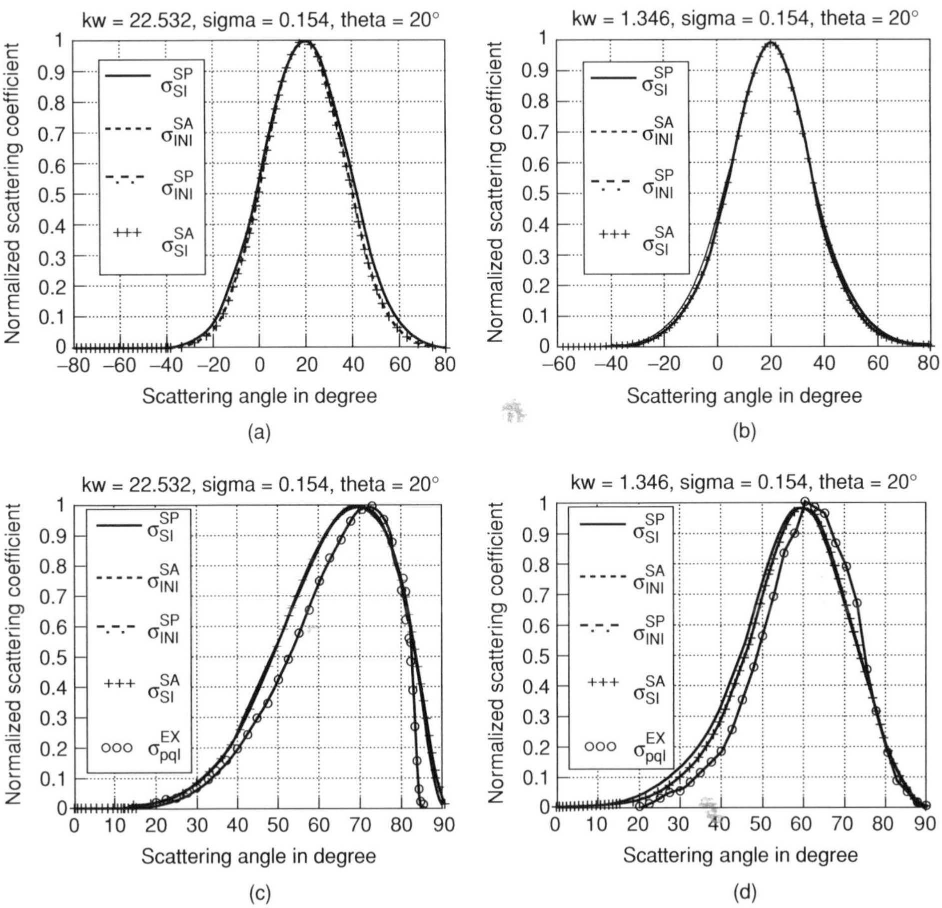

Figure 14.101 Comparison of models with the experimental data. (From Bourlier *et al.* [II.110]. Reproduced by permission of © 2001 Taylor & Francis)

to the fact that the experimental surface height autocorrelation function is not perfectly Gaussian. Indeed, it is observed to have an oscillatory behaviour, which becomes significant at grazing angles [II.112].

The surface parameters of Figure 14.101(d) are exactly the same as those found in Figure 14.101(b) and $|\theta| = 70°$. Similar to the case of Figure 14.101(c), the results deviate from the experimental data, but they are correct for grazing angles.

Thus, the simulations executed by Bourlier *et al.* [II.110] showed that the shadowing effect is important at near grazing incidence and scattering angles and when the surface slope variance becomes large. From an isotropic perfectly conducting surface, the offered models were compared with experimental data of Reference [II.111]. According to the scattering angle, the results of Bourlier *et al.* deviate from the experimental data when the scattering angles are smaller than the specular direction. Above the specular direction, the shadowing effect involved a nonsymmetrical behaviour and the models deviated less than those obtained without a shadow. Although the shadowing function was included, this deviation may be explained by the fact that the criterion (14.30) is not verified for grazing incidence angles. The results

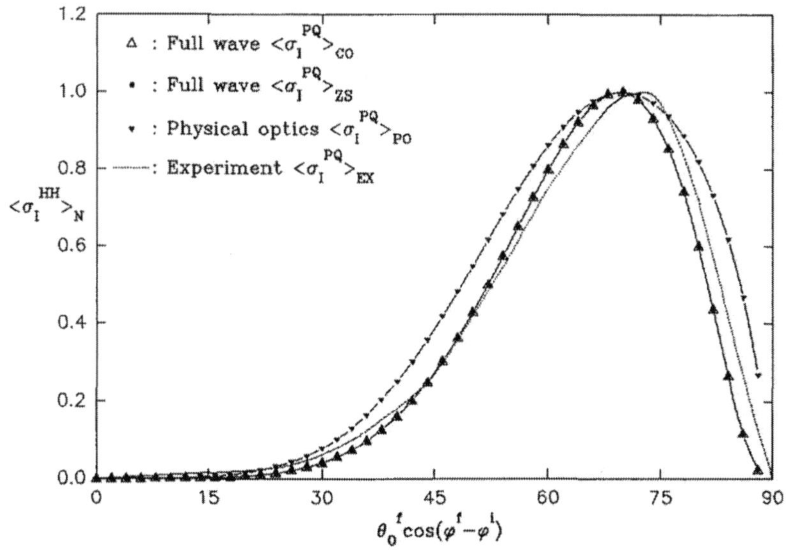

Figure 14.102 Bistatic scattering cross-section from a perfectly conducting 2D surface for horizontally polarized waves. (From Bahar and Lee [II.112]. Reproduced by permission of © 1996 Taylor & Francis)

computed from the scalar approximation were weakly smaller than those determined with the stationary phase.

In Reference [II.113], Hsieh and Fung developed a surface scattering model to separate the single and multiple scattering (see Figure 14.103) for studying the contributions to the scattering coefficients from randomly rough surfaces. With the consideration of multiple scattering, the comparisons between the model prediction and the measured data of bistatic scattering in the laboratory were shown to be in good agreement.

The model developed by Hsieh and Fung [II.113] was used to predict the scattering behaviour of bistatic multiple scattering from randomly very rough surfaces. For the purpose of illustration Hsieh and Fung assumed a Gaussian height distribution. To study the contribution of multiple scattering they rearrange and split the scattering coefficient into two parts: single and multiple scattering coefficients.

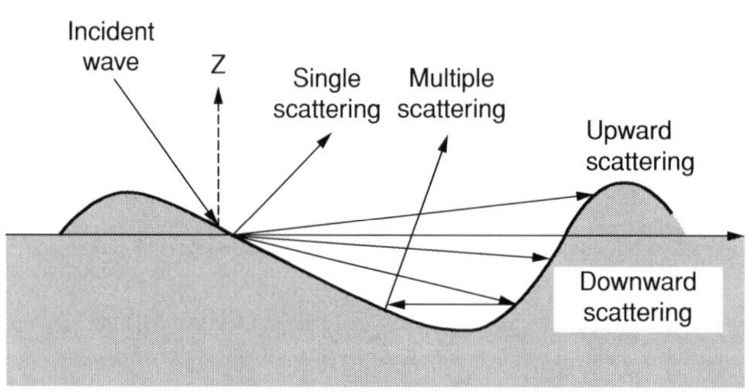

Figure 14.103 Single and multiple (upward and downward) scattering from a randomly rough surface. (From Hsieh and Fung [II.113]. Reproduced by permission of © 2003 IEE)

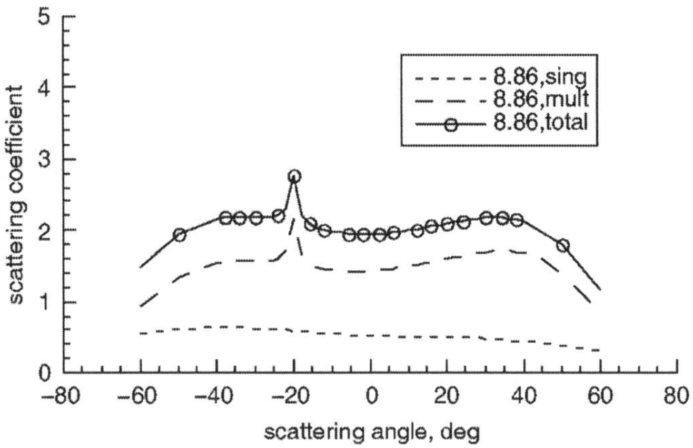

Figure 14.104 Comparison of VV polarized single, multiple and total scattering coefficients from random very rough surfaces with a normalized surface height deviation of 6.283, surface correlation length 8.86 and RMS surface slope 1.0 (incident angle = 20°). (From Hsieh and Fung [II.113]. Reproduced by permission of © 2003 IEE)

The multiple scattering model may be viewed, according to Hsieh and Fung, as a correction to the single scattering model for both the high- and low-frequency regions. To meet the real measurement, shadowing was used in Reference [II.113] to modify the scattering coefficient. These shadowing functions were based on the assumption of the geometric optics model. Hsieh and Fung believed that the shadowing correction is extremely important for the large scatter angle and for a very rough surface with a large slope.

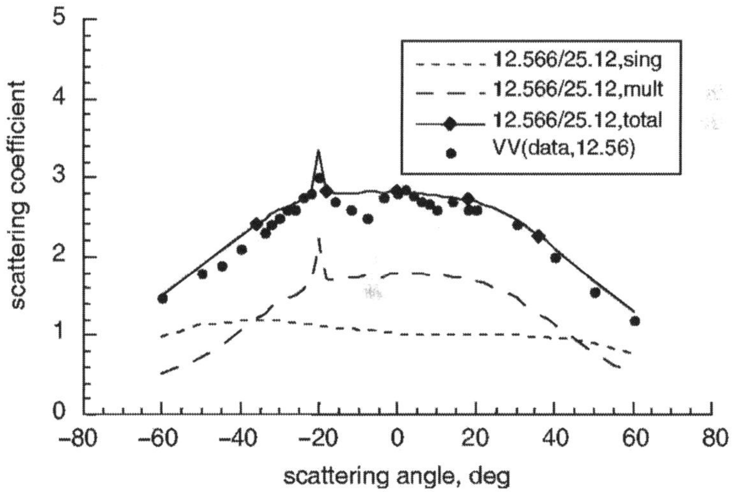

Figure 14.105 Comparison of single, multiple and total scattering patterns with measured data from random very rough surfaces with a normalized surface height deviation of 12.566, surface correlation length 25.12 and RMS surface slope 0.707 (incident angle = 20°). (From Hsieh and Fung [II.113]. Reproduced by permission of © 2003 IEE)

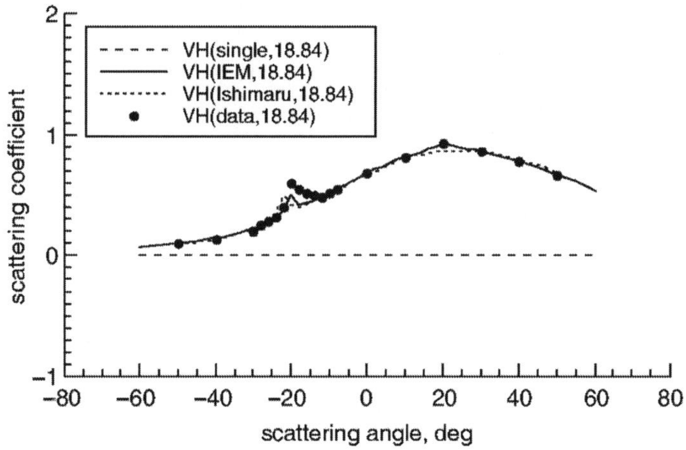

Figure 14.106 Comparison of single, multiple and total scattering patterns with measured data from random very rough surfaces with a normalized surface height deviation of 6.283, surface correlation length 18.84 and RMS surface slope 0.47 (incident angle = 20°). (From Hsieh and Fung [II.113]. Reproduced by permission of © 2003 IEE)

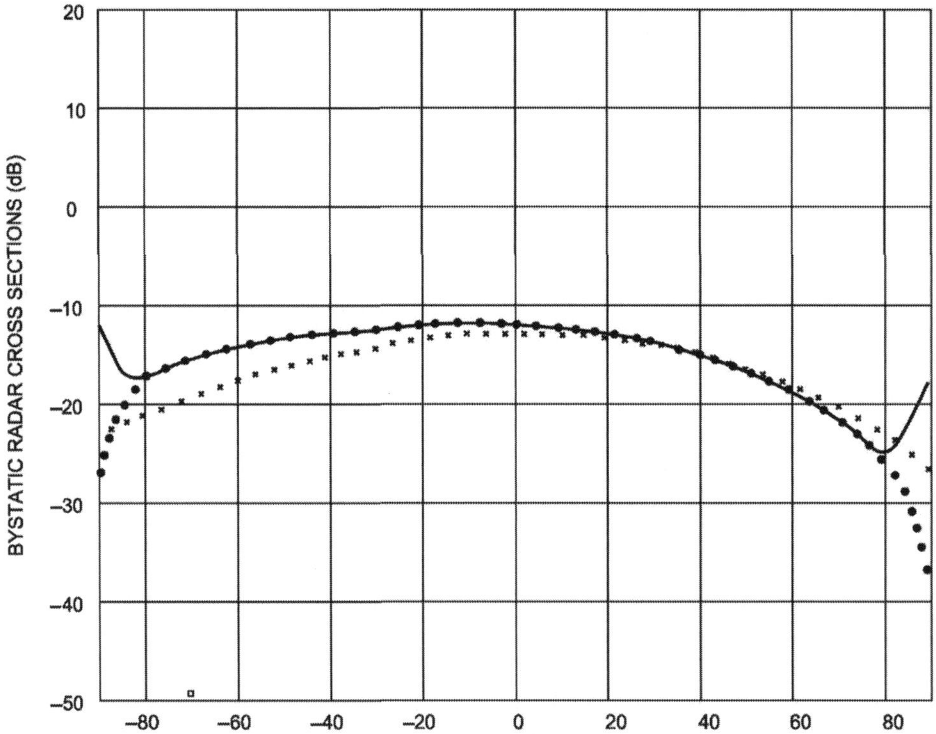

Figure 14.107 Comparison of the two-scale, small perturbation, Kirchhoff approximation bistatic radar cross-sections for the roughness parameters $k\sigma = 0.10$ and $kL = 1.40$, $\varphi = 0$ (From Khenchaf *et al.* [II.103]. Reproduced by permission of © 1996 IEEE)

Figure 14.104 shows the comparisons of VV polarized single, multiple and total scattering coefficients from random very rough surfaces with a normalized surface height deviation of 6.283 and surface correlation length 8.86. To evaluate the multiple contribution for electromagnetic wave scattering from randomly very rough surfaces, Hsieh and Fung showed the comparisons of the like- and cross-polarized single, multiple and total bistatic scattering coefficients with the measured data.

The predictions were compared in Reference [II.113] with the laboratory scattering measurements from the statistically known isotropic rough surfaces. The multiangle and multipolarization measurements were taken from a Gaussian distributed, Gaussian correlated and perfectly conducting rough surface. Comparisons of model prediction of the like- and cross-polarized bistatic single, multiple and total scattering with measurements are shown in Figure 14.105 and those for cross-polarized bistatic scattering are shown in Figure 14.106. The incident angle was 20° from the perfect conducting rough surfaces.

With consideration of multiple surface scattering, the excellent agreement of like-polarized scattering can be seen in Figure 14.105. In Figure 14.106, the comparison was made with cross-polarized measurements from a randomly rough surface. Without the contribution of multiple scattering, the excellent agreement between the model prediction and the measured data cannot be obtained according to Hsieh and Fung: the cross-polarized multiple scattering is extremely important, especially in the plane of incidence, where the single cross-polarized scattering reaches zero.

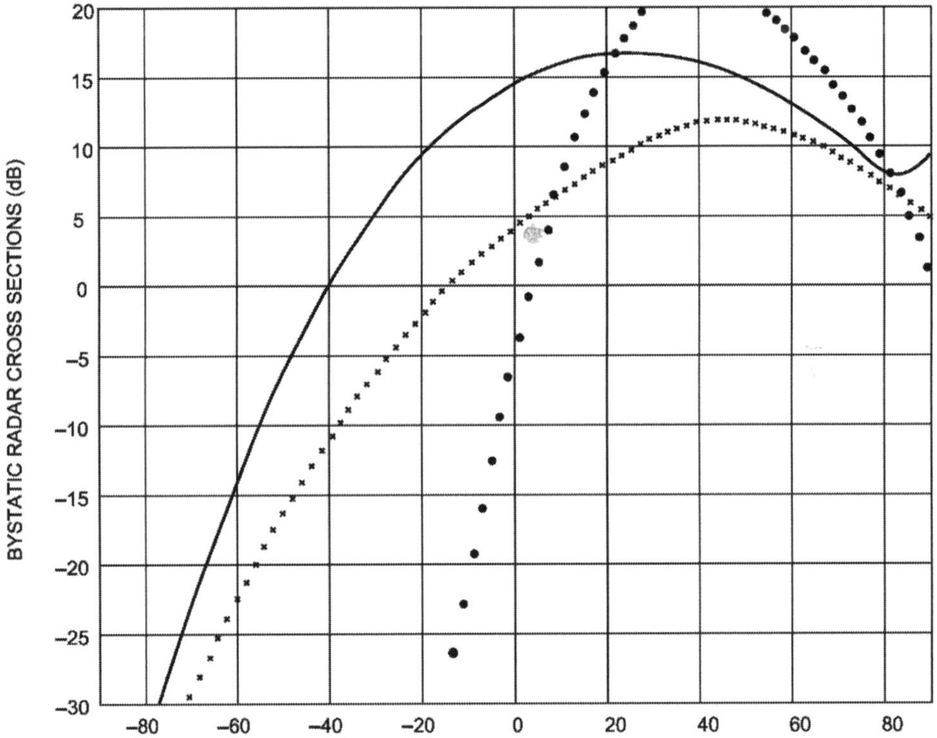

Figure 14.108 Comparison of the two-scale, small perturbation, Kirchhoff approximation bistatic radar cross-sections for the roughness parameters $k\sigma = 1.29$ and $kL = 7.31$, $\varphi = 0$. (From Khenchaf *et al.* [II.103]. Reproduced by permission of © 1996 IEEE)

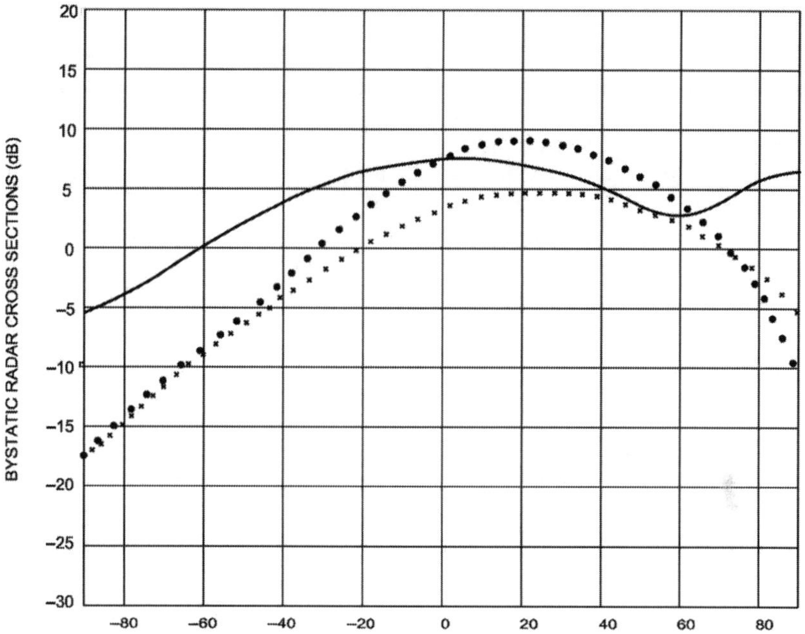

Figure 14.109 Comparison of the two-scale, small perturbation, Kirchhoff approximation bistatic radar cross-sections for the roughness parameters $k\sigma = 0.67$ and $kL = 2.80$, $\varphi = 0$. (From Khenchaf *et al.* [II.103]. Reproduced by permission of © 1996 IEEE)

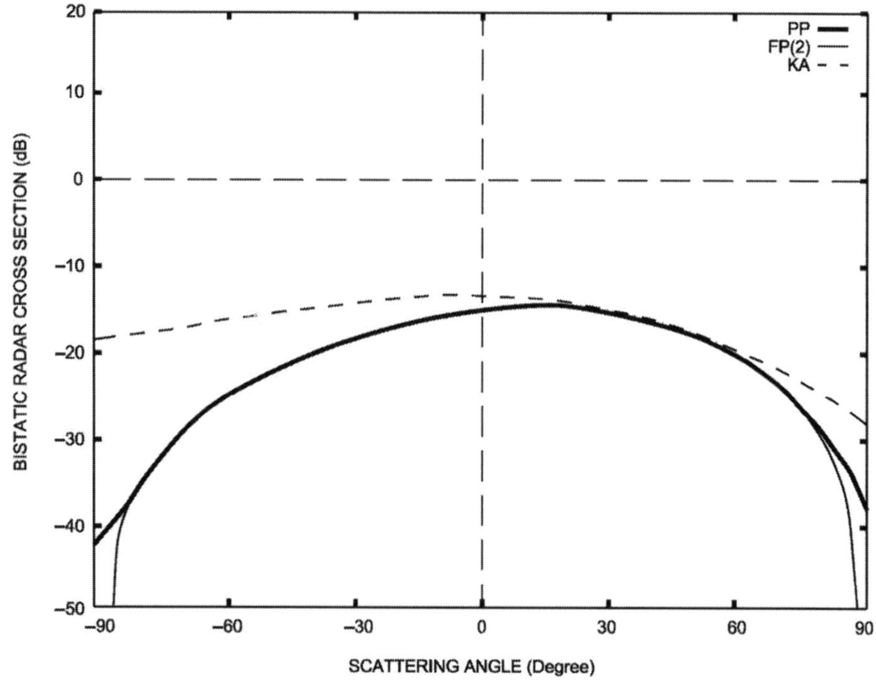

Figure 14.110 Comparison of the second-order field perturbation, phase perturbation and Kirchhoff approximation bistatic radar cross-sections for the roughness parameters $k\sigma = 0.10$ and $kL = 1.40$. (From Khenchaf *et al.* [II.103]. Reproduced by permission of © 1996 IEEE)

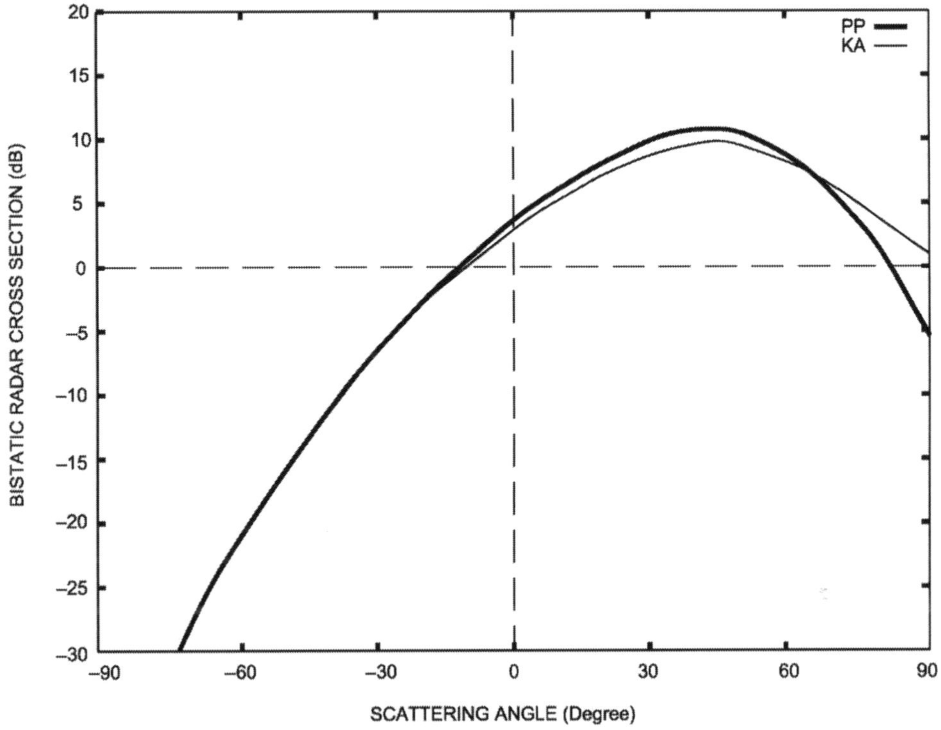

Figure 14.111 Comparison of the phase perturbation and Kirchhoff approximation bistatic radar cross sections for the roughness parameters $k\sigma = 1.29$ and $kL = 7.31$, $\varphi = 0$. (From Khenchaf *et al.* [II.103]. Reproduced by permission of © 1996 IEEE)

14.2.3 Sea Surface

Numerical results for the bistatic radar cross-section of the sea surface ($T = 20$ °C, $S = 35$ ppm) with a Gaussian roughness spectrum are presented in Reference [II.103]. A moderate incident angle of 45° was used for each example and the cross-sections are computed over the range of scattering angles from −90° to 90° (−90° to 0° is the back-scattering case; 0° to 90° means forward-scattering) in the plane of incidence, as shown in Figures 14.107 to 14.109 (••, small perturbation model; –, two-scale model; ××, scalar approximation). The results are compared with those given by Broschat *et al.* [II.114] for the Dirichlet problem for 2D surfaces with a 2D isotropic Gaussian spectrum (see Figures. 14.110 to 14.112).

The two-scale model for the sea surface scattering analysis was used in Reference [II.110]. The small scale that characterizes the capillary waves and the scattering coefficient was evaluated with, for example, the small perturbation method. The high scale corresponding to the gravity waves and the scattering coefficient was computed with the stationary phase. Assuming that both of these scales are independent, the bistatic cross-section is then evaluated in Reference [II.110] from summing both of these terms.

Figure 14.113 represents the normalized bistatic incoherent scattering coefficients, $\sigma_{\mathrm{INI}}^{\mathrm{SP}}$, and $\sigma_{\mathrm{USI}}^{\mathrm{SP}}$ (the index US means that the shadow is ignored) for a two-dimensional sea surface versus the scattering angle. The wind directions were $\phi = 0°$, 45° and 90° and different sea states, $u_{10} = 5$ and 10 m/s, were examined. Figure 14.113(a) is shown for $u_{10} = 5$ m/s and $|\theta| = 40°$, Figure 14.113(b) for $u_{10} = 5$ m/s and $|\theta| = 85°$, Figure 14.113(c) for $u_{10} = 10$ m/s and $|\theta| = 40°$, and Figure 14.113(d) for $u_{10} = 10$ m/s

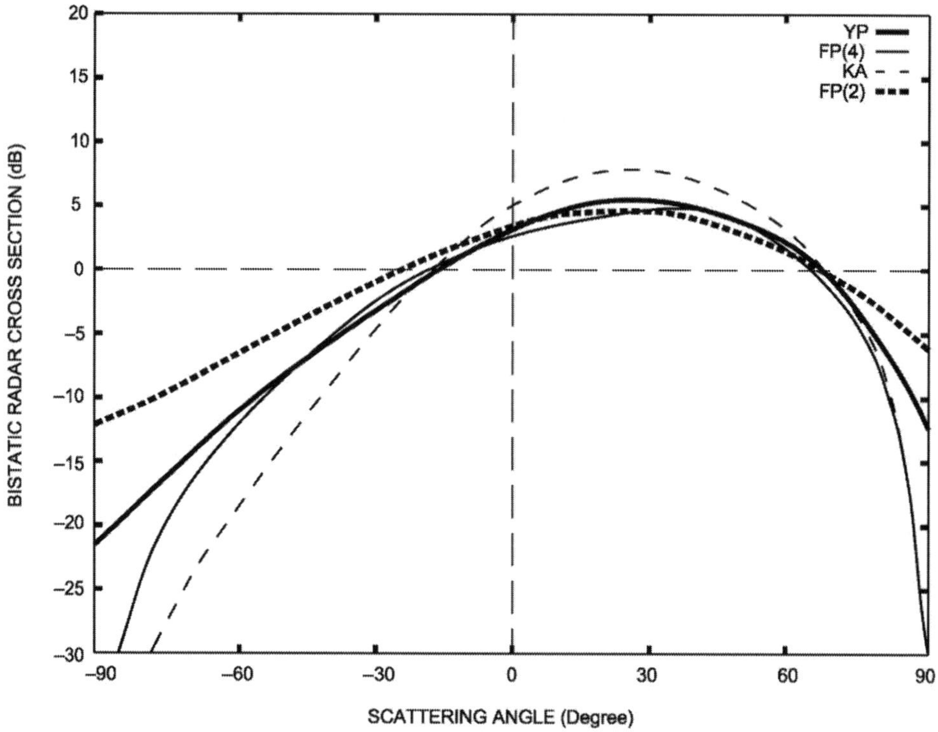

Figure 14.112 Comparison of the phase perturbation, Kirchhoff approximation and second- and fourth-order field perturbation bistatic radar cross-sections for the roughness parameters $k\sigma = 0.67$ and $kL = 2.80$, $\varphi = 0$. (From Khenchaf *et al.* [II.103]. Reproduced by permission of © 1996 IEEE)

and $|\theta| = 85°$. The normalization was obtained from the maximum computed with $\phi = 90°$. The electromagnetic wavelength was equal to $\lambda = 3$ cm (X-band). The surface relative complex permittivity was equal to 56 – 38j, which corresponds, according to Reference [II.110], to temperature and the salinity of the sea surface of 20 °C and 35 g/l respectively, with $\lambda = 3$ cm.

In Figures 14.113(a) and 14.113(b), the wind speed $u_{10} = 5$ m/s involves the surface RMS slopes $\sigma_x = 0.0775$, 0.0638 and 0.0462 with respect to $\phi = 0°$, 45° and 90°, $\omega = 0.162$ m, $k_0\omega = 34.00$ and $|\theta| = 40°$ and 85°. As ϕ increases, the surface slope variance decreases causing the bandwidth to decrease, whereas the maximum enhances the bandwidth. For a near grazing incidence angle, Bourlier *et al.* [II.110] saw that the shadowing effect is not negligible.

In Figures 14.113(c) and 14.113(d), the parameters are the same as in Figures 14.113(a) and 14.113(b) respectively, with $u_{10} = 10$ m/s. This means that $\sigma_x = 0.0784, 0.0644$ and 0.0463, $\omega = 0.658$ m and $k_0\omega = 137.89$. Although $k_0\omega$ is greater than in Figures 14.113(a) and 14.113(b), the behaviour is similar to a weak diminution of the maximum because the slope variance is greater. This fact comes from the fact that $k_0\omega$ is so large that the geometrical optics approximation is valid and the scattering coefficient then becomes independent of $k_0\omega$ [II.110].

In Figure 14.114, for a two-dimensional sea surface, the bistatic incoherent scattering coefficient $\sigma_{\mathrm{INI}}^{\mathrm{SP}}$ is represented versus the scattering angle with $\phi = 0°$, $u_{10} = 10$ m/s, $|\theta| = 80°$ according to the frequency $f = 1$, 5 and 10 GHz and polarization VV and HH. Since the sea surface is a dispersive medium, its permittivity depends on the frequency. With a temperature and salinity of the sea surface of 20 °C and 35 g/l, Bourlier *et al.* used ε_r ($f = 1$, 5 and 10 GHz) = 72 – 85j, 67 – 41j and 56 – 38j. As

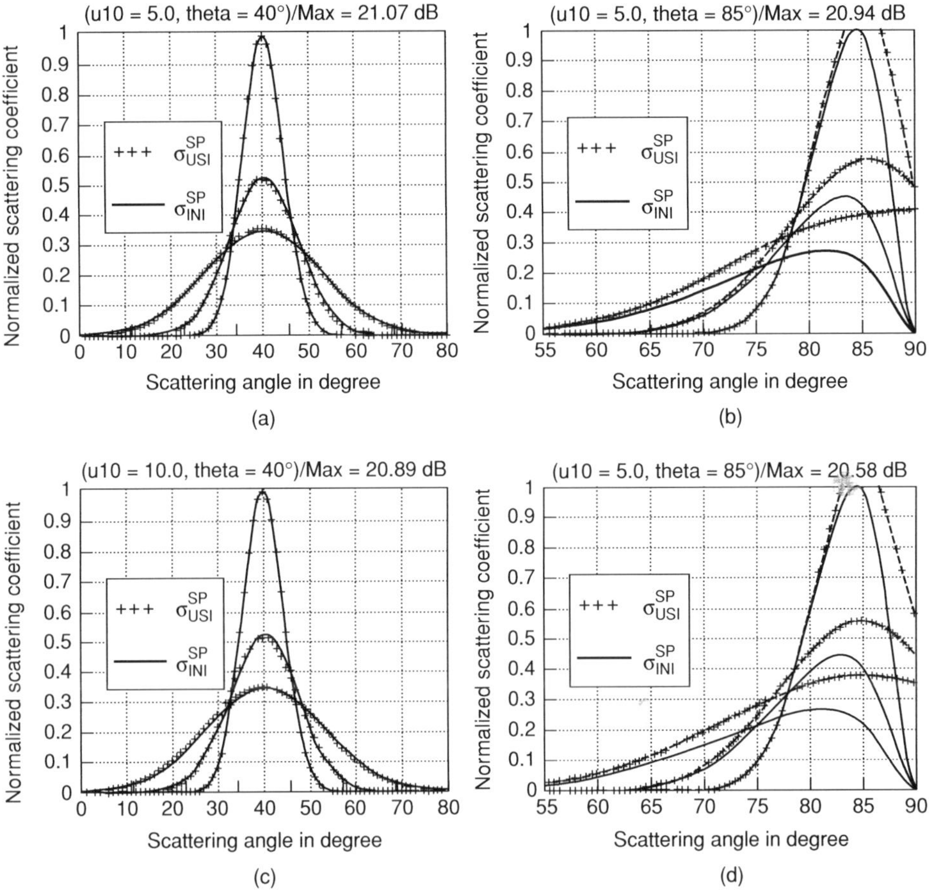

Figure 14.113 Normalized bistatic incoherent scattering coefficients, $\sigma_{\mathrm{INI}}^{\mathrm{SP}}$ and $\sigma_{\mathrm{USI}}^{\mathrm{SP}}$, for a two-dimensional sea surface versus the scattering angle with wind directions $\phi = 0°$, $45°$ and $90°$. (From Bourlier *et al.* [II.110]. Reproduced by permission of © 2001 Taylor & Francis)

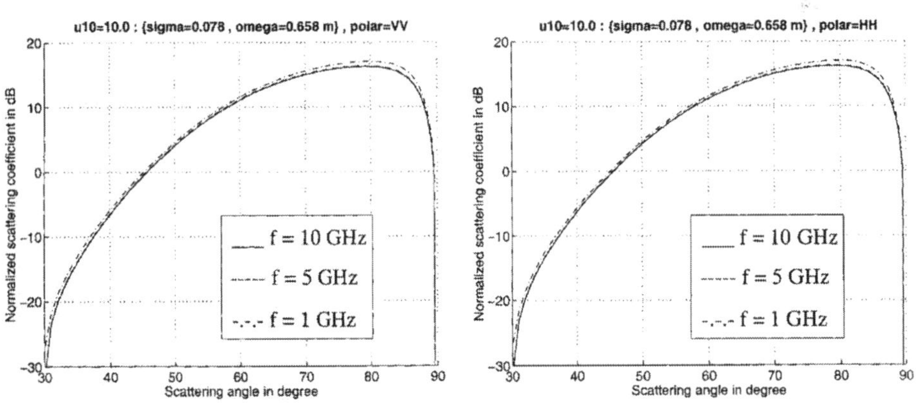

Figure 14.114 Bistatic incoherent scattering coefficient $\sigma_{\mathrm{INI}}^{\mathrm{SP}}$ for a two-dimensional sea surface versus the scattering angle with $\phi = 0°$, $u_{10} = 10$ m/s, $|\theta| = 80°$ and according to the polarizations: (a) VV and (b) HH, with $f = 10$ GHz (full curve), $f = 5$GHz (broken curve) and $f = 1$ GHz (chain curve). (From Bourlier *et al.* [II.110]. Reproduced by permission of © 2001 Taylor & Francis)

depicted in Figure 14.113, the maximum of the scattering coefficient decreases when $k_0\omega$, proportional to the wind speed u_{10}, increases. Therefore, in Figure 14.114, since $k_0\omega = \{137.89, 68.94, 13.79\}$ the incoherent component should increase with the frequency f. However, the sea permittivity decreases with frequency as the Fresnel coefficients in VV and HH polarization increase when the frequency decreases. Both of these phenomena are then in opposition and allow Bourlier *et al.* to explain that the scattering coefficient observed in Figure 14.114 increases when the frequency decreases. Bourlier *et al.* also noted that the scattering coefficient varies weakly with the polarization.

Summary of Part II

For bistatic, multistatic and multisite radars, the transmitter and receiver are not co-located. The configuration of a bistatic radar assumes the presence of one transmitting station and one receiving station, separated in space. From a system theory point of view, the bistatic radar may be considered as an elementary cell of a multistatic radar system. However, the bistatic radar has an independent importance of its own. Selected problems of bistatic radar theory have been stated in this part.

A bistatic radar may be stationary, shipborne, airborne and space-based. This part has been devoted solely to stationary bistatic radars. It is important that the transmitter and receiver locations for a stationary bistatic radar are not changed during the operation process.

The widely known phenomenon of electromagnetic wave scattering is fundamental for bistatic radar. Some of the basic concepts and results of electromagnetic wave scattering theory were reviewed. The bases of electromagnetism have been stated and some key bistatic radar concepts, such as cross-sections and scattering matrices, have been presented.

An important question within the theory of a bistatic radar concerns the choice of the system of coordinates and the system of reference. This question has been studied in detail. Important problems such as resolution, maximum range and effective area have also been discussed.

A significant portion of this part has been devoted to signal models: for an understanding of the bistatic radar system, it is important to obtain the relationships between electromagnetic waves reradiated by a target and the radar signal parameters. Chapter 14 was dedicated to reviewing the methods of analysis of bistatic scattering problems.

Abbreviations

AF	ambiguity function
AMP	Antenna Modeling Program
BOR	body of revolution
BRCS	bistatic radar cross-section
BSCS	bistatic scattering cross-section
CFD	computational fluid dynamics
CMT	current marching technique
EEC	equivalent edge current
EFIE	electrical field integral equation
EM	electromagnetic
EMW	electromagnetic wave
FDTD	finite differen time-domain
FEM	finite element method
FSR	forward-scattering radar
FVTD	finite volume time-domain
GDT	geometrical diffraction theory
GPS	global positioning system
GTD	geometrical theory of diffraction
HF	high frequency
HH	horizontal polarization for both (transmitting and receiving) antennas
HIM	hybrid iterative method
IBC	impedance boundary conditions
JRC	Joint Research Centre
MBET	monostatic–bistatic equivalence theorem
MFIE	magnetic field integral equation
MOM	method of moments
MRCS	monostatic radar cross-section
MW	microwave

Bistatic Radar: Principles and Practice Edited by M. Cherniakov

NEC	numerical electromagnetic code
PDF	probability density function
PE	parabolic equation
PEC	perfect electric conductors
PO	physical optics
PTD	physical theory of diffraction
RBRCS	relative bistatic radar cross-section
RCS	radar cross-section
RCSC	radar cross-section coefficient
RF	radio frequency
RMS	root mean square
SAR	synthetic aperture radar
SCS	scattering cross-section
SCSC	scattering cross-section coefficient
SIE	surface integral equation
SPM	small perturbation method
TE	transverse electric
TM	transverse magnetic
VV	vertical polarization for both (transmitting and receiving) antennas

Variables

a	large main axis of a polarizing ellipse
a	radius of a sphere
A	effective area of any antenna
A	constant
A_0	Huynen's parameter
A	vector potential
$A(\rho)$	spatial function of autocorrelation
A_{R}	effective area of the receiving antenna
b	small main axis of a polarizing ellipse
B	Huynen's parameter
B	constant
B_0	Huynen's parameter
$\mathbf{B}$	magnetic induction
c	speed of light in vacuum
C	Huynen's parameter
C	constant
d	thickness of skin
$\mathrm{d}\sigma^{\mathrm{sc}}$	differential scattering cross-section
D	fractal dimension of a surface
D	overall dimension of the receiving antenna
D	Huynen's parameter
D	constant
$\mathbf{D}$	electrical displacement
$\boldsymbol{e}_h$	ort, i.e. unit vectors in a direction of increase of coordinate h
$\boldsymbol{e}_{\mathrm{H}}$	ort in the horizontal direction
$\boldsymbol{e}_q$	ort, i.e. unit vectors in a direction of increase of coordinate q
$\boldsymbol{e}_{\mathrm{s}}$	unit vector in the direction of a Poynting vector
$\boldsymbol{e}_{\mathrm{s}}^{\mathrm{in}}$	unit vector in the direction of a Poynting vector of an incident wave

$\boldsymbol{e}_{\mathrm{s}}^{\mathrm{sc}}$	unit vector in the direction of a Poynting vector of a scattering wave
$\boldsymbol{e}_{\upsilon}$	ort, i.e. unit vectors in a direction of increase of coordinate υ
$\boldsymbol{e}_{\mathrm{V}}$	ort in the vertical direction
$\boldsymbol{e}_{x}$	unit ort of the coordinate axis OX
$\boldsymbol{e}_{y}$	unit ort of the coordinate axis OY
E	Huynen's parameter
E	intensity of an electrical field
E	Jones vector
E^{in}	incident Jones vector
E^{sc}	scattered Jones vector
E^{in}	intensity of an incident electrical field
E_{0}^{in}	constant amplitude of intensity of an incident electrical field
$\mathbf{E}^{\mathrm{sc}}$	intensity of a scattering electrical wave
$\mathbf{E}_{0}$	complex amplitude of intensity of an electrical field, $\mathbf{E}$
$\mathbf{E}_{0x}$ and $\mathbf{E}_{0y}$	two orthogonal components of $\mathbf{E}_{0}$
f	frequency
F	Huynen's parameter
g	directivity factor of any antenna
g_{T}	directivity factor of the transmitting antenna
G	gain coefficient (factor) of any antenna
G	Huynen's parameter
$G(\boldsymbol{k})$	spatial power spectrum of eminences
$G(\boldsymbol{r}, \boldsymbol{r}')$	Green's function
G_{R}	gain coefficient for the receiving antenna
G_{T}	gain coefficient for the transmitting antenna
h	coordinate
H	Huynen's parameter
H	Hurst parameter of a surface
$H(\boldsymbol{r}, \boldsymbol{r}')$	solution of the Helmholtz homogeneous equation
$\mathbf{H}$	intensity of a magnetic field
$\mathbf{H}^{\mathrm{in}}$	intensity of an incident magnetic field
$\mathbf{H}_{0}^{\mathrm{in}}$	constant amplitude of intensity of an incident magnetic field
$\mathbf{H}^{\mathrm{sc}}$	intensity of a scattering magnetic wave
I	Stokes parameter
$\mathbf{I}$	Stokes vector
$i_{\Gamma}(t)$	current, which is induced in the target by the field $\dot{u}_{\Gamma}(t)$
$\mathbf{J}$	density of a current
$\mathbf{J}_{\mathrm{s}}$	surface current density
k	wave number
k_{I}	imaginary part of a complex wave number
k_{R}	real part of a complex wave number
$\dot{k}$	complex wave number
$\mathbf{K}$	the Kennaugh matrix
L	baseline of the bistatic radar
L	autocorrelation length
L_{0}	topothesy of a surface
L_{e}	effective bistatic base
L_{P1}	losses that appear during wave propagation in the area of the transmitter–target
L_{P2}	losses which appear during wave propagation in the area of the target-receiver

L_s	system losses
$\boldsymbol{n}$	normal (unit vector)
N	north pole of the Poincare sphere
P_R	signal power at the input of the receiver
P_T	output power of the transmitting antenna
P_{Rmin}	sensitivity of the radar receiver
q	polarizing relation
q	coordinate
q_S	surface charge
q^{in}	polarization ratio for the incident wave
q^{sc}	polarization ratio for the scattering wave
Q	Stokes parameter
r	spherical coordinate
$\boldsymbol{r}$	radius vector of the point in space
$\boldsymbol{r}$	vector, which is directed from the phase centre of the receiving antenna to the point of target
R	receiver
R	range sum (total range)
R	range of the target in the monostatic radar case
$\mathbf{R}$	vector, which is directed from the phase centre of the transmitting antenna to the point of target
R_{max}	maximum range of monostatic radar
$R_\perp$ and $R_\parallel$	Fresnel coefficients
R_1	range of the target relative to the transmitter
R_2	range of the target relative to the receiver
S	surface
S	south pole of the Poincare sphere
S	Poynting's vector
$s(t)$	signal
$s_T(t)$	probing signal
$\dot{s}_0(t)$	complex signal of unit power
t	current time
T	transmitter
$\mathbf{T}$	Jones matrix
$T_\perp$ and $T_\parallel$	Fresnel coefficients
U	Stokes parameter
$U(t)$	law of the amplitude modulation
$\dot{U}(t)$	complex envelope of the complex signal of unit power
$\dot{u}(t, \boldsymbol{r})$	field of the spherical wave
$\dot{u}_R(t)$	complex envelope of the field in the phase centre of the receiving antenna
$\dot{u}_\Gamma(t)$	complex field intensity of a wave, which reaches the target
v	coordinate
V	volume
V	Stokes parameter
$\mathbf{V}$	vector of speed
$\mathbf{V}_r$	radial component of the vector of speed
$\mathbf{V}_t$	tangential component of the vector of speed
${}_\Gamma V_r$	module of the radial constituent of the target velocity relative to the phase centre of the receiving antenna

${}_{\Gamma}V_{t}$	module of the tangential constituent of the target velocity relative to the phase centre of the receiving antenna
W	electromagnetic energy
x	Cartesian coordinate
x_A	Cartesian coordinates of any point A of space
x_R	Cartesian coordinate of the receiver
x_T	Cartesian coordinate of the transmitter
y	Cartesian coordinate
y_A	Cartesian coordinates of any point A of space
y_R	Cartesian coordinate of the receiver
y_T	Cartesian coordinate of the transmitter
z	Cartesian coordinate
z_A	Cartesian coordinates of any point A of space
z_R	Cartesian coordinate of the receiver
z_T	Cartesian coordinate of the transmitter
α	spectral parameter of a surface
α, β, and γ	direction cosines of axis OZ'
β	bistatic angle
Γ	target (scattering body, object, heterogeneity of medium)
δ	angle of losses in the dielectric
δ_{hv}	the Kronecker symbol
$\delta(\boldsymbol{x})$	Dirac delta function
$\Delta\vartheta$	phase shift
$\Delta\varphi$	difference of phases
ε	permittivity
ε_0	dielectric constant
$\dot{\varepsilon}$	complex dielectric permittivity
$\zeta(\rho)$	height of a rough surface above a point ρ of a plane XOY
$\bar{\zeta}$	average height of roughness on a surface
η_A	efficiency factor of the antenna
θ	spherical angle coordinate
θ_R	spherical angle coordinate of the receiver rather than the target
${}_{R}\theta$	look angle of the receiver
θ_T	spherical angle coordinate of the transmitter rather than the target
${}_{T}\theta$	look angle of the transmitter
θ_{1Br}	Brewster angle
ϑ	elevation
ι	aberration angle
λ	wavelength
λ_0	wavelength
μ	permeability
μ_0	magnetic constant
Ξ	energy flux
$\boldsymbol{\Pi}$	Hertz vector
Π_R	power flux density of the wave reradiated by a target
Π_Γ	power flux density of the direct electromagnetic wave in the vicinity of a target
ρ	volumetric density of a charge
ρ	a difference of the path of the useful signal and the path of the direct wave from the transmitter to the receiver
ρ	effective target range

$\boldsymbol{\rho}$	vector, which is directed from the phase centre of the transmitting antenna to the phase centre of the receiving antenna
$\boldsymbol{\rho}$	radius vector of a point on a coordinate plane XOY
ρ_{m}	radius of the mth Fresnel region
ρ_{pol}	polarization efficiency (polarization match factor)
ρ_0	target range equal to the length of the median for the bistatic triangle
ρ_1	radius of the first Fresnel region
σ	active conductivity
σ	radar cross-section of a target
σ	root mean square meaning of roughness on a surface
σ_{b}	bistatic RCS of a target
σ_{m}	monostatic RCS of a target
σ_{r}	relative bistatic radar cross-section
σ^{sc}	bistatic scattering cross-section of a target
σ_0^{sc}	scattering cross-section coefficient (scattering cross-section of the unit area of the target)
σ_0	radar cross-section coefficient
Σ	motionless system of reference
$\boldsymbol{\Sigma}$	Sinclair matrix
$\boldsymbol{\Sigma}_{\mathrm{S}}$	symmetric Sinclair matrix
$\boldsymbol{\Sigma}_{\mathrm{SS}}$	skew-symmetric Sinclair matrix
Σ'	moving system of reference
τ	target vector
ς	energy potential (ratio of power of the probing signal to receiver sensitivity)
$\upsilon(t)$	law of the angle modulation
$\dot{\Upsilon}$	complex conductivity of a medium
ϕ	spherical angle coordinate
ϕ	angle between the bistatic bisector and target velocity vector
ϕ_{R}	spherical angle coordinate of the receiver rather than the target
ϕ_{T}	spherical angle coordinate of the transmitter rather than the target
Φ	scalar potential
$\Phi(\mathbf{R})$	amplifier coefficient of the transmitting antenna
χ	casual phase having uniform distribution in the range $(-\pi, \pi]$
ψ	any Cartesian component of intensity of an electrical field or intensity of a magnetic field
ψ	set of 2×2 complex Pauli matrices
ω	circular frequency
$\Omega_{\mathrm{D}}(\boldsymbol{\rho})$	Doppler frequency shift of the signal in the receiving antenna aperture
$\aleph$	relative coefficient

Part III

Forward-Scattering Radars

A.B. Blyakhman, A.G. Ryndyk, A.V. Myakinkov

15

Basic Principles of Forward-Scattering Radars

15.1 FORWARD-SCATTER RADAR CROSS-SECTION

Bistatic radars based on the principle of forward-scattering, or forward-scattering radars, are designed to detect and track targets moving in the narrow region stretched along the baseline, which connects the transmit and receive sides. In this region, the RCS of targets increases sharply due to the so-called 'forward-scatter effect'. The topology of the forward-scattering radar (FSR) is shown in Figure 15.1, where Tr is a transmit position, Re is a receive position, x, y, z are Cartesian coordinates, with the origin corresponding to the phase centre of the receiving antenna, b is the base of the system, Tg is a target, $\boldsymbol{V}$ is a target velocity vector, φ is an array of trajectory inclination towards the baseline in the horizontal plane, γ is the angle of trajectory inclination towards the horizontal plane, β_{B} is a bistatic angle, α is a target azimuth and β is its elevation.

In addition, the neighbourhood of a target location is shown separately in Figure 15.1 to define some of the angle relationships. The local Cartesian coordinate system (x', y', z') is collinear to the system (x, y, z) with the origin corresponding to the current target location. Poynting vectors of incident wave and wave scattered in the direction of the receive side are denoted by $\boldsymbol{S}_{\mathrm{inc}}$ and $\boldsymbol{S}_{\mathrm{tg}}$ respectively. The diffraction angle is denoted by α_{D}, while its projections on to the horizontal and vertical planes are denoted by α_{h} and α_{v} respectively. Obviously, the bistatic angle, β_{B}, and diffraction angle, α_{D}, are concatenated by the relationship $\alpha_{\mathrm{D}} = \pi - \beta_{\mathrm{B}}$.

In the first part of the book the definition of the target RCS was presented on the basis of a radar equation analysis. Let us consider the RCS on the basis of the incident and reflected waves electric fields intensity [III.1–III.3]. Expressing the RCS in terms of the electric field intensity gives:

$$\sigma = 4\pi R^2 \left(\frac{|E_{\mathrm{rc}}|^2}{|E_{\mathrm{inc}}|^2} \right), \tag{15.1}$$

where E_{inc} and E_{rc} are the electric field intensity of the incident wave in the target location and the electric field intensity at the receive side respectively; R is the target range from the receive side. The above definition is true for both monostatic and bistatic radar configurations. In the latter case, the radar cross-section of the target is called the bistatic RCS or BRCS. The RCS of most real targets essentially

Bistatic Radars: Principles and Practice Edited by M. Cherniakov

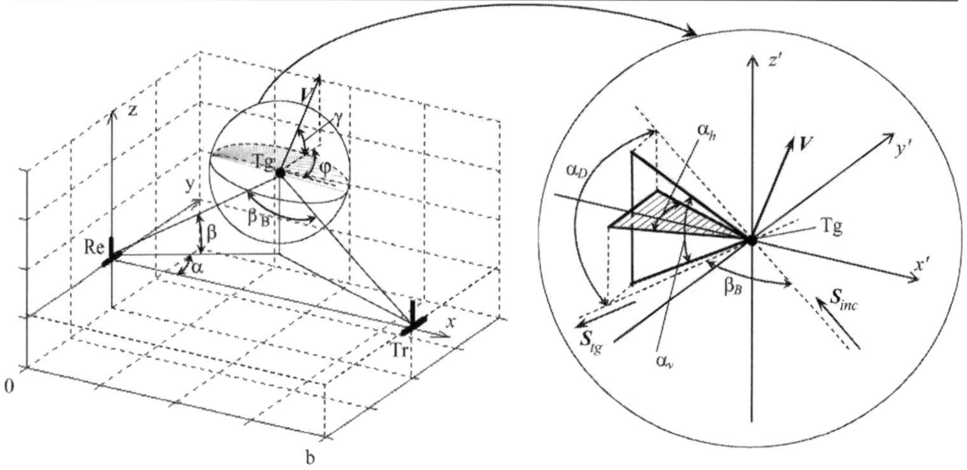

Figure 15.1 FSR topology

depends on the corresponding bistatic angle. In general, the bistatic RCS, σ_B, differs from its monostatic counterpart, σ_0.

Consider the process of forming the electromagnetic field scattered by a target in the bistatic case. This field consists of two components. First of all, any real target creates a self-scattering field that is produced by the currents induced on the surface of the target by the incident wave. An intensity of the electric component of the self-scattering field is designated as $\boldsymbol{E}_S$.

It is known from electromagnetic fields and waves theory [III.4] that if an absolute black body of finite size, which is larger or comparable with the wavelength, λ, is placed in the way of wave propagation, then a scattering field or shadow field appears behind the body. This field appears as a result of a primary field disturbance (the shadowing part of the incident wave front). Designating an electric component of the shadow field as $\boldsymbol{E}_{SH}$, the full scattered field is then determined by the sum

$$\boldsymbol{E}_{\Sigma} = \boldsymbol{E}_S + \boldsymbol{E}_{SH}. \tag{15.2}$$

The targets with the dimensions comparable to and bigger than the wavelength have both shadow and self-scattering fields that are separated in space. The shadow field is concentrated in the narrow solid angle near $\beta_B = 180^\circ$ and is called the forward-scattering (FS) field. In general, in this region the self-scattering field is much weaker than the shadow field. Because of this, analysing the scattering field in the FS region allows the effect of the currents on the surface of the target to be neglected, i.e. to consider a real target to be an absolute black body. Similarly, by considering a scattering field at moderate β_B, the shadow field can be neglected.

It is obvious that when calculating the RCS of a target in the forward-scattering condition or the FS RCS, it is necessary for the target model to be significantly different from the model used in the monostatic case. Consider an absolute black body located in the way of plane wave propagation (Figure 15.2).

If the size of an absolute black body is much larger than the wavelength, the shadow region is formed just behind the body. The shadow region is characterized by the fact that the field in this region is equal to zero; in particular, the electric component is equal to zero, i.e. $\boldsymbol{E} = 0$. Such a situation can be described mathematically in the following way: in the shadow region there is a certain compensation field $\boldsymbol{E}_C = -\boldsymbol{E}_{inc}$, which is equal in amplitude and opposite in sign to the field of the incident wave. It is very important that the cross-section of the shadow region is determined only by the shadow contour or silhouette of the target, and does not depend upon its three-dimensional shape. Thus a certain planar cophasal aperture emitting the field, $\boldsymbol{E}_C$, the shape of which is determined by the shadow contour of the

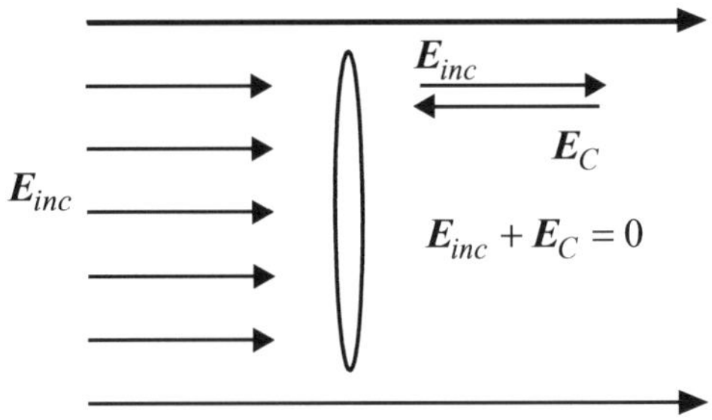

Figure 15.2 Shadow field forming

body, is formed just behind the black body whose shadow size is much larger than the wavelength. This aperture forms a shadow FS field in the distant receiving point.

When the size of the target is much larger than the wavelength, the self-scattering field of the target in the shadow region can be neglected. Hence, in an approximation of physical optics, the FS field is not dependent upon the three-dimensional shape of the target and is completely determined by the target shadow contour as well as the field of the absolute black body.

For a quantitative assessment of the FS RCS, Babinet's principle discussed in Part II can be used. According to this principle, the diffraction pattern from an opaque body is identical to that from a hole of the same size and shape, except for the overall forward beam intensity. The field arising from the incident wave diffraction on the opening coincides (to within a sign) with the shadow field of the initial screen [III.2, III.5, III.6]. Let us consider the shadow field, $\boldsymbol{E}_{SH}$, in the distant receiving point for the case when the incident wave is plane. This field corresponds to the radiation field of a planar cophasal aperture, which is perpendicular to the direction of the incident wave propagation. The aperture form is defined by the projection of the target shadow contour on to the plane, which is perpendicular to the direction of propagation of the incident wave.

Figure 15.3 shows the cophasal aperture, A_{SH}, which forms the shadow field in the distant receiving point Re. In Figure 15.3, ρ is the radius vector of an arbitrary point M of the aperture, A_{SH}, $\boldsymbol{r}_0$ is a unit vector oriented towards the receive side, x, y, z are Cartesian coordinates with an origin in the conditional centre of the aperture, θ_x, θ_y are arguments of direction cosines of the unit vector $\boldsymbol{r}_0$, $\boldsymbol{R} = \{R_x, R_y, R_z\}$ is the radius vector of the receiving point and R_x, R_y, R_z are projections of the vector $\boldsymbol{R}$. Hence, assuming that aperture A_{SH} presents the target while point Re presents the phase centre of the receiving antenna, magnitude $R = |\boldsymbol{R}|$ of vector $\boldsymbol{R}$ presents the target range.

In view of the 'radiating' aperture, A_{SH}, the target FS RCS for a distant receiving point at bistatic angles, close to 180°, is determined by the following expression:

$$\sigma_F(\boldsymbol{r}_0) = 4\pi R^2 \left(\frac{|E_{SH}|^2}{|E_{inc}|^2}\right) = \frac{4\pi}{\lambda^2}\left|\int_{A_{SH}} \exp\left[j\left(\frac{2\pi}{\lambda}\right)\boldsymbol{\rho r}_0\right] dS\right|^2, \tag{15.3}$$

where E_{SH} is the electric field intensity of the shadow field, λ is the wavelength of the transmitter and dS is the surface element of the cophasal aperture. It follows from Equation (15.3) that when $\boldsymbol{r}_0 \perp \boldsymbol{\rho}$, the FS RCS reaches the maximum. It can be seen from Figure 15.3 that in this case $\beta_B = 180°$ or $\alpha_D = 0$. Considering the FS RCS as a function of the diffraction angle $\sigma_F = \sigma_F(\alpha_D)$, gives

$$\sigma_F(0) = 4\pi\left(\frac{S_A}{\lambda}\right)^2, \tag{15.4}$$

where S_A is the area of aperture, A_{SH}. For clarity Equation (15.4) will be presented as $\sigma_F(0) = G_A S_A$, where $G_A = 4\pi S_A/\lambda^2$ is the directive gain of the co-phasal aperture A_{SH} with an area S_A. Thus, the

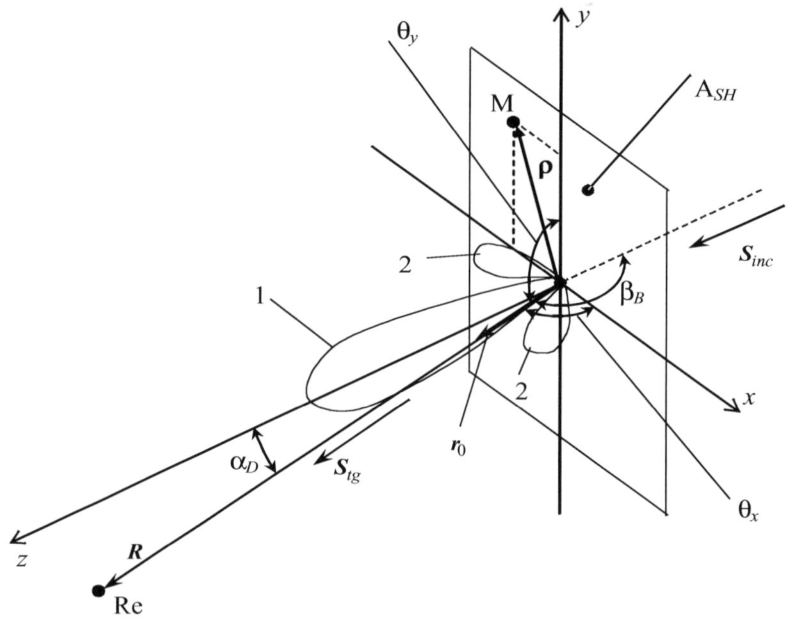

Figure 15.3 Creation of a co-phasal aperture shadow field at a distant point

FS RCS $\sigma_F(0)$ is larger than the geometrical square of the aperture A_{SH} in G_A times. If $S_A \gg \lambda^2$ then $G_A \gg 1$, and hence $\sigma_F(0) \gg S_A$. As a result the FS RCS happens to be much larger than the monostatic RCS. For example, for an ideally conductive full sphere of radius $r_S = 20\lambda$, the monostatic RCS and BRCS (at $\beta_B < 140$–$150°$) is $\sigma_0 = \sigma_B = S_A = \pi r_S^2 = 400\,\pi\lambda^2$. According to Equation (15.4), the FS RCS is $\sigma_F\,(0) = 4\pi(\pi r_S^2/\lambda)^2 = 64\pi^3 \times 10^4\lambda^2$. Therefore, $\sigma_F\,(0)$ is $1600\pi^2$ times or 42 dB bigger than σ_0.

The sharp increase in the magnitude of the target RCS in the FS region is called the forward-scatter effect. This effect makes it possible to improve the power budget of the radar. It is very important to mention here that it is impossible to reduce the target's FS RCS by applying antiradar coating or using other measures that are used to reduce the monostatic RCS. Moving on to scattering on the ideally conducting bodies, it is known that applying even ideally absorbent coating on the metallic body it is possible to reduce the value of its total scattered power by only 3 dB [III.6].

However, the values of the FS RCS σ_F, determined by Equation (15.4), hold true only in the narrow sector around the FSR baseline. The directional pattern of the considered co-phasal aperture usually has a lobe structure. In Figure 15.3 the main- and sidelobes are marked by 1 and 2 correspondingly. Therefore, in terms of the directional pattern structure, deviation of the bistatic angle β_B from 180° or deviation of the diffraction angle, α_D, from zero should not be more than half the width of the mainlobe of the directional pattern of the 'radiating' aperture, A_{SH}. It is known that its half-power level width is $\Delta\theta \approx \lambda/l_A$ in radians, where l_A is a linear aperture size. For example, for $r_S = 20\lambda$ sphere, the FS RCS decreases by 3 dB when α_D deviates from zero by only $\pm 0.75°$.

Nevertheless, an essential gain in the RCS may remain in the sidelobe region of the directional pattern of the aperture, A_{SH}, leading to the excess of $\sigma_F(\alpha_D)$ over σ_0. This gain can be evaluated.

Now let us obtain an expression for an approximate calculation of the FS RCS as a function of the horizontal and vertical diffraction angles, assuming that the target size considerably exceeds the emitted wavelength. The diffraction angle is understood to be the angle between the rays, passing from the transmit side towards the target and from the target towards the receive side. Horizontal and vertical projections of this angle are horizontal, α_h, and vertical, α_v, diffraction angles respectively. The corresponding angle relationships are illustrated in Figure 15.1.

It can be seen from Figure 15.3 that $\boldsymbol{r}_0 = \{\cos\theta_x;\ \cos\theta_y;\ \sin\theta_x\}$ and $\boldsymbol{\rho} = \{x, y, 0\}$. Passing on to Cartesian coordinates in Equation (15.3) gives

$$\sigma_F(\mathbf{r}_0) = \frac{4\pi}{\lambda^2}\left|\iint_{A_{SH}} \exp\left[j\frac{2\pi}{\lambda}\left(x\cos\theta_x + y\cos\theta_y\right)\right] dx\,dy\right|^2. \tag{15.5}$$

From Figure 15.3, the Poynting vector of an electromagnetic wave illuminating the target is collinear to axis z. Hence, in the region of small diffraction angles arguments of direction cosines θ_x, θ_y are concatenated with the horizontal and vertical diffraction angles, α_h, α_v, by the expressions $\alpha_h \approx \pi/2 - \theta_x$ and $\alpha_v \approx \pi/2 - \theta_y$. In the domain of small diffraction angles, on the assumption of a plane wave normally illuminating the shadow contour of the target, then

$$\cos\theta_x = \sin\left(\frac{\pi}{2} - \theta_x\right) \approx \sin\alpha_h \approx \alpha_h, \qquad \cos\theta_y = \sin\left(\frac{\pi}{2} - \theta_y\right) \approx \sin\alpha_v \approx \alpha_v. \tag{15.6}$$

In view of Equation (15.6), Equation (15.3) takes the form

$$\begin{aligned}\sigma_F(\alpha_h, \alpha_v) &\approx \frac{4\pi}{\lambda^2}\left|\iint_{A_{SH}} \exp\left[jk_W(x\alpha_h + y\alpha_v)\right] dx\,dy\right|^2\\ &= \frac{4\pi}{\lambda^2}\left|\int_{x_{min}}^{x_{max}} \exp(jk_W x\alpha_h)\,dx \int_{y_2(x)}^{y_1(x)} \exp(jk_W y\alpha_v)\,dy\right|^2\\ &= \frac{4\pi}{\lambda^2}\left|\int_{x_{min}}^{x_{max}} \exp(jk_W x\alpha_h)\,dx \frac{1}{jk_W\alpha_v}\left\{\exp\left[jk_W\alpha_v y_1(x)\right] - \exp\left[jk_W\alpha_v y_2(x)\right]\right\} dx\right|^2,\end{aligned} \tag{15.7}$$

where $k_W = 2\pi/\lambda$ is a wave number and $y_1(x)$ and $y_2(x)$ are the upper and the lower bounds of the co-phasal aperture.

In References [III.7] and [III.8] a simple method of the approximate calculation of an integral defined by Equation (15.7) is considered. The upper and the lower bounds of the co-phasal aperture are approximated by piecewise-linear segments (Figure 15.4). The integral in Equation (15.7) is then evaluated as the sum of integrals by the separate aperture segments, approximated by trapeziums. The

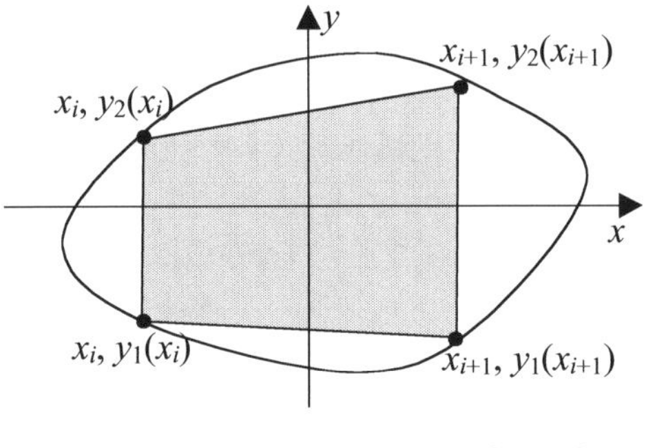

Figure 15.4 Approximating segment in the form of a trapezium

final analytic expression for the evaluation of the FS RCS takes the form

$$\begin{aligned}
\sigma_F(\alpha_h, \alpha_v) &= \frac{4\pi}{\lambda^2}\left|\frac{-1}{k_W^2\alpha_v}\sum_{i=1}^{N}(I_{1i} - I_{2i})\right|^2,\\
I_{1i} &= \frac{1}{\alpha_h + \alpha_v b_1}\exp(jk_W\alpha_v a_1)\,[\exp(jk_W\{\alpha_h x_{i+1} + \alpha_v[y_1(x_{i+1}) - y_1(x_i)]\})\\
&\quad - \exp(jk_W x_i\alpha_h)],\\
I_{1i} &= \frac{1}{\alpha_h + \alpha_v b_2}\exp(jk_W\alpha_v a_2)[\exp(jk_W\{\alpha_h x_{i+1} + \alpha_v(y_2(x_{i+1}) - y_2(x_i)]\})\\
&\quad - \exp(jk_W x_i\alpha_h)],\\
a_1 &= y_1(x_i),\\
b_1 &= \frac{y_1(x_{i+1}) - y_1(x_i)}{x_{i+1} - x_i},\\
a_2 &= y_2(x_i),\\
b_2 &= \frac{y_2(x_{i+1}) - y_2(x_i)}{x_{i+1} - x_i}.
\end{aligned} \tag{15.8}$$

The calculation results of the FS RCS for cruise missile type and helicopter type targets are presented in Figure 15.5. Dependence of the FS RCS upon the target azimuth, $\sigma_F(\alpha)$, is presented in dB with respect to m^2. It was assumed that the target was following a preassigned trajectory at a constant altitude, h. When the target altitude is low, the range of target azimuth values, $\alpha \approx \alpha_h$, corresponds to the forward-scatter RCS region, which is in the order of $\pi/6$. A solid line on these figures (curves 1) corresponds to the emitted signal wavelength $\lambda = 0.3$ m; a dashed line (curves 2) is for $\lambda = 1$ m.

After simultaneous analysis of the maximum FS RCS values and the width of the target observation angular sector, where the FS RCS still essentially exceeds its monostatic counterpart, it is possible to conclude that the range $\lambda = 0.3$–0.8 m is optimum for FSR applications against the discussed targets.

The monostatic RCS of the cruise missile in the specified range is about 0.1–0.2 m^2 and that of the helicopter is in the order of 2–3 m^2 [15.9]. Thus, the calculation shows considerable (by 30–50 dB) energy gain of the FSR in comparison with the monostatic radar in the FS region. The main lobe width of the FS RCS could be from 1–2° to about 10° depending on the target shadow contour size and transmitter wavelength. However, a significant FS RCS increase also takes place in the region of sidelobes up to diffraction angle values of about 10–15°. This allows an effective FS detection of targets in the angle sector of about ±15°.

The accuracy of the proposed method of the FS RCS calculation depends upon a number of factors, such as the relationship between target size and wavelength, the number of trapezium segments used for integral calculation and the range of diffraction angles. Solving the problems of FS target detection and tracking, the precise shape of the FS RCS is not so important. Here the qualitative conclusions concerning the maximal value of the FS RCS, its mainlobe width and sidelobe level are much more important. However, results obtained via an experimental FSR prove that the considered method of calculation gives sufficiently adequate estimates of the FS RCS in dependence on the diffraction angle inside the FSR coverage. The experimental results mentioned above are described in Chapter 19.

15.2 ADVANTAGES AND PROBLEMS OF THE FSR

The basic advantage of the FSR is the essential increase in the power budget in the directions close to a given baseline. It was shown in the previous section that when a bistatic angle exceeds an angle of 150°–160°, a sharp increase in the target radar RCS is observed and, hence, the intensity of an echo signal also increases at the receive side [III.7, III.10–III.14]. The key feature of the forward-scatter effect

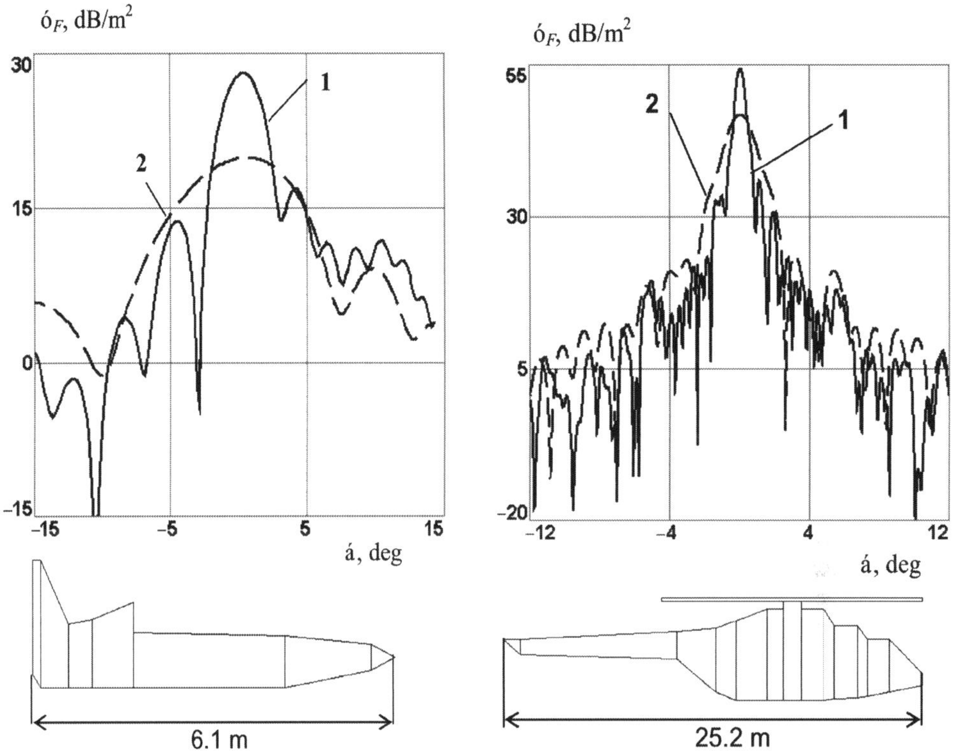

Figure 15.5 FS RCS as a function of the target azimuth: (a) cruise missile with the trajectory parameters: $x_{CR} = 15$ km; $\varphi = 45°$; $\gamma = 0$; $h = 1$ km; (b) helicopter with the trajectory parameters: $x_{CR} = 20$ km; $\varphi = 90°$; $\gamma = 0$; $h = 300$ m

for radar applications is its fundamental independence on the radio absorbent coating of a target, as well as a target's particular shape.

Another advantage of a forward-scattered signal is a reduction in its fluctuations. In a monostatic radar the phase interference of reflections from two and more main scatterers or scattering centres causes noticeable distortions of an echo signal. This happens in the most widespread practice of radar, when the size of a target exceeds the wavelength of radiation. As a result, with regards to the radar, even small variations of an aspect angle of the target give rise to sharp fluctuations of an echo signal. The echo signal burst modulation caused by this worsens the radar's characteristics of target detection and tracking. This effect was considered for monostatic radars in the Part I of the book.

In the FSR this effect is essentially reduced [III.7, III.10]. This is explained by the fact that in the FSR the magnitude and phase of the signal scattered by a target is defined by the shadow contour [III.4–III.7], which insignificantly varies even with appreciable changes in the target aspect angle. Another explanation of this FSR feature is presented in Reference [III.15]. A two-point target model is considered to analyse the influence of a target's natural swinging upon the fluctuations of the echo signal. It is shown that in the case of a conventional radar, a target's natural swinging essentially influences the phase difference between signals returned from considered reflection points. This leads to significant fluctuations of the echo signal. At the same time, in the FSR the influence of a target's natural swinging upon the mentioned phase difference is negligibly small. Hence, echo signal fluctuations are insignificant.

As in any other real system, the FSR has not only advantages but also inherent drawbacks and limitations. The main disadvantages are listed below:

- The FS effect occurs only within a relatively small solid angle along the baseline. Therefore, an increase in the RCS will not be observed if the target does not pass near the baseline.
- The observed Doppler frequency shift of a signal will be reduced to zero when the target crosses the baseline. Therefore, the radar loses its ability to measure range when the target crosses the baseline.
- All ground returns towards the baseline will act on the receiver practically simultaneously with the target echo. This can hamper selection of target signals against a background of passive interference.

It is necessary to add at least two technical problems to the above list of fundamental problems:

- A very strong direct signal is present at the input of the FSR receiver. This may exceed the receiver dynamic range and put it into the saturation mode.
- In general, the FSR requires synchronization separated by a long-distance transmitter and the receiver position, which may cause essential technical problems.

These problems are some of the main reasons why the FSR is not widely used at present. Let us consider possible ways of overcoming, or at least ameliorating, the drawbacks listed above.

Development of radar technology has demonstrated that those systems which are capable of not only target detection but also target position measurement have far better prospects. A number of bistatic systems with coordinate measurement capability [III.13, III.16, III.17] are known. Measurement of target coordinates via the FSR is performed by two stages. The first stage is signal processing. At this stage both target detection and measurement of target return primary parameters are executed. Primary parameters can be, for example, the target return Doppler frequency shift, the target return angle of arrival, the transmitter–target–receiver range sum and the rate of change of this sum. Using measurements of primary parameters, target trajectory parameters are to be estimated by a certain algorithm at the second stage. Trajectory parameters may be current Cartesian coordinates of the target, its velocity and heading. It is important to select methods of obtaining a necessary set of the primary parameters properly, which allows transferring the potential advantages of the FSR into a practical domain and bypassing their drawbacks where possible.

First of all, the type of probing signal has to be selected and, in general, two main approaches are possible. The first approach involves using an impulse probing signal, while the second uses a continuous one. In the next chapters of this part the second approach is preferred, and the explanations for this shall be given.

Let us carry out a simplified analysis of a two-dimensional FSR. Two-dimensional radar allows measurement of the horizontal target coordinates, x, y, and their derivatives, V_x, V_y, but does not allow measurement of the target altitude, h. Assume that the impulse signal is used for the measurement of the transmitter–target–receiver range sum. The range sum is obtained unambiguously from measurement of the difference between delays of the target return and the direct transmitter signal. In the considered case, the delay in the target return with respect to the direct transmitter signal measured is small due to the FS effect region being stretched along the baseline. Hence, the probing signal bandwidth should be sufficiently wide.

Let us estimate the band of signal power spectrum necessary to provide the required range resolution. In Figure 15.6 the target is shown in the plane (x, y) at a distance y_0 from the baseline of the 2D FSR. It is also equidistant from the transmitter and the receiver; thus $R_R = R_T$. In this case a delay, τ, of the scattered signal with respect to the probing signal is maximal. Assuming the considered geometrical relationships, the value of the delay, τ, is

$$\tau = \frac{b}{c}\left[\left(1 + \frac{4y_0^2}{b^2}\right)^{1/2} - 1\right], \tag{15.9}$$

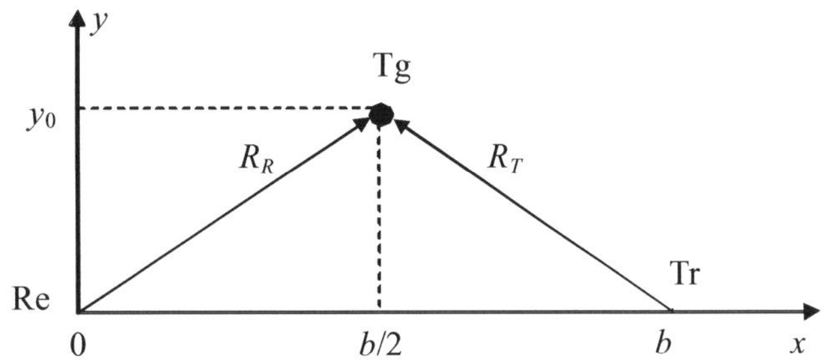

Figure 15.6 Two-dimensional FSR

where c is the speed of light. When $y_0/b \ll 1$, $\tau \cong 2y_0^2/(bc)$. For measurement of this delay with satisfactory accuracy, the duration of the optimum processed echo signal should be essentially less than the delays themselves. Accordingly, a frequency band of the power spectrum, Δ_f, is

$$\Delta_f \geq \frac{bc}{2y^2}. \tag{15.10}$$

For example, to measure the range sum when $y_0 = 10^3$ m, the signal bandwidth should be $\Delta_f > 6$ MHz and for $y_0 = 10^2$ m it should be $\Delta_f > 600$ MHz.

Therefore, the ranging signal with a band of 100 MHz and more is required for the target range tracking up to a distance of several hundreds of metres from a baseline. It can be seen that the use of an impulse probing signal leads to the need for the development and implementation of a transmitter and receiver operating with a very wide band signal. The technical realization of such a radar is much more complicated in comparison with the case of a continuous wave (CW) FSR. In addition, the required pulse transmitter power is much bigger in comparison with the averaged transmitter power of the CW FSR.

The factors discussed above, as well as those that will be considered later, are favourable for the CW FSR and, as such, it is only this type of radar that will be considered in this chapter. An additional set of factors, listed below, are also sufficiently significant to cause the CW FSR to be chosen specifically for further consideration:

1. In the FSR, the mode of an external coherence is easily realized by combining the signal scattered by the target and a direct signal that is 40–60 dB more powerful than an echo signal at the receive side. The Doppler beats formed in this way are simply separated after usual envelope detection.
2. The spectral analysis of a quasi-harmonic CW echo signal allows the implementation of optimum filtering of a useful signal as well, as effective suppression of interference and a direct signal. Simultaneously the Doppler frequency shift could be estimated with a high accuracy by means of standard computational methods. Thus, synchronization of receive and transmit sides is not required in the CW FSR.
3. The required transmitting power level in the CW FSR with a base length of 40 km is about 10 W. Such a transmitting power level could easily be achieved in the CW radar.

Thus, the problem of power budget limitation and precise synchronization requirements could be overcome by means of quasi-harmonic CW signal utilization in the FSR. However, the problem of coordinate measurement remains, and its effective solution requires utilizing the advantages of the FSR.

In this part the FSR based on the following principles will be considered:

1. A quasi-harmonic CW signal is used as a ranging signal at the transmit side.
2. A low-gain antenna simultaneously illuminating all the FS region is used at the transmit side.

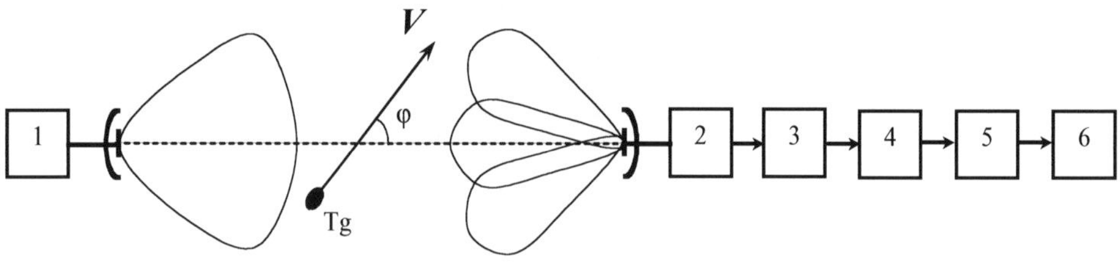

Figure 15.7 Block diagram of a CW FSR

3. A monopulse antenna is used at the receive side.
4. Several beams of the receiving antenna and appropriate receiver channels are used to provide parallel signal processing within the FSR coverage.
5. Spectral analysis in the Doppler frequency shift domain.
6. Tracking the target by a vector of primary (measured) parameters f_D, α and β.
7. Functional transformation of the measured parameters into spatial Cartesian coordinates x, y, z and tracking the target in these coordinates.

A simplified block diagram of the FSR constructed on the principles listed above is shown in Figure 15.7 [III.13, III.14], where the following are designated as: 1, a transmit side (position); 2, analogue receivers in receiving channels of each directional pattern partial beam; 3, amplitude detectors (ADs) of receiving channels; 4, analogue-to-digital converters (ADCs); 5, module of digital signal processing; 6, tracking module; Tg, target; and ***V***, vector of the target velocity. In such a system, the Doppler frequency shift measured at the output of the amplitude detectors and target return angle of arrival (AOA) measured by the amplitude monopulse method are used as the primary (measured) parameters.

Let us consider the basic features of an FSR of this type.

15.3 COVERAGE OF THE FSR

Let us estimate a detection coverage area of the FSR for the case when the echo signal from a moving target is received against the background of the receiver's thermal noise and reflections from the underlying surface. Coverage is defined as a region of space where the detection probability, D, is not less than a certain defined value while the false alarm rate, F, does not exceed the assigned value.

The coverage calculation is based on the radar equation for a bistatic system [III.1, III.2, III.5]:

$$Q = \frac{P_S}{P_0} = \frac{P_T G_T G_R \lambda^2 \eta F_T^2(\alpha, \beta) F_R^2(\alpha, \beta)}{(2\pi)^3 P_0 R_T^2 R_R^2} \sigma_B(\alpha_v, \alpha_h). \tag{15.11}$$

In Equation (15.11) P_S is the scattered signal power at the receive side; P_0 is the power corresponding to the threshold signal; $Q = P_S/P_0$ is the excess of the received signal over a threshold signal; P_T is the average transmitter power; G_T and G_R are the gains of the transmitting and the receiving antennas; λ is the wavelength; η is the generalized loss; $\sigma_B(\alpha_v, \alpha_h)$ is the target bistatic RCS as a function of diffraction angles in the vertical (α_v) and horizontal (α_h) planes; $F_T(\alpha, \beta)$ and $F_R(\alpha, \beta)$ are the directional patterns of the transmitting and the receiving antennas which include possible deformation due to the surface reflections; and R_T and R_R are the transmitter–target range and the target–receiver range.

To solve the problem of constructing a coverage diagram, the threshold value of the signal-to-noise ratio (SNR) for a given false alarm rate requires determination. An example of a coverage calculation for the FSR with a given set of parameters is presented below.

The false alarm rate, F, is determined by the total number of resolution cells, N_{RC}, in the observation space. $F \cong 1/N_{RC}$ for the case of not more than one false alarm in the observation space and $N_{RC} \gg 1$. If the observation is performed in a time interval, T, in the azimuth sector, $\Delta\alpha$, and in range of the Doppler frequency shift, Δf, then the total number of independent resolution cells $N_{RC} = n_T n_\alpha n_f$, where n_T, n_α and n_f are numbers of resolution cells in the domains of time, azimuth and Doppler frequency. The order of magnitude for these factors will be estimated.

It can be shown [III.18] that the maximum possible Doppler shift of the signal returned by a target is defined by the expression

$$f_{D\,max} = \frac{2V_{max}}{\lambda}\cos\left(\frac{\beta_{B\,min}}{2}\right), \tag{15.12}$$

where V_{max} is the maximum possible target velocity and $\beta_{B\,min}$ is the minimum value of the bistatic angle within the coverage. From Equation (15.12), it can be seen that f_D is at a maximum when the target orthogonally crosses the baseline in the middle.

If the detection of targets moving at speeds close to the sonic barrier is required, then $V \leq 1200$ km/h. On the basis of the above BRCS estimates, it can be assumed that the minimum bistatic angle, at which the forward-scatter effect is still present, is approximately $\beta_{B\,min} \approx 150°$. Then, according to Equation (15.12), when $\lambda \geq 0.3$ m, $f_D \leq 500$ Hz. The frequency resolution, Δf_{res}, is determined by the duration of the primary measurement interval, T. Thus, when $T \approx 1$ s, $\Delta f_{res} \approx 1$ Hz and $n_f \approx 500$.

The forward-scattering radar is intended for the detection of the targets that are difficult to detect using monostatic radars. Therefore, such a radar is expected to provide not only a high detection probability in the order of $D = 0.9$ but also a low false alarm rate. It must be not more than one false alarm per day. Thus, for a primary measurement interval of $T = 1$ s, $n_T = 24 \times 3600 \cong 86\,400$.

Finally, the value of n_α is determined by the beamwidth $\Delta\theta_\alpha$ of the receive position. Since the forward-scatter RCS sector (coverage sector of the receive position) is about $\Delta\alpha \approx 30°$ and $\Delta\theta_\alpha = 5°$–$10°$, $n_\alpha \leq 10$.

Thus, $N_{RC} = 500 \times 10 \times 86\,400 \cong 432 \times 10^6$ and the false alarm rate of the FSR is in the order of $F \cong 10^{-8}$–10^{-9}. The threshold value of the SNR corresponding to the detection of a single nonfluctuating signal with $D = 0.9$ and $F = 10^{-8}$ is $q_0 = 14$ dB [III.1].

The receiver internal noise power is equal to $n_0 = k_B T_K n_{NF}\,\Delta f_{opt}$, where $k_B = 1.38 \times 10^{-23}$ is the Boltzmann constant, n_{NF} is noise factor, T_K is the absolute temperature and Δf_{opt} is the optimum processing band. With $n_{NF} = 3$, $T_K = 290$ K and $\Delta f_{opt} \approx 1$ Hz, we obtain $n_0 \cong 1.2 \times 10^{-20}$ W. The parameters of the FSR and their values used in the coverage calculations are given in Table 15.1.

Figure 15.8 shows the directivity patterns (in terms of field intensity) for the transmitting and receiving antennas in view of a 10° upward mechanical tilt of the receiving antenna curtain, adopted in the calculations. Reflections from the rough spherical surface of the Earth are, by some means, taken into

Table 15.1 FSR parameters for coverage calculation

Parameter	Value
P_T	6 W
G_T	11 dB
G_R	23 dB
η_T (transmission loss)	– 2 dB
η_R (reception loss)	– 1.5 dB
η_{prc} (processing loss)	– 6.5 dB
n_0	1.2×10^{-20} W
q_0	14 dB
λ	0.714 m

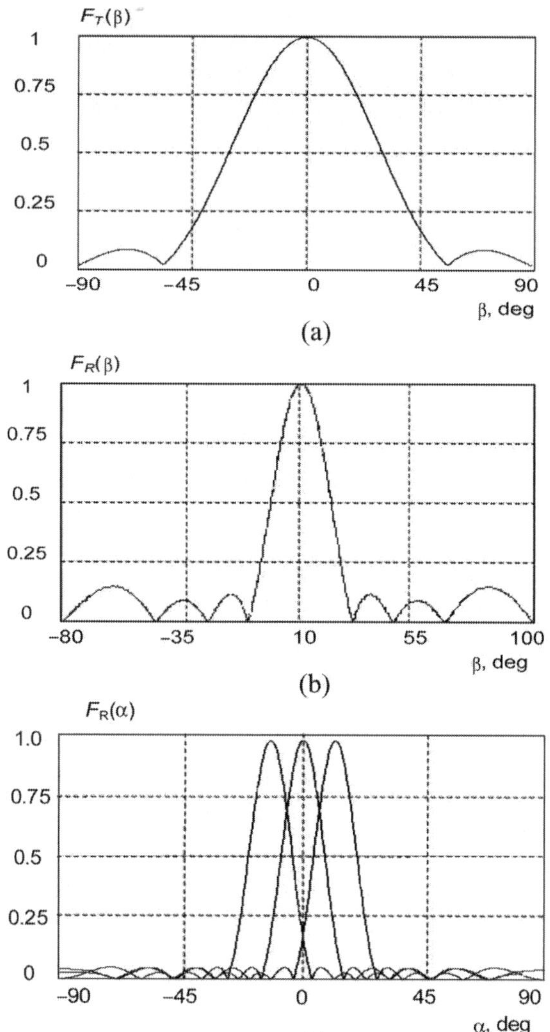

Figure 15.8 Transmitting and receiving antenna directional patterns : (a) transmitting antenna elevation pattern; (b) receive antenna elevation pattern; (c) receive antenna azimuth plane pattern

account in the coverage calculation [III.1, III.19]. Besides, when performing the complex summation of the direct wave and the reflected wave, it is necessary to take into account a path difference between the direct and the reflected rays, both on the 'transmitter–target' and on the 'target–receiver' paths. The calculation was performed with the following assumptions: a root mean square (RMS) deviation of the Earth surface irregularity $\sigma_S \approx 0.5$ m, a moderately wet soil with a relative permittivity $\varepsilon = 10$ and a specific electric conductivity $\delta = 10\,(\Omega\,\mathrm{m})^{-1}$.

The targets considered were assumed to be flying at a constant altitude. The flight trajectory inclination with respect to the baseline was taken as $\varphi = 90°$. The least favourable conditions for the detection of small-size targets take place at flight altitudes where the two-way product pattern determining the spatial field distribution has interference minima.

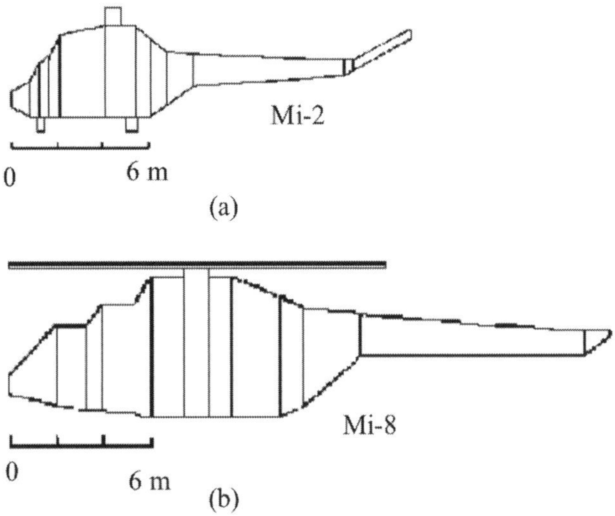

Figure 15.9 Shadow contours of selected targets

Figure 15.9 shows shadow contours of two different helicopters, Mi-2 and Mi-8. These contours were used in the calculation of the FSR coverage. They are divided into a number of trapezoidal sections to evaluate the estimation of the FS RCS corresponding to the method considered above.

Figures 15.10 and 15.11 show horizontal cross-sections of the FSR coverage against selected airborne targets for different altitudes from 30 up to 1000 m. The coverage diagrams depicted in the figures are topology maps of the excess of the SNR over the threshold q_0. The gradation of the SNR is shown on the grey scale below each figure.

It can be seen from Figures 15.10 and 15.11, that the radar coverage sections have essentially different structures at different altitudes. For each target location in coordinates (x, y) there are more favourable and less favourable heights in respect to the SNR. Thus, the least favourable conditions for the detection of small-size targets take place at flight altitudes where the product of the transmitting and receiving antenna patterns determining the spatial field distribution has interference minima. The FS RCS estimation is correct within the FS range, which is limited by diffraction angles $\alpha_{v,h} \leq 30°–40°$ (or bistatic angles $\beta_B > 140°–150°$). Obviously, the SNR will not exhibit a sharp difference from the calculated value beyond the borders of the above-mentioned range of diffraction angles. Nevertheless, a specific change of field intensity in the space outside $\alpha_{v,h} \leq 30°–40°$ needs to be defined more exactly.

The calculation demonstrates a high power level in the forward-scatter RCS region. The level of echo signals from small-size targets, such as an Mi-2 helicopter, exceeds the receiver internal noise by 45–60 dB with a microwave radiation power of 1–3 W, which ensures a detection probability close to unity with a very low false alarm rate.

Figure 15.12 is a vertical section of the radar coverage along the baseline up to a height of about 10 km for a helicopter-type target. The presented figures provide a full characterization of the configuration and size of detection coverage for a given target.

Thus, the following main conclusions to be drawn in this section are that FSR coverage presents a region of complex configurations stretched along the baseline. Since in the FS region the FS RCS of air targets exceeds the monostatic RCS by 30–50 dB, a sufficiently small enough CW transmitter power (less than 10 W) provides a very high probability (close to unity) of target detection in the whole FS region and a very low probability of a false alarm (about 10^{-6}).

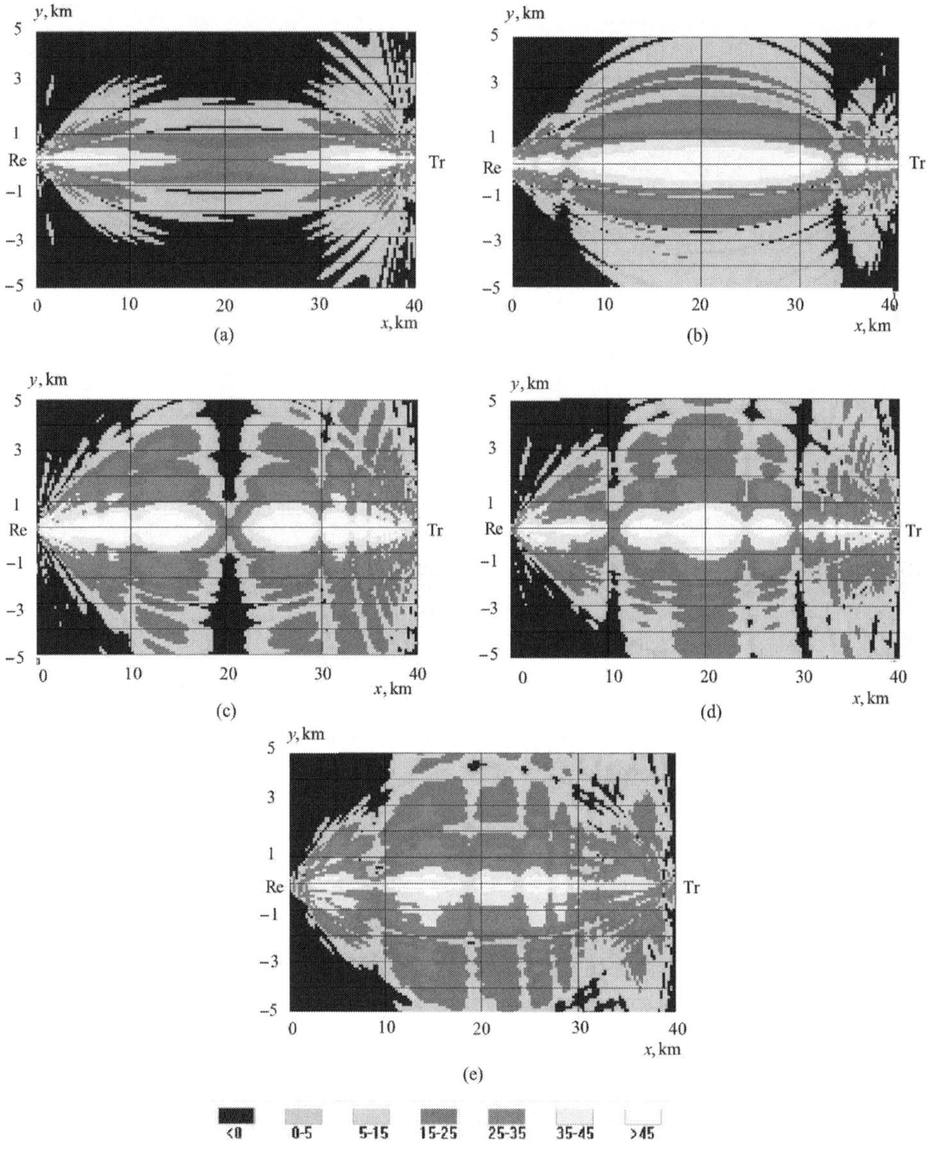

Figure 15.10 SNR over threshold q_0. (dB) for an Mi-2 helicopter: (a) $h = 30$ m; (b) $h = 100$ m; (c) $h = 325$ m; (d) $h = 500$ m, (e) $h = 1000$ m

15.4 CHARACTERISTICS OF THE INTERFERENTIAL SIGNAL

Disregarding the possible interferences and noise, the signal at the input of the receiving antenna, $u_R(t)$, can be presented as a sum of the target return and the direct signal:

$$u_R(t) = U_{dir}\cos(\omega_0 t + \varphi_0) + U_{tg}(t)\cos[\omega_0 t + \varphi_D(t) + \varphi_x], \tag{15.13}$$

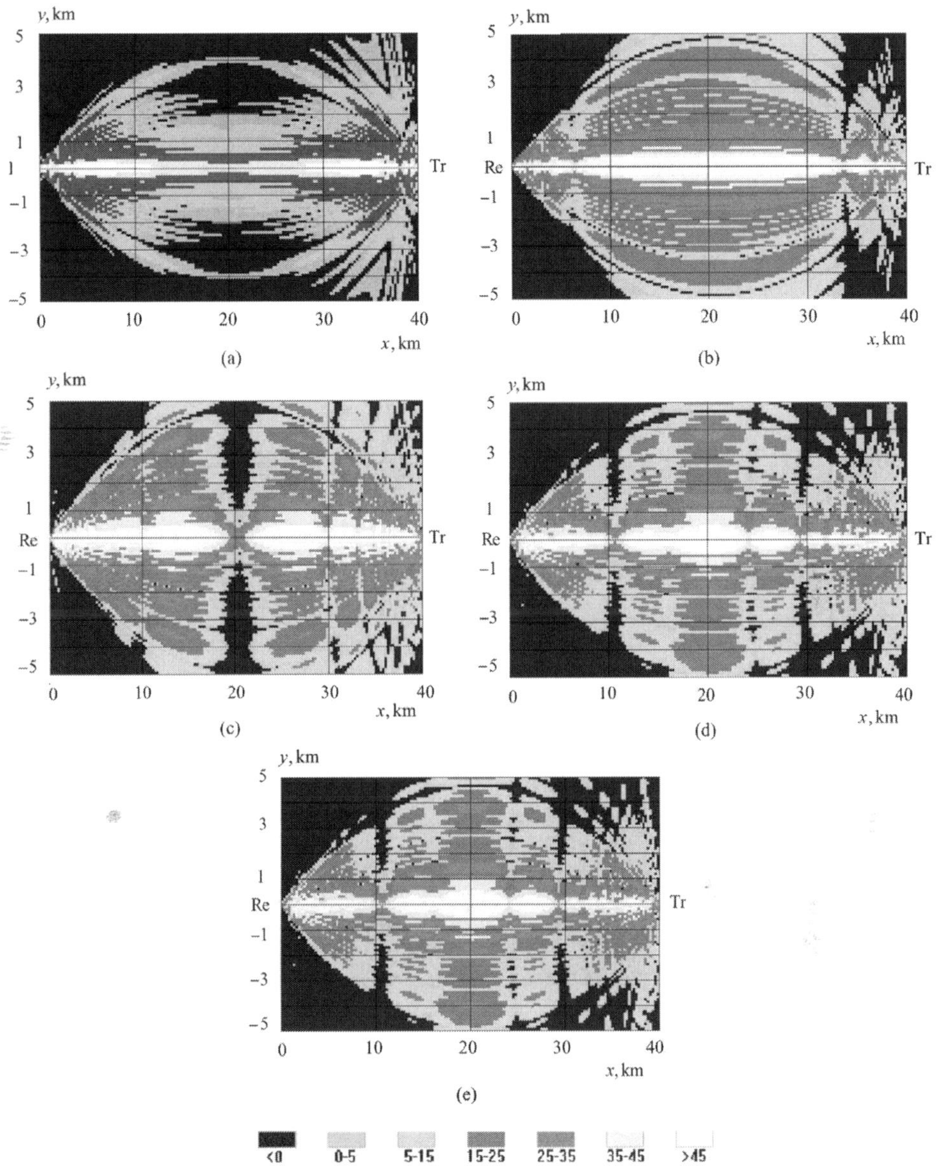

Figure 15.11 SNR over threshold q_0. (dB) for an Mi-8 helicopter: (a) $h = 30$ m; (b) $h = 100$ m; (c) $h = 325$m; (d) $h = 500$ m; (*e*) $h = 1000$ m

where $t \in (0; T_{\text{OBS}})$, $\omega_0 = 2\pi f_0$, $\varphi_{\text{D}}(t) = \int_0^t 2\pi f_{\text{D}}(t)\,dt$ and T_{OBS} is the full target observation time. On the right side of Equation (15.13), the first term of the sum represents the direct non-modulated signal of the transmitter with an amplitude, U_{dir}, a carrier frequency, f_0, and an initial phase, φ_0. The second term of the sum represents the signal returned by the target with unknown values of amplitude, $U_{\text{tg}}(t)$, Doppler shift, $f_{\text{D}}(t)$, and phase, φ_x; T_{OBS} is the full target observation time. The target returned signal amplitude, $U_{\text{tg}}(t)$, depends on the scattering properties of the target determined by its RCS and its location. The

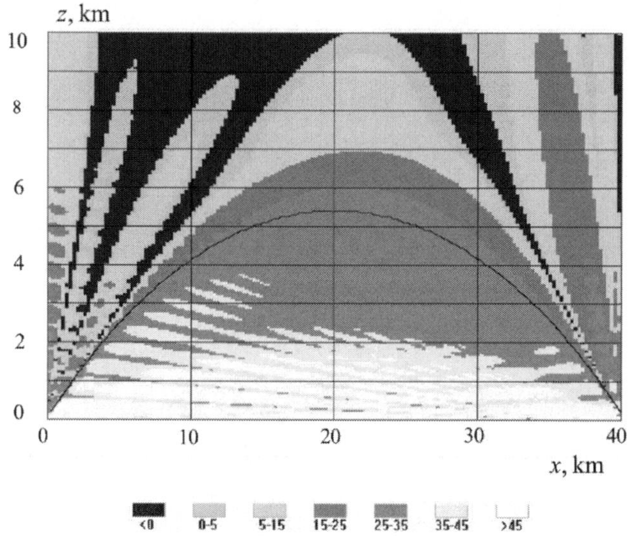

Figure 15.12 SNR over threshold q_0 (dB) for a vertical section along the baseline for an Mi-2 helicopter

power of the target signal may be obtained from the radar equation (15.11) for bistatic systems. The target RCS depends on its aspect angle and angular position, which could be defined via horizontal and vertical diffraction angles. Hence the amplitude of the target signal, $U_{tg}(t)$, is a function of time. First the characteristics of the interferential signal for small, in comparison with T_{OBS}, time intervals will be examined. The amplitude, $U_{tg}(t) = U_{tg}$, at such an interval is assumed to be constant.

The basic parameters of an interferential signal (15.13) are Doppler shift, f_D, target returned signal amplitude, U_{tg}, and its direction of arrival α, β (Figure 15.1). Let us examine the relation of these parameters to the target characteristics and trajectory.

When $U_{tg} = Constant$ and $U_{dir} \gg U_{tg}$, the resultant oscillation defined by Equation (15.13) can be expressed in the form of a biharmonic signal:

$$u_R(t) = U(t)\cos[\omega_0 t + \Theta(t)], \tag{15.14}$$

where

$$\begin{aligned} U(t) &= \sqrt{U_{dir}^2 + U_{tg}^2 + 2U_{dir}U_{tg}\cos[\omega_D(t)t + \varphi_x - \varphi_0]} \\ &\approx U_{dir} + U_{tg}\cos[\omega_D(t)t + \varphi_x - \varphi_0], \end{aligned} \tag{15.15}$$

$$\begin{aligned} \Theta(t) &= \operatorname{arctg}\left\{\frac{(U_{tg}/U_{dir})\sin[\omega_D(t)t + \varphi_x - \varphi_0]}{1 + (U_{tg}/U_{dir})\cos[\omega_D(t)t + \varphi_x - \varphi_0]}\right\} \\ &\approx \operatorname{arctg}\left\{\frac{U_{tg}}{U_{dir}}\sin[\omega_D(t)t + \varphi_x - \varphi_0]\right\} \approx \frac{U_{tg}}{U_{dir}}\sin(\omega_D(t)t + \varphi_x - \varphi_0). \end{aligned}$$

It is seen from Equation (15.15) that the amplitude and phase of the sum signal are oscillations with a current frequency defined by the target return Doppler shift. It was noted above, and will be shown in detail later, that target tracking in the CW FSR can be based on the measurements of the Doppler shift of the echo signal. The simplest way to measure the Doppler shift is to obtain the Doppler frequency

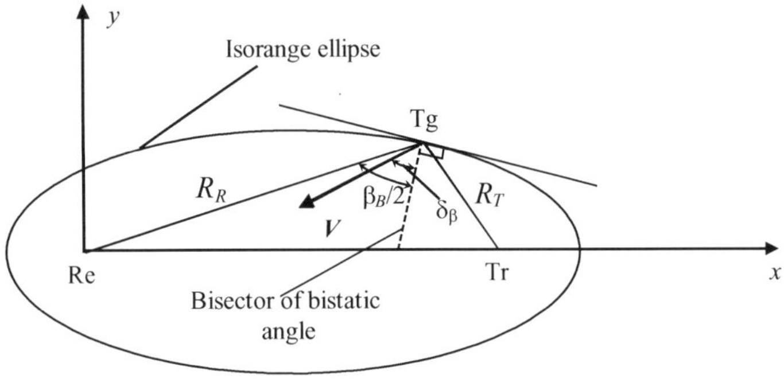

Figure 15.13 2D FSR geometry

signal from the interferential signal by means of an amplitude detector. In this case, there is no need for a coherent reference signal at the receive side.

In Figure 15.13, the 2D FSR geometry illustrates the Doppler relationships considered below. The Doppler shift of the signal reflected by the target and received by the bistatic radar is defined as [III.1, III.5, III.20]

$$f_D(t) = -\frac{1}{\lambda}\frac{dL(t)}{dt} = -\frac{1}{\lambda}\frac{d\,[R_T(t) + R_R(t)]}{dt}, \tag{15.16}$$

where $R_T(t)$ and $R_R(t)$ are the transmitter–target and target–receiver ranges respectively; $L(t) = R_T(t) + R_R(t)$ is the range sum.

It is seen from Equation (15.16) that there is no Doppler shift if the range sum does not vary, or $L(t) = constant$, i.e. when the target is moving along the ellipsoidal surface of equal range sums. The foci of the equal range sum ellipsoids are the phase centres of the receiving and the transmitting antennas. Projections of equal range sum ellipsoids on to the plane xOy are ellipses with the same foci. It is clear that the Doppler shift is equal to zero when the target is on the baseline, which is also a degenerate ellipse of the equal range sum. Furthermore, this range sum is equal to the length of the baseline. The change of Doppler shift is maximal when the target moves in the orthogonal direction with respect to the surface of equal range sums, or in the direction of the bisector of the bistatic angle.

Target detection and tracking in the radar under consideration is performed within a narrow region stretching along the baseline, where the forward-scatter effect is present. Therefore, the target velocity vector, $\boldsymbol{V}$, can be assumed to be constant and the target trajectory to be a straight line within the FSR coverage.

Let us find an expression for the Doppler frequency shift in the case of uniform rectilinear motion of the target in space (Figure 15.1). With (x_0, y_0, z_0) denoting the coordinates of the point of target entry into the coverage of the FSR, γ and φ denoting the angles determining the target heading (Figure 15.1) and V denoting the constant target velocity, equations can be written for the target motion along the relevant coordinate axes:

$$\begin{aligned} x(t) &= x_0 + V_x t, \\ y(t) &= y_0 + V_y t, \\ z(t) &= z_0 + V_z t, \end{aligned} \tag{15.17}$$

where $V_x = V_{hor}\cos\varphi$, $V_y = V_{hor}\sin\varphi$, $V_z = V\sin\gamma$ are the projections of target velocity on to the coordinate axes; $V_{hor} = V\cos\gamma$ is the projection of the target velocity vector on to the plane xOy. In this case, the expressions for the target–receiver range and the transmitter–target range take the form

$$R_R(t) = \sqrt{(x_0 + V_x t)^2 + (y_0 + V_y t)^2 + (z_0 + V_z t)^2},$$
$$R_T(t) = \sqrt{[d - (x_0 + V_x t)]^2 + (y_0 + V_y t)^2 + (z_0 + V_z t)^2}. \quad (15.18)$$

Substituting Equations (15.18) into (15.16) gives:

$$f_D(t) = -\frac{1}{\lambda}\left\{\frac{(x_0 + V_x t)V_x + (y_0 + V_y t)V_y + (z_0 + V_z t)V_z}{\sqrt{(x_0 + V_x t)^2 + (y_0 + V_y t)^2 + (z_0 + V_z t)^2}} + \frac{(y_0 + V_y t)V_y - [b - (x_0 + V_x t)] + (z_0 + V_z t)V_z}{\sqrt{[b - (x_0 + V_x t)]^2 + (y_0 + V_y t)^2 + (z_0 + V_z t)^2}}\right\}. \quad (15.19)$$

For a 2D FSR ($z = 0$, $V_z = 0$), Equation (15.19) can be rewritten as

$$f_D(t) = -\frac{1}{\lambda}\left\{\frac{(x_0 + V_x t)V_x + (y_0 + V_y t)V_y}{\sqrt{(x_0 + V_x t)^2 + (y_0 + V_y t)^2}} + \frac{(y_0 + V_y t)V_y - [b - (x_0 + V_x t)]}{\sqrt{[b - (x_0 + V_x t)]^2 + (y_0 + V_y t)^2}}\right\}. \quad (15.20)$$

The resultant equation can be rearranged in the form [III.5, III.18]:

$$f_D(t) = -\frac{2V}{\lambda}\cos\delta_\beta(t)\cos\left[\frac{\beta_B(t)}{2}\right]. \quad (15.21)$$

With a given maximum possible target velocity, V_{max}, the maximum possible Doppler shift, in accordance with Equation (15.21), is

$$f_{D\,max} = \frac{2V_{max}}{\lambda}\cos\left(\frac{\beta_{B\,min}}{2}\right). \quad (15.22)$$

It follows from Equations (15.21) and (15.22) that the points of the maximum Doppler shift reside on a circular arc corresponding to equal values of the bistatic angle $\beta_{B\,min}$. Besides, the target is expected to move along the bisector of the angle $\beta_{B\,min}$ (the first cosine factor in Equation (15.21) is equal to unity).

In Figure 15.14 dependencies of the maximum Doppler shift, $f_{D\,max}$, upon the minimum value of the bistatic angle, $\beta_{B\,min}$, are shown. These are plotted using Equation (15.22) for the cases of different target velocities, V_{max}. Graph 1 corresponds to the target velocity of $V = 550$ m/s, graph 2 corresponds to $V = 400$ m/s, graph 3 to 200 m/s and graph 4 to 50 m/s. In all cases the wavelength is assumed to be equal, $\lambda = 1$ m. Analysing such plots with a given minimum value of the bistatic angle within the radar coverage, the maximum possible Doppler shift can be found that, determines the Doppler range and the passband of the receiver.

Figure 15.16 presents the Doppler shift against the y coordinate for the rectilinear target trajectories shown in Figure 15.15. The graphs were plotted for a system with the following parameters: $b = 40$ km, $\lambda = 1$ m, and $\beta_{B\,min} = 150°$. The target motion is set in the plane xOy and the target velocity is $V = 200$ m/s. It is obvious from Figure 15.16 that the plots under consideration are practically linear. With linear trajectories, the plots of the Doppler shift against time will also be close to linear.

Analysis of the graphs given in Figure 15.16, as well as graphs plotted for other linear target trajectories, indicates that:

- For targets moving rectilinearly and with a constant velocity, the dependence of Doppler frequency on time $f_D(t)$ is, in most cases, quasi-linear.
- Dependence $f_D(t)$ remains quasi-linear with target velocities varying over a wide range.

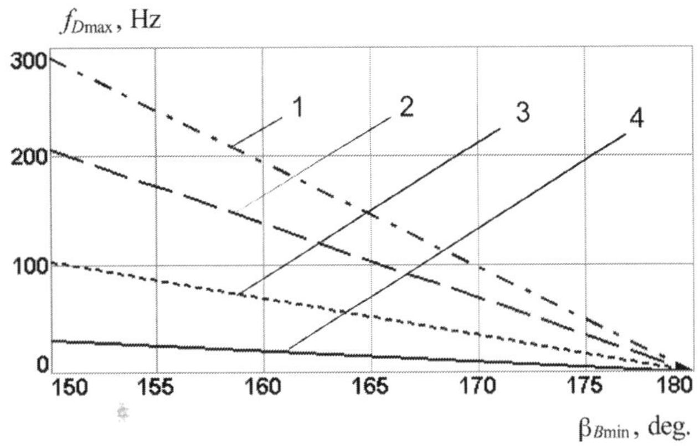

Figure 15.14 Maximum Doppler shift, $f_{D\ max}$, plotted against the width of the visibility area bounded by circular arcs corresponding to $\beta_{B\ min}$

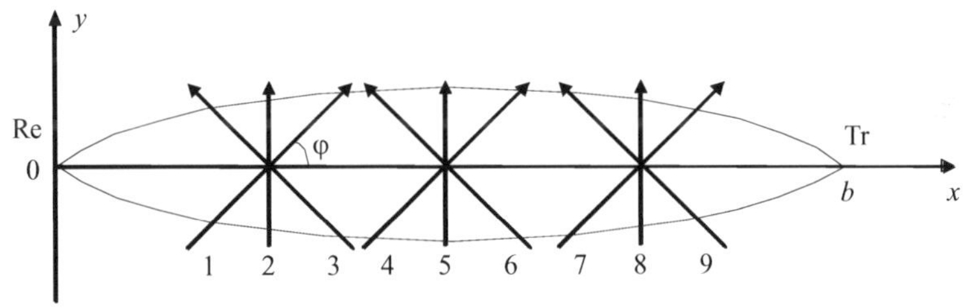

Figure 15.15 Examples of target trajectories: 1, $x_{CR}/b = 0.25$, $\varphi = 45°$; 2, $x_{CR}/b = 0.25$, $\varphi = 90°$; 3, $x_{CR}/b = 0.25$, $\varphi = 135°$; 4, $x_{CR}/b = 0.5$, $\varphi = 45°$; 5, $x_{CR}/b = 0.5$, $\varphi = 90°$; 6, $x_{CR}/b = 0.5$, $\varphi = 135°$; 7, $x_{CR}/b = 0.75$, $\varphi = 45°$; 8, $x_{CR}/b = 0.75$, $\varphi = 90°$; 9, $x_{CR}/b = 0.75$, $\varphi = 135°$

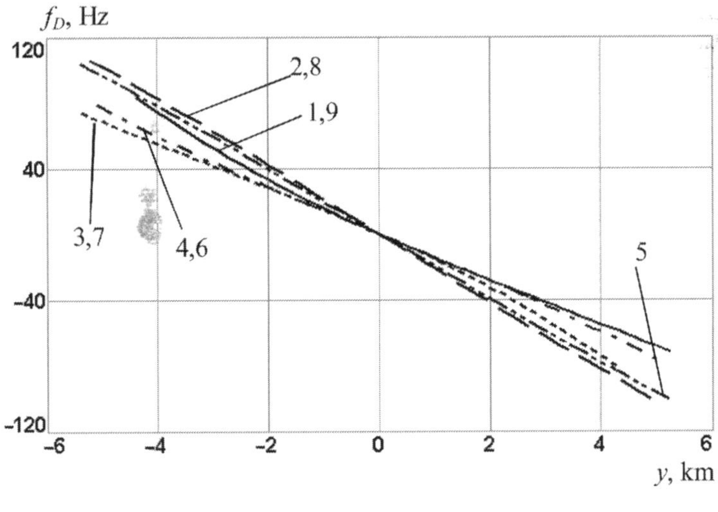

Figure 15.16 Frequency, $f_D(t)$, plotted against the current coordinate, $y(t)$, for targets moving along linear trajectories. Numbers 1, 2 , . . . , 9 correspond to those used in Figure 15.15

- The maximal Doppler shift corresponds to the instant of target detection on the boundary of the FSR coverage.
- The rate of Doppler shift change is higher for target trajectories closer to the transmitter or the receiver.

Examination of the Doppler shift dependence on target altitude has demonstrated that the quasi-linear behaviour of this dependence remains unchanged with increasing flight altitudes. Furthermore, at the instant when the target flies across the baseline, its Doppler shift may differ from zero. Its sign and value at the moment of crossing the baseline are determined by the inclination of the target trajectory with respect to the baseline and by target altitude.

The absolute rate of Doppler shift change can be found approximately from the quasi-linear frequency variation condition and is defined as

$$V_{\rm f} \approx \frac{f_{\rm D\,max}}{\Delta t_1}, \tag{15.23}$$

where Δt_1 is the time of the target flight from the first observed point where $f_{\rm D} = f_{\rm D\,max}$ to the baseline.

If a limit is set on the minimum range from the crossing point on the baseline to the receiving or transmitting positions, as well as on the minimum bistatic angle value, $\beta_{\rm B\,min}$, both the maximum possible Doppler shift, $f_{\rm D\,max}$, and its maximum rate of change, $V_{\rm f\,max}$, can then be estimated. These quantities have an effect on the choice of receiver passband and the parameters of space–time signal processing (in particular, the span of the primary measurement interval T).

The target azimuth (Figure 15.1) is defined as

$$\alpha(t) = \operatorname{arctg}\frac{y(t)}{x(t)}, \tag{15.24}$$

where $x(t), y(t)$ are the current coordinates of the target. In the case of a target moving along a rectilinear trajectory with a constant velocity, in view of Equation (15.17), Equation (15.24) takes the form

$$\alpha(t) = \operatorname{arctg}\frac{y_0 + V_y t}{x_0 + V_x t}. \tag{15.25}$$

Figure 15.17 shows the plots for the azimuth, $\alpha(t)$, versus time for the same trajectories and conditions as the plots for the Doppler shift in Figure 15.16.

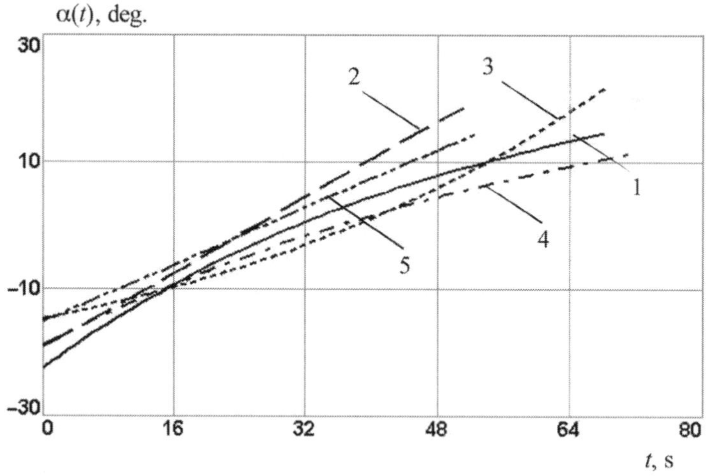

Figure 15.17 Plots for the azimuth versus time for rectilinear target trajectories. Numbers 1, 2, . . . , 5 correspond to those used in Figure 15.15

Analysis of these plots shows the following:

- The plot of $\alpha(t)$ is close to linear for the model of rectilinear and uniform target motion.
- The rate of azimuth change is higher for the target trajectories closer to the receiver.
- The highest rate of azimuth change takes place when the target moves orthogonally relative to the baseline ($\varphi = 90°$).

The maximum azimuth change rate is

$$V_{\alpha\,\max} \approx \frac{\alpha_{\max}}{\Delta t_{1\,\min}}, \tag{15.26}$$

where $\alpha_{\max} = \alpha(0)$ is the maximum possible target azimuth for the system under consideration (this is determined by the minimum range from the target trajectory to the receiving position and the width of the coverage) and $\Delta t_{1\min}$ is the minimum time of the target motion from the moment of detection until the moment of crossing the baseline. The maximum rate of azimuth change will correspond to a target that is at the boundary of the coverage at the moment of detection and follows a trajectory of the minimum range from the receiving position with an inclination towards the baseline $\varphi = 90°$. Furthermore, $\Delta t_{1\,\min} = y_0 / V_{y\,\max}$ since, in this case, $V_{y\,\max} = V_{\max}$. Analysis of the values of $\alpha_{\max}$ and $V_{\alpha\,\max}$ specified for a particular system under development allows their effect on the parameters and procedure of signal processing to be determined, in particular the number of space channels, the beamwidth of the antennas and the primary measurement interval, T.

16

Measurement of Target Coordinates in a 2D FSR

16.1 MEASUREMENT OF PRIMARY PARAMETERS

The development of a two-dimensional continuous-wave forward-scattering radar (Figure 15.7) is based on measuring the target return Doppler frequency shift and the target azimuth. Using measurements of these primary parameters in subsequent processing, trajectories of the target motion can be determined within the coverage area of the FSR. We will consider the principles of primary measurements in the CW FSR.

Due to the target motion, Doppler shift and, hence, the frequency of a signal at the input of the amplitude detector (AD) will change. In this case one of the major problems for subsequent trajectory processing is choosing an interval of primary measurements. The interval of primary measurements, T, should be chosen on the basis of the following requirements:

- A change in the target azimuth during the T interval is permissible in terms of the dynamic error.
- There should be sufficient time of coherent integration of the signal for its reliable detection and the measurement of its primary parameters.
- A sufficient amount of primary measurements (the angle α and frequency f_{D}) is required for construction of a target trajectory at a certain moment of time.
- A required rate of radar information updating is needed.

In the first approximation, the law of frequency change can be considered to be linear. Therefore, the optimum signal detection in the measurement interval should be implemented by means of filters according to the linear law of Doppler frequency shift. However, the value of the average Doppler frequency and the frequency rate of change in the measurement interval are not known beforehand. Therefore, the structure of optimum processing should be multichannel both in Doppler frequency and in the rate of its change. Thus, the required number of filters can appear excessively large. For example, an estimate will be made of the quantity of filters required for covering all of the Doppler frequency bands, assuming that the deviation of the frequency in the interval of primary measurement is known. In this case the number of filters can be estimated in the approximate form:

$$N_{\mathrm{f}} \cong \frac{f_{\mathrm{D\,max}}}{f_{\mathrm{cor}}} \cong 0.833 f_{\mathrm{D\,max}} T, \tag{16.1}$$

Bistatic Radars: Principles and Practice Edited by M. Cherniakov

where $f_{\mathrm{D\,max}}$ is the highest possible value of Doppler frequency shift of a signal of any target for the given system and $f_{\mathrm{cor}} = 1.2/T$ is the width of the high correlation region of the signal ambiguity function in the frequency domain [III.2, III.21]. (Note that Equation (16.1) assumes the rounding of results when the resultant number is not integer.) When the target velocity, $V_{\max}$, is the highest possible, the value $f_{\mathrm{D\,max}}$ is determined from Equation (15.22). For example, if $V_{\max} = 200$ m/s, $\lambda = 1$ m and $\beta_{\mathrm{B\,min}} = 150°$, $N_{\mathrm{f}} = 104$. Taking into account that the target return Doppler frequency shift rate of change is also unknown, the required number of matched filters may be several times greater.

Let us consider the possibility of reducing the total number of matched filters to minimal, when multichannelling by frequency deviation is to be absent. The Doppler frequency deviation in the interval of measurement can be ignored when the signal base insignificantly differs from unity:

$$B_{\mathrm{S}} = \Delta f_{\mathrm{D}} T = V_{\mathrm{f\,max}} T^2 \approx 1, \tag{16.2}$$

where $\Delta f_{\mathrm{D}} = V_{\mathrm{f\,max}}\ T$ for a signal with the maximal Doppler frequency rate of change, $V_{\mathrm{f\,max}}$. From Equation (16.2), the value of the primary measurement interval should satisfy

$$T \approx \sqrt{\frac{1}{V_{\mathrm{f\,max}}}}. \tag{16.3}$$

Thus, in the case when the Doppler frequency deviation during interval T of the primary measurements is not taken into account, multichannelling by frequency deviations is absent, and processing is reduced to the spectral analysis of the output voltage of the scheme 'amplitude detector–notch filter'. In general, to decrease the influence of the Doppler frequency shift deviation at the T interval, this interval should ideally be shorter.

On the other hand, the FSR possesses resolution in space of the Doppler frequency shift. Hence, it is desirable to improve this resolution by way of increasing interval T. Let us consider a compromise between these cases. This means that a value of the primary measurement interval is selected to provide values of the parameter B_{S} in the range of 1–4 in dependence on the target velocity. For example, in the case of the FSR with a base length $b = 40$ km and a wavelength of $\lambda = 0.5$ m the values $T = 0.25$–1 s are appropriate.

The target return AOA measurement in the block diagram shown in Figure 15.7 is performed using the monopulse method by the ratio of signal amplitudes in partial beams of the antenna directional pattern. The quantity of partial beams is determined depending on the width of the main lobe of an antenna directional pattern, the width of the radar coverage and the required accuracy of the angle, α, measurement.

With the known shape of the directional pattern of the receiving antenna, the difference pattern is formed to estimate the target return AOA by the relative difference of signal amplitudes in partial beams:

$$D_{\mathrm{dif}}(\alpha) = \frac{F_{\mathrm{R1}}(\alpha) - F_{\mathrm{R2}}(\alpha)}{F_{\mathrm{R1}}(\alpha) + F_{\mathrm{R2}}(\alpha)}, \tag{16.4}$$

where $F_{\mathrm{R1,2}}(\alpha)$ are receiving antenna directional patterns in the direction of the adjacent partial beams.

Figure 16.1 shows the directional pattern of two partial beams and the corresponding difference pattern. It is assumed that the ith beam of the directional pattern in the direction α_{0i} is described by

$$\begin{aligned} F_{\mathrm{R}}(\alpha) &= w(\alpha)\frac{\sin\left[2\pi c_1(\alpha - \alpha_{0i})\right]}{2\pi c_1(\alpha - \alpha_{0i})}, \\ w(\alpha) &= 0.54 + 0.46\cos\left[2\pi c_2(\alpha - \alpha_{0i})\right], \end{aligned} \tag{16.5}$$

where $w(\alpha)$ is a weight function and $c_{1,2}$ are constant factors. It is seen from Figure 16.1 that the difference pattern has a unique area of unambiguous AOA measurement. This area is located between two zero points of the directional patterns of partial beams. The mentioned zero points are shown in Figure 16.1 as well as the level k_{cr} of the beam crossing (in Figure 16.1, $k_{\mathrm{cr}} = 0.707$). In this area the difference

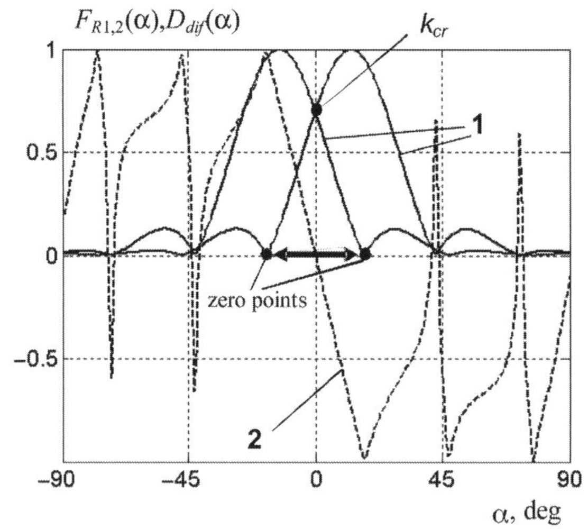

Figure 16.1 Formation of the difference pattern: 1, beams of the directional pattern; 2, the difference pattern

pattern is close to linear in form. If the number of beams is more than two, a similar difference pattern is created for each pair of adjacent beams.

In Figure 16.2, the block diagram of signal processing in the receiver of the FSR is shown where the spectral analysis on the basis of the fast Fourier transform (FFT) and measurement of the AOA by the monopulse method are used to obtain primary parameter measurements. Analogue processing is not considered in detail, but its main stages will briefly be described. After filtering and amplification in the RF unit, amplitude detection is performed in the receiving channels corresponding to each partial beam of the antenna pattern to obtain the Doppler signal. Notched filters (NFs) are then used to reject constant components caused by a direct transmitter signal, while lowpass filters (LPFs) provide a required sampling rate of the ADC. After analogue-to-digital conversion, signal samples are registered in the buffers (Buf) of the size N.

After filling the buffers with new samples the fast Fourier transform is performed on the obtained arrays of samples. Magnitudes of these complex FFT samples are calculated. In each channel samples exceeding an adaptive threshold are then found. For each detected target the channel with the maximal response is chosen. The number of FFT samples in this channel can be considered as the first rough estimate of the Doppler frequency shift. Improved estimation of the Doppler frequency shift is obtained

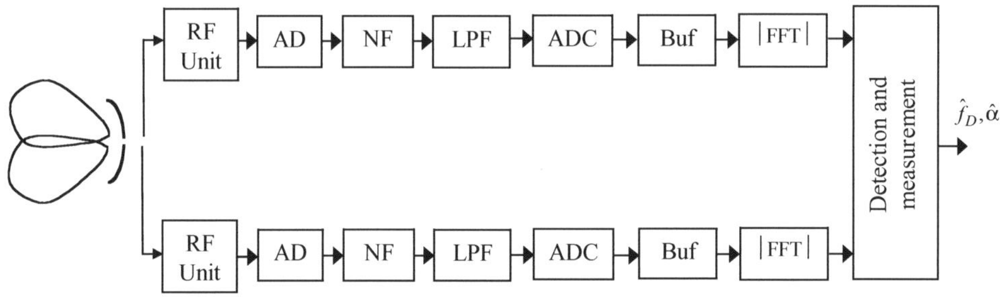

Figure 16.2 FSR signal processing block diagram

from the ratio of sample magnitudes in three neighbouring Doppler filters (FFT samples). For this purpose various methods of interpolation can be used. One of the simplest is to approximate magnitudes of three successive samples of the FFT by parabola and to calculate the coordinate of its maximum.

When measuring the wave AOA by the amplitude monopulse method, the value of the angle is estimated by the ratio of amplitudes of FFT samples corresponding to the same target, in the channels corresponding to various beams of a directional pattern. For this purpose the calculated or previously measured difference pattern corresponding to a directional pattern of the receiving antenna is used.

In conclusion, it should be noted that the considered processing algorithm using an amplitude detector to obtain the envelope of interferential signal, strictly speaking, is not optimal. The signal-to-noise ratio for a Doppler frequency signal can be increased using processing algorithms where the Doppler frequency signal is obtained not by means of the amplitude detector but by means of synchronous detecting in two quadrature channels. It is not difficult to demonstrate that such a processing structure allows a 3 dB gain to be achieved in the SNR. In the presence of a powerful direct signal and the target return, the amplitude detector acts as a synchronous detector relevant to the Doppler signal. The signal processing procedure is the equivalent to one of two of the quadrature down conversions and the output signal corresponds to one quadrature component of the target return complex envelope. Since the amplitude detector performs a target return synchronous detection, a quasi-harmonic Doppler shift frequency signal (real envelope) is extracted without any nonlinear distortions.

However, synchronous quadrature processing is difficult to realize due to the absence of a reference signal. It is for this reason that the use of signal processing with the amplitude detector is the most interesting, as the receiver and the transmitter operate independently.

16.2 COORDINATE MEASUREMENT ALGORITHM BASED ON THE MAXIMUM LIKELIHOOD METHOD

This section is dedicated to target tracking in the FSR via the maximum likelihood method. The main goal is to obtain accuracy of maximum likelihood estimations of target coordinates. This accuracy is calculated theoretically. The mathematical modelling results are also presented.

16.2.1 Mathematical Model of Measuring Process; Maximum Likelihood Estimation of Trajectory Parameters

It is assumed in the observation model that a target has velocity $\boldsymbol{V}$ along the linear trajectory, with an inclination angle, φ, towards the baseline and coordinate x_{CR} of the baseline crossing point (Figure 16.3). Since the two-dimensional FSR is being considered, it is supposed that the target moves in the plane (x, y), i.e has zero altitude, $h = 0$. It is assumed that n consecutive measurements of Doppler frequency

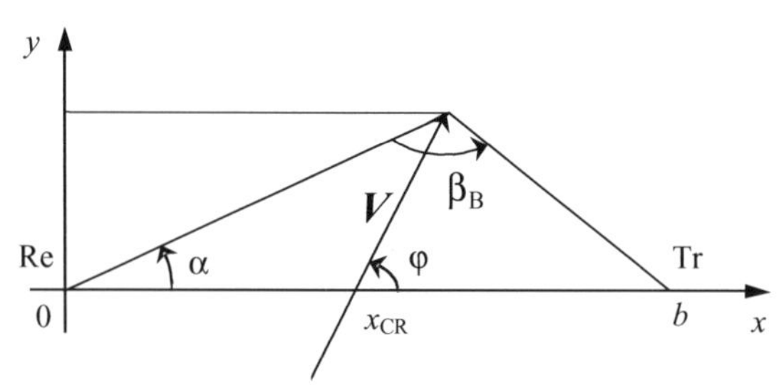

Figure 16.3 Target motion model

shift, f_{D}, and azimuth, α, form the primary measurement vector

$$z_n = \left[\hat{f}_{\mathrm{D}1}, \hat{\alpha}_1, \hat{f}_{\mathrm{D}2}, \hat{\alpha}_2, \ldots, \hat{f}_{\mathrm{D}n}, \hat{\alpha}_n\right]^{\mathrm{T}}. \tag{16.6}$$

In this case, a nonlinear system of equations concatenating values of primary parameters with values of trajectory parameters in the two-dimensional system (at zero flight altitude) in accordance with Equations (15.20), (15.24) and (15.25) appears as

$$f_{\mathrm{D}i}(\boldsymbol{x}_n) = -\frac{1}{\lambda}\left[\frac{x_i V_x + y_i V_y}{\sqrt{x_i^2 + y_i^2}} + \frac{y_i V_y + (b - x_i) V_x}{\sqrt{y_i^2 + (b - x_i)^2}}\right],$$
$$\alpha_i(\boldsymbol{x}_n) = \mathrm{arctg}\frac{y_i}{x_i}, \tag{16.7}$$

where $\boldsymbol{x}_n = \left[x_n, y_n, V_x, V_y\right]^{\mathrm{T}}$ is a vector of the target trajectory parameters at the time moment, t_{n}; x_{n}, y_{n} are values of coordinates; V_{x}, V_{y} are the rates of change of these coordinates; $x_i = x_n - V_x(n - i)T$ and $y_i = y_n - V_y(n - i)T$, where T is the interval of primary measurements; b is the base of the system; and λ is the wavelength.

A model of the measuring process is based on Equations (16.7). Statistical dependence of the measured values of the primary parameters on the target trajectory parameters is described by a measurement equation. It can be written in the following compact form:

$$z_n = \boldsymbol{h}_n(\boldsymbol{x}_n) + \Delta z_n, \tag{16.8}$$

where $\boldsymbol{h}_n(\boldsymbol{x}_n) = [f_{\mathrm{D}1}(\boldsymbol{x}_n), \alpha_1(\boldsymbol{x}_n), f_{\mathrm{D}2}(\boldsymbol{x}_n), \alpha_2(\boldsymbol{x}_n), \ldots, f_{\mathrm{D}n}(\boldsymbol{x}_n), \alpha_n(\boldsymbol{x}_n)]^{\mathrm{T}}$ is a nonlinear vector function with elements defined by Equation (16.7) and Δz_n is a vector of random primary measurement errors.

For further consideration, it is supposed that primary measurement errors are Gaussian with zero mean and variances, $\sigma_{\mathrm{f}}^2, \sigma_{\alpha}^2$, which are considered to be the same for all the moments of time, unless otherwise stipulated. It is also supposed that the measurements of the primary parameters are mutually independent at any time shift, as well as the measurements of every parameter being independent at different moments of observation.

The maximum likelihood estimation of the tracking vector, $\hat{\boldsymbol{x}}_n$, is obtained from maximizing the conditional probability density of the primary measurement vector, $p(z_n | \boldsymbol{x}_n)$. Using the accepted Gaussian model of primary measurement errors, the optimum estimation of the vector of trajectory parameters can be found as

$$\hat{\boldsymbol{x}}_n = \mathrm{argmin}\, \varphi_n(\boldsymbol{x}_n), \tag{16.9}$$

where $\varphi_n(\boldsymbol{x}_n) = [z_n - \boldsymbol{h}(\boldsymbol{x}_n)]^{\mathrm{T}} \mathbf{G}_n [z_n - \boldsymbol{h}(\boldsymbol{x}_n)]$ and $\mathbf{G}_n = \left[M\left\{\Delta z_n \Delta z_n^{\mathrm{T}}\right\}\right]^{-1}$ is an inverse correlation matrix of the primary measurement error vector, Δz_n.

16.2.2 Potential Accuracy of Trajectory Parameters Measurement

Let us evaluate the potential accuracy of the estimation of the trajectory parameter vector, $\boldsymbol{x}_n$, via the primary measurement vector, z_n. When using the maximum likelihood method, the potential accuracy is determined by the informational Fisher matrix. For the observation process of the introduced model, the elements of the Fisher matrix of References [III.5], and [III.22] are defined as

$$\mathbf{J}_{lk}(\boldsymbol{x}_n) = -M\left\{\frac{\partial^2 \ln p(z_n | \boldsymbol{x}_n)}{\partial x_l\, \partial x_k}\right\} = \frac{1}{\sigma_{\mathrm{f}}^2}\sum_{i=1}^{n}\frac{\partial f_i(\boldsymbol{x})}{\partial x_l}\frac{\partial f_i(\boldsymbol{x})}{\partial x_k} + \frac{1}{\sigma_{\alpha}^2}\sum_{i=1}^{n}\frac{\partial \alpha_i(\boldsymbol{x})}{\partial x_l}\frac{\partial \alpha_i(\boldsymbol{x})}{\partial x_k}, \tag{16.10}$$

where $x_l, x_k (l, k = 1, 4)$ are elements of the vector $\boldsymbol{x}_n$ ($x_1 = x_n$, $x_2 = y_n$, $x_3 = V_x$, $x_4 = V_y$) and $M\{\cdot\}$ is a sign of statistical averaging.

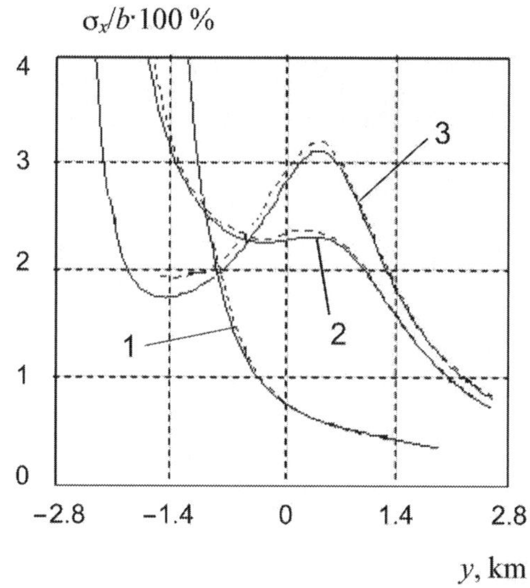

Figure 16.4 RMS deviation of the x coordinate measurement [III.23]

Variances $\sigma_x^2, \sigma_y^2, \sigma_{Vx}^2, \sigma_{Vy}^2$ of estimations of trajectory parameters can be calculated as diagonal elements of a matrix inverse to the Fisher matrix. In Figure 16.4, dependences of normalized RMS deviations of estimations of coordinate x as functions of the value of coordinate y are shown by solid lines. These are calculated for $\sigma_f = 1$ Hz and $\sigma_\alpha = 0.5°$. Three different target trajectories with parameters $x_{CR}/b = 0.25$, $\varphi = 90°$ (curve 1), $x_{CR}/b = 0.5$, $\varphi = 135°$ (curve 2) and $x_{CR}/b = 0.5$, $\varphi = 60°$ (curve 3) were considered. For all the cases, it was assumed that $b = 40$ km, $V = 200$ m/s, $\lambda = 1$ m and $T = 0.25$ s. FSR coverage was assumed to be limited by the values of $y = \pm 2$ km.

Analysis of the plots (Figure 16.4) shows that a monotonous reduction of RMS deviations of errors of coordinate x estimation takes place while the observation time grows (at small values of the angle α they practically coincide with RMS deviations of target range estimation errors), but an insignificant rise on the baseline is possible. The degradation in accuracy of the estimation of coordinate x at the moment of crossing the baseline can be explained by the fact that the Doppler frequency shift on the baseline does not depend on x. The large errors at the initial segment of the trajectory can be explained by the lack of measurements of the primary parameters. At the moment the target leaves the coverage, RMS deviation of the error of estimation of coordinate x becomes sufficiently small, about 0.5–1 % of the length of the baseline under the conditions accepted above.

The calculation of RMS deviation of the error of estimation of coordinate y has shown that it is insignificant in comparison with σ_x and does not have a significant effect on the accuracy of a target position estimation.

16.2.3 Iterative Algorithm of a Coordinate Estimation

In practice, iterative algorithms are used in order to obtain the solution of the minimization problem (16.9), giving the maximal likelihood estimation of the vector of trajectory parameters. The use of the Gauss–Newton algorithm gives [III.14, III.23, III.24]

$$\hat{\mathbf{x}}_n^{i+1} = \hat{\mathbf{x}}_n^i + k_C(\mathbf{H}_n^T\mathbf{G}_n\mathbf{H}_n)^{-1}\mathbf{H}_n^T\mathbf{G}_n[z_n - \boldsymbol{h}_n(\hat{\mathbf{x}}_n^i)]. \tag{16.11}$$

Here $\hat{\boldsymbol{x}}_n^i$ is an estimate of the vector of trajectory parameters on the ith iteration, $\mathbf{H}_n = \partial \boldsymbol{h}_n(\boldsymbol{x}_n)/\partial x|_{\boldsymbol{x}_n=\hat{\boldsymbol{x}}_n^i}$ is a derivative matrix and k_C is the parameter determining the rate of convergence.

In a practical application of the given iterative algorithm, the key problem is to choose an initial approximation, $\hat{\boldsymbol{x}}_n^0$. Since the minimized function, $\varphi_n(\boldsymbol{x}_n)$, is essentially nonlinear, the application of the Gauss–Newton algorithm guarantees finding the local minimum, which does not always coincide with the absolute minimum. An insufficiently accurate initial approximation, $\hat{\boldsymbol{x}}_n^0$, can lead to significant errors of estimation of the trajectory parameter vector. Consider the following possible approach of defining an initial approximation.

16.2.4 Evaluation of the Initial Approximation

Suppose there are two pairs of primary measurements, (f_{D1}, α_1) and (f_{Dn}, α_n), taken at a time interval, $(n-1)T$. From Equation (16.7) a system can be deduced of four nonlinear equations concatenating primary parameters f_{D1}, α_1, f_{Dn} and α_n and trajectory parameters x_n, y_n, V_x and V_y. Taking into account the fact that the values of the bistatic angle, β_B, are close to 180°, this system can be approximately presented as

$$
\begin{aligned}
f_{Dn} &= -\frac{V_y}{\lambda}\left(\alpha_n + \frac{y_n}{b - x_n}\right),\\
\alpha_n &= \frac{y_n}{x_n},\\
f_{D1} &= -\frac{V_y}{\lambda}\left(\alpha_1 + \frac{y_1}{b - x_1}\right),\\
\alpha_1 &= \frac{y_1}{x_1}.
\end{aligned}
\qquad (16.12)
$$

System (16.12) has a unique analytical solution defined by

$$
\begin{aligned}
x_n &= \frac{\left(\lambda f_{Dn} f_{D1}\Delta t + b\alpha_1^2 f_{Dn} - b\alpha_n\alpha_1 f_{D1}\right)}{u},\\
y_n &= \alpha_n x_n,\\
V_x &= \frac{b\left[f_{Dn}\alpha_1^2 + f_{D1}\alpha_n^2 - \alpha_n\alpha_1(f_{D1} + f_{Dn})\right]}{u\Delta t},\\
V_y &= -\frac{\lambda f_{Dn} f_{D1}(\alpha_1 - \alpha_n)}{u},
\end{aligned}
\qquad (16.13)
$$

where $u = \lambda f_{Dn} f_{D1}\Delta t/b + f_{Dn}\alpha_1^2 - f_{D1}\alpha_n^2$. Substitution of estimations of primary parameters $f_{D1}, \alpha_1, f_{Dn}, \alpha_n$ into Equations (16.13) allows an initial approximation to be obtained for the trajectory parameters vector, $\hat{\boldsymbol{x}}_n^0$.

The initial and final measurements of the Doppler frequency shift, f_D, and azimuth, α, at the initial section of the trajectory can be used as parameter values $(f_{D1}, \alpha_1, f_{Dn}, \alpha_n)$ in Equations (16.13). Better results can be achieved if an initial approximation is found using all primary measurements obtained. Consider the vector of first N primary measurements, z_N. Using the least squares method, from the available sample collection an approximation can be obtained of the Doppler shift, f_D, and azimuth, α, as a function of time in the form of a polynomial of a given power. A linear approximation yields

$$
\begin{aligned}
\tilde{f}_D(t) &= a_f t + b_f,\\
\tilde{\alpha}(t) &= a_\alpha t + b_\alpha,
\end{aligned}
\qquad (16.14)
$$

where

$$a_{\mathrm{f}} = \frac{\sum_{i=1}^{N} (t_i - \hat{m}_{\mathrm{t}})(\hat{f}_i - \hat{m}_{\mathrm{f}})}{\sum_{i=1}^{N} (t_i - \hat{m}_{\mathrm{t}})^2}, \qquad b_{\mathrm{f}} = \hat{m}_{\mathrm{f}} - a_{\mathrm{f}}\hat{m}_{\mathrm{t}},$$

$$a_{\alpha} = \frac{\sum_{i=1}^{N} (t_i - \hat{m}_{\mathrm{t}})(\hat{\alpha}_i - \hat{m}_{\alpha})}{\sum_{i=1}^{N} (t_i - \hat{m}_{\mathrm{t}})^2}, \qquad b_{\alpha} = \hat{m}_{\alpha} - a_{\mathrm{f}}\hat{m}_{\mathrm{t}}, \tag{16.15}$$

$$\hat{m}_{\mathrm{t}} = \frac{1}{N}\sum_{i=1}^{N} t_i, \quad \hat{m}_{\mathrm{f}} = \frac{1}{N}\sum_{i=1}^{N} \hat{f}_i, \qquad \hat{m}_{\alpha} = \frac{1}{N}\sum_{i=1}^{N} \hat{\alpha}_i.$$

Substituting $f_{\mathrm{D}1}, \alpha_1, f_{\mathrm{D}n}, \alpha_n$ in Equations (16.13) for estimates $\tilde{f}_{\mathrm{D}1}, \tilde{\alpha}_1, \tilde{f}_{\mathrm{D}N}, \tilde{\alpha}_N$, the sought initial approximation is obtained.

A generalization of the proposed method will now be discussed. For any two pairs of $(\tilde{f}_{\mathrm{D}i}, \tilde{\alpha}_i)$, $(\tilde{f}_{\mathrm{D}j}, \tilde{\alpha}_j)$, $i, j = \overline{1, N}, i < j$, from Equations (16.12) and (16.13) an estimate can be obtained of the state vector at instant t_j. Extrapolating it to the moment of time, t_N, a first approximation, $\hat{\boldsymbol{x}}_N^0(i, j)$, is obtained. Its accuracy is dependent on the value of the likelihood function at the point $\hat{\boldsymbol{x}}_N^0(i, j)$. Finally, it is assumed that

$$\hat{\boldsymbol{x}}_N^0 = \arg\min_{\forall(i,j)} \varphi_N\left(\hat{\boldsymbol{x}}_N^0(i, j)\right). \tag{16.16}$$

For all $n > N$, an initial approximation of $\hat{\boldsymbol{x}}_n^0$ can be obtained through extrapolation of the updated value of the trajectory parameter vector $\hat{\boldsymbol{x}}_{n-1}$ from the previous, $(n-1)$th, instant of time to the current instant, i.e. $\hat{x}_n^0 = \hat{x}_{n-1} + \hat{V}_{x,n-1}$, $\hat{y}_n^0 = \hat{y}_{n-1} + \hat{V}_{y,n-1}$, $\hat{V}_{x,n}^0 = \hat{V}_{x,n-1}$ and $\hat{V}_{y,n}^0 = \hat{V}_{y,n-1}$. The assessment of the initial approximation accuracy for iterative algorithm performance and of the algorithm itself was carried out by the method of mathematical modelling. The error probability density was taken as Gaussian, and the averaging was performed in terms of 1000 independent implementations.

Figure 16.5 shows the plots of RMS deviations of errors in estimating the x coordinate of the initial approximation $\hat{\boldsymbol{x}}_N^0$ against y coordinate values. Curves marked by numbers 1 to 3 are obtained for the trajectories with the same parameters used to obtain the graphs in Figure 16.4. Solid lines denote where the initial approximation using extreme (first and last) measurements has been derived, while the dotted lines are from Equation (16.16). It is seen from the plots that when the initial approximation is determined by the extreme points and the values of y are close to zero, there is a sharp degradation in accuracy. This fact is accounted for by low sensitivity of the Doppler shift, f_{D}, to a variation of the coordinate x in the vicinity of the baseline. The generalized method is practically free of this disadvantage. At the same time, the computational cost of implementing this method is considerably higher than that of the method of extreme points. As demonstrated by the calculations, for the trajectories and target velocities in question, the choice of $N > N_{\min} \approx 16$ ensures the required accuracy of the initial approximation. It is not productive to select a large value of N, since this results in a longer time spent on forming the initial estimate of the trajectory parameters vector, which is not always suitable.

Using the iterative algorithm (16.11) and initial approximation (16.16), the estimations of the target coordinate measurement accuracy was obtained by the mathematical modelling. In Figure 16.4 the dotted lines present the results of the modelling. It is seen from the comparison of the theoretical curves and curves obtained during modelling that the considered iterative algorithm provides close to potential target tracking accuracy.

16.2.5 Parametric Ambiguity Elimination

Equation (16.11), which specifies the iterative procedure for refining the estimate of the trajectory parameters vector $\boldsymbol{x}_n$, contains $\mathbf{G}_n$, an inverse correlation matrix of primary measurement errors. This

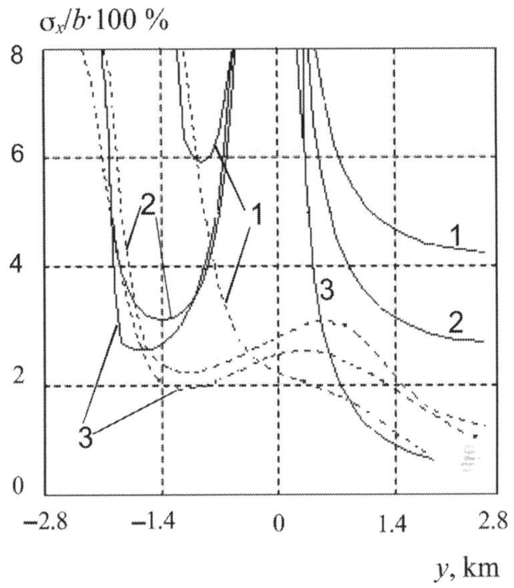

Figure 16.5 RMS deviations of the x coordinate first approximation errors [III.23]

can be exactly defined only when values of RMS deviations of measurement errors of primary parameters, $\sigma_{\mathrm{f}n}$ and $\sigma_{\alpha\mathrm{n}}$, are known. As a rule, the exact values of RMS deviations of measurement errors are not known. In this case, they can be replaced by estimations of the corresponding quantities [III.14, III.17, III.23, III.25].

A natural simplification of the problem would be to adopt a hypothesis about the measurement errors being constant and independent, i.e. $\forall t_i \sigma_{\mathrm{f}i} = \sigma_{\mathrm{f}},\ \sigma_{\alpha i} = \sigma_{\alpha}$. Furthermore, if some average estimates of variances of primary measurement errors $\hat{\sigma}_{\mathrm{f}}, \hat{\sigma}_{\alpha}$ are previously established for the system, these estimates can be used to set the matrix $\mathbf{G}_n$. In this case, the correlation matrix and, consequently, the matrix $\mathbf{G}_n$ of size $2n \times 2n$ take the diagonal form: $\mathbf{G}_n = \mathrm{diag}[1/\hat{\sigma}_{\mathrm{f}}^2, 1/\hat{\sigma}_{\alpha}^2]$.

If the average estimates of RMS deviations of measurement errors are not known, one possible solution would be to ignore the variance uncertainty. As a result, the matrix $\mathbf{G}_n$ is disregarded in the likelihood function and in the iterative algorithm (16.11). In this case, the maximum likelihood method under consideration goes over to the least squares method [III.22] with a unit weight matrix. The corresponding expressions take the form

$$\varphi(\boldsymbol{x}_n) = [z_n - \boldsymbol{h}_n(\boldsymbol{x}_n)]^{\mathrm{T}} [z_n - \boldsymbol{h}_n(\boldsymbol{x}_n)], \tag{16.17}$$

$$\hat{\boldsymbol{x}}_n^{i+1} = \hat{\boldsymbol{x}}_n^i + k_{\mathrm{C}} \left(\mathbf{H}_n^{\mathrm{T}} \mathbf{H}_n\right)^{-1} \mathbf{H}_n^{\mathrm{T}} \left[z_n - \boldsymbol{h}_n\left(\hat{\boldsymbol{x}}_n^i\right)\right]. \tag{16.18}$$

In this case, a considerably lower coordinate measurement accuracy can be achieved. With $i \to \infty$, from Equation (16.18),

$$\left(\mathbf{H}_n^{\mathrm{T}} \mathbf{H}_n\right)^{-1} \mathbf{H}_n^{\mathrm{T}} \left[z_n - \boldsymbol{h}_n\left(\hat{\boldsymbol{x}}_{n,\mathrm{opt}}^i\right)\right] = 0, \tag{16.19}$$

where $\hat{\boldsymbol{x}}_{n,\mathrm{opt}}$ is the optimum (most accurate in the considered conditions) estimate obtainable subject to algorithm convergence.

Let us represent z_n and $\hat{\boldsymbol{x}}_{n,\text{opt}}$ in Equation (16.19) as the sums of their values in the absence of measurement errors $z_n = [f_{\text{D}1}, \alpha_1, f_{\text{D}2}, \alpha_2, \ldots, f_{\text{D}n}, \alpha_n]^{\text{T}}$, $\boldsymbol{x}_{n,\text{opt}} = [x_{n,\text{opt}}, y_{n,\text{opt}}, V_{x,\text{opt}}, V_{y,\text{opt}}]^{\text{T}}$ and errors caused by primary measurement errors $\Delta z_n = [\Delta f_{\text{D}1}, \Delta\alpha_1, \Delta f_{\text{D}2}, \Delta\alpha_2, \ldots, \Delta f_{\text{D}n}, \Delta\alpha_n]^{\text{T}}$, $\Delta\boldsymbol{x}_n = [\Delta x_n, \Delta y_n, \Delta V_x, \Delta V_y]^{\text{T}}$. Assuming that these errors are small, we obtain from Equation (16.19) after expanding the nonlinear function $\boldsymbol{h}_n(\boldsymbol{x}_n)$ in a series by the powers $\Delta\boldsymbol{x}_n$ and rejecting terms of the second order of smallness:

$$\Delta\boldsymbol{x}_n = \mathbf{R}_n \Delta z_n = \left(\mathbf{H}_n^{\text{T}}\mathbf{H}_n\right)^{-1}\mathbf{H}_n^{\text{T}}\Delta z_n. \tag{16.20}$$

Hence, the correlation matrix of the errors of measurement of trajectory parameters, which defines the potential accuracy for implementation of algorithm (16.18), will be equal to [III.22]

$$\mathbf{K}_n^{\text{LS}} = \mathbf{M}\left\{\Delta\boldsymbol{x}_n \Delta\boldsymbol{x}_n^{\text{T}}\right\} = \mathbf{R}_n\mathbf{M}\left\{\Delta z_n \Delta z_n^{\text{T}}\right\}\mathbf{R}_n^{\text{T}}, \tag{16.21}$$

where $\mathbf{R}_n = \left(\mathbf{H}_n^{\text{T}}\mathbf{H}_n\right)^{-1}\mathbf{H}_n^{\text{T}}$, $\mathbf{H}_n = \partial\boldsymbol{h}(\boldsymbol{x})/\partial\boldsymbol{x}|_{\boldsymbol{x}=\boldsymbol{x}_n^i}$ and $\mathbf{M}\{\Delta z_n \Delta z_n^{\text{T}}\}$ is the correlation matrix of primary measurement errors.

It should be noted that Equation (16.21) defines the accuracy of estimation of the trajectory parameters vector $\boldsymbol{x}_n$ from the observation vector z_n using the least squares method in the case when the measurement errors are small and the vector function, $\boldsymbol{h}_n(\boldsymbol{x}_n)$, can be linearized in the vicinity of the true trajectory of the target.

Let us compare the potential accuracies for previously known and previously unknown correlation matrices of primary measurement errors. The solid lines (curves 1) in Figure 16.6 represent RMS deviations of the x coordinate measurement with a previously known correlation matrix of errors for target trajectories with parameters (a) $x_{\text{CR}}/b = 0.25$, $\varphi = 90°$; (b) $x_{\text{CR}}/b = 0.5$, $\varphi = 60°$. The curves have been plotted as functions of the coordinate y at the moment of measurement n. It was assumed for all cases that $V = 200$ m/s, $\lambda = 1$ m and $b = 40$ km. The RMS deviations of primary measurement errors were taken in the calculations to be equal to $\sigma_{\text{f}} = 1$ Hz and $\sigma_\alpha = 0.5°$, and as independent for all moments of time ($\mathbf{G}_n$ is a diagonal matrix). The dashed lines (curves 2) in Figure 16.6 show the

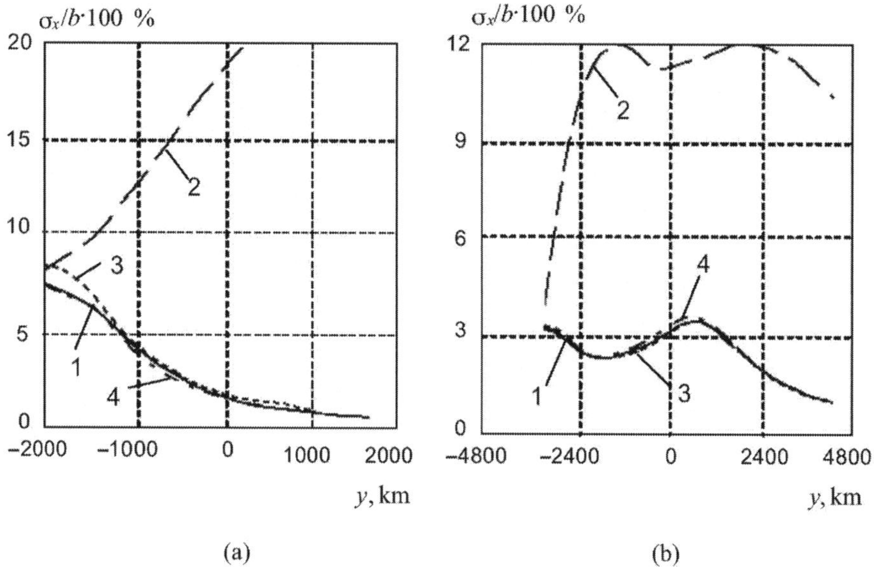

Figure 16.6 Normalized RMS deviations of the x coordinate measurement errors versus the current y coordinate

normalized RMS deviations of x coordinate measurement errors with a previously unknown correlation matrix of errors ($\mathbf{G}_n$ is a unit matrix). Target trajectories and radar parameters remain the same.

A comparison of solid and dashed curves shows that estimates that are optimal by the least squares criterion (when $\mathbf{G}_n$ is a unit matrix) provide essentially lower target position measurement accuracy in comparison with maximum likelihood estimates.

One way to eliminate previously discussed ambiguity is to estimate the variance of primary measurement errors during the trajectory processing and to use the values thus found instead of the unknown true values of elements of matrix $\mathbf{G}_n$. According to the measurement model considered, the elements of the observation vector, z_n, have a Gaussian distribution with $\boldsymbol{h}_n(\boldsymbol{x}_n)$ as a mean and with constant variances, $\sigma_{\rm f}^2$ and σ_α^2. This allows the following estimates of unknown variances to be obtained for n measurements:

$$\hat{\sigma}_{\text{f}n}^2 = \frac{1}{n}\sum_{i=1}^{n}\left[\hat{f}_{\text{D}i} - f_{\text{D}i}(\hat{\boldsymbol{x}}_n)\right]^2, \qquad \hat{\sigma}_{\alpha n}^2 = \frac{1}{n}\sum_{i=1}^{n}\left[\hat{\alpha}_i - \alpha_i(\hat{\boldsymbol{x}}_n)\right]^2. \tag{16.22}$$

An essential feature of Equation (16.22) is that the exact value of the trajectory parameter vector is unknown, which predetermines substitution of its estimated value into the equation. Hence, to find estimates of variances $\hat{\sigma}_{\text{f}n}^2$ and $\hat{\sigma}_{\alpha n}^2$ by formula (16.22), it is necessary to know the estimate of the trajectory parameter vector, $\hat{\boldsymbol{x}}_n$. The required estimate can be obtained from Equation (16.11) provided that, for calculation of the matrix $\mathbf{G}_n$, the estimates of $\hat{\sigma}_{\text{f}(n-1)}^2$ and $\hat{\sigma}_{\alpha(n-1)}^2$ found at the previous stage were used as variances of measurements of the Doppler shift, f_{D}, and the angular coordinate, α.

It is noteworthy that knowledge of the variances of the measurements of primary parameters is not required for the first approximation, which permits the initial estimates of $\hat{\sigma}_{\text{f}}^2$ and $\hat{\sigma}_\alpha^2$ to be obtained from the initial estimate of the trajectory parameter vector.

A method for obtaining $\hat{\sigma}_{\text{f}n}^2$ and $\hat{\sigma}_{\alpha n}^2$ that does not require knowledge of the initial estimate of the trajectory parameter vector can also be considered. This is based on the replacement of the unknown functions $f_{\text{D}}(t)$ and $\alpha(t)$ with their linear approximations (16.14). In this case, the estimated variances of the primary measurement errors are

$$\hat{\sigma}_{\text{f}n}^2 = \frac{1}{n}\sum_{i=1}^{n}\left[\hat{f}_{\text{D}i} - \tilde{f}_{\text{D}}(t_i)\right]^2, \qquad \hat{\sigma}_{\alpha n}^2 = \frac{1}{n}\sum_{i=1}^{n}\left[\hat{\alpha}_i - \tilde{\alpha}(t_i)\right]^2. \tag{16.23}$$

Let us evaluate the accuracy of the target position measurement using the maximum likelihood method for different ways of *a priori* ambiguity elimination with respect to primary measurement errors. For a comparison of methods use will be made of the plots of base length-normalized RMS deviations of x coordinate measurement errors versus the current coordinate y.

In Figure 16.6 the dotted and dashed-and-dotted lines (curves 3 and 4) present discussed plots constructed by the method of mathematical modelling. Target trajectories and radar parameters were the same as those used to construct the corresponding theoretical plots for the scenarios when *a priori* ambiguity is absent and *a priori* ambiguity is ignored. The plots in dotted lines were constructed with the use of algorithm (16.22) for obtaining the estimates of $\hat{\sigma}_{\text{f}n}^2$ and $\hat{\sigma}_{\alpha n}^2$; the dashed-and-dotted plots were constructed with the use of variance estimates obtained according to Equations (16.23). Estimations of RMS deviations are obtained by averaging the results of 1000 independent experiments. The initial approximation was formed on the basis of the first six measurements using the method of linearization of plots for primary parameters (Equations (16.12) and (16.13)).

For the target trajectories considered, the use of estimates of variances of primary measurement errors, $\hat{\sigma}_{\text{f}n}^2$ and $\hat{\sigma}_{\alpha n}^2$, obtained by Equations (16.22) or by (16.23) provides practically the same accuracy of coordinate$x$ measurement. Moreover, the estimates thus obtained are practically as accurate as those calculated with previously known σ_{f}^2 and σ_α^2 values, except for the initial section.

For uniform and rectilinear target motions, it is more practical to use the method based on Equations (16.23) for the calculation of $\hat{\sigma}_{\text{f}n}^2$ and $\hat{\sigma}_{\alpha n}^2$, as it is rather easy to implement. At the same time, this method presupposes that the functions $f_D(t)$ and, $\alpha(t)$ are close to linear. The RMS deviations of primary measurement errors calculated by Equation (16.22) are free of such limitations and can be used with the nonlinear functions $f_D(t)$ and $\alpha(t)$.

Simulation of the iterative algorithm (16.11) shows that the main contribution to the correction of the initial approximation comes from the first iteration with $k_C = 1$. This circumstance allows a single iteration to be used in the practical implementation of algorithm (16.11).

At the same time, it has been shown experimentally that the use of an iterative algorithm to obtain approximate estimates of the tracking vector does not always yield accuracy close to its potential. First of all, a prerequisite for obtaining close-to-potential accuracy is the knowledge of the correlation matrix of primary measurement errors. In practice, it demands that these errors are to be estimated in the process of tracking. In some cases, using this method results in the formation of false trajectories.

Trajectories will be considered to be 'false' if coordinate measurement errors increase while tracking inside radar coverage. The probability of false trajectory formation essentially depends on the form of the true trajectory and the magnitude of primary measurement errors. For example, in the cases considered, false trajectories were initiated for the trajectory with parameters $\varphi = 135^\circ$, $x_{CR}/b = 0.5$. For example, when $\sigma_f = 1$ Hz and $\sigma_\alpha = 0.5^\circ$, the number of false trajectories per 100 was approximately one. When measurement errors were doubled, the number of false trajectories amounted to 10. For trajectories with $\varphi = 90^\circ$, the values of occurrence of false trajectories are less by an order, while for a trajectory with $\varphi = 60^\circ$, they are 0.5 and 4 % respectively.

In conclusion, it can be stated that using the considered iterative target tracking algorithm, on the basis of the maximum likelihood method in the CW FSR, provides a high enough accuracy of moving target tracking. Close agreement between the theoretical calculations and mathematical modelling results takes place. It is shown that parametrical ambiguity [III.26], with respect to the correlation matrix of errors of primary measurements, can be eliminated during tracking.

16.3 EXTRAPOLATION ALGORITHM OF THE TARGET COORDINATE MEASUREMENT

16.3.1 Basic Relationships

When a target is moving in the azimuth plane, its position can be defined as the coordinates of the cross point of the radius vector, $\boldsymbol{r}$, and the position ellipse with foci corresponding to the phase centres of the receiving (Re) and transmitting (Tr) antennas. The sum of the focal radii, meanwhile, is equal to the sum of ranges to the target, $L = R_T + R_R$ (Figure 16.7). With a known range sum, L, and angle, α,

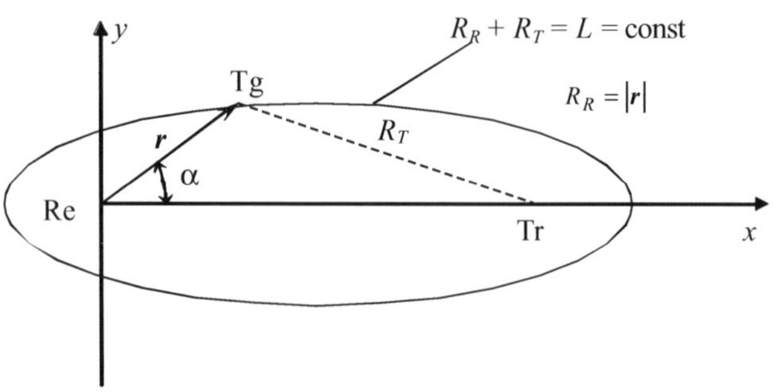

Figure 16.7 Target position determination

the distance to the target, r (here and later it is supposed that the distance to the target, r, corresponds to the distance from the receive side, R_R, if another is not specified), is given by [III.13, III.18, III.26]

$$r = \frac{L^2(t) - b^2}{2[L^2(t) - b\cos\alpha(t)]}. \tag{16.24}$$

In the CW FSR, the range sum cannot be measured directly. To solve this problem, a method was proposed in References [III.13] and [III.26], which permits $L(t)$ estimation indirectly from measurements of the Doppler frequency shift, f_D. The method is based on the well-known relationship between these parameters:

$$f_D(t) = -\frac{1}{\lambda}\frac{dL(t)}{dt}, \tag{16.25}$$

where λ is the signal wavelength. Integration of the signal's Doppler frequency shift (16.25) at the time interval of the target observation yields the range sum at the current instant of time:

$$L(t) = L_0 - \lambda\int_{t_0}^{t} f_D(t)\,dt. \tag{16.26}$$

Where L_0 is the initial value of $L(t)$ at the time instant of target detection, t_0, i.e. $L(t_0)$. At the time t_{CR}, when the target crosses the baseline, the Doppler frequency shift $f_D(t_{CR}) = 0$ and the value of $L(t_{CR})$ is known and equal to the base length, b. This allows the value of L_0 to be found from Equation (16.26) as

$$L_0 = b + \lambda\int_{t_0}^{t_{CR}} f_D(t)\,dt. \tag{16.27}$$

From Equations (16.26) and (16.27) equations are obtained for the calculation of L both before and after the target crosses the baseline [III.13, III.26]:

$$\begin{aligned} L(t) &= b + \lambda\int_{t}^{t_{CR}} f_D(t)\,dt, \quad t \le t_{CR}, \\ L(t) &= b - \lambda\int_{t_{CR}}^{t} f_D(t)\,dt, \quad t > t_{CR}. \end{aligned} \tag{16.28}$$

Thus, the knowledge of the baseline crossing time, t_{CR}, and the dependence of the Doppler shift, $f_D(t)$, allows determination of the range sum using Equations (16.28), both before the baseline crossing time (on a time interval (t, t_{CR})) and after this time (on a time interval (t_{CR}, t)). The range to the target can then be found using Equation (16.24). Therefore, if the azimuth is measured, target coordinates can be calculated.

16.3.2 Features of Tracking before Baseline Crossing

The use of the method discussed in the previous section allows an estimation to be made of the location of the target moving along an arbitrary trajectory. Thus, the entire trajectory can be divided into two parts, and the application of this method has distinctive features in both parts.

The first part is the path after the target crosses the baseline, i.e. at $t > t_{CR}$. Here the estimation, $\hat{L}$, of the range sum can be obtained by direct integration of the receiving signal Doppler frequency shift estimation, $\hat{f}_D(t)$, at an interval (t_{CR}, t).

The second part is the initial path up to the point where the target crosses the baseline at a given time, t_{CR}. During this part of the trajectory the estimation, $\hat{L}$, cannot be obtained directly, as the baseline crossing time and the dependence, $f_D(t)$, at an interval from the moment of observation to the moment t_{CR} are unknown. There are two possible approaches to define target positions on this part of the trajectory. The first is to define the positions only after the target crosses the baseline. This approach excludes the estimation of the target location in real time. Due to this, the second approach seems more appropriate.

This allows target coordinate estimation at the initial trajectory part before the baseline crossing time, i.e. in real time. To provide such an estimation, extrapolation of dependence, $f_D(t)$, from the current observation at the initial trajectory part to the baseline crossing time is performed. The extrapolated dependences, $f_D^e(t)$, are then used in Equations (16.28) [III.13, III.26].

For an approximate evaluation of the baseline crossing time, t_{CR}, by the target, it is generally possible to use both extrapolation of the Doppler frequency shift, $f_D(t)$, and the angular coordinate, $\alpha(t)$, which tend to be zero at this time instant. If a target moves along a linear trajectory with a constant velocity, functions $f_D(t)$ and $\alpha(t)$ are close to linear when the azimuth, α, has small values. This allows representation of these functions with a sufficiently high degree of accuracy by a polynomial approximation with a small number of coefficients. It is then possible to extrapolate them over the trajectory path before crossing the baseline. It is necessary to note that, in this case, rectilinear motion with a constant velocity can be considered as typical for air targets. It is not likely that the target would fulfil a manoeuver during the short time period of the FS region crossing.

The calculation of the coefficients of an approximating polynomial can be made according to criterion of the least squares. Using this criterion, the vector of coefficients of the approximating polynomial, for the case of discrete measurement intervals, is obtained from the expression [III.13, III.26]

$$\boldsymbol{c}_n = (\mathbf{P}_n^{\mathrm{T}}\mathbf{P}_n)^{-1}\mathbf{P}_n^{\mathrm{T}} f_D^{(n)} = \mathbf{B}\boldsymbol{f}_D^{(N)}, \tag{16.29}$$

where $\boldsymbol{f}_D^{(N)} = [\hat{f}_{D(n-N+1)}, \ldots, \hat{f}_{D(n-1)}, \hat{f}_{Dn}]^{\mathrm{T}}$ is the N-dimensional vector of the measurements of the Doppler frequency shift; $\boldsymbol{c}_n = [\hat{c}_{0,n}, \hat{c}_{1,n}, \ldots, \hat{c}_{M,n}]^{\mathrm{T}}$ is the M-dimensional vector of coefficients of an approximating polynomial; N is the number of primary measurements used to make the approximation; $\hat{f}_{Dn}$ is the Doppler shift estimation in the target observation time $t_n = nT$, where T is the primary measurement interval; n is the current measurement number; M is the order of the approximating polynomial ($N > M$); $(\,)^{\mathrm{T}}$ is the transposition sign; $\mathbf{B} = (\mathbf{P}_n^{\mathrm{T}}\mathbf{P}_n)^{-1}\mathbf{P}_n^{\mathrm{T}}$; and

$$\mathbf{P}_n = \begin{bmatrix} 1, & \ldots, & t_{n-N+1}^{M-1}, & t_{n-N+1}^{M} \\ \ldots & \ldots & \ldots & \ldots \\ 1, & \ldots, & t_{n-1}^{M-1}, & t_{n-1}^{M} \\ 1, & \ldots, & t_{n}^{M-1}, & t_{n}^{M} \end{bmatrix}.$$

Taking into account the above, the extrapolated dependence $f_D^e(t)$ for $t < t_{CR}$ can be presented as

$$f_D^e(t) = \boldsymbol{c}_n^{\mathrm{T}}\boldsymbol{t} = \hat{c}_{0,n} + \cdots + \hat{c}_{M-1,n}t^{M-1} + \hat{c}_{M,n}t^{M}, \tag{16.30}$$

where $\boldsymbol{t} = [1, t, \ldots, t^{M-1}, t^{M}]^{\mathrm{T}}$. Substitution of the extrapolating function (16.30) into Equations (16.28), instead of $f_D(t)$ at $t < t_{CR}$, allows a range sum L estimation to be obtained on the initial part sections of the trajectory before the target crosses the baseline ($t_n < t_{CR}$):

$$\hat{L}_n = b + \lambda \sum_{i=0}^{M} \hat{c}_{i,n} \frac{(\hat{t}_{CR,n})^{i+1} - t_n^{i+1}}{i+1}. \tag{16.31}$$

The value $\hat{t}_{CR,n}$ in Equation (16.31) represents the estimation of the baseline crossing time. To find this estimation, the extrapolated dependence, $f_D^e(t)$, has to be equated to zero. Then, the estimation $\hat{t}_{CR,n}$ can be found as a solution of the following equation:

$$\boldsymbol{c}_n^{\mathrm{T}}\boldsymbol{t} = [\mathbf{B}\boldsymbol{f}_D^{e}]^{\mathrm{T}}\boldsymbol{t} = 0. \tag{16.32}$$

16.3.3 Tracking Accuracy Estimation

If the considered extrapolation method of target coordinate estimation is used, the errors of the target location measurement consist of two components. The first component is the fluctuation errors caused

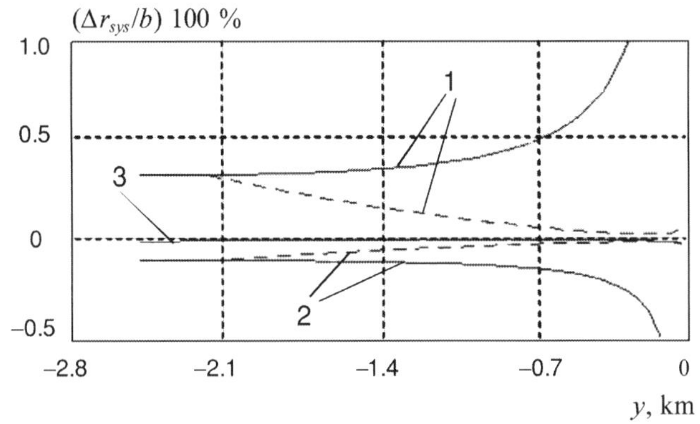

Figure 16.8 Systematic errors of target tracking with the extrapolation method [III.27]

by random errors of measurement of the primary parameters, while the second is the systematic errors arising because of extrapolation inaccuracy.

To define the systematic error component, the N-dimensional vector of the measurements of the Doppler shift, $\hat{\boldsymbol{f}}_{\mathrm{D}}^{(N)}$, will be replaced in Equation (16.29) by the vector of the true values of the Doppler shift, $\boldsymbol{f}_{\mathrm{D}}^{(N)}$. We will find the range sum estimation from Equation (16.31) in view of estimations $\hat{c}_n$ and $\hat{t}_{\mathrm{CR},n}$ calculated after the noted replacement. After substitution of the obtained estimation, $\hat{L}_n$, and the true value of the azimuth, α, into Equation (16.24) the range estimation, $\hat{r}_n$, can be found. We are then able to calculate the systematic error component of the measurement error, Δr_{sys}, as a difference of the obtained estimation, $\hat{r}_n$, and the true value of the range, r_n.

Figure 16.8 shows the dependences normalized to the baseline b systematic error of the range determination upon the value of the coordinate y, calculated for a straight-line target trajectory. The following trajectory parameters were assumed: $\varphi = 90°$, $x_{\mathrm{CR}}/b = 0.5$ and $V = 200$ m/s. The linear ($M = 1$, curves 1), quadratic ($M = 2$, curves 2) and cubic ($M = 3$, curve 3) extrapolations for all measurements ($N = n$, solid lines) and a fixed number of last measurements of Doppler frequency shift ($N = 10$, dashed lines) were applied. Diagrams are calculated for the case of the following radar parameters: $b = 40$ km, $\lambda = 1$ m and $T = 0.25$ s; coverage has been limited to minimal bistatic angle values, $\beta_{\mathrm{B\,min}} = 165°$, which correspond to the coordinate y of the target coming into the coverage $y = 2633$ m; and the initial values of the target coordinates were calculated at $n = 4$.

The graphs plotted in Figure 16.8 show that the use of linear and quadratic extrapolation of the Doppler frequency shift laws of change, $f_{\mathrm{D}}(t)$, provides the values of the systematic errors of the target range measurement, Δr_{sys}, not exceeding several per cent of the base length during almost the whole trajectory part before the target crosses the baseline. On approaching the baseline, the considered systematic errors grow (solid line curves). This growth is explained by the growth in sensitivity of the range sum estimation accuracy to the errors of Doppler frequency shift estimations. This sensitivity is essentially strong in the region of small values of the azimuth, α. Hence, the influence of the Doppler frequency shift extrapolation errors upon the errors of target range estimation grows stronger on approaching the baseline.

Extrapolation errors do not tend to zero on approaching the baseline if the extrapolation is made by the results of an approximation of all measurements obtained. The reason for this is that the extrapolated dependence does not coincide in accuracy with the dependence, $f_{\mathrm{D}}(t)$. If only the last N measurements of the Doppler frequency shift are used for extrapolation, the error, Δr_{sys}, monotonously decreases (dashed curves). With cubic extrapolation the systematic error is practically absent along the whole path (curve 3).

Let us obtain the estimations of the coordinate measurement fluctuation errors at the trajectory path before crossing the baseline. It is assumed that these fluctuation errors of coordinate measurement are small enough, while the estimations of measuring the primary parameters, α and f_D, are supposed to be unbiased, independent and distributed accordingly to Gaussian law. In Equation (16.31), estimations of the approximating polynomial coefficients, $c_{i,n}$, and the crossing time, $t_{CR,n}$, could be presented as the sums $\hat{c}_{i,n} = \hat{c}^c_{i,n} + \Delta c_{i,n}$ and $\hat{t}_{CR,n} = \hat{t}^c_{CR,n} + \Delta t_{CR,n}$ of their values, $\hat{c}^c_{i,n}$ and $\hat{t}^c_{CR,n}$ without fluctuation errors and deviations, $\Delta c_{i,n}$ and Δt_{CR}, caused by action of fluctuation errors of the primary measurement. Thus, the dependence $L_n(\Delta c_{i,n}, \Delta t_{CR})$, defined by Equation (16.31), is developed as a series in $\Delta c_{i,n}$ and Δt_{CR}, taking into account only linear members. Then, excepting systematic error, the fluctuation component of the error of measurement of the range sum before crossing the baseline is obtained

$$\Delta L_n \approx \lambda \sum_{i=0}^{M} (\Delta c_{i,n}) \frac{(\hat{t}^c_{CR,n})^{i+1} - t_n^{i+1}}{i+1} = \lambda \Delta \boldsymbol{c}_n^T \boldsymbol{\eta}_n^c, \tag{16.33}$$

where

$$\Delta \boldsymbol{c}_n = \boldsymbol{B} \Delta \boldsymbol{f}_D^{(n)}, \tag{16.34}$$

where $\Delta \boldsymbol{f}_D^{(n)} = [\Delta f_{D,(n-N+1)}, \ldots, \Delta f_{D,(n-1)}, \Delta f_{D,n}]^T$ is the Doppler frequency shift measurement error vector and; $\boldsymbol{\eta}_n^c = [\eta^c_{0,n}, \eta^c_{1,n}, \ldots, \eta^c_{M,n}]^T$ and $\boldsymbol{\eta}^c_{i,n} = \{[(\hat{t}^c_{CR,n})^{i+1} - t_n^{i+1}]/(i+1)\} i = \overline{0, M}$.

Then, developing the function $r(\alpha, L)$, defined by Equation (16.24), as a series of deviations in $\Delta\alpha$ and ΔL, and using only linear members, the expression is obtained for the fluctuation component of the range measurement:

$$\Delta r_n \approx \rho_{L,n} \Delta L_n + \rho_{\alpha,n} \Delta \alpha_n, \tag{16.35}$$

where the components $\rho_{L,n}$ and $\rho_{\alpha,n}$ determine the sensitivity of the range measurement to measurement errors of parameters L and α:

$$\rho_{\alpha,n} = \left(\frac{\partial r}{\partial \alpha} \right)_{\substack{L=L_n^c \\ \alpha=\alpha_n}} = \frac{b \sin \alpha_n}{2\left(L_n^c - b\cos\alpha_n\right)^2} \left[\left(L_n^c\right)^2 - b^2 \right], \tag{16.36}$$

$$\rho_{L,n} = \left(\frac{\partial r}{\partial L} \right)_{\substack{L=L_n^c \\ \alpha=\alpha_n}} = \frac{\left(L_n^c\right)^2 - 2L_n^c b \sin\alpha_n + b^2}{2\left(L_n^c - b\cos\alpha_n\right)^2}, \tag{16.37}$$

where α_n^c and L_n^c are values of parameters in the absence of fluctuation noise.

Taking Equation (16.35) and the independence of the measurements of the azimuth and the range sum into account, the variance of the range measurement error at the observation time t_n gives

$$\sigma^2_{r,n} = \rho^2_{L,n} \sigma^2_{L,n} + \rho^2_{\alpha,n} \sigma^2_{\alpha,n}, \tag{16.38}$$

where $\sigma^2_{\alpha,n}$ is the variance of the azimuth estimation error and $\sigma^2_{L,n}$ is the variance of the range sum estimation error. Using Equations (16.33) and (16.34), it is possible to obtain the following expression:

$$\sigma^2_{L,n} = \lambda^2 (\eta_n^c)^T \mathbf{B} \mathbf{K}_f \mathbf{B}^T \eta_n^c. \tag{16.39}$$

Where $\mathbf{K}_f$ is a correlation matrix of the Doppler shift measurement errors.

After the target crosses the baseline ($t > t_{CR}$), the range sum estimation $\hat{L}(t)$ is determined according to Equations (16.28) by direct integration of the target return Doppler shift estimation, $\hat{f}_D(t)$. When estimations of the primary parameters, $\hat{f}_D(t)$ and $\hat{\alpha}(t)$, are unbiased, the range sum estimation is also unbiased. The range sum estimation error in the case of discrete measurements of the Doppler frequency shift, taking into account Equations (16.28), appears as

$$\Delta L_N = -\lambda T \sum_{i=0}^{N} \Delta f_{\mathrm{D},i}. \tag{16.40}$$

From Equation (16.40) the following expression is obtained for the variance of the range sum estimation error after the baseline is crossed:

$$\sigma_{L,N}^2 = \lambda^2 T^2 \sum_{i=0}^{N} \sigma_{\mathrm{f},i}^2. \tag{16.41}$$

Where $\sigma_{\mathrm{f},i}^2$ is a variance of the ith order after the baseline crossing time estimation of the Doppler shift and N is the number of measurements. Having substituted Equation (16.41) into (16.38), it is possible to obtain the expression for the calculation of the variance of the range estimation error after the baseline crossing. Since $= r\cos(\alpha)$, in the FS region, i.e. at small values of α, the variance of the estimation error of coordinate x of the target $\sigma_{x,n}^2$ practically coincides with the variance of the range estimation error $\sigma_{r,n}^2$.

Figure 16.9 shows the dependences of RMS deviations $\sigma_{x,n}$ of the fluctuation error, normalized to the base length, upon the value of coordinate y for the same linear trajectory of the target ($\varphi = 90°$, $x_{\mathrm{CR}}/b = 0.5$ and $V = 200$ m/s) and parameters of the system that were adopted for calculation of patterns given in Figure 16.8. Root mean square deviations of primary measuring errors were accepted to be equal to $\sigma_{\mathrm{f}} = 1$ Hz and $\sigma_\alpha = 0.5°$ for all measurements.

In Figure 16.9, curves marked by 1 correspond to the case of linear extrapolation ($M = 1$), performed using all Doppler shift measurements, obtained before the current time t_n ($N = n$, solid line) and using the last N measurements ($N = 10$, dashed line). Curves marked by 2 and 3 are obtained for the case of quadratic ($M = 2$) and cubic ($M = 3$) extrapolation accordingly performed using all Doppler shift measurements. The curve marked by 4 corresponds to the potential accuracy of the maximum likelihood estimation of coordinate x, plotted by the calculation of the elements of the Fisher information matrix. Potential accuracy is determined for the case when the current target coordinates are estimated using all measurements of the Doppler shift, f_{D}, and azimuth, α, from the first moment of target detection in the radar coverage.

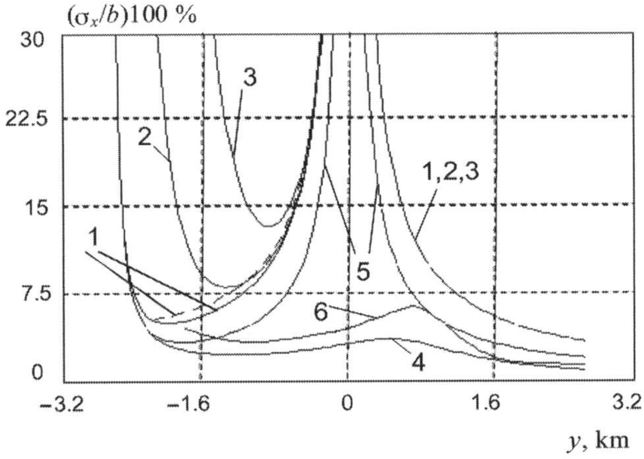

Figure 16.9 RMS deviations of the target x coordinate estimation [III.27]

From Figure 16.9 it is evident that the coordinate x (or range r) estimation error has the largest values, both at the beginning of the trajectory and in the region near to the baseline. Large values of estimation errors at the start of the trajectory are caused by small amounts of measurements of the primary parameters. The growth in error at the baseline crossing point is a consequence of the increase of the sensitivity coefficients, $\rho_{\alpha,n}$ and $\rho_{L,n}$. This increase takes place when the azimuth, α, approaches zero values.

From a comparison of curves 1, 2 and 3 it follows that an increase in the extrapolating polynomial degree results in a significant increase in fluctuation errors of coordinate determination. A significant increase in fluctuation errors is observed at the initial section of the trajectory; this is due to the fact that the duration of the time interval to the baseline crossing time is maximal here, and extrapolation errors can be large. Therefore, in spite of the fact that a reduction in the extrapolating polynomial degree can lead to an increase in systematic errors of coordinate measurement, the linear extrapolation looks the most suitable. The limitation in the number of primary measurements, N, used for extrapolation in all cases results in an increase in fluctuation error of coordinates measurements (the dashed line curve marked by 1 in Figure 16.9 corresponds to the case of $N = 10$). However, the reduction of N is required to provide a monotonous decrease of systematic error in the target coordinate estimation, when linear extrapolation is used (Figure 16.8, dashed lines). Limiting the number of primary measurements used for linear extrapolation allows, algorithm efficiency to be retained in the case of nonlinear target trajectories.

It should be noted that for a linear trajectory of the target, parameter $\alpha(t)$ is also close to the linear function of time. In this case, further improvement in the accuracy of the current target position estimation could be achieved by means of linear smoothing of azimuth measurements. To realize such smoothing, linear polynomial coefficients are computed using the last K azimuth measurements. To compute polynomial coefficients, the least squares method may be used. The efficiency of azimuth measurement smoothing can be illustrated by curve 5 in Figure 16.9, plotted for the case of linear smoothing of azimuth measurements using $K = 10$ consecutive measurements, while other conditions are equal to the conditions for curve 1.

The peculiarity of the RMS deviation, $\sigma_{x,n}$, behaviour is its unlimited increase in the target range estimation error in the region close to the baseline. To eliminate this disadvantage, it is possible to use smoothing of the target trajectory coordinate estimations. In Figure 16.9, curve 6 presents a dependence of normalized RMS deviations of the coordinate x estimation error upon the coordinate y value, for the case where linear smoothing of coordinate x measurements is performed using the last 32 measurements. This is obtained by a mathematical modelling method via an average of 1000 independent realizations. Other conditions were the same, as in the calculations of curve 5. At small values of the azimuth, $\alpha(|\alpha| < 0.3^\circ)$, linear extrapolation of the target trajectory has been performed using all previous coordinate measurements. Comparing curve 6 with the other curves, it is possible to conclude that the use of trajectory smoothing allows a reduction of RMS deviations of the target range measurements when the target is located on the baseline. The acceptable values of these RMS deviations of the target range are close to the RMS deviations corresponding to the potential accuracy of the maximum likelihood target coordinate estimations. The RMS deviation of measurement error of coordinate $y = r \sin \alpha$ is substantially less than $\sigma_{x,n}$. In addition, they have no increase at $y = 0$ and do not exert a significant influence on the accuracy of the target location definition.

Results of the calculations carried out for various linear trajectories show that in a range of changes of parameters, $30^\circ < \varphi < 150^\circ, 210^\circ < \varphi < 330^\circ$ and $0.1 < x_{CR}/b < 0.9$, cited assertions remain valid. Although the numerical results may change two to three times, depending on the trajectory inclination towards the base and the distance of the target from the FSR receiver.

Mathematical modelling has been carried out to estimate the influence of a target flight altitude upon the accuracy of the target horizontal coordinate measurement. The results of modelling show that this influence is negligible for flight altitudes that do not exceed 3–5 % of the base length depending on the trajectory. At $b = 40$ km this corresponds to flight altitudes up to 1.2–2.0 km. For higher altitudes, the tracking algorithm should be modified to take the target elevation angle into account.

In conclusion, it can be stated that applying the tracking algorithm based on the preliminary range sum estimation in a 2D CW FSR provides an estimation of the target coordinates using the target return Doppler shift and the target azimuth measurements. In the case where the target moves with constant velocity along a linear trajectory, the coordinate estimation accuracy is close to the potential accuracy of the maximum likelihood estimation, both before and after the baseline is crossed. In practical terms, the target range estimation errors may be several per cent of the baseline length before the target will be out of FSR coverage.

17

Coordinate Measurement in a 3D FSR

17.1 SYSTEMATIC ERRORS OF TARGET TRACKING IN A 2D FSR

In the previous chapter algorithms of the target coordinate measurement in a 2D FSR were considered. These algorithms have been built with the assumption that the target altitude is equal to zero. At the same time, a nonzero target altitude leads to systematic errors of target tracking in horizontal coordinates. In other words, the estimations of target coordinates are biased when its altitude is nonzero. This is explained by the fact that the target return Doppler frequency shift depends upon the target altitude. However, this dependence could not have been taken into account in the 2D FSR because the target elevation is not measured.

To estimate the influence of the target altitude upon the discussed systematic errors, a 2D CW FSR is considered using the iterative algorithm (16.11) of the target coordinate measurement on the basis of the maximum likelihood method. It should be remembered that in the considered type of FSR, estimations of target coordinates are obtained using primary measurements of the target return Doppler frequency shift and the target azimuth.

Estimations of systematic errors of the target coordinate measurement via the considered 2D FSR have been obtained by mathematical modelling. As used earlier, the FSR has a base length $b = 40$ km and a transmitter wavelength $\lambda = 1$ m. The target is assumed to move along linear trajectories with different parameters, such as the altitude, h, the angle of trajectory inclination towards the baseline, φ, and the coordinate of the baseline crossing point, x_{CR}. Diagrams of the dependence of normalized systematic errors of coordinate x estimation upon the coordinate y values are shown in Figure 17.1. Curve 1 corresponds to the systematic errors in two-dimensional radar during a target flight at zero altitude $h = 0$, curve 2 corresponds to the systematic error at $h = 1000$ m and curve 3 is for $h = 2000$ m.

From the diagrams given in Figure 17.1, it is seen that systematic errors of the target coordinate estimation in a 2D FSR grow with an increase in the target flight altitude. For specified parameters of the system at altitudes not exceeding 1 km, this error is insignificant. A further increase of altitude leads to an essential increase in systematic errors. These errors can achieve several per cent of the base length at $h = 2000$ m. However, a significant growth in the coordinate estimation error, caused by an altitude increase, is not observed for all the trajectories of the target flight. When the targets move along the trajectories, crossing the baseline near its middle, or the trajectory inclination angle is close to 90°

Bistatic Radars: Principles and Practice Edited by M. Cherniakov

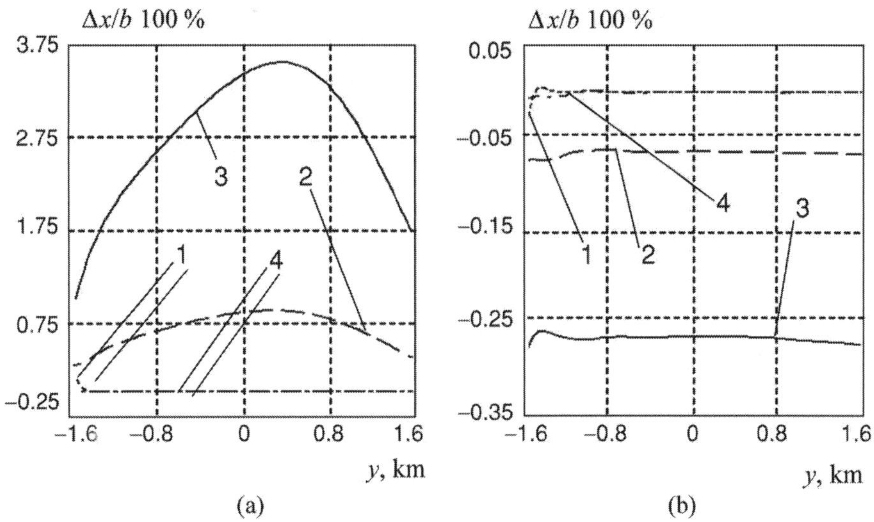

Figure 17.1 Normalized systematic error of a coordinate x estimation in a 2D FSR, where $T = 0.25$ s, $\lambda = 1$ m, $V = 200$ m/s and $b = 40\,000$ m: (a) $x_{CR}/b = 0.25$, $\varphi = 45°$; (b) $x_{CR}/b = 0.25$, $\varphi = 90°$

(normal trajectory), the systematic error grows weakly with an altitude increase. This is explained by the dependence of the Doppler shift on coordinate y for noted trajectory types, which is close to the analogous dependence for trajectories with zero altitude. In fact, for trajectories crossing the baseline in the middle (at any inclination angle) or at a right inclination angle (at any point), the Doppler shift is equal to zero at the moment of crossing the baseline irrespective of the target flight altitude. This phenomenon is illustrated in Figure 17.1(b). The specific feature of most of the obtained results is that systematic errors have maximal values in the region close to the baseline, but at the moment when the target leaves radar coverage, systematic errors fall to rather small values. The given conclusions, as the calculation shows, are correct for systematic errors of estimation of both plane coordinates, as well as of their derivatives (target velocity components) at different trajectories of the target.

Note that variances in random errors of primary parameter measurements were assigned to be small to obtain estimations of systematic errors. As was shown earlier, in the considered CW FSR the values of the primary parameters (the Doppler frequency shift and the azimuth) and trajectory parameters (target Cartesian coordinates and their derivatives) are concatenated with the system of nonlinear equations (16.7). Thus, the considered iterative algorithm of target coordinate measurements solves the problem of the nonlinear estimation of the trajectory parameter vector by the vector of the primary parameters.

Due to this nonlinearity, the increase in the variances of primary measurement errors leads to an increase in both variances and mean values of coordinate estimation errors. Therefore, coordinate estimations become more strongly biased with the growth of fluctuation primary measurement errors.

Since fluctuation errors of primary measurements are defined by the signal-to-noise ratio, the presented results of systematic error estimation correspond to the case of a large SNR, or even to the absence of noise. At the same time, as the modelling results show, in the conditions where noise is present the considered systematic errors can be several times larger.

The following conclusions may thus be drawn from this section. By increasing the target flight altitude, the estimations of target coordinates in the azimuth plane, obtained via a 2D FSR, can be significantly biased or, in other words, can include large systematic errors. To eliminate these systematic errors it is necessary to use measurements of the third coordinate, the elevation angle of the target. In this case it is possible not only to eliminate the discussed systematic errors but also to obtain an estimation of the target flight altitude. These possibilities are to be realized in a three-dimensional FSR. Algorithms of target tracking in a 3D FSR are themselves a subject for further consideration.

17.2 ITERATIVE COORDINATE ESTIMATION ALGORITHM FOR A 3D FSR

17.2.1 Mathematical Model of the Measuring Process

Consider a three-dimensional forward-scattering radar [III.17, III.28]. The 3D FSR topology is shown in Figure 17.2. Most of the designations in Figure 17.2 coincide with similar designations in Figure 15.1, although the target motion model in Figure 17.2 is more detailed. In particular, AB is the target trajectory, CD is the trajectory projection on to the horizontal plane (x, y), x_{CR} is the coordinate of target flight over the baseline and h is the target altitude.

In the 3D FSR, the Doppler frequency shift, f_D, azimuth, α, and elevation angle, β, are the measured primary parameters. A primary measurement vector looks like

$$z_n = \left[\hat{f}_{D1}(\boldsymbol{x}_n), \hat{\alpha}_1(\boldsymbol{x}_n), \hat{\beta}_1(\boldsymbol{x}_n), \hat{f}_{D2}(\boldsymbol{x}_n), \hat{\alpha}_2(\boldsymbol{x}_n), \hat{\beta}_2(\boldsymbol{x}_n), \ldots, \hat{f}_{Dn}(\boldsymbol{x}_n), \hat{\alpha}_n(\boldsymbol{x}_n), \hat{\beta}_n(\boldsymbol{x}_n)\right]^{\mathrm{T}}. \tag{17.1}$$

Where $\hat{f}_{Di}(\boldsymbol{x}_n)$, $\hat{\alpha}_i(\boldsymbol{x}_n)$ and $\hat{\beta}_i(\boldsymbol{x}_n)$ are estimations at the moment of time, $t_i = iT$, corresponding to the ith interval of primary measurements $((i-1)T, iT)$, $i \in [1, n]$; $t_n = nT$ is the moment of observation; and $\boldsymbol{x}_n$ is the vector of target trajectory parameters.

It is assumed, as earlier, that the target follows a linear trajectory at constant altitude, h, with constant velocity, V. In this case,

$$\boldsymbol{x}_n = [x_n, y_n, h, V_x, V_y]^{\mathrm{T}}, \tag{17.2}$$

where x_n, y_n, h, V_x and V_y, are the values of the Cartesian coordinates and their rates of change at the moment of observation, t_n. The processing goal is to estimate the vector of trajectory parameters, $\boldsymbol{x}_n$, by the primary measurement vector, z_n.

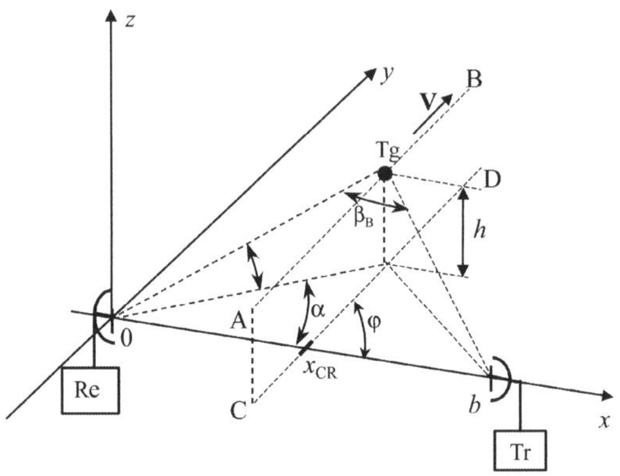

Figure 17.2 Model of the target motion in a 3D FSR

Taking into account Equation (15.19), the system of equations concatenating the exact values of measured primary parameters with the exact values of the trajectory parameters is written as follows:

$$
\begin{aligned}
f_D(\boldsymbol{x}_n) &= -\frac{1}{\lambda}\left(\frac{[x_n - V_x(n-i)T]\,V_x + \left[y_n - V_y(n-i)T\right]V_y}{\sqrt{[x_n - V_x(n-i)T]^2 + \left[y_n - V_y(n-i)T\right]^2 + h^2}}\right.\\
&\left.+\frac{\left[y_n - V_y(n-i)T\right]V_y - \{b - [x_n - V_x(n-i)T]\}\,V_x}{\sqrt{\{b - [x_n - V_x(n-i)T]\}^2 + \left[y_n - V_y(n-i)T\right]^2 + h^2}}\right),\\
\alpha_i(\boldsymbol{x}_n) &= \operatorname{arctg}\frac{y_n - V_y(n-i)T}{x_n - V_x(n-i)T},\\
\beta_i(\boldsymbol{x}_n) &= \operatorname{arctg}\frac{h}{\sqrt{[x_n - V_x(n-i)T]^2 + \left[y_n - V_y(n-i)T\right]^2}}.
\end{aligned}
\tag{17.3}
$$

Similar to the case of the 2D FSR, it is assumed that the primary measurement errors are Gaussian with zero mean and known RMS deviations, σ_f, σ_α, σ_β, considered to be identical for all moments of time. It is also supposed that the measurements of the primary parameters are independent mutually, as well as for different observation moments.

17.2.2 Potential Accuracy

If the maximum likelihood method is used, the potential accuracy of the target coordinate measurement is determined by the Fisher information matrix. For the observation model mentioned above, elements of the Fisher matrix [III.5, III.22, III.23] are defined by the following equation:

$$
\begin{aligned}
\mathbf{J}_{lk}(\boldsymbol{x}_n) &= -\mathrm{M}\left\{\frac{\partial^2 \ln p(z_n|\boldsymbol{x}_n)}{\partial x_l\,\partial x_k}\right\}\\
&= \frac{1}{\sigma_\mathrm{f}^2}\sum_{i=1}^{n}\left[\frac{\partial f_{\mathrm{D}i}(\boldsymbol{x}_n)}{\partial x_l}\,\frac{\partial f_{\mathrm{D}i}(\boldsymbol{x}_n)}{\partial x_k}\right] + \frac{1}{\sigma_\alpha^2}\sum_{i=1}^{n}\left[\frac{\partial \alpha_i(\boldsymbol{x}_n)}{\partial x_l}\,\frac{\partial \alpha_i(\boldsymbol{x}_n)}{\partial x_k}\right]\\
&+\frac{1}{\sigma_\beta^2}\sum_{i=1}^{n}\left[\frac{\partial \beta_i(\boldsymbol{x}_n)}{\partial x_l}\,\frac{\partial \beta_i(\boldsymbol{x}_n)}{\partial x_k}\right],
\end{aligned}
\tag{17.4}
$$

where $p(z_n|\boldsymbol{x}_n)$ is the posterior probability density of the primary measurement vector z_n; x_l and x_k are elements of the vector $\boldsymbol{x}_n$; $l, k \in [1, 5]$ ($x_1 = x_n$, $x_2 = y_n$, $x_3 = h$, $x_4 = V_x$, $x_5 = V_y$); and $\mathbf{M}\{\cdot\}$ is a symbol of statistical averaging.

Using Equation (17.4), we find the variances, $\sigma_{x,n}^2$, $\sigma_{y,n}^2$, $\sigma_{h,n}^2$, $\sigma_{Vx,n}^2$ and $\sigma_{Vy,n}^2$, of errors of trajectory parameter estimations as diagonal elements of the matrix, inverse to the Fisher matrix. The diagrams of dependence of normalized RMS deviations of coordinate x estimation errors on coordinate y are shown in Figure 17.3. In the calculations, the system parameters were accepted to be equal: the primary measurement interval $T = 0.25$ s, base length $b = 40$ km, wavelength of the transmitter $\lambda = 1$ m, and RMS deviations of the primary measuring errors $\sigma_f = 1$ Hz, $\sigma_\alpha = \sigma_\beta = 0.5°$. The coverage of the FSR was limited by values of coordinate $y = \pm 2000$ m. Target trajectory parameters were assigned to be: (a) $x_\mathrm{CR}/b = 0.25$, $\varphi = 45°$ and (b) $x_\mathrm{CR}/b = 0.75$, $\varphi = 45°$. In addition, in each case the target was assigned to be moving at two different altitudes. In Figure 17.3, diagrams 1 correspond to the potential accuracy when the altitude of the target flight $h = 0$ and diagrams 2 are for $h = 2000$ m.

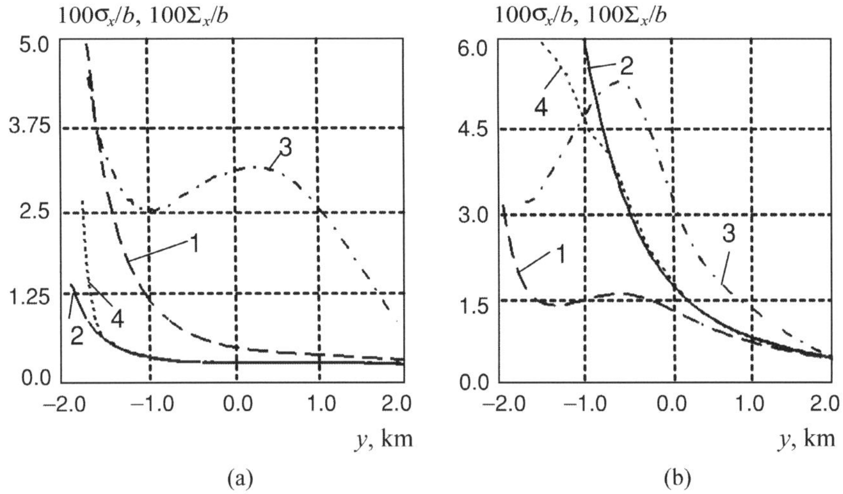

Figure 17.3 Dependence of coordinate x measurement errors on the value of coordinate y

Using Equations (17.3) and (17.4) it is clear that the potential accuracy of the estimation of the horizontal target coordinates in 2D and 3D radars at $h = 0$ is identical. Therefore, the diagrams constructed for zero altitude correspond to the potential accuracy in the 2D FSR.

The ratio of the potential accuracy of the coordinate x measurement in two- and three-dimensional radar depends on the target trajectory parameters. When $h > 0$, the potential accuracy of the target coordinate estimation in the 3D FSR is higher if the target moves near to the receive side, but the accuracy in the 2D FSR is higher if the target moves near to the transmit side. This feature has two causes. First, the rate of change of primary parameters depends on the target location between the transmit and receive sides. This is higher when the target is near to the receiver. Second, in the 3D FSR, the length of the primary measurement vector, z_n, is increased in comparison with the 2D due to the measurements of the target elevation angle added. Therefore, the Fisher matrix includes additional components. These components determine the derivatives of the elevation angle by elements of the trajectory parameter vector, $\boldsymbol{x}_n$. Therefore, in the region close to the receiver the second factor intensifies the first one to provide a higher accuracy in the 3D FSR, while the inverse situation takes place in the proximity of the transmitter region.

Errors of the coordinate y estimation are small in comparison with errors of the coordinate x estimation and do not affect the accuracy of the target position estimation. As the calculation shows, potential accuracies of the target horizontal coordinate estimation in the 2D FSR at $h = 0$ and 3D FSR at $h > 0$ differ essentially only in the segment of the target trajectory before the moment of crossing the baseline. After the baseline is crossed, potential accuracies of the target horizontal coordinate estimation are close in both types of radar.

The accuracy of the altitude estimation in the 3D FSR has also been evaluated. In Figure 17.4, the dependence of normalized RMS deviations of altitude estimation errors on the value of coordinate y in the 3D FSR is shown. Curves 1 correspond to the target flight altitude of $h = 1000$ m and curves 2 to $h = 2000$ m. Curves corresponding to the potential accuracy are shown by dashed and solid lines, while curves obtained by mathematical modelling are shown by chain and dotted lines.

By increasing the altitude, the accuracy of the altitude evaluation falls. This can be easily explained from the viewpoint of geometry. At an equal horizontal range, the linear size of the error region of the altitude estimation increases with the growth in altitude.

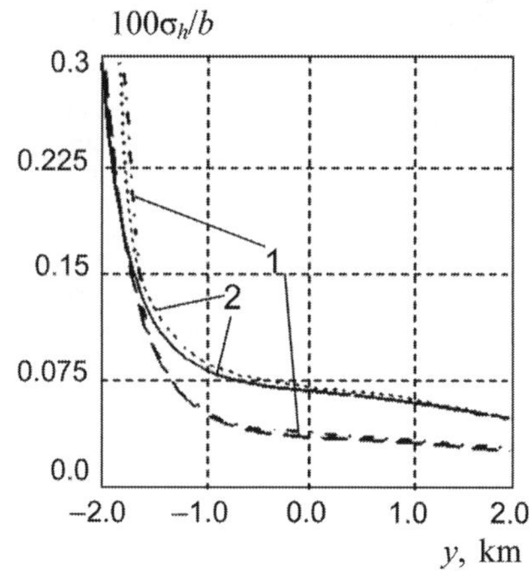

Figure 17.4 Normalized RMS deviation of the target altitude estimation

17.2.3 Iterative Algorithm of the Coordinate Estimation

An iterative algorithm used to obtain the maximum likelihood estimation of the trajectory parameter vector in the 3D FSR is formally similar to the algorithm in the 2D FSR and appears as

$$\hat{\boldsymbol{x}}_n^{i+1} = \hat{\boldsymbol{x}}_n^i + k_{\mathrm{C}}\left(\mathbf{H}_n^{\mathrm{T}}\mathbf{G}_n\mathbf{H}_n\right)^{-1}\mathbf{H}_n^{\mathrm{T}}\mathbf{G}_n\left[z_n - \boldsymbol{h}_n\left(\hat{\boldsymbol{x}}_n^i\right)\right], \tag{17.5}$$

where $\hat{\boldsymbol{x}}_n^i$ is the estimate of the parameter vector on the ith iteration, $\mathbf{H}_n = \partial \boldsymbol{h}_n(\boldsymbol{x})/\partial \boldsymbol{x}|_{\boldsymbol{x}=\hat{\boldsymbol{x}}_n^i}$ is the derivative matrix, $\boldsymbol{h}_n(\boldsymbol{x}_n) = [f_{\mathrm{D1}}(\boldsymbol{x}_n), \alpha_1(\boldsymbol{x}_n), \beta_1(\boldsymbol{x}_n), f_{\mathrm{D2}}(\boldsymbol{x}_n), \alpha_2(\boldsymbol{x}_n), \beta_2(\boldsymbol{x}_n), \ldots, f_{\mathrm{D}n}(\boldsymbol{x}_n), \alpha_n(\boldsymbol{x}_n), \beta_n(\boldsymbol{x}_n)]^{\mathrm{T}}$ is the nonlinear vector function determined by dependences (17.3) and k_{C} is the parameter determining the rate of convergence.

In this case, the first approximation of the vector of trajectory parameters has to be found taking into account the target elevation angle. As in the two-dimensional case, two triplets of values of primary parameters $(f_{\mathrm{D1}}, \alpha_1, \beta_1)$ and $(f_{\mathrm{D}n}, \alpha_n, \beta_n)$ are considered, taken at the time interval $\Delta t = (n-1)T$, but now elevation angle values are also used. Using these triplets of primary parameters, a system of six nonlinear equations concatenating the values of the primary parameters and the trajectory parameters can be obtained from Equation (17.3). Taking into account the fact that the bistatic angle, β_{B}, is close to $180°$, the proposed system can be presented in a linearized aspect:

$$\begin{aligned} f_{\mathrm{D}n} &= -\frac{V_y}{\lambda}\left(\alpha_n + \frac{y_n}{b - x_n}\right), \\ \alpha_n &= \frac{y_n}{x_n}, \\ \beta_n &= \frac{h}{\sqrt{x_n^2 + y_n^2}}, \\ f_{\mathrm{D1}} &= -\frac{V_y}{\lambda}\left(\alpha_1 + \frac{y_1}{b - x_1}\right), \\ \alpha_1 &= \frac{y_1}{x_1}, \\ \beta_1 &= \frac{h}{\sqrt{x_1^2 + y_1^2}}. \end{aligned} \tag{17.6}$$

The system obtained always has the unique solution determined by the expressions

$$
\begin{aligned}
x_n &= \left(\lambda f_{\mathrm{D}n} f_{\mathrm{D}1} \Delta t + b\alpha_1^2 f_{\mathrm{D}n} - b\alpha_n \alpha_1 f_{\mathrm{D}1}\right)/u, \\
y_n &= \alpha_n x_n, \\
h &= \sqrt{x_n^2 + y_n^2}\,\beta_n, \\
V_x &= b\left[f_{\mathrm{D}n}\alpha_1^2 + f_{\mathrm{D}1}\alpha_n^2 - \alpha_n\alpha_1\left(f_{\mathrm{D}1} + f_{\mathrm{D}n}\right)\right]/(u\Delta t), \\
V_y &= -\lambda f_{\mathrm{D}n} f_{\mathrm{D}1}\left(\alpha_1 - \alpha_n\right)/u,
\end{aligned}
\tag{17.7}
$$

where $u = \lambda f_{\mathrm{D}n} f_{\mathrm{D}1} \Delta t/b + f_{\mathrm{D}n}\alpha_1^2 - f_{\mathrm{D}1}\alpha_n^2$.

If an initial approximation is found according to Equations (17.7), systematic errors arise due to the linearization of system (17.3), and these errors increase at a reduction of the bistatic angle [III.23]. Note that the representation of the dependence of the Doppler shift on the trajectory parameters in the linearized aspect does not take into account the fact that when the target flight altitude is nonzero, the Doppler shift is not equal to zero at the moment of flight above the baseline in general. Moreover, an account of this altitude-caused effect is possible only by keeping a nonlinearity of system (17.3). This fact results in the occurrence of additional systematic errors in the calculation of the first approximation of trajectory parameters. Calculations of the first approximation of coordinate x, executed according to the technique mentioned above, with the use of Equation (17.7) have shown that when the altitude grows, the systematic errors of estimation of the first approximation can both essentially increase and slightly decrease. The first is explained by ignoring the dependence of the Doppler shift on altitude, while the second can be explained by the greater proximity of the time dependence of the Doppler shift at $h > 0$ to linear for some trajectories. Some increase in the systematic errors of the first approximation does not result in loss of efficiency of the constructed algorithm on the basis of the maximum likelihood method.

In the analysis of the 2D FSR it was noted that it is sufficient to use a single iteration in most cases for the minimization of computing expenses in algorithm (17.5). Moreover, to reduce the effect of random errors of the primary measurements on the accuracy of the first approximation, it is suggested that the first approximation can be found using all the previously obtained primary measurements. This method is also applicable in the 3D FSR. As in the 2D FSR considered earlier, the approximation of the dependence of the Doppler shift, f_{D}, azimuth, α, and elevation angle, β, on time can be obtained by the least squares method. The polynomial of an assigned degree (but usually a linear polynomial) is used. Then, substituting smoothed estimations $\tilde{f}_{\mathrm{D}1}$, $\tilde{f}_{\mathrm{D}n}$, $\tilde{\alpha}_1$, $\tilde{\alpha}_n$, $\tilde{\beta}_1$ and $\tilde{\beta}_n$ in Equation (17.7), instead of $f_{\mathrm{D}1}, f_{\mathrm{D}n}, \alpha_1, \alpha_n, \beta_1$ and β_n, the first approximation is obtained.

As noted earlier, the method of first approximation computing can be exposed to subsequent optimization selecting from all pairs of triplets, $(\tilde{f}_{\mathrm{D}i}, \tilde{\alpha}_i, \tilde{\beta}_i)$, $(\tilde{f}_{\mathrm{D}j}, \tilde{\alpha}_j, \tilde{\beta}_j)$, $i, j \in [1, N]$, $i < j$, a pair that satisfies the condition

$$
\hat{\boldsymbol{x}}_N^0 = \underset{\forall(i,j)}{\arg\min}\, \varphi_N\left(\hat{\boldsymbol{x}}_N^0(i, j)\right), \tag{17.8}
$$

where N is the number of primary measurement samples on which the evaluation of the first approximation is found. Estimation of the vector of trajectory parameters for the moment t_N is found by means of extrapolation of the estimation for the moment t_j. For all $n > N$ the first approximation can be obtained by extrapolation of the maximum likelihood estimation of the trajectory parameter vector from the previous $(n-1)$th moment of time.

17.2.4 Mathematical Modelling Results

The accuracy of the target coordinate estimation in the 3D FSR, by means of the constructed iterative tracking algorithm, has been evaluated by the method of mathematical modelling. Evaluation of the vector of trajectory parameters was carried out by means of algorithm (17.5), and the first approximation was found accordingly to Equation (17.8). It was assumed that the trajectory was unique, and primary measurements were associated with the trajectory without errors. The statistics of primary measurement errors were accepted as Gaussian and the averaging was made on 1000 independent realizations.

The diagrams of dependence of normalized total errors of the coordinate x estimation on coordinate y for two different trajectories, obtained by the method of mathematical modelling, are plotted in Figure 17.3. Curve 3 corresponds to the 2D FSR, while curve 4 is for the 3D FSR. In both cases the altitude of the target flight amounted to $h = 2000$ m. System parameters, trajectories and RMS deviations of primary measuring errors are equal to those used in the calculation of potential accuracy. Total errors of the coordinate x estimation $\Sigma_{x,n}$ were calculated by the following equation:

$$\Sigma_{x,n} = \sqrt{\hat{\sigma}_{x,n}^2 + \hat{\varepsilon}_{x,n}^2}, \tag{17.9}$$

where $\hat{\sigma}_{x,n}$ and $\hat{\varepsilon}_{x,n}$ are estimations of RMS deviations and bias of the coordinate x measurement in the nth moment of time.

From the diagrams given in Figure 17.3, it is seen that total errors of the target coordinate estimation in the 2D FSR essentially exceed the values of RMS deviations, corresponding to potential accuracy. This is explained by the fact that systematic errors make a significant (and with the increase of altitude, prevailing) contribution to the values of total errors of the coordinate estimation in the 2D FSR. At the same time, the values of total errors in the 3D FSR appear to be close to the values of RMS deviation, corresponding to potential accuracy, as the systematic component of the coordinate definition error in the 3D FSR is practically absent. The diagrams of dependence of systematic errors of the coordinate x estimation on coordinate y in the 3D FSR, obtained by mathematical modelling, are given in Figure 17.1 (curves 4). It is seen that these systematic errors are very small.

The dependences of normalized RMS deviations of altitude estimation errors in the 3D FSR on coordinate y, obtained by the method of mathematical simulation, are pictured in Figure 17.4 by dash–dot and dotted lines. Curves 1 and 2 correspond to the altitude of flight $h = 1000$ m and $h = 2000$ m respectively. System parameters and trajectories are similar to those that have been described in the calculation of potential accuracy. It can be seen from Figure 17.4, that accuracies of altitude definition, obtained by mathematical simulation, appear close to their potential.

Note that, similar to the case of the 2D FSR, the correlation matrix of primary measurement errors should be evaluated to reduce the coordinate estimation errors. Modelling results of the trajectory processing algorithm, with the preliminary evaluation of RMS deviations of primary measuring errors, have shown that the accuracy of the target coordinate estimation worsens insignificantly in comparison with the case when the correlation matrix of primary measurement errors is known beforehand.

In conclusion, the main advantages of the 3D FSR in comparison to the 2D one can be listed.

1. Estimations of the target horizontal coordinates obtained via the 3D FSR are free of systematic errors due to information about target elevation measurements being used. Thus, the achievable accuracy of target tracking in the 3D FSR is practically close to its potential accuracy and corresponds to maximum likelihood estimations.
2. The 3D FSR allows measurement of the target altitude with high accuracy. The presented mathematical modelling results prove the corresponding theoretical statements.

17.3 EXTRAPOLATION TRACKING ALGORITHM FOR A 3D FSR

The extrapolation tracking algorithm with a preliminary range sum estimation was considered earlier. It could be used in the 3D FSR, but it has its peculiarities. The range sum transmitter–target–receiver $L(t)$ in the 3D FSR can also be determined on the basis of Equations (16.28), where the moment t_{CR} should be considered as the time of target flight above the baseline instead of the baseline crossing time in the two-dimensional radar. However, considering the three-dimensional radar, both the flight above the baseline and the baseline crossing are used as convertible terms.

Estimates of current target coordinates can be obtained using the measurements of the range sum, $\hat{L}$, azimuth, $\hat{\alpha}$, and elevation angle, $\hat{\beta}$, from the following expressions:

$$\begin{aligned}\hat{x} &= \frac{\left(\hat{L}^2 - b^2\right)\cos\hat{\beta}\cos\hat{\alpha}}{2\left(\hat{L} - b\cos\hat{\alpha}\cos\hat{\beta}\right)},\\ \hat{y} &= \frac{\left(\hat{L}^2 - b^2\right)\cos\hat{\beta}\sin\hat{\alpha}}{2\left(\hat{L} - b\cos\hat{\alpha}\cos\hat{\beta}\right)},\\ \hat{z} &= \frac{\left(\hat{L}^2 - b^2\right)\sin\hat{\beta}}{2\left(\hat{L} - b\cos\hat{\alpha}\cos\hat{\beta}\right)}.\end{aligned} \tag{17.10}$$

If the target moves in the horizontal plane (x, y), the value of the constant of integration L_0 in expressions (16.26) to (16.28) is exactly known at the moment of crossing the baseline and is equal to the base length, b. If the altitude of the target is distinct from zero, *a priori* ambiguity takes place in respect of the constant of integration, L_0. If the base length is used as an approximate value of the constant of integration, L_0, systematic errors of the target coordinate definition occur, increasing with the growth in altitude. In the case of a horizontal trajectory, these systematic errors can be substantially compensated. Some ways of compensating for the systematic errors in determining the range sum caused by *a priori* ambiguity in respect of the constant of integration L_0 may be considered.

If we understand the term 'range sum', L, as meaning the sum of horizontal ranges, then the condition $L(t_{CR}) = b$ remains correct. In this case errors of the horizontal range estimation are caused only by the fact that the Doppler shift depends upon the target altitude. This results in the occurrence of systematic errors in the range calculation.

The calculations show that, by assuming a linear model of the target motion, the following approximate relation takes place:

$$f_D(t, h) \approx f_D(t, 0) + f_D(t_{CR}, h), \tag{17.11}$$

where h is the constant altitude of the target flight. For a reduction of high-altitude errors in the calculations by Equation (16.28), it is possible to use the corrected values, $f_D(t, h) - f_D(t_{CR}, h)$, instead of the measured values, $f_D(t, h)$. The value $f_D(t_{CR}, h)$, can be measured at the time t_{CR} of flight above the baseline or can be obtained by means of extrapolation of the Doppler shift, $f_D(t)$, with measurements on an initial segment of the trajectory up to the time t_{CR}.

Note that in the case of a nonzero target altitude, it is necessary to extrapolate both the Doppler shift and the azimuth, as the estimation of the time, t_{CR}, of flight above the baseline can be obtained only when the azimuth is equal to zero. Thus, an estimation of the time of flight above the baseline can be found using the linear extrapolation of the target azimuth law of change:

$$\hat{t}_{CR,n} = -\frac{b_{\alpha,n}}{a_{\alpha,n}}, \tag{17.12}$$

where $a_{\alpha,n}$, $b_{\alpha,n}$ are the coefficients of the linear polynomial, $\alpha^e(t) = a_{\alpha,n}t + b_{\alpha,n}$, approximating a dependence of the target azimuth on time.

Estimation of the Doppler shift value in the expected moment when the target flies above the baseline is then found by

$$\hat{f}_{\mathrm{D}}(t_{\mathrm{CR}}, h)\,\hat{f}_{\mathrm{D}} = a_{\mathrm{f},n}\hat{t}_{\mathrm{CR},n} + b_{\mathrm{f},n}, \tag{17.13}$$

where $a_{\mathrm{f},n}$, $b_{\mathrm{f},n}$ are the coefficients of the linear polynomial $f_{\mathrm{D}}^{\mathrm{e}}(t) = a_{\mathrm{f},n}t + b_{\mathrm{f},n}$ approximating the dependence of the Doppler shift on time. Using the estimation of the Doppler shift at the baseline crossing time obtained by Equation (17.13), the dependence of the Doppler shift is corrected by means of this value subtraction. Finally, calculations are executed corresponding to Equations (16.28) to obtain the range sum estimation.

The analysis shows that the efficiency of the suggested method of compensation of systematic errors is only high enough at the trajectory section after the baseline is crossed. This is when the estimation of the time of flight above the baseline by means of extrapolation of the azimuth variation law is not required. To provide the efficiency of this method in the trajectory section before the baseline crossing point, a very high accuracy of primary measurements is required. This is explained by the strong increase in random errors of the range sum estimation. The increase in random errors of the range sum estimation is caused by the random error of the Doppler shift extrapolation up to the expected time of flight above the baseline. This random error of the Doppler shift extrapolation is integrated as a constant component over the whole observation interval. As a result, before crossing the baseline, a correction in the Doppler shift measurements does not provide significant reduction of total errors when determining the range sum, in comparison with the case when the target altitude is not taken into account.

For compensation of systematic errors when determining the range sum, before the baseline is crossed, a more effective method of the sequential iterative estimation of the constant of integration, L_0, exists. In the first stage, the constant of integration, L_0, is expected to be equal to the base length, b. Estimates of the range sum and target coordinates by Equations (16.28) and (17.10) are found.

The value of the elevation angle is extrapolated to the moment of flight above the baseline:

$$\hat{\beta}(\hat{t}_{\mathrm{CR},n}) = a_{\beta,n}\hat{t}_{\mathrm{CR},n} + b_{\beta,n}, \tag{17.14}$$

where $a_{\beta,n}$ and $b_{\beta,n}$ are the coefficients of a linear polynomial approximating the dependence of the elevation angle on time and $\hat{t}_{\mathrm{CR},n}$ is the estimation of the moment of flight above the baseline, found from Equation (17.12). Lastly, estimation of the constant of integration,L_0, for the next stage is found:

$$\hat{L}_{0,n+1} = \sqrt{\hat{z}_n^2\left[1 + \frac{1}{\sin^2\hat{\beta}(\hat{t}_{\mathrm{CR},n})}\right]} + \sqrt{\hat{z}_n^2 + \left\{b - \sqrt{\hat{z}_n^2\left[1 + \frac{1}{\sin^2\hat{\beta}(\hat{t}_{\mathrm{CR},n})}\right]}\right\}^2}. \tag{17.15}$$

Estimation of the constant of integration, (17.15), is used for the definition of the range sum and target coordinates in the next stage of tracking. The procedure is then repeated. As simulation results show, the systematic error of definition of the range sum, caused by prior ambiguity with respect to the constant of integration, decreases to acceptably small values. However, the described approach is effective only for the case of trajectories close to horizontal, when the target altitude varies weakly during motion and it is possible to assume $z(t) \approx h = \text{constant}$.

A detailed analysis of statistical characteristics of errors of the target coordinate definition via the extrapolation algorithm in the 2D FSR was given in Section 16.3. Most conclusions are also correct with respect to the 3D FSR.

In conclusion, taking into account the features of the 3D FSR mentioned above, associated with systematic errors of the target coordinate definition at nonzero altitudes, an appropriate application of the extrapolation tracking algorithm needs to be defined with a preliminary estimation of the range sum in the 3D FSR. Thus, it seems to be appropriate in the 3D FSR to obtain an initial approximation of the coordinate estimates. The obtained initial approximation is to be used in the more accurate iterative algorithm on the basis of the maximum likelihood method. In other words, in the problem of the initial approximation calculation, the algorithm considered here presents an alternative to the method of linearization of the nonlinear equation system, concatenating the values of primary parameters and trajectory parameters.

18

3D FSR with an Array Antenna

18.1 INTRODUCTION

We have previously considered a bistatic forward scattering radar realization wherein the measurement of the Doppler shift was carried out after amplitude detection of the received interferential signal. The measurement of angular coordinates was performed by the amplitude monopulse method. Magnitudes of Doppler signals in the beams of the antenna pattern are compared to measure the target azimuth.

Such a system is relatively easy to create, but has a number of limitations. In the considered radar, extraction of the Doppler signal is performed by amplitude detection. Thus, the relationship between the gains of the beams depends on the level of the direct transmitter signal. This is explained by the direct transmitter signal level determining the gains of the amplitude detectors in the receive channels of the corresponding beams. The direct transmitter signal in the beam directed towards the transmitter is too strong, which may result in saturation. At the same time, in the other beams the level of the direct transmitter signal is determined by the level of sidelobes of the antenna directional pattern and therefore has a random nature. This leads to signal angle measurement errors and reduces the accuracy of the target position measurement. Moreover, it is difficult to ensure acceptable functioning of the beam directed towards the transmitter because of its saturation.

Modern radars are usually designed for operation with array antennas [III.29, III.30]. The application of phased arrays in monostatic radar systems possesses a number of indisputable advantages. As far as bistatic forward-scattering radars are concerned, utilization of a phased array antenna is essentially complicated due to an absence of a coherent reference signal on the receive side. To ensure coherent processing, it is necessary to send a transmitted (reference) signal via a dedicated communication channel, or to extract a reference waveform from the direct signal. Each solution is rather complex in terms of practical realization. At the same time, the FSR has a unique peculiarity which makes it possible to use space–time processing on the basis of array antenna with beamforming after amplitude detection in the receive channels [III.31–III.33].

Direct signal levels are the same in all channels of the array antenna and consequently all the receiving channels have an equal gain. At the input of the amplitude detectors in all channels, a strong direct signal and a relatively weak target return are present. In these conditions the amplitude detector acts as a parametric synchronous detector, which is a linear device relevant to the weak waveform, i.e.

Bistatic Radars: Principles and Practice Edited by M. Cherniakov

target return. Thus at the amplitude detector output, a quasi-harmonic Doppler signal could be selected by an appropriate filter. Using this approach, there is also no need for a reference signal on the receive side. Due to this, the radar's technical implementation is significantly simplified in comparison with the realization of the radars built on the basis of coherent array antenna with beam-forming at RF (or equivalent processing at video frequency (VF) without phase information loss).

18.2 SPACE–TIME PROCESSING ALGORITHM

Let us consider a three-dimensional forward scattering radar space–time processing algorithm on the basis of two orthogonal linear equispaced array antennas with beam-forming after amplitude detection in receive channels of the array antenna elements. To simplify the analysis, it is supposed that the elements of both receiving linear array antennas are omnidirectional. The general structure of the considered FSR is shown in Figure 18.1.

The FSR structure presented in Figure 18.1, in a broad sense, corresponds to general FSR topology (Figure 15.1). The major particularity is the receiving antenna structure. This consists of two linear array antennas with omnidirectional elements marked by 1 (azimuth array antenna) and 2 (elevation array antenna). It can be seen that the azimuth array antenna is placed in the horizontal plane (x, y), while the elevation array antenna is placed in the vertical plane (x, z), but inclined towards the plane (y, z). The reasons for this will be discussed later in this chapter.

Firstly, let us consider the azimuth array antenna. The functional block diagram of the azimuth array antenna signal processing algorithm is presented in Figure 18.2, where AD are amplitude detectors, NF are notch filters, LPF are lowpass filters and ADC are analogue-to-digital converters. The block diagram of the elevation array antenna signal processing is similar.

All devices included in the receiving channel of any single element of the array antenna, before the unit of digital space–time processing (STP), are specified as the array antenna receiving channel. The unit of digital STP then provides beam-forming, filtering, target detection and measurement of the primary parameters of the target return. Thus, the output data of the digital STP unit are estimates of the target return Doppler frequency shift, $\hat{f}_D$, and target azimuth, $\hat{\alpha}$.

Input signals undergo filtering, amplification and amplitude detection in each array antenna receiving channel. As mentioned above, in the presence of a powerful direct transmitter signal, the Doppler signal

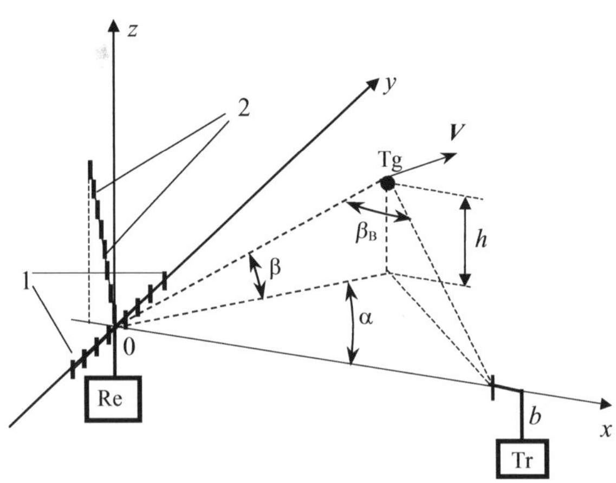

Figure 18.1 Structure of the 3D FSR with an array antenna

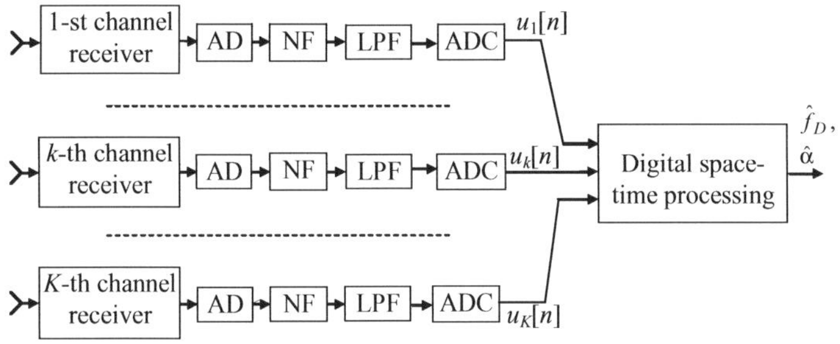

Figure 18.2 Functional block diagram of CW FSR signal processing

occurs at the output of an AD without any nonlinear distortions. This signal represents one quadrature component of the target return complex envelope.

It is assumed that the normal to the azimuth array antenna aperture is pointed directly at the transmitting antenna phase centre. Figure 18.3 illustrates the process of the target return reception by the azimuth array when the target azimuth is equal to α using, as an example, a three-element array. In Figure 18.3, 1, 2 and 3 are the azimuth array antenna receiving channels, 4 is the beam-forming unit, $\mathbf{S}_{\text{tg,hor}}$ is the projection of the target scattered electromagnetic wave Poynting vector upon the horizontal plane (x, y) and $\mathbf{S}_{\text{dir}}$ is the Poynting vector of the direct transmitter signal wave.

Since the direct transmitter signal wave Poynting vector is normal to the antenna aperture, there is no phase shift between direct signal samples in different channels. Then, the phase shift of the Doppler signal samples between the different channels is defined only by the phase shift of target return samples.

Taking this into account, the following equation gives the process at the AD output in the kth receiving channel of the azimuth array antenna (or the kth azimuth channel in short notation) [III.31–III.33]:

$$u_{\text{hor},k}(t) = U_0 + U_{\text{S}} \cos\left[\varphi_{\text{D}}(t) + 2\pi \frac{kd}{\lambda} \sin\alpha(t) + \varphi_x\right] + \xi_k(t), \tag{18.1}$$

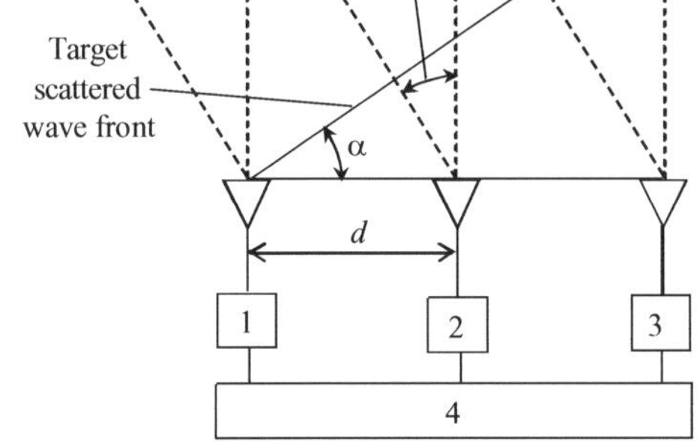

Figure 18.3 Azimuth array antenna angular relationships

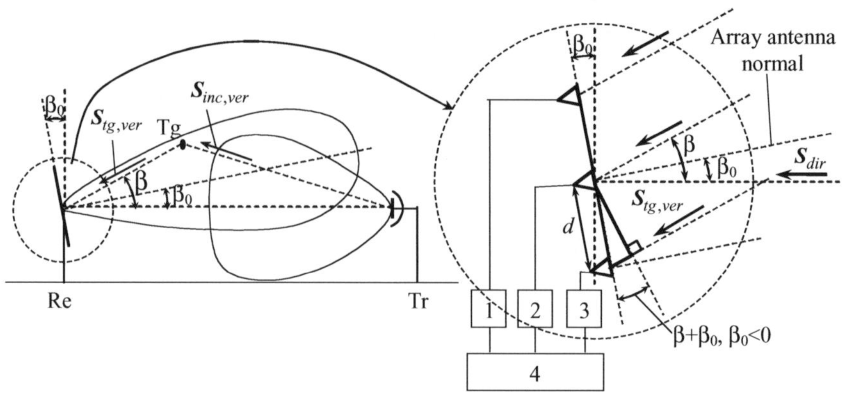

Figure 18.4 Elevation array antenna angular relationships

where U_0 is the constant component caused by the direct transmitter signal presence, $\varphi_D(t)$ is the phase shift coursed by the target return Doppler shift, U_S and φ_x are the amplitude and random phase of the quasi-harmonic Doppler signal in the kth azimuth channel correspondingly, $2\pi(kd/\lambda)\sin[\alpha(t)]$ is the phase shift between the first and the kth azimuth channel, determined by the target azimuth, d is the constant array pitch, λ is the transmitter wavelength, α is the target azimuth and $\xi_k(t)$ is the kth channel noise.

The phase shift, $\varphi_D(t)$, is determined by the target return Doppler frequency shift, $f_D(t)$, as follows:

$$\varphi_D(t) = 2\pi \int_0^t f_D(t)\, dt. \tag{18.2}$$

The phase shift between the elevation angle antenna array elements is determined by the target elevation angle, $\beta(t)$. However, to calculate this phase shift account must be taken of the fact that the elevation array antenna has an inclination, β_0, towards the vertical to reduce the ground surface clutter. Figure 18.4 illustrates this feature. It shows a three-element elevation array antenna. A target return reception where a direct transmitter signal is present is considered. Since the baseline concatenating the phase centres of the transmitting and receiving antennas is lower than normal to the elevation array antenna, it is supposed that angle $\beta_0 < 0$. This corresponds to the elevation of the antenna array normal towards the azimuth plane by the angle $|\beta_0|$.

In Figure 18.4 $\mathbf{S}_{\text{inc,ver}}$ and $\mathbf{S}_{\text{tg,ver}}$ are the vertical plane projections of the Poynting vectors of incidence on to the target and scattered by the target electromagnetic waves correspondingly. Digits 1, 2, 3 and 4 denote the elevation array antenna receiving channels (elevation channels, briefly) and the elevation beam-forming unit correspondingly. From Figure 18.4, the phase relationships between the direct transmitter signal and the target return can be obtained. The angle between the normal to the array antenna aperture and the target return direction of arrival in the vertical plane is equal to $(\beta + \beta_0)$, $\beta_0 < 0$. Thus, the phase shift, $\Delta\varphi_{\text{tg}}$, between the target return samples in the adjacent array elements is defined by $\Delta\varphi_{\text{tg}} = 2\pi(d/\lambda)\sin(\beta + \beta_0)$. Similarly, the angle between the normal to the array antenna aperture and the direct transmitter signal direction of arrival is equal to β_0. Thus, the phase shift, $\Delta\varphi_{\text{dir}}$, between the samples of the direct transmitter signal in the adjacent elevation channels is $\Delta\varphi_{\text{dir}} = 2\pi(d/\lambda)\sin\beta_0$. At the outputs of the elevation channels, the phase shift of the Doppler signal samples is defined by the difference between the phase shift of the target return and the phase shift of the direct signal. Hence, the signal at the AD output in the kth elevation channel is given by

$$u_{\text{ver},k}(t) = U_0 + U_S \cos\left(\varphi_D(t) + 2\pi\frac{kd}{\lambda}\{\sin[\beta(t) + \beta_0] - \sin\beta_0\} + \varphi_x\right) + \xi_k(t). \tag{18.3}$$

It can be seen from Equations (18.1) and (18.3) that at the outputs of the AD in receiving channels of azimuth and elevation arrays, the constant component, U_0, is present. This component is caused by the direct transmitter signal. Since the direct transmitter signal is much stronger (by 30–60 dB) than the target return, this constant component is similarly much stronger than the Doppler signal. To suppress the constant component, U_0, notch filters are used (Figure 18.2). The analogue NF included before ADC allows a reduction in dynamic range requirements, in comparison with the case of a digital NF application.

Figure 18.2 also shows lowpass filters. These are used to specify the ADC minimal sampling rate, which should have an order of maximum Doppler frequency. Thus, after analogue signal processing and analogue-to-digital conversion, discrete samples of processes at the outputs of the azimuth and elevation channels are given by

$$\begin{aligned} u_{\mathrm{hor},k}[n] &= U_{\mathrm{S}} \cos\left(\varphi_{\mathrm{D}}[n] + 2\pi \frac{kd}{\lambda} \sin\alpha[n] + \varphi_x\right) + \xi_k[n], \\ u_{\mathrm{ver},k}[n] &= U_{\mathrm{S}} \cos\left\{\varphi_{\mathrm{D}}[n] + 2\pi \frac{kd}{\lambda}[\sin(\beta[n] + \beta_0) - \sin\beta_0] + \varphi_x\right\} + \xi_k[n], \end{aligned} \tag{18.4}$$

where $u_{\mathrm{hor(ver)},k}[n] = u_{\mathrm{hor(ver)},k}(t = n\Delta t)$ and Δt is the sampling interval.

We will now consider features of the processing of signals at the outputs of the azimuth and elevation channels. It is seen from Equation (18.4) that at a given time instant the aperture amplitude distributions of the array antennas represent space harmonics [III.31–III.34]. Since the considered antenna is an array, we have discrete samples of discussed space harmonics. Similar to the common term of 'sampling rate' in the time domain, we can consider the term of 'space sampling rate'. The space sampling rate is specified as the value inversed to the separation between elements in the array, d. In addition, the mentioned separation between the elements of the array, d, is termed the 'array pitch'. The frequencies of space harmonics are determined by the azimuth, α, and the elevation angle, β, of the target. To provide a baseband beam-forming, a discrete Fourier transform (DFT) should be applied to an array of the signal samples at the outputs of the receive channels.

Each sample of the discussed DFT can be considered as an output sample of the corresponding space filter. A space filter is defined as a certain functional transform. Its input is a vector of signal samples from the outputs of the array antenna receiving channels. The output of the space filter is scalar, presenting the value of the space spectrum at the corresponding space frequency. Thus, the considered DFT is equal to the array of outputs of the corresponding space filters.

We will describe the selectivity of the considered space filters by their directional characteristics. In the present case, we are dealing with beam-forming after amplitude detection, where information about the phase of the RF signal is lost. Therefore, the term 'directional characteristic' is used instead of the traditional directional pattern or antenna pattern, to underline the noted particularity of this space processing. Thus, we will define the directional characteristic, $U_{\mathrm{SP}}(\alpha)$, of a space filter as a dependence of its response amplitude upon the target return angle of arrival. Such a characteristic can be obtained with the assumption that an electromagnetic wave scattered by the target with the azimuth position, α, comes to the input of the array antenna.

Figures 18.5(a) and 18.5(b) present the directional characteristics, $U_{\mathrm{SP}}(\alpha)$, for the zero ($i = 0$) and the second ($i = 2$) space filters of the azimuth antenna array correspondingly, where i is the number of space filter. They are plotted under the condition that the target moves along a previously known trajectory. The directional characteristics of elevation array space filters are analogous functions of the target elevation angle, β.

It is obvious from Figure 18.5 that the directional characteristics obtained by the method described above are fast-oscillating and do not allow the target return AOA to be measured. The oscillations are caused by signal phase incursion at the output of the AD resulting from the target return Doppler shift.

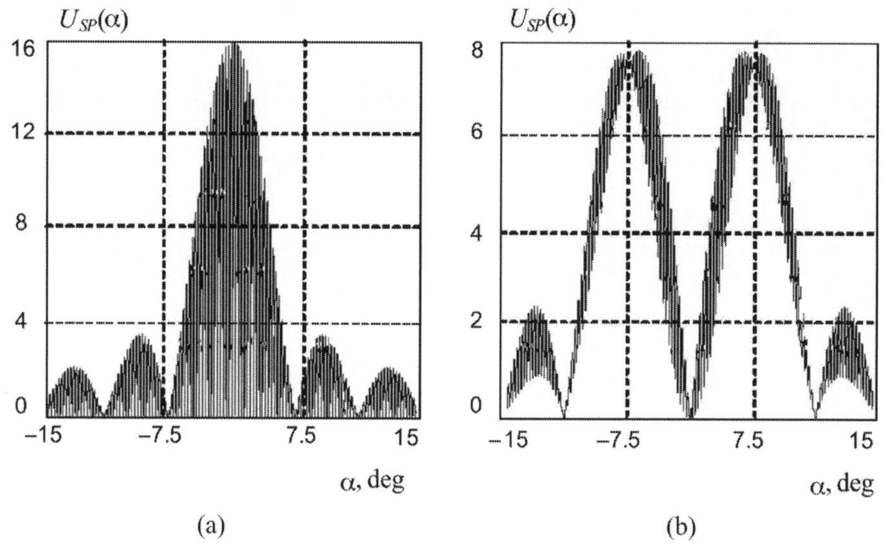

Figure 18.5 Directional characteristics of space filters: (a) $i = 0$; (b) $i = 2$

Thus, the time variation of the phase shift, $\varphi_D[n]$, leads to a change of phase distribution in the array elements. In other words, the initial phases of space harmonics, in fact, change with a rate equal to the Doppler frequency shift of the target return. Since the considered space harmonics are obtained after the AD, they are real functions of the space coordinate, and the magnitude of each space filter response is significantly dependent on the harmonic initial phase. This dependence is especially strong when the space harmonic frequency is small. This corresponds to small values of the target return AOA. In this case only a few periods (or even less than one period) of space harmonics are sampled by the array antenna elements. Exactly this fact leads to the mentioned phase dependence for the case of real harmonics.

To eliminate these oscillations of the directional characteristics of space filters, it is necessary to move from space processing, independent from time processing, to space–time processing [III.34, III.35], i.e. to take a 2D DFT of the discrete signal array. This will be a two-dimensional array of signal samples from the AD outputs in the array antenna channels. These samples are obtained during the primary measurement interval. Figure 18.6 shows the directional characteristics calculated for the same filters (for $i = 0$ and $i = 2$), but using the space–time processing algorithm.

In Figure 18.6, the dotted lines correspond to the envelopes of the values of the 2D DFT in the azimuth channels for each primary measurement interval. These envelopes are smooth curves with shapes determined by the frequency characteristics of the corresponding DFT filters, since transition to the 2D DFT provides averaging of the space spectrum over the primary measurement interval. The primary measurement interval was taken to be equal to $T = 1$ s, and the number of signal samples in the receiving channel of each antenna element over the primary measurements interval was adopted to be $N = 2048$.

Thus, in the considered system it is necessary to use 2D STP of signals from the outputs of the azimuth and elevation channels in order to measure the target return AOA and its Doppler frequency shift. The delimiting of space and time processing in this system is not possible because of the peculiarities of beam-forming after amplitude detection.

From the above, it can be concluded that quasi-optimal STP of a 2D array of samples in the receive channels of the azimuth or elevation arrays $u[k,n]$, $k \in [0, K-1]$, $n \in [0, N-1]$ can be realized with

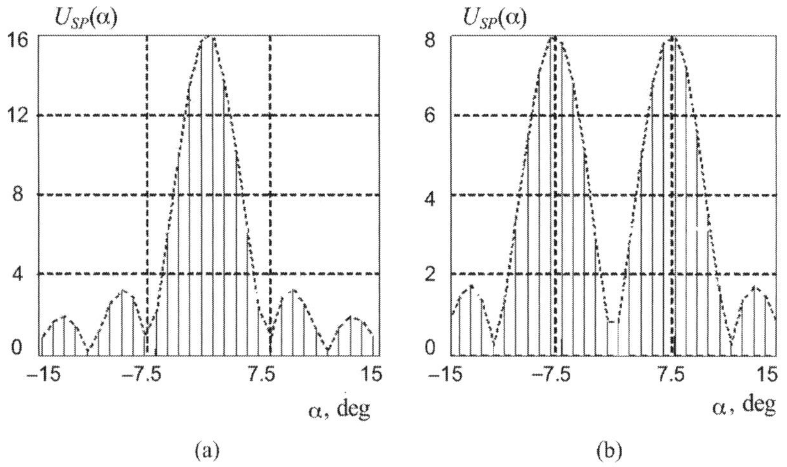

Figure 18.6 Directional characteristics of space filters with the use of space–time processing

the use of a 2D DFT defined by

$$S(i, m) = \sum_{n=0}^{N-1} \sum_{k=0}^{K-1} u[k, n] \, e^{-j(2\pi/K)ki} e^{-j(2\pi/N)nm}, \tag{18.5}$$

where $i \in [0, K-1]$, $m \in [0, N-1]$, K is the number of array elements, and N is the number of samples per primary measurement interval.

According to Equation (18.5), the modulus of the function $S(i,m)$ is at a maximum when the equalities $(d/\lambda)\sin\alpha = \pm i/K$, $(d/\lambda)[\sin(\beta + \beta_0) - \sin\beta_0] = \pm i/K$ and $f_D = \pm m/N$ are satisfied. The position of the maximum can be used to estimate the primary parameters, i.e. the target return Doppler shift and the angle of arrival. However, this leads to ambiguity in respect to the signs of Doppler frequency shift and the target return AOA, because the magnitude of the 2D DFT of a real signal possesses central symmetry. This means that it has equal values in diagonally opposite quadrants. The graphs of function $|S(i, m)|$ are plotted in Figure 18.7(a) for the cases when the Doppler shift and the AOA have the same and different signs (curves 1 and 2 respectively). In addition, the location of the maximums of function $|S(i, m)|$ associated with the plane (f_D, f_{SP}) is shown in Figure 18.7(b), where $f_{SP} = \sin\alpha/\lambda$ is the space frequency.

However, from the positions of the maximums of the 2D DFT the relationship between the signs of the target return angle of arrival and the Doppler shift can be inferred. To explain this, let us consider quadrants marked by I and IV (Figure 18.7(b)). One of the two maximums of $|S(i, m)|$ is always located in one of these quadrants. If it is located in quadrant I, the Doppler frequency shift and the AOA have the same signs, but if it is located in quadrant IV, the signs of the Doppler frequency shift and the AOA are opposite.

Considering the central symmetry of the two-dimensional spectrum allows the use of only one-half of the samples of the 2D DFT. For example, using only the samples corresponding to the positive Doppler frequency domain ($f_D > 0$), the absolute values of the primary parameters can be estimated, as well as their mutual signs.

It follows from the above that, sign ambiguity of the azimuth angle α can be cancelled without any difficulties during tracking [III.31–III.33]. A positive Doppler shift corresponds to the target motion in the direction to the baseline. In this case, the Doppler frequency shift reduces with time. A negative

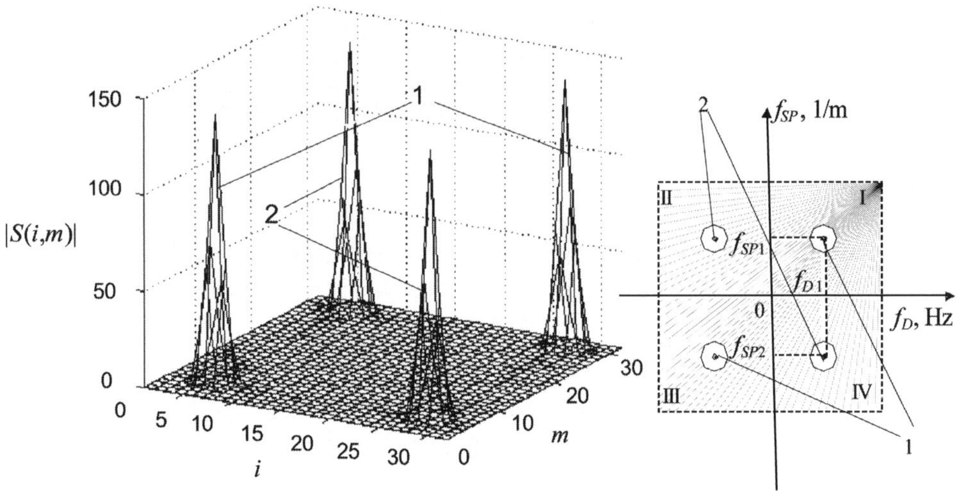

Figure 18.7 2D DFT of a real signal: $N = 32$, $K = 32$ [III.33]

Doppler shift corresponds to the target motion out of the baseline. In this case, the absolute value of the Doppler frequency shift increases. Therefore, the measurement of the first derivative of the absolute value of the Doppler shift, performed during tracking, allows the Doppler shift sign to be obtained. It was shown that, with a known sign of Doppler shift, the sign of the azimuth, α, is determined taking into account the maximum position of the 2D DFT.

It is important to mention here that the errors in determining the sign of the Doppler shift from the sign of the first derivative of its modulus as a function of time can lead to anomalous errors in angular measurement. However, simple estimates show that the probability of such errors is small enough. To illustrate this, a case is considered where the measurements of the Doppler frequency shift are independent for different measurement intervals. Furthermore, the Gaussian distribution of measurement error is assumed. Thus, in the case of a linear variation of the Doppler frequency shift, the probability of false decision making in respect to the Doppler frequency shift sign is defined by the expression

$$F_{\mathrm{DFS}} = \begin{cases} P\{\hat{a}_n > 0\} = 1 - \Phi\left(\dfrac{-M\{\hat{a}_n\}}{\sigma_{an}}\right), & f > 0, \\ P\{\hat{a}_n < 0\} = \Phi\left(\dfrac{-M\{\hat{a}_n\}}{\sigma_{an}}\right), & f < 0. \end{cases} \tag{18.6}$$

Here $\Phi(x) = (1/\sqrt{2\pi}) \int_{-\infty}^{x} e^{-t^2/2}\, dt$ is the cumulative distribution function for zero mean and unity variance, $M\{\hat{a}_n\}$ and σ_{an} are the mean value and RMS deviation of the estimate respectively and $\hat{a}_n$ is the Doppler shift modulus rate of change obtained from n measurements.

For real values of system and target motion parameters, with only two measurements of frequency available, the probability of incorrect sign determination is negligibly small. For example, for a target moving in the azimuth plane with a constant velocity $V = 200$ m/s along a rectilinear trajectory crossing the baseline at an angle $\varphi = 45°$ and at a distance of 15 km from the receiving position, the base being equal to 40 km, the probability of discussed errors does not exceed 10^{-2}.

Measurement ambiguity is present in both the azimuth and elevation channels. However, since the elevation of the the target is always positive, it suffices to eliminate ambiguity with respect to the target azimuth.

18.3 PRIMARY MEASUREMENT CHARACTERISTICS

When a 2D DFT is used, evaluation of the primary parameters is derived from the number of the beam, i_0, and the number of the Doppler filter, m_0, corresponding to a maximal sample of function $|S(i,m)|$ (Figure 18.7). The following algorithm can be used to obtain more accurate measurements of the target return Doppler frequency shift and its angle of arrival.

The main principle is to use a parabolic approximation of the two-dimensional spectrum in the neighbourhood of the maximal sample. Figure 18.8 illustrates the essence of such an approximation.

At first, the maximal sample of the function $|S(i,m)|$ is found. It is denoted by $|S(i_0,m_0)|$. Using this sample and two adjacent samples, $|S(i_0,m_0-1)|$ and $|S(i_0,m_0+1)|$, it is then possible to calculate coefficients a_1 and a_2 of the approximating quadratic polynomial. This polynomial describes a parabola with the position of the vertex $x_{max} = -a_1/(2a_2)$. Using this position of the parabola vertex, an estimation is obtained of the Doppler frequency shift, $\hat{f}_D = (x_{max}/N)f_S$, where f_S is the sampling rate. Similarly, using samples $|S(i_0-1,m_0)|$ and $|S(i_0+1,m_0)|$, an estimation is obtained of the space frequency, $\hat{f}_{SP}$. To obtain an AOA estimation, however, it is necessary to make the transform $\hat{\alpha} = \arcsin(\hat{f}_{SP}\lambda)$.

The instrumental accuracy of the Doppler frequency shift and AOA measurements can be additionally improved by way of time and aperture weighting of samples before spectral analysis. Furthermore, to improve the accuracy of the measurement of the target return AOA, it is worthwhile to increase the number of spectral components analysed in terms of spatial frequency [III.36]. For this purpose, zero padding should be used. This means that the array of samples from the outputs of amplitude detectors in the receive channels is complemented with zero samples. This leads to an increase in the number of space filters (formed beams) and improves accuracy of AOA measurement by the 2D DFT.

Statistical characteristics of errors in primary measurements, performed using the proposed STP algorithm, were examined by the method of mathematical modelling. The target was assumed to move along known trajectories at constant altitude, h (Figure 18.1). The target trajectory in this case is parallel to the horizontal plane (x, y). Thus, it is worthwhile to obtain dependences of primary measurement errors upon the y coordinate of the target. Discrete signals at the outputs of the azimuth and elevation channels were modelled according to Equations (18.4). The simulation of the measuring algorithm was carried out with the assigned signal-to-noise ratio, q, at the outputs of the amplitude detectors. The receiver noise was modelled as white Gaussian noise. The signal initial phase was assumed to be uniformly distributed over the interval $[-\pi, \pi]$.

Figure 18.9 presents plots of Doppler shift measurement errors versus the y coordinate for trajectories with different parameters. Curve 1 shows an estimate of the bias, Δf_B, of the Doppler frequency shift measurement. Curve 2 corresponds to an estimate of RMS deviations, σ_f, of the Doppler frequency

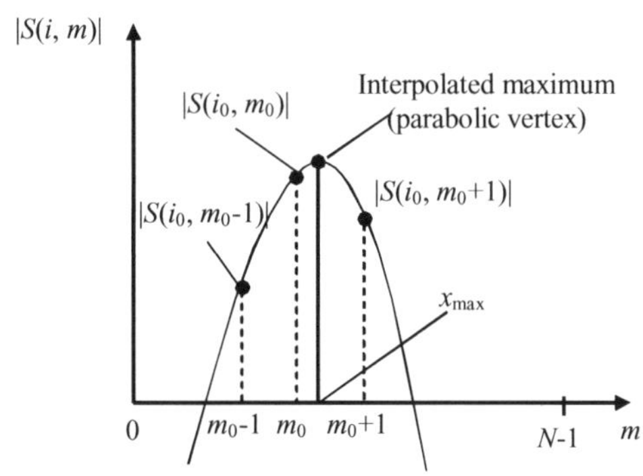

Figure 18.8 Parabolic approximation of the spectrum

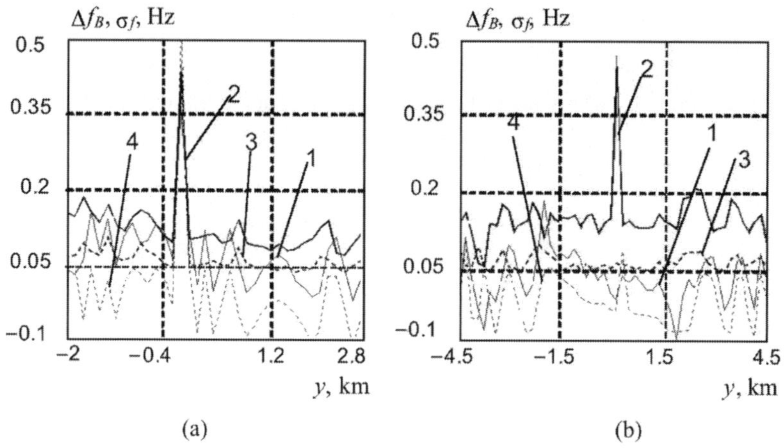

Figure 18.9 Bias and RMS deviations of Doppler frequency shift measurement errors [III.33]: (a) $x_{CR}/b = 0.25$, $\varphi = 45°$; (b) $x_{CR}/b = 0.5$, $\varphi = 90°$

shift measurement errors for the SNR at the outputs of the amplitude detectors equal to $q = -20$ dB. Curve 3 is the same for $q = -14$ dB. Figure 18.10 shows analogous plots for azimuth measurement errors and Figure 18.11 for the elevation angle measurement errors. Similarly to Figure 18.9, curves denoted by 1 present an estimation of bias, while those denoted by 2 and 3 are estimations of RMS deviations for $q = -20$ dB and $q = -14$ dB respectively. The averaging was performed over the results of 100 independent experiments. FSR coverage was limited by the minimum value of the bistatic angle, $\beta_{B\,min} = 150°$. The target was assumed to move at the altitude $h = 1000$ m, with velocity $V = 200$ m/s. The following values of the radar parameters were adopted: the primary measurement interval, $T = 1$ s, wavelength $\lambda = 1$ m, the number of samples per primary measurement interval for each space channel $N = 2048$ (which corresponds to the number of Doppler filters with a bandwidth of 1 Hz), the number of array elements $K = 16$ and the number of space filters (beams) formed $M_{SF} = 32$.

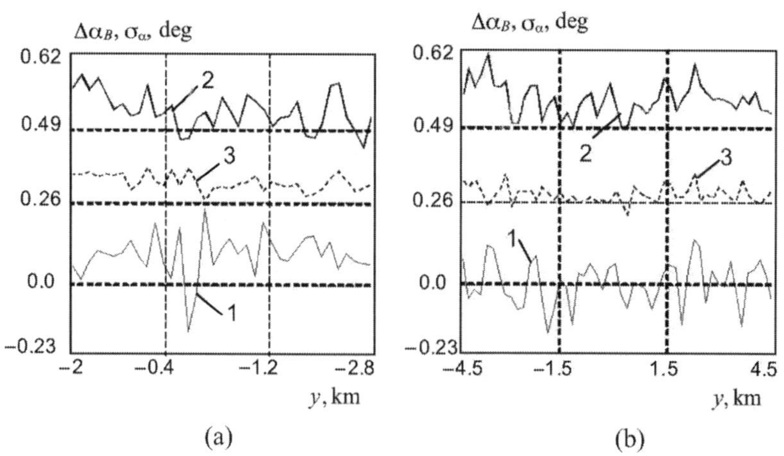

Figure 18.10 Bias and RMS deviations of azimuth measurement errors [III.33]: (a) $x_{CR}/b = 0.25$, $\varphi = 45°$; (b) $x_{CR}/b = 0.5$, $\varphi = 90°$

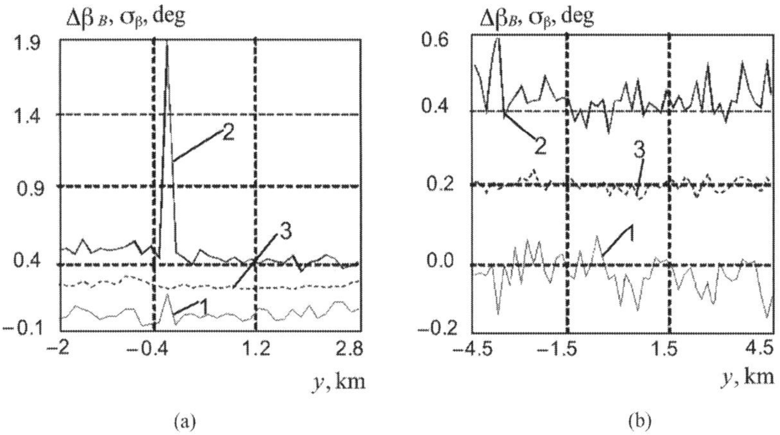

Figure 18.11 Bias and RMS deviations of the measurement errors of the elevation angle: (a) $x_{CR}/b = 0.25$, $\varphi = 45°$; (b) $x_{CR}/b = 0.5$, $\varphi = 90°$

Analysis of the above curves yields the following inferences. The bias of primary measurement errors for each point of the trajectory actually represents instrumental measurement errors. When using the described STP algorithm, the instrumental primary measurement errors in most cases do not exceed $\Delta f_B \approx 0.1$ Hz, $\Delta\alpha_B \approx \Delta\beta_B \approx 0.1°$.

As demonstrated by the simulation results, the magnitude of the Doppler shift measurement error practically does not depend on the random initial phase of the Doppler signal on the measurement interval, and is only determined by the law of Doppler shift change. Dependence of the error on the initial phase is observed solely with small values of Doppler frequency shift, when the measurement interval contains a small number of Doppler signal periods. In view of the above, it is clear that in the absence of additive noise, the variance of the random errors in the Doppler frequency shift measurement is practically equal to zero (except for the narrow region near zero frequency), even in the presence of the random initial phase of the signal, which is independent for each realization. The total Doppler frequency shift measurement error will, in that case, be determined by the instrumental errors corresponding to the bias of measurements for each point of a given trajectory. The magnitude of variance of the random errors in the Doppler frequency shift measurement is only dependent on the SNR at the output of the amplitude detectors. The random initial phase has a considerable effect on the variance of the errors in the Doppler shift measurement only with small values of Doppler shift in the vicinity of the baseline crossing point.

Similar conclusions remain true for the influence of the SNR and random initial phase on the accuracy of the target return AOA measurement. As shown above, the random initial phase considerably affects the instantaneous space spectrum of a real two-dimensional signal because the array antenna aperture size corresponds to just a few incomplete periods (or even less than one period) of a real space harmonic. However, with sufficiently large values of Doppler shift, the space spectrum is averaged over the primary measurement interval. Therefore, as in the case with errors in the Doppler shift measurement, the random initial phase affects the variance of errors of angle estimates only in the region of small values of the AOA and Doppler shift (near the baseline).

A certain increase in measurement errors is possible if the values of the Doppler shift and angles are close to zero (Figures 18.9–18.11). This is caused by the central-axis symmetry of a two-dimensional amplitude spectrum of a real two-dimensional signal. If the 2D DFT maximums are in zero-tuned filters in both Doppler and space dimensions the Doppler shift and angle can be measured with the accuracy determined by the space–time filter bandpass in dimensions of Doppler and space frequencies. In this case, the relations between the amplitude responses of space–time filters are not useful when obtaining a precise estimation of 2D DFT maximum coordinates.

As seen from the graphs in Figure 18.9, the averaged value of the errors in the Doppler shift measurement by the trajectory is not zero. This indicates that the Doppler shift estimates obtained are biased. The bias of estimation of the Doppler frequency shift arises from the target return AOA variation during the primary measurement interval. To explain this effect qualitatively, it will be assumed that the Doppler shift is constant on the primary measurement interval, while the target return AOA has linear variations. The expression for the instantaneous phase of the signal in the kth azimuth channel then takes the form

$$\varphi(t) = 2\pi f_{\mathrm{D}} t + \frac{2\pi k d}{\lambda} \sin\left[\alpha_0 + \Delta\alpha_{\mathrm{r}}(t - t_0)\right], \tag{18.7}$$

where α_0 is the initial angle value at the current primary measurement interval, $\Delta\alpha_{\mathrm{r}}$ is the angle rate of change and t_0 is the initial moment of the beginning of the current measurement interval.

Let us write the sine of the sum in Equation (18.7) in the form

$$\sin\left[\alpha_0 + \Delta\alpha_{\mathrm{r}}(t - t_0)\right] = \sin\alpha_0 \cos\Delta\alpha_{\mathrm{r}}(t - t_0) + \cos\alpha_0 \sin\Delta\alpha_{\mathrm{r}}(t - t_0).$$

Taking the fact into account that $\Delta\alpha_{\mathrm{r}}(t - t_0) \ll 1$, $\Delta\alpha_{\mathrm{r}}^2(t - t_0)^2 \ll \Delta\alpha_{\mathrm{r}}(t - t_0)$, we pass on to equivalent infinitesimals and neglect all terms with powers larger than the first:

$$\varphi(t) = 2\pi f_{\mathrm{D}} t + \frac{2\pi k d}{\lambda}\left[\sin\alpha_0 + \cos\alpha_0 \Delta\alpha_{\mathrm{r}}(t - t_0)\right]. \tag{18.8}$$

The phase rate of change will then be defined by a certain equivalent frequency:

$$f_{\mathrm{eq}} = \frac{1}{2\pi}\frac{\mathrm{d}\varphi(t)}{\mathrm{d}t} = f_{\mathrm{D}} + k\frac{d}{\lambda}\Delta\alpha_{\mathrm{r}}\cos\alpha_0, \tag{18.9}$$

where the second term defines the signal frequency offset in the kth azimuth channel relative to the Doppler shift:

$$\Delta f_{\mathrm{off}} = k\frac{d}{\lambda}\Delta\alpha_{\mathrm{r}}\cos\alpha_0. \tag{18.10}$$

This offset depends on the channel number, k, the initial value of the target return AOA on the current measurement interval, α_0, and the angle rate of change, $\Delta\alpha_{\mathrm{r}}$. When the primary measurement estimates are obtained by calculating a 2D DFT, the frequency offset averaged over the receive channels of the elements of the array antenna (18.10) results in a bias of the Doppler shift estimate.

To compensate to this Doppler shift estimation bias, it is suggested that the following value is to be subtracted from the Doppler shift estimation:

$$f_{\mathrm{cor}} = k_{\mathrm{m}}\frac{d}{\lambda}\Delta\alpha_{\mathrm{r}}\cos\alpha_0, \tag{18.11}$$

where k_{m} is the number of the middle azimuth channel. The thin dashed-and-dotted lines in Figure 18.9 (curves 4) present the bias of the Doppler shift measurement error plotted using the correction (18.11). Analysis shows that Doppler shift estimates obtained with the use of correction are practically unbiased.

The results of mathematical modelling show that the discussed STP algorithm provides a sufficiently high accuracy of Doppler shift and angle of arrival estimation. The considered SNR, $q = -20$ to -14 dB, at the outputs of the AD in the receiving channels, for the case of analysed target trajectories, approximately corresponds to the threshold SNR, at which there are no anomalous measurement errors. Therefore, it can be inferred that the system with the above parameters has RMS deviations of primary measurement errors not exceeding $\sigma_{\mathrm{f}} = 0.5$ Hz and $\sigma_\alpha = \sigma_\beta = 0.5°$. Such primary measurement accuracy, as shown in previous sections, allows target coordinates to be estimated with acceptable accuracy.

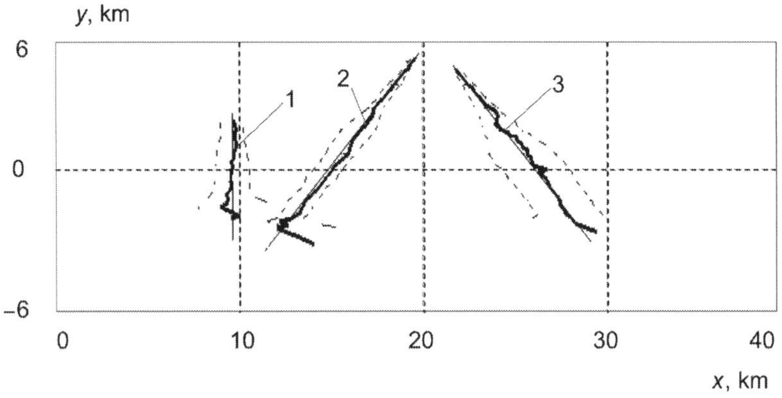

Figure 18.12 Results of combined modelling of STP and tracking in a 3D FSR [III.33]

Presented in Figure 18.12 are target trajectories plotted on the basis of primary measurements obtained according to the considered STP algorithm. Moving target coordinates were calculated with the use of the iterative algorithm on the basis of the maximum likelihood method considered in earlier sections. The thin lines in Figure 18.12 denote true trajectories of the targets, while the thick line denotes trajectories plotted during processing. The thin dashed-and-dotted lines show the areas of x coordinate errors, limited by error values $\pm 3\sigma_X$. Root mean square deviations of x coordinate errors were estimated over 100 independent simulations. As shown earlier, the radar parameters were supposed to equal the following: primary measurement interval $T = 1$ s, wavelength $\lambda = 1$ m and the SNR at the outputs of the AD in receiving channels of azimuth and elevation arrays $q = -20$ dB. Trajectories presented in Figure 18.12 have the following parameters: curve 1 corresponds to $x_{CR} = 10$ km, $\varphi = 90°$, $V = 200$ m/s; curve 2 corresponds to $x_{CR} = 15$ km, $\varphi = 45°$, $V = 200$ m/s; and curve 3 is for $x_{CR} = 27$ km, $\varphi = 135°$, $V = 200$ m/s. For all cases the target altitude was equal to $h = 2000$ m.

The obtained results of simultaneous modelling of the considered STP algorithm, and the iterative tracking algorithm discussed earlier, prove the efficiency of the 3D FSR with an array antenna of a considered type.

19

FSR Design and Experimental Investigation

19.1 INTRODUCTION

Since the early stages of its study, the unique properties of the forward-scatter effect have stimulated interest in its practical application. We will not consider the first experiments with FSR pertaining to the first half of the twentieth century; in these experiments the FS effect was demonstrated, but not specifically studied. The object-oriented research into the FSR began in earnest in the 1950s [III.30], when model experiments demonstrated a sharp increase in the target FS RCS and its independence from the target material in the FS region. The model experiments confirmed the existence of the FS effect and the theory of forward-scattering, but they did not suggest any methods of its practical utilization.

Later on, we will consider the results of the full-scale experiments with the FSR involving aircrafts of different sizes and types [III.7, III.10, III.14].

19.2 EXPERIMENTAL FSR

The pilot radar for forward-scattering experiments was developed in accordance with the principles listed in Section 15.2. The block diagram of the experimental FSR is presented in Figure 19.1. A simplified structure of the radar has already been given above (in Section 15.2); therefore we will now discuss the main components of the radar in more detail [III.10].

In Figure 19.1, PAA is the phased array antenna, RFA is the RF amplifier, RF WBA is the RF wideband amplifier, IFA is the intermediate-frequency (IF) amplifier, AD is the amplitude detector, NF is the notch filter, LPF is the lowpass filter and ADC is the analogue-to-digital converter.

The transmitter was implemented on the basis of a standard 400 MHz radio station with a power output of about 6 W (Figure 19.2). Thus, the transmitted signal of the FSR is a continuous narrowband quasi-harmonic waveform. The antenna of the radio station (Figure 19.3) was mounted on a mast at a height of 20–25 m and formed a single beam with a –3 dB width of 60° and 40° in the azimuth and elevation planes respectively. Due to losses in the feeder, the radiated microwave power level was 1–2 W.

Bistatic Radars: Principles and Practice Edited by M. Cherniakov

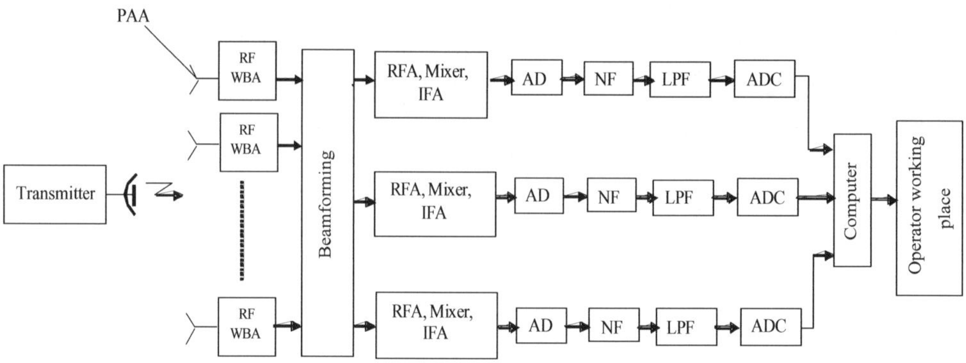

Figure 19.1 Experimental FSR block diagram

The receiving antenna is presented in Figure 19.4. It was mounted on a similar mast at a height of 25–30 m. It was a planar phased array with dimensions of about 5 × 2.5 m. The directional pattern of the antenna had three azimuth-plane partial beams with a −3 dB width of approximately 12° in the azimuth plane and 20° in the elevation plane.

The partial beam patterns measured in the experiment are given in Figure 19.5. The numbers of beams 1, 2 and 3 are shown. The small container behind the antenna curtain accommodates the RF parts of the receiving channels corresponding to the partial beams. These are located immediately on the antenna to minimize the loss inherent in the forming of partial channels and the loss in the feeders, which conduct the echo signal received on the antenna beams to the processing equipment.

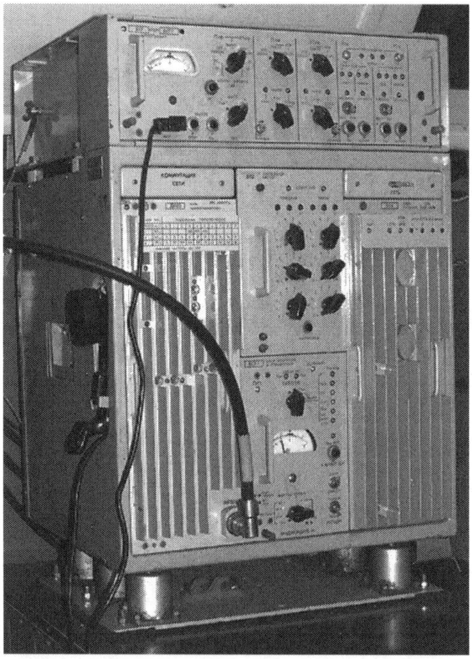

Figure 19.2 Radio station R415V (transmitter)

Figure 19.3 Antenna of the transmit position

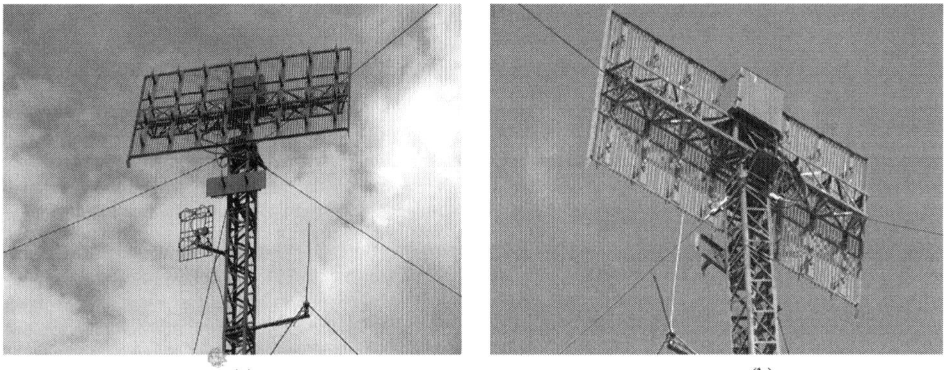

(a) (b)

Figure 19.4 Antenna of the receive position: (a) receiving antenna and communication antenna (front view); (b) receiving antenna and receiver container (rear view)

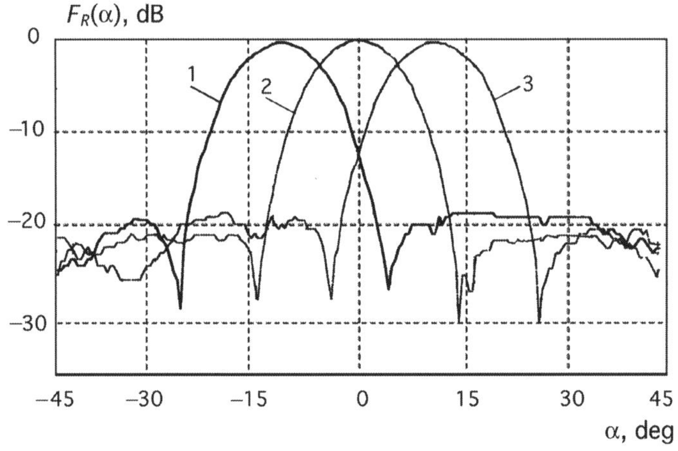

Figure 19.5 Partial beams of receiving antenna directional pattern in azimuth plane

Figure 19.6 Processing cabinet at the receive position

It is seen from Figure 19.4 that the antenna curtain has a tilt of about 10° in the vertical plane, which corresponds roughly to half the beamwidth in the elevation plane. The antenna tilt makes it possible to maximize the coverage of the radar in the vertical plane and to minimize the level of the underlying surface clutter, preserving the capability for detection of low-flying objects. The subsequent analogue and digital processing of the echo signals is carried out by the equipment accommodated inside the cabin (Figure 19.6).

Time processing is identical in the receiving channels of three partial beams and includes amplification, amplitude detection of signals, clutter rejection and coherent integration of the useful Doppler signal. As discussed in the previous chapter, in the presence of a powerful direct transmitter signal (by 30–60 dB stronger than the target return), amplitude detectors perform synchronous detection of the target return. Thus, nonlinear distortions of the Doppler signal are absent. This circumstance is essential since it determines the possibility of further optimal time processing (coherent integration of the Doppler signal) and optimal space processing (measuring the target return AOA). The linear processing requirement leads to the necessity of a high dynamic range of receiving channels. This is required to avoid saturation in the receiving channels in the presence of a direct signal. In the present case, the dynamic range was as high as 60 dB.

After amplitude detection, notch filters are used to suppress the constant component caused by the presence of a direct transmitter signal. Also, lowpass filters are used to exclude spectral foldover after the sampling. The processes at LPF outputs are converted into a digital code by the ADC. In the pilot radar, a 10-bit ADC was implemented with a sampling rate of 500 Hz. The bit capacity in the ADC is governed by the dynamic range of the receiving channel. Since a large constant component caused by the strong direct signal is suppressed by the NF, the ADC dynamic range requirement is significantly reduced. Therefore, the considered 10-bit conversion is appropriate. The requirement of the sampling rate is defined by the maximal Doppler frequency shift of the target return. In the present case, the selected sampling rate was appropriate for the targets flying at near-sonic speeds.

Digital codes from the receiving channels of the beams were entered on the computer. The following tasks were performed by means of a regular computer. Multichannel filtering in the Doppler frequency domain was realized on the basis of FFT processing. Previous Hamming weighting was used. The use of the FFT provides target echo signal integration with simultaneous clutter rejection. The most beneficial conditions for this kind of processing are achieved by transmitting a continuous harmonic waveform. For the analysis of the underlying surface spectrum, the Doppler spectrum analyser was connected prior to the clutter rejection stage.

After detection of the target return spectrum components, measurement of the primary parameters (the Doppler frequency shift and AOA) was performed according to the techniques described earlier in Section 16.1. Using the measurements of these primary parameters, the trajectory parameters were estimated via an iterative algorithm on the basis of the maximum likelihood method considered in Section 16.2.

19.3 EXPERIMENTAL CONDITIONS

Full-scale experiments on a full-featured FSR were carried out from September 1997 to September 1998 in the vicinity of the city of Nizhniy Novgorod (Russia). The processed results of these experiments are presented in this section.

The transmitter was positioned at a distance of $b = 39.3$ km from the receive side. The height of the receiving antenna was $h_R = 30$ m and that of the transmitting antenna $h_T = 22$ m relative to ground level. The type of terrain within the coverage of the forward-scattering radar was flat woodland. The level of the underlying surface clutter and atmospheric clutter was estimated over a lengthy period encompassing all seasons and a variety of weather conditions.

19.4 CLUTTER LEVEL AND CLUTTER SPECTRUM ESTIMATION

As noted above, analysis of underlying surface clutter and atmospheric clutter was performed with a spectrum analyser. It was connected to the output of the AD in the receiving channel of the antenna pattern partial beam. Figure 19.7 shows spectrograms for the underlying surface clutter obtained in the central beam of the antenna pattern (Figure 19.5). In the side beams, the spectrum of clutter virtually does not exceed the intrinsic noise level.

Consecutive spectrum realizations were registered at an interval of 1 minute. The amplitude scale on the display is linear. The clutter spectrum width is approximately 0.3–0.5 Hz. The level of clutter is 10–20 dB higher than the intrinsic noise level of the receiving channel. Such characteristics of the clutter were observed in the period from September 1997 to early April 1998.

A notch filter was used to eliminate clutter and the direct transmitter signal. There was no clutter at the output of the NF. Thus, the use of the NF the proved to be an expedient and effective solution.

Shown in Figure 19.8 are the clutter spectrum patterns registered in April and May 1998. The spectrograms taken on the days of noticeable atmospheric activity and nonhomogeneity showed that the spectrum could vary significantly on certain days and even within a single day. An essential increase of

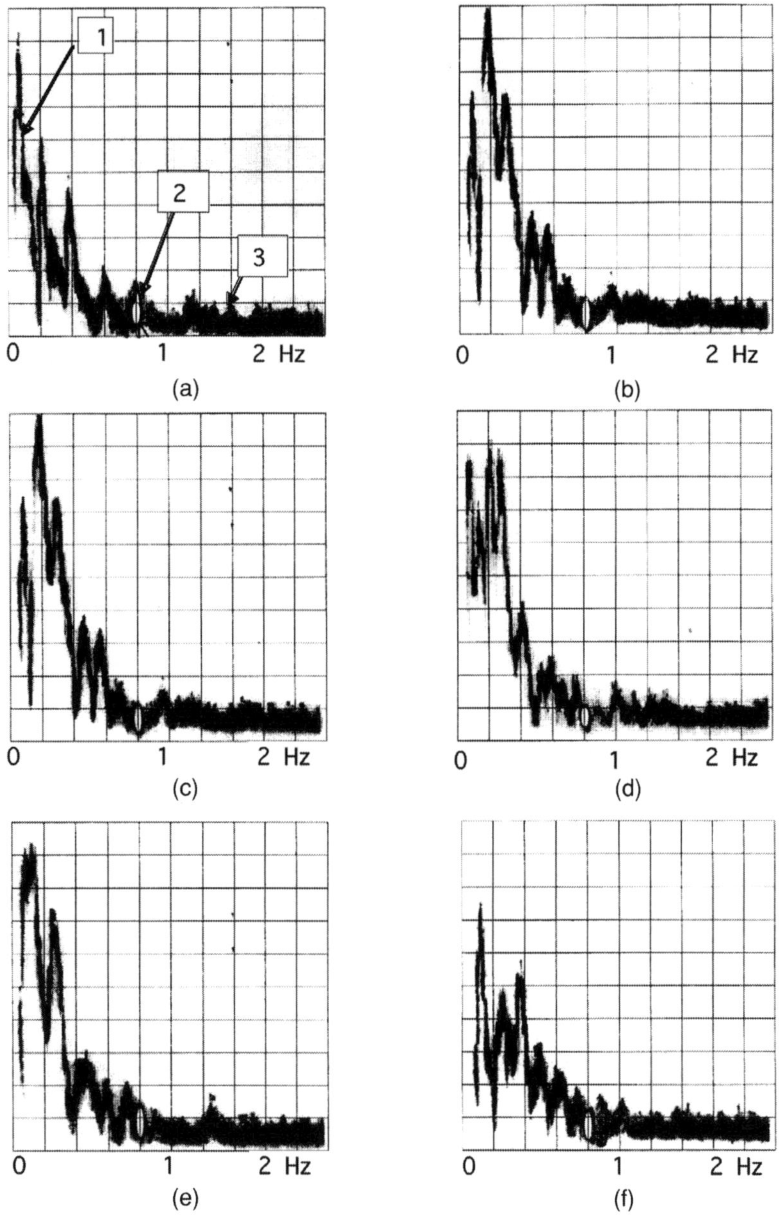

Figure 19.7 Spectrum of underlying surface clutter, linear scale (November 1997): 1, clutter; 2, 1 Hz mark; 3, intrinsic noise

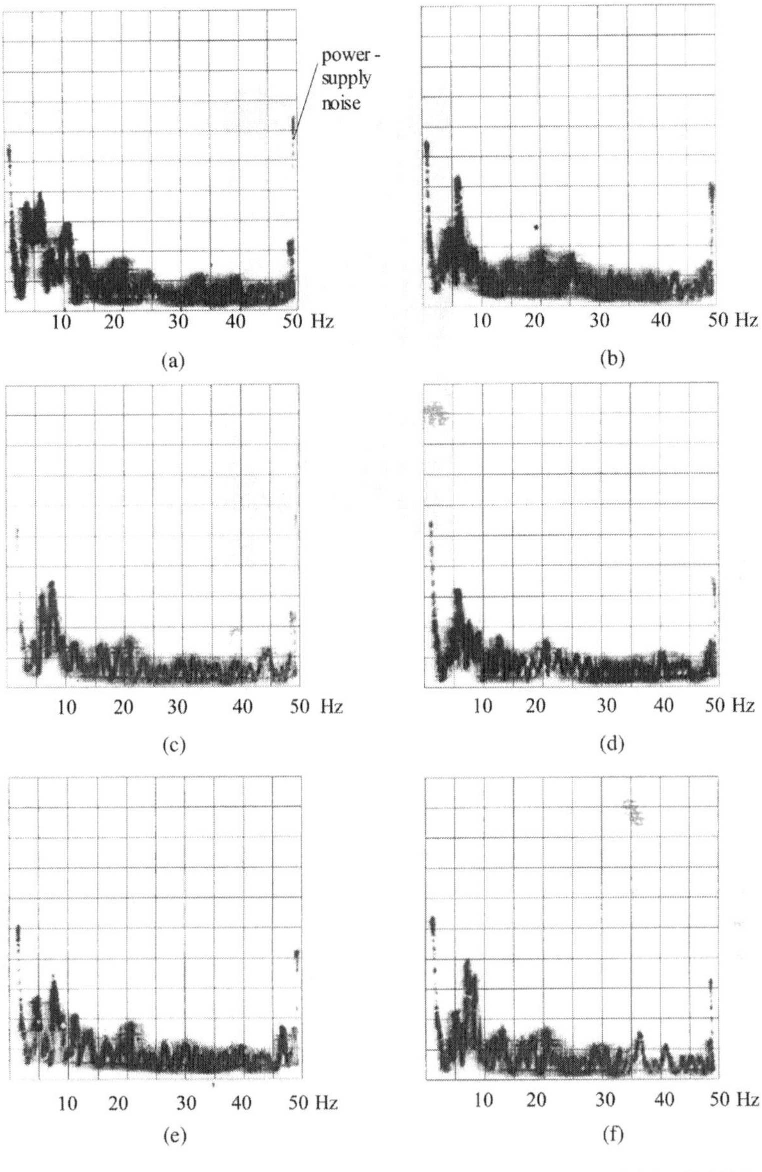

Figure 19.8 Spectrum of underlying surface clutter, linear scale (23 April 1998)

the clutter spectrum width up to 10–25 Hz was observed on such days. The level of the clutter was about 10–30 dB over the noise level.

The sharp increase in clutter power and the widening of its Doppler spectrum is caused by considerable activity of atmospheric processes in April and May. This seasonal activity is characteristic for this region. Great temperature differences between night and day, essentially soil moisture and air humidity result in the formation of large atmospheric irregularities. These irregularities ascend and travel fast carried by the wind. The irregularities form specific clutter caused by forward-scattering. The band of such clutter is much wider than the band of underlying surface clutter. This kind of clutter was only observed over a short period of the year.

19.5 DETECTION OF AIRBORNE TARGETS

19.5.1 Experimental Conditions

The experiments with air targets were carried out on a full-scale basis. In all the operating modes and characteristics, the radar and the air objects corresponded to real radar operating scenarios. The underlying surface in the FSR coverage was represented by an undulating terrain predominantly covered by forest – the Central Russian Upland in the region of Nizhniy Novgorod. The observations were conducted in all seasons, under various weather conditions including snow and rain.

The airborne targets used were light propeller-driven planes, Yak-52 and M-101T ('Gzhel'), helicopters, Mi-2 and Mi-8, a small-sized jet, L-29, as well as motor hang-gliders. The typical flight trajectories are shown in Figure 19.9. The inclination of the trajectories towards the FSR baseline was varied from 90° to 30°. Target altitudes were from 30 m to 3000 m. The aircraft fly at velocities from 70 km/h (motor hang-glider) to 550 km/h (L-29).

Trajectories 1, 2 and 3 correspond to targets flying orthogonally to the baseline at a range of 13 km, 20 km and 27 km from the receiver respectively. Trajectories 4, 5, 6 and 7 are for targets flying slantwise, at an angle of about 30° towards the baseline. The targets flew at altitudes from 30 m to 3000 m at the following speeds:

Mi-2 and Mi-8 helicopters	from 130 to 200 km/h;
Yak-52 airplane	from 210 to 270 km/h;
L-29 airplane	from 440 to 530 km/h;
Motor hang-glider	from 70 to 90 km/h.

A satellite navigation system was used to obtain reference coordinates. GPS receivers were mounted on board the aircraft and at the receive side. They provided an ensured universal time signal reception as well as a target coordinate measurement and data registration.

19.5.2 Echo Signals from Airborne Targets

It was explained in Section 15.4 that the target return in a CW FSR is a continuous wave with a Doppler shift varying in accordance with a slightly nonlinear law. The amplitude of the echo signal varies in accordance with the target FS RCS depending on diffraction angles. Hence, the waveform at the NF output is a frequency-modulated (chirp) signal of variable amplitude. This was specifically confirmed by experiments [III.10, III.37, III.38]. Figures 16.10 and 16.11, present Doppler signals of all the aircraft involved in the sorties, which were recorded at the output of the NF.

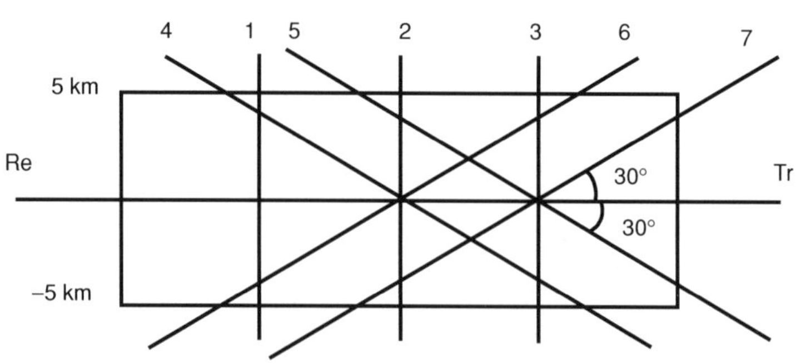

Figure 19.9 Target trajectories in the coverage of FSR

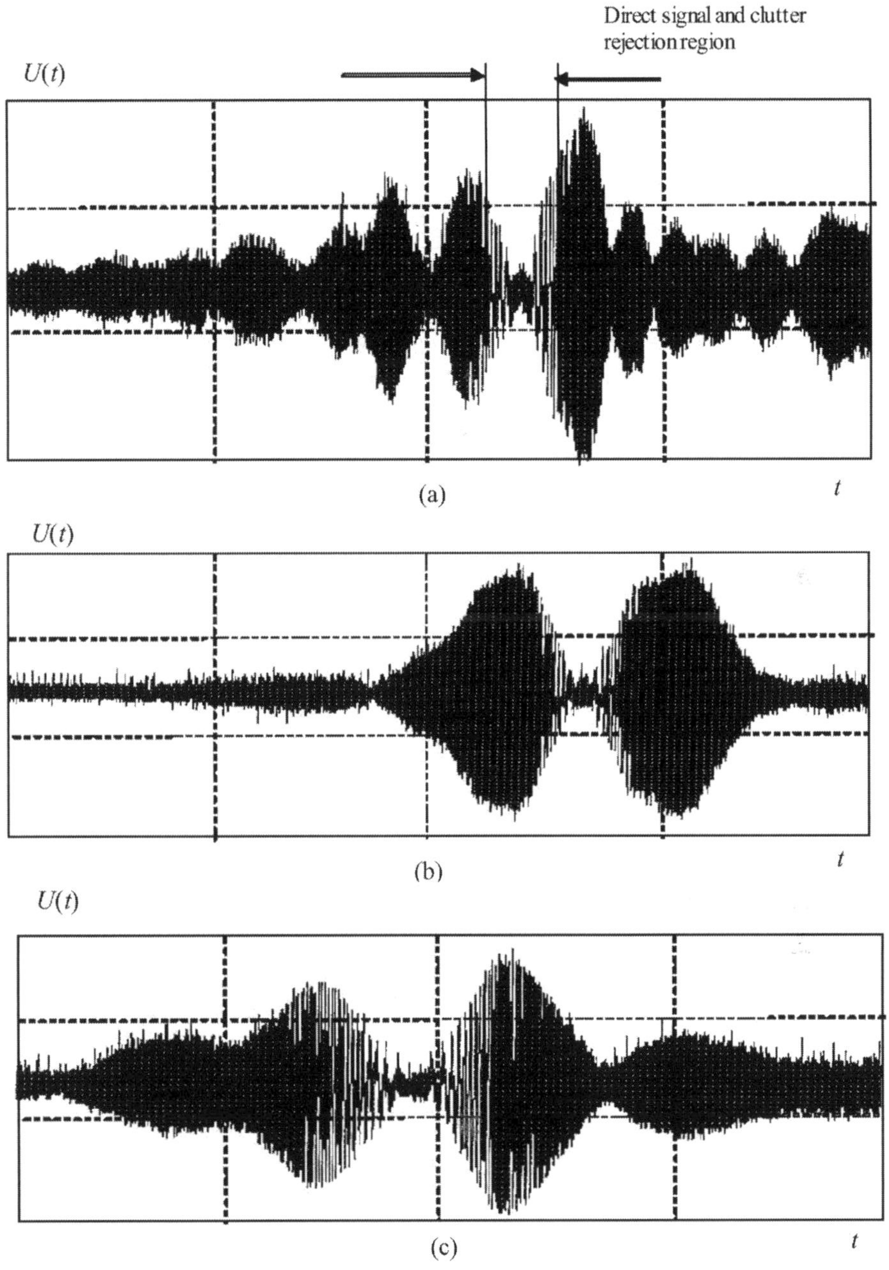

Figure 19.10 Doppler signal: (a) Mi-8 helicopter, (b) Mi-2 helicopter, (c) Yak-52 airplane

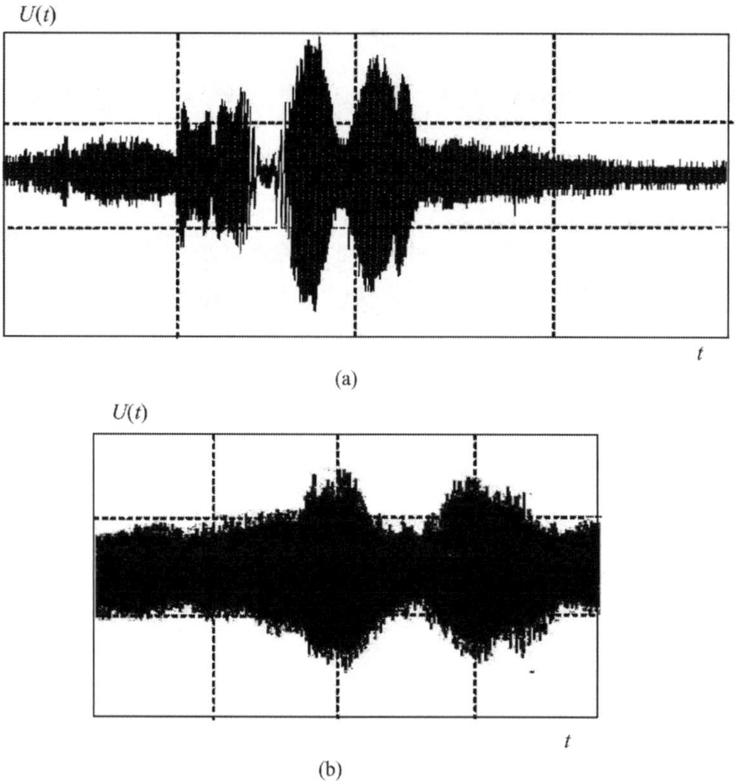

Figure 19.11 Doppler signal: (a) L-29 airplane, (b) motor hang-glider

The envelope of the Doppler signal well describes the structure of the FS RCS, as was shown in Chapter 15, which is similar to a shadow aperture directional pattern containing a mainlobe and sidelobes. These pictures show the possibility in principle of target identification.

Figure 19.12 shows Doppler signals in the receiving channels of three partial beams. The numbers of the beams marked by 1, 2 and 3 correspond to the numbers shown in Figure 19.5. Differences can be seen in the amplitudes of the Doppler signals in the channels and their frequency modulation.

19.5.3 FS RCS

Experiments for target FS RCS estimation are an unquestionably essential part of the FSR study. The objective of the experiments was to measure the FS RCS of the targets and to compare measurement results with the results obtained earlier by the theoretical analysis. Since the FS RCS is dependent upon the target shadow contour, targets with different sizes and shapes of shadow contour were involved in the experiment.

As the trajectory parameters were known beforehand, the horizontal and vertical diffraction angles were also known for each trajectory point. Taking into account the known parameters of the experimental FSR, the SNR at the receiver output was calculated as a function of the diffraction angles α_h and α_v. Experimental the FSR parameters are presented in Table 19.1.

A GPS device installed on board an aircraft provided sufficiently accurate knowledge of the true target trajectory. Thus, the current target position was known with high accuracy. This allowed the

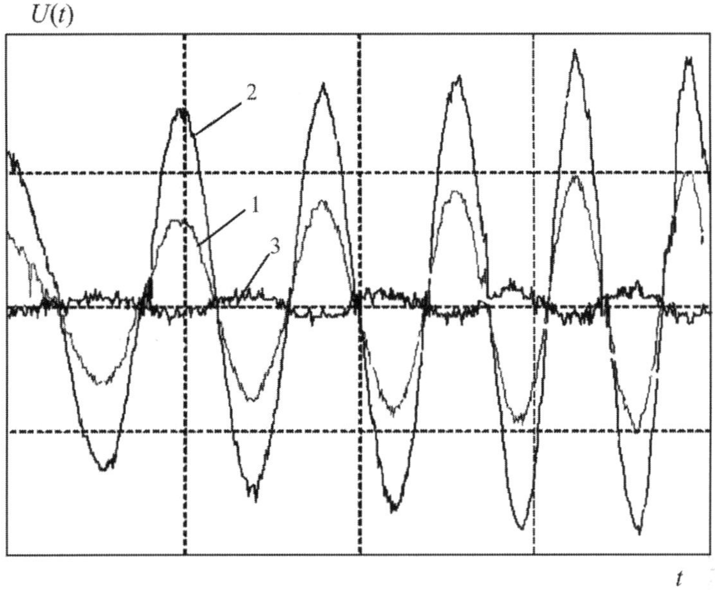

Figure 19.12 Structure of echo signals in the three partial channels

possibility of comparing the FS RCS calculated theoretically with those measured in the experiment for the corresponding target positions.

The experiment was aimed at estimating the level of Doppler signal for an air target flying across the radar baseline at different altitudes. Figure 19.13 shows the SNR q measurements (curves 1 and 2) and the theoretical calculation results (curves 3). Graphics are plotted against the horizontal diffraction angle, α_h, with a fixed vertical diffraction angle, α_v.

The curve discontinuity in the region close to zero diffraction angles corresponds to the baseline crossing. In this region, the target return Doppler frequency shift is close to zero. In this case, the Doppler signal is in the rejection band of the NF corresponding to the band of underlying surface clutter. Knowledge of the frequency response of the NF allows the level of echo signals in this region to be restored (curve 2, circlets).

It follows from the radar equation that the measured values of the SNR are practically defined by the corresponding values of the FS RCS. Hence, the measured values of the SNR could be considered

Table 19.1 Experimental FSR specification

Parameter	Value
Wavelength (m)	0.77
Radiated power (W)	≈ 2
Base length (m)	39 300
Transmitting antenna beam width (horizontal/vertical) (deg)	60/40
Transmitting antenna mounting height (m)	22
Gain (dB)	11
Receiving antenna beam width (horizontal/vertical) (deg)	36–40/20
Receiving antenna mounting height (m)	30
Receiving antenna gain	23

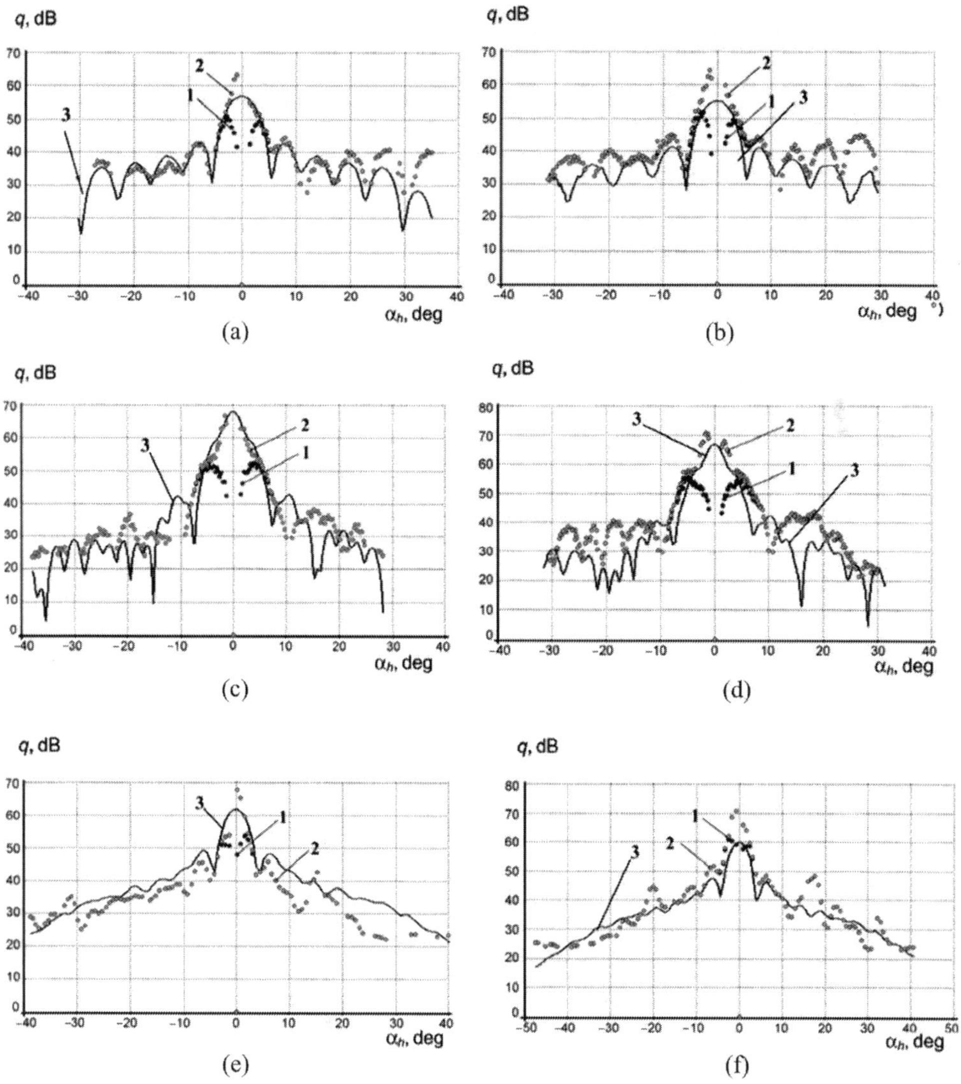

Figure 19.13 SNR measurement and calculation results: curves 1 and 2, measurement results; curve 3, theoretical calculation results; for a Yak-52 airplane flying at an altitude of (a) 100 m and (b) 200 m; for an Mi-2 helicopter flying at an altitude of (c) 100 m and (d) 200 m; for an L-29 jet airplane flying at an altitude of (e) 100 and (f) 200 m

as experimentally obtained FS RCS values. Of course, the quasi-constant coefficient of proportionality defined by FSR parameters takes place. Comparison of theoretical and experimental data demonstrates good agreement between theory and experiment. It is seen from Figure 19.13 that a certain discrepancy in the form of sidelobes between the theoretical and experiment dependences could be present. This is caused by a number of factors. First, when calculating the FS RCS, it is very difficult to accurately take into account the shape of the target shadow contour for all aspect angles. Second, the form of the receiving antenna directional pattern sidelobes affects the form of experimental curves. Also, GPS data indicate that the air target did not maintain the assigned flight altitude. This leads to considerable change in the gains of the receiving and transmitting antennas due to ground influence.

However, theoretical and experimental results for the mainlobe of the FS RCS are practically the same, as well as the envelopes of sidelobes in the range of diffraction angles up to 30–40°. It is known that within the frequency range of the FSR, the value of monostatic RCS is 0.4–0.7 m^2 for a Yak-52 light aircraft, 2–3 m^2 for an Mi-2 helicopter and 0.6–1 m^2 for an L-29 small-sized jet. Thus, the full-scale experiments proved an increase of 3–4 orders in the FS RCS of different targets in comparison to their monostatic RCS due to the forward-scatter effect.

19.5.4 Radar Coverage

FS coverage is the most important radar characteristic. It defines the area where the radar accomplishes its main mission: target detection with the desired probability.

One of the experimental goals was to estimate FSR coverage and to compare the obtained estimations with the results of theoretical calculation. Experimental estimations of FSR coverage were obtained in the following way. Using GPS data and measurements of the target return amplitude, the experimental dependences of the SNR upon the target plane coordinates for different target trajectories were obtained. The target was assumed to be inside FSR coverage until the measured SNR exceeded the assigned threshold. Segments of target trajectories satisfying this condition are shown in Figure 19.14 as light paths on the grey background corresponding to the theoretical FSR coverage. Theoretical coverage was evaluated correspondingly to the technique described in Section 15.3. Digits 1 to 5 are the numbers of paths, while the arrows show the directions of target motion along the trajectories. The presented results demonstrate that theory is in good agreement with experiment.

19.5.5 Measurement of Primary Coordinates and Trajectory Parameters

In the considered FSR, the target trajectory parameters (Cartesian coordinates and their derivatives) are determined using the measurements of the primary parameters (the target return Doppler frequency shift and the AOA). It has been shown above that the accuracy of the primary measurements determines the accuracy of target tracking.

In an experimental FSR, air target trajectories were estimated using the iterative algorithm considered earlier, on the basis of the maximum likelihood method. A first approximation of the trajectory parameter vector was obtained from the first six primary measurements. The primary measurement interval was assigned to be equal to $T = 1$ s. The Doppler frequency shift and AOA measurement variances were estimated during target tracking. The true trajectories were obtained using the data provided by the onboard GPS equipment.

Figure 19.15 shows the results of a target tracking experiment. The solid lines correspond to true trajectories. For known true trajectories, functions $f_D(y)$ and $\alpha(y)$ have been calculated. Dotted lines represent the measurements of the primary parameters and the estimations of the coordinates obtained experimentally.

Thus, the experiment validated the suitability of the maximum likelihood algorithm for establishing trajectory parameters with sufficient accuracy in real time.

19.6 CONCLUSION

The full-scale experimental studies of FSR show:

1. The CW FSR has the main traditional functions of radar systems, such as target detection and tracking.
2. The size and shape of FSR coverage were estimated, where the length is equal to the base length, while the width reaches 30 % of the length.

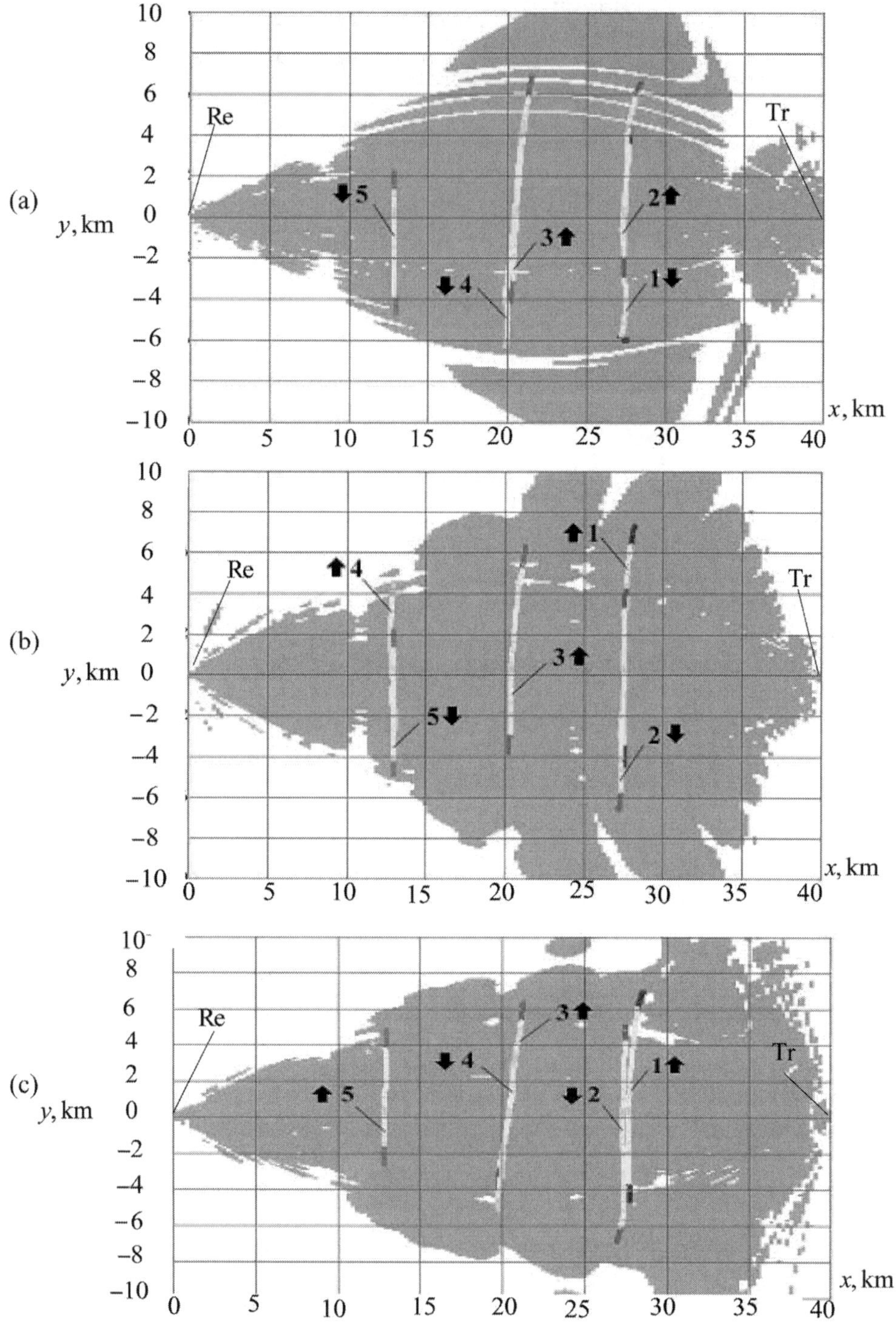

Figure 19.14 Coverage estimation results for an Mi-2 helicopter. Flight altitudes: (a) 100 m, (b) 500 m, (c) 1000 m. Curves 1 to 5 are numbers of flight paths on 10/02/98

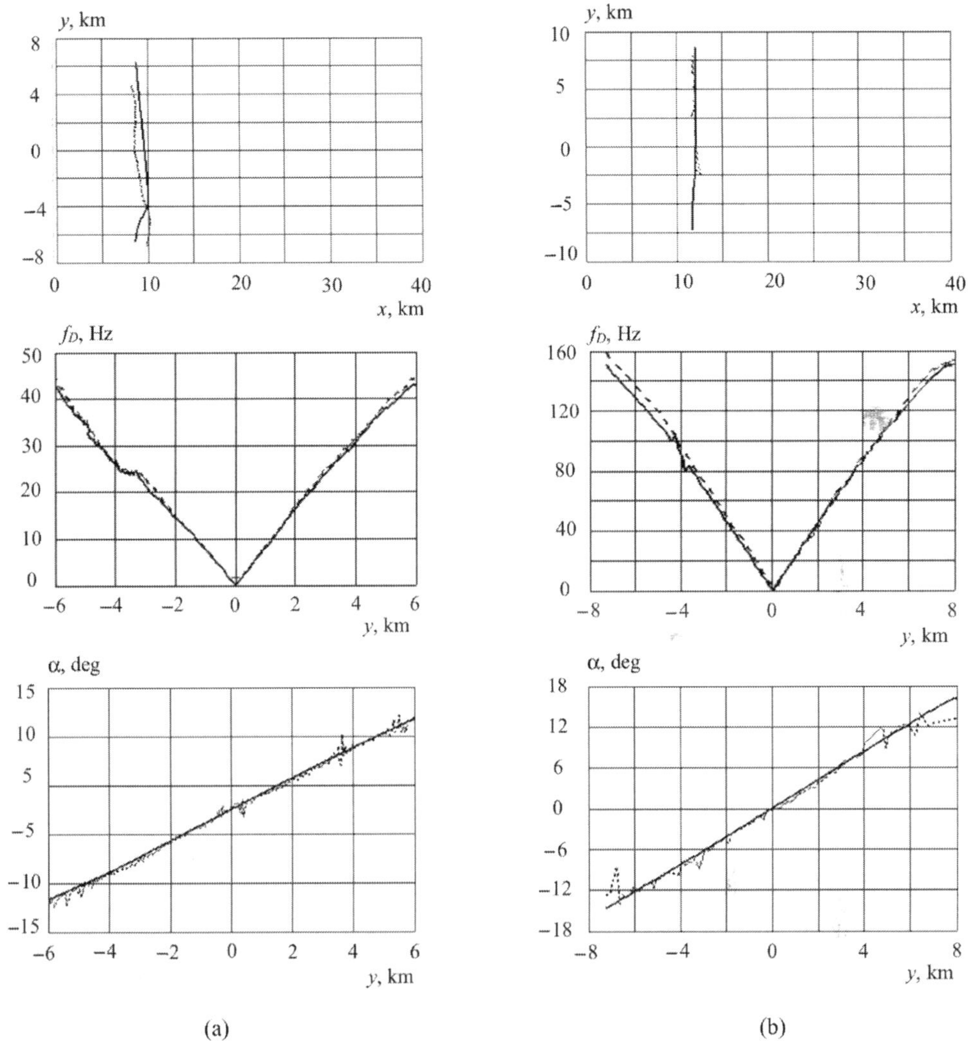

Figure 19.15 Results of tracking experiments [III.14]: (a) Mi-8 helicopter, altitude $H = 200$ m, velocity $V = 44$ m/s; (b) L-29 jet airplane, altitude $H = 200$ m, velocity $V = 148$ m/s

3. The target FS RCS increases by 30–50 dB in the FS region in comparison with its equivalent monostatic counterpart and depends only on the shape of the object's shadow contour.
4. The level of underlying surface clutter and atmospheric clutter in the FSR is relatively small (below 30 dB in respect to the intrinsic noise). Also, the clutter spectrum width is small enough. Thus, clutter cancellation could be realized without deterioration in the characteristics of moving target detection and tracking.
5. The developed algorithms for the measurement of trajectory parameters ensure real-time target tracking.

Summary of Part III

Theory and practice of the CW FSR were discussed in this part. An attempt has been made to cover a broad spectrum of problems related to this radar type.

In Chapter 15, the basic principles of the FSR were considered. First of all, the calculation of the target FS RCS was presented. It was shown that considerable (by 30–50 dB) energy gain of the FSR, in comparison with the monostatic radar, in the FS region takes place. It is very important that the FS RCS depends only on the target shadow contour. The obtained results of the FS RCS calculation have become a departure point for further investigations.

Using these results, the FSR coverage was evaluated on the basis of radar equation analysis. It presents the region stretched along the baseline of the FSR. This region lies inside the azimuth sector by about $\pm 15^{\circ}$. In this sector the FSR provides very effective detection of targets, including those using stealth technology. CW provides the possibility of using a simple transmitter with low power. Thus, a 10 W transmitter provides high probability of air target detection in an FS region of about 40 km long, keeping the probability of false alarms low. In such radar, the Doppler frequency shift and AOA can be used as the primary parameters of the target return to be measured. It was shown that for the linear target motion model these parameters vary by quasi-linear laws when the target is in FSR coverage.

In Chapter 16, the problems of target coordinate measurement in a 2D FSR were considered and the generalized structure of the signal processing algorithm was discussed. It was shown that the Doppler signal can be obtained without nonlinear distortions via amplitude detection, due to the presence of a powerful direct transmitter signal. The Doppler frequency shift measurement can be implemented on the basis of the FFT, while the AOA can be measured by the monopulse method. The measurements of these primary parameters are used further for target tracking. The potential accuracy of the maximum-likelihood estimations of target coordinates was evaluated by means of the informational Fisher matrix analysis. For the parameters of the FSR and targets considered in this part, RMS deviations of the coordinate measurement errors were about 1–2 % of the base length along the baseline and approximately 10 times less in the orthogonal direction.

Two tracking algorithms for the 2D FSR were considered: an iterative algorithm based on the maximum likelihood method and an extrapolation algorithm with preliminary estimation of the range sum. Both algorithms were simulated. A prior ambiguity in the correlation matrix of the primary parameter measurement errors was taken into account for the iterative algorithm. This matrix was estimated during

Bistatic Radars: Principles and Practice Edited by M. Cherniakov

target tracking. Simulation results demonstrated good agreement with potential accuracy. It was shown that the use of corresponding linear smoothing techniques allows us to get close to potential tracking accuracy via the extrapolation algorithm.

Chapter 17 concerns the tracking algorithms for the 3D FSR. It was shown that using the 2D approach, the coordinate estimations in the horizontal plane contain strong systematic errors for the nonzero target altitude. This error occurs due to the Doppler frequency dependency on the target altitude. To exclude these systematic errors, the iterative tracking algorithm was modified. Estimations of the target elevation angle were included in the vector of the primary parameters and the target altitude was added to the vector of trajectory parameters. Modelling results proved that estimations of target horizontal coordinates obtained via the 3D FSR are free of systematic errors. Thus, practically achievable accuracy of target tracking in the 3D FSR is close to the potential accuracy. The potential accuracy of the target horizontal coordinates estimation in the 3D FSR is approximately the same as in the 2D radar. Also, the 3D FSR allows measurement of the target altitude with high accuracy, i.e. RMS deviations are about 0.1 % of the base length.

In Chapter 18, an STP algorithm for the 3D FSR with an array antenna was considered. It was shown that, in the FSR, array antenna beam-forming could be performed after amplitude detection in the receiving channels of the array elements. The main advantages of the proposed technique are as follows:

1. In the presence of a strong direct transmitter signal, the amplitude detector in the array receiving channels acts as a linear synchronous detector and there is no need for a coherent reference signal.
2. Direct signal levels are the same in all receiving channels of the array antenna and consequently they have an equal gain. Thus, the direct signal does not affect the accuracy of the target angular coordinate measurement, in contrast to the monopulse angular measurement with RF beam-forming.
3. The requirements of the receiving channels phase tuning are essentially reduced in comparison with beam-forming at RF.
4. Simulation results have shown that the proposed STP algorithm in the 3D FSR provides sufficiently high accuracy of Doppler shift and angle of arrival estimations. At an SNR of -20 to -14 dB at the AD outputs, RMS deviations of Doppler frequency shift measurement errors do not exceed the values of 0.5 Hz, while RMS deviations of AOA measurement errors are less then 0.5°. As shown in Chapter 16 and 17, such primary measurement estimation allows target coordinates to be estimated with an acceptable accuracy.

In Chapter 19, the results of the full-scale experiment with the FSR were described. The main goal of this experiment was to prove the major theoretical statements formulated in all previous chapters. The following results were obtained:

1. Experimental measurements of target FS RCS and FSR coverage demonstrated good agreement with theoretical calculations.
2. The measurement of target coordinates during the experimentation proved the efficiency of the proposed signal processing and tracking algorithms.
3. Experimental characteristics of ground surface and precipitation clutter were obtained. It was shown that nearly full clutter cancellation could be achieved by filtering in the frequency domain.

Abbreviations

2D	two-dimensional
3D	three-dimensional
AD	amplitude detector
ADC	analogue-to-digital converter
AOA	angle of arrival
BRCS	bistatic radar cross-section
CW	continuous wave
DFT	discrete Fourier transform
FFT	fast Fourier transform
FS	forward-scattering
FSR	forward-scattering radar
LPF	lowpass filter
NF	notched filter
RCS	radar cross-section
Re	receiver
RF	radio frequency
RFA	radio frequency amplifier
RMS	root mean square
SNR	signal-to-noise ratio
STP	space–time processing
Tg	target
Tr	transmitter

Bistatic Radars: Principles and Practice Edited by M. Cherniakov

Variables

$a_f, a_\alpha, b_f, b_\alpha$	coefficients of linear approximating polynomials
A_{SH}	cophasal aperture
b	base length
B_S	spreading ratio of the chirp waveform
c	speed of light
$\boldsymbol{c}_n$	vector of polynomial coefficients
d	array pitch
dS	surface element
D	detection probability
$D_{dif}(\alpha)$	difference pattern
$\mathbf{E}_C$	electric field intensity of the compensation field
$\mathbf{E}_{inc}$	electric field intensity of the incident wave
$\mathbf{E}_{rc}$	electric field intensity at the receive side
$\mathbf{E}_S$	electric field intensity of the self-scattering field
$\mathbf{E}_{SH}$	electric field intensity of the shadow field
$\mathbf{E}_\Sigma$	electric field intensity of the full scattered field
f_{cor}	width of the high-correlation region
f_D	target return Doppler frequency shift
$f_D^e(t)$	extrapolated function of the Doppler frequency shift
$\boldsymbol{f}_D^{(N)}$	vector of Doppler shift measurements
f_{eq}	equivalent frequency
f_S	sampling rate
f_{SP}	space frequency
F	false alarm rate
F_{DFS}	probability of false decision making in respect to the Doppler shift sign
$F_T(\alpha, \beta), F_R(\alpha, \beta)$	directional patterns of transmitting and receiving antennas
G_A	cophasal aperture gain
$\mathbf{G}_n$	inverse correlation matrix of primary measurement errors

G_R	receiving antenna gain
G_T	transmitting antenna gain
h	target altitude
h_T, h_R	height of transmitting and receiving antennas
$\boldsymbol{h}(\boldsymbol{x}_n)$	nonlinear vector function
$\boldsymbol{H}_n$	derivative matrix
$\mathbf{J}(\boldsymbol{x}_n)$	Fisher matrix
k_B	Boltzman constant
k_C	parameter determining the rate of iterative algorithm convergence
k_W	wave number
K	number of array elements
$\mathbf{K}_f$	correlation matrix of the Doppler shift measurement error vector
l_A	linear aperture size
L	range sum
M	polynomial degree
M_{SF}	number of space filters
n_{NF}	noise factor
n_T, n_α and n_f	numbers of resolution cells in domains of time, azimuth and Doppler frequency
n_0	receiver internal noise power
N	number of signal samples or measurements of primary parameters
N_f	number of Doppler filters
N_{RC}	total number of resolution cells
$p(z_n/\boldsymbol{x}_n)$	conditional probability density of primary measurement vector
P_S	scattered signal power at the receive side
P_T	average transmitter power
P_0	power corresponding to the threshold signal
q	signal-to-noise ratio
q_0	threshold signal-to-noise ratio
Q	excess of the received signal over a threshold signal
$\boldsymbol{r}_0$	unit vector pointed towards the receive side
$\boldsymbol{r}$	target radius vector
R	target range
R_R	target–receiver range
R_T	transmitter–target range
S_A	aperture area
$\mathbf{S}_{dir}$	Poynting vector of the direct transmitter signal wave
$\mathbf{S}_{inc}$	Poynting vector of the incident wave
$\mathbf{S}_{tg}$	Poynting vector of the target scattered wave
t	time
t_{CR}	baseline crossing time
t_0	initial moment of the current primary measurement interval
T	primary measurement interval
T_K	absolute temperature
T_{OBS}	full target observation time
$u_{hor,k}[n]$	nth sample at the output of the amplitude detector in the reception channel of the kth horizontal array element
$u_R(t)$	signal at the input of the receiving antenna
$u_{ver,k}[n]$	nth sample at the output of the amplitude detector in the reception channel of the kth elevation array element

U_{dir}	amplitude of the direct transmitted signal
$U(t)$	amplitude of the interference signal
U_{S}	signal amplitude
$U_{\text{SP}}(\alpha)$	directional characteristic of the space filter
$U_{\text{tg}}(t)$	amplitude of the target return at the input of the receiving antenna
U_0	constant component caused by the presence of the direct transmitter signal
$\mathbf{V}$	target velocity vector
V_{f}	Doppler shift rate of change
$V_{\text{f max}}$	maximal Doppler shift rate of change
V_{hor}	projection of the target velocity on to the horizontal plane
V_{max}	maximal target velocity
V_{α}	azimuth rate of change
$V_{\alpha\,\text{max}}$	maximal azimuth rate of change
$w(\alpha)$	weight function
x, y, z	Cartesian coordinates of the target
x_{CR}	base crossing point coordinate
$\boldsymbol{x}_n$	trajectory parameter vector
z_n	primary measurement vector
α	target azimuth
α_{D}	diffraction angle
$\alpha_{\text{h,v}}$	projections of diffraction angle on to horizontal and vertical planes
α_0	initial angle value at primary measurement interval
β	target elevation angle
β_{B}	bistatic angle
γ	angle of trajectory inclination towards the base in the vertical plane
Δ	sampling interval
Δ_{f}	frequency band of a power spectrum
$\Delta\alpha$	azimuth sector of the FS region
$\Delta\alpha_{\text{B}}$	bias of the azimuth estimation
$\Delta\alpha_{\text{r}}$	angle rate of change
$\Delta\beta_{\text{B}}$	bias of the elevation angle estimation
$\Delta\theta$	beam width
Δf	range of the Doppler frequency shift
Δf_{B}	bias of the Doppler frequency shift estimation
Δf_{D}	Doppler frequency deviation
Δf_{off}	frequency offset of the signal
Δf_{opt}	optimum processing band
Δf_{res}	Doppler frequency resolution
ΔL_{n}	fluctuation error of the range sum estimation
Δr_{sys}	systematic error of the range estimation
Δr_n	fluctuation error of the range estimation
Δt_1	time of target flight from the first observed point to the baseline
δ	specific earth conductivity
ε	relative permittivity
ε_x	bias of coordinate x estimation
η	generalized loss
θ_x, θ_y	arguments of directional cosines of the unit vector
$\Theta(t)$	phase of the interference signal

λ	wavelength
$\xi_k[n]$	samples of receiver noise in reception channel of k-th array element
$\boldsymbol{\rho}$	cophasal aperture point radius vector
$\rho_{L,n}$, $\rho_{\alpha,n}$	sensitivity coefficients
σ	radar cross-section
σ_{B}	bistatic radar cross-section
$\sigma_f^2, \sigma_\alpha^2, \sigma_\beta^2$	variances of primary measuring errors
σ_{F}	forward scatter radar cross-section
$\sigma_{r,n}^2$	variance of range estimation fluctuation error at nth time
σ_{S}	root mean square value of Earth surface irregularities
σ_0	monostatic radar cross-section
$\Sigma_{x,n}$	total errors of coordinate x estimation
τ	time delay
φ	angle of trajectory inclination towards the base in horizontal plane
$\varphi_{\mathrm{D}}[n]$	phase shift caused by the target return Doppler shift
$\varphi_n\left(\vec{x}_n\right)$	argument of the Gaussian likelihood function
φ_x	target return random phase
φ_0	initial phase of direct signal
$\Phi(x)$	cumulative distribution function for zero mean and unity variance
ω_0	carrier of a transmitted signal $2\pi f_0$

References

PART I

I.1. Blake, L.V. (1990) Prediction of radar range, in Radar Handbook, 2nd edn, (ed. M. Skolnik), McGraw-Hill, New York.

I.2. Nathanson, F.E. (1990) *Radar Design Principles*, 2nd edn, McGraw-Hill, New York.

I.3. Skolnik, M. (2001) *Introduction to Radar Systems*, 3rd edn, McGraw-Hill, New York.

I.4. Crispin Jr, J.W. and Siegel, K.M. (eds) (1968) *Methods of Radar Cross Section Analysis*, Academic Press, New York.

I.5. Knott, E.F., Shaeffer, J.F. and Tuley, M.T. (1985) *Radar Cross Section*, Artech House, Norwood, Massachusetts.

I.6. Chernyi, F.B. (1972) Radiowaves propagation, Moscow, Sovetskoe Radio (in Russian).

I.7. Shirman, Ya.D. (ed.) (1970) *Theoretical Fundamentals of Radar*, Sovetskoye Radio, Moscow [in Russian].

I.8. Bakulev, P.A. and Veitsel, V.A. (eds) (1996) *A Book of Problems to the Course 'Theoretical Fundamentals of Radar Systems'*, Radio i Svyaz, Moscow [in Russian].

I.9. Gonorovsky, I.S. (1996) *Radio Circuits and Signals*, Radio i Svyaz, Moscow [in Russian].

I.10. Levin, B.R. (1968) *Theoretical Fundamentals of Statistical Radiotechnika*. Sovetskoe Radio, Moscow [in Russian].

I.11. Korn, G.A. and Korn, T.M. (1961) *Mathematical Handbook for Scientists and Engineers*, McGraw-Hill, New York.

I.12. Rabiner, L.R. and Gold, B. (1975) *Theory and Application of Digital Signal Processing*, Prentice-Hall, Inc., Englewood Cliffs, New Jersey.

I.13. Oppenheim, A.V. (ed.) (1978) *Applications of Digital Signal Processing*, Prentice-Hall, Inc., Englewood Cliffs, New Jersey.

I.14. Morgan, D.P. (1985) *Surface-Wave Devices for Signal Processing*, Elsevier Science Publishers, Amsterdam.

I.15. Farnett, E.C. and Stevens, G.H. (1990) Pulse compression radar, in *Radar Handbook*, 2nd edn (ed. M. Skolnik), McGraw-Hill, New York.

Bistatic Radars: Principles and Practice Edited by M. Cherniakov

I.16. Smirnov, N.I. (1973) Properties and characteristics of noise like signals, in *Noise Like Signals in Telecommunication Systems* (ed. V.B. Pestryakov), Moscow, Sovetskoe Radio [in Russian].
I.17. Blake, L.V. (1953) The effective number of pulses per beamwidth for a scanning radar, *Proc. IRE*, **41**, 770–4.
I.18. Barton, D.K. (1969) Simple procedures for radar detection calculations, *IEEE Trans.*, **AES-5**(5), 837–46.
I.19. Harmer, J.D. and O'Hara, W.S. (1961) Some advances in CW radar techniques, in Proceedings of the National Conference on *Military Electron*, Washington DC, pp. 311–23.
I.20. O'Hara, F.J. and Moore, G.M. (1963) A high performance CW receiver using Feed-Thru nulling, *Microwave J.*, **6**, 63–71.
I.21. Peebles Jr, P.Z. (1998) *Radar Principles*, John Wiley & Sons, Inc., New York.
I.22. Konovalov, G.F. (2003) *Radio Automatic*, Radiotechnic, Moscow [in Russian].
I.23. Barton, D.K. (1988) *Radar System Analysis*, Prentice-Hall Inc., Englewood Cliffs, New Jersey.
I.24. Barton, D.K. and Ward, H.R. (1969) *Handbook of Radar Measurement*, Prentice-Hall Inc., Englewood Cliffs, New Jersey.
I.25. Howard, D.D. (1990) Tracking radar, in *Radar Handbook*, 2nd edn (ed. M. Skolnik), McGraw-Hill, New York.
I.26. Stutzman, W.L. (1998) Estimating directivity and gain of antennas, *IEEE AP Magazine*, **40**, 7–11.
I.27. Silver, S. (1949) *Microwave Antenna Theory and Design*, Vol. 12 of the M.I.T. Radiation Laboratory Series, McGraw-Hill, New York.
I.28. Josefson. L.G. (1971) A broad-band twist reflector, *IEEE Trans.*, **AP-19**, 552–4.
I.29. Lewis, B.L. and Shelton, J.P. (1980) Mirror scan antenna technology, Record of the IEEE International Radar Conference, Arlington, Virginia, pp. 279–83.
I.30. Hall Jr, W.P. and Nordmeyer, R.D. (1991) Active element, phased array radar: affordable performance for the 1990s, in Proceedings of the IEEE National Telesystems Conference, Atlanta, Georgia, pp. 193–7.
I.31. Cheston, T.C. and Frank, J. (1990) Phased array antennas, in *Radar Handbook*, 2nd edn (ed. M. Skolnik), McGraw-Hill, New York.
I.32. Miller, C.J. (1964) Minimizing the effects of phase quantization in an electronically scanned array, in Proceedings of the Symposium on *Electrically Scanned Array Techniques and Applications*, Rome Air Development Center Technical Documentary Report RADC-TDR-64-225, Vol.1, pp. 17–38.
I.33. Liberty, J.C. and Rappaport, T.C. (1999) *Smart Antennas for Wireless Communications:IS-95 and Third Generation CDMA Applications*, Prentice-Hall Inc., Englewood Cliffs, New Jersey.
I.34. Hudson, J.E. (1981) *Adaptive Array Principles*, Peter Peregrinus Ltd, Loughborough.
I.35. Monzingo, R.A. and Miller, T.W. (1980) *Introduction to Adaptive Arrays*, Hughes Aircraft Company, Fillerton, California, A Wiley Interscience Publication, John Wiley & Sons, Inc., New York.
I.36. Haykin, S. (2001) *Adaptive Filter Theory*, 4th edn, Prentice-Hall Inc., Englewood Cliffs, New York.
I.37. Sayed, A.H. (2003) *Fundamentals of Adaptive Filtering*, John Wiley & Sons, Inc., Hoboken, New Jersey.
I.38. Steyskal, H., Shore, R.A., Haupt, R.L. (1986) Methods for null control and their effects on the radiation pattern. *IEEE Trans.*, **AP-34**(3), 404–9.
I.39. Frost, O.L. (1972) An algorithm for linearly constrained adaptive array processing, *Proc. IEEE*, **60**, 926–35.
I.40. Dzhigan, V.I. (2004) Multichannel RLS- and fast RLS-algorithms of adaptive filtering. *Achievements in Modern Radio Electronics*, **11**, 48–77 [in Russian].
I.41. Bakrunov, O.A., Nezlin, D.V. and Sizov, V.I. Chistyukhin V.V. (1997) Aerological Radiotheodolite. Russian Federation Commitee for patents and trademarks (Rospatent). Certificat N 4387, June 16, 1997 (in Russian).

PART II

II.1. Nathanson, F.E., Reilly, J.P. and Cohen M.N. (1999) *Radar Design Principles: Signal Processing and the Environment*, 2nd edn, SciTech Publishing, New York.

II.2. Peebles Jr, P.Z. (1998) *Radar Principles*, John Wiley & Sons, Inc., New York.

II.3. Caspers, J.M. (1970) Bistatic and multistatic radar, in *Radar Handbook* (ed. M.I. Skolnik), Chapter 36, McGraw-Hill, New York.

II.4. Chernyak, V.S. (1998) *Fundamentals of Multisite Radar Systems. Multistatic Radar and Multiradar Systems*, Gordon and Breach Science Publishers, Amsterdam.

II.5. Willis, N.J. (1990) Bistatic radar, in *Radar Handbook* (ed. M.I. Skolnik), Chapter 25, McGraw-Hill, New York.

II.6. Kremer, I.Ya. (ed.) (1984) *Prostranstvenno-vremennaya obrabotka signalov (Space-Time Signal Processing)*, Radio i Svyaz, Moscow [in Russian].

II.7. Aver'yanov, V.Ya. (1978) *Raznesennye radiolokatsionnye stantsiy i sistemy (Radar Stations and Systems with Space Diversity)*, Nauka i Tekhnika, Belarus, [in Russian].

II.8. Kontorov, D.S. and Golubev-Novozhilov, V.S. (1971) *Vvedenie v radiolokatsionnuyu sistemoteckniku (Introduction to Radar System Engineering)*, Sovietskoye Radio, Moscow [in Russian].

II.9. Hershey, J.E. and Schroeder, J.E. (1990) Bistatic radar without transmitted signal – covert altimetry concept, *IEEE Trans.*, **AES-26**(5), 732–6.

II.10. Finkel'shteyn, M.I. (1983) *Osnovy radiolokatsii (Fundamentals of Radar)*, Radio i Svyaz, Moscow [in Russian].

II.11. Germond, A-L., Pottier, E. and Saillard, J. (1997) Foundations of bistatic radar polarimetry theory, in Proceedings of the International Conference on *Radar 97*, October 1997, pp. 833–7.

II.12. Shirman, Ya.D. (ed.) (2001) *Computer Simulation of Aerial Target Radar Scattering, Recognition, Detection and Tracking*. Artech House, Boston.

II.13. Knott, E.F., Shaeffer, J.E. and Tuly, M.T. (1993) *Radar Cross Section*, Artech House, Norwood, Massachusetts.

II.14. Beckmann, P. and Spizzichino, A. (1963) *The Scattering of Electromagnetic Waves from Rough Surfaces*, Pergamon Press, Ltd, Oxford.

II.15. Sayles, R.S. and Thomas, T.R. (1978) Surface topography as a nonstationary random process, *Nature*, **271** (5644), 431–4.

II.16. Berry, M.V. and Hannay, J.H. (1978) Topography of random surfaces, *Nature*, **273** (5663), 573.

II.17. Kobak, V.O. (1975) *Radiolokatsionnye otrazhateli (Radar Reflectors)*, Sovietskoye Radio, Moscow [in Russian].

II.18. Chung-Chi Cha, M.J. and Starszewski, E. (1988) An RCS analysis of generic airborne vehicles depending on frequency and bistatic angle, in Proceedings of the IEEE National Radar Conference, pp. 214–9.

II.19. Ewell, W. and Zehner, S.P. (1980) Bistatic radar cross section of ship targets, *IEEE Journal of Oceanic Engineering*, **OE-5** (4), 211–5.

II.20. Ewell, G.W. and Zehner, S.P. (1982) Bistatic sea clutter return near grazing incidence, in Proceedings of the IEE International Radar Conference, pp. 188–192.

II.21. Willis, N.J. (1991) *Bistatic Radar*, Artech House, Norwood, Massachusetts.

II.22. Knotts, M.E., Michel, T.R. and O'Donnell, K.A. (1993) Comparison of theory and experiment in light scattering from a randomly rough surface, *J. Opt. Sac. Amr. A*, **10**(5), 928–41.

II.23. Mott, H. (1992) *Antennas for Radar and Communications: A Polarimetric Approach*, John Wiley & Sons, Inc., New York.

II.24. Fung, A. (1994) *Microwave Scattering and Emission Models and Their Applications*, Artech House, Norwood, Massachusetts.

II.25. Cloude, S.R. and Pottier, E. (1996) A review of target decomposition theorems in radar polarimetry, *IEEE Trans.*, **GRS-34**, 498–518.
II.26. Huynen, J.R. (1970) Phenomenological theory of radar targets, PhD Dissertation, PQR Press.
II.27. Shaeffer, J.E., Hom, K.W., Baucke, R.G., Cooper, B.A. and Talcott Jr, N.A. (1996) Bistatic *k*-space imaging for electromagnetic prediction codes for scattering and antennas, NASA Technical Paper 3569.
II.28. Menzel, D.H. (ed.) (1955) *Fundamental Formulas of Physics*, Prentice-Hall, New York.
II.29. Jackson, M.C. (1986) The geometry of bistatic systems, *IEE Proc.*, **F 133**(7), 604–12.
II.30. Tsao, T., Slamani, M., Varshney, P., Weiner, D., Schwarzlander, H. and Borek, S. (1997) Ambiguity function for a bistatic radar, *IEEE Trans.*, **AES-33**(3), 1041–51.
II.31. Varshney, P.K., *et al.* (1995) Analysis of ambiguity function for bistatic radar, Technical Report RL-TR-95-23, Rome Laboratory, Rome, New York.
II.32. Shirman, Ya.D. (ed.) (1970) *Teoreticheskiye osnovy radiolokatsii (Radar Fundamentals)*, Sovietskoye Radio, Moscow [in Russian].
II.33. Dulevich, V.E. (ed.) (1978) *Teoreticheskiye osnovy radiolokatsii (Radar Fundamentals)*, Sovietskoye Radio, Moscow [in Russian].
II.34. Dunsmore, M.R.B. (1993) Bistatic radars, in *Advanced Radar Techniques and Systems* (ed. G. Galati), Chapter 11, Peter Peregrinus, Houghborough.
II.35. Cherniakov, M., Salous, M., Jancovic, P., Abdullah, R. and Kostylev, V. (2005) Forward scattering radar for ground targets detection and recognition, in Proceedings of the 2nd Annual Technical DTC Conference. Edinburgh.
II.36. Cherniakov, M., Salous, M., Kostylev, V. and Abdullah, R. (2005) Analysis of forward scattering radar for ground target detection, in Proceedings of the 2nd European Radar Conference, pp. 145–8.
II.37. Born, M. and Wolf, E. (1965) *Principles of Optics*, 3rd edn., Pergamon Press, New York.
II.38. Abramowitz, M. and Stegun, I.A. (eds) (1964) *Handbook of Mathematical Functions with Formulas, Graphs, and Mathematical Tables*, Applied Mathematics Series 55, National Bureau of Standards, Washington, DC.
II.39. Luke, Y.L. (1975) *Mathematical Functions and Their Approximations*, Academic Press, New York.
II.40. Soroko, L.M. (1971) *Osnovy golografii i kogerentnoy optiki (Fundamentals of Holography and Coherent Optics)* Nauka, Moscow [in Russian].
II.41. Jackson, J.D. (1998) *Classical Electrodynamics*, 3rd edn, John Wiley & Sons, Inc., New York.
II.42. Stratton, J.A. (1941) *Electromagnetic Theory*, McGraw-Hill, New York.
II.43. Kay, I.W. (1968) Some Basic Concepts and Results from Electromagnetic Theory, in *Methods of Radar Cross Section Analysis* (eds J.W. Crispin and K.M. Seigel), Chapter 1, Academic Press, New York.
II.44. Wainstein, L.A. (1988) *Elektromagnitnye volny (Electromagnetic waves),* Radio i Svyaz, Moscow [in Russian].
II.45. Shtager, E.A. (1986) *Rasseyanie radiovoln na telakh slozhnoy formi (Wave Dissipation on Bodies of Complex Configuration)*, Radio i Svyaz, Moscow [in Russian].
II.46. Ruck, G.T., Barrick, D.E. and Stuart, W.D. (1970) Theory, in *Radar Cross Section Handbook* (eds G.T. Ruck, D.E. Barrick, W.D. Stuart and C.K Krichbaum) Chapter 2, Plenum, New York.
II.47. Ulaby, F.T., Moore, R.K. and Fung, A.K. (1986) *Microwave Remote Sensing, Active and Passive*, Artech House, Boston, Massachusetts.
II.48. Gasiòn, O.S. (1994) Modelos de reflectividad radar de superficies terrestres y marinas, Projecte final de carrera, UPC, Barcelona.
II.49. Marchand, R.T. and Brown, G.S. (1998) Inferring rough surface parameters from average scattering data using approximate scattering models. 1. Gaussian spectrum, *Radio Science*, **33**(4), 821–34.

II.50. Valenzuela, A. (1978) Theories for the interaction of electromagnetic and oceanic waves – a review, *Boundary-Layer Meteorology*, **13**, 61–85.

II.51. Ufimtsev, P.Ya. (1962) *Metod kraevyh voln v fizicheskoy teorii difraktsii* (*Method of Edge Waves in the Physical Theory of Diffraction*), Sovetskoe Radio, Moscow [in Russian]. (This book was translated into English by US. Air Force, Foreign Technology. Division Wright-Patterson AFB, Ohio, 1971: Technical. Report. AD N 733203 DTIC, Cameron Station, Alexandria, VA 22304-6145.)

II.52. Yu, J.S. (1970) Radar cross section of a thin plate near grazing incidence. *IEEE Trans.*, **AP-18**(5), 711.

II.53. James, G.L. (1976) *Geometrical Theory of Diffraction for Electromagnetic Waves*, Peter Peregrinus, London.

II.54. Fock, V.A. (1970) *Problemy difraktsii i rasprostraneniya electromagnitnyh voln* (*Problems of Diffraction and Propagation of Electromagnetic Waves*), Sovetskoe Radio, Moscow [in Russian].

II.55. Rayleigh, L. (1896) *The Theory of Sound*, 3rd edn. Macmillan, London.

II.56. Rice, S.O. (1951) Reflection of electromagnetic waves from slightly rough surfaces, *Commun. Pure Appl. Math.*, **4**, 351–78.

II.57. Mitzner, K.M. (1964) Effect of small irregularities on electromagnetic scattering from an interface of arbitrary shape, *J. Math. Phys.*, **5**, 1776–86.

II.58. Hecht, E. and Zajac. A. (1974) *Optics*, Addison Wesley, Reading Massachusetts.

II.59. Zavorotny, V. and Voronovich, A. (1998) Modeling of GPS signal scattering for ocean-surface remote sensing using various techniques, Presented at the IDS Workshop at JPL, pp. 1–3.

II.60. Snoeij, P., Swart, P.J.F. and Unal, C.M.H. Study on the response of the radar echo from the ocean surface to the surface wind vector at frequencies between 1 and 18 GHz, Final Report for ESA under Contract 8722/89/F/FL(SC).

II.61. Eigel Jr, R.L., Collins, P.J., Terzuoli Jr, A.J., Nesti, G. and Fortuny, J. (2000) Bistatic scattering characterization of complex objects, *IEEE Trans.*, **GRS-38** (5), 2078–92.

II.62. Harrington, R.F. (1967) Matrix methods for field problems, *Proc. IEEE*, **55** (2), 136–49.

II.63. Harrington, R.F. (1993) *Field Computation by Moment Methods* (ed. D.G. Dudley), IEEE/OUP, New York.

II.64. Burke, G.F. and Poggio, A.J. (1981) Numerical electromagnetic code – (NEC) – method of moments, Technical Document 116, prepared for Naval Electronic Systems Command (ELEX 3041), San Diego, California.

II.65. Mei, K.K. (1974) Unimoment method of solving antennas and scattering problem, *IEEE Trans.*, **AP-22**(6), 760–6.

II.66. Kvam, T.M. and Mei, K.K. (1981) The internal fields of a layered radome excited by a plane wave, Digest of papers presented at the 1981 International IEEE AP-S Symposium, Vol. 2, pp. 608–11.

II.67. Peterson, A.F. and Mittra, R, (1985) Method of conjugate gradients for the numerical solution of large body electromagnetic scattering problems, *J. Opt. Soc. Am. A*, **2** (6), 971–7.

II.68. Bennett, C.L. and Weeks, W.L. (1970) Transient scattering from conducting cylinders, *IEEE Trans.*, **AP-18**(5).

II.69. Taflove, A. and Umashankar, K. (1983) Radar cross section of general three-dimensional scatterers, *IEEE Trans.*, **EMC-24**(4), 433–40.

II.70. Shankar, V., Hall, W.F. and Mohammadian, A.H. (1989) A time-domain differential solver for electromagnetic scattering problems, *Proc. IEEE,* **77**(5), 709–21.

II.71. Zaporozhets, A.A. and Levy, M.F. (1999) Bistatic RCS calculations with the vector parabolic equation method, *IEEE Trans.*, **AP-47**(11), 1688–96.

II.72. Levy, M.F. and Borsboom, P.-.P (1996) Radar cross-section computations using the parabolic equation method, *Electronics Lett.*, **32** (6), 1234–6.

II.73. Levy, M.F., Borsboom. P.-P., Zaporozhets. A.A. and Zebic-Le Hyaric, A. (1996) RCS calculations with the parabolic wave equation, in Proceedings of the AGARD Conference, Ankara, Turkey, Vol. 583, pp. 5.1–5.9.

II.74. Levy, M.F. and Zaporozhets, A.A. (1998) Target scattering calculations with the parabolic equation method, *J. Acoust. Soc. Am.*, **103**(2), 735–41.
II.75. Bowman, J.J., Senior, T.B.A. and Uslenghi, P.L.E. (1987) *Electromagnetic Scattering by Simple Shapes*. Hemisphere, New York.
II.76. Borzì, G. (2004) Trigonometric approximations for the computation of radar cross sections, *IEEE Trans.*, **AP-52**(6), 1596–602.
II.77. Alfonzetti, S. and Borzì, G. (2000) A fast method for computation of the radar cross section, *IEEE Trans.*, **M-36**(4), 921–84.
II.78. Medgyesi-Mitschanc, L.N. and Wang D.-S. (1989) Hybrid methods for analysis of complex scatterers, *Proc. IEEE*, **77**(5), 770–9.
II.79. Burnside, W.D., Yu, C.L. and Marhefka, R.J. (1975) A technique to combine the geometric theory of diffraction and the moment method, *IEEE Trans.*, **AP-23**, 551–8.
II.80. Knott, E.F. and Senior, T.B.A. (1973) Non-specula radar cross section study, Report AFAL-TR-73-2. The University of Michigan, Michigan.
II.81. Knott, E.F. (1977) RCS reduction of dihedral corners, *IEEE Trans.*, **AP-25**, 406–9.
II.82. Wang, D.-S. (1987) Limits and validity of the impedance boundary condition for penetrable surfaces, *IEEE Trans.*, **AP-35**, 453–7.
II.83. Wang, D.-S. and Medgyesi-Mitschang, L.N. (1985) Electromagnetic scattering from finite circular and elliptic cones, *IEEE Trans.*, **AP-33**, 488–97.
II.84. Wang Hao, Xu Minjia, Wang Chen, Hubing Todd (2003) Impedance boundary conditions in a hybrid FEM/MOM formulation, *IEEE Trans.*, **EC-45**(2), 198–206.
II.85. Schuh, M.J., Woo, A.C. and Simon, M.P. (1994) The monostatic/bistatic approximation, *IEEE Antennas and Propagation Magazine*, **36**(4), 76–8.
II.86. Woo, A.C., Wang, H.T.G., Schuh, M.J. and Sanders, M.L. (1993) Benchmark radar targets for the validation of computational electromagnetics programs, *IEEE Antennas and Propagation Magazine*, **35**(1).
II.87. Crispin Jr, J.W., Goodrich, R.F. and Siegel, K.M. (1959) A theoretical method for the calculation of the radar cross sections of aircraft and missiles, The University of Michigan Radiation Laboratory, prepared for the Air Force Cambridge Research Center.
II.88. Crispin, J.W. and Siegel, K.M. (1968) *Methods of Radar Cross Section Analysis*, Academic Press New York.
II.89. Kell, R.E. (1965) On the derivation of the bistatic RCS from monostatic measurements, *Proc. IEEE*, **53**(8), 983–8.
II.90. Skolnik, M.J. (1961) An analysis of bistatic radar, *IRE Trans.*, **ANE-8**(1), 19–27.
II.91. Crispin Jr, J.W. and Maffet, A.L. (1965) Radar cross section estimation for complex shapes, *Proc. IEEE*, **53**(8), 972–82.
II.92. Zasada, N. (1980) Multistage radar system for aircraft defense, *Signal*, **34**(8), 65–75.
II.93. Fawcette, J. (1978) Bistatic radar may be a 'sanctuary' in space, *Electronic Warfare/Defence Electronics*, **10**(1), 84–6.
II.94. Glaser, J.I. (1984) Bistatic radar hold promise for future systems, *MSN*, **16**(11), 119–36.
II.95. Wu, T.K. (1989) Radar cross section of arbitrarily shaped bodies of revolution, *Proc. IEEE*, **77**(5), 735–40.
II.96. Mautz, J.R. and Harrington, R.F. (1979) Electromagnetic scattering from a homogeneous material body of revolution, *AEU*, **33**, 71.
II.97. Penno, R.P., Thiele, G.A. and Pasala, K.M. (1989) Scattering from a perfectly conducting cube, *Proc. IEEE*, **77**(5), 815–23.
II.98. Yaghjian, A.D. (1984) Broadside radar cross section of the perfectly conducting cube, Report RADC-TR-84-11, Rome Air Development Center, Air Force Systems Command, Criffiss Air Force Base, New York.
II.99. McCahan, R.V. (1983) Scattering experiments at the Ipswich Electromagnetic Measurements

Facility: backscatter from metal cubes, RADC Report, Rome Air Development Center, Air Force Systems Command, Criffiss Air Force Base, New York.

II.100. Pouliguen, P., Lucas, L., Muller, F., Quete, S. and Terret, C. (2002) Calculation and analysis of electromagnetic scattering by helicopter rotating blades, *IEEE Trans.*, **AP-50**(10), 1396–408.

II.101. Robillard, P.E. and Herrett, R.H. (1976) Circular polarized static RCS patterns for bistatic scattering – I (8-deg flat back cones), Report R76–85, Delco Electronics, CMC, Santa Barbara Operations, Goleta, California.

II.102. Glaser, J.I. (1989) Some results in the bistatic radar cross section (RCS) of complex objects, *Proc. IEEE*, **77**(5), 639–48.

II.103. Khenchaf, A., Daout, F. and Saillard, J. (1996) The two-scale model for random rough surface scattering, in Proceedings of the MTS/IEEE 'OCEANS'96, Prospects for the 21st Century' Conference, pp. 887–91.

II.104. Khenchaf, A., Daout. F. and Saillard, J. (1995) Bistatic polarized scattering by rough surface, in JIPR Third International Workshop on *Radar Polarimetry*, pp. 21–3.

II.105. Bahar, E. and Lee, B.S. (1995) Full wave vertically polarized bistatic radar cross section for random rough surfaces – comparison with experimental and numerical results, *IEEE Trans.*, **AP-43**(2), 205–13.

II.106. Sanchez-Gil, J.A. and Nieto-Vesperinas, M. (1991) Light scattering from random rough dielectric surfaces, *J. Opt. Soc. Am. A*, **8**(8), 1270–86.

II.107. Hsieh, C.-Y. (2001) Study of out of plane bistatic scattering from rough dielectric surfaces, *IEE Proc.-Microwave Antennas Propagation*, **148**(3), 157–62.

II.108. Hauck, B., Ulaby, F. and Deroo, R. (1998) Polarimctric bistatic-measurement facility for point and distributed targets, *IEEE Antenna Propagation Magazine*, **40**(1).

II.109. Sakurai, H., Ohki, M., Motojima, K. and Kozaki, S. (2004) Scattering of Gaussian beam from a hemispherical boss on a conducting plane, *IEEE Trans.*, **AP-52**(3), 892–4.

II.110. Bourlier, C., Berginc, G. and Saillard, J. (2001) Bistatic scattering coefficient from one- and two-dimensional random surfaces using the stationary phase and scalar approximation with shadowing effect: comparisons with experiments and application to the sea surface, *Waves in Random Media*, November, 119–47.

II.111. O'Donnell, K.A. and Mendez, E.R. (1987) Experimental study of scattering from characterized random surfaces, *J. Opt. Soc. Am. A*, **4**(7) 1194–205.

II.112. Bahar, E. and Lee, B.S. (1996) Radar scatter cross sections for two dimensional random rough surfaces – full wave solution and comparisons with experiments, *Waves in Random Media*, June, 1–23.

II.113. Hsieh, C.-Y. and Fung, A.K. (2003) Bistatic multiple scattering from randomly rough surfaces, *IEE Proc.-Microwave. Antennas Propagation*, **150**(4), 214–8.

II.114. Broschat, S.L., Thorsos, E.I. and Ishimaru, A. (1989) The phase perturbation technique versus an exact numerical method for random rough surface scattering, *J. Electromagnetic Waves Applic.*, **3**, 237–56.

II.115. Ross, R.A. (1966) Radar cross section of rectangular flat plates as a function of aspect angle, *IEEE Trans.*, **AP-14**(3), 329–35.

PART III

III.1. Skolnik, M.I. (1990) *Radar Handbook*, 2nd edn, McGraw-Hill, New York.

III.2. Shirman, Y.D. (ed.), Golikov, V.N., Busygin, I.N., *et al.* (1970) *Theoretical Foundations of Radiolocation*, Sovetskoe Radio, Moscow [in Russian].

III.3. Kobak, V.O. (1975) *Radar Reflectors*, Sovetskoe Radio, Moscow [in Russian].

III.4. Balanis, C.A. (1989) *Advanced Engineering Electromagnetic*, John Wiley & Sons, Ltd, Chichester.

III.5. Chernyak, V.S. (1998) *Fundamentals of Multisite Radar Systems*, Gordon and Breach Scientific Publishers, the Netherlands.

III.6. Ufimcev, P.Y. (1989) Black bodies and shadow radiation, *Radiotekhnika i elektronika*, **35**(12), 2519–27 [in Russian].

III.7. Blyakhman, A.B. and Runova, I.A. (1999) Forward scattering radiolocation bistatic RCS and target detection, in Proceedings of the IEEE Radar Conference, Waltham, Massachusetts, 20–22, April 1999, pp. 203–8.

III.8. Chapursky, V.V. (1989) Forward scattering inverse hologram spectrum and complex object radar cross section calculation, *Izvestia vuzov 'Radioelectronika'*, **32**(7), 75–7 [in Russian].

III.9. Skolnik, M.I. (2001) *Introduction to Radar Systems*, 3rd edn, McGraw-Hill, New York.

III.10. Blyakhman, A.B. (1998) Forward scattering bistatic radar, in proceedings of the PIERS Workshop on *Advances in Radar Methods*, Italy, July 1998, pp. 107–13.

III.11. Blyakhman, A.B. and Runova, I.A. (2001) Bistatic radar cross section and the detection of objects from their forward scatter. *J. Communications Technology and Electronics*, **46**(4), 393–401.

III.12. Glaser, J.I. (1984) Bistatic radar hold promise for future systems, *Microwave Systems News*, **16**(11), 119–36.

III.13. Blyakhman, A.B., Ryndyk, A.G. and Kovalev, F.N. (1998) The equipment for target movement parameter measurement, Russian Federation Patent 2124220 with priority of 29.10.97, registered 27.12.98.

III.14. Blyakhman, A.B., Ryndyk, A.G. and Sidorov, S.B. (2000) Forward scattering radar moving objects coordinates measurement, in Proceedings of the IEEE 2000 International Radar Conference, Alexandria, USA, 7–12 May, 2000, p. 678.

III.15. Cherniakov, M., Abdullah, R., Jančovič P. and Chapursky, V. (2006) Automatic ground target classification using FSR, *IEEE Proc., Radar, Sonar Navigation* (in press).

III.16. Glaser, J. (1985) Bistatic RCS of complex objects near forward scatter. *IEEE Trans. Aerospace and Electronic Systems*, **AES-21** (1), January, 70–8.

III.17. Blyakhman, A.B., Myakinkov, A.V. and Ryndyk, A.G. (2005) Tracking algorithm for three-dimensional bistatic forward scattering radar with weighting of primary measurements, in proceedings of the 2nd European Radar Conference, Paris, pp. 153–6.

III.18. Willis, N.I. (1995) *Bistatic Radar*, Technology Service Corporation, Silver Spring.

III.19. Cherniy, B. (1972) *Propagation of Radio Waves*, Sovetskoe Radio, Moscow [in Russian].

III.20. Howland, P.E. (1994) A passive metric radar using a transmitter of opportunity, in Proceedings of the IEEE International Radar Conference, Paris, pp. 370–5.

III.21. Cook, C.E. and Bernfeld M. (1967) *Radar Signals. An Introduction to Theory and Application*, Academic Press, New York and London.

III.22. Sage, A.P. and Melse, J.L. (1972) *Estimation Theory with Application to Communication and Control*. McGraw-Hill, New York.

III.23. Ryndyk, A.G., Sidorov, S.B., Blyakhman, A.B. and Kovalev, F.N. (1999) Accuracy of estimating the target coordinates by the maximum likelihood method using a transmissive radar, J. *Communications Technology and Electronics*, **44** (12), 1322–6.

III.24. Blyakhman, A.B., Kovalev, F.N., Ryndyk, A.G. and Sidorov, S.B. (2000) The equipment for target movement parameter measurement, Russian Federation Patent 2154840 with priority of 23.09.99, registered 20.08.2000f.

III.25. Myakinkov, A.V. (2003) Forward scattering radar tracking algorithm with estimation of primary measurement accuracy variation, in Proceedings of the IInd International Science Technical Conference on *Physics and Applications of Wave Processes*, Samara, Russia, September 2003, p. 166 [in Russian].

III.26. Repin, V.G. and Tartakovsky, G.P. (1977) *Statistical Synthesis under Conditions of A-Priory Ambiguity and Adaptation of Informational Systems*, Sovetskoe Radio, Moscow [in Russian].

III.27. Blyakhman, A.B., Kovalev, F.N. and Ryndyk, A.G. (2001) Target location method in forward scattering bistatic radar, *Radiotehnika*, 1, 4–15 [in Russian].
III.28. Blyakhman, A.B., Myakinkov, A.V. and Ryndyk, A.G. (2004) Algorithm of target tracking for three-dimensional bistatic forward scattering radar, in Proceedings of the IVth International Radar Symposium, Warsaw, Poland, 17–21, May 2004, pp. 309–24.
III.29. Wirth, W.D. (2001) *Radar Techniques Using Array Antennas*, IEE Radar, Sonar, Navigation and Avionics Series.
III.30. Monzingo, R.A. and Miller, T.W. (1980) *Introduction to Adaptive Arrays*, John Wiley & Sons, Ltd, Chiehester.
III.31. Blyakhman, A.B., Myakinkov, A.V. and Ryndyk, A.G. (2002) Phased antenna arrays in bistatic forward scattering radar system, in Progress in Electromagnetics Research Symposium Proceedings, Boston, Massachusetts, 1 July 2002, p. 163.
III.32. Blyakhman, A.B., Myakinkov, A.V. and Ryndyk A.G. (2004) Space-time signal processing in a bistatic forward-scattering radar system with an array antenna, *J. Communications Technology and Electronics*, **49** (6), 664–9.
III.33. Myakinkov, A.V. and Ryndyk, A.G. (2003) Space-time processing in three-dimensional forward scattering radar, in Proceedings of the IVth International Conference on *Antenna Theory and Techniques*, Sevastopol, Ukraine, 9–12 September 2003, pp. 355–8.
III.34. Korostelev, A.A. (1987) *Space–Time Theory of Radio Systems*, Radio i Svyaz, Moscow, [in Russian].
III.35. Klemm, R. (2002) *Principles of Space–Time Processing*, IEE Publisher, Series 12.
III.36. Marple Jr, S.L. (1987) *Digital Spectral Analysis with Applications*, Prentice-Hall, Englewood Cliffs, New Jersey.
III.37. Glaser, J.I. (1989) Some Results in the Bistatic Radar Cross-Section of Complex Objects, *IEEE Trans.*, **77**(5), May 1989.
III.38. Chapurskiy, V. and Sablin, V. (2000) SISAR: shadow inverse synthetic aperture radiolocation, in IEEE 2000 International Radar Conference, Alexandria, USA, May 2000, pp. 322–328.

Index

C

Printed and bound by CPI Group (UK) Ltd, Croydon, CR0 4YY

09/12/2025

14787605-0001